Universitext

Universitext

Universitext is a series of textbooks that presents material from a wide variety of mathematical disciplines at master's level and beyond. The books, often well class-tested by their author, may have an informal, personal even experimental approach to their subject matter. Some of the most successful and established books in the series have evolved through several editions, always following the evolution of teaching curricula, to very polished texts.

Thus as research topics trickle down into graduate-level teaching, first textbooks written for new, cutting-edge courses may make their way into *Universitext.*

For further volumes:
www.springer.com/series/223

Vladimir A. Zorich

Mathematical Analysis I

Second Edition

Vladimir A. Zorich
Department of Mathematics
Moscow State University
Moscow, Russia

Translators:
Roger Cooke (first English edition translated from the 4th Russian edition)
Burlington, Vermont, USA
and
Octavio Paniagua T. (Appendices A–F and new problems of the 6th Russian edition)
Berlin, Germany

Original Russian edition: Matematicheskij Analiz (Part I, 6th corrected edition, Moscow, 2012) MCCME (Moscow Center for Continuous Mathematical Education Publ.)

ISSN 0172-5939 ISSN 2191-6675 (electronic)
Universitext
ISBN 978-3-662-48790-7 ISBN 978-3-662-48792-1 (eBook)
DOI 10.1007/978-3-662-48792-1

Library of Congress Control Number: 2016930048

Mathematics Subject Classification (2010): 26-01, 26Axx, 26Bxx, 42-01

Springer Heidelberg New York Dordrecht London

Printed on acid-free paper

Springer is part of Springer Science+Business Media (www.springer.com)

Prefaces

Preface to the Second English Edition

Science has not stood still in the years since the first English edition of this book was published. For example, Fermat's last theorem has been proved, the Poincaré conjecture is now a theorem, and the Higgs boson has been discovered. Other events in science, while not directly related to the contents of a textbook in classical mathematical analysis, have indirectly led the author to learn something new, to think over something familiar, or to extend his knowledge and understanding. All of this additional knowledge and understanding end up being useful even when one speaks about something apparently completely unrelated.[1]

In addition to the original Russian edition, the book has been published in English, German, and Chinese. Various attentive multilingual readers have detected many errors in the text. Luckily, these are local errors, mostly misprints. They have assuredly all been corrected in this new edition.

But the main difference between the second and first English editions is the addition of a series of appendices to each volume. There are six of them in the first and five of them in the second. So as not to disturb the original text, they are placed at the end of each volume. The subjects of the appendices are diverse. They are meant to be useful to students (in mathematics and physics) as well as to teachers, who may be motivated by different goals. Some of the appendices are surveys, both prospective and retrospective. The final survey contains the most important conceptual achievements of the whole course, which establish connections between analysis and other parts of mathematics as a whole.

[1] There is a story about Erdős, who, like Hadamard, lived a very long mathematical and human life. When he was quite old, a journalist who was interviewing him asked him about his age. Erdős replied, after deliberating a bit, "I remember that when I was very young, scientists established that the Earth was two billion years old. Now scientists assert that the Earth is four and a half billion years old. So, I am approximately two and a half billion years old."

I was happy to learn that this book has proven to be useful, to some extent, not only to mathematicians, but also to physicists, and even to engineers from technical schools that promote a deeper study of mathematics.

It is a real pleasure to see a new generation that thinks bigger, understands more deeply, and is able to do more than the generation on whose shoulders it grew.

Moscow, Russia V. Zorich
2015

Preface to the First English Edition

An entire generation of mathematicians has grown up during the time between the appearance of the first edition of this textbook and the publication of the fourth edition, a translation of which is before you. The book is familiar to many people, who either attended the lectures on which it is based or studied out of it, and who now teach others in universities all over the world. I am glad that it has become accessible to English-speaking readers.

This textbook consists of two parts. It is aimed primarily at university students and teachers specializing in mathematics and natural sciences, and at all those who wish to see both the rigorous mathematical theory and examples of its effective use in the solution of real problems of natural science.

Note that Archimedes, Newton, Leibniz, Euler, Gauss, Poincaré, who are held in particularly high esteem by us, mathematicians, were more than mere mathematicians. They were scientists, natural philosophers. In mathematics resolving of important specific questions and development of an abstract general theory are processes as inseparable as inhaling and exhaling. Upsetting this balance leads to problems that sometimes become significant both in mathematical education and in science in general.

The textbook exposes classical analysis as it is today, as an integral part of the unified Mathematics, in its interrelations with other modern mathematical courses such as algebra, differential geometry, differential equations, complex and functional analysis.

Rigor of discussion is combined with the development of the habit of working with real problems from natural sciences. The course exhibits the power of concepts and methods of modern mathematics in exploring specific problems. Various examples and numerous carefully chosen problems, including applied ones, form a considerable part of the textbook. Most of the fundamental mathematical notions and results are introduced and discussed along with information, concerning their history, modern state and creators. In accordance with the orientation toward natural sciences, special attention is paid to informal exploration of the essence and roots of the basic concepts and theorems of calculus, and to the demonstration of numerous, sometimes fundamental, applications of the theory.

For instance, the reader will encounter here the Galilean and Lorentz transforms, the formula for rocket motion and the work of nuclear reactor, Euler's theorem on homogeneous functions and the dimensional analysis of physical quantities, the Legendre transform and Hamiltonian equations of classical mechanics, elements of hydrodynamics and the Carnot's theorem from thermodynamics, Maxwell's equations, the Dirac delta-function, distributions and the fundamental solutions, convolution and mathematical models of linear devices, Fourier series and the formula for discrete coding of a continuous signal, the Fourier transform and the Heisenberg uncertainty principle, differential forms, de Rham cohomology and potential fields, the theory of extrema and the optimization of a specific technological process, numerical methods and processing the data of a biological experiment, the asymptotics of the important special functions, and many other subjects.

Within each major topic the exposition is, as a rule, inductive, sometimes proceeding from the statement of a problem and suggestive heuristic considerations concerning its solution, toward fundamental concepts and formalisms. Detailed at first, the exposition becomes more and more compressed as the course progresses. Beginning *ab ovo* the book leads to the most up-to-date state of the subject.

Note also that, at the end of each of the volumes, one can find the list of the main theoretical topics together with the corresponding simple, but nonstandard problems (taken from the midterm exams), which are intended to enable the reader both determine his or her degree of mastery of the material and to apply it creatively in concrete situations.

More complete information on the book and some recommendations for its use in teaching can be found below in the prefaces to the first and second Russian editions.

Moscow, Russia
2003

V. Zorich

Preface to the Sixth Russian Edition

On my own behalf and on behalf of future readers, I thank all those, living in different countries, who had the possibility to inform the publisher or me personally about errors (typos, errors, omissions), found in Russian, English, German and Chinese editions of this textbook.

As it turned out, the book has been also very useful to physicists; I am very happy about that. In any case, I really seek to accompany the formal theory with meaningful examples of its application both in mathematics and outside of it.

The sixth edition contains a series of appendices that may be useful to students and lecturers. Firstly, some of the material is actually real lectures (for example, the transcription of two introductory survey lectures for students of first and third semesters), and, secondly, this is some mathematical information (sometimes of current interest, such as the relation between multidimensional geometry and the theory of probability), lying close to the main subject of the textbook.

Moscow, Russia
2011

V. Zorich

Preface to the Second Russian Edition

In this second edition of the book, along with an attempt to remove the misprints that occurred in the first edition,[2] certain alterations in the exposition have been made (mainly in connection with the proofs of individual theorems), and some new problems have been added, of an informal nature as a rule.

The preface to the first edition of this course of analysis (see below) contains a general description of the course. The basic principles and the aim of the exposition are also indicated there. Here I would like to make a few remarks of a practical nature connected with the use of this book in the classroom.

Usually both the student and the teacher make use of a text, each for his own purposes.

At the beginning, both of them want most of all a book that contains, along with the necessary theory, as wide a variety of substantial examples of its applications as possible, and, in addition, explanations, historical and scientific commentary, and descriptions of interconnections and perspectives for further development. But when preparing for an examination, the student mainly hopes to see the material that will be on the examination. The teacher likewise, when preparing a course, selects only the material that can and must be covered in the time alloted for the course.

In this connection, it should be kept in mind that the text of the present book is noticeably more extensive than the lectures on which it is based. What caused

[2]No need to worry: in place of the misprints that were corrected in the plates of the first edition (which were not preserved), one may be sure that a host of new misprints will appear, which so enliven, as Euler believed, the reading of a mathematical text.

this difference? First of all, the lectures have been supplemented by essentially an entire problem book, made up not so much of exercises as substantive problems of science or mathematics proper having a connection with the corresponding parts of the theory and in some cases significantly extending them. Second, the book naturally contains a much larger set of examples illustrating the theory in action than one can incorporate in lectures. Third and finally, a number of chapters, sections, or subsections were consciously written as a supplement to the traditional material. This is explained in the sections "On the introduction" and "On the supplementary material" in the preface to the first edition.

I would also like to recall that in the preface to the first edition I tried to warn both the student and the beginning teacher against an excessively long study of the introductory formal chapters. Such a study would noticeably delay the analysis proper and cause a great shift in emphasis.

To show what in fact can be retained of these formal introductory chapters in a realistic lecture course, and to explain in condensed form the syllabus for such a course as a whole while pointing out possible variants depending on the student audience, at the end of the book I give a list of problems from the midterm exam, along with some recent examination topics for the first two semesters, to which this first part of the book relates. From this list the professional will of course discern the order of exposition, the degree of development of the basic concepts and methods, and the occasional invocation of material from the second part of the textbook when the topic under consideration is already accessible for the audience in a more general form.

In conclusion I would like to thank colleagues and students, both known and unknown to me, for reviews and constructive remarks on the first edition of the course. It was particularly interesting for me to read the reviews of A.N. Kolmogorov and V.I. Arnol'd. Very different in size, form, and style, these two have, on the professional level, so many inspiring things in common.

Moscow, Russia V. Zorich
1997

From the Preface to the First Russian Edition

The creation of the foundations of the differential and integral calculus by Newton and Leibniz three centuries ago appears even by modern standards to be one of the greatest events in the history of science in general and mathematics in particular.

Mathematical analysis (in the broad sense of the word) and algebra have intertwined to form the root system on which the ramified tree of modern mathematics is supported and through which it makes its vital contact with the nonmathematical sphere. It is for this reason that the foundations of analysis are included as a necessary element of even modest descriptions of so-called higher mathematics; and it is probably for that reason that so many books aimed at different groups of readers are devoted to the exposition of the fundamentals of analysis.

This book has been aimed primarily at mathematicians desiring (as is proper) to obtain thorough proofs of the fundamental theorems, but who are at the same time interested in the life of these theorems outside of mathematics itself.

The characteristics of the present course connected with these circumstances reduce basically to the following:

In the Exposition Within each major topic the exposition is as a rule inductive, sometimes proceeding from the statement of a problem and suggestive heuristic considerations toward its solution to fundamental concepts and formalisms.

Detailed at first, the exposition becomes more and more compressed as the course progresses.

An emphasis is placed on the efficient machinery of smooth analysis. In the exposition of the theory I have tried (to the extent of my knowledge) to point out the most essential methods and facts and avoid the temptation of a minor strengthening of a theorem at the price of a major complication of its proof.

The exposition is geometric throughout wherever this seemed worthwhile in order to reveal the essence of the matter.

The main text is supplemented with a rather large collection of examples, and nearly every section ends with a set of problems that I hope will significantly complement even the theoretical part of the main text. Following the wonderful precedent of Pólya and Szegő, I have often tried to present a beautiful mathematical result or an important application as a series of problems accessible to the reader.

The arrangement of the material was dictated not only by the architecture of mathematics in the sense of Bourbaki, but also by the position of analysis as a component of a unified mathematical or, one should rather say, natural-science/mathematical education.

In Content This course is being published in two books (Part 1 and Part 2).

The present Part 1 contains the differential and integral calculus of functions of one variable and the differential calculus of functions of several variables.

In differential calculus we emphasize the role of the differential as a linear standard for describing the local behavior of the variation of a variable. In addition to numerous examples of the use of differential calculus to study functional relations (monotonicity, extrema) we exhibit the role of the language of analysis in writing simple differential equations – mathematical models of real-world phenomena and the substantive problems connected with them.

We study a number of such problems (for example, the motion of a body of variable mass, a nuclear reactor, atmospheric pressure, motion in a resisting medium) whose solution leads to important elementary functions. Full use is made of the language of complex variables; in particular, Euler's formula is derived and the unity of the fundamental elementary functions is shown.

The integral calculus has consciously been explained as far as possible using intuitive material in the framework of the Riemann integral. For the majority of applications, this is completely adequate.[3] Various applications of the integral are pointed out, including those that lead to an improper integral (for example, the work involved in escaping from a gravitational field, and the escape velocity for the Earth's gravitational field) or to elliptic functions (motion in a gravitational field in the presence of constraints, pendulum motion).

The differential calculus of functions of several variables is very geometric. In this topic, for example, one studies such important and useful consequences of the implicit function theorem as curvilinear coordinates and local reduction to canonical form for smooth mappings (the rank theorem) and functions (Morse's lemma), and also the theory of extrema with constraint.

Results from the theory of continuous functions and differential calculus are summarized and explained in a general invariant form in two chapters that link up naturally with the differential calculus of real-valued functions of several variables. These two chapters open the second part of the course. The second book, in which we also discuss the integral calculus of functions of several variables up to the general Newton–Leibniz–Stokes formula thus acquires a certain unity.

We shall give more complete information on the second book in its preface. At this point we add only that, in addition to the material already mentioned, it contains information on series of functions (power series and Fourier series included), on integrals depending on a parameter (including the fundamental solution, convolution, and the Fourier transform), and also on asymptotic expansions (which are usually absent or insufficiently presented in textbooks).

We now discuss a few particular problems.

On the Introduction I have not written an introductory survey of the subject, since the majority of beginning students already have a preliminary idea of differential and integral calculus and their applications from high school, and I could hardly claim to write an even more introductory survey. Instead, in the first two chapters I bring the former high-school student's understanding of sets, functions, the use of logical symbolism, and the theory of a real number to a certain mathematical completeness.

This material belongs to the formal foundations of analysis and is aimed primarily at the mathematics major, who may at some time wish to trace the logical structure of the basic concepts and principles used in classical analysis. Mathematical analysis proper begins in the third chapter, so that the reader who wishes to get effective machinery in his hands as quickly as possible and see its applications can in general begin a first reading with Chap. 3, turning to the earlier pages whenever something seems nonobvious or raises a question which hopefully I also have thought of and answered in the early chapters.

On the Division of Material The material of the two books is divided into chapters numbered continuously. The sections are numbered within each chapter sepa-

[3]The "stronger" integrals, as is well known, require fussier set-theoretic considerations, outside the mainstream of the textbook, while adding hardly anything to the effective machinery of analysis, mastery of which should be the first priority.

rately; subsections of a section are numbered only within that section. Theorems, propositions, lemmas, definitions, and examples are written in italics for greater logical clarity, and numbered for convenience within each section.

On the Supplementary Material Several chapters of the book are written as a natural extension of classical analysis. These are, on the one hand, Chaps. 1 and 2 mentioned above, which are devoted to its formal mathematical foundations, and on the other hand, Chaps. 9, 10, and 15 of the second part, which give the modern view of the theory of continuity, differential and integral calculus, and finally Chap. 19, which is devoted to certain effective asymptotic methods of analysis.

The question as to which part of the material of these chapters should be included in a lecture course depends on the audience and can be decided by the lecturer, but certain fundamental concepts introduced here are usually present in any exposition of the subject to mathematicians.

In conclusion, I would like to thank those whose friendly and competent professional aid has been valuable and useful to me during the work on this book.

The proposed course was quite detailed, and in many of its aspects it was coordinated with subsequent modern university mathematics courses – such as, for example, differential equations, differential geometry, the theory of functions of a complex variable, and functional analysis. In this regard my contacts and discussions with V.I. Arnol'd and the especially numerous ones with S.P. Novikov during our joint work with the so-called "experimental student group in natural-science/mathematical education" in the Department of Mathematics at MSU, were very useful to me.

I received much advice from N.V. Efimov, chair of the Section of Mathematical Analysis in the Department of Mechanics and Mathematics at Moscow State University.

I am also grateful to colleagues in the department and the section for remarks on the mimeographed edition of my lectures.

Student transcripts of my recent lectures which were made available to me were valuable during the work on this book, and I am grateful to their owners.

I am deeply grateful to the official reviewers L.D. Kudryavtsev, V.P. Petrenko, and S.B. Stechkin for constructive comments, most of which were taken into account in the book now offered to the reader.

Moscow, Russia
1980

V. Zorich

Contents

Chapter 1
Some General Mathematical Concepts and Notation

1.1 Logical Symbolism

1.1.1 Connectives and Brackets

The language of this book, like the majority of mathematical texts, consists of ordinary language and a number of special symbols from the theories being discussed. Along with the special symbols, which will be introduced as needed, we use the common symbols of mathematical logic $\neg$, $\wedge$, $\vee$, $\Rightarrow$, and $\Leftrightarrow$ to denote respectively negation (*not*) and the logical connectives *and*, *or*, *implies*, and *is equivalent to*.[1]

For example, take three statements of independent interest:

L. *If the notation is adapted to the discoveries..., the work of thought is marvelously shortened.* (G. Leibniz)[2]

P. *Mathematics is the art of calling different things by the same name.* (H. Poincaré).[3]

G. *The great book of nature is written in the language of mathematics.* (Galileo).[4]

Then, according to the notation given above, Table 1.1 relates L, P, G.

---

[1] The symbol & is often used in logic in place of $\wedge$. Logicians more often write the implication symbol $\Rightarrow$ as $\rightarrow$ and the relation of logical equivalence as $\longleftrightarrow$ or $\leftrightarrow$. However, we shall adhere to the symbolism indicated in the text so as not to overburden the symbol $\rightarrow$, which has been traditionally used in mathematics to denote passage to the limit.

[2] G.W. Leibniz (1646–1716) – outstanding German scholar, philosopher, and mathematician to whom belongs the honor, along with Newton, of having discovered the foundations of the infinitesimal calculus.

[3] H. Poincaré (1854–1912) – French mathematician whose brilliant mind transformed many areas of mathematics and achieved fundamental applications of it in mathematical physics.

[4] Galileo Galilei (1564–1642) – Italian scholar and outstanding scientific experimenter. His works lie at the foundation of the subsequent physical concepts of space and time. He is the father of modern physical science.

V.A. Zorich, *Mathematical Analysis I*, Universitext,
DOI 10.1007/978-3-662-48792-1_1

Table 1.1

Notation	Meaning
$L \Rightarrow P$	L implies P
$L \Leftrightarrow P$	L is equivalent to P
$((L \Rightarrow P) \wedge (\neg P)) \Rightarrow (\neg L)$	If P follows from L and P is false, then L is false
$\neg((L \Leftrightarrow G) \vee (P \Leftrightarrow G))$	G is not equivalent either to L or to P

We see that it is not always reasonable to use only formal notation, avoiding colloquial language.

We remark further that parentheses are used in the writing of complex statements composed of simpler ones, fulfilling the same syntactical function as in algebraic expressions. As in algebra, in order to avoid the overuse of parentheses one can make a convention about the order of operations. To that end, we shall agree on the following order of priorities for the symbols:

$$\neg, \quad \wedge, \quad \vee, \quad \Rightarrow, \quad \Leftrightarrow .$$

With this convention the expression $\neg A \wedge B \vee C \Rightarrow D$ should be interpreted as $(((\neg A) \wedge B) \vee C) \Rightarrow D$, and the relation $A \vee B \Rightarrow C$ as $(A \vee B) \Rightarrow C$, not as $A \vee (B \Rightarrow C)$.

We shall often give a different verbal expression to the notation $A \Rightarrow B$, which means that A implies B, or, what is the same, that B follows from A, saying that B is a *necessary criterion* or *necessary condition* for A and A in turn is a *sufficient condition* or *sufficient criterion* for B, so that the relation $A \Leftrightarrow B$ can be read in any of the following ways:

A is necessary and sufficient for B;
A holds when B holds, and only then;
A if and only if B;
A is equivalent to B.

Thus the notation $A \Leftrightarrow B$ means that A implies B and simultaneously B implies A.

The use of the conjunction *and* in the expression $A \wedge B$ requires no explanation.

It should be pointed out, however, that in the expression $A \vee B$ the conjunction *or* is not exclusive, that is, the statement $A \vee B$ is regarded as true if at least one of the statements A and B is true. For example, let x be a real number such that $x^2 - 3x + 2 = 0$. Then we can write that the following relation holds:

$$\left(x^2 - 3x + 2 = 0\right) \Leftrightarrow (x = 1) \vee (x = 2).$$

1.1.2 Remarks on Proofs

A typical mathematical proposition has the form $A \Rightarrow B$, where A is the assumption and B the conclusion. The proof of such a proposition consists of constructing a chain $A \Rightarrow C_1 \Rightarrow \cdots \Rightarrow C_n \Rightarrow B$ of implications, each element of which is either an axiom or a previously proved proposition.[5]

In proofs we shall adhere to the classical rule of inference: if A is true and $A \Rightarrow B$, then B is also true.

In proof by contradiction we shall also use the law of excluded middle, by virtue of which the statement $A \vee \neg A$ (A or not-A) is considered true independently of the specific content of the statement A. Consequently we simultaneously accept that $\neg(\neg A) \Leftrightarrow A$, that is, double negation is equivalent to the original statement.

1.1.3 Some Special Notation

For the reader's convenience and to shorten the writing, we shall agree to denote the end of a proof by the symbol $\Box$.

We also agree, whenever convenient, to introduce definitions using the special symbol := (equality by definition), in which the colon is placed on the side of the object being defined.

For example, the notation

$$\int_a^b f(x)\,\mathrm{d}x := \lim_{\lambda(P)\to 0} \sigma(f; P, \xi)$$

defines the left-hand side in terms of the right-hand side, whose meaning is assumed to be known.

Similarly, one can introduce abbreviations for expressions already defined. For example

$$\sum_{i=1}^{n} f(\xi_i)\Delta x_i =: \sigma(f; P, \xi)$$

introduces the notation $\sigma(f; P, \xi)$ for the sum of special form on the left-hand side.

1.1.4 Concluding Remarks

We note that here we have spoken essentially about notation only, without analyzing the formalism of logical deductions and without touching on the profound questions of truth, provability, and deducibility, which form the subject matter of mathematical logic.

How are we to construct mathematical analysis if we have no formalization of logic? There may be some consolation in the fact that we always know more than

[5] The notation $A \Rightarrow B \Rightarrow C$ will be used as an abbreviation for $(A \Rightarrow B) \wedge (B \Rightarrow C)$.

we can formalize at any given time, or perhaps we should say we know how to do more than we can formalize. This last sentence may be clarified by the well-known proverb of the centipede who forgot how to walk when asked to explain exactly how it dealt with so many legs.

The experience of all the sciences convinces us that what was considered clear or simple and unanalyzable yesterday may be subjected to reexamination or made more precise today. Such was the case (and will undoubtedly be the case again) with many concepts of mathematical analysis, the most important theorems and machinery of which were discovered in the seventeenth and eighteenth centuries, but which acquired its modern formalized form with a unique interpretation that is probably responsible for its being generally accessible, only after the creation of the theory of limits and the fully developed theory of real numbers needed for it in the nineteenth century.

This is the level of the theory of real numbers from which we shall begin to construct the whole edifice of analysis in Chap. 2.

As already noted in the preface, those who wish to make a rapid acquaintance with the basic concepts and effective machinery of differential and integral calculus proper may begin immediately with Chap. 3, turning to particular places in the first two chapters only as needed.

1.1.5 Exercises

We shall denote true assertions by the symbol 1 and false ones by 0. Then to each of the statements $\neg A$, $A \wedge B$, $A \vee B$, and $A \Rightarrow B$ one can associate a so-called *truth table*, which indicates its truth or falsehood depending on the truth of the statements A and B. These tables are a formal definition of the logical operations $\neg$, $\wedge$, $\vee$, $\Rightarrow$. Here they are:

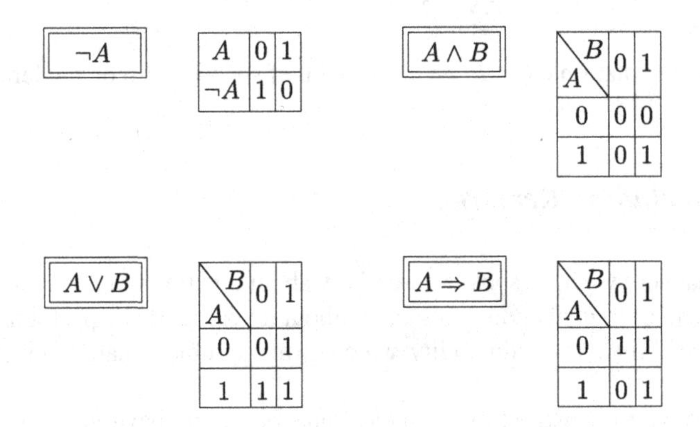

$\neg A$

A	0	1
$\neg A$	1	0

$A \wedge B$

$A \backslash B$	0	1
0	0	0
1	0	1

$A \vee B$

$A \backslash B$	0	1
0	0	1
1	1	1

$A \Rightarrow B$

$A \backslash B$	0	1
0	1	1
1	0	1

1. Check whether all of these tables agree with your concept of the corresponding logical operation. (In particular, pay attention to the fact that if A is false, then the implication $A \Rightarrow B$ is always true.)

2. Show that the following simple, but very useful relations, which are widely used in mathematical reasoning, are true:

a) $\neg(A \wedge B) \Leftrightarrow \neg A \vee \neg B$;
b) $\neg(A \vee B) \Leftrightarrow \neg A \wedge \neg B$;
c) $(A \Rightarrow B) \Leftrightarrow (\neg B \Rightarrow \neg A)$;
d) $(A \Rightarrow B) \Leftrightarrow (\neg A \vee B)$;
e) $\neg(A \Rightarrow B) \Leftrightarrow A \wedge \neg B$.

1.2 Sets and Elementary Operations on Them

1.2.1 The Concept of a Set

Since the late nineteenth and early twentieth centuries the most universal language of mathematics has been the language of set theory. This is even manifest in one of the definitions of mathematics as the science that studies different structures (relations) on sets.[6]

"We take a *set* to be an assemblage of definite, perfectly distinguishable objects of our intuition or our thought into a coherent whole." Thus did Georg Cantor,[7] the creator of set theory, describe the concept of a set.

Cantor's description cannot, of course, be considered a definition, since it appeals to concepts that may be more complicated than the concept of a set itself (and in any case, have not been defined previously). The purpose of this description is to explain the concept by connecting it with other concepts.

The basic assumptions of Cantorian (or, as it is generally called, "naive") set theory reduce to the following statements.

1^0. A set may consist of any distinguishable objects.
2^0. A set is unambiguously determined by the collection of objects that comprise it.
3^0. Any property defines the set of objects having that property.

If x is an object, P is a property, and $P(x)$ denotes the assertion that x has property P, then the class of objects having the property P is denoted $\{x \mid P(x)\}$. The objects that constitute a class or set are called the *elements* of the class or set.

The set consisting of the elements $x_1, \dots, x_n$ is usually denoted $\{x_1, \dots, x_n\}$. Wherever no confusion can arise we allow ourselves to denote the one-element set $\{a\}$ simply as a.

The words "class", "family", "totality", and "collection" are used as synonyms for "set" in naive set theory.

---

[6]Bourbaki, N. "The architecture of mathematics" in: N. Bourbaki, *Elements of the history of mathematics*, translated from the French by John Meldrum, Springer, New York, 1994.

[7]G. Cantor (1845–1918) – German mathematician, the creator of the theory of infinite sets and the progenitor of set-theoretic language in mathematics.

The following examples illustrate the application of this terminology:

- the set of letters “a” occurring in the word “I”;
- the set of wives of Adam;
- the collection of ten decimal digits;
- the family of beans;
- the set of grains of sand on the Earth;
- the totality of points of a plane equidistant from two given points of the plane;
- the family of sets;
- the set of all sets.

The variety in the possible degree of determinacy in the definition of a set leads one to think that a set is, after all, not such a simple and harmless concept.

And in fact the concept of the set of all sets, for example, is simply contradictory.

Proof Indeed, suppose that for a set M the notation $P(M)$ means that M is not an element of itself.

Consider the class $K = \{M \mid P(M)\}$ of sets having property P.

If K is a set either $P(K)$ or $\neg P(K)$ is true. However, this dichotomy does not apply to K. Indeed, $P(K)$ is impossible; for it would then follow from the definition of K that K contains K as an element, that is, that $\neg P(K)$ is true; on the other hand, $\neg P(K)$ is also impossible, since that means that K contains K as an element, which contradicts the definition of K as the class of sets that do not contain themselves as elements.

Consequently K is not a set. □

This is the classical paradox of Russell,[8] one of the paradoxes to which the naive conception of a set leads.

In modern mathematical logic the concept of a set has been subjected to detailed analysis (with good reason, as we see). However, we shall not go into that analysis. We note only that in the current axiomatic set theories a set is defined as a mathematical object having a definite collection of properties.

The description of these properties constitutes an axiom system. The core of axiomatic set theory is the postulation of rules by which new sets can be formed from given ones. In general any of the current axiom systems is such that, on the one hand, it eliminates the known contradictions of the naive theory, and on the other hand it provides freedom to operate with specific sets that arise in different areas of mathematics, most of all, in mathematical analysis understood in the broad sense of the word.

Having confined ourselves for the time being to remarks on the concept of a set, we pass to the description of the set-theoretic relations and operations most commonly used in analysis.

Those wishing a more detailed acquaintance with the concept of a set should study Sect. 1.4.2 in the present chapter or turn to the specialized literature.

---

[8]B. Russell (1872–1970) – British logician, philosopher, sociologist and social activist.

Fig. 1.1

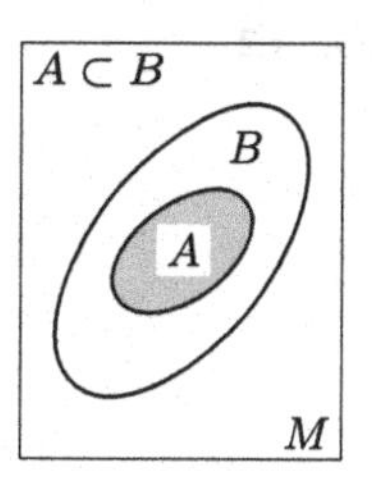

1.2.2 The Inclusion Relation

As has already been pointed out, the objects that comprise a set are usually called the *elements* of the set. We tend to denote sets by uppercase letters and their elements by the corresponding lowercase letters.

The statement, "x is an element of the set X" is written briefly as

$$x \in X \quad (\text{or } X \ni x),$$

and its negation as

$$x \notin X \quad (\text{or } X \not\ni x).$$

When statements about sets are written, frequent use is made of the logical operators $\exists$ ("there exists" or "there are") and $\forall$ ("every" or "for any") which are called the *existence* and *generalization* quantifiers respectively.

For example, the string $\forall x\ ((x \in A) \Leftrightarrow (x \in B))$ means that for any object x the relations $x \in A$ and $x \in B$ are equivalent. Since a set is completely determined by its elements, this statement is usually written briefly as

$$A = B,$$

read "A equals B", and means that the sets A and B are the same.

Thus two sets are *equal* if they consist of the same elements.

The negation of equality is usually written as $A \neq B$.

If every element of A is an element of B, we write $A \subset B$ or $B \supset A$ and say that A is a *subset* of B or that B contains A or that B includes A. In this connection the relation $A \subset B$ between sets A and B is called the *inclusion relation* (Fig. 1.1).

Thus

$$(A \subset B) := \forall x\ \big((x \in A) \Rightarrow (x \in B)\big).$$

If $A \subset B$ and $A \neq B$, we shall say that the inclusion $A \subset B$ is *strict* or that A is a *proper* subset of B.

Using these definitions, we can now conclude that

$$(A = B) \Leftrightarrow (A \subset B) \wedge (B \subset A).$$

Fig. 1.2

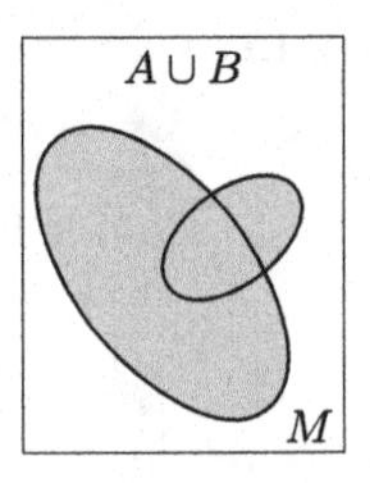

If M is a set, any property P distinguishes in M the subset

$$\{x \in M \mid P(x)\}$$

consisting of the elements of M that have the property.

For example, it is obvious that

$$M = \{x \in M \mid x \in M\}.$$

On the other hand, if P is taken as a property that no element of the set M has, for example, $P(x) := (x \neq x)$, we obtain the set

$$\varnothing = \{x \in M \mid x \neq x\},$$

called the *empty* subset of M.

1.2.3 Elementary Operations on Sets

Let A and B be subsets of a set M.

a. The *union* of A and B is the set

$$A \cup B := \{x \in M \mid (x \in A) \vee (x \in B)\},$$

consisting of precisely the elements of M that belong to at least one of the sets A and B (Fig. 1.2).

b. The *intersection* of A and B is the set

$$A \cap B := \{x \in M \mid (x \in A) \wedge (x \in B)\},$$

formed by the elements of M that belong to both sets A and B (Fig. 1.3).

c. The *difference* between A and B is the set

$$A \backslash B := \{x \in M \mid (x \in A) \wedge (x \notin B)\},$$

consisting of the elements of A that do not belong to B (Fig. 1.4).

Fig. 1.3

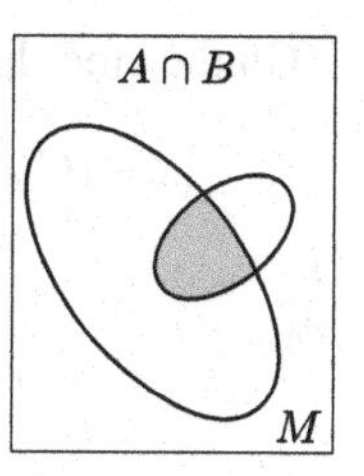

Fig. 1.4

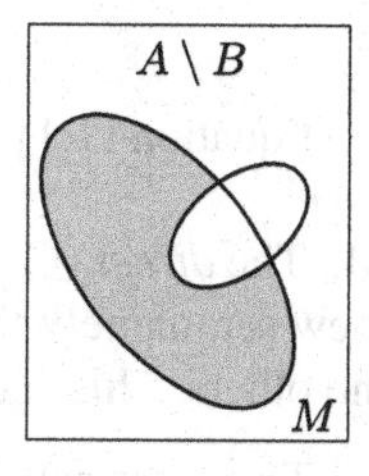

Fig. 1.5

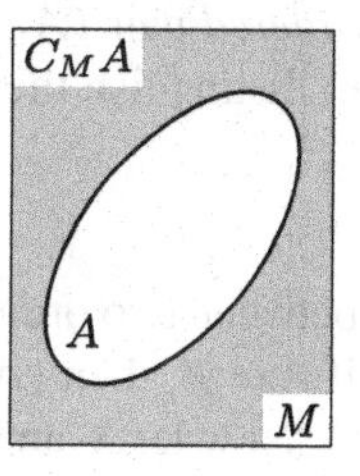

The difference between the set M and one of its subsets A is usually called the *complement* of A in M and denoted $C_M A$, or CA when the set in which the complement of A is being taken is clear from the context (Fig. 1.5).

Example As an illustration of the interaction of the concepts just introduced, let us verify the following relations (the so-called de Morgan[9] rules):

$$C_M(A \cup B) = C_M A \cap C_M B, \tag{1.1}$$

$$C_M(A \cap B) = C_M A \cup C_M B. \tag{1.2}$$

Proof We shall prove the first of these equalities by way of example:

$$\big(x \in C_M(A \cup B)\big) \Rightarrow \big(x \notin (A \cup B)\big) \Rightarrow \big((x \notin A) \wedge (x \notin B)\big) \Rightarrow$$
$$\Rightarrow (x \in C_M A) \wedge (x \in C_M B) \Rightarrow \big(x \in (C_M A \cap C_M B)\big).$$

Thus we have established that

$$C_M(A \cup B) \subset (C_M A \cap C_M B). \tag{1.3}$$

---

[9] A. de Morgan (1806–1871) – Scottish mathematician.

On the other hand,

$$\begin{aligned}\big(x \in (C_M A \cap C_M B)\big) \Rightarrow \big((x \in C_M A) \wedge (x \in C_M B)\big) \Rightarrow \\ \Rightarrow \big((x \notin A) \wedge (x \notin B)\big) \Rightarrow \big(x \notin (A \cup B)\big) \Rightarrow \\ \Rightarrow \big(x \in C_M(A \cup B)\big),\end{aligned}$$

that is,

$$(C_M A \cap C_M B) \subset C_M(A \cup B). \tag{1.4}$$

Equation (1.1) follows from (1.3) and (1.4). □

d. The *direct* (*Cartesian*) *product of sets*. For any two sets A and B one can form a new set, namely the pair $\{A, B\} = \{B, A\}$, which consists of the sets A and B and no others. This set has two elements if $A \neq B$ and one element if $A = B$.

This set is called the *unordered pair* of sets A and B, to be distinguished from the *ordered* pair (A, B) in which the elements are endowed with additional properties to distinguish the first and second elements of the pair $\{A, B\}$. The equality

$$(A, B) = (C, D)$$

between two ordered pairs means by definition that $A = C$ and $B = D$. In particular, if $A \neq B$, then $(A, B) \neq (B, A)$.

Now let X and Y be arbitrary sets. The set

$$X \times Y := \big\{(x, y) \mid (x \in X) \wedge (y \in Y)\big\},$$

formed by the ordered pairs (x, y) whose first element belongs to X and whose second element belongs to Y, is called the *direct* or *Cartesian product* of the sets X and Y (in that order!).

It follows obviously from the definition of the direct product and the remarks made above about the ordered pair that in general $X \times Y \neq Y \times X$. Equality holds only if $X = Y$. In this last case we abbreviate $X \times X$ as X^2.

The direct product is also called the Cartesian product in honor of Descartes,[10] who arrived at the language of analytic geometry in terms of a system of coordinates independently of Fermat.[11] The familiar system of Cartesian coordinates in the plane makes this plane precisely into the direct product of two real axes. This familiar object shows vividly why the Cartesian product depends on the order of the factors. For example, different points of the plane correspond to the pairs (0, 1) and (1, 0).

---

[10] R. Descartes (1596–1650) – outstanding French philosopher, mathematician and physicist who made fundamental contributions to scientific thought and knowledge.

[11] P. Fermat (1601–1665) – remarkable French mathematician, a lawyer by profession. He was one of the founders of a number of areas of modern mathematics: analysis, analytic geometry, probability theory, and number theory.

In the ordered pair $z = (x_1, x_2)$, which is an element of the direct product $Z = X_1 \times X_2$ of the sets X_1 and X_2, the element x_1 is called the *first projection* of the pair z and denoted $\mathrm{pr}_1\, z$, while the element x_2 is the *second projection* of z and is denoted $\mathrm{pr}_2\, z$.

By analogy with the terminology of analytic geometry, the projections of an ordered pair are often called the (first and second) *coordinates* of the pair.

1.2.4 Exercises

In Exercises 1, 2, and 3 the letters A, B, and C denote subsets of a set M.

1. Verify the following relations.

a) $(A \subset C) \wedge (B \subset C) \Leftrightarrow ((A \cup B) \subset C)$;
b) $(C \subset A) \wedge (C \subset B) \Leftrightarrow (C \subset (A \cap B))$;
c) $C_M(C_M A) = A$;
d) $(A \subset C_M B) \Leftrightarrow (B \subset C_M A)$;
e) $(A \subset B) \Leftrightarrow (C_M A \supset C_M B)$.

2. Prove the following statements.

a) $A \cup (B \cup C) = (A \cup B) \cup C =: A \cup B \cup C$;
b) $A \cap (B \cap C) = (A \cap B) \cap C =: A \cap B \cap C$;
c) $A \cap (B \cup C) = (A \cap B) \cup (A \cap C)$;
d) $A \cup (B \cap C) = (A \cup B) \cap (A \cup C)$.

3. Verify the connection (duality) between the operations of union and intersection:

a) $C_M(A \cup B) = C_M A \cap C_M B$;
b) $C_M(A \cap B) = C_M A \cup C_M B$.

4. Give geometric representations of the following Cartesian products.

a) The product of two line segments (a rectangle).
b) The product of two lines (a plane).
c) The product of a line and a circle (an infinite cylindrical surface).
d) The product of a line and a disk (an infinite solid cylinder).
e) The product of two circles (a torus).
f) The product of a circle and a disk (a solid torus).

5. The set $\Delta = \{(x_1, x_2) \in X^2 \mid x_1 = x_2\}$ is called the *diagonal* of the Cartesian square X^2 of the set X. Give geometric representations of the diagonals of the sets obtained in parts a), b), and e) of Exercise 4.

6. Show that

a) $(X \times Y = \varnothing) \Leftrightarrow (X = \varnothing) \vee (Y = \varnothing)$, and if $X \times Y \neq \varnothing$, then
b) $(A \times B \subset X \times Y) \Leftrightarrow (A \subset X) \wedge (B \subset Y)$,
c) $(X \times Y) \cup (Z \times Y) = (X \cup Z) \times Y$,

d) $(X \times Y) \cap (X' \times Y') = (X \cap X') \times (Y \cap Y')$.
Here $\varnothing$ denotes the empty set, that is, the set having no elements.

7. By comparing the relations of Exercise 3 with relations a) and b) from Exercise 2 of Sect. 1.1, establish a correspondence between the logical operators $\neg$, $\wedge$, $\vee$ and the operations C, $\cap$, and $\cup$ on sets.

1.3 Functions

1.3.1 The Concept of a Function (Mapping)

We shall now describe the concept of a functional relation, which is fundamental both in mathematics and elsewhere.

Let X and Y be certain sets. We say that there is a *function* defined on X with values in Y if, by virtue of some rule f, to each element $x \in X$ there corresponds an element $y \in Y$.

In this case the set X is called the *domain of definition* of the function. The symbol x used to denote a general element of the domain is called the *argument* of the function, or the *independent variable*. The element $y_0 \in Y$ corresponding to a particular value $x_0 \in X$ of the argument x is called the *value* of the function at x_0, or the value of the function at the value $x = x_0$ of its argument, and is denoted $f(x_0)$. As the argument $x \in X$ varies, the value $y = f(x) \in Y$, in general, varies depending on the values of x. For that reason, the quantity $y = f(x)$ is often called the *dependent variable*.

The set

$$f(X) := \{y \in Y \mid \exists x \; ((x \in X) \wedge (y = f(x)))\}$$

of values assumed by a function on elements of the set X will be called the *set of values* or the *range* of the function.

The term "function" has a variety of useful synonyms in different areas of mathematics, depending on the nature of the sets X and Y: *mapping, transformation, morphism, operator, functional*. The commonest is *mapping*, and we shall also use it frequently.

For a function (mapping) the following notations are standard:

$$f : X \to Y, \quad X \xrightarrow{f} Y.$$

When it is clear from the context what the domain and range of a function are, one also uses the notation $x \mapsto f(x)$ or $y = f(x)$, but more frequently a function in general is simply denoted by the single symbol f.

Two functions f_1 and f_2 are considered *identical* or *equal* if they have the same domain X and at each element $x \in X$ the values $f_1(x)$ and $f_2(x)$ are the same. In this case we write $f_1 = f_2$.

If $A \subset X$ and $f : X \to Y$ is a function, we denote by $f|A$ or $f|_A$ the function $\varphi : A \to Y$ that agrees with f on A. More precisely, $f|_A(x) := \varphi(x)$ if $x \in A$. The function $f|_A$ is called the *restriction* of f to A, and the function $f : X \to Y$ is called an *extension* or a *continuation* of φ to X.

We see that it is sometimes necessary to consider a function $\varphi : A \to Y$ defined on a subset A of some set X while the range $\varphi(A)$ of φ may also turn out be a subset of Y that is different from Y. In this connection, we sometimes use the term *domain of departure* of the function to denote any set X containing the domain of a function, and *domain of arrival* to denote any subset of Y containing its range.

Thus, defining a function (mapping) involves specifying a triple (X, Y, f), where

X is the set being mapped, or domain of the function;
Y is the set into which the mapping goes, or a domain of arrival of the function;
f is the rule according to which a definite element $y \in Y$ is assigned to each element $x \in X$.

The asymmetry between X and Y that appears here reflects the fact that the mapping goes from X to Y, and not the other direction.

Now let us consider some examples of functions.

Example 1 The formulas $l = 2\pi r$ and $V = \frac{4}{3}\pi r^3$ establish functional relationships between the circumference l of a circle and its radius r and between the volume V of a ball and its radius r. Each of these formulas provides a particular function $f : \mathbb{R}_+ \to \mathbb{R}_+$ defined on the set $\mathbb{R}_+$ of positive real numbers with values in the same set.

Example 2 Let X be the set of inertial coordinate systems and $c : X \to \mathbb{R}$ the function that assigns to each coordinate system $x \in X$ the value $c(x)$ of the speed of light *in vacuo* measured using those coordinates. The function $c : X \to \mathbb{R}$ is constant, that is, for any $x \in X$ it has the same value c. (This is a fundamental experimental fact.)

Example 3 The mapping $G : \mathbb{R}^2 \to \mathbb{R}^2$ (the direct product $\mathbb{R}^2 = \mathbb{R} \times \mathbb{R} = \mathbb{R}_t \times \mathbb{R}_x$ of the time axis $\mathbb{R}_t$ and the spatial axis $\mathbb{R}_x$) into itself defined by the formulas

$$x' = x - vt,$$
$$t' = t,$$

is the classical Galilean transformation for transition from one inertial coordinate system (x, t) to another system (x', t') that is in motion relative to the first at speed v.

The same purpose is served by the mapping $L : \mathbb{R}^2 \to \mathbb{R}^2$ defined by the relations

$$x' = \frac{x - vt}{\sqrt{1 - (\frac{v}{c})^2}},$$

$$t' = \frac{t - (\frac{v}{c^2})x}{\sqrt{1 - (\frac{v}{c})^2}}.$$

This is the well-known (one-dimensional) *Lorentz*[12] *transformation*, which plays a fundamental role in the special theory of relativity. The speed c is the speed of light.

Example 4 The *projection* $\mathrm{pr}_1 : X_1 \times X_2 \to X_1$ defined by the correspondence $X_1 \times X_2 \ni (x_1, x_2) \overset{\mathrm{pr}_1}{\longmapsto} x_1 \in X_1$ is obviously a function. The second projection $\mathrm{pr}_2 : X_1 \times X_2 \to X_2$ is defined similarly.

Example 5 Let $\mathcal{P}(M)$ be the set of subsets of the set M. To each set $A \in \mathcal{P}(M)$ we assign the set $C_M A \in \mathcal{P}(M)$, that is, the complement to A in M. We then obtain a mapping $C_M : \mathcal{P}(M) \to \mathcal{P}(M)$ of the set $\mathcal{P}(M)$ into itself.

Example 6 Let $E \subset M$. The real-valued function $\chi_E : M \to \mathbb{R}$ defined on the set M by the conditions $(\chi_E(x) = 1 \text{ if } x \in E) \wedge (\chi_E(x) = 0 \text{ if } x \in C_M E)$ is called the *characteristic function* of the set E.

Example 7 Let $M(X; Y)$ be the set of mappings of the set X into the set Y and x_0 a fixed element of X. To any function $f \in M(X; Y)$ we assign its value $f(x_0) \in Y$ at the element x_0. This relation defines a function $F : M(X; Y) \to Y$. In particular, if $Y = \mathbb{R}$, that is, Y is the set of real numbers, then to each function $f : X \to \mathbb{R}$ the function $F : M(X; \mathbb{R}) \to \mathbb{R}$ assigns the number $F(f) = f(x_0)$. Thus F is a function defined on functions. For convenience, such functions are called *functionals*.

Example 8 Let Γ be the set of curves lying on a surface (for example, the surface of the earth) and joining two given points of the surface. To each curve $\gamma \in \Gamma$ one can assign its length. We then obtain a function $F : \Gamma \to \mathbb{R}$ that often needs to be studied in order to find the shortest curve, or as it is called, the *geodesic* between the two given points on the surface.

Example 9 Consider the set $M(\mathbb{R}; \mathbb{R})$ of real-valued functions defined on the entire real line $\mathbb{R}$. After fixing a number $a \in \mathbb{R}$, we assign to each function $f \in M(\mathbb{R}; \mathbb{R})$ the function $f_a \in M(\mathbb{R}; \mathbb{R})$ connected with it by the relation $f_a(x) = f(x + a)$. The function $f_a(x)$ is usually called the *translate* or *shift* of the function f by a. The mapping $A : M(\mathbb{R}; \mathbb{R}) \to M(\mathbb{R}; \mathbb{R})$ that arises in this way is called the *translation* of *shift operator*. Thus the operator A is defined on functions and its values are also functions $f_a = A(f)$.

---

[12]H.A. Lorentz (1853–1928) – outstanding Dutch theoretical physicist. Poincaré called these transformations *Lorentz transformations* in honor of Lorentz, who stimulated the research of symmetries in Maxwell's equations. They were used by Einstein in 1905 in the formulation of his theory of special relativity.

This last example might seem artificial if not for the fact that we encounter real operators at every turn. Thus, any radio receiver is an operator $f \overset{F}{\longmapsto} \hat{f}$ that transforms electromagnetic signals f into acoustic signals $\hat{f}$; any of our sensory organs is an operator (transformer) with its own domain of definition and range of values.

Example 10 The position of a particle in space is determined by an ordered triple of numbers (x, y, z) called its spatial coordinates. The set of all such ordered triples can be thought of as the direct product $\mathbb{R} \times \mathbb{R} \times \mathbb{R} = \mathbb{R}^3$ of three real lines $\mathbb{R}$.

A particle in motion is located at some point of the space $\mathbb{R}^3$ having coordinates $(x(t), y(t), z(t))$ at each instant t of time. Thus the motion of a particle can be interpreted as a mapping $\gamma : \mathbb{R} \to \mathbb{R}^3$, where $\mathbb{R}$ is the time axis and $\mathbb{R}^3$ is three-dimensional space.

If a system consists of n particles, its configuration is defined by the position of each of the particles, that is, it is defined by an ordered set $(x_1, y_1, z_1; x_2, y_2, z_2; \dots; x_n, y_n, z_n)$ consisting of $3n$ numbers. The set of all such ordered sets is called the *configuration space* of the system of n particles. Consequently, the configuration space of a system of n particles can be interpreted as the direct product $\mathbb{R}^3 \times \mathbb{R}^3 \times \cdots \times \mathbb{R}^3 = \mathbb{R}^{3n}$ of n copies of $\mathbb{R}^3$.

To the motion of a system of n particles there corresponds a mapping $\gamma : \mathbb{R} \to \mathbb{R}^{3n}$ of the time axis into the configuration space of the system.

Example 11 The potential energy U of a mechanical system is connected with the mutual positions of the particles of the system, that is, it is determined by the configuration that the system has. Let Q be the set of possible configurations of a system. This is a certain subset of the configuration space of the system. To each position $q \in Q$ there corresponds a certain value $U(q)$ of the potential energy of the system. Thus the potential energy is a function $U : Q \to \mathbb{R}$ defined on a subset Q of the configuration space with values in the domain $\mathbb{R}$ of real numbers.

Example 12 The kinetic energy K of a system of n material particles depends on their velocities. The total mechanical energy of the system E, defined as $E = K + U$, that is, the sum of the kinetic and potential energies, thus depends on both the configuration q of the system and the set of velocities v of its particles. Like the configuration q of the particles in space, the set of velocities v, which consists of n three-dimensional vectors, can be defined as an ordered set of $3n$ numbers. The ordered pairs (q, v) corresponding to the states of the system form a subset Φ in the direct product $\mathbb{R}^{3n} \times \mathbb{R}^{3n} = \mathbb{R}^{6n}$, called the *phase space* of the system of n particles (to be distinguished from the configuration space $\mathbb{R}^{3n}$).

The total mechanical energy of the system is therefore a function $E : \Phi \to \mathbb{R}$ defined on the subset Φ of the phase space $\mathbb{R}^{6n}$ and assuming values in the domain $\mathbb{R}$ of real numbers.

In particular, if the system is isolated, that is, no external forces are acting on it, then by the law of conservation of energy, at each point of the set Φ of states of the system the function E will have the same value $E_0 \in \mathbb{R}$.

Fig. 1.6

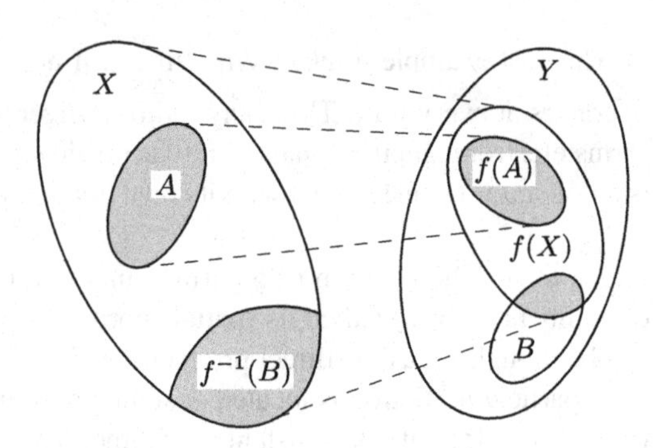

1.3.2 Elementary Classification of Mappings

When a function $f : X \to Y$ is called a mapping, the value $f(x) \in Y$ that it assumes at the element $x \in Y$ is usually called the *image* of x.

The *image* of a set $A \subset X$ under the mapping $f : X \to Y$ is defined as the set

$$f(A) := \big\{y \in Y \mid \exists x \ \big((x \in A) \wedge \big(y = f(x)\big)\big)\big\}$$

consisting of the elements of Y that are images of elements of A.

The set

$$f^{-1}(B) := \big\{x \in X \mid f(x) \in B\big\}$$

consisting of the elements of X whose images belong to B is called the *pre-image* (or *complete pre-image*) of the set $B \subset Y$ (Fig. 1.6).

A mapping $f : X \to Y$ is said to be

surjective (a mapping of X *onto* Y) if $f(X) = Y$;
injective (or an *imbedding* or *injection*) if for any elements x_1, x_2 of X

$$\big(f(x_1) = f(x_2)\big) \Rightarrow (x_1 = x_2),$$

that is, distinct elements have distinct images;
bijective (or a *one-to-one correspondence*) if it is both surjective and injective.

If the mapping $f : X \to Y$ is bijective, that is, it is a one-to-one correspondence between the elements of the sets X and Y, there naturally arises a mapping

$$f^{-1} : Y \to X,$$

defined as follows: if $f(x) = y$, then $f^{-1}(y) = x$ that is, to each element $y \in Y$ one assigns the element $x \in X$ whose image under the mapping f is y. By the surjectivity of f there exists such an element, and by the injectivity of f, it is unique. Hence the mapping f^{-1} is well-defined. This mapping is called the *inverse* of the original mapping f.

Fig. 1.7

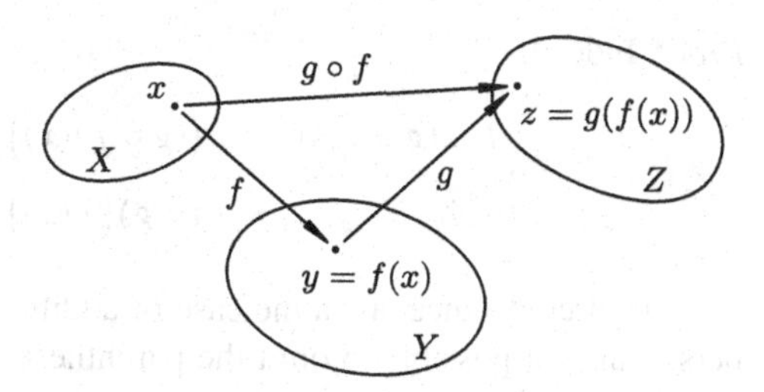

It is clear from the construction of the inverse mapping that $f^{-1} : Y \to X$ is itself bijective and that its inverse $(f^{-1})^{-1} : X \to Y$ is the same as the original mapping $f : X \to Y$.

Thus the property of two mappings of being inverses is reciprocal: if f^{-1} is inverse for f, then f is inverse for f^{-1}.

We remark that the symbol $f^{-1}(B)$ for the pre-image of a set $B \subset Y$ involves the symbol f^{-1} for the inverse function; but it should be kept in mind that the pre-image of a set is defined for any mapping $f : X \to Y$, even if it is not bijective and hence has no inverse.

1.3.3 Composition of Functions and Mutually Inverse Mappings

The operation of composition of functions is on the one hand a rich source of new functions and on the other hand a way of resolving complex functions into simpler ones.

If the mappings $f : X \to Y$ and $g : Y \to Z$ are such that one of them (in our case g) is defined on the range of the other (f), one can construct a new mapping

$$g \circ f : X \to Z,$$

whose values on elements of the set X are defined by the formula

$$(g \circ f)(x) := g\big(f(x)\big).$$

The compound mapping $g \circ f$ so constructed is called the *composition* of the mapping f and the mapping g (in that order!).

Figure 1.7 illustrates the construction of the composition of the mappings f and g.

You have already encountered the composition of mappings many times, both in geometry, when studying the composition of rigid motions of the plane or space, and in algebra in the study of "complicated" functions obtained by composing the simplest elementary functions.

The operation of composition sometimes has to be carried out several times in succession, and in this connection it is useful to note that it is associative, that is,

$$h \circ (g \circ f) = (h \circ g) \circ f.$$

Proof Indeed,

$$h \circ (g \circ f)(x) = h\big((g \circ f)(x)\big) = h\big(g\big(f(x)\big)\big) =$$
$$= (h \circ g)\big(f(x)\big) = \big((h \circ g) \circ f\big)(x). \qquad \square$$

This circumstance, as in the case of addition and multiplication of several numbers, makes it possible to omit the parentheses that prescribe the order of the pairings.

If all the terms of a composition $f_n \circ \dots \circ f_1$ are equal to the same function f, we abbreviate it to f^n.

It is well known, for example, that the square root of a positive number a can be computed by successive approximations using the formula

$$x_{n+1} = \frac{1}{2}\left(x_n + \frac{a}{x_n}\right),$$

starting from any initial approximation $x_0 > 0$. This none other than the successive computation of $f^n(x_0)$, where $f(x) = \frac{1}{2}(x + \frac{a}{x})$. Such a procedure, in which the value of the function computed at the each step becomes its argument at the next step, is called a *recursive* procedure. Recursive procedures are widely used in mathematics.

We further note that even when both compositions $g \circ f$ and $f \circ g$ are defined, in general

$$g \circ f \neq f \circ g.$$

Indeed, let us take for example the two-element set $\{a, b\}$ and the mappings $f : \{a, b\} \to a$ and $g : \{a, b\} \to b$. Then it is obvious that $g \circ f : \{a, b\} \to b$ while $f \circ g : \{a, b\} \to a$.

The mapping $f : X \to X$ that assigns to each element of X the element itself, that is $x \overset{f}{\longmapsto} y$, will be denoted e_X and called the *identity mapping* on X.

Lemma

$$(g \circ f = e_X) \Rightarrow (g \textit{ is surjective}) \wedge (f \textit{ is injective}).$$

Proof Indeed, if $f : X \to Y$, $g : Y \to X$, and $g \circ f = e_X : X \to X$, then

$$X = e_X(X) = (g \circ f)(X) = g\big(f(X)\big) \subset g(Y)$$

and hence g is surjective.

Further, if $x_1 \in X$ and $x_2 \in X$, then

$$(x_1 \neq x_2) \Rightarrow \big(e_X(x_1) \neq e_X(x_2)\big) \Rightarrow \big((g \circ f)(x_1) \neq (g \circ f)(x_2)\big) \Rightarrow$$
$$\Rightarrow \big(g\big(f(x_1)\big)\big) \neq g\big(f(x_2)\big) \Rightarrow \big(f(x_1) \neq f(x_2)\big),$$

and therefore f is injective. $\square$

Using the operation of composition of mappings one can describe mutually inverse mappings.

Proposition *The mappings* $f : X \to Y$ *and* $g : Y \to X$ *are bijective and mutually inverse to each other if and only if* $g \circ f = e_X$ *and* $f \circ g = e_Y$.

Proof By the lemma the simultaneous fulfillment of the conditions $g \circ f = e_X$ and $f \circ g = e_Y$ guarantees the surjectivity and injectivity, that is, the bijectivity, of both mappings.

These same conditions show that $y = f(x)$ if and only if $x = g(y)$. □

In the preceding discussion we started with an explicit construction of the inverse mapping. It follows from the proposition just proved that we could have given a less intuitive, yet more symmetric definition of mutually inverse mappings as those mappings that satisfy the two conditions $g \circ f = e_X$ and $f \circ g = e_Y$. (In this connection, see Exercise 6 at the end of this section.)

1.3.4 Functions as Relations. The Graph of a Function

In conclusion we return once again to the concept of a function. We note that it has undergone a lengthy and rather complicated evolution.

The term *function* first appeared in the years from 1673 to 1692 in works of G. Leibniz (in a somewhat narrower sense, to be sure). By the year 1698 the term had become established in a sense close to the modern one through the correspondence between Leibniz and Johann Bernoulli.[13] (The letter of Bernoulli usually cited in this regard dates to that same year.)

Many great mathematicians have participated in the formation of the modern concept of functional dependence.

A description of a function that is nearly identical to the one given at the beginning of this section can be found as early as the work of Euler (mid-eighteenth century) who also introduced the notation $f(x)$. By the early nineteenth century it had appeared in the textbooks of S. Lacroix.[14] A vigorous advocate of this concept of a function was N.I. Lobachevskii,[15] who noted that "a comprehensive view of

---

[13]Johann Bernoulli (1667–1748) – one of the early representatives of the distinguished Bernoulli family of Swiss scholars; he studied analysis, geometry and mechanics. He was one of the founders of the calculus of variations. He gave the first systematic exposition of the differential and integral calculus.

[14]S.F. Lacroix (1765–1843) – French mathematician and educator (professor at the École Normale and the École Polytechnique, and member of the Paris Academy of Sciences).

[15]N.I. Lobachevskii (1792–1856) – great Russian scholar, to whom belongs the credit – shared with the great German scientist C.F. Gauss (1777–1855) and the outstanding Hungarian mathematician J. Bólyai (1802–1860) – for having discovered the non-Euclidean geometry that bears his name.

theory admits only dependence relationships in which the numbers connected with each other are understood as if they were *given as a single unit*."[16] It is this idea of precise definition of the concept of a function that we are about to explain.

The description of the concept of a function given at the beginning of this section is quite dynamic and reflects the essence of the matter. However, by modern canons of rigor it cannot be called a definition, since it uses the concept of a correspondence, which is equivalent to the concept of a function. For the reader's information we shall show here how the definition of a function can be given in the language of set theory. (It is interesting that the concept of a relation, to which we are now turning, preceded the concept of a function, even for Leibniz.)

a. Relations

Definition 1 A *relation* $\mathcal{R}$ is any set of ordered pairs (x, y).

The set X of first elements of the ordered pairs that constitute $\mathcal{R}$ is called the *domain of definition* of $\mathcal{R}$, and the set Y of second elements of these pairs the *range of values* of $\mathcal{R}$.

Thus, a relation can be interpreted as a subset $\mathcal{R}$ of the direct product $X \times Y$. If $X \subset X'$ and $Y \subset Y'$, then of course $\mathcal{R} \subset X \times Y \subset X' \times Y'$, so that a given relation can be defined as a subset of different sets.

Any set containing the domain of definition of a relation is called a *domain of departure* for that relation. A set containing the region of values is called a *domain of arrival* of the relation.

Instead of writing $(x, y) \in \mathcal{R}$, we often write $x\mathcal{R}y$ and say that x *is connected with* y *by the relation* $\mathcal{R}$.

If $\mathcal{R} \subset X^2$, we say that the relation $\mathcal{R}$ is *defined on* X.

Let us consider some examples.

Example 13 The diagonal

$$\Delta = \{(a, b) \in X^2 \mid a = b\}$$

is a subset of X^2 defining the relation of equality between elements of X. Indeed, $a\Delta b$ means that $(a, b) \in \Delta$, that is, $a = b$.

Example 14 Let X be the set of lines in a plane.

Two lines $a \in X$ and $b \in X$ will be considered to be in the relation $\mathcal{R}$, and we shall write $a\mathcal{R}b$, if b is parallel to a. It is clear that this condition distinguishes a set $\mathcal{R}$ of pairs (a, b) in X^2 such that $a\mathcal{R}b$. It is known from geometry that the relation of parallelism between lines has the following properties:

$a\mathcal{R}a$ (reflexivity);

[16]Lobachevskii, N.I. *Complete Works*, Vol. 5, Moscow–Leningrad: Gostekhizdat, 1951, p. 44 (Russian).

$a\mathcal{R}b \Rightarrow b\mathcal{R}a$ (symmetry);
$(a\mathcal{R}b) \wedge (b\mathcal{R}c) \Rightarrow a\mathcal{R}c$ (transitivity).

A relation $\mathcal{R}$ having the three properties just listed, that is, reflexivity,[17] symmetry, and transitivity, is usually called an *equivalence relation.* An equivalence relation is denoted by the special symbol $\sim$, which in this case replaces the letter $\mathcal{R}$. Thus, in the case of an equivalence relation we shall write $a \sim b$ instead of $a\mathcal{R}b$ and say that a is *equivalent* to b.

Example 15 Let M be a set and $X = \mathcal{P}(M)$ the set of its subsets. For two arbitrary elements a and b of $X = \mathcal{P}(M)$, that is, for two subsets a and b of M, one of the following three possibilities always holds: a is contained in b; b is contained in a; a is not a subset of b and b is not a subset of a.

As an example of a relation $\mathcal{R}$ on X^2, consider the relation of inclusion for subsets of M, that is, make the definition

$$a\mathcal{R}b := (a \subset b).$$

This relation obviously has the following properties:

$a\mathcal{R}a$ (reflexivity);
$(a\mathcal{R}b) \wedge (b\mathcal{R}c) \Rightarrow a\mathcal{R}c$ (transitivity);
$(a\mathcal{R}b) \wedge (b\mathcal{R}a) \Rightarrow a\Delta b$, that is, $a = b$ (antisymmetry).

A relation between pairs of elements of a set X having these three properties is usually called a *partial ordering* on X. For a partial ordering relation on X, we often write $a \preceq b$ and say that *b follows a*.

If the condition

$$\forall a\, \forall b\, \big((a\mathcal{R}b) \vee (b\mathcal{R}a)\big)$$

holds in addition to the last two properties defining a partial ordering relation, that is, any two elements of X are comparable, the relation $\mathcal{R}$ is called an *ordering*, and the set X with the ordering defined on it is said to be *linearly ordered.*

The origin of this term comes from the intuitive image of the real line $\mathbb{R}$ on which a relation $a \leq b$ holds between any pair of real numbers.

b. Functions and Their Graphs

A relation $\mathcal{R}$ is said to be *functional* if

$$(x\mathcal{R}y_1) \wedge (x\mathcal{R}y_2) \Rightarrow (y_1 = y_2).$$

---

[17]For the sake of completeness it is useful to note that a relation $\mathcal{R}$ is *reflexive* if its domain of definition and its range of values are the same and the relation $a\mathcal{R}a$ holds for any element a in the domain of $\mathcal{R}$.

A functional relation is called a *function*.

In particular, if X and Y are two sets, not necessarily distinct, a relation $\mathcal{R} \subset X \times Y$ between elements x of X and y of Y is a *functional* relation on X if for every $x \in X$ there exists a unique element $y \in Y$ in the given relation to x, that is, such that $x\mathcal{R}y$ holds.

Such a functional relation $\mathcal{R} \subset X \times Y$ is a *mapping from X into Y*, or a *function from X into Y*.

We shall usually denote functions by the letter f. If f is a function, we shall write $y = f(x)$ or $x \overset{f}{\longmapsto} y$, as before, rather than xfy, calling $y = f(x)$ the *value* of f at x or the *image* of x under f.

As we now see, assigning an element $y \in Y$ "corresponding" to $x \in X$ in accordance with the "rule" f, as was discussed in the original description of the concept of a function, amounts to exhibiting for each $x \in X$ the unique $y \in Y$ such that xfy, that is, $(x, y) \in f \subset X \times Y$.

The *graph* of a function $f : X \to Y$, as understood in the original description, is the subset Γ of the direct product $X \times Y$ whose elements have the form $(x, f(x))$. Thus

$$\Gamma := \{(x, y) \in X \times Y \mid y = f(x)\}.$$

In the new description of the concept of a function, in which we define it as a subset $f \subset X \times Y$, of course, there is no longer any difference between a function and its graph.

We have exhibited the theoretical possibility of giving a formal set-theoretic definition of a function, which reduces essentially to identifying a function and its graph. However, we do not intend to confine ourselves to that way of defining a function. At times it is convenient to define a functional relation analytically, at other times by giving a table of values, and at still other times by giving a verbal description of a process (algorithm) making it possible to find the element $y \in Y$ corresponding to a given $x \in X$. With each method of presenting a function it is meaningful to ask how the function could have been defined using its graph. This problem can be stated as the problem of constructing the graph of the function. Defining numerical-valued functions by a good graphical representation is often useful because it makes the basic qualitative properties of the functional relation visualizable. One can also use graphs (nomograms) for computations; but, as a rule, only in cases where high precision is not required. For precise computations we do use the table definition of a function, but more often we use an algorithmic definition that can be implemented on a computer.

1.3.5 Exercises

1. The *composition* $\mathcal{R}_2 \circ \mathcal{R}_1$ of the relations $\mathcal{R}_1$ and $\mathcal{R}_2$ is defined as follows:

$$\mathcal{R}_2 \circ \mathcal{R}_1 := \{(x, z) \mid \exists y\ (x\mathcal{R}_1 y \wedge y\mathcal{R}_2 z)\}.$$

In particular, if $\mathcal{R}_1 \subset X \times Y$ and $\mathcal{R}_2 \subset Y \times Z$, then $\mathcal{R} = \mathcal{R}_2 \circ \mathcal{R}_1 \subset X \times Z$, and

$$x\mathcal{R}z := \exists y \big((y \in Y) \wedge (x\mathcal{R}_1 y) \wedge (y\mathcal{R}_2 z)\big).$$

a) Let Δ_X be the diagonal of X^2 and Δ_Y the diagonal of Y^2. Show that if the relations $\mathcal{R}_1 \subset X \times Y$ and $\mathcal{R}_2 \subset Y \times X$ are such that $(\mathcal{R}_2 \circ \mathcal{R}_1 = \Delta_X) \wedge (\mathcal{R}_1 \circ \mathcal{R}_2 = \Delta_Y)$, then both relations are functional and define mutually inverse mappings of X and Y.

b) Let $\mathcal{R} \subset X^2$. Show that the condition of transitivity of the relation $\mathcal{R}$ is equivalent to the condition $\mathcal{R} \circ \mathcal{R} \subset \mathcal{R}$.

c) The relation $\mathcal{R}' \subset Y \times X$ is called the *transpose* of the relation $\mathcal{R} \subset X \times Y$ if $(y\mathcal{R}'x) \Leftrightarrow (x\mathcal{R}y)$.

Show that a relation $\mathcal{R} \subset X^2$ is antisymmetric if and only if $\mathcal{R} \cap \mathcal{R}' \subset \Delta_X$.

d) Verify that any two elements of X are connected (in some order) by the relation $\mathcal{R} \subset X^2$ if and only if $\mathcal{R} \cup \mathcal{R}' = X^2$.

2. Let $f : X \to Y$ be a mapping. The pre-image $f^{-1}(y) \subset X$ of the element $y \in Y$ is called the *fiber* over y.

a) Find the fibers for the following mappings:

$$\mathrm{pr}_1 : X_1 \times X_2 \to X_1, \qquad \mathrm{pr}_2 : X_1 \times X_2 \to X_2.$$

b) An element $x_1 \in X$ will be considered to be connected with an element $x_2 \in X$ by the relation $\mathcal{R} \subset X^2$, and we shall write $x_1\mathcal{R}x_2$ if $f(x_1) = f(x_2)$, that is, x_1 and x_2 both lie in the same fiber.

Verify that $\mathcal{R}$ is an equivalence relation.

c) Show that the fibers of a mapping $f : X \to Y$ do not intersect one another and that the union of all the fibers is the whole set X.

d) Verify that any equivalence relation between elements of a set makes it possible to represent the set as a union of mutually disjoint equivalence classes of elements.

3. Let $f : X \to Y$ be a mapping from X into Y. Show that if A and B are subsets of X, then

a) $(A \subset B) \Rightarrow (f(A) \subset f(B)) \not\Rightarrow (A \subset B)$.
b) $(A \neq \varnothing) \Rightarrow (f(A) \neq \varnothing)$,
c) $f(A \cap B) \subset f(A) \cap f(B)$,
d) $f(A \cup B) = f(A) \cup f(B)$;

if A' and B' are subsets of Y, then

e) $(A' \subset B') \Rightarrow (f^{-1}(A') \subset f^{-1}(B'))$,
f) $f^{-1}(A' \cap B') = f^{-1}(A') \cap f^{-1}(B')$,
g) $f^{-1}(A' \cup B') = f^{-1}(A') \cup f^{-1}(B')$;

if $Y \supset A' \supset B'$, then

h) $f^{-1}(A' \backslash B') = f^{-1}(A') \backslash f^{-1}(B')$,

i) $f^{-1}(C_Y A') = C_X f^{-1}(A')$;

and for any $A \subset X$ and $B' \subset Y$

j) $f^{-1}(f(A)) \supset A$,
k) $f(f^{-1}(B')) \subset B'$.

4. Show that the mapping $f: X \to Y$ is

a) surjective if and only if $f(f^{-1}(B')) = B'$ for every set $B' \subset Y$;
b) bijective if and only if

$$\big(f^{-1}(f(A)) = A\big) \wedge \big(f\big(f^{-1}(B')\big) = B'\big)$$

for every set $A \subset X$ and every set $B' \subset Y$.

5. Verify that the following statements about a mapping $f: X \to Y$ are equivalent:

a) f is injective;
b) $f^{-1}(f(A)) = A$ for every $A \subset X$;
c) $f(A \cap B) = f(A) \cap f(B)$ for any two subsets A and B of X;
d) $f(A) \cap f(B) = \varnothing \Leftrightarrow A \cap B = \varnothing$;
e) $f(A \backslash B) = f(A) \backslash f(B)$ whenever $X \supset A \supset B$.

6. a) If the mappings $f: X \to Y$ and $g: Y \to X$ are such that $g \circ f = e_X$, where e_X is the identity mapping on X, then g is called a *left inverse* of f and f a *right inverse* of g. Show that, in contrast to the uniqueness of the inverse mapping, there may exist many one-sided inverse mappings.

Consider, for example, the mappings $f: X \to Y$ and $g: Y \to X$, where X is a one-element set and Y a two-element set, or the mappings of sequences given by

$$(x_1, \dots, x_n, \dots) \overset{f_a}{\longmapsto} (a, x_1, \dots, x_n, \dots),$$

$$(y_2, \dots, y_n, \dots) \overset{g}{\longleftarrow\!\!\shortmid} (y_1, y_2, \dots, y_n, \dots).$$

b) Let $f: X \to Y$ and $g: Y \to Z$ be bijective mappings. Show that the mapping $g \circ f: X \to Z$ is bijective and that $(g \circ f)^{-1} = f^{-1} \circ g^{-1}$.

c) Show that the equality

$$(g \circ f)^{-1}(C) = f^{-1}\big(g^{-1}(C)\big)$$

holds for any mappings $f: X \to Y$ and $g: Y \to Z$ and any set $C \subset Z$.

d) Verify that the mapping $F: X \times Y \to Y \times X$ defined by the correspondence $(x, y) \mapsto (y, x)$ is bijective. Describe the connection between the graphs of mutually inverse mappings $f: X \to Y$ and $f^{-1}: Y \to X$.

7. a) Show that for any mapping $f: X \to Y$ the mapping $F: X \to X \times Y$ defined by the correspondence $x \overset{F}{\longmapsto} (x, f(x))$ is injective.

b) Suppose a particle is moving at uniform speed on a circle Y; let X be the time axis and $x \overset{f}{\longmapsto} y$ the correspondence between the time $x \in X$ and the position

$y = f(x) \in Y$ of the particle. Describe the graph of the function $f : X \to Y$ in $X \times Y$.

8. a) For each of the examples 1–12 considered in Sect. 1.3 determine whether the mapping defined in the example is surjective, injective, or bijective or whether it belongs to none of these classes.

b) Ohm's law $I = V/R$ connects the current I in a conductor with the potential difference V at the ends of the conductor and the resistance R of the conductor. Give sets X and Y for which some mapping $O : X \to Y$ corresponds to Ohm's law. What set is the relation corresponding to Ohm's law a subset of?

c) Find the mappings G^{-1} and L^{-1} inverse to the Galilean and Lorentz transformations.

9. a) A set $S \subset X$ is *stable* with respect to a mapping $f : X \to X$ if $f(S) \subset S$. Describe the sets that are stable with respect to a shift of the plane by a given vector lying in the plane.

b) A set $I \subset X$ is *invariant* with respect to a mapping $f : X \to X$ if $f(I) = I$. Describe the sets that are invariant with respect to rotation of the plane about a fixed point.

c) A point $p \in X$ is a *fixed point* of a mapping $f : X \to X$ if $f(p) = p$. Verify that any composition of a shift, a rotation, and a similarity transformation of the plane has a fixed point, provided the coefficient of the similarity transformation is less than 1.

d) Regarding the Galilean and Lorentz transformations as mappings of the plane into itself for which the point with coordinates (x, t) maps to the point with coordinates (x', t'), find the invariant sets of these transformations.

10. Consider the steady flow of a fluid (that is, the velocity at each point of the flow does not change over time). In time t a particle at point x of the flow will move to some new point $f_t(x)$ of space. The mapping $x \mapsto f_t(x)$ that arises thereby on the points of space occupied by the flow depends on time and is called the *mapping after time t*. Show that $f_{t_2} \circ f_{t_1} = f_{t_1} \circ f_{t_2} = f_{t_1+t_2}$ and $f_t \circ f_{-t} = e_X$.

1.4 Supplementary Material

1.4.1 The Cardinality of a Set (Cardinal Numbers)

The set X is said to be *equipollent* to the set Y if there exists a bijective mapping of X onto Y, that is, a point $y \in Y$ is assigned to each $x \in X$, the elements of Y assigned to different elements of X are different, and every point of Y is assigned to some point of X.

Speaking fancifully, each element $x \in X$ has a seat all to itself in Y, and there are no vacant seats $y \in Y$.

It is clear that the relation $X\mathcal{R}Y$ thereby introduced is an *equivalence relation*. For that reason we shall write $X \sim Y$ instead of $X\mathcal{R}Y$, in accordance with our earlier convention.

The relation of equipollence partitions the collection of all sets into classes of mutually equivalent sets. The sets of an equivalence class have the same number of elements (they are equipollent), and sets from different equivalence classes do not.

The class to which a set X belongs is called the *cardinality* of X, and also the *cardinal* or *cardinal number* of X. It is denoted $\operatorname{card} X$. If $X \sim Y$, we write $\operatorname{card} X = \operatorname{card} Y$.

The idea behind this construction is that it makes possible a comparison of the numbers of elements in sets without resorting to an intermediate count, that is, without measuring the number by comparing it with the natural numbers $\mathbb{N} = \{1, 2, 3, \ldots\}$. Doing the latter, as we shall soon see, is sometimes not even theoretically possible.

The cardinal number of a set X is said to be *not larger* than the cardinal number of a set Y, and we write $\operatorname{card} X \leq \operatorname{card} Y$, if X is equipollent to some subset of Y.

Thus,

$$(\operatorname{card} X \leq \operatorname{card} Y) := \exists Z \subset Y \ (\operatorname{card} X = \operatorname{card} Z).$$

If $X \subset Y$, it is clear that $\operatorname{card} X \leq \operatorname{card} Y$. It turns out, however, that the relation $X \subset Y$ does not exclude the inequality $\operatorname{card} Y \leq \operatorname{card} X$, even when X is a proper subset of Y.

For example, the correspondence $x \mapsto \frac{x}{1-|x|}$ is a bijective mapping of the interval $-1 < x < 1$ of the real axis $\mathbb{R}$ onto the entire axis.

The possibility of being equipollent to a proper subset of itself is a characteristic of infinite sets that Dedekind[18] even suggested taking as the definition of an infinite set. Thus a set is called *finite* (in the sense of Dedekind) if it is not equipollent to any proper subset of itself; otherwise, it is called *infinite*.

Just as the relation of inequality orders the real numbers on a line, the inequality just introduced orders the cardinal numbers of sets. To be specific, one can prove that the relation just constructed has the following properties:

1^0 $(\operatorname{card} X \leq \operatorname{card} Y) \wedge (\operatorname{card} Y \leq \operatorname{card} Z) \Rightarrow (\operatorname{card} X \leq \operatorname{card} Z)$ (obvious).

2^0 $(\operatorname{card} X \leq \operatorname{card} Y) \wedge (\operatorname{card} Y \leq \operatorname{card} X) \Rightarrow (\operatorname{card} X = \operatorname{card} Y)$ (the Schröder–Bernstein theorem[19]).

3^0 $\forall X \, \forall Y \ (\operatorname{card} X \leq \operatorname{card} Y) \vee (\operatorname{card} Y \leq \operatorname{card} X)$ (Cantor's theorem).

Thus the class of cardinal numbers is linearly ordered.

---

[18] R. Dedekind (1831–1916) – German algebraist who took an active part in the development of the theory of a real number. He was the first to propose the axiomatization of the set of natural numbers usually called the Peano axiom system after G. Peano (1858–1932), the Italian mathematician who formulated it somewhat later.

[19] F. Bernstein (1878–1956) – German mathematician, a student of G. Cantor. E. Schröder (1841–1902) – German mathematician.

We say that the cardinality of X is *less* than the cardinality of Y and write $\operatorname{card} X < \operatorname{card} Y$, if $\operatorname{card} X \leq \operatorname{card} Y$ but $\operatorname{card} X \neq \operatorname{card} Y$. Thus $(\operatorname{card} X < \operatorname{card} Y) := (\operatorname{card} X \leq \operatorname{card} Y) \wedge (\operatorname{card} X \neq \operatorname{card} Y)$.

As before, let $\varnothing$ be the empty set and $\mathcal{P}(X)$ the set of all subsets of the set X. Cantor made the following discovery:

Theorem $\operatorname{card} X < \operatorname{card} \mathcal{P}(X)$.

Proof The assertion is obvious for the empty set, so that from now on we shall assume $X \neq \varnothing$.

Since $\mathcal{P}(X)$ contains all one-element subsets of X, $\operatorname{card} X \leq \operatorname{card} \mathcal{P}(X)$.

To prove the theorem it now suffices to show that $\operatorname{card} X \neq \operatorname{card} \mathcal{P}(X)$ if $X \neq \varnothing$.

Suppose, contrary to the assertion, that there exists a bijective mapping $f : X \to \mathcal{P}(X)$. Consider the set $A = \{x \in X : x \notin f(x)\}$ consisting of the elements $x \in X$ that do not belong to the set $f(x) \in \mathcal{P}(X)$ assigned to them by the bijection. Since $A \in \mathcal{P}(X)$, there exists $a \in X$ such that $f(a) = A$. For the element a the relation $a \in A$ is impossible by the definition of A, and the relation $a \notin A$ is impossible, also by the definition of A. We have thus reached a contradiction with the law of excluded middle. □

This theorem shows in particular that if infinite sets exist, then even "infinities" are not all the same.

1.4.2 Axioms for Set Theory

The purpose of the present subsection is to give the interested reader a picture of an axiom system that describes the properties of the mathematical object called a *set* and to illustrate the simplest consequences of those axioms.

1^0. (Axiom of extensionality) *Sets A and B are equal if and only if they have the same elements.*

This means that we ignore all properties of the object known as a "set" except the property of having elements. In practice it means that if we wish to establish that $A = B$, we must verify that $\forall x\, ((x \in A) \Leftrightarrow (x \in B))$.

2^0. (Axiom of separation) *To any set A and any property P there corresponds a set B whose elements are those elements of A, and only those, having property P.*

More briefly, it is asserted that if A is a set, then $B = \{x \in A \mid P(x)\}$ is also a set.

This axiom is used very frequently in mathematical constructions, when we select from a set the subset consisting of the elements having some property.

For example, it follows from the axiom of separation that there exists an empty subset $\varnothing_X = \{x \in X \mid x \neq x\}$ in any set X. By virtue of the axiom of extensionality we conclude that $\varnothing_X = \varnothing_Y$ for all sets X and Y, that is, the empty set is unique. We denote this set by $\varnothing$.

It also follows from the axiom of separation that if A and B are sets, then $A \setminus B = \{x \in A \mid x \notin B\}$ is also a set. In particular, if M is a set and A a subset of M, then $C_M A$ is also a set.

3^0. (Union axiom) *For any set M whose elements are sets there exists a set $\bigcup M$, called the* union *of M and consisting of those elements and only those that belong to some element of M.*

If we use the phrase "family of sets" instead of "a set whose elements are sets" the axiom of union assumes a more familiar sound: there exists a set consisting of the elements of the sets in the family. Thus, a union of sets is a set, and $x \in \bigcup M \Leftrightarrow \exists X\, ((X \in M) \wedge (x \in X))$.

When we take account of the axiom of separation, the union axiom makes it possible to define the *intersection of the set M* (or family of sets) as the set

$$\bigcap M := \Big\{x \in \bigcup M \mid \forall X\, \big((X \in M) \Rightarrow (x \in X)\big)\Big\}.$$

4^0. (Pairing axiom) *For any sets X and Y there exists a set Z such that X and Y are its only elements.*

The set Z is denoted $\{X, Y\}$ and is called the *unordered pair* of sets X and Y. The set Z consists of one element if $X = Y$.

As we have already pointed out, the *ordered* pair (X, Y) differs from the unordered pair by the presence of some property possessed by one of the sets in the pair. For example, $(X, Y) := \{\{X, X\}, \{X, Y\}\}$.

Thus, the unordered pair makes it possible to introduce the ordered pair, and the ordered pair makes it possible to introduce the direct product of sets by using the axiom of separation and the following important axiom.

5^0. (Power set axiom) *For any set X there exists a set $\mathcal{P}(X)$ having each subset of X as an element, and having no other elements.*

In short, there exists a set consisting of all the subsets of a given set.

We can now verify that the ordered pairs (x, y), where $x \in X$ and $y \in Y$, really do form a set, namely

$$X \times Y := \big\{p \in \mathcal{P}\big(\mathcal{P}(X) \cup \mathcal{P}(Y)\big) \mid \big(p = (x, y)\big) \wedge (x \in X) \wedge (y \in Y)\big\}.$$

Axioms 1^0–5^0 limit the possibility of forming new sets. Thus, by Cantor's theorem (which asserts that $\operatorname{card} X < \operatorname{card} \mathcal{P}(X)$) there is an element in the set $\mathcal{P}(X)$ that does not belong to X. Therefore the "set of all sets" does not exist. And it was precisely on this "set" that Russell's paradox was based.

In order to state the next axiom we introduce the concept of the *successor* X^+ of the set X. By definition $X^+ = X \cup \{X\}$. More briefly, the one-element set $\{X\}$ is adjoined to X.

Further, a set is called *inductive* if the empty set is one of its elements and the successor of each of its elements also belongs to it.

6^0. (Axiom of infinity) *There exist inductive sets.*

When we take Axioms 1^0–4^0 into account, the axiom of infinity makes it possible to construct a standard model of the set $\mathbb{N}_0$ of natural numbers (in the sense of von Neumann),[20] by defining $\mathbb{N}_0$ as the intersection of all inductive sets, that is, the smallest inductive set. The elements of $\mathbb{N}_0$ are

$$\varnothing, \quad \varnothing^+ = \varnothing \cup \{\varnothing\} = \{\varnothing\}, \qquad \{\varnothing\}^+ = \{\varnothing\} \cup \{\{\varnothing\}\}, \quad \ldots,$$

which are a model for what we denote by the symbols $0, 1, 2, \ldots$ and call the natural numbers.

7^0. (Axiom of replacement) *Let $\mathcal{F}(x, y)$ be a statement (more precisely, a formula) such that for every x_0 in the set X there exists a unique object y_0 such that $\mathcal{F}(x_0, y_0)$ is true. Then the objects y for which there exists an element $x \in X$ such that $\mathcal{F}(x, y)$ is true form a set.*

We shall make no use of this axiom in our construction of analysis.

Axioms 1^0–7^0 constitute the axiom system known as the Zermelo–Fraenkel axioms.[21]

To this system another axiom is usually added, one that is independent of Axioms 1^0–7^0 and used very frequently in analysis.

8^0. (Axiom of choice) *For any family of nonempty and mutually nonintersecting sets there exists a set C such that for each set X in the family $X \cap C$ consists of exactly one element.*

In other words, from each set of the family one can choose exactly one representative in such a way that the representatives chosen form a set C.

The axiom of choice, known as Zermelo's axiom in mathematics, has been the subject of heated debates among specialists.

1.4.3 Remarks on the Structure of Mathematical Propositions and Their Expression in the Language of Set Theory

In the language of set theory there are two basic, or *atomic* types of mathematical statements: the assertion $x \in A$, that an object x is an element of a set A, and the

[20] J. von Neumann (1903–1957) – American mathematician who worked in functional analysis, the mathematical foundations of quantum mechanics, topological groups, game theory, and mathematical logic. He was one of the leaders in the creation of the first computers.

[21] E. Zermelo (1871–1953) – German mathematician. A. Fraenkel (1891–1965) – German (later, Israeli) mathematician.

assertion $A = B$, that the sets A and B are identical. (However, when the axiom of extensionality is taken into account, the second statement is a combination of statements of the first type: $(x \in A) \Leftrightarrow (x \in B)$.)

A complex statement or logical formula can be constructed from atomic statements by means of logical operators – the connectors $\neg$, $\wedge$, $\vee$ $\Rightarrow$ and the quantifiers $\forall$, $\exists$ – by use of parentheses ( ). When this is done, the formation of any statement, no matter how complicated, reduces to carrying out the following elementary logical operations:

a) forming a new statement by placing the negation sign before some statement and enclosing the result in parentheses;

b) forming a new statement by substituting the necessary connectors $\wedge$, $\vee$, and $\Rightarrow$ between two statements and enclosing the result in parentheses.

c) forming the statement "for every object x property P holds", (written as $\forall x\ P(x)$) or the statement "there exists an object x having property P" (written as $\exists x\ P(x)$).

For example, the cumbersome expression

$$\exists x \ \big(P(x) \wedge \big(\forall y \ \big(P(y) \Rightarrow (y = x)\big)\big)\big)$$

means that there exists an object having property P and such that if y is any object having this property, then $y = x$. In brief: there exists a unique object x having property P. This statement is usually written $\exists! x\ P(x)$, and we shall use this abbreviation.

To simplify the writing of a statement, as already pointed out, one attempts to omit as many parentheses as possible while retaining the unambiguous interpretation of the statement. To this end, in addition to the priority of the operators $\neg$, $\wedge$, $\vee$, $\Rightarrow$ mentioned earlier, we assume that the symbols in a formula are most strongly connected by the symbols $\in$, $=$, then $\exists$, $\forall$, and then the connectors $\neg$, $\wedge$, $\vee$, $\Rightarrow$.

Taking account of this convention, we can now write

$$\exists! x \ P(x) := \exists x \ \big(P(x) \wedge \forall y \ \big(P(y) \Rightarrow y = x\big)\big).$$

We also make the following widely used abbreviations:

$$(\forall x \in X)\ P := \forall x \ \big(x \in X \Rightarrow P(x)\big),$$
$$(\exists x \in X)\ P := \exists x \ \big(x \in X \wedge P(x)\big),$$
$$(\forall x > a)\ P := \forall x \ \big(x \in \mathbb{R} \wedge x > a \Rightarrow P(x)\big),$$
$$(\exists x > a)\ P := \exists x \ \big(x \in \mathbb{R} \wedge x > a \wedge P(x)\big).$$

Here $\mathbb{R}$, as always, denotes the set of real numbers.

Taking account of these abbreviations and the rules a), b), c) for constructing complex statements, we can, for example, give an unambiguous expression

$$\Big(\lim_{x \to A} f(x) = a\Big) := \forall \varepsilon > 0 \ \exists \delta > 0 \ \forall x \in \mathbb{R} \ \big(0 < |x - a| < \delta \Rightarrow \big|f(x) - A\big| < \varepsilon\big)$$

of the fact that the number A is the limit of the function $f : \mathbb{R} \to \mathbb{R}$ at the point $a \in \mathbb{R}$.

For us perhaps the most important result of what has been said in this subsection will be the rules for forming the negation of a statement containing quantifiers.

The negation of the statement "for some x, $P(x)$ is true" means that "for any x, $P(x)$ is false", while the negation of the statement "for any x, $P(x)$ is true" means that "there exists an x such that $P(x)$ is false".

Thus,

$$\neg \exists x\ P(x) \Leftrightarrow \forall x\ \neg P(x),$$
$$\neg \forall x\ P(x) \Leftrightarrow \exists x\ \neg P(x).$$

We recall also (see the exercises in Sect. 1.1) that

$$\neg(P \wedge Q) \Leftrightarrow \neg P \vee \neg Q,$$
$$\neg(P \vee Q) \Leftrightarrow \neg P \wedge \neg Q,$$
$$\neg(P \Rightarrow Q) \Leftrightarrow P \wedge \neg Q.$$

On the basis of what has just been said, one can conclude, for example, that

$$\neg\big((\forall x > a)P\big) \Leftrightarrow (\exists x > a)\neg P.$$

It would of course be wrong to express the right-hand side of this last relation as $(\exists x \le a)\neg P$.

Indeed,

$$\begin{aligned}\neg\big((\forall x > a)P\big) &:= \neg\big(\forall x\ \big(x \in \mathbb{R} \wedge x > a \Rightarrow P(x)\big)\big) \Leftrightarrow \\ &\Leftrightarrow \exists x\ \big(\neg\big(x \in \mathbb{R} \wedge x > a \Rightarrow P(x)\big)\big) \Leftrightarrow \\ &\Leftrightarrow \exists x\ \big((x \in \mathbb{R} \wedge x > a) \wedge \neg P(x)\big) =: (\exists x > a)\neg P.\end{aligned}$$

If we take into account the structure of an arbitrary statement mentioned above, we can now use the negations just constructed for the simplest statements to form the negation of any particular statement.

For example,

$$\neg\Big(\lim_{x\to a} f(x) = A\Big) \Leftrightarrow \exists \varepsilon > 0\ \forall \delta > 0\ \exists x\ \in \mathbb{R}$$
$$\big(0 < |x - a| < \delta \wedge \big|f(x) - A\big| \ge \varepsilon\big).$$

The practical importance of the rule for forming a negation is connected, in particular, with the method of proof by contradiction, in which the truth of a statement P is deduced from the fact that the statement $\neg P$ is false.

1.4.4 Exercises

1. a) Prove the equipotence of the closed interval $\{x \in \mathbb{R} \mid 0 \le x \le 1\}$ and the open interval $\{x \in \mathbb{R} \mid 0 < x < 1\}$ of the real line $\mathbb{R}$ both using the Schröder–Bernstein theorem and by direct exhibition of a suitable bijection.

b) Analyze the following proof of the Schröder–Bernstein theorem:

$$(\operatorname{card} X \le \operatorname{card} Y) \wedge (\operatorname{card} Y \le \operatorname{card} X) \Rightarrow (\operatorname{card} X = \operatorname{card} Y).$$

Proof It suffices to prove that if the sets X, Y, and Z are such that $X \supset Y \supset Z$ and $\operatorname{card} X = \operatorname{card} Z$, then $\operatorname{card} X = \operatorname{card} Y$. Let $f : X \to Z$ be a bijection. A bijection $g : X \to Y$ can be defined, for example, as follows:

$$g(x) = \begin{cases} f(x), & \text{if } x \in f^n(X) \backslash f^n(Y) \text{ for some } n \in \mathbb{N}, \\ x & \text{otherwise.} \end{cases}$$

Here $f^n = f \circ \cdots \circ f$ is the nth iteration of the mapping f and $\mathbb{N}$ is the set of natural numbers. □

2. a) Starting from the definition of a pair, verify that the definition of the direct product $X \times Y$ of sets X and Y given in Sect. 1.4.2 is unambiguous, that is, the set $\mathcal{P}(\mathcal{P}(X) \cup \mathcal{P}(Y))$ contains all ordered pairs (x, y) in which $x \in X$ and $y \in Y$.

b) Show that the mappings $f : X \to Y$ from one given set X into another given set Y themselves form a set $M(X, Y)$.

c) Verify that if $\mathcal{R}$ is a set of ordered pairs (that is, a relation), then the first elements of the pairs belonging to $\mathcal{R}$ (like the second elements) form a set.

3. a) Using the axioms of extensionality, pairing, separation, union, and infinity, verify that the following statements hold for the elements of the set $\mathbb{N}_0$ of natural numbers in the sense of von Neumann:

1^0 $x = y \Rightarrow x^+ = y^+$;
2^0 $(\forall x \in \mathbb{N}_0)\ (x^+ \ne \varnothing)$;
3^0 $x^+ = y^+ \Rightarrow x = y$;
4^0 $(\forall x \in \mathbb{N}_0)\ (x \ne \varnothing \Rightarrow (\exists y \in \mathbb{N}_0)\ (x = y^+))$.

b) Using the fact that $\mathbb{N}_0$ is an inductive set, show that the following statements hold for any of its elements x and y (which in turn are themselves sets):

1^0 $\operatorname{card} x \le \operatorname{card} x^+$;
2^0 $\operatorname{card} \varnothing < \operatorname{card} x^+$;
3^0 $\operatorname{card} x < \operatorname{card} y \Leftrightarrow \operatorname{card} x^+ < \operatorname{card} y^+$;
4^0 $\operatorname{card} x < \operatorname{card} x^+$;
5^0 $\operatorname{card} x < \operatorname{card} y \Rightarrow \operatorname{card} x^+ \le \operatorname{card} y$;

6^0 $x = y \Leftrightarrow \operatorname{card} x = \operatorname{card} y$;
7^0 $(x \subset y) \vee (x \supset y)$.

c) Show that in any subset X of $\mathbb{N}_0$ there exists a (minimal) element x_m such that $(\forall x \in X)\ (\operatorname{card} x_m \leq \operatorname{card} x)$. (If you have difficulty doing so, come back to this problem after reading Chap. 2.)

4. We shall deal only with sets. Since a set consisting of different elements may itself be an element of another set, logicians usually denote all sets by uppercase letters. In the present exercise, it is very convenient to do so.

a) Verify that the statement

$$\forall x\ \exists y\ \forall z\ \big(z \in y \Leftrightarrow \exists w\ (z \in w \wedge w \in x)\big)$$

expresses the axiom of union, according to which y is the union of the sets belonging to x.

b) State which axioms of set theory are represented by the following statements:

$$\forall x\ \forall y\ \forall z\ \big((z \in x \Leftrightarrow z \in y) \Leftrightarrow x = y\big),$$
$$\forall x\ \forall y\ \exists z\ \forall v\ \big(v \in z \Leftrightarrow (v = x \vee v = y)\big),$$
$$\forall x\ \exists y\ \forall z\ \big(z \in y \Leftrightarrow \forall u\ (u \in z \Rightarrow u \in x)\big),$$
$$\exists x\ \big(\forall y\ \big(\neg \exists z\ (z \in y) \Rightarrow y \in x\big) \wedge$$
$$\wedge \forall w\ \big(w \in x \Rightarrow \forall u\ \big(\forall v\ \big(v \in u \Leftrightarrow (v = w \vee v \in w)\big) \Rightarrow u \in x\big)\big)\big).$$

c) Verify that the formula

$$\forall z\ \big(z \in f \Rightarrow \big(\exists x_1\ \exists y_1\ \big(x_1 \in x \wedge y_1 \in y \wedge z = (x_1, y_1)\big)\big)\big) \wedge$$
$$\wedge \forall x_1\ \big(x_1 \in x \Rightarrow \exists y_1\ \exists z\ \big(y_1 \in y \wedge z = (x_1, y_1) \wedge z \in f\big)\big) \wedge$$
$$\wedge \forall x_1\ \forall y_1\ \forall y_2\ \big(\exists z_1\ \exists z_2\ \big(z_1 \in f \wedge z_2 \in f \wedge z_1 = (x_1, y_1) \wedge z_2 = (x_1, y_2)\big) \Rightarrow$$
$$\Rightarrow y_1 = y_2\big)$$

imposes three successive restrictions on the set f: f is a subset of $x \times y$; the projection of f on x is equal to x; to each element x_1 of x there corresponds exactly one y_1 in y such that $(x_1, y_1) \in f$.

Thus what we have here is a definition of a mapping $f : x \to y$.

This example shows yet again that the formal expression of a statement is by no means always the shortest and most transparent in comparison with its expression in ordinary language. Taking this circumstance into account, we shall henceforth use logical symbolism only to the extent that it seems useful to us to achieve greater compactness or clarity of exposition.

5. Let $f : X \to Y$ be a mapping. Write the logical negation of each of the following statements:

a) f is surjective;
b) f is injective;
c) f is bijective.

6. Let X and Y be sets and $f \subset X \times Y$. Write what it means to say that the set f is not a function.

Chapter 2
The Real Numbers

Mathematical theories, as a rule, find uses because they make it possible to transform one set of numbers (the initial data) into another set of numbers constituting the intermediate or final purpose of the computations. For that reason numerical-valued functions occupy a special place in mathematics and its applications. These functions (more precisely, the so-called differentiable functions) constitute the main object of study of classical analysis. But, as you may already have sensed from your school experience, and as will soon be confirmed, any description of the properties of these functions that is at all complete from the point of view of modern mathematics is impossible without a precise definition of the set of real numbers, on which these functions operate.

Numbers in mathematics are like time in physics: everyone knows what they are, and only experts find them hard to understand. This is one of the basic mathematical abstractions, which seems destined to undergo significant further development. A very full separate course could be devoted to this subject. At present we intend only to unify what is basically already known to the reader about real numbers from high school, exhibiting as axioms the fundamental and independent properties of numbers. In doing this, our purpose is to give a precise definition of real numbers suitable for subsequent mathematical use, paying particular attention to their property of completeness or continuity, which contains the germ of the idea of passage to the limit – the basic nonarithmetical operation of analysis.

2.1 The Axiom System and Some General Properties of the Set of Real Numbers

2.1.1 Definition of the Set of Real Numbers

Definition 1 A set $\mathbb{R}$ is called the set of *real numbers* and its elements are *real numbers* if the following list of conditions holds, called the axiom system of the real numbers.

V.A. Zorich, *Mathematical Analysis I*, Universitext,
DOI 10.1007/978-3-662-48792-1_2

(I) (AXIOMS FOR ADDITION) *An operation*

$$+ : \mathbb{R} \times \mathbb{R} \to \mathbb{R},$$

(the operation of addition) is defined, assigning to each ordered pair (x, y) *of elements* x, y *of* $\mathbb{R}$ *a certain element* $x + y \in \mathbb{R}$*, called the* sum *of* x *and* y*. This operation satisfies the following conditions:*

1_+. *There exists a neutral, or identity element* 0 *(called zero) such that*

$$x + 0 = 0 + x = x$$

for every $x \in \mathbb{R}$.

2_+. *For every element* $x \in \mathbb{R}$ *there exists an element* $-x \in \mathbb{R}$ *called the* negative *of* x *such that*

$$x + (-x) = (-x) + x = 0.$$

3_+. *The operation* $+$ *is associative, that is, the relation*

$$x + (y + z) = (x + y) + z$$

holds for any elements x, y, z *of* $\mathbb{R}$.

4_+. *The operation* $+$ *is commutative, that is,*

$$x + y = y + x$$

for any elements x, y *of* $\mathbb{R}$.

If an operation is defined on a set G satisfying Axioms 1_+, 2_+, and 3_+, we say that a *group structure* is defined on G or that G is a *group*. If the operation is called addition, the group is called an *additive* group. If it is also known that the operation is commutative, that is, condition 4_+ holds, the group is called *commutative* or *Abelian*.[1]

Thus, Axioms 1_+–4_+ assert that $\mathbb{R}$ is an additive abelian group.

(II) (AXIOMS FOR MULTIPLICATION) *An operation*

$$\bullet : \mathbb{R} \times \mathbb{R} \to \mathbb{R},$$

(the operation of multiplication) is defined, assigning to each ordered pair (x, y) *of elements* x, y *of* $\mathbb{R}$ *a certain element* $x \cdot y \in \mathbb{R}$*, called the* product *of* x *and* y*. This operation satisfies the following conditions:*

---

[1] N.H. Abel (1802–1829) – outstanding Norwegian mathematician, who proved that the general algebraic equation of degree higher than four cannot be solved by radicals.

$1_\bullet$. *There exists a neutral, or identity element* $1 \in \mathbb{R}\backslash 0$ *(called one) such that*

$$x \cdot 1 = 1 \cdot x = x$$

for every $x \in \mathbb{R}$.

$2_\bullet$. *For every element* $x \in \mathbb{R}\backslash 0$ *there exists an element* $x^{-1} \in \mathbb{R}$, *called the* inverse *or reciprocal of* x, *such that*

$$x \cdot x^{-1} = x^{-1} \cdot x = 1.$$

$3_\bullet$. *The operation* $\bullet$ *is associative, that is, the relation*

$$x \cdot (y \cdot z) = (x \cdot y) \cdot z$$

holds for any elements x, y, z *of* $\mathbb{R}$.

$4_\bullet$. *The operation* $\bullet$ *is commutative, that is,*

$$x \cdot y = y \cdot x$$

for any elements x, y *of* $\mathbb{R}$.

We remark that with respect to *the* operation of multiplication the set $\mathbb{R}\backslash 0$, as one can verify, is a (*multiplicative*) group.

(I, II) (THE CONNECTION BETWEEN ADDITION AND MULTIPLICATION) *Multiplication is distributive with respect to addition, that is*

$$(x + y)z = xz + yz$$

for all $x, y, z \in \mathbb{R}$.

We remark that by the commutativity of multiplication, this equality continues to hold if the order of the factors is reversed on either side.

If two operations satisfying these axioms are defined on a set G, then G is called a *field*.

(III) (ORDER AXIOMS) *Between elements of* $\mathbb{R}$ *there is a relation* $\le$, *that is, for elements* $x, y \in \mathbb{R}$ *one can determine whether* $x \le y$ *or not. Here the following conditions must hold*:

$0_\le$. $\forall x \in \mathbb{R}\ (x \le x)$.

$1_\le$. $(x \le y) \wedge (y \le x) \Rightarrow (x = y)$.

$2_\le$. $(x \le y) \wedge (y \le z) \Rightarrow (x \le z)$.

$3_\le$. $\forall x \in \mathbb{R}\ \forall y \in \mathbb{R}\ (x \le y) \vee (y \le x)$.

The relation $\le$ on $\mathbb{R}$ is called *inequality*.

A set on which there is a relation between pairs of elements satisfying Axioms $0_\le$, $1_\le$, and $2_\le$, as you know, is said to be *partially ordered*. If in addition

Axiom $3_{\le}$ holds, that is, any two elements are comparable, the set is *linearly ordered*. Thus the set of real numbers is linearly ordered by the relation of inequality between elements.

(I, III) (THE CONNECTION BETWEEN ADDITION AND ORDER ON $\mathbb{R}$) *If x, y, z are elements of $\mathbb{R}$, then*

$$(x \le y) \Rightarrow (x + z \le y + z).$$

(II, III) (THE CONNECTION BETWEEN MULTIPLICATION AND ORDER ON $\mathbb{R}$) *If x and y are elements of $\mathbb{R}$, then*

$$(0 \le x) \wedge (0 \le y) \Rightarrow (0 \le x \cdot y).$$

(IV) (THE AXIOM OF COMPLETENESS (CONTINUITY)) *If X and Y are nonempty subsets of $\mathbb{R}$ having the property that $x \le y$ for every $x \in X$ and every $y \in Y$, then there exists $c \in \mathbb{R}$ such that $x \le c \le y$ for all $x \in X$ and $y \in Y$.*

We now have a complete list of axioms such that any set on which these axioms hold can be considered a concrete realization or *model* of the real numbers.

This definition does not formally require any preliminary knowledge about numbers, and from it "by turning on mathematical thought" we should, again formally, obtain as theorems all the other properties of real numbers. On the subject of this axiomatic formalism we would like to make a few informal remarks.

Imagine that you had not passed from the stage of adding apples, cubes, or other named quantities to the addition of abstract natural numbers; you had not studied the measurement of line segments and arrived at rational numbers; you did not know the great discovery of the ancients that the diagonal of a square is incommensurable with its side, so that its length cannot be a rational number, that is, that irrational numbers are needed; you did not have the concept of "greater" or "smaller" that arises in the process of measurement; you did not picture order to yourself using, for example, the real line. If all these preliminaries had not occurred, the axioms just listed would not be perceived as the outcome of intellectual progress; they would seem at the very least a strange, and in any case arbitrary, fruit of the imagination.

In relation to any abstract system of axioms, at least two questions arise immediately.

First, are these axioms consistent? That is, does there exist a set satisfying all the conditions just listed? This is the problem of *consistency* of the axioms.

Second, does the given system of axioms determine the mathematical object uniquely? That is, as the logicians would say, is the axiom system *categorical*? Here uniqueness must be understood as follows. If two people A and B construct models independently, say of number systems $\mathbb{R}_A$ and $\mathbb{R}_B$, satisfying the axioms, then a bijective correspondence can be established between the systems $\mathbb{R}_A$ and $\mathbb{R}_B$, say $f : \mathbb{R}_A \to \mathbb{R}_B$, preserving the arithmetic operations and the order, that is,

$$f(x + y) = f(x) + f(y),$$

$$f(x \cdot y) = f(x) \cdot f(y),$$

$$x \le y \Leftrightarrow f(x) \le f(y).$$

In this case, from the mathematical point of view, $\mathbb{R}_A$ and $\mathbb{R}_B$ are merely distinct but equally valid realizations (models) of the real numbers (for example, $\mathbb{R}_A$ might be the set of infinite decimal fractions and $\mathbb{R}_B$ the set of points on the real line). Such realizations are said to be *isomorphic* and the mapping f is called an *isomorphism*. The result of this mathematical activity is thus not about any particular realization, but about each model in the class of isomorphic models of the given axiom system.

We shall not discuss the questions posed above, but instead confine ourselves to giving informative answers to them.

A positive answer to the question of consistency of an axiom system is always of a hypothetical nature. In relation to numbers it has the following appearance: Starting from the axioms of set theory that we have accepted (see Sect. 1.4.2), one can construct the set of natural numbers, then the set of rational numbers, and finally the set $\mathbb{R}$ of real numbers satisfying all the properties listed.

The question of the categoricity of the axiom system for the real numbers can be established. Those who wish to do so may obtain it independently by solving Exercises 23 and 24 at the end of this section.

2.1.2 Some General Algebraic Properties of Real Numbers

We shall show by examples how the known properties of numbers can be obtained from these axioms.

a. Consequences of the Addition Axioms

1^0. *There is only one zero in the set of real numbers.*

Proof If 0_1 and 0_2 are both zeros in $\mathbb{R}$, then by definition of zero,

$$0_1 = 0_1 + 0_2 = 0_2 + 0_1 = 0_2. \qquad \square$$

2^0. *Each element of the set of real numbers has a unique negative.*

Proof If x_1 and x_2 are both negatives of $x \in \mathbb{R}$, then

$$x_1 = x_1 + 0 = x_1 + (x + x_2) = (x_1 + x) + x_2 = 0 + x_2 = x_2. \qquad \square$$

Here we have used successively the definition of zero, the definition of the negative, the associativity of addition, again the definition of the negative, and finally, again the definition of zero.

3^0. *In the set of real numbers* $\mathbb{R}$ *the equation*

$$a + x = b$$

has the unique solution

$$x = b + (-a).$$

Proof This follows from the existence and uniqueness of the negative of every element $a \in \mathbb{R}$:

$$\begin{aligned}(a + x = b) &\Leftrightarrow \big((x + a) + (-a) = b + (-a)\big) \Leftrightarrow \\ &\Leftrightarrow \big(x + \big(a + (-a)\big) = b + (-a)\big) \Leftrightarrow \big(x + 0 = b + (-a)\big) \Leftrightarrow \\ &\Leftrightarrow \big(x = b + (-a)\big).\end{aligned}$$ □

The expression $b + (-a)$ can also be written as $b - a$. This is the shorter and more common way of writing it, to which we shall adhere.

b. Consequences of the Multiplication Axioms

1^0. *There is only one multiplicative unit in the real numbers.*
2^0. *For each* $x \neq 0$ *there is only one reciprocal* x^{-1}.
3^0. *For* $a \in \mathbb{R}\backslash 0$, *the equation* $a \cdot x = b$ *has the unique solution* $x = b \cdot a^{-1}$.

The proofs of these propositions, of course, merely repeat the proofs of the corresponding propositions for addition (except for a change in the symbol and the name of the operation); they are therefore omitted.

c. Consequences of the Axiom Connecting Addition and Multiplication

Applying the additional axiom (I, II) connecting addition and multiplication, we obtain further consequences.

1^0. *For any* $x \in \mathbb{R}$

$$x \cdot 0 = 0 \cdot x = 0.$$

Proof

$$\big(x \cdot 0 = x \cdot (0 + 0) = x \cdot 0 + x \cdot 0\big) \Rightarrow \big(x \cdot 0 = x \cdot 0 + \big(-(x \cdot 0)\big) = 0\big).$$ □

From this result, incidentally, one can see that if $x \in \mathbb{R}\backslash 0$, then $x^{-1} \in \mathbb{R}\backslash 0$.

2^0. $(x \cdot y = 0) \Rightarrow (x = 0) \vee (y = 0)$.

Proof If, for example, $y \neq 0$, then by the uniqueness of the solution of the equation $x \cdot y = 0$ for x, we find $x = 0 \cdot y^{-1} = 0$. □

3^0. *For any* $x \in \mathbb{R}$

$$-x = (-1) \cdot x.$$

Proof $x + (-1) \cdot x = (1 + (-1)) \cdot x = 0 \cdot x = x \cdot 0 = 0$, and the assertion now follows from the uniqueness of the negative of a number. □

4^0. *For any* $x \in \mathbb{R}$

$$(-1)(-x) = x.$$

Proof This follows from 3^0 and the uniqueness of the negative of $-x$. □

5^0. *For any* $x \in \mathbb{R}$

$$(-x) \cdot (-x) = x \cdot x.$$

Proof

$$(-x)(-x) = \big((-1) \cdot x\big)(-x) = \big(x \cdot (-1)\big)(-x) = x\big((-1)(-x)\big) = x \cdot x.$$

Here we have made successive use of the preceding propositions and the commutativity and associativity of multiplication. □

d. Consequences of the Order Axioms

We begin by noting that the relation $x \leq y$ (read "x is less than or equal to y") can also be written as $y \geq x$ ("y is greater than or equal to x"); when $x \neq y$, the relation $x \leq y$ is written $x < y$ (read "x is less than y") or $y > x$ (read "y is greater than x"), and is called *strict inequality*.

1^0. *For any x and y in $\mathbb{R}$ precisely one of the following relations holds*:

$$x < y, \qquad x = y, \qquad x > y.$$

Proof This follows from the definition of strict inequality just given and Axioms $1_{\leq}$ and $3_{\leq}$. □

2^0. *For any* $x, y, z \in \mathbb{R}$

$$(x < y) \wedge (y \leq z) \Rightarrow (x < z),$$

$$(x \leq y) \wedge (y < z) \Rightarrow (x < z).$$

Proof We prove the first assertion as an example. By Axiom $2_{\leq}$, which asserts that the inequality relation is transitive, we have

$$(x \leq y) \wedge (y < z) \Leftrightarrow (x \leq y) \wedge (y \leq z) \wedge (y \neq z) \Rightarrow (x \leq z).$$

It remains to be verified that $x \neq z$. But if this were not the case, we would have

$$(x \leq y) \wedge (y < z) \Leftrightarrow (z \leq y) \wedge (y < z) \Leftrightarrow (z \leq y) \wedge (y \leq z) \wedge (y \neq z).$$

By Axiom $1_{\leq}$ this relation would imply

$$(y = z) \wedge (y \neq z),$$

which is a contradiction. □

e. Consequences of the Axioms Connecting Order with Addition and Multiplication

If in addition to the axioms of addition, multiplication, and order, we use Axioms (I, III) and (II, III), which connect the order with the arithmetic operations, we can obtain, for example, the following propositions.

1^0. *For any* $x, y, z, w \in \mathbb{R}$

$$(x < y) \Rightarrow (x + z) < (y + z),$$

$$(0 < x) \Rightarrow (-x < 0),$$

$$(x \leq y) \wedge (z \leq w) \Rightarrow (x + z) \leq (y + w),$$

$$(x \leq y) \wedge (z < w) \Rightarrow (x + z < y + w).$$

Proof We shall verify the first of these assertions.

By definition of strict inequality and the axiom (I, III) we have

$$(x < y) \Rightarrow (x \leq y) \Rightarrow (x + z) \leq (y + z).$$

It remains to be verified that $x + z \neq y + z$. Indeed,

$$\big((x + z) = (y + z)\big) \Rightarrow \big(x = (y + z) - z = y + (z - z) = y\big),$$

which contradicts the assumption $x < y$. □

2^0. *If* $x, y, z \in \mathbb{R}$, *then*

$$(0 < x) \wedge (0 < y) \Rightarrow (0 < xy),$$

$$(x < 0) \wedge (y < 0) \Rightarrow (0 < xy),$$

$$(x < 0) \wedge (0 < y) \Rightarrow (xy < 0),$$
$$(x < y) \wedge (0 < z) \Rightarrow (xz < yz),$$
$$(x < y) \wedge (z < 0) \Rightarrow (yz < xz).$$

Proof We shall verify the first of these assertions. By definition of strict inequality and the axiom (II, III) we have

$$(0 < x) \wedge (0 < y) \Rightarrow (0 \le x) \wedge (0 \le y) \Rightarrow (0 \le xy).$$

Moreover, $0 \neq xy$ since, as already shown,

$$(x \cdot y = 0) \Rightarrow (x = 0) \vee (y = 0).$$

Let us further verify, for example, the third assertion:

$$\begin{aligned}(x < 0) \wedge (0 < y) &\Rightarrow (0 < -x) \wedge (0 < y) \Rightarrow \\ &\Rightarrow \big(0 < (-x) \cdot y\big) \Rightarrow \big(0 < \big((-1) \cdot x\big)y\big) \Rightarrow \\ &\Rightarrow \big(0 < (-1) \cdot (xy)\big) \Rightarrow \big(0 < -(xy)\big) \Rightarrow (xy < 0).\end{aligned}$$ □

The reader is now invited to prove the remaining relations independently and also to verify that if nonstrict inequality holds in one of the parentheses on the left-hand side, then the inequality on the right-hand side will also be nonstrict.

3^0. $0 < 1$.

Proof We know that $1 \in \mathbb{R} \backslash 0$, that is $0 \neq 1$. If we assume $1 < 0$, then by what was just proved,

$$(1 < 0) \wedge (1 < 0) \Rightarrow (0 < 1 \cdot 1) \Rightarrow (0 < 1).$$

But we know that for any pair of numbers $x, y \in \mathbb{R}$ exactly one of the possibilities $x < y$, $x = y$, $x > y$ actually holds. Since $0 \neq 1$ and the assumption $1 < 0$ implies the relation $0 < 1$, which contradicts it, the only remaining possibility is the one in the statement of the proposition. □

4^0. $(0 < x) \Rightarrow (0 < x^{-1})$ and $(0 < x) \wedge (x < y) \Rightarrow (0 < y^{-1}) \wedge (y^{-1} < x^{-1})$.

Proof Let us verify the first of these assertions. First of all, $x^{-1} \neq 0$. Assuming $x^{-1} < 0$, we obtain

$$\big(x^{-1} < 0\big) \wedge (0 < x) \Rightarrow \big(x \cdot x^{-1} < 0\big) \Rightarrow (1 < 0).$$

This contradiction completes the proof. □

We recall that numbers larger than zero are called *positive* and those less than zero *negative*.

Thus we have shown, for example, that 1 is a positive number, that the product of a positive and a negative number is a negative number, and that the reciprocal of a positive number is also positive.

2.1.3 The Completeness Axiom and the Existence of a Least Upper (or Greatest Lower) Bound of a Set of Numbers

Definition 2 A set $X \subset \mathbb{R}$ is said to be *bounded above* (resp. *bounded below*) if there exists a number $c \in \mathbb{R}$ such that $x \leq c$ (resp. $c \leq x$) for all $x \in X$.

The number c in this case is called an *upper bound* (resp. *lower bound*) of the set X. It is also called a *majorant* (resp. *minorant*) of X.

Definition 3 A set that is bounded both above and below is called *bounded*.

Definition 4 An element $a \in X$ is called the *largest* or *maximal* (resp. *smallest* or *minimal*) element of X if $x \leq a$ (resp. $a \leq x$) for all $x \in X$.

We now introduce some notation and at the same time give a formal expression to the definition of maximal and minimal elements:

$$(a = \max X) := \big(a \in X \wedge \forall x \in X\ (x \leq a)\big),$$
$$(a = \min X) := \big(a \in X \wedge \forall x \in X\ (a \leq x)\big).$$

Along with the notation $\max X$ (read "the maximum of X") and $\min X$ (read "the minimum of X") we also use the respective expressions $\max_{x\in X} x$ and $\min_{x\in X} x$.

It follows immediately from the order Axiom $1_{\leq}$ that if there is a maximal (resp. minimal) element in a set of numbers, it is the only one.

However, not every set, not even every bounded set, has a maximal or minimal element.

For example, the set $X = \{x \in \mathbb{R} \mid 0 \leq x < 1\}$ has a minimal element. But, as one can easily verify, it has no maximal element.

Definition 5 The smallest number that bounds a set $X \subset \mathbb{R}$ from above is called the *least upper bound* (or the *exact upper bound*) of X and denoted $\sup X$ (read "the supremum of X") or $\sup_{x\in X} x$.

This is the basic concept of the present subsection. Thus

$$(s = \sup X) := \forall x \in X\ \big((x \leq s) \wedge \big(\forall s' < s\ \exists x' \in X\ \big(s' < x'\big)\big)\big).$$

The expression in the first set of parentheses on the right-hand side here says that s is an upper bound for X; the expression in the second set says that s is the smallest number having this property. More precisely, the expression in the second set of parentheses asserts that any number smaller than s is not an upper bound of X.

The concept of the *greatest lower bound* (or *exact lower bound*) of a set X is introduced similarly as the largest of the lower bounds of X.

Definition 6

$$(i = \inf X) := \forall x \in X \left((i \le x) \wedge \left(\forall i' > i \; \exists x' \in X \; \left(x' < i'\right)\right)\right).$$

Along with the notation $\inf X$ (read "the infimum of X") one also uses the notation $\inf_{x \in X} x$ for the greatest lower bound of X.

Thus we have given the following definitions:

$$\sup X := \min\left\{c \in \mathbb{R} \mid \forall x \in X \; (x \le c)\right\},$$
$$\inf X := \max\left\{c \in \mathbb{R} \mid \forall x \in X \; (c \le x)\right\}.$$

But we said above that not every set has a minimal or maximal element. Therefore the definitions we have adopted for the least upper bound and greatest lower bound require an argument, provided by the following lemma.

Lemma (The least upper bound principle) *Every nonempty set of real numbers that is bounded from above has a unique least upper bound.*

Proof Since we already know that the minimal element of a set of numbers is unique, we need only verify that the least upper bound exists.

Let $X \subset \mathbb{R}$ be a given set and $Y = \{y \in \mathbb{R} \mid \forall x \in X \; (x \le y)\}$. By hypothesis, $X \ne \varnothing$ and $Y \ne \varnothing$. Then, by the completeness axiom there exists $c \in \mathbb{R}$ such that $\forall x \in X \; \forall y \in Y \; (x \le c \le y)$. The number c is therefore both a majorant of X and a minorant of Y. Being a majorant of X, c is an element of Y. But then, as a minorant of Y, it must be the minimal element of Y. Thus $c = \min Y = \sup X$. □

Naturally the existence and uniqueness of the greatest lower bound of a nonempty set of numbers that is bounded from below is analogous, that is, the following proposition holds.

Lemma (X *nonempty and bounded below*) $\Rightarrow$ ($\exists! \inf X$).

We shall not take time to give the proof.

We now return to the set $X = \{x \in \mathbb{R} \mid 0 \le x < 1\}$. By the lemma just proved it must have a least upper bound. By the very definition of the set X and the definition of the least upper bound, it is obvious that $\sup X \le 1$.

To prove that $\sup X = 1$ it is thus necessary to verify that for any number $q < 1$ there exists $x \in X$ such that $q < x$; simply put, this means merely that there are numbers between q and 1. This of course, is also easy to prove independently (for example, by showing that $q < 2^{-1}(q+1) < 1$), but we shall not do so at this point, since such questions will be discussed systematically and in detail in the next section.

As for the greatest lower bound, it always coincides with the minimal element of a set, if such an element exists. Thus, from this consideration alone we have $\inf X = 0$ in the present example.

Other, more substantive examples of the use of the concepts introduced here will be encountered in the next section.

2.2 The Most Important Classes of Real Numbers and Computational Aspects of Operations with Real Numbers

2.2.1 *The Natural Numbers and the Principle of Mathematical Induction*

a. Definition of the Set of Natural Numbers

The numbers of the form $1, 1+1, (1+1)+1$, and so forth are denoted respectively by $1, 2, 3, \dots$ and so forth and are called *natural numbers*.

Such a definition will be meaningful only to one who already has a complete picture of the natural numbers, including the notation for them, for example in the decimal system of computation.

The continuation of such a process is by no means always unique, so that the ubiquitous "and so forth" actually requires a clarification provided by the fundamental principle of mathematical induction.

Definition 1 A set $X \subset \mathbb{R}$ is *inductive* if for each number $x \in X$, it also contains $x+1$.

For example, $\mathbb{R}$ is an inductive set; the set of positive numbers is also inductive.

The intersection $X = \bigcap_{\alpha \in A} X_\alpha$ of any family of inductive sets X_α, if not empty, is an inductive set.

Indeed,

$$\left(x \in X = \bigcap_{\alpha \in A} X_\alpha\right) \Rightarrow \big(\forall \alpha \in A\ (x \in X_\alpha)\big) \Rightarrow$$

$$\Rightarrow \big(\forall \alpha \in A\ \big((x+1) \in X_\alpha\big)\big) \Rightarrow \left((x+1) \in \bigcap_{\alpha \in A} X_\alpha = X\right).$$

We now adopt the following definition.

Definition 2 The set of *natural numbers* is the smallest inductive set containing 1, that is, the intersection of all inductive sets that contain 1.

The set of natural numbers is denoted $\mathbb{N}$; its elements are called *natural numbers*.

From the set-theoretic point of view it might be more rational to begin the natural numbers with 0, that is, to introduce the set of natural numbers as the smallest inductive set containing 0; however, it is more convenient for us to begin numbering with 1.

The following fundamental and widely used principle is a direct corollary of the definition of the set of natural numbers.

b. The Principle of Mathematical Induction

If a subset E of the set of natural numbers $\mathbb{N}$ is such that $1 \in E$ and together with each number $x \in E$, the number $x+1$ also belongs to E, then $E = \mathbb{N}$.

Thus,

$$(E \subset \mathbb{N}) \wedge (1 \in E) \wedge \big(x \in E \Rightarrow (x+1) \in E\big) \Rightarrow E = \mathbb{N}.$$

Let us illustrate this principle in action by using it to prove several useful properties of the natural numbers that we will be using constantly from now on.

1^0. *The sum and product of natural numbers are natural numbers.*

Proof Let $m, n \in \mathbb{N}$; we shall show that $(m+n) \in \mathbb{N}$. We denote by E the set of natural numbers n for which $(m+n) \in \mathbb{N}$ for all $m \in \mathbb{N}$. Then $1 \in E$ since $(m \in \mathbb{N}) \Rightarrow ((m+1) \in \mathbb{N})$ for any $m \in \mathbb{N}$. If $n \in E$, that is, $(m+n) \in \mathbb{N}$, then $(n+1) \in E$ also, since $(m+(n+1)) = ((m+n)+1) \in \mathbb{N}$. By the principle of induction, $E = \mathbb{N}$, and we have proved that addition does not lead outside of $\mathbb{N}$.

Similarly, taking E to be the set of natural numbers n for which $(m \cdot n) \in \mathbb{N}$ for all $m \in \mathbb{N}$, we find that $1 \in E$, since $m \cdot 1 = m$, and if $n \in E$, that is, $m \cdot n \in \mathbb{N}$, then $m \cdot (n+1) = mn + m$ is the sum of two natural numbers, which belongs to $\mathbb{N}$ by what was just proved above. Thus $(n \in E) \Rightarrow ((n+1) \in E)$, and so by the principle of induction $E = \mathbb{N}$. □

2^0. $(n \in \mathbb{N}) \wedge (n \neq 1) \Rightarrow ((n-1) \in \mathbb{N})$.

Proof Consider the set E consisting of all real numbers of the form $n-1$, where n is a natural number different from 1; we shall show that $E = \mathbb{N}$. Since $1 \in \mathbb{N}$, it follows that $2 := (1+1) \in \mathbb{N}$ and hence $1 = (2-1) \in E$.

If $m \in E$, then $m = n-1$, where $n \in \mathbb{N}$; then $m+1 = (n+1)-1$, and since $n+1 \in \mathbb{N}$, we have $(m+1) \in E$. By the principle of induction we conclude that $E = \mathbb{N}$. □

3^0. *For any $n \in \mathbb{N}$ the set $\{x \in \mathbb{N} \mid n < x\}$ contains a minimal element, namely*

$$\min\{x \in \mathbb{N} \mid n < x\} = n+1.$$

Proof We shall show that the set E of $n \in \mathbb{N}$ for which the assertion holds coincides with $\mathbb{N}$.

We first verify that $1 \in E$, that is,

$$\min\{x \in \mathbb{N} \mid 1 < x\} = 2.$$

We shall also verify this assertion by the principle of induction. Let

$$M = \{x \in \mathbb{N} \mid (x = 1) \vee (2 \le x)\}.$$

By definition of M we have $1 \in M$. Then if $x \in M$, either $x = 1$, in which case $x + 1 = 2 \in M$, or else $2 \le x$, and then $2 \le (x + 1)$, and once again $(x + 1) \in M$. Thus $M = \mathbb{N}$, and hence if $(x \ne 1) \wedge (x \in \mathbb{N})$, then $2 \le x$, that is, indeed $\min\{x \in \mathbb{N} \mid 1 < x\} = 2$. Hence $1 \in E$.

We now show that if $n \in E$, then $(n + 1) \in E$.

We begin by remarking that if $x \in \{x \in \mathbb{N} \mid n + 1 < x\}$, then

$$(x - 1) = y \in \{y \in \mathbb{N} \mid n < y\}.$$

For, by what has already been proved, every natural number is at least as large as 1; therefore $(n + 1 < x) \Rightarrow (1 < x) \Rightarrow (x \ne 1)$, and then by the assertion in 2^0 we have $(x - 1) = y \in \mathbb{N}$.

Now let $n \in E$, that is, $\min\{y \in \mathbb{N} \mid n < y\} = n + 1$. Then $x - 1 \ge y \ge n + 1$ and $x \ge n + 2$. Hence,

$$\big(x \in \{x \in \mathbb{N} \mid n + 1 < x\}\big) \Rightarrow (x \ge n + 2)$$

and consequently, $\min\{x \in \mathbb{N} \mid n + 1 < x\} = n + 2$, that is, $(n + 1) \in E$.

By the principle of induction $E = \mathbb{N}$, and 3^0 is now proved. □

As immediate corollaries of 2^0 and 3^0 above, we obtain the following properties (4^0, 5^0, and 6^0) of the natural numbers.

4^0. $(m \in \mathbb{N}) \wedge (n \in \mathbb{N}) \wedge (n < m) \Rightarrow (n + 1 \le m)$.

5^0. *The number* $(n + 1) \in \mathbb{N}$ *is the immediate successor of the number* $n \in \mathbb{N}$; *that is, if* $n \in \mathbb{N}$, *there are no natural numbers* x *satisfying* $n < x < n + 1$.

6^0. *If* $n \in \mathbb{N}$ *and* $n \ne 1$, *then* $(n - 1) \in \mathbb{N}$ *and* $(n - 1)$ *is the immediate predecessor of* n *in* $\mathbb{N}$; *that is, if* $n \in \mathbb{N}$, *there are no natural numbers* x *satisfying* $n - 1 < x < n$.

We now prove one more property of the set of natural numbers.

7^0. *In any nonempty subset of the set of natural numbers there is a minimal element.*

Proof Let $M \subset \mathbb{N}$. If $1 \in M$, then $\min M = 1$, since $\forall n \in \mathbb{N}\ (1 \le n)$.

Now suppose $1 \notin M$, that is, $1 \in E = \mathbb{N} \backslash M$. The set E must contain a natural number n such that all natural numbers not larger than n belong to E, but

$(n+1) \in M$. If there were no such n, the set $E \subset \mathbb{N}$, which contains 1, would contain along with each of its elements n, the number $(n+1)$ also; by the principle of induction, it would therefore equal $\mathbb{N}$. But the latter is impossible, since $\mathbb{N} \backslash E = M \neq \varnothing$.

The number $(n+1)$ so found must be the smallest element of M, since there are no natural numbers between n and $n+1$, as we have seen. □

2.2.2 *Rational and Irrational Numbers*

a. The Integers

Definition 3 The union of the set of natural numbers, the set of negatives of natural numbers, and zero is called the set of *integers* and is denoted $\mathbb{Z}$.

Since, as has already been proved, addition and multiplication of natural numbers do not take us outside $\mathbb{N}$, it follows that these same operations on integers do not lead outside of $\mathbb{Z}$.

Proof Indeed, if $m, n \in \mathbb{Z}$, either one of these numbers is zero, and then the sum $m+n$ equals the other number, so that $(m+n) \in \mathbb{Z}$ and $m \cdot n = 0 \in \mathbb{Z}$, or both numbers are non-zero. In the latter case, either $m, n \in \mathbb{N}$ and then $(m+n) \in \mathbb{N} \subset \mathbb{Z}$ and $(m \cdot n) \in \mathbb{N} \subset \mathbb{Z}$, or $(-m), (-n) \in \mathbb{N}$ and then $m \cdot n = ((-1)m)((-1)n) \in \mathbb{N}$ or $(-m), n \in \mathbb{N}$ and then $(-m \cdot n) \in \mathbb{N}$, that is, $m \cdot n \in \mathbb{Z}$, or, finally, $m, -n \in \mathbb{N}$ and then $(-m \cdot n) \in \mathbb{N}$ and once again $m \cdot n \in \mathbb{Z}$. □

Thus $\mathbb{Z}$ is an Abelian group with respect to addition. With respect to multiplication $\mathbb{Z}$ is not a group, nor is $\mathbb{Z} \backslash 0$, since the reciprocals of the integers are not in $\mathbb{Z}$ (except the reciprocals of 1 and -1).

Proof Indeed, if $m \in \mathbb{Z}$ and $m \neq 0, 1$, then assuming first that $m \in \mathbb{N}$, we have $0 < 1 < m$, and, since $m \cdot m^{-1} = 1 > 0$, we must have $0 < m^{-1} < 1$ (see the consequences of the order axioms in the previous subsection). Thus $m^{-1} \notin \mathbb{Z}$. The case when m is a negative integer different from -1 reduces immediately to the one already considered. □

When $k = m \cdot n^{-1} \in \mathbb{Z}$ for two integers $m, n \in \mathbb{Z}$, that is, when $m = k \cdot n$ for some $k \in \mathbb{Z}$, we say that m is *divisible* by n or a *multiple* of n, or that n is a *divisor* of m.

The divisibility of integers reduces immediately via suitable sign changes, that is, through multiplication by -1 when necessary, to the divisibility of the corresponding natural numbers. In this context it is studied in number theory.

We recall without proof the so-called fundamental theorem of arithmetic, which we shall use in studying certain examples.

A number $p \in \mathbb{N}$, $p \neq 1$, is *prime* if it has no divisors in $\mathbb{N}$ except 1 and p.

The fundamental theorem of arithmetic *Each natural number admits a representation as a product*

$$n = p_1 \cdots p_k,$$

where $p_1, \dots, p_k$ are prime numbers. This representation is unique except for the order of the factors.

Numbers $m, n \in \mathbb{Z}$ are said to be *relatively prime* if they have no common divisors except 1 and –1.

It follows in particular from this theorem that if the product $m \cdot n$ of relatively prime numbers m and n is divisible by a prime p, then one of the two numbers is also divisible by p.

b. The Rational Numbers

Definition 4 Numbers of the form $m \cdot n^{-1}$, where $m, n \in \mathbb{Z}$, are called *rational.*

We denote the set of rational numbers by $\mathbb{Q}$.

Thus, the ordered pair (m, n) of integers defines the rational number $q = m \cdot n^{-1}$ if $n \neq 0$.

The number $q = m \cdot n^{-1}$ can also be written as a quotient[2] of m and n, that is, as a so-called rational fraction $\frac{m}{n}$.

The rules you learned in school for operating with rational numbers in terms of their representation as fractions follow immediately from the definition of a rational number and the axioms for real numbers. In particular, "the value of a fraction is unchanged when both numerator and denominator are multiplied by the same nonzero integer", that is, the fractions $\frac{mk}{nk}$ and $\frac{m}{n}$ represent the same rational number. In fact, since $(nk)(k^{-1}n^{-1}) = 1$, that is $(n \cdot k)^{-1} = k^{-1} \cdot n^{-1}$, we have $(mk)(nk)^{-1} = (mk)(k^{-1}n^{-1}) = m \cdot n^{-1}$.

Thus the different ordered pairs (m, n) and (mk, nk) define the same rational number. Consequently, after suitable reductions, any rational number can be presented as an ordered pair of relatively prime integers.

On the other hand, if the pairs (m_1, n_1) and (m_2, n_2) define the same rational number, that is, $m_1 \cdot n_1^{-1} = m_2 \cdot n_2^{-1}$, then $m_1 n_2 = m_2 n_1$, and if, for example, m_1 and n_1 are relatively prime, it follows from the corollary of the fundamental theorem of arithmetic mentioned above that $n_2 \cdot n_1^{-1} = m_2 \cdot m_1^{-1} = k \in \mathbb{Z}$.

We have thus demonstrated that two ordered pairs (m_1, n_1) and (m_2, n_2) define the same rational number if and only if they are proportional. That is, there exists an integer $k \in \mathbb{Z}$ such that, for example, $m_2 = km_1$ and $n_2 = kn_1$.

---

[2]The notation $\mathbb{Q}$ comes from the first letter of the English word *quotient*, which in turn comes from the Latin *quota*, meaning the unit part of something, and *quat*, meaning *how many*.

c. The Irrational Numbers

Definition 5 The real numbers that are not rational are called *irrational.*

The classical example of an irrational real number is $\sqrt{2}$, that is, the number $s \in \mathbb{R}$ such that $s > 0$ and $s^2 = 2$. By the Pythagorean theorem, the irrationality of $\sqrt{2}$ is equivalent to the assertion that the diagonal and side of a square are incommensurable.

Thus we begin by verifying that *there exists a real number* $s \in \mathbb{R}$ *whose square equals* 2, and then that $s \notin \mathbb{Q}$.

Proof Let X and Y be the sets of positive real numbers such that $\forall x \in X\ (x^2 < 2)$, $\forall y \in Y\ (2 < y^2)$. Since $1 \in X$ and $2 \in Y$, it follows that X and Y are nonempty sets.

Further, since $(x < y) \Leftrightarrow (x^2 < y^2)$ for positive numbers x and y, every element of X is less than every element of Y. By the completeness axiom there exists $s \in \mathbb{R}$ such that $x \le s \le y$ for all $x \in X$ and all $y \in Y$.

We shall show that $s^2 = 2$.

If $s^2 < 2$, then, for example, the number $s + \frac{2-s^2}{3s}$, which is larger than s, would have a square less than 2. Indeed, we know that $1 \in X$, so that $1^2 \le s^2 < 2$, and $0 < \Delta := 2 - s^2 \le 1$. It follows that

$$\left(s + \frac{\Delta}{3s}\right)^2 = s^2 + 2 \cdot \frac{\Delta}{3s} + \left(\frac{\Delta}{3s}\right)^2 < s^2 + 3 \cdot \frac{\Delta}{3s} < s^2 + 3 \cdot \frac{\Delta}{3s} = s^2 + \Delta = 2.$$

Consequently, $(s + \frac{\Delta}{3s}) \in X$, which is inconsistent with the inequality $x \le s$ for all $x \in X$.

If $2 < s^2$, then the number $s - \frac{s^2-2}{3s}$, which is smaller than s, would have a square larger than 2. Indeed, we know that $2 \in Y$, so that $2 < s^2 \le 2^2$ or $0 < \Delta := s^2 - 2 < 3$ and $0 < \frac{\Delta}{3} < 1$. Hence,

$$\left(s - \frac{\Delta}{3s}\right)^2 = s^2 - 2 \cdot \frac{\Delta}{3s} + \left(\frac{\Delta}{3s}\right)^2 > s^2 - 3 \cdot \frac{\Delta}{3s} = s^2 - \Delta = 2,$$

and we have now contradicted the fact that s is a lower bound of Y.

Thus the only remaining possibility is that $s^2 = 2$.

Let us show, finally, that $s \notin Q$. Assume that $s \in \mathbb{Q}$ and let $\frac{m}{n}$ be an irreducible representation of s. Then $m^2 = 2 \cdot n^2$, so that m^2 is divisible by 2 and therefore m also is divisible by 2. But, if $m = 2k$, then $2k^2 = n^2$, and for the same reason, n must be divisible by 2. But this contradicts the assumed irreducibility of the fraction $\frac{m}{n}$. □

We have worked hard just now to prove that there exist irrational numbers. We shall soon see that in a certain sense nearly all real numbers are irrational. It will be shown that the cardinality of the set of irrational numbers is larger than that of the set of rational numbers and that in fact the former equals the cardinality of the set of real numbers.

Among the irrational numbers we make a further distinction between the so-called algebraic irrational numbers and the transcendental numbers.

A real number is called *algebraic* if it is the root of an algebraic equation

$$a_0x^n + \cdots + a_{n-1}x + a_n = 0$$

with rational (or equivalently, integer) coefficients.

Otherwise the number is called *transcendental*.

We shall see that the cardinality of the set of algebraic numbers is the same as that of the set of rational numbers, while the cardinality of the set of transcendental numbers is the same as that of the set of real numbers. For that reason the difficulties involved in exhibiting specific transcendental numbers – more precisely, proving that a given number is transcendental – seem at first sight paradoxical and unnatural.

For example, it was not proved until 1882 that the classical geometric number π is transcendental,[3] and one of the famous Hilbert[4] problems was to prove the transcendence of the number α^β, where α is algebraic, $(\alpha > 0) \wedge (\alpha \neq 1)$ and β is an irrational algebraic number (for example, $\alpha = 2$, $\beta = \sqrt{2}$).

2.2.3 The Principle of Archimedes

We now turn to the principle of Archimedes,[5] which is important in both its theoretical aspect and the application of numbers in measurement and computations. We shall prove it using the completeness axiom (more precisely, the least-upper-bound principle, which is equivalent to the completeness axiom). In other axiom systems for the real numbers this fundamental principle is frequently included in the list of axioms.

We remark that the propositions that we have proved up to now about the natural numbers and the integers have made no use at all of the completeness axiom. As

[3]The number π equals the ratio of the circumference of a circle to its diameter in Euclidean geometry. That is the reason this number has been conventionally denoted since the eighteenth century, following Euler by π, which is the initial letter of the Greek word περιφέρια – *periphery* (circumference). The transcendence of π was proved by the German mathematician F. Lindemann (1852–1939). It follows in particular from the transcendence of π that it is impossible to construct a line segment of length π with compass and straightedge (the problem of rectification of the circle), and also that the ancient problem of squaring the circle cannot be solved with compass and straightedge.

[4]D. Hilbert (1862–1943) – outstanding German mathematician who stated 23 problems from different areas of mathematics at the 1900 International Congress of Mathematicians in Paris. These problems came to be known as the "Hilbert problems". The problem mentioned here (Hilbert's seventh problem) was given an affirmative answer in 1934 by the Soviet mathematician A.O. Gel'fond (1906–1968) and the German mathematician T. Schneider (1911–1989).

[5]Archimedes (287–212 BCE) – brilliant Greek scholar, about whom Leibniz, one of the founders of analysis said, "When you study the works of Archimedes, you cease to be amazed by the achievements of modern mathematicians."

will be seen below, the principle of Archimedes essentially reflects the properties of the natural numbers and integers connected with completeness. We begin with these properties.

1^0. *Any nonempty subset of natural numbers that is bounded from above contains a maximal element.*

Proof If $E \subset \mathbb{N}$ is the subset in question, then by the least-upper-bound lemma, $\exists! \sup E = s \in \mathbb{R}$. By definition of the least upper bound there is a natural number $n \in E$ satisfying the condition $s - 1 < n \leq s$. But then, $n = \max E$, since a natural number that is larger than n must be at least $n + 1$, and $n + 1 > s$. □

Corollaries 2^0 *The set of natural numbers is not bounded above.*

Proof Otherwise there would exist a maximal natural number. But $n < n + 1$. □

3^0. *Any nonempty subset of the integers that is bounded from above contains a maximal element.*

Proof The proof of 1^0 can be repeated verbatim, replacing $\mathbb{N}$ with $\mathbb{Z}$. □

4^0. *Any nonempty subset of integers that is bounded below contains a minimal element.*

Proof One can, for example, repeat the proof of 1^0, replacing $\mathbb{N}$ by $\mathbb{Z}$ and using the greatest-lower-bound principle instead of the least-upper-bound principle.

Alternatively, one can pass to the negatives of the numbers ("change signs") and use what has been proved in 3^0. □

5^0. *The set of integers is unbounded above and unbounded below.*

Proof This follows from 3^0 and 4^0, or directly from 2^0. □

We can now state the principle of Archimedes.

6^0. (The principle of Archimedes). *For any fixed positive number h and any real number x there exists a unique integer k such that $(k - 1)h \leq x < kh$.*

Proof Since $\mathbb{Z}$ is not bounded above, the set $\{n \in \mathbb{Z} \mid \frac{x}{h} < n\}$ is a nonempty subset of the integers that is bounded below. Then (see 4^0) it contains a minimal element k, that is $(k - 1) \leq x/h < k$. Since $h > 0$, these inequalities are equivalent to those given in the statement of the principle of Archimedes. The uniqueness of $k \in \mathbb{Z}$ satisfying these two inequalities follows from the uniqueness of the minimal element of a set of numbers (see Sect. 2.1.3). □

And now some corollaries:

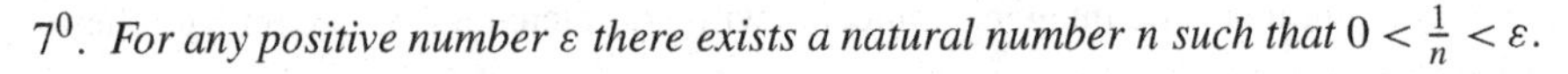

7^0. *For any positive number* ε *there exists a natural number* n *such that* $0 < \frac{1}{n} < \varepsilon$.

Proof By the principle of Archimedes there exists $n \in \mathbb{Z}$ such that $1 < \varepsilon \cdot n$. Since $0 < 1$ and $0 < \varepsilon$, we have $0 < n$. Thus $n \in \mathbb{N}$ and $0 < \frac{1}{n} < \varepsilon$. □

8^0. *If the number* $x \in \mathbb{R}$ *is such that* $0 \le x$ *and* $x < \frac{1}{n}$ *for all* $n \in \mathbb{N}$, *then* $x = 0$.

Proof The relation $0 < x$ is impossible by virtue of 7^0. □

9^0. *For any numbers* $a, b \in \mathbb{R}$ *such that* $a < b$ *there is a rational number* $r \in \mathbb{Q}$ *such that* $a < r < b$.

Proof Taking account of 7^0, we choose $n \in \mathbb{N}$ such that $0 < \frac{1}{n} < b - a$. By the principle of Archimedes we can find a number $m \in \mathbb{Z}$ such that $\frac{m-1}{n} \le a < \frac{m}{n}$. Then $\frac{m}{n} < b$, since otherwise we would have $\frac{m-1}{n} \le a < b \le \frac{m}{n}$, from which it would follow that $\frac{1}{n} \ge b - a$. Thus $r = \frac{m}{n} \in \mathbb{Q}$ and $a < \frac{m}{n} < b$. □

10^0. *For any number* $x \in \mathbb{R}$ *there exists a unique integer* $k \in \mathbb{Z}$ *such that* $k \le x < k + 1$.

Proof This follows immediately from the principle of Archimedes. □

The number k just mentioned is denoted $[x]$ and is called the *integer part* of x. The quantity $\{x\} := x - [x]$ is called the *fractional part* of x. Thus $x = [x] + \{x\}$, and $\{x\} \ge 0$.

2.2.4 The Geometric Interpretation of the Set of Real Numbers and Computational Aspects of Operations with Real Numbers

a. The Real Line

In relation to real numbers we often use a descriptive geometric language connected with a fact that you know in general terms from school. By the axioms of geometry there is a one-to-one correspondence $f : \mathbb{L} \to \mathbb{R}$ between the points of a line $\mathbb{L}$ and the set $\mathbb{R}$ of real numbers. Moreover this correspondence is connected with the rigid motions of the line. To be specific, if T is a parallel translation of the line $\mathbb{L}$ along itself, there exists a number $t \in \mathbb{R}$ (depending only on T) such that $f(T(x)) = f(x) + t$ for each point $x \in \mathbb{L}$.

The number $f(x)$ corresponding to a point $x \in \mathbb{L}$ is called the *coordinate* of x. In view of the one-to-one nature of the mapping $f : \mathbb{L} \to \mathbb{R}$, the coordinate of a point is often called simply a point. For example, instead of the phrase "let us take the point whose coordinate is 1" we say "let us take the point 1". Given the correspondence $f : \mathbb{L} \to \mathbb{R}$, we call the line $\mathbb{L}$ the *coordinate axis* or the *number axis* or the *real line.* Because f is bijective, the set $\mathbb{R}$ itself is also often called the real line and its points are called points of the real line.

As noted above, the bijective mapping $f : \mathbb{L} \to \mathbb{R}$ that defines coordinates on $\mathbb{L}$ has the property that under a parallel translation T the coordinates of the images of points of the line $\mathbb{L}$ differ from the coordinates of the points themselves by a number $t \in \mathbb{R}$, the same for every point. For this reason f is determined completely by specifying the point that is to have coordinate 0 and the point that is to have coordinate 1, or more briefly, by the point 0, called the *origin*, and the point 1. The closed interval determined by these points is called the *unit interval.* The direction determined by the ray with origin at 0 containing 1 is called the positive direction and a motion in that direction (from 0 to 1) is called a motion from left to right. In accordance with this convention, 1 lies to the right of 0 and 0 to the left of 1.

Under a parallel translation T that moves the origin x_0 to the point $x_1 = T(x_0)$ with coordinate 1, the coordinates of the images of all points are one unit larger than those of their pre-images, and therefore we locate the point $x_2 = T(x_1)$ with coordinate 2, the point $x_3 = T(x_2)$ with coordinate $3, \dots$, and the point $x_{n+1} = T(x_n)$ with coordinate $n + 1$, as well as the point $x_{-1} = T^{-1}(x_0)$ with coordinate $-1, \dots$, the point $x_{-n-1} = T^{-1}(x_{-n})$ with coordinate $-n - 1$. In this way we obtain all points with integer coordinates $m \in \mathbb{Z}$.

Knowing how to double, triple, ... the unit interval, we can use Thales' theorem to partition this interval into n congruent subintervals. By taking the subinterval having an endpoint at the origin, we find that the coordinate of its other end, which we denote by x, satisfies the equation $n \cdot x = 1$, that is, $x = \frac{1}{n}$. From this we find all points with rational coordinates $\frac{m}{n} \in \mathbb{Q}$.

But there still remain points of $\mathbb{L}$, since we know there are intervals incommensurable with the unit interval. Each such point, like every other point of the line, divides the line into two rays, on each of which there are points with integer or rational coordinates. (This is a consequence of the original geometric principle of Archimedes.) Thus a point produces a partition, or, as it is called, a *cut* of $\mathbb{Q}$ into two nonempty sets X and Y corresponding to the rational points (points with rational coordinates) on the left-hand and right-hand rays. By the axiom of completeness, there is a number c that separates X and Y, that is, $x \le c \le y$ for all $x \in X$ and all $y \in Y$. Since $X \cup Y = \mathbb{Q}$, it follows that $\sup X = s = i = \inf Y$. For otherwise, $s < i$ and there would be a rational number between s and i lying neither in X nor in Y. Thus $s = i = c$. This uniquely determined number c is assigned to the corresponding point of the line.

The assignment of coordinates to points of the line just described provides a visualizable model for both the order relation in $\mathbb{R}$ (hence the term "linear ordering") and for the axiom of completeness or continuity in $\mathbb{R}$, which in geometric language means that there are no "holes" in the line $\mathbb{L}$, which would separate it into two pieces having no points in common. (Such a separation could only come about by use of some point of the line $\mathbb{L}$.)

We shall not go into further detail about the construction of the mapping $f : \mathbb{L} \to \mathbb{R}$, since we shall invoke the geometric interpretation of the set of real numbers only for the sake of visualizability and perhaps to bring into play the reader's very useful geometric intuition. As for the formal proofs, just as before, they will rely either on the collection of facts we have obtained from the axioms for the real numbers or directly on the axioms themselves.

Geometric language, however, will be used constantly.

We now introduce the following notation and terminology for the number sets listed below:

$]a, b[:= \{x \in \mathbb{R} \mid a < x < b\}$ is the open interval ab;

$[a, b] := \{x \in \mathbb{R} \mid a \le x \le b\}$ is the closed interval ab;

$]a, b] := \{x \in \mathbb{R} \mid a < x \le b\}$ is the half-open interval ab containing b;

$[a, b[:= \{x \in \mathbb{R} \mid a \le x < b\}$ is the half-open interval ab containing a.

Definition 6 Open, closed, and half-open intervals are called *numerical intervals* or simply *intervals*. The numbers determining an interval are called its *endpoints*.

The quantity $b–a$ is called the *length* of the interval ab. If I is an interval, we shall denote its length by $|I|$. (The origin of this notation will soon become clear.)

The sets

$$]a, +\infty[:= \{x \in \mathbb{R} \mid a < x\}, \qquad]-\infty, b[:= \{x \in \mathbb{R} \mid x < b\}$$

$$[a, +\infty[:= \{x \in \mathbb{R} \mid a \le x\}, \qquad]-\infty, b] := \{x \in \mathbb{R} \mid x \le b\}$$

and $]-\infty, +\infty[:= \mathbb{R}$ are conventionally called *unbounded intervals* or *infinite intervals*.

In accordance with this use of the symbols $+\infty$ (read "plus infinity") and $-\infty$ (read "minus infinity") it is customary to denote the fact that the numerical set X is not bounded above (resp. below), by writing $\sup X = +\infty$ ($\inf X = -\infty$).

Definition 7 An open interval containing the point $x \in \mathbb{R}$ will be called a *neighborhood* of this point.

In particular, when $\delta > 0$, the open interval $]x - \delta, x + \delta[$ is called the δ-*neighborhood* of x. Its length is 2δ.

The distance between points $x, y \in \mathbb{R}$ is measured by the length of the interval having them as endpoints.

So as not to have to investigate which of the points is "left" and which is "right", that is, whether $x < y$ or $y < x$ and whether the length is $y - x$ or $x - y$, we can use the useful function

$$|x| = \begin{cases} x & \text{when } x > 0, \\ 0 & \text{when } x = 0, \\ -x & \text{when } x < 0, \end{cases}$$

which is called the *modulus* or *absolute value* of the number.

Definition 8 The *distance* between $x, y \in \mathbb{R}$ is the quantity $|x - y|$.

The distance is nonnegative and equals zero only when the points x and y are the same. The distance from x to y is the same as the distance from y to x, since $|x - y| = |y - x|$. Finally, if $z \in \mathbb{R}$, then $|x - y| \leq |x - z| + |z - y|$. That is, the so-called *triangle inequality* holds.

The triangle inequality follows from a property of the absolute value that is also called the triangle inequality (since it can be obtained from the preceding triangle inequality by setting $z = 0$ and replacing y by $-y$). To be specific, *the inequality*

$$|x + y| \leq |x| + |y|$$

holds for any numbers x and y, and equality holds only when the numbers x and y are both negative or both positive.

Proof If $0 \leq x$ and $0 \leq y$, then $0 \leq x + y$, $|x + y| = x + y$, $|x| = x$, and $|y| = y$, so that equality holds in this case.

If $x \leq 0$ and $y \leq 0$, then $x + y \leq 0$, $|x + y| = -(x + y) = -x - y$, $|x| = -x$, $|y| = -y$, and again we have equality.

Now suppose one of the numbers is negative and the other positive, for example, $x < 0 < y$. Then either $x < x + y \leq 0$ or $0 \leq x + y < y$. In the first case $|x + y| < |x|$, and in the second case $|x + y| < |y|$, so that in both cases $|x + y| < |x| + |y|$. □

Using the principle of induction, one can verify that

$$|x_1 + \cdots + x_n| \leq |x_1| + \cdots + |x_n|,$$

and equality holds if and only if the numbers $x_1, \ldots, x_n$ are all nonnegative or all nonpositive.

The number $\frac{a+b}{2}$ is often called the *midpoint* or *center* of the interval with endpoints a and b, since it is equidistant from the endpoints of the interval.

In particular, a point $x \in \mathbb{R}$ is the center of its δ-neighborhood $]x - \delta, x + \delta[$ and all points of the δ-neighborhood lie at a distance from x less than δ.

b. Defining a Number by Successive Approximations

In measuring a real physical quantity, we obtain a number that, as a rule, changes when the measurement is repeated, especially if one changes either the method of making the measurement or the instrument used. Thus the result of measurement is usually an approximate value of the quantity being sought. The quality or precision of a measurement is characterized, for example, by the magnitude of the possible discrepancy between the true value of the quantity and the value obtained for it by measurement. When this is done, it may happen that we can never exhibit the exact value of the quantity (if it exists theoretically). Taking a more constructive position, however, we may (or should) consider that we know the desired quantity completely if we can measure it with any preassigned precision. Taking this position is tantamount to identifying the number with a sequence[6] of more and more precise approximations by numbers obtained from measurement. But every measurement is a finite set of comparisons with some standard or with a part of the standard commensurable with it, so that the result of the measurement will necessarily be expressed in terms of natural numbers, integers, or, more generally, rational numbers. Hence theoretically the whole set of real numbers can be described in terms of sequences of rational numbers by constructing, after due analysis, a mathematical copy or, better expressed, a model of what people do with numbers who have no notion of their axiomatic description. The latter add and multiply the approximate values rather than the values being measured, which are unknown to them. (To be sure, they do not always know how to say what relation the result of these operations has to the result that would be obtained if the computations were carried out with the exact values. We shall discuss this question below.)

Having identified a number with a sequence of approximations to it, we should then, for example, add the sequences of approximate values when we wish to add two numbers. The new sequence thus obtained must be regarded as a new number, called the sum of the first two. But is it a number? The subtlety of the question resides in the fact that not every randomly constructed sequence is the sequence of arbitrarily precise approximations to some quantity. That is, one still has to learn how to determine from the sequence itself whether it represents some number or not. Another question that arises in the attempt to make a mathematical copy of operations with approximate numbers is that different sequences may be approximating sequences for the same quantity. The relation between sequences of approximations defining a number and the numbers themselves is approximately the same as that between a point on a map and an arrow on the map indicating the point. The arrow determines the point, but the point determines only the tip of the arrow, and does not exclude the use of a different arrow that may happen to be more convenient.

A precise description of these problems was given by Cauchy,[7] who carried out the entire program of constructing a model of the real numbers, which we have only sketched. One may hope that after you study the theory of limits you will be able to repeat these constructions independently of Cauchy.

What has been said up to now, of course, makes no claim to mathematical rigor. The purpose of this informal digression has been to direct the reader's attention to the theoretical possibility that more than one natural model of the real numbers may exist. I have also tried to give a picture of the relation of numbers to the world around us and to clarify the fundamental role of natural and rational numbers. Finally, I wished to show that approximate computations are both natural and necessary.

---

[6]If n is the number of the measurement and x_n the result of that measurement, the correspondence $n \mapsto x_n$ is simply a function $f : \mathbb{N} \to \mathbb{R}$ of a natural-number argument, that is, by definition a *sequence* (in this case a sequence of numbers). Section 3.1 is devoted to a detailed study of numerical sequences.

[7]A. Cauchy (1789–1857) – French mathematician, one of the most active creators of the language of mathematics and the machinery of classical analysis.

The next part of the present section is devoted to simple but important estimates of the errors that arise in arithmetic operations on approximate quantities. These estimates will be used below and are of independent interest.

We now give precise statements.

Definition 9 If x is the exact value of a quantity and $\tilde{x}$ a known approximation to the quantity, the numbers

$$\Delta(\tilde{x}) := |x - \tilde{x}|$$

and

$$\delta(\tilde{x}) := \frac{\Delta(\tilde{x})}{|\tilde{x}|}$$

are called respectively the *absolute* and *relative* error of approximation by $\tilde{x}$. The relative error is not defined when $\tilde{x} = 0$.

Since the value x is unknown, the values of $\Delta(\tilde{x})$ and $\delta(\tilde{x})$ are also unknown. However, one usually knows some upper bounds $\Delta(\tilde{x}) < \Delta$ and $\delta(\tilde{x}) < \delta$ for these quantities. In this case we say that the absolute or relative error does not exceed Δ or δ respectively. In practice we need to deal only with estimates for the errors, so that the quantities Δ and δ themselves are often called the *absolute* and *relative* errors. But we shall not do this.

The notation $x = \tilde{x} \pm \Delta$ means that $\tilde{x} - \Delta \le x \le \tilde{x} + \Delta$.

For example,

gravitational constant	$G = (6.672598 \pm 0.00085) \times 10^{-11}\ \mathrm{N \cdot m^2/kg^2}$,
speed of light *in vacuo*	$c = 299\,792\,458\ \mathrm{m/s}$ (exactly),
Planck's constant	$h = (6.6260755 \pm 0.0000040) \times 10^{-34}\ \mathrm{J \cdot s}$,
charge of an electron	$e = (1.60217733 \pm 0.00000049) \times 10^{-19}\ \mathrm{C}$,
rest mass of an electron	$m_e = (9.1093897 \pm 0.0000054) \times 10^{-31}\ \mathrm{kg}$.

The main indicator of the precision of a measurement is the relative error in approximation, usually expressed as a percent.

Thus in the examples just given the relative errors are at most (in order):

$$13 \times 10^{-5}; \quad 0; \quad 6 \times 10^{-7}; \quad 31 \times 10^{-8}; \quad 6 \times 10^{-7}$$

or, as percents of the measured values,

$$13 \times 10^{-3}\ \%; \quad 0\ \%; \quad 6 \times 10^{-5}\ \%; \quad 31 \times 10^{-6}\ \%; \quad 6 \times 10^{-5}\ \%.$$

We now estimate the errors that arise in arithmetic operations with approximate quantities.

Proposition *If*

$$|x - \tilde{x}| = \Delta(\tilde{x}), \qquad |y - \tilde{y}| = \Delta(\tilde{y}),$$

then

$$\Delta(\tilde{x}+\tilde{y}) := \big|(x+y)-(\tilde{x}+\tilde{y})\big| \le \Delta(\tilde{x})+\Delta(\tilde{y}), \tag{2.1}$$

$$\Delta(\tilde{x}\cdot\tilde{y}) := |x\cdot y-\tilde{x}\cdot\tilde{y}| \le |\tilde{x}|\Delta(\tilde{y})+|\tilde{y}|\Delta(\tilde{x})+\Delta(\tilde{x})\cdot\Delta(\tilde{y}); \tag{2.2}$$

if, in addition,

$$y \ne 0, \qquad \tilde{y}\ne 0 \quad \textit{and} \quad \delta(\tilde{y}) = \frac{\Delta(\tilde{y})}{|\tilde{y}|} < 1,$$

then

$$\Delta\left(\frac{\tilde{x}}{\tilde{y}}\right) := \left|\frac{x}{y}-\frac{\tilde{x}}{\tilde{y}}\right| \le \frac{|\tilde{x}|\Delta(\tilde{y})+|\tilde{y}|\Delta(\tilde{x})}{\tilde{y}^2}\cdot\frac{1}{1-\delta(\tilde{y})}. \tag{2.3}$$

Proof Let $x=\tilde{x}+\alpha$ and $y=\tilde{y}+\beta$. Then

$$\begin{aligned}
\Delta(\tilde{x}+\tilde{y}) &= \big|(x+y)-(\tilde{x}+\tilde{y})\big| = |\alpha+\beta| \le |\alpha|+|\beta| = \Delta(\tilde{x})+\Delta(\tilde{y}),\\
\Delta(\tilde{x}\cdot\tilde{y}) &= |xy-\tilde{x}\cdot\tilde{y}| = \big|(\tilde{x}+\alpha)(\tilde{y}+\beta)-\tilde{x}\cdot\tilde{y}\big| =\\
&= |\tilde{x}\beta+\tilde{y}\alpha+\alpha\beta| \le |\tilde{x}||\beta|+|\tilde{y}||\alpha|+|\alpha\beta| =\\
&= |\tilde{x}|\Delta(\tilde{y})+|\tilde{y}|\Delta(\tilde{x})+\Delta(\tilde{x})\cdot\Delta(\tilde{y}),\\
\Delta\left(\frac{\tilde{x}}{\tilde{y}}\right) &= \left|\frac{x}{y}-\frac{\tilde{x}}{\tilde{y}}\right| = \left|\frac{x\tilde{y}-y\tilde{x}}{y\tilde{y}}\right| =\\
&= \left|\frac{(\tilde{x}+\alpha)\tilde{y}-(\tilde{y}+\beta)\tilde{x}}{\tilde{y}^2}\right|\cdot\left|\frac{1}{1+\beta/\tilde{y}}\right| \le \frac{|\tilde{x}||\beta|+|\tilde{y}||\alpha|}{\tilde{y}^2}\cdot\frac{1}{1-\delta(\tilde{y})} =\\
&= \frac{|\tilde{x}|\Delta(\tilde{y})+|\tilde{y}|\Delta(\tilde{x})}{\tilde{y}^2}\cdot\frac{1}{1-\delta(\tilde{y})}.
\end{aligned}$$

□

These estimates for the absolute errors imply the following estimates for the relative errors:

$$\delta(\tilde{x}+\tilde{y}) \le \frac{\Delta(\tilde{x})+\Delta(\tilde{y})}{|\tilde{x}+\tilde{y}|}, \tag{2.1'}$$

$$\delta(\tilde{x}\cdot\tilde{y}) \le \delta(\tilde{x})+\delta(\tilde{y})+\delta(\tilde{y})\cdot\delta(\tilde{y}), \tag{2.2'}$$

$$\delta\left(\frac{\tilde{x}}{\tilde{y}}\right) \le \frac{\delta(\tilde{x})+\delta(\tilde{y})}{1-\delta(\tilde{y})}. \tag{2.3'}$$

In practice, when working with sufficiently good approximations, we have $\Delta(\tilde{x})\cdot\Delta(\tilde{y})\approx 0$, $\delta(\tilde{x})\cdot\delta(\tilde{y})\approx 0$, and $1-\delta(\tilde{y})\approx 1$, so that one can use the following simplified and useful, but formally incorrect, versions of formulas (2.2), (2.3), (2.2′), and (2.3′):

$$\Delta(\tilde{x}\cdot\tilde{y}) \le |\tilde{x}|\Delta(\tilde{y})+|\tilde{y}|\Delta(\tilde{x}),$$

$$\Delta\left(\frac{\tilde{x}}{\tilde{y}}\right) \le \frac{|\tilde{x}|\Delta(\tilde{y}) + \tilde{y}\Delta(\tilde{x})}{\tilde{y}^2},$$

$$\delta(\tilde{x}\cdot\tilde{y}) \le \delta(\tilde{x}) + \delta(\tilde{y}),$$

$$\delta\left(\frac{\tilde{x}}{\tilde{y}}\right) \le \delta(\tilde{x}) + \delta(\tilde{y}).$$

Formulas (2.3) and (2.3′) show that it is necessary to avoid dividing by a number that is near zero and also to avoid using rather crude approximations in which $\tilde{y}$ or $1-\delta(\tilde{y})$ is small in absolute value.

Formula (2.1′) warns against adding approximate quantities if they are close to each other in absolute value but opposite in sign, since then $|\tilde{x}+\tilde{y}|$ is close to zero.

In all these cases, the errors may increase sharply.

For example, suppose your height has been measured twice by some device, and the precision of the measurement is ± 0.5 cm. Suppose a sheet of paper was placed under your feet before the second measurement. It may nevertheless happen that the results of the measurement are as follows: $H_1 = (200 \pm 0.5)$ cm and $H_2 = (199.8 \pm 0.5)$ cm respectively.

It does not make sense to try to find the thickness of the paper in the form of the difference $H_2 - H_1$, from which it would follow only that the thickness of the paper is not larger than 0.8 cm. That would of course be a crude reflection (if indeed one could even call it a "reflection") of the true situation.

However, it is worthwhile to consider another more hopeful computational effect through which comparatively precise measurements can be carried out with crude devices. For example, if the device just used for measuring your height was used to measure the thickness of 1000 sheets of the same paper, and the result was (20 ± 0.5) cm, then the thickness of one sheet of paper is (0.02 ± 0.0005) cm, which is (0.2 ± 0.005) mm, as follows from formula (2.1).

That is, with an absolute error not larger than 0.005 mm, the thickness of one sheet is 0.2 mm. The relative error in this measurement is at most 0.025 or 2.5 %.

This idea can be developed and has been proposed, for example, as a way of detecting a weak periodic signal amid the larger random static usually called white noise.

c. The Positional Computation System

It was stated above that every real number can be presented as a sequence of rational approximations. We now recall a method, which is important when it comes to computation, for constructing in a uniform way a sequence of such rational approximations for every real number. This method leads to the positional computation system.

Lemma *If a number $q > 1$ is fixed, then for every positive number $x \in \mathbb{R}$ there exists a unique integer $k \in \mathbb{Z}$ such that*

$$q^{k-1} \le x < q^k.$$

Proof We first verify that the set of numbers of the form q^k, $k \in \mathbb{N}$, is not bounded above. If it were, it would have a least upper bound s, and by definition of the least upper bound, there would be a natural number $m \in \mathbb{N}$ such that $\frac{s}{q} < q^m \le s$. But then $s < q^{m+1}$, so that s could not be an upper bound of the set.

Since $1 < q$, it follows that $q^m < q^n$ when $m < n$ for all $m, n \in \mathbb{Z}$. Hence we have also shown that for every real number $c \in \mathbb{R}$ there exists a natural number $N \in \mathbb{N}$ such that $c < q^n$ for all $n > N$.

It follows that for any $\varepsilon > 0$ there exists $M \in \mathbb{N}$ such that $\frac{1}{q^m} < \varepsilon$ for all natural numbers $m > M$.

Indeed, it suffices to set $c = \frac{1}{\varepsilon}$ and $N = M$; then $\frac{1}{\varepsilon} < q^m$ when $m > M$.

Thus the set of integers $m \in \mathbb{Z}$ satisfying the inequality $x < q^m$ for $x > 0$ is bounded below. It therefore has a minimal element k, which obviously will be the one we are seeking, since, for this integer, $q^{k-1} \le x < q^k$.

The uniqueness of such an integer k follows from the fact that if $m, n \in \mathbb{Z}$ and, for example, $m < n$, then $m \le n - 1$. Hence if $q > 1$, then $q^m \le q^{n-1}$.

Indeed, it can be seen from this remark that the inequalities $q^{m-1} \le x < q^m$ and $q^{n-1} \le x < q^n$, which imply $q^{n-1} \le x < q^m$, are incompatible if $m \ne n$. □

We shall use this lemma in the following construction. Fix $q > 1$ and take an arbitrary positive number $x \in \mathbb{R}$. By the lemma we find a unique number $p \in \mathbb{Z}$ such that

$$q^p \le x < q^{p+1}. \tag{2.4}$$

Definition 10 The number p satisfying (2.4) is called the *order of x in the base q* or (when q is fixed) simply the *order of x*.

By the principle of Archimedes, we find a unique natural number $\alpha_p \in \mathbb{N}$ such that

$$\alpha_p q^p \le x < \alpha_p q^p + q^p. \tag{2.5}$$

Taking (2.4) into account, one can assert that $\alpha_p \in \{1, \dots, q-1\}$.

All of the subsequent steps in our construction will repeat the step we are about to take, starting from relation (2.5).

It follows from relation (2.5) and the principle of Archimedes that there exists a unique number $\alpha_{p-1} \in \{0, 1, \dots, q-1\}$ such that

$$\alpha_p q^p + \alpha_{p-1} q^{p-1} \le x < \alpha_p q^p + \alpha_{p-1} q^{p-1} + q^{p-1}. \tag{2.6}$$

If we have made n such steps, obtaining the relation

$$\begin{aligned} &\alpha_p q^p + \alpha_{p-1} q^{p-1} + \dots + \alpha_{p-n} q^{p-n} \le \\ &\quad \le x < \alpha_p q^p + \alpha_{p-1} q^{p-1} + \dots + \alpha_{p-n} q^{p-n} + q^{p-n}, \end{aligned}$$

then by the principle of Archimedes there exists a unique number $\alpha_{p-n-1} \in \{0, 1, \dots, q-1\}$ such that

$$\alpha_p q^p + \dots + \alpha_{p-n} q^{p-n} + \alpha_{p-n-1} q^{p-n-1} \le$$
$$\le x < \alpha_p q^p + \dots + \alpha_{p-n} q^{p-n} + \alpha_{p-n-1} q^{p-n-1} + q^{p-n-1}.$$

Thus we have exhibited an algorithm by means of which a sequence of numbers $\alpha_p, \alpha_{p-1}, \dots, \alpha_{p-n}, \dots$ from the set $\{0, 1, \dots, q-1\}$ is placed in correspondence with the positive number x. Less formally, we have constructed a sequence of rational numbers of the special form

$$r_n = \alpha_p q^p + \dots + \alpha_{p-n} q^{p-n}, \tag{2.7}$$

and such that

$$r_n \le x < r_n + \frac{1}{q^{n-p}}. \tag{2.8}$$

In other words, we construct better and better approximations from below and from above to the number x using the special sequence (2.7). The symbol $\alpha_p \dots \alpha_{p-n} \dots$ is a code for the entire sequence $\{r_n\}$. To recover the sequence $\{r_n\}$ from this symbol it is necessary to indicate the value of p, the order of x.

For $p \ge 0$ it is customary to place a period or comma after α_0; for $p < 0$, the convention is to place $|p|$ zeros left of α_p and a period or comma right of the leftmost zero (we recall that $\alpha_p \ne 0$).

For example, when $q = 10$,

$$123.45 := 1 \times 10^2 + 2 \times 10^1 + 3 \times 10^0 + 4 \times 10^{-1} + 5 \times 10^{-2},$$
$$0.00123 := 1 \times 10^{-3} + 2 \times 10^{-4} + 3 \times 10^{-5};$$

and when $q = 2$,

$$1000.001 := 1 \cdot 2^3 + 1 \cdot 2^{-3}.$$

Thus the value of a digit in the symbol $\alpha_p \dots \alpha_{p-n} \dots$ depends on the position it occupies relative to the period or comma.

With this convention, the symbol $\alpha_p \dots \alpha_0 \dots$ makes it possible to recover the whole sequence of approximations.

It can be seen by inequalities (2.8) (verify this!) that different sequences $\{r_n\}$ and $\{r'_n\}$, and therefore different symbols $\alpha_p \dots \alpha_0 \dots$ and $\alpha'_p \dots \alpha'_0 \dots$, correspond to different numbers x and x'.

We now answer the question whether some real number $x \in \mathbb{R}$ corresponds to every symbol $\alpha_p \dots \alpha_0 \dots$. The answer turns out to be negative.

We remark that by virtue of the algorithm just described for obtaining the numbers $\alpha_{p-n} \in \{0, 1, \dots, q-1\}$ successively, it cannot happen that all these numbers from some point on are equal to $q-1$.

Indeed, if

$$r_n = \alpha_p q^p + \cdots + \alpha_{p-k} q^{p-k} + (q-1) q^{p-k-1} + \cdots + (q-1) q^{p-n}$$

for all $n > k$, that is,

$$r_n = r_k + \frac{1}{q^{k-p}} - \frac{1}{q^{n-p}}, \tag{2.9}$$

then by (2.8) we have

$$r_k + \frac{1}{q^{k-p}} - \frac{1}{q^{n-p}} \le x < r_k + \frac{1}{q^{k-p}}.$$

Then for any $n > k$

$$0 < r_k + \frac{1}{q^{k-p}} - x < \frac{1}{q^{n-p}},$$

which, as we know from 8^0 above, is impossible.

It is also useful to note that if at least one of the numbers $\alpha_{p-k-1}, \ldots, \alpha_{p-n}$ is less than $q-1$, then instead of (2.9) we can write

$$r_n < r_k + \frac{1}{q^{k-p}} - \frac{1}{q^{n-p}}$$

or, what is the same

$$r_n + \frac{1}{q^{n-p}} < r_k + \frac{1}{q^{k-p}}. \tag{2.10}$$

We can now prove that any symbol $\alpha_n \ldots \alpha_0 \ldots$, composed of the numbers $\alpha_k \in \{0, 1, \ldots, q-1\}$, and in which there are numbers different from $q-1$ with arbitrarily large indices, corresponds to some number $x \ge 0$.

Indeed, from the symbol $\alpha_p \ldots \alpha_{p-n} \ldots$ let us construct the sequence $\{r_n\}$ of the form (2.7). By virtue of the relations $r_0 \le r_1 \le r_n \le \cdots$, taking account of (2.9) and (2.10), we have

$$r_0 \le r_1 \le \cdots \le \cdots < \cdots \le r_n + \frac{1}{q^{n-p}} \le \cdots \le r_1 + \frac{1}{q^{1-p}} \le r_0 + \frac{1}{q^{-p}}. \tag{2.11}$$

The strict inequalities in this last relation should be understood as follows: every element of the left-hand sequence is less than every element of the right-hand sequence. This follows from (2.10).

If we now take $x = \sup_{n \in \mathbb{N}} r_n (= \inf_{n \in \mathbb{N}} (r_n + q^{-(n-p)}))$, then the sequence $\{r_n\}$ will satisfy conditions (2.7) and (2.8), that is, the symbol $\alpha_p \ldots \alpha_{p-n} \ldots$ corresponds to the number $x \in \mathbb{R}$.

Thus, we have established a one-to-one correspondence between the positive numbers $x \in \mathbb{R}$ and symbols of the form $\alpha_p \ldots \alpha_0, \ldots$ if $p \ge 0$ or $\underbrace{0, 0 \ldots 0}_{|p| \text{ zeros}} \alpha_p \ldots$

if $p < 0$. The symbol assigned to x is called the *q-ary representation of* x; the numbers that occur in the symbol are called its *digits*, and the position of a digit relative to the period is called its *rank*.

We agree to assign to a number $x < 0$ the symbol for the positive number $-x$, prefixed by a negative sign. Finally, we assign the symbol $0.0\dots0\dots$ to the number 0.

In this way we have constructed the *positional q-ary system of writing real numbers*.

The most useful systems are the decimal system (in common use) and for technical reasons the binary system (in electronic computers). Less common, but also used in some parts of computer engineering are the ternary and octal systems.

Formulas (2.7) and (2.8) show that if only a finite number of digits are retained in the q-ary expression of x (or, if we wish, we may say that the others are replaced with zeros), then the absolute error of the resulting approximation (2.7) for x does not exceed one unit in the last rank retained.

This observation makes it possible to use the formulas obtained in Paragraph b to estimate the errors that arise when doing arithmetic operations on numbers as a result of replacing the exact numbers by the corresponding approximate values of the form (2.7).

This last remark also has a certain theoretical value. To be specific, if we identify a real number x with its q-ary expression, as was suggested in Paragraph b, once we have learned to perform arithmetic operations directly on the q-ary symbols, we will have constructed a new model of the real numbers, seemingly of greater value from the computational point of view.

The main problems that need to be solved in this direction are the following:

To two q-ary symbols it is necessary to assign a new symbol representing their sum. It will of course be constructed one step at a time. To be specific, by adding more and more precise rational approximations of the original numbers, we shall obtain rational approximations corresponding to their sum. Using the remark made above, one can show that as the precision of the approximations of the terms increases, we shall obtain more and more q-ary digits of the sum, which will then not vary under subsequent improvements in the approximation.

This same problem needs to be solved with respect to multiplication.

Another, less constructive, route for passing from rational numbers to all real numbers is due to Dedekind.

Dedekind identifies a real number with a cut in the set $\mathbb{Q}$ of rational numbers, that is, a partition of $\mathbb{Q}$ into two disjoint sets A and B such that $a < b$ for all $a \in A$ and all $b \in B$. Under this approach to real numbers our axiom of completeness (continuity) becomes a well-known theorem of Dedekind. For that reason the axiom of completeness in the form we have given it is sometimes called Dedekind's axiom.

To summarize, in the present section we have exhibited the most important classes of numbers. We have shown the fundamental role played by the natural and rational numbers. It has been shown how the basic properties of these numbers follow from the axiom system[8] we have adopted. We have given a picture of various models of the set of real numbers. We have discussed the computational aspects of the theory of real numbers: estimates of the errors arising during arithmetical operations with approximate magnitudes, and the q-ary positional computation system.

2.2.5 Problems and Exercises

1. Using the principle of induction, show that

a) the sum $x_1+\cdots+x_n$ of real numbers is defined independently of the insertion of parentheses to specify the order of addition;

b) the same is true of the product $x_1\cdots x_n$;

c) $|x_1+\cdots+x_n| \le |x_1|+\cdots+|x_n|$;

d) $|x_1\cdots x_n| = |x_1|\cdots|x_n|$;

e) $((m,n\in\mathbb{N})\wedge(m<n)) \Rightarrow ((n-m)\in\mathbb{N})$;

f) $(1+x)^n \ge 1+nx$ for $x>-1$ and $n\in\mathbb{N}$, equality holding only when $n=1$ or $x=0$ (*Bernoulli's inequality*);

g) $(a+b)^n = a^n + \frac{n}{1!}a^{n-1}b + \frac{n(n-1)}{2!}a^{n-2}b^2 + \cdots + \frac{n(n-1)\cdots 2}{(n-1)!}ab^{n-1} + b^n$ (*Newton's binomial formula*).

2. a) Verify that $\mathbb{Z}$ and $\mathbb{Q}$ are inductive sets.

b) Give examples of inductive sets different from $\mathbb{N}$, $\mathbb{Z}$, $\mathbb{Q}$, and $\mathbb{R}$.

3. Show that an inductive set is not bounded above.

4. a) Show that an inductive set is infinite (that is, equipotent with one of its subsets different from itself).

b) The set $E_n = \{x\in\mathbb{N} \mid x\le n\}$ is finite. (We denote card E_n by n.)

5. (The *Euclidean algorithm*) a) Let $m,n\in\mathbb{N}$ and $m>n$. Their greatest common divisor $(\gcd(m,n) = d\in\mathbb{N})$ can be found in a finite number of steps using the following algorithm of Euclid involving successive divisions with remainder.

$$\begin{aligned} m &= q_1 n + r_1 \quad (r_1 < n),\\ n &= q_2 r_1 + r_2 \quad (r_2 < r_1),\\ r_1 &= q_3 r_2 + r_3 \quad (r_3 < r_2),\\ &\ \ \vdots\\ r_{k-1} &= q_{k+1} r_k + 0. \end{aligned}$$

Then $d = r_k$.

[8]It was stated by Hilbert in almost the form given above at the turn of the twentieth century. See for example Hilbert, D. *Foundations of Geometry*, Chap. III, §13. (Translated from the second edition of *Grundlagen der Geometrie*, La Salle, Illinois: Open Court Press, 1971. This section was based on Hilbert's article "Über den Zahlbegriff" in *Jahresbericht der deutschen Mathematikervereinigung* **8** (1900).)

b) If $d = \gcd(m, n)$, one can choose numbers $p, q \in \mathbb{Z}$ such that $pm + qn = d$; in particular, if m and n are relatively prime, then $pm + qn = 1$.

6. Try to give your own proof of the fundamental theorem of arithmetic (Paragraph a in Sect. 2.2.2).

7. Show that if the product $m \cdot n$ of natural numbers is divisible by a prime p, that is, $m \cdot n = p \cdot k$, where $k \in \mathbb{N}$, then either m or n is divisible by p.

8. It follows from the fundamental theorem of arithmetic that the set of prime numbers is infinite.

9. Show that if the natural number n is not of the form k^m, where $k, m \in \mathbb{N}$, then the equation $x^m = n$ has no rational roots.

10. Show that the expression of a rational number in any q-ary computation system is periodic, that is, starting from some rank it consists of periodically repeating groups of digits.

11. Let us call an irrational number $\alpha \in \mathbb{R}$ *well approximated* by rational numbers if for any natural numbers $n, N \in \mathbb{N}$ there exists a rational number $\frac{p}{q}$ such that $|\alpha - \frac{p}{q}| < \frac{1}{Nq^n}$.

a) Construct an example of a well-approximated irrational number.

b) Prove that a well-approximated irrational number cannot be algebraic, that is, it is transcendental (*Liouville's theorem*).[9]

12. Knowing that $\frac{m}{n} := m \cdot n^{-1}$ by definition, where $m \in \mathbb{Z}$ and $n \in \mathbb{N}$, derive the "rules" for addition, multiplication, and division of fractions, and also the condition for two fractions to be equal.

13. Verify that the rational numbers $\mathbb{Q}$ satisfy all the axioms for real numbers except the axiom of completeness.

14. Adopting the geometric model of the set of real numbers (the real line), show how to construct the numbers $a + b$, $a - b$, ab, and $\frac{a}{b}$ in this model.

15. a) Illustrate the axiom of completeness on the real line.

b) Prove that the least-upper-bound principle is equivalent to the axiom of completeness.

16. a) If $A \subset B \subset \mathbb{R}$, then $\sup A \le \sup B$ and $\inf A \ge \inf B$.

b) Let $\mathbb{R} \supset X \ne \varnothing$ and $\mathbb{R} \supset Y \ne \varnothing$. If $x \le y$ for all $x \in X$ and all $y \in Y$, then X is bounded above, Y is bounded below, and $\sup X \le \inf Y$.

c) If the sets X, Y in b) are such that $X \cup Y = \mathbb{R}$, then $\sup X = \inf Y$.

d) If X and Y are the sets defined in c), then either X has a maximal element or Y has a minimal element. (*Dedekind's theorem.*)

e) (Continuation.) Show that Dedekind's theorem is equivalent to the axiom of completeness.

17. Let $A + B$ be the set of numbers of the form $a + b$ and $A \cdot B$ the set of numbers of the form $a \cdot b$, where $a \in A \subset \mathbb{R}$ and $b \in B \subset \mathbb{R}$. Determine whether it is always true that

---

[9] J. Liouville (1809–1882) – French mathematician, who wrote on complex analysis, geometry, differential equations, number theory, and mechanics.

a) $\sup(A+B) = \sup A + \sup B$,
b) $\sup(A \cdot B) = \sup A \cdot \sup B$.

18. Let $-A$ be the set of numbers of the form $-a$, where $a \in A \subset \mathbb{R}$. Show that $\sup(-A) = -\inf A$.

19. a) Show that for $n \in \mathbb{N}$ and $a > 0$ the equation $x^n = a$ has a positive root (denoted $\sqrt[n]{a}$ or $a^{1/n}$).

b) Verify that for $a > 0$, $b > 0$, and $n, m \in \mathbb{N}$

$$\sqrt[n]{ab} = \sqrt[n]{a} \cdot \sqrt[n]{b} \quad \text{and} \quad \sqrt[n]{\sqrt[m]{a}} = \sqrt[n \cdot m]{a}.$$

c) $(a^{\frac{1}{n}})^m = (a^m)^{\frac{1}{n}} =: a^{m/n}$ and $a^{1/n} \cdot a^{1/m} = a^{1/n+1/m}$.
d) $(a^{m/n})^{-1} = (a^{-1})^{m/n} =: a^{-m/n}$.
e) Show that for all $r_1, r_2 \in \mathbb{Q}$

$$a^{r_1} \cdot a^{r_2} = a^{r_1+r_2} \quad \text{and} \quad \left(a^{r_1}\right)^{r_2} = a^{r_1 r_2}.$$

20. a) Show that the inclusion relation is a partial ordering relation on sets (but not a linear ordering!).

b) Let A, B, and C be sets such that $A \subset C$, $B \subset C$, $A \backslash B \neq \varnothing$, and $B \backslash A \neq \varnothing$. We introduce a partial ordering into this triple of sets as in a). Exhibit the maximal and minimal elements of the set $\{A, B, C\}$. (Pay attention to the non-uniqueness!)

21. a) Show that, just like the set $\mathbb{Q}$ of rational numbers, the set $\mathbb{Q}(\sqrt{n})$ of numbers of the form $a + b\sqrt{n}$, where $a, b \in \mathbb{Q}$ and n is a fixed natural number that is not the square of any integer, is an ordered set satisfying the principle of Archimedes but not the axiom of completeness.

b) Determine which axioms for the real numbers do not hold for $\mathbb{Q}(\sqrt{n})$ if the standard arithmetic operations are retained in $\mathbb{Q}(\sqrt{n})$ but order is defined by the rule $(a + b\sqrt{n} \leq a' + b'\sqrt{n}) := ((b < b') \vee ((b = b') \wedge (a \leq a')))$. Will $\mathbb{Q}(\sqrt{n})$ now satisfy the principle of Archimedes?

c) Order the set $\mathbb{P}[x]$ of polynomials with rational or real coefficients by specifying that

$$P_m(x) = a_0 + a_1 x + \cdots + a_m x^m \succ 0, \quad \text{if } a_m > 0.$$

d) Show that the set $\mathbb{Q}(x)$ of rational fractions

$$R_{m,n} = \frac{a_0 + a_1 x + \cdots + a_m x^m}{b_0 + b_1 x + \cdots + b_n x^n}$$

with coefficients in $\mathbb{Q}$ or $\mathbb{R}$ becomes an ordered field, but not an Archimedean ordered field, when the order relation $R_{m,n} \succ 0$ is defined to mean $a_m b_n > 0$ and the usual arithmetic operations are introduced. This means that the principle of Archimedes cannot be deduced from the other axioms for $\mathbb{R}$ without using the axiom of completeness.

22. Let $n \in \mathbb{N}$ and $n > 1$. In the set $E_n = \{0, 1, \dots, n-1\}$ we define the sum and product of two elements as the remainders when the usual sum and product in $\mathbb{R}$ are divided by n. With these operations defined on it, the set E_n is denoted $\mathbb{Z}_n$.

a) Show that if n is not a prime number, then there are nonzero numbers m, k in $\mathbb{Z}_n$ such that $m \cdot k = 0$. (Such numbers are called *zero divisors*.) This means that in $\mathbb{Z}_n$ the equation $a \cdot b = c \cdot b$ does not imply that $a = c$, even when $b \neq 0$.

b) Show that if p is prime, then there are no zero divisors in $\mathbb{Z}_p$ and $\mathbb{Z}_p$ is a field.

c) Show that, no matter what the prime p, $\mathbb{Z}_p$ cannot be ordered in a way consistent with the arithmetic operations on it.

23. Show that if $\mathbb{R}$ and $\mathbb{R}'$ are two models of the set of real numbers and $f : \mathbb{R} \to \mathbb{R}'$ is a mapping such that $f(x+y) = f(x) + f(y)$ and $f(x \cdot y) = f(x) \cdot f(y)$ for any $x, y \in \mathbb{R}$, then

a) $f(0) = 0'$;

b) $f(1) = 1'$ if $f(x) \not\equiv 0'$, which we shall henceforth assume;

c) $f(m) = m'$ where $m \in \mathbb{Z}$ and $m' \in \mathbb{Z}'$, and the mapping $f : \mathbb{Z} \to \mathbb{Z}'$ is injective and preserves the order.

d) $f(\frac{m}{n}) = \frac{m'}{n'}$, where $m, n \in \mathbb{Z}$, $n \neq 0$, $m', n' \in \mathbb{Z}'$, $n' \neq 0'$, $f(m) = m'$, $f(n) = n'$. Thus $f : \mathbb{Q} \to \mathbb{Q}'$ is a bijection that preserves order.

e) $f : \mathbb{R} \to \mathbb{R}'$ is a bijective mapping that preserves order.

24. On the basis of the preceding exercise and the axiom of completeness, show that the axiom system for the set of real numbers determines it completely up to an isomorphism (method of realizing it), that is, if $\mathbb{R}$ and $\mathbb{R}'$ are two sets satisfying these axioms, then there exists a one-to-one correspondence $f : \mathbb{R} \to \mathbb{R}'$ that preserves the arithmetic operations and the order: $f(x+y) = f(x) + f(y)$, $f(x \cdot y) = f(x) \cdot f(y)$, and $(x \le y) \Leftrightarrow (f(x) \le f(y))$.

25. A number x is represented on a computer as

$$x = \pm q^p \sum_{n=1}^{k} \frac{\alpha_n}{q^n},$$

where p is the order of x and $M = \sum_{n=1}^{k} \frac{\alpha_n}{q^n}$ is the mantissa of the number x ($\frac{1}{q} \le M < 1$).

Now a computer works only with a certain range of numbers: for $q = 2$ usually $|p| \le 64$, and $k = 35$. Evaluate this range in the decimal system.

26. a) Write out the (6×6) multiplication table for multiplication in base 6.

b) Using the result of a), multiply "columnwise" in the base-6 system

$$\begin{array}{c} (532)_6 \\ \times \\ \underline{(145)_6} \end{array}$$

and check your work by repeating the computation in the decimal system.

c) Perform the "long" division

$$(1301)_6 \underline{|(25)_6}$$

and check your work by repeating the computation in the decimal system.

d) Perform the "columnwise" addition

$$\begin{array}{c}(4052)_6\\ \times\\ \underline{(3125)_6}\end{array}$$

27. Write $(100)_{10}$ in the binary and ternary systems.

28. a) Show that along with the unique representation of an integer as

$$(\alpha_n \alpha_{n-1} \dots \alpha_0)_3,$$

where $\alpha_i \in \{0, 1, 2\}$, it can also be written as

$$(\beta_n \beta_{n-1} \dots \beta_0)_3,$$

where $\beta \in \{-1, 0, 1\}$.

b) What is the largest number of coins from which one can detect a counterfeit in three weighings with a pan balance, if it is known in advance only that the counterfeit coin differs in weight from the other coins?

29. What is the smallest number of questions to be answered "yes" or "no" that one must pose in order to be sure of determining a 7-digit telephone number?

30. a) How many different numbers can one define using 20 decimal digits (for example, two ranks with 10 possible digits in each)? Answer the same question for the binary system. Which system does a comparison of the results favor in terms of efficiency?

b) Evaluate the number of different numbers one can write, having at one's disposal n digits of a q-ary system. (Answer: $q^{n/q}$.)

c) Draw the graph of the function $f(x) = x^{n/x}$ over the set of natural-number values of the argument and compare the efficiency of the different systems of computation.

2.3 Basic Lemmas Connected with the Completeness of the Real Numbers

In this section we shall establish some simple useful principles, each of which could have been used as the axiom of completeness in our construction of the real numbers.[10]

[10] See Problem 4 at the end of this section.

We have called these principles basic lemmas in view of their extensive application in the proofs of a wide variety of theorems in analysis.

2.3.1 The Nested Interval Lemma (Cauchy–Cantor Principle)

Definition 1 A function $f : \mathbb{N} \to X$ of a natural-number argument is called a *sequence* or, more fully, a *sequence of elements of* X.

The value $f(n)$ of the function f corresponding to the number $n \in \mathbb{N}$ is often denoted x_n and called the nth term of the sequence.

Definition 2 Let $X_1, X_2, \dots, X_n, \dots$ be a sequence of sets. If $X_1 \supset X_2 \supset \dots \supset X_n \supset \cdots$, that is $X_n \supset X_{n+1}$ for all $n \in \mathbb{N}$, we say the sequence is *nested.*

Lemma (Cauchy–Cantor) *For any nested sequence $I_1 \supset I_2 \supset \dots \supset I_n \supset \cdots$ of closed intervals, there exists a point $c \in \mathbb{R}$ belonging to all of these intervals.*

If in addition it is known that for any $\varepsilon > 0$ there is an interval I_k whose length $|I_k|$ is less than ε, then c is the unique point common to all the intervals.

Proof We begin by remarking that for any two closed intervals $I_m = [a_m, b_m]$ and $I_n = [a_n, b_n]$ of the sequence we have $a_m \le b_n$. For otherwise we would have $a_n \le b_n < a_m \le b_m$, that is, the intervals I_m and I_n would be mutually disjoint, while one of them (the one with the larger index) is contained in the other.

Thus the numerical sets $A = \{a_m \mid m \in \mathbb{N}\}$ and $B = \{b_n \mid n \in \mathbb{N}\}$ satisfy the hypotheses of the axiom of completeness, by virtue of which there is a number $c \in \mathbb{R}$ such that $a_m \le c \le b_n$ for all $a_m \in A$ and all $b_n \in B$. In particular, $a_n \le c \le b_n$ for all $n \in \mathbb{N}$. But that means that the point c belongs to all the intervals I_n.

Now let c_1 and c_2 be two points having this property. If they are different, say $c_1 < c_2$, then for any $n \in \mathbb{N}$ we have $a_n \le c_1 < c_2 \le b_n$, and therefore $0 < c_2 - c_1 < b_n - a_n$, so that the length of an interval in the sequence cannot be less than $c_2 - c_1$. Hence if there are intervals of arbitrarily small length in the sequence, their common point is unique. □

2.3.2 The Finite Covering Lemma (Borel–Lebesgue Principle, or Heine–Borel Theorem)

Definition 3 A system $S = \{X\}$ of sets X is said to *cover* a set Y if $Y \subset \bigcup_{X \in S} X$, (that is, if every element $y \in Y$ belongs to at least one of the sets X in the system S).

A subset of a set $S = \{X\}$ that is a system of sets will be called a *subsystem* of S. Thus a subsystem of a system of sets is itself a system of sets of the same type.

Lemma (Borel–Lebesgue[11]) *Every system of open intervals covering a closed interval contains a finite subsystem that covers the closed interval.*

Proof Let $S = \{U\}$ be a system of open intervals U that cover the closed interval $[a, b] = I_1$. If the interval I_1 could not be covered by a finite set of intervals of the system S, then, dividing I_1 into two halves, we would find that at least one of the two halves, which we denote by I_2, does not admit a finite covering. We now repeat this procedure with the interval I_2, and so on.

In this way a nested sequence $I_1 \supset I_2 \supset \cdots \supset I_n \supset \cdots$ of closed intervals arises, none of which admit a covering by a finite subsystem of S. Since the length of the interval I_n is $|I_n| = |I_1| \cdot 2^{-n}$, the sequence $\{I_n\}$ contains intervals of arbitrarily small length (see the lemma in Paragraph c of Sect. 2.2.4). But the nested interval theorem implies that there exists a point c belonging to all of the intervals I_n, $n \in \mathbb{N}$. Since $c \in I_1 = [a, b]$ there exists an open interval $]\alpha, \beta[= U \in S$ containing c, that is, $\alpha < c < \beta$. Let $\varepsilon = \min\{c - \alpha, \beta - c\}$. In the sequence just constructed, we find an interval I_n such that $|I_n| < \varepsilon$. Since $c \in I_n$ and $|I_n| < \varepsilon$, we conclude that $I_n \subset U =]\alpha, \beta[$. But this contradicts the fact that the interval I_n cannot be covered by a finite set of intervals from the system. □

2.3.3 The Limit Point Lemma (Bolzano–Weierstrass Principle)

We recall that we have defined a *neighborhood* of a point $x \in \mathbb{R}$ to be an open interval containing the point and the δ-*neighborhood* about x to be the open interval $]x - \delta, x + \delta[$.

Definition 4 A point $p \in \mathbb{R}$ is a *limit point* of the set $X \subset \mathbb{R}$ if every neighborhood of the point contains an infinite subset of X.

This condition is obviously equivalent to the assertion that every neighborhood of p contains at least one point of X different from p itself. (Verify this!)

We now give some examples.

If $X = \{\frac{1}{n} \in \mathbb{R} \mid n \in \mathbb{N}\}$, the only limit point of X is the point $0 \in \mathbb{R}$.

For an open interval $]a, b[$ every point of the closed interval $[a, b]$ is a limit point, and there are no others.

For the set $\mathbb{Q}$ of rational numbers every point of $\mathbb{R}$ is a limit point; for, as we know, every open interval of the real numbers contains rational numbers.

---

[11]É. Borel (1871–1956) and H. Lebesgue (1875–1941) – well-known French mathematicians who worked in the theory of functions.

Lemma (Bolzano–Weierstrass[12]) *Every bounded infinite set of real numbers has at least one limit point.*

Proof Let X be the given subset of $\mathbb{R}$. It follows from the definition of boundedness that X is contained in some closed interval $I \subset \mathbb{R}$. We shall show that at least one point of I is a limit point of X.

If such were not the case, then each point $x \in I$ would have a neighborhood $U(x)$ containing either no points of X or at most a finite number. The totality of such neighborhoods $\{U(x)\}$ constructed for the points $x \in I$ forms a covering of I by open intervals $U(x)$. By the finite covering lemma we can extract a system $U(x_1), \dots, U(x_n)$ of open intervals that cover I. But, since $X \subset I$, this same system also covers X. However, there are only finitely many points of X in $U(x_i)$, and hence only finitely many in their union. That is, X is a finite set. This contradiction completes the proof. □

2.3.4 Problems and Exercises

1. Show that

a) if I is any system of nested closed intervals, then

$$\sup\{a \in \mathbb{R} \mid [a, b] \in I\} = \alpha \le \beta = \inf\{b \in \mathbb{R} \mid [a, b] \in I\}$$

and

$$[\alpha, \beta] = \bigcap_{[a,b] \in I} [a, b];$$

b) if I is a system of nested open intervals $]a, b[$ the intersection $\bigcap_{]a,b[\in I}]a, b[$ may happen to be empty.

Hint : $]a_n, b_n[=]0, \frac{1}{n}[$.

2. Show that

a) from a system of closed intervals covering a closed interval it is not always possible to choose a finite subsystem covering the interval;

b) from a system of open intervals covering an open interval it is not always possible to choose a finite subsystem covering the interval;

c) from a system of closed intervals covering an open interval it is not always possible to choose a finite subsystem covering the interval.

[12]B. Bolzano (1781–1848) – Czech mathematician and philosopher. K. Weierstrass (1815–1897) – German mathematician who devoted a great deal of attention to the logical foundations of mathematical analysis.

3. Show that if we take only the set $\mathbb{Q}$ of rational numbers instead of the complete set $\mathbb{R}$ of real numbers, taking a closed interval, open interval, and neighborhood of a point $r \in \mathbb{Q}$ to mean respectively the corresponding subsets of $\mathbb{Q}$, then none of the three lemmas proved above remains true.
4. Show that we obtain an axiom system equivalent to the one already given if we take as the axiom of completeness

a) the Bolzano–Weierstrass principle

or

b) the Borel–Lebesgue principle (Heine–Borel theorem).

Hint: The principle of Archimedes and the axiom of completeness in the earlier form both follow from a).

c) Replacing the axiom of completeness by the Cauchy–Cantor principle leads to a system of axioms that becomes equivalent to the original system if we also postulate the principle of Archimedes. (See Problem 21 in Sect. 2.2.2.)

2.4 Countable and Uncountable Sets

We now make a small addition to the information about sets that was provided in Chap. 1. This addition will be useful below.

2.4.1 Countable Sets

Definition 1 A set X is *countable* if it is equipollent with the set $\mathbb{N}$ of natural numbers, that is, $\operatorname{card} X = \operatorname{card} \mathbb{N}$.

Proposition

a) *An infinite subset of a countable set is countable.*
b) *The union of the sets of a finite or countable system of countable sets is a countable set.*

Proof a) It suffices to verify that every infinite subset E of $\mathbb{N}$ is equipollent with $\mathbb{N}$. We construct the needed bijective mapping $f : \mathbb{N} \to E$ as follows. There is a minimal element of $E_1 := E$, which we assign to the number $1 \in \mathbb{N}$ and denote $e_1 \in E$. The set E is infinite, and therefore $E_2 := E_1 \backslash e_1$ is nonempty. We assign the minimal element of E_2 to the number 2 and call it $e_2 \in E_2$. We then consider $E_3 := E \backslash \{e_1, e_2\}$, and so forth. Since E is an infinite set, this construction cannot terminate at any finite step with index $n \in \mathbb{N}$. As follows from the principle of induction, we assign in this way a certain number $e_n \in E$ to each $n \in \mathbb{N}$. The mapping $f : \mathbb{N} \to E$ is obviously injective.

It remains to verify that it is surjective, that is, $f(\mathbb{N}) = E$. Let $e \in E$. The set $\{n \in \mathbb{N} \mid n \le e\}$ is finite, and hence the subset of it $\{n \in E \mid n \le e\}$ is also finite. Let k be the number of elements in the latter set. Then by construction $e = e_k$.

b) If $X_1, \dots, X_n, \dots$ is a countable system of sets and each set $X_m = \{x_m^1, \dots, x_m^n, \dots\}$ is itself countable, then since the cardinality of the set $X = \bigcup_{n\in\mathbb{N}} X_n$, which consists of the elements x_m^n where $m, n \in \mathbb{N}$, is not less than the cardinality of each of the sets X_m, it follows that X is an infinite set. The element $x_m^n \in X_m$ can be identified with the pair (m, n) of natural numbers that defines it. Then the cardinality of X cannot be greater than the cardinality of the set of all such ordered pairs. But the mapping $f : \mathbb{N} \times \mathbb{N} \to \mathbb{N}$ given by the formula $(m, n) \mapsto \frac{(m+n-2)(m+n-1)}{2} + m$, as one can easily verify, is bijective. (It has a visualizable meaning: we are enumerating the points of the plane with coordinates (m, n) by successively passing from points of one diagonal on which $m + n$ is constant to the points of the next such diagonal, where the sum is one larger.)

Thus the set of ordered pairs (m, n) of natural numbers is countable. But then $\operatorname{card} X \le \operatorname{card} \mathbb{N}$, and since X is an infinite set we conclude on the basis of a) that $\operatorname{card} X = \operatorname{card} \mathbb{N}$. □

It follows from the proposition just proved that any subset of a countable set is either finite or countable. If it is known that a set is either finite or countable, we say it is *at most countable*. (An equivalent expression is $\operatorname{card} X \le \operatorname{card} \mathbb{N}$.)

We can now assert, in particular, that *the union of an at most countable family of at most countable sets is at most countable.*

Corollaries

1) $\operatorname{card} \mathbb{Z} = \operatorname{card} \mathbb{N}$.
2) $\operatorname{card} \mathbb{N}^2 = \operatorname{card} \mathbb{N}$.
   (*This result means that the direct product of countable sets is countable.*)
3) $\operatorname{card} \mathbb{Q} = \operatorname{card} \mathbb{N}$, *that is, the set of rational numbers is countable.*

Proof A rational number $\frac{m}{n}$ is defined by an ordered pair (m, n) of integers. Two pairs (m, n) and (m', n') define the same rational number if and only if they are proportional. Thus, choosing as the unique pair representing each rational number the pair (m, n) with the smallest possible positive integer denominator $n \in \mathbb{N}$, we find that the set $\mathbb{Q}$ is equipollent to some infinite subset of the set $\mathbb{Z} \times \mathbb{Z}$. But $\operatorname{card} \mathbb{Z}^2 = \operatorname{card} \mathbb{N}$ and hence $\operatorname{card} \mathbb{Q} = \operatorname{card} \mathbb{N}$. □

4) *The set of algebraic numbers is countable.*

Proof We remark first of all that the equality $\mathbb{Q} \times \mathbb{Q} = \operatorname{card} \mathbb{N}$ implies, by induction, that $\operatorname{card} \mathbb{Q}^k = \operatorname{card} \mathbb{N}$ for every $k \in \mathbb{N}$.

An element $r \in \mathbb{Q}^k$ is an ordered set $(r_1, \dots, r_k)$ of k rational numbers.

An algebraic equation of degree k with rational coefficients can be written in the reduced form $x^k + r_1 x^{k-1} + \dots + r_k = 0$, where the leading coefficient is 1. Thus there are as many different algebraic equations of degree k as there are different ordered sets $(r_1, \dots, r_k)$ of rational numbers, that is, a countable set.

The algebraic equations with rational coefficients (of arbitrary degree) also form a countable set, being a countable union (over degrees) of countable sets. Each such equation has only a finite number of roots. Hence the set of algebraic numbers is at most countable. But it is infinite, and hence countable. □

2.4.2 *The Cardinality of the Continuum*

Definition 2 The set $\mathbb{R}$ of real numbers is also called the *number continuum*,[13] and its cardinality the *cardinality of the continuum.*

Theorem (Cantor) $\operatorname{card} \mathbb{N} < \operatorname{card} \mathbb{R}$.

This theorem asserts that the infinite set $\mathbb{R}$ has cardinality greater than that of the infinite set $\mathbb{N}$.

Proof We shall show that even the closed interval $[0, 1]$ is an uncountable set.

Assume that it is countable, that is, can be written as a sequence $x_1, x_2, \dots, x_n, \dots$. Take the point x_1 and on the interval $[0, 1] = I_0$ fix a closed interval of positive length I_1 not containing the point x_1. In the interval I_1 construct an interval I_2 not containing x_2. If the interval I_n has been constructed, then, since $|I_n| > 0$, we construct in it an interval I_{n+1} so that $x_{n+1} \notin I_{n+1}$ and $|I_{n+1}| > 0$. By the nested set lemma, there is a point c belonging to all of the intervals $I_0, I_1, \dots, I_n, \dots$. But this point of the closed interval $I_0 = [0, 1]$ by construction cannot be any point of the sequence $x_1, x_2, \dots, x_n, \dots$. □

Corollaries

1) $\mathbb{Q} \neq \mathbb{R}$, *and so irrational numbers exist.*
2) *There exist transcendental numbers, since the set of algebraic numbers is countable.*

(*After solving Exercise* 3 *below, the reader will no doubt wish to reinterpret this last proposition, stating it as follows*: Algebraic numbers are occasionally encountered among the real numbers.)

At the very dawn of set theory the question arose whether there exist sets of cardinality between countable sets and sets having cardinality of the continuum, and the conjecture was made, known as the continuum hypothesis, *that there are no intermediate cardinalities.*

---

[13]From the Latin *continuum*, meaning *continuous*, or *solid.*

The question turned out to involve the deepest parts of the foundations of mathematics. It was definitively answered in 1963 *by the American mathematician P. Cohen. Cohen proved that the continuum hypothesis is undecidable by showing that neither the hypothesis nor its negation contradicts the standard axiom system of set theory, so that the continuum hypothesis can be neither proved nor disproved within that axiom system. This situation is very similar to the way in which Euclid's fifth postulate on parallel lines is independent of the other axioms of geometry.*

2.4.3 Problems and Exercises

1. Show that the set of real numbers has the same cardinality as the points of the interval $]-1, 1[$.

2. Give an explicit one-to-one correspondence between

a) the points of two open intervals;
b) the points of two closed intervals;
c) the points of a closed interval and the points of an open interval;
d) the points of the closed interval $[0, 1]$ and the set $\mathbb{R}$.

3. Show that

a) every infinite set contains a countable subset;
b) the set of even integers has the same cardinality as the set of all natural numbers;
c) the union of an infinite set and an at most countable set has the same cardinality as the original infinite set;
d) the set of irrational numbers has the cardinality of the continuum;
e) the set of transcendental numbers has the cardinality of the continuum.

4. Show that

a) the set of increasing sequences of natural numbers $\{n_1 < n_2 < \cdots\}$ has the same cardinality as the set of fractions of the form $0.\alpha_1\alpha_2\ldots$;
b) the set of all subsets of a countable set has cardinality of the continuum.

5. Show that

a) the set $\mathcal{P}(X)$ of subsets of a set X has the same cardinality as the set of all functions on X with values 0, 1, that is, the set of mappings $f : X \to \{0, 1\}$;
b) for a finite set X of n elements, $\operatorname{card} \mathcal{P}(X) = 2^n$;
c) taking account of the results of Exercises 4b) and 5a), one can write $\operatorname{card} \mathcal{P}(X) = 2^{\operatorname{card} X}$, and, in particular, $\operatorname{card} \mathcal{P}(\mathbb{N}) = 2^{\operatorname{card} \mathbb{N}} = \operatorname{card} \mathbb{R}$;
d) for any set X

$$\operatorname{card} X < 2^{\operatorname{card} X}, \quad \text{in particular,} \quad n < 2^n \quad \text{for any } n \in \mathbb{N}.$$

Hint: See Cantor's theorem in Sect. 1.4.1.

6. Let $X_1, \dots, X_n$ be a finite system of finite sets. Show that

$$\begin{aligned}\operatorname{card}\left(\bigcup_{i=1}^{m} X_i\right) = \sum_{i_1} \operatorname{card} X_{i_1} - \sum_{i_1<i_2} \operatorname{card}(X_{i_1} \cap X_{i_2}) + \\ + \sum_{i_1<i_2<i_3} \operatorname{card}(X_{i_1} \cap X_{i_2} \cap X_{i_3}) - \dots + \\ + (-1)^{m-1} \operatorname{card}(X_1 \cap \dots \cap X_m),\end{aligned}$$

the summation extending over all sets of indices from 1 to m satisfying the inequalities under the summation signs.

7. On the closed interval $[0, 1] \subset \mathbb{R}$ describe the sets of numbers $x \in [0, 1]$ whose ternary representation $x = 0.\alpha_1\alpha_2\alpha_3 \dots$, $\alpha_i \in \{0, 1, 2\}$, has the property:

a) $\alpha_1 \neq 1$;
b) $(\alpha_1 \neq 1) \wedge (\alpha_2 \neq 1)$;
c) $\forall i \in \mathbb{N}\ (\alpha_i \neq 1)$ (the *Cantor set*).

8. (Continuation of Exercise 7.) Show that

a) the set of numbers $x \in [0, 1]$ whose ternary representation does not contain 1 has the same cardinality as the set of all numbers whose binary representation has the form $0.\beta_1\beta_2 \dots$;

b) the Cantor set has the same cardinality as the closed interval $[0, 1]$.

Chapter 3
Limits

In discussing the various aspects of the concept of a real number we remarked in particular that in measuring real physical quantities we obtain sequences of approximate values with which one must then work.

Such a state of affairs immediately raises at least the following three questions:

1) What relation does the sequence of approximations so obtained have to the quantity being measured? We have in mind the mathematical aspect of the question, that is, we wish to obtain an exact expression of what is meant in general by the expression "sequence of approximate values" and the extent to which such a sequence describes the value of the quantity. Is the description unambiguous, or can the same sequence correspond to different values of the measured quantity?
2) How are operations on the approximate values connected with the same operations on the exact values, and how can we characterize the operations that can legitimately be carried out by replacing the exact values with approximate ones?
3) How can one determine from a sequence of numbers whether it can be a sequence of arbitrarily precise approximations of the values of some quantity?

The answer to these and related questions is provided by the concept of the limit of a function, one of the fundamental concepts of analysis.

We begin our discussion of the theory of limits by considering the limit of a function of a natural-number argument (a sequence), in view of the fundamental role played by these functions, as already explained, and also because all the basic facts of the theory of limits can actually be clearly seen in this simplest situation.

3.1 The Limit of a Sequence

3.1.1 Definitions and Examples

We recall the following definition.

Definition 1 A function $f : \mathbb{N} \to X$ whose domain of definition is the set of natural numbers is called a *sequence*.

V.A. Zorich, *Mathematical Analysis I*, Universitext,
DOI 10.1007/978-3-662-48792-1_3

The values $f(n)$ of the function f are called the *terms* of the sequence. It is customary to denote them by a symbol for an element of the set into which the mapping goes, endowing each symbol with the corresponding index of the argument. Thus, $x_n := f(n)$. In this connection the sequence itself is denoted $\{x_n\}$, and also written as $x_1, x_2, \dots, x_n, \dots$. It is called a *sequence in* X or a *sequence of elements of* X.

The element x_n is called the *nth term of the sequence*.

Throughout the next few sections we shall be considering only sequences $f : \mathbb{N} \to \mathbb{R}$ of real numbers.

Definition 2 A number $A \in \mathbb{R}$ is called the *limit of the numerical sequence* $\{x_n\}$ if for every neighborhood $V(A)$ of A there exists an index N (depending on $V(A)$) such that all terms of the sequence having index larger than N belong to the neighborhood $V(A)$.

We shall give an expression in formal logic for this definition below, but we first point out another common formulation of the definition of the limit of a sequence.

A number $A \in \mathbb{R}$ is called the *limit of the sequence* $\{x_n\}$ if for every $\varepsilon > 0$ there exists an index N such that $|x_n - A| < \varepsilon$ for all $n > N$.

The equivalence of these two statements is easy to verify (verify it!) if we remark that any neighborhood $V(A)$ of A contains some ε-neighborhood of the point A.

The second formulation of the definition of a limit means that no matter what precision $\varepsilon > 0$ we have prescribed, there exists an index N such that the absolute error in approximating the number A by terms of the sequence $\{x_n\}$ is less than ε as soon as $n > N$.

We now write these formulations of the definition of a limit in the language of symbolic logic, agreeing that the expression "$\lim_{n\to\infty} x_n = A$" is to mean that A is the limit of the sequence $\{x_n\}$. Thus

$$\boxed{\left(\lim_{n\to\infty} x_n = A\right) := \forall V(A)\ \exists N \in \mathbb{N}\ \forall n > N\ \big(x_n \in V(A)\big)}$$

and respectively

$$\left(\lim_{n\to\infty} x_n = A\right) := \forall \varepsilon > 0\ \exists N \in \mathbb{N}\ \forall n > N\ \big(|x_n - A| < \varepsilon\big).$$

Definition 3 If $\lim_{n\to\infty} x_n = A$, we say that the sequence $\{x_n\}$ *converges* to A or *tends* to A and write $x_n \to A$ as $n \to \infty$.

A sequence having a limit is said to be *convergent*. A sequence that does not have a limit is said to be *divergent*.

Let us consider some examples.

Example 1 $\lim_{n\to\infty} \frac{1}{n} = 0$, since $|\frac{1}{n} - 0| = \frac{1}{n} < \varepsilon$ when $n > N = [\frac{1}{\varepsilon}]$.[1]

---

[1] We recall that $[x]$ is the integer part of the number x. (See Corollaries 7^0 and 10^0 of Sect. 2.2.)

Example 2 $\lim_{n\to\infty} \frac{n+1}{n} = 1$, since $|\frac{n+1}{n} - 1| = \frac{1}{n} < \varepsilon$ if $n > [\frac{1}{\varepsilon}]$.

Example 3 $\lim_{n\to\infty}(1 + \frac{(-1)^n}{n}) = 1$, since $|(1 + \frac{(-1)^n}{n}) - 1| = \frac{1}{n} < \varepsilon$ when $n > [\frac{1}{\varepsilon}]$.

Example 4 $\lim_{n\to\infty} \frac{\sin n}{n} = 0$, since $|\frac{\sin n}{n} - 0| \le \frac{1}{n} < \varepsilon$ for $n > [\frac{1}{\varepsilon}]$.

Example 5 $\lim_{n\to\infty} \frac{1}{q^n} = 0$ if $|q| > 1$.

Let us verify this last assertion using the definition of the limit. As was shown in Paragraph c of Sect. 2.2.4, for every $\varepsilon > 0$ there exists $N \in \mathbb{N}$ such that $\frac{1}{|q|^N} < \varepsilon$. Since $|q| > 1$, we shall have $|\frac{1}{q^n} - 0| < \frac{1}{|q|^n} < \frac{1}{|q|^N} < \varepsilon$ for $n > N$, and the condition in the definition of the limit is satisfied.

Example 6 The sequence $1, 2, \frac{1}{3}, 4, \frac{1}{5}, 6, \frac{1}{7}, \ldots$ whose nth term is $x_n = n^{(-1)^n}$, $n \in \mathbb{N}$, is divergent.

Proof Indeed, if A were the limit of this sequence, then, as follows from the definition of limit, any neighborhood of A would contain all but a finite number of terms of the sequence.

A number $A \neq 0$ cannot be the limit of this sequence; for if $\varepsilon = \frac{|A|}{2} > 0$, all the terms of the sequence of the form $\frac{1}{2k+1}$ for which $\frac{1}{2k+1} < \frac{|A|}{2}$ lie outside the ε-neighborhood of A.

But the number 0 also cannot be the limit, since, for example, there are infinitely many terms of the sequence lying outside the 1-neighborhood of 0. □

Example 7 One can verify similarly that the sequence $1, -1, +1, -1, \ldots$, for which $x_n = (-1)^n$, has no limit.

3.1.2 Properties of the Limit of a Sequence

a. General Properties

We assign to this group the properties possessed not only by numerical sequences, but by other kinds of sequences as well, as we shall see below, although at present we shall study these properties only for numerical sequences.

A sequence assuming only one value will be called a *constant* sequence.

Definition 4 If there exists a number A and an index N such that $x_n = A$ for all $n > N$, the sequence $\{x_n\}$ will be called *ultimately constant*.

Definition 5 A sequence $\{x_n\}$ is *bounded* if there exists M such that $|x_n| < M$ for all $n \in \mathbb{N}$.

Theorem 1 a) *An ultimately constant sequence converges.*

b) *Any neighborhood of the limit of a sequence contains all but a finite number of terms of the sequence.*

c) *A convergent sequence cannot have two different limits.*

d) *A convergent sequence is bounded.*

Proof a) If $x_n = A$ for $n > N$, then for any neighborhood $V(A)$ of A we have $x_n \in V(A)$ when $n > N$, that is, $\lim_{n\to\infty} x_n = A$.

b) This assertion follows immediately from the definition of a convergent sequence.

c) This is the most important part of the theorem. Let $\lim_{n\to\infty} x_n = A_1$ and $\lim_{n\to\infty} x_n = A_2$. If $A_1 \neq A_2$, we fix nonintersecting neighborhoods $V(A_1)$ and $V(A_2)$ of A_1 and A_2. These neighborhoods might be, for example, the δ-neighborhoods of A_1 and A_2 for $\delta < \frac{1}{2}|A_1 - A_2|$. By definition of limit we find indices N_1 and N_2 such that $x_n \in V(A_1)$ for all $n > N_1$ and $x_n \in V(A_2)$ for all $n > N_2$. But then for $N = \max\{N_1, N_2\}$ we have $x_n \in V(A_1) \cap V(A_2)$. But this is impossible, since $V(A_1) \cap V(A_2) = \varnothing$.

d) Let $\lim_{n\to\infty} x_n = A$. Setting $\varepsilon = 1$ in the definition of a limit, we find N such that $|x_n - A| < 1$ for all $n > N$. Then for $n > N$ we have $|x_n| < |A| + 1$. If we now take $M > \max\{|x_1|, \dots, |x_n|, |A| + 1\}$ we find that $|x_n| < M$ for all $n \in \mathbb{N}$. □

b. Passage to the Limit and the Arithmetic Operations

Definition 6 If $\{x_n\}$ and $\{y_n\}$ are two numerical sequences, their *sum*, *product*, and *quotient* (in accordance with the general definition of sum, product, and quotient of functions) are the sequences

$$\{(x_n + y_n)\}, \quad \{(x_n \cdot y_n)\}, \quad \left\{\left(\frac{x_n}{y_n}\right)\right\}.$$

The quotient, of course, is defined only when $y_n \neq 0$ for all $n \in \mathbb{N}$.

Theorem 2 *Let* $\{x_n\}$ *and* $\{y_n\}$ *be numerical sequences. If* $\lim_{n\to\infty} x_n = A$ *and* $\lim_{n\to\infty} y_n = B$, *then*

a) $\lim_{n\to\infty}(x_n + y_n) = A + B$;
b) $\lim_{n\to\infty}(x_n \cdot y_n) = A \cdot B$;
c) $\lim_{n\to\infty} \frac{x_n}{y_n} = \frac{A}{B}$, *provided* $y_n \neq 0$ $(n = 1, 2, \dots)$ *and* $B \neq 0$.

Proof As an exercise we use the estimates for the absolute errors that arise under arithmetic operations with approximate values of quantities, which we already know (Sect. 2.2.4).

Set $|A - x_n| = \Delta(x_n)$, $|B - y_n| = \Delta(y_n)$. Then for case a) we have

$$\bigl|(A + B) - (x_n + y_n)\bigr| \le \Delta(x_n) + \Delta(y_n).$$

Suppose $\varepsilon > 0$ is given. Since $\lim_{n\to\infty} x_n = A$, there exists N' such that $\Delta(x_n) < \varepsilon/2$ for all $n > N'$. Similarly, since $\lim_{n\to\infty} y_n = B$, there exists N'' such that $\Delta(y_n) < \varepsilon/2$ for all $n > N''$. Then for $n > \max\{N', N''\}$ we shall have

$$\bigl|(A+B) - (x_n + y_n)\bigr| < \varepsilon,$$

which, by definition of limit, proves assertion a).

b) We know that

$$\bigl|(A\cdot B) - (x_n \cdot y_n)\bigr| \le |x_n|\Delta(y_n) + |y_n|\Delta(x_n) + \Delta(x_n)\cdot\Delta(y_n).$$

Given $\varepsilon > 0$ find numbers N' and N'' such that

$$\forall n > N' \quad \left(\Delta(x_n) < \min\left\{1, \frac{\varepsilon}{3(|B|+1)}\right\}\right),$$

$$\forall n > N'' \quad \left(\Delta(y_n) < \min\left\{1, \frac{\varepsilon}{3(|A|+1)}\right\}\right).$$

Then for $n > N = \max\{N', N''\}$ we shall have

$$|x_n| < |A| + \Delta(x_n) < |A| + 1,$$

$$|y_n| < |B| + \Delta(y_n) < |B| + 1,$$

$$\Delta(x_n)\cdot\Delta(y_n) < \min\left\{1, \frac{\varepsilon}{3}\right\}\cdot\min\left\{1, \frac{\varepsilon}{3}\right\} \le \frac{\varepsilon}{3}.$$

Hence for $n > N$ we have

$$|x_n|\Delta(y_n) < \bigl(|A|+1\bigr)\cdot\frac{\varepsilon}{3(|A|+1)} < \frac{\varepsilon}{3},$$

$$|y_n|\Delta(x_n) < \bigl(|B|+1\bigr)\cdot\frac{\varepsilon}{3(|B|+1)} < \frac{\varepsilon}{3},$$

$$\Delta(x_n)\cdot\Delta(y_n) < \frac{\varepsilon}{3},$$

and therefore $|AB - x_n y_n| < \varepsilon$ for $n > N$.

c) We use the estimate

$$\left|\frac{A}{B} - \frac{x_n}{y_n}\right| \le \frac{|x_n|\Delta(y_n) + |y_n|\Delta(x_n)|}{y_n^2}\cdot\frac{1}{1-\delta(y_n)},$$

where $\delta(y_n) = \frac{\Delta(y_n)}{|y_n|}$.

For a given $\varepsilon > 0$ we find numbers N' and N'' such that

$$\forall n > N' \quad \left(\Delta(x_n) < \min\left\{1, \frac{\varepsilon|B|}{8}\right\}\right),$$

$$\forall n > N'' \quad \left(\Delta(y_n) < \min\left\{\frac{|B|}{4}, \frac{\varepsilon \cdot B^2}{16(|A|+1)}\right\}\right).$$

Then for $n > \max\{N', N''\}$ we shall have

$$|x_n| < |A| + \Delta(x_n) < |A| + 1,$$

$$|y_n| > |B| - \Delta(y_n) > |B| - \frac{|B|}{4} > \frac{|B|}{2},$$

$$\frac{1}{|y_n|} < \frac{2}{|B|},$$

$$0 < \delta(y_n) = \frac{\Delta(y_n)}{|y_n|} < \frac{|B|/4}{|B|/2} = \frac{1}{2},$$

$$1 - \delta(y_n) > \frac{1}{2},$$

and therefore

$$|x_n| \cdot \frac{1}{y_n^2} \Delta(y_n) < \big(|A| + 1\big) \cdot \frac{4}{B^2} \cdot \frac{\varepsilon \cdot B^2}{16(|A|+1)} = \frac{\varepsilon}{4},$$

$$\left|\frac{1}{y_n}\right| \Delta(x_n) < \frac{2}{|B|} \cdot \frac{\varepsilon |B|}{8} = \frac{\varepsilon}{4},$$

$$0 < \frac{1}{1 - \delta(y_n)} < 2,$$

and consequently

$$\left|\frac{A}{B} - \frac{x_n}{y_n}\right| < \varepsilon \quad \text{when } n > N. \qquad \square$$

Remark The statement of the theorem admits another, less constructive method of proof that is probably known to the reader from the high-school course in the rudiments of analysis. We shall mention this method when we discuss the limit of an arbitrary function. But here, when considering the limit of a sequence, we wished to call attention to the way in which bounds on the errors in the result of an arithmetic operations can be used to set permissible bounds on the errors in the values of quantities on which an operation is carried out.

c. Passage to the Limit and Inequalities

Theorem 3 a) *Let* $\{x_n\}$ *and* $\{y_n\}$ *be two convergent sequences with* $\lim_{n\to\infty} x_n = A$ *and* $\lim_{n\to\infty} y_n = B$. *If* $A < B$, *then there exists an index* $N \in \mathbb{N}$ *such that* $x_n < y_n$ *for all* $n > N$.

b) *Suppose the sequences* $\{x_n\}$, $\{y_n\}$, *and* $\{z_n\}$ *are such that* $x_n \leq y_n \leq z_n$ *for all* $n > N \in \mathbb{N}$. *If the sequences* $\{x_n\}$ *and* $\{z_n\}$ *both converge to the same limit, then the sequence* $\{y_n\}$ *also converges to that limit.*

Proof a) Choose a number C such that $A < C < B$. By definition of limit, we can find numbers N' and N'' such that $|x_n - A| < C - A$ for all $n > N'$ and $|y_n - B| < B - C$ for all $n > N''$. Then for $n > N = \max\{N', N''\}$ we shall have $x_n < A + C - A = C = B - (B - C) < y_n$.

b) Suppose $\lim_{n\to\infty} x_n = \lim_{n\to\infty} z_n = A$. Given $\varepsilon > 0$ choose N' and N'' such that $A - \varepsilon < x_n$ for all $n > N'$ and $z_n < A + \varepsilon$ for all $n > N''$. Then for $n > N = \max\{N', N''\}$ we shall have $A - \varepsilon < x_n \leq y_n \leq z_n < A + \varepsilon$, which says $|y_n - A| < \varepsilon$, that is $A = \lim_{n\to\infty} y_n$. □

Corollary *Suppose* $\lim_{n\to\infty} x_n = A$ *and* $\lim_{n\to\infty} y_n = B$. *If there exists N such that for all n > N we have*

a) $x_n > y_n$, *then* $A \geq B$;
b) $x_n \geq y_n$, *then* $A \geq B$;
c) $x_n > B$, *then* $A \geq B$;
d) $x_n \geq B$, *then* $A \geq B$.

Proof Arguing by contradiction, we obtain the first two assertions immediately from part a) of the theorem. The third and fourth assertions are the special cases of the first two obtained when $y_n \equiv B$. □

It is worth noting that strict inequality may become equality in the limit. For example $\frac{1}{n} > 0$ for all $n \in \mathbb{N}$, yet $\lim_{n\to\infty} \frac{1}{n} = 0$.

3.1.3 Questions Involving the Existence of the Limit of a Sequence

a. The Cauchy Criterion

Definition 7 A sequence $\{x_n\}$ is called a *fundamental* or *Cauchy* sequence[2] if for any $\varepsilon > 0$ there exists an index $N \in \mathbb{N}$ such that $|x_m - x_n| < \varepsilon$ whenever $n > N$ and $m > N$.

Theorem 4 (Cauchy's convergence criterion) *A numerical sequence converges if and only if it is a Cauchy sequence.*

[2]Bolzano introduced Cauchy sequences in an attempt to prove, without having at his disposal a precise concept of a real number, that a fundamental sequence converges. Cauchy gave a proof, taking the nested interval principle, which was later justified by Cantor, as obvious.

Proof Suppose $\lim_{n\to\infty} x_n = A$. Given $\varepsilon > 0$, we find an index N such that $|x_n - A| < \frac{\varepsilon}{2}$ for $n > N$. Then if $m > N$ and $n > N$, we have $|x_m - x_n| \le |x_m - A| + |x_n - A| < \frac{\varepsilon}{2} + \frac{\varepsilon}{2} = \varepsilon$, and we have thus verified that the sequence is a Cauchy sequence.

Now let $\{x_k\}$ be a fundamental sequence. Given $\varepsilon > 0$, we find an index N such that $|x_m - x_k| < \frac{\varepsilon}{3}$ when $m \ge N$ and $k \ge N$. Fixing $m = N$, we find that for any $k > N$

$$x_N - \frac{\varepsilon}{3} < x_k < x_N + \frac{\varepsilon}{3}, \tag{3.1}$$

but since only a finite number of terms of the sequence have indices not larger than N, we have shown that a fundamental sequence is bounded.

For $n \in \mathbb{N}$ we now set $a_n := \inf_{k\ge n} x_k$, and $b_n := \sup_{k\ge n} x_k$.

It is clear from these definitions that $a_n \le a_{n+1} \le b_{n+1} \le b_n$ (since the greatest lower bound does not decrease and the least upper bound does not increase when we pass to a smaller set). By the nested interval principle, there is a point A common to all of the closed intervals $[a_n, b_n]$.

Since

$$a_n \le A \le b_n$$

for any $n \in \mathbb{N}$ and

$$a_n = \inf_{k\ge n} x_k \le x_k \le \sup_{k\ge n} x_k = b_k$$

for $k \ge n$, it follows that

$$|A - x_k| \le b_n - a_n. \tag{3.2}$$

But it follows from Eq. (3.1) that

$$x_N - \frac{\varepsilon}{3} \le \inf_{k\ge n} x_k = a_n \le b_n = \sup_{k\ge n} x_k \le x_N + \frac{\varepsilon}{3}$$

for $n > N$, and therefore

$$b_n - a_n \le \frac{2\varepsilon}{3} < \varepsilon \tag{3.3}$$

for $n > m$. Comparing Eqs. (3.2) and (3.3), we find that

$$|A - x_k| < \varepsilon,$$

for any $k > N$, and we have proved that $\lim_{k\to\infty} x_k = A$. □

Example 8 The sequence $(-1)^n$ $(n = 1, 2, \ldots)$ has no limit, since it is not a Cauchy sequence. Even though this fact is obvious, we shall give a formal verification. The negation of the statement that $\{x_n\}$ is a Cauchy sequence is the following:

$$\exists \varepsilon > 0\ \forall N \in \mathbb{N}\ \exists n > N\ \exists m > N\ \big(|x_m - x_n| \ge \varepsilon\big),$$

that is, there exists $\varepsilon > 0$ such that for any $N \in \mathbb{N}$ two numbers n, m larger than N exist for which $|x_m - x_n| \geq \varepsilon$.

In our case it suffices to set $\varepsilon = 1$. Then for any $N \in \mathbb{N}$ we shall have $|x_{N+1} - x_{N+2}| = |1 - (-1)| = 2 > 1 = \varepsilon$.

Example 9 Let

$$x_1 = 0.\alpha_1, \quad x_2 = 0.\alpha_1\alpha_2, \quad x_3 = 0.\alpha_1\alpha_2\alpha_3, \quad \dots, \quad x_n = 0.\alpha_1\alpha_2\dots\alpha_n, \quad \dots$$

be a sequence of finite binary fractions in which each successive fraction is obtained by adjoining a 0 or a 1 to its predecessor. We shall show that such a sequence always converges. Let $m > n$. Let us estimate the difference $x_m - x_n$:

$$|x_m - x_n| = \left| \frac{\alpha_{n+1}}{2^{n+1}} + \dots + \frac{\alpha_m}{2^m} \right| \leq$$

$$\leq \frac{1}{2^{n+1}} + \dots + \frac{1}{2^m} = \frac{(\frac{1}{2})^{n+1} - (\frac{1}{2})^{m+1}}{1 - \frac{1}{2}} < \frac{1}{2^n}.$$

Thus, given $\varepsilon > 0$, if we choose N so that $\frac{1}{2^N} < \varepsilon$, we obtain the estimate $|x_m - x_n| < \frac{1}{2^n} < \frac{1}{2^N} < \varepsilon$ for all $m > n > N$, which proves that the sequence $\{x_n\}$ is a Cauchy sequence.

Example 10 Consider the sequence $\{x_n\}$, where

$$x_n = 1 + \frac{1}{2} + \dots + \frac{1}{n}.$$

Since

$$|x_{2n} - x_n| = \frac{1}{n+1} + \dots + \frac{1}{n+n} > n \cdot \frac{1}{2n} = \frac{1}{2},$$

for all $n \in \mathbb{N}$, the Cauchy criterion implies immediately that this sequence does not have a limit.

b. A Criterion for the Existence of the Limit of a Monotonic Sequence

Definition 8 A sequence $\{x_n\}$ is *increasing* if $x_n < x_{n+1}$ for all $n \in \mathbb{N}$, *nondecreasing* if $x_n \leq x_{n+1}$ for all $n \in \mathbb{N}$, *nonincreasing* if $x_n \geq x_{n+1}$ for all $n \in \mathbb{N}$, and *decreasing* if $x_n > x_{n+1}$ for all $n \in \mathbb{N}$. Sequences of these four types are called *monotonic* sequences.

Definition 9 A sequence $\{x_n\}$ is *bounded above* if there exists a number M such that $x_n < M$ for all $n \in \mathbb{N}$.

Theorem 5 (Weierstrass) *In order for a nondecreasing sequence to have a limit it is necessary and sufficient that it be bounded above.*

Proof The fact that any convergent sequence is bounded was proved above under general properties of the limit of a sequence. For that reason only the sufficiency assertion is of interest.

By hypothesis the set of values of the sequence $\{x_n\}$ is bounded above and hence has a least upper bound $s = \sup_{n\in\mathbb{N}} x_n$.

By definition of the least upper bound, for every $\varepsilon > 0$ there exists an element $x_N \in \{x_n\}$ such that $s - \varepsilon < x_N \leq s$. Since the sequence $\{x_n\}$ is nondecreasing, we now find that $s - \varepsilon < x_N \leq x_n \leq s$ for all $n > N$. That is, $|s - x_n| = s - x_n < \varepsilon$. Thus we have proved that $\lim_{n\to\infty} x_n = s$. □

Of course an analogous theorem can be stated and proved for a nonincreasing sequence that is bounded below. In this case $\lim_{n\to\infty} x_n = \inf_{n\in\mathbb{N}} x_n$.

Remark The boundedness from above (resp. below) of a nondecreasing (resp. nonincreasing) sequence is obviously equivalent to the boundedness of that sequence.

Let us consider some useful examples.

Example 11 $\lim_{n\to\infty} \frac{n}{q^n} = 0$ if $q > 1$.

Proof Indeed, if $x_n = \frac{n}{q^n}$, then $x_{n+1} = \frac{n+1}{nq} x_n$ for $n \in \mathbb{N}$. Since $\lim_{n\to\infty} \frac{n+1}{nq} = \lim_{n\to\infty}(1 + \frac{1}{n})\frac{1}{q} = \lim_{n\to\infty}(1 + \frac{1}{n}) \cdot \lim_{n\to\infty} \frac{1}{q} = 1 \cdot \frac{1}{q} = \frac{1}{q} < 1$, there exists an index N such that $\frac{n+1}{nq} < 1$ for $n > N$. Thus we shall have $x_{n+1} < x_n$ for $n > N$, so that the sequence will be monotonically decreasing from index N on. As one can see from the definition of a limit, a finite set of terms of a sequence has no effect the convergence of a sequence or its limit, so that it now suffices to find the limit of the sequence $x_{N+1} > x_{N+2} > \cdots$.

The terms of this sequence are positive, that is, the sequence is bounded below. Therefore it has a limit.

Let $x = \lim_{n\to\infty} x_n$. It now follows from the relation $x_{n+1} = \frac{n+1}{nq} x_n$ that

$$x = \lim_{n\to\infty} (x_{n+1}) = \lim_{n\to\infty} \left(\frac{n+1}{nq} x_n\right) = \lim_{n\to\infty} \frac{n+1}{nq} \cdot \lim_{n\to\infty} x_n = \frac{1}{q} x,$$

from which we find $(1 - \frac{1}{q})x = 0$, and so $x = 0$. □

Corollary 1

$$\lim_{n\to\infty} \sqrt[n]{n} = 1.$$

Proof By what was just proved, for a given $\varepsilon > 0$ there exists $N \in \mathbb{N}$ such that $1 \leq n < (1 + \varepsilon)^n$ for all $n > N$. Then for $n > N$ we obtain $1 \leq \sqrt[n]{n} < 1 + \varepsilon$ and hence $\lim_{n\to\infty} \sqrt[n]{n} = 1$. □

Corollary 2

$$\lim_{n\to\infty} \sqrt[n]{a} = 1 \quad \textit{for any } a > 0.$$

Proof Assume first that $a \geq 1$. For any $\varepsilon > 0$ there exists $N \in \mathbb{N}$ such that $1 \leq a < (1+\varepsilon)^n$ for all $n > N$, and we then have $1 \leq \sqrt[n]{a} < 1+\varepsilon$ for all $n > N$, which says $\lim_{n\to\infty} \sqrt[n]{a} = 1$.

For $0 < a < 1$, we have $1 < \frac{1}{a}$, and then

$$\lim_{n\to\infty} \sqrt[n]{a} = \lim_{n\to\infty} \frac{1}{\sqrt[n]{\frac{1}{a}}} = \frac{1}{\lim_{n\to\infty} \sqrt[n]{\frac{1}{a}}} = 1.$$

□

Example 12 $\lim_{n\to\infty} \frac{q^n}{n!} = 0$; here q is any real number, $n \in \mathbb{N}$, and $n! := 1 \cdot 2 \cdot \ldots \cdot n$.

Proof If $q = 0$, the assertion is obvious. Further, since $|\frac{q^n}{n!}| = \frac{|q|^n}{n!}$, it suffices to prove the assertion for $q > 0$. Reasoning as in Example 11, we remark that $x_{n+1} = \frac{q}{n+1} x_n$. Since the set of natural numbers is not bounded above, there exists an index N such that $0 < \frac{q}{n+1} < 1$ for all $n > N$. Then for $n > N$ we shall have $x_{n+1} < x_n$, and since the terms of the sequence are positive, one can now guarantee that the limit $\lim_{n\to\infty} x_n = x$ exists. But then

$$x = \lim_{n\to\infty} x_{n+1} = \lim_{n\to\infty} \frac{q}{n+1} x_n = \lim_{n\to\infty} \frac{q}{n+1} \cdot \lim_{n\to\infty} x_n = 0 \cdot x = 0.$$

□

c. The Number e

Example 13 Let us prove that the limit $\lim_{n\to\infty}(1+\frac{1}{n})^n$ exists.

In this case the limit is a number denoted by the letter e, after Euler. This number is just as central to analysis as the number 1 to arithmetic or π to geometry. We shall revisit it many times for a wide variety of reasons.

We begin by verifying the following inequality, sometimes called Jacob Bernoulli's inequality:[3]

$$(1+\alpha)^n \geq 1 + n\alpha \quad \text{for } n \in \mathbb{N} \text{ and } \alpha > -1.$$

Proof The assertion is true for $n = 1$. If it holds for $n \in \mathbb{N}$, then it must also hold for $n+1$, since we then have

$$(1+\alpha)^{n+1} = (1+\alpha)(1+\alpha)^n \geq (1+\alpha)(1+n\alpha) =$$
$$= 1 + (n+1)\alpha + n\alpha^2 \geq 1 + (n+1)\alpha.$$

[3] Jacob (James) Bernoulli (1654–1705) – Swiss mathematician, a member of the famous Bernoulli family of scholars. He was one of the founders of the calculus of variations and probability theory.

By the principle of induction the assertion is true for all $n \in \mathbb{N}$.

Incidentally, the computation shows that strict inequality holds if $\alpha \neq 0$ and $n > 1$. □

We now show that the sequence $y_n = (1+\frac{1}{n})^{n+1}$ is decreasing.

Proof Let $n \geq 2$. Using Bernoulli's inequality, we find that

$$\frac{y_{n-1}}{y_n} = \frac{(1+\frac{1}{n-1})^n}{(1+\frac{1}{n})^{n+1}} = \frac{n^{2n}}{(n^2-1)^n} \cdot \frac{n}{n+1} = \left(1+\frac{1}{n^2-1}\right)^n \frac{n}{n+1} \geq$$

$$\geq \left(1+\frac{n}{n^2-1}\right)\frac{n}{n+1} > \left(1+\frac{1}{n}\right)\frac{n}{n+1} = 1.$$

Since the terms of the sequence are positive, the limit $\lim_{n\to\infty}(1+\frac{1}{n})^{n+1}$ exists. But we then have

$$\lim_{n\to\infty}\left(1+\frac{1}{n}\right)^n = \lim_{n\to\infty}\left(1+\frac{1}{n}\right)^{n+1}\left(1+\frac{1}{n}\right)^{-1} =$$

$$= \lim_{n\to\infty}\left(1+\frac{1}{n}\right)^{n+1} \cdot \lim_{n\to\infty}\frac{1}{1+\frac{1}{n}} = \lim_{n\to\infty}\left(1+\frac{1}{n}\right)^{n+1}.$$ □

Thus we make the following definition:

Definition 10

$$\boxed{e := \lim_{n\to\infty}\left(1+\frac{1}{n}\right)^n.}$$

d. Subsequences and Partial Limits of a Sequence

Definition 11 If $x_1, x_2, \dots, x_n, \dots$ is a sequence and $n_1 < n_2 < \dots < n_k < \dots$ an increasing sequence of natural numbers, then the sequence $x_{n_1}, x_{n_2}, \dots, x_{n_k}, \dots$ is called a *subsequence* of the sequence $\{x_n\}$.

For example, the sequence $1, 3, 5, \dots$ of positive odd integers in their natural order is a subsequence of the sequence $1, 2, 3, \dots$. But the sequence $3, 1, 5, 7, 9, \dots$ it not a subsequence of this sequence.

Lemma 1 (Bolzano–Weierstrass) *Every bounded sequence of real numbers contains a convergent subsequence.*

Proof Let E be the set of values of the bounded sequence $\{x_n\}$. If E is finite, there exists a point $x \in E$ and a sequence $n_1 < n_2 < \cdots$ of indices such that $x_{n_1} = x_{n_2} = \cdots = x$. The subsequence $\{x_{n_k}\}$ is constant and hence converges.

If E is infinite, then by the Bolzano–Weierstrass principle it has a limit point x. Since x is a limit point of E, one can choose $n_1 \in \mathbb{N}$ such that $|x_{n_1} - x| < 1$. If $n_k \in \mathbb{N}$ have been chosen so that $|x_{n_k} - x| < \frac{1}{k}$, then, because x is a limit point of E, there exists $n_{k+1} \in \mathbb{N}$ such that $n_k < n_{k+1}$ and $|x_{n_{k+1}} - x| < \frac{1}{k+1}$.

Since $\lim_{k\to\infty} \frac{1}{k} = 0$, the sequence $x_{n_1}, x_{n_2}, \ldots, x_{n_k}, \ldots$ so constructed converges to x. □

Definition 12 We shall write $x_n \to +\infty$ and say that the sequence $\{x_n\}$ *tends to positive infinity* if for each number c there exists $N \in \mathbb{N}$ such that $x_n > c$ for all $n > N$.

Let us write this and two analogous definitions in logical notation:

$$(x_n \to +\infty) := \forall c \in \mathbb{R}\ \exists N \in \mathbb{N}\ \forall n > N\ (c < x_n),$$

$$(x_n \to -\infty) := \forall c \in \mathbb{R}\ \exists N \in \mathbb{N}\ \forall n > N\ (x_n < c),$$

$$(x_n \to \infty) := \forall c \in \mathbb{R}\ \exists N \in \mathbb{N}\ \forall n > N\ \big(c < |x_n|\big).$$

In the last two cases we say that the sequence $\{x_n\}$ *tends to negative infinity* and *tends to infinity* respectively.

We remark that a sequence may be unbounded and yet not tend to positive infinity, negative infinity, or infinity. An example is $x_n = n^{(-1)^n}$.

Sequences that tend to infinity will not be considered convergent.

It is easy to see that these definitions enable us to supplement Lemma 1, stating it in a slightly different form.

Lemma 2 *From each sequence of real numbers one can extract either a convergent subsequence or a subsequence that tends to infinity.*

Proof The new case here occurs when the sequence $\{x_n\}$ is not bounded. Then for each $k \in \mathbb{N}$ we can choose $n_k \in \mathbb{N}$ such that $|x_{n_k}| > k$ and $n_k < n_{k+1}$. We then obtain a subsequence $\{x_{n_k}\}$ that tends to infinity. □

Let $\{x_k\}$ be an arbitrary sequence of real numbers. If it is bounded below, one can consider the sequence $i_n = \inf_{k\ge n} x_k$ (which we have already encountered in proving the Cauchy convergence criterion). Since $i_n \le i_{n+1}$ for any $n \in \mathbb{N}$, either the sequence $\{i_n\}$ has a finite limit $\lim_{n\to\infty} i_n = l$, or $i_n \to +\infty$.

Definition 13 The number $l = \lim_{n\to\infty} \inf_{k\ge n} x_k$ is called the *inferior limit* of the sequence $\{x_k\}$ and denoted $\underline{\lim}_{k\to\infty} x_k$ or $\liminf_{k\to\infty} x_k$. If $i_n \to +\infty$, it is said that the inferior limit of the sequence equals positive infinity, and we write $\underline{\lim}_{k\to\infty} x_k = +\infty$ or $\liminf_{k\to\infty} x_k = +\infty$. If the original sequence $\{x_k\}$ is not bounded below, then we shall have $i_n = \inf_{k\geq n} x_k = -\infty$ for all n. In that case we say that the inferior limit of the sequence equals negative infinity and write $\underline{\lim}_{k\to\infty} x_k = -\infty$ or $\liminf_{k\to\infty} x_k = -\infty$.

Thus, taking account of all the possibilities just enumerated, we can now write down briefly the definition of the inferior limit of a sequence $\{x_k\}$:

$$\boxed{\underline{\lim}_{k\to\infty} x_k := \lim_{n\to\infty} \inf_{k\geq n} x_k.}$$

Similarly by considering the sequence $s_n = \sup_{k\geq n} x_k$, we arrive at the definition of the superior limit of the sequence $\{x_k\}$:

Definition 14

$$\boxed{\overline{\lim}_{k\to\infty} x_k := \lim_{n\to\infty} \sup_{k\geq n} x_k.}$$

We now give several examples:

Example 14 $x_k = (-1)^k$, $k \in \mathbb{N}$:

$$\underline{\lim}_{k\to\infty} x_k = \lim_{n\to\infty} \inf_{k\geq n} x_k = \lim_{n\to\infty} \inf_{k\geq n} (-1)^k = \lim_{n\to\infty} (-1) = -1,$$

$$\overline{\lim}_{k\to\infty} x_k = \lim_{n\to\infty} \sup_{k\geq n} x_k = \lim_{n\to\infty} \sup_{k\geq n} (-1)^k = \lim_{n\to\infty} 1 = 1.$$

Example 15 $x_k = k^{(-1)^k}$, $k \in \mathbb{N}$:

$$\underline{\lim}_{k\to\infty} k^{(-1)^k} = \lim_{n\to\infty} \inf_{k\geq n} k^{(-1)^k} = \lim_{n\to\infty} 0 = 0,$$

$$\overline{\lim}_{k\to\infty} k^{(-1)^k} = \lim_{n\to\infty} \sup_{k\geq n} k^{(-1)^k} = \lim_{n\to\infty} (+\infty) = +\infty.$$

Example 16 $x_k = k$, $k \in \mathbb{N}$:

$$\underline{\lim}_{k\to\infty} k = \lim_{n\to\infty} \inf_{k\geq n} k = \lim_{n\to\infty} n = +\infty,$$

$$\overline{\lim}_{k\to\infty} k = \lim_{n\to\infty} \sup_{k\geq n} k = \lim_{n\to\infty} (+\infty) = +\infty.$$

Example 17 $x_k = \frac{(-1)^k}{k}$, $k \in \mathbb{N}$:

$$\underline{\lim}_{k\to\infty} \frac{(-1)^k}{k} = \lim_{n\to\infty} \inf_{k\geq n} \frac{(-1)^k}{k} = \lim_{n\to\infty} \begin{Bmatrix} -\frac{1}{n}, & \text{if } n = 2m+1 \\ -\frac{1}{n+1}, & \text{if } n = 2m \end{Bmatrix} = 0,$$

$$\overline{\lim_{k\to\infty}} \frac{(-1)^k}{k} = \lim_{n\to\infty} \sup_{k\geq n} \frac{(-1)^k}{k} = \lim_{n\to\infty} \left\{ \begin{array}{ll} \frac{1}{n}, & \text{if } n = 2m \\ \frac{1}{n+1}, & \text{if } n = 2m+1 \end{array} \right\} = 0.$$

Example 18 $x_k = -k^2, k \in \mathbb{N}$:

$$\underline{\lim_{k\to\infty}} \left(-k^2\right) = \lim_{n\to\infty} \inf_{k\geq n} \left(-k^2\right) = -\infty.$$

Example 19 $x_k = (-1)^k k, k \in \mathbb{N}$:

$$\underline{\lim_{k\to\infty}} (-1)^k k = \lim_{n\to\infty} \inf_{k\geq n} (-1)^k k = \lim_{n\to\infty} (-\infty) = -\infty,$$

$$\overline{\lim_{k\to\infty}} (-1)^k k = \lim_{n\to\infty} \sup_{k\geq n} (-1)^k k = \lim_{n\to\infty} (+\infty) = +\infty.$$

To explain the origin of the terms "superior" and "inferior" limit of a sequence, we make the following definition.

Definition 15 A number (or the symbol $-\infty$ or $+\infty$) is called a *partial limit* of a sequence, if the sequence contains a subsequence converging to that number.

Proposition 1 *The inferior and superior limits of a bounded sequence are respectively the smallest and largest partial limits of the sequence.*[4]

Proof Let us prove this, for example, for the inferior limit $i = \underline{\lim}_{k\to\infty} x_k$. What we know about the sequence $i_n = \inf_{k\geq n} x_k$ is that it is nondecreasing and that $\lim_{n\to\infty} i_n = i \in \mathbb{R}$. For the numbers $n \in \mathbb{N}$, using the definition of the greatest lower bound, we choose by induction numbers $k_n \in \mathbb{N}$ such that $k_n < k_{n+1}$ and $i_{k_n} \leq x_{k_n} < i_{k_n} + \frac{1}{n}$. (Taking i_1 we find k_1; taking i_{k_1+1} we find k_2, etc.) Since $\lim_{n\to\infty} i_n = \lim_{n\to\infty}(i_n + \frac{1}{n}) = i$, we can assert, by properties of limits, that $\lim_{n\to\infty} x_{k_n} = i$. We have thus proved that i is a partial limit of the sequence $\{x_k\}$. It is the smallest partial limit since for every $\varepsilon > 0$ there exists $n \in \mathbb{N}$ such that $i - \varepsilon < i_n$, that is $i - \varepsilon < i_n = \inf_{k\geq n} x_k \leq x_k$ for any $k \geq n$.

The inequality $i - \varepsilon < x_k$ for $k > n$ means that no partial limit of the sequence can be less than $i - \varepsilon$. But $\varepsilon > 0$ is arbitrary, and hence no partial limit can be less than i.

The proof for the superior limit is of course analogous. □

We now remark that if a sequence is not bounded below, then one can select a subsequence of it tending to $-\infty$. But in this case we also have $\underline{\lim}_{k\to\infty} x_k = -\infty$, and we can make the convention that the inferior limit is once again the smallest partial limit. The superior limit may be finite; if so, by what has been proved it

[4]Here we are assuming the natural relations $-\infty < x < +\infty$ between the symbols $-\infty$, $+\infty$ and numbers $x \in \mathbb{R}$.

must be the largest partial limit. But it may also be infinite. If $\overline{\lim}_{k\to\infty} x_k = +\infty$, then the sequence is also unbounded from above, and one can select a subsequence tending to $+\infty$. Finally, if $\overline{\lim}_{k\to\infty} x_k = -\infty$, which is also possible, this means that $\sup_{k\geq n} x_k = s_n \to -\infty$, that is, the sequence $\{x_n\}$ itself tends to $-\infty$, since $s_n \geq x_n$. Similarly, if $\underline{\lim}_{k\to\infty} x_k = +\infty$, then $x_k \to +\infty$.

Taking account of what has just been said we deduce the following proposition.

Proposition 1′ *For any sequence, the inferior limit is the smallest of its partial limits and the superior limit is the largest of its partial limits.*

Corollary 3 *A sequence has a limit or tends to negative or positive infinity if and only if its inferior and superior limits are the same.*

Proof The cases when $\underline{\lim}_{k\to\infty} x_k = \overline{\lim}_{k\to\infty} x_k = +\infty$ or $\underline{\lim}_{k\to\infty} x_k = \overline{\lim}_{k\to\infty} x_k = -\infty$ have been investigated above, and so we may assume that $\underline{\lim}_{k\to\infty} x_k = \overline{\lim}_{k\to\infty} x_k = A \in \mathbb{R}$. Since $i_n = \inf_{k\geq n} x_k \leq x_n \leq \sup_{k\geq n} x_k = s_n$ and by hypothesis $\lim_{n\to\infty} i_n = \lim_{n\to\infty} s_n = A$, we also have $\lim_{n\to\infty} x_n = A$ by properties of limits. □

Corollary 4 *A sequence converges if and only if every subsequence of it converges.*

Proof The inferior and superior limits of a subsequence lie between those of the sequence itself. If the sequence converges, its inferior and superior limits are the same, and so those of the subsequence must also be the same, proving that the subsequence converges. Moreover, the limit of the subsequence must be the same as that of the sequence itself.

The converse assertion is obvious, since the subsequence can be chosen as the sequence itself. □

Corollary 5 *The Bolzano–Weierstrass Lemma in its restricted and wider formulations follows from Propositions* 1 *and* 1′ *respectively.*

Proof Indeed, if the sequence $\{x_k\}$ is bounded, then the points $i = \underline{\lim}_{k\to\infty} x_k$ and $s = \overline{\lim}_{k\to\infty} x_k$ are finite and, by what has been proved, are partial limits of the sequence. Only when $i = s$ does the sequence have a unique limit point. When $i < s$ there are at least two.

If the sequence is unbounded on one side or the other, there exists a subsequence tending to the corresponding infinity. □

Concluding Remarks We have carried out all three points of the program outlined at the beginning of this section (and even gone beyond it in some ways). We have given a precise definition of the limit of a sequence, proved that the limit is unique, explained the connection between the limit operation and the structure of the set of real numbers, and obtained a criterion for convergence of a sequence.

We now study a special type of sequence that is frequently encountered and very useful – a series.

3.1.4 Elementary Facts About Series

a. The Sum of a Series and the Cauchy Criterion for Convergence of a Series

Let $\{a_n\}$ be a sequence of real numbers. We recall that the sum $a_p + a_{p+1} + \cdots + a_q$, $(p \leq q)$ is denoted by the symbol $\sum_{n=p}^{q} a_n$. We now wish to give a precise meaning to the expression $a_1 + a_2 + \cdots + a_n + \cdots$, which expresses the sum of all the terms of the sequence $\{a_n\}$.

Definition 16 The expression $a_1 + a_2 + \cdots + a_n + \cdots$ is denoted by the symbol $\sum_{n=1}^{\infty} a_n$ and usually called a *series* or an *infinite series* (in order to emphasize its difference from the sum of a finite number of terms).

Definition 17 The elements of the sequence $\{a_n\}$, when regarded as elements of the series, are called the *terms of the series*. The element a_n is called the *nth term.*

Definition 18 The sum $s_n = \sum_{k=1}^{n} a_k$ is called the *partial sum of the series*, or, when one wishes to exhibit its index, the *nth partial sum of the series.*[5]

Definition 19 If the sequence $\{s_n\}$ of partial sums of a series converges, we say the series is *convergent*. If the sequence $\{s_n\}$ does not have a limit, we say the series is *divergent.*

Definition 20 The limit $\lim_{n\to\infty} s_n = s$ of the sequence of partial sums of the series, if it exists, is called the *sum of the series.*

It is in this sense that we shall henceforth understand the expression

$$\sum_{n=1}^{\infty} a_n = s.$$

Since convergence of a series is equivalent to convergence of its sequence of partial sums $\{s_n\}$, applying the Cauchy convergence criterion to the sequence $\{s_n\}$ yields the following theorem.

Theorem 6 (The Cauchy convergence criterion for a series) *The series $a_1 + \cdots + a_n + \cdots$ converges if and only if for every $\varepsilon > 0$ there exists $N \in \mathbb{N}$ such that the inequalities $m \geq n > N$ imply $|a_n + \cdots + a_m| < \varepsilon$.*

Corollary 6 *If only a finite number of terms of a series are changed, the resulting new series will converge if the original series did and diverge if it diverged.*

---

[5]Thus we are actually defining a series to be an ordered pair $(\{a_n\}, \{s_n\})$ of sequences connected by the relation $(s_n = \sum_{k=1}^{n} a_k)$ for all $n \in \mathbb{N}$.

Proof For the proof it suffices to assume that the number N in the Cauchy convergence criterion is larger than the largest index among the terms that were altered. □

Corollary 7 *A necessary condition for convergence of the series* $a_1 + \cdots + a_n + \cdots$ *is that the terms tend to zero as* $n \to \infty$, *that is, it is necessary that* $\lim_{n\to\infty} a_n = 0$.

Proof It suffices to set $m = n$ in the Cauchy convergence criterion and use the definition of the limit of a sequence. □

Here is another proof: $a_n = s_n - s_{n-1}$, and, given that $\lim_{n\to\infty} s_n = s$, we have $\lim_{n\to\infty} a_n = \lim_{n\to\infty}(s_n - s_{n-1}) = \lim_{n\to\infty} s_n - \lim_{n\to\infty} s_{n-1} = s - s = 0$.

Example 20 The series $1 + q + q^2 + \cdots + q^n + \cdots$ is often called the *geometric series*. Let us investigate its convergence.

Since $|q^n| = |q|^n$, we have $|q^n| \geq 1$ when $|q| \geq 1$, and in this case the necessary condition for convergence is not met.

Now suppose $|q| < 1$. Then

$$s_n = 1 + q + \cdots + q^{n-1} = \frac{1 - q^n}{1 - q}$$

and $\lim_{n\to\infty} s_n = \frac{1}{1-q}$, since $\lim_{n\to\infty} q^n = 0$ if $|q| < 1$.

Thus the series $\sum_{n=1}^{\infty} q^{n-1}$ converges if and only if $|q| < 1$, and in that case its sum is $\frac{1}{1-q}$.

Example 21 The series $1 + \frac{1}{2} + \cdots + \frac{1}{n} + \cdots$ is called the *harmonic series*, since each term from the second on is the harmonic mean of the two terms on either side of it (see Exercise 6 at the end of this section).

The terms of the series tend to zero, but the sequence of partial sums

$$s_n = 1 + \frac{1}{2} + \cdots + \frac{1}{n},$$

as was shown in Example 10, diverges. This means that in this case $s_n \to +\infty$ as $n \to \infty$.

Thus the harmonic series diverges.

Example 22 The series $1 - 1 + 1 - \cdots + (-1)^{n+1} + \cdots$ diverges, as can be seen both from the sequence of partial sums $1, 0, 1, 0, \ldots$ and from the fact that the terms do not tend to zero.

If we insert parentheses and consider the new series

$$(1 - 1) + (1 - 1) + \cdots,$$

whose terms are the sums enclosed in parentheses, this new series does converge, and its sum is obviously zero.

If we insert the parentheses in a different way and consider the series

$$1 + (-1 + 1) + (-1 + 1) + \cdots,$$

the result is a convergent series with sum 1.

If we move all the terms that are equal to -1 in the original series two places to the right, we obtain the series

$$1 + 1 - 1 + 1 - 1 + 1 - \cdots,$$

we can then, by inserting parentheses, arrive at the series

$$(1 + 1) + (-1 + 1) + (-1 + 1) + \cdots,$$

whose sum equals 2.

These observations show that the usual laws for dealing with finite sums can in general not be extended to series.

There is nevertheless an important type of series that can be handled exactly like finite sums, as we shall see below. These are the so-called *absolutely convergent* series. They are the ones we shall mainly work with.

b. Absolute Convergence. The Comparison Theorem and Its Consequences

Definition 21 The series $\sum_{n=1}^{\infty} a_n$ is *absolutely* convergent if the series $\sum_{n=1}^{\infty} |a_n|$ converges.

Since $|a_n + \cdots + a_m| \leq |a_n| + \cdots |a_m|$, the Cauchy convergence criterion implies that an absolutely convergent series converges.

The converse of this statement is generally not true, that is, absolute convergence is a stronger requirement than mere convergence, as one can show by an example.

Example 23 The series $1 - 1 + \frac{1}{2} - \frac{1}{2} + \frac{1}{3} - \frac{1}{3} + \cdots$, whose partial sums are either $\frac{1}{n}$ or 0, converges to 0.

At the same time, the series of absolute values of its terms

$$1 + 1 + \frac{1}{2} + \frac{1}{2} + \frac{1}{3} + \frac{1}{3} + \cdots$$

diverges, as follows from the Cauchy convergence criterion, just as in the case of the harmonic series:

$$\left|\frac{1}{n+1} + \frac{1}{n+1} + \cdots + \frac{1}{n+n} + \frac{1}{n+n}\right| =$$

$$= 2\left(\frac{1}{n+1} + \cdots + \frac{1}{n+n}\right) > 2n \cdot \frac{1}{n+n} = 1.$$

To learn how to determine whether a series converges absolutely or not, it suffices to learn how to investigate the convergence of series with nonnegative terms. The following theorem holds.

Theorem 7 (Criterion for convergence of series of nonnegative terms) *A series $a_1 + \cdots + a_n + \cdots$ whose terms are nonnegative converges if and only if the sequence of partial sums is bounded above.*

Proof This follows from the definition of convergence of a series and the criterion for convergence of a nondecreasing sequence, which the sequence of partial sums is, in this case: $s_1 \le s_2 \le \cdots \le s_n \le \cdots$. □

This criterion implies the following simple theorem, which is very useful in practice.

Theorem 8 (Comparison theorem) *Let $\sum_{n=1}^{\infty} a_n$ and $\sum_{n=1}^{\infty} b_n$ be two series with nonnegative terms. If there exists an index $N \in \mathbb{N}$ such that $a_n \le b_n$ for all $n > N$, then the convergence of the series $\sum_{n=1}^{\infty} b_n$ implies the convergence of $\sum_{n=1}^{\infty} a_n$, and the divergence of $\sum_{n=1}^{\infty} a_n$ implies the divergence of $\sum_{n=1}^{\infty} b_n$.*

Proof Since a finite number of terms has no effect on the convergence of a series, we can assume with no loss of generality that $a_n \le b_n$ for every index $n \in \mathbb{N}$. Then $A_n = \sum_{k=1}^{n} a_k \le \sum_{k=1}^{n} b_k = B_n$. If the series $\sum_{n=1}^{\infty} b_n$ converges, then the sequence $\{B_n\}$, which is nondecreasing, tends to a limit B. But then $A_n \le B_n \le B$ for all $n \in \mathbb{N}$, and so the sequence A_n of partial sums of the series $\sum_{n=1}^{\infty} a_n$ is bounded. By the criterion for convergence of a series with nonnegative terms (Theorem 7), the series $\sum_{n=1}^{\infty} a_n$ converges.

The second assertion of the theorem follows from what has just been proved through proof by contradiction. □

Example 24 Since $\frac{1}{n(n+1)} < \frac{1}{n^2} < \frac{1}{(n-1)n}$ for $n \ge 2$, we conclude that the series $\sum_{n=1}^{\infty} \frac{1}{n^2}$ and $\sum_{n=1}^{\infty} \frac{1}{n(n+1)}$ converge or diverge together.

But the latter series can be summed directly, by observing that $\frac{1}{k(k+1)} = \frac{1}{k} - \frac{1}{k+1}$, and therefore $\sum_{k=1}^{n} \frac{1}{k(k+1)} = 1 - \frac{1}{n+1}$. Hence $\sum_{n=1}^{\infty} \frac{1}{n(n+1)} = 1$. Consequently the series $\sum_{n=1}^{\infty} \frac{1}{n^2}$ converges. It is interesting that $\sum_{n=1}^{\infty} \frac{1}{n^2} = \frac{\pi^2}{6}$, as will be proved below.

Example 25 It should be observed that the comparison theorem applies only to series with nonnegative terms. Indeed, if we set $a_n = -n$ and $b_n = 0$, for example, we have $a_n < b_n$ and the series $\sum_{n=1}^{\infty} b_n$ converges while $\sum_{n=1}^{\infty} a_n$ diverges.

Corollary 8 (The Weierstrass M-test for absolute convergence) *Let $\sum_{n=1}^{\infty} a_n$ and $\sum_{n=1}^{\infty} b_n$ be series. Suppose there exists an index $N \in \mathbb{N}$ such that $|a_n| \le b_n$ for all $n > N$. Then a sufficient condition for absolute convergence of the series $\sum_{n=1}^{\infty} a_n$ is that the series $\sum_{n=1}^{\infty} b_n$ converge.*

Proof In fact, by the comparison theorem the series $\sum_{n=1}^{\infty}|a_n|$ will then converge, and that is what is meant by the absolute convergence of $\sum_{n=1}^{\infty}a_n$. □

This important sufficiency test for absolute convergence is often stated briefly as follows: *If the terms of a series are majorized* (in absolute value) *by the terms of a convergent numerical series, then the original series converges absolutely.*

Example 26 The series $\sum_{n=1}^{\infty}\frac{\sin n}{n^2}$ converges absolutely, since $|\frac{\sin n}{n^2}|\le\frac{1}{n^2}$ and the series $\sum_{n=1}^{\infty}\frac{1}{n^2}$ converges, as we saw in Example 24.

Corollary 9 (Cauchy's test) *Let $\sum_{n=1}^{\infty}a_n$ be a given series and $\alpha=\overline{\lim}_{n\to\infty}\sqrt[n]{|a_n|}$. Then the following are true:*

a) *if $\alpha<1$, the series $\sum_{n=1}^{\infty}a_n$ converges absolutely;*
b) *if $\alpha>1$, the series $\sum_{n=1}^{\infty}a_n$ diverges;*
c) *there exist both absolutely convergent and divergent series for which $\alpha=1$.*

Proof a) If $\alpha<1$, we can choose $q\in\mathbb{R}$ such that $\alpha<q<1$. Fixing q, by definition of the superior limit, we find $N\in\mathbb{N}$ such that $\sqrt[n]{|a_n|}<q$ for all $n>N$. Thus we shall have $|a_n|<q^n$ for $n>N$, and since the series $\sum_{n=1}^{\infty}q^n$ converges for $|q|<1$, it follows from the comparison theorem or from the Weierstrass criterion that the series $\sum_{n=1}^{\infty}a_n$ converges absolutely.

b) Since α is a partial limit of the sequence $\{\sqrt[n]{|a_n|}\}$ (Proposition 1), there exists a subsequence $\{a_{n_k}\}$ such that $\lim_{n\to\infty}\sqrt[n_k]{|a_{n_k}|}=\alpha$. Hence if $\alpha>1$, there exists $K\in\mathbb{N}$ such that $|a_{n_k}|>1$ for all $k>K$, and so the necessary condition for convergence ($a_n\to 0$) does not hold for the series $\sum_{n=1}^{\infty}a_n$. It therefore diverges.

c) We already know that the series $\sum_{n=1}^{\infty}\frac{1}{n}$ diverges and $\sum_{n=1}^{\infty}\frac{1}{n^2}$ converges (absolutely, since $|\frac{1}{n^2}|=\frac{1}{n^2}$). At the same time, $\overline{\lim}_{n\to\infty}\sqrt[n]{\frac{1}{n}}=\lim_{n\to\infty}\frac{1}{\sqrt[n]{n}}=1$ and $\overline{\lim}_{n\to\infty}\sqrt[n]{\frac{1}{n^2}}=\lim_{n\to\infty}\frac{1}{\sqrt[n]{n^2}}=\lim_{n\to\infty}(\frac{1}{\sqrt[n]{n}})^2=1$. □

Example 27 Let us investigate the values of $x\in\mathbb{R}$ for which the series

$$\sum_{n=1}^{\infty}\bigl(2+(-1)^n\bigr)^n x^n$$

converges.

We compute $\alpha=\overline{\lim}_{n\to\infty}\sqrt[n]{|(2+(-1)^n)^n x^n|}=|x|\,\overline{\lim}_{n\to\infty}|2+(-1)^n|=3|x|$. Thus for $|x|<\frac{1}{3}$ the series converges and even absolutely, while for $|x|>3$ the series diverges. The case $|x|=\frac{1}{3}$ requires separate consideration. In the present case that is an elementary task, since for $|x|=\frac{1}{3}$ and n even ($n=2k$), we have $(2+(-1)^{2k})^{2k}x^{2k}=3^{2k}(\frac{1}{3})^{2k}=1$. Therefore the series diverges, since it does not fulfill the necessary condition for convergence.

Corollary 10 (d'Alembert's[6] test) *Suppose the limit* $\lim_{n\to\infty}|\frac{a_{n+1}}{a_n}| = \alpha$ *exists for the series* $\sum_{n=1}^{\infty} a_n$. *Then,*

a) *if* $\alpha < 1$, *the series* $\sum_{n=1}^{\infty} a_n$ *converges absolutely;*
b) *if* $\alpha > 1$, *the series* $\sum_{n=1}^{\infty} a_n$ *diverges;*
c) *there exist both absolutely convergent and divergent series for which* $\alpha = 1$.

Proof a) If $\alpha < 1$, there exists a number q such that $\alpha < q < 1$. Fixing q and using properties of limits, we find an index $N \in \mathbb{N}$ such that $|\frac{a_{n+1}}{a_n}| < q$ for $n > N$. Since a finite number of terms has no effect on the convergence of a series, we shall assume without loss of generality that $|\frac{a_{n+1}}{a_n}| < q$ for all $n \in \mathbb{N}$.

Since

$$\left|\frac{a_{n+1}}{a_n}\right| \cdot \left|\frac{a_n}{a_{n-1}}\right| \cdots \left|\frac{a_2}{a_1}\right| = \left|\frac{a_{n+1}}{a_1}\right|,$$

we find that $|a_{n+1}| \le |a_1| \cdot q^n$. But the series $\sum_{n=1}^{\infty} |a_1|q^n$ converges (its sum is obviously $\frac{|a_1|q}{1-q}$), so that the series $\sum_{n=1}^{\infty} a_n$ converges absolutely.

b) If $\alpha > 1$, then from some index $N \in \mathbb{N}$ on we have $|\frac{\alpha_{n+1}}{a_n}| > 1$, that is, $|a_n| < |a_{n+1}|$, and the condition $a_n \to 0$, which is necessary for convergence, does not hold for the series $\sum_{n=1}^{\infty} a_n$.

c) As in the case of Cauchy's test, the series $\sum_{n=1}^{\infty} \frac{1}{n}$ and $\sum_{n=1}^{\infty} \frac{1}{n^2}$ provide examples. □

Example 28 Let us determine the values of $x \in \mathbb{R}$ for which the series

$$\sum_{n=1}^{\infty} \frac{1}{n!} x^n$$

converges.

For $x = 0$ it obviously converges absolutely.

For $x \neq 0$ we have $\lim_{n\to\infty} |\frac{a_{n+1}}{a_n}| = \lim_{n\to\infty} \frac{|x|}{n+1} = 0$.

Thus, this series converges absolutely for every value of $x \in \mathbb{R}$.

Finally, let us consider another special, but frequently encountered class of series, namely those whose terms form a monotonic sequence. For such series we have the following necessary and sufficient condition:

Proposition 2 (Cauchy) *If* $a_1 \ge a_2 \ge \cdots \ge 0$, *the series* $\sum_{n=1}^{\infty} a_n$ *converges if and only if the series* $\sum_{k=0}^{\infty} 2^k a_{2^k} = a_1 + 2a_2 + 4a_4 + 8a_8 + \cdots$ *converges.*

Proof Since

$$a_2 \le a_2 \le a_1,$$

[6] J.L. d'Alembert (1717–1783) – French scholar specializing in mechanics. He was a member of the group of *philosophers* who wrote the *Encyclopédie*.

$$
\begin{aligned}
2a_4 &\le a_3 + a_4 \le 2a_2,\\
4a_8 &\le a_5 + a_6 + a_7 + a_8 \le 4a_4,\\
&\vdots\\
2^n a_{2^{n+1}} &\le a_{2^n+1} + \cdots + a_{2^{n+1}} \le 2^n a_{2^n},
\end{aligned}
$$

by adding these inequalities, we find

$$
\frac{1}{2}(S_{n+1} - a_1) \le A_{2^{n+1}} - a_1 \le S_n,
$$

where $A_k = a_1 + \cdots + a_k$ and $S_n = a_1 + 2a_2 + \cdots + 2^n a_{2^n}$ are the partial sums of the two series in question. The sequences $\{A_k\}$ and $\{S_n\}$ are nondecreasing, and hence from these inequalities one can conclude that they are either both bounded above or both unbounded above. Then, by the criterion for convergence of series with nonnegative terms, it follows that the two series indeed converge or diverge together. □

This result implies a useful corollary.

Corollary *The series* $\sum_{n=1}^{\infty} \frac{1}{n^p}$ *converges for* $p > 1$ *and diverges for* $p \le 1$.[7]

Proof If $p \ge 0$, the proposition implies that the series converges or diverges simultaneously with the series

$$
\sum_{k=0}^{\infty} 2^k \frac{1}{(2^k)^p} = \sum_{k=0}^{\infty} (2^{1-p})^k,
$$

and a necessary and sufficient condition for the convergence of this series is that $q = 2^{1-p} < 1$, that is, $p > 1$.

If $p \le 0$, the divergence of the series $\sum_{n=1}^{\infty} \frac{1}{n^p}$ is obvious, since all the terms of the series are not smaller than 1. □

The importance of this corollary is that the series $\sum_{n=1}^{\infty} \frac{1}{n^p}$ is often used as a comparison series to study the convergence of other series.

c. The Number e as the Sum of a Series

To conclude our study of series we return once again to the number e and obtain a series the provides a very convenient way of computing it.

---

[7] Up to now in this book the number n^p has been defined formally only for rational values of p, so that for the moment the reader is entitled to take this proposition as applying only to values of p for which n^p is defined.

We shall use Newton's binomial formula to expand the expression $(1+\frac{1}{n})^n$. Those who are unfamiliar with this formula from high school and have not solved part g) of Exercise 1 in Sect. 2.2 may omit the present appendix on the number e with no loss of continuity and return to it after studying Taylor's formula, of which Newton's binomial formula may be regarded as a special case.

We know that $\mathrm{e}=\lim_{n\to\infty}(1+\frac{1}{n})^n$.

By Newton's binomial formula

$$\left(1+\frac{1}{n}\right)^n = 1+\frac{n}{1!}\frac{1}{n}+\frac{n(n-1)}{2!}\frac{1}{n^2}+\cdots+ \\ +\frac{n(n-1)\cdots(n-k+1)}{k!}\frac{1}{n^k}+\cdots+\frac{1}{n^n}= \\ =1+1+\frac{1}{2!}\left(1-\frac{1}{n}\right)+\cdots+\frac{1}{k!}\left(1-\frac{1}{n}\right)\left(1-\frac{2}{n}\right)\times\cdots\times \\ \times\left(1-\frac{k-1}{n}\right)+\cdots+\frac{1}{n!}\left(1-\frac{1}{n}\right)\cdots\left(1-\frac{n-1}{n}\right).$$

Setting $(1+\frac{1}{n})^n=e_n$ and $1+1+\cdots\frac{1}{2!}+\cdots+\frac{1}{n!}=s_n$, we thus have $e_n<s_n$ $(n=1,2,\dots)$.

On the other hand, for any fixed k and $n\geq k$, as can be seen from the same expansion, we have

$$1+1+\frac{1}{2!}\left(1-\frac{1}{n}\right)+\cdots+\frac{1}{k!}\left(1-\frac{1}{n}\right)\cdots\left(1-\frac{k-1}{n}\right)<e_n.$$

As $n\to\infty$ the left-hand side of this inequality tends to s_k and the right-hand side to e. We can now conclude that $s_k\leq \mathrm{e}$ for all $k\in\mathbb{N}$.

But then from the relations

$$e_n<s_n\leq \mathrm{e}$$

we find that $\lim_{n\to\infty}s_n=\mathrm{e}$.

In accordance with the definition of the sum of a series, we can now write

$$\boxed{\mathrm{e}=1+\frac{1}{1!}+\frac{1}{2!}+\cdots+\frac{1}{n!}+\cdots.}$$

This representation of the number e is very well adapted for computation.

Let us estimate the difference $\mathrm{e}-s_n$:

$$0<\mathrm{e}-s_n=\frac{1}{(n+1)!}+\frac{1}{(n+2)!}+\cdots= \\ =\frac{1}{(n+1)!}\left[1+\frac{1}{n+2}+\frac{1}{(n+2)(n+3)}+\cdots\right]<$$

$$< \frac{1}{(n+1)!}\left[1 + \frac{1}{n+2} + \frac{1}{(n+2)^2} + \cdots\right] =$$

$$= \frac{1}{(n+1)!}\frac{1}{1-\frac{1}{n+2}} = \frac{n+2}{n!(n+1)^2} < \frac{1}{n!n}.$$

Thus, in order to make the absolute error in the approximation of e by s_n less than, say 10^{-3}, it suffices that $\frac{1}{n!n} < \frac{1}{1000}$. This condition is already satisfied by s_6.

Let us write out the first few decimal digits of e:

$$\mathrm{e} = 2.7182818284590\ldots.$$

This estimate of the difference $\mathrm{e} - s_n$ can be written as the equality

$$\mathrm{e} = s_n + \frac{\theta_n}{n!n}, \quad \text{where } 0 < \theta_n < 1.$$

It follows immediately from this representation of e that it is irrational. Indeed, if we assume that $\mathrm{e} = \frac{p}{q}$, where $p, q \in \mathbb{N}$, then the number $q!\mathrm{e}$ must be an integer, while

$$q!\mathrm{e} = q!\left(s_q + \frac{\theta_q}{q!q}\right) = q! + \frac{q!}{1!} + \frac{q!}{2!} + \cdots + \frac{q!}{q!} + \frac{\theta_q}{q},$$

and then the number $\frac{\theta_q}{q}$ would have to be an integer, which is impossible.

For the reader's information we note that e is not only irrational, but also transcendental.

3.1.5 Problems and Exercises

1. Show that a number $x \in \mathbb{R}$ is rational if and only if its q-ary expression in any base q is periodic, that is, from some rank on it consists of periodically repeating digits.

2. A ball that has fallen from height h bounces to height qh, where q is a constant coefficient $0 < q < 1$. Find the time that elapses until it comes to rest and the distance it travels through the air during that time.

3. We mark all the points on a circle obtained from a fixed point by rotations of the circle through angles of n radians, where $n \in \mathbb{Z}$ ranges over all integers. Describe all the limit points of the set so constructed.

4. The expression

$$n_1 + \cfrac{1}{n_2 + \cfrac{1}{n_3 + \cfrac{}{\ddots \cfrac{1}{n_{k-1} + \frac{1}{n_k}}}}}$$

where $n_k \in \mathbb{N}$, is called a finite *continued fraction*, and the expression

$$n_1 + \cfrac{1}{n_2 + \cfrac{1}{n_3 + \ddots}}$$

is called an infinite continued fraction. The fractions obtained from a continued fraction by omitting all its elements from a certain one on are called the *convergents*. The value assigned to an infinite continued fraction is the limit of its convergents. Show that:

a) Every rational number $\frac{m}{n}$, where $m, n \in \mathbb{N}$ can be expanded in a unique manner as a continued fraction:

$$\frac{m}{n} = q_1 + \cfrac{1}{q_2 + \cfrac{1}{q_3 + \ddots \cfrac{1}{q_{n-1} + \frac{1}{q_n}}}},$$

assuming that $q_n \neq 1$ for $n > 1$.

Hint: The numbers $q, \ldots, q_n$, called the *incomplete quotients* or *elements*, can be obtained from the Euclidean algorithm

$$m = n \cdot q_1 + r_1,$$
$$n = r_1 \cdot q_2 + r_2,$$
$$r_1 = r_2 \cdot q_3 + r_3,$$
$$\vdots$$

by writing it in the form

$$\frac{m}{n} = q_1 + \frac{1}{n/r_1} = q_1 + \cfrac{1}{q_2 + \ddots}.$$

b) The convergents $R_1 = q_1$, $R_2 = q_1 + \frac{1}{q_2}, \ldots$ satisfy the inequalities

$$R_1 < R_3 < \cdots < R_{2k-1} < \frac{m}{n} < R_{2k} < R_{2k-2} < \cdots < R_2.$$

c) The numerators P_k and denominators Q_k of the convergents are formed according to the following rule:

$$P_k = P_{k-1}q_k + P_{k-2}, \qquad P_2 = q_1 q_2, \qquad P_1 = q_1,$$
$$Q_k = Q_{k-1}q_k + Q_{k-2}, \qquad Q_2 = q_2, \qquad Q_1 = 1.$$

d) The difference of successive convergents can be computed from the formula

$$R_k - R_{k-1} = \frac{(-1)^k}{Q_k Q_{k-1}} \quad (k > 1).$$

e) Every infinite continued fraction has a determinate value.
f) The value of an infinite continued fraction is irrational.
g)

$$\frac{1+\sqrt{5}}{2} = 1 + \cfrac{1}{1 + \cfrac{1}{1 + \ddots}}.$$

h) The Fibonacci numbers $1, 1, 2, 3, 5, 8, \dots$ (that is, $u_n = u_{n-1} + u_{n-2}$ and $u_1 = u_2 = 1$), which are obtained as the denominators of the convergents in g), are given by the formula

$$u_n = \frac{1}{\sqrt{5}}\left[\left(\frac{1+\sqrt{5}}{2}\right)^n - \left(\frac{1-\sqrt{5}}{2}\right)^n\right].$$

i) The convergents $R_k = \frac{P_k}{Q_k}$ in g) are such that $\left|\frac{1+\sqrt{5}}{2} - \frac{P_k}{Q_k}\right| > \frac{1}{Q_k^2\sqrt{5}}$. Compare this result with the assertions of Exercise 11 in Sect. 2.2.

5. Show that

a) the equality

$$1 + \frac{1}{1!} + \frac{1}{2!} + \cdots + \frac{1}{n!} + \frac{1}{n!n} = 3 - \frac{1}{1 \cdot 2 \cdot 2!} - \cdots - \frac{1}{(n-1) \cdot n \cdot n!}$$

holds for $n \geq 2$;
b) $e = 3 - \sum_{n=0}^{\infty} \frac{1}{(n+1)(n+2)(n+2)!}$;
c) for computing the number e approximately the formula $e \approx 1 + \frac{1}{1!} + \frac{1}{2!} + \cdots + \frac{1}{n!} + \frac{1}{n!n}$ is much better than the original formula $e \approx 1 + \frac{1}{1!} + \frac{1}{2!} + \cdots + \frac{1}{n!}$. (Estimate the errors, and compare the result with the value of e given on p. 103.)

6. If a and b are positive numbers and p an arbitrary nonzero real number, then the *mean of order* p of the numbers a and b is the quantity

$$S_p(a, b) = \left(\frac{a^p + b^p}{2}\right)^{\frac{1}{p}}.$$

In particular for $p = 1$ we obtain the arithmetic mean of a and b, for $p = 2$ their square-mean, and for $p = -1$ their harmonic mean.

a) Show that the mean $S_p(a, b)$ of any order lies between the numbers a and b.

b) Find the limits of the sequences

$$\{S_n(a,b)\}, \quad \{S_{-n}(a,b)\}.$$

7. Show that if $a > 0$, the sequence $x_{n+1} = \frac{1}{2}(x_n + \frac{a}{x_n})$ converges to the square root of a for any $x_1 > 0$.

Estimate the rate of convergence, that is, the magnitude of the absolute error $|x_n - \sqrt{a}| = |\Delta_n|$ as a function of n.

8. Show that

a)

$$S_0(n) = 1^0 + \cdots + n^0 = n,$$

$$S_1(n) = 1^1 + \cdots + n^1 = \frac{n(n+1)}{2} = \frac{1}{2}n^2 + \frac{1}{2}n,$$

$$S_2(n) = 1^2 + \cdots + n^2 = \frac{n(n+1)(2n+1)}{6} = \frac{1}{3}n^3 + \frac{1}{2}n^2 + \frac{1}{6}n,$$

$$S_3(n) = \frac{n^2(n+1)^2}{4} = \frac{1}{4}n^4 + \frac{1}{2}n^3 + \frac{1}{4}n^2,$$

and in general that

$$S_k(n) = a_{k+1}n^{k+1} + \cdots + a_1 n + a_0$$

is a polynomial in n of degree $k+1$.

b) $\lim_{n\to\infty} \frac{S_k(n)}{n^{k+1}} = \frac{1}{k+1}$.

3.2 The Limit of a Function

3.2.1 Definitions and Examples

Let E be a subset of $\mathbb{R}$ and a a limit point of E. Let $f : E \to \mathbb{R}$ be a real-valued function defined on E.

We wish to write out what it means to say that the value $f(x)$ of the function f approaches some number A as the point $x \in E$ approaches a. It is natural to call such a number A the limit of the values of the function f, or the limit of f as x tends to a.

Definition 1 We shall say (following Cauchy) that the function $f : E \to \mathbb{R}$ *tends to* A *as* x *tends to* a, or that A *is the limit of* f *as* x *tends to* a, if for every $\varepsilon > 0$ there exists $\delta > 0$ such that $|f(x) - A| < \varepsilon$ for every $x \in E$ such that $0 < |x-a| < \delta$.

In logical symbolism these conditions are written as

$$\forall \varepsilon > 0\ \exists \delta > 0\ \forall x \in E\ \big(0 < |x-a| < \delta \Rightarrow \big|f(x) - A\big| < \varepsilon\big).$$

If A is the limit of $f(x)$ as x tends to a in the set E, we write $f(x) \to A$ as $x \to a$, $x \in E$, or $\lim_{x\to a, x\in E} f(x) = A$. Instead of the expression $x \to a$, $x \in E$, we shall as a rule use the shorter notation $E \ni x \to a$, and instead of $\lim_{x\to a, x\in E} f(x)$ we shall write $\lim_{E\ni x\to a} f(x)$.

Example 1 Let $E = \mathbb{R}\backslash 0$, and $f(x) = x \sin \frac{1}{x}$. We shall verify that

$$\lim_{E\ni x\to 0} x \sin \frac{1}{x} = 0.$$

Indeed, for a given $\varepsilon > 0$ we choose $\delta = \varepsilon$. Then for $0 < |x| < \delta = \varepsilon$, taking account of the inequality $|x \sin \frac{1}{x}| \le |x|$, we shall have $|x \sin \frac{1}{x}| < \varepsilon$.

Incidentally, one can see from this example that a function $f : E \to \mathbb{R}$ may have a limit as $E \ni x \to a$ without even being defined at the point a itself. This is exactly the situation that most often arises when limits must be computed; and, if you were paying attention, you may have noticed that this circumstance is taken into account in our definition of limit, where we wrote the strict inequality $0 < |x - a|$.

We recall that a *neighborhood* of a point $a \in \mathbb{R}$ is any open interval containing the point.

Definition 2 A *deleted neighborhood* of a point is a neighborhood of the point from which the point itself has been removed.

If $U(a)$ denotes a neighborhood of a, we shall denote the corresponding deleted neighborhood by $\overset{\circ}{U}(a)$.

The sets

$$U_E(a) := E \cap U(a)$$
$$\overset{\circ}{U}_E(a) := E \cap \overset{\circ}{U}(a)$$

will be called respectively a *neighborhood of a in E* and a *deleted neighborhood of a in E*.

If a is a limit point of E, then $\overset{\circ}{U}_E(a) \neq \varnothing$ for every neighborhood $U(a)$.

If we temporarily adopt the cumbersome symbols $\overset{\circ}{U}{}^{\delta}_E(a)$ and $V^{\varepsilon}_{\mathbb{R}}(A)$ to denote the deleted δ-neighborhood of a in E and the ε-neighborhood of A in $\mathbb{R}$, then Cauchy's so-called "ε–δ-definition" of the limit of a function can be rewritten as

$$\Big(\lim_{E\ni x\to a} f(x) = A\Big) := \forall V^{\varepsilon}_{\mathbb{R}}(A)\ \exists \overset{\circ}{U}{}^{\delta}_E(a)\ \Big(f\big(\overset{\circ}{U}{}^{\delta}_E(a)\big) \subset V^{\varepsilon}_{\mathbb{R}}(A)\Big).$$

This expression says that A is the limit of the function $f : E \to \mathbb{R}$ as x tends to a in the set E if for every ε-neighborhood $V^{\varepsilon}_{\mathbb{R}}(A)$ of A there exists a deleted neighborhood $\overset{\circ}{U}{}^{\delta}_E(a)$ of a in E whose image $f(\overset{\circ}{U}{}^{\delta}_E(a))$ under the mapping $f : E \to \mathbb{R}$ is entirely contained in $V^{\varepsilon}_{\mathbb{R}}(A)$.

Taking into account that every neighborhood of a point on the real line contains a symmetric neighborhood (a δ-neighborhood) of the same point, we arrive at the following expression for the definition of a limit, which we shall take as our main definition:

Definition 3

$$\Bigl(\lim_{E\ni x\to a} f(x)=A\Bigr):=\forall V_{\mathbb{R}}(A)\ \exists \overset{\circ}{U}_E(a)\ \bigl(f\bigl(\overset{\circ}{U}_E(a)\bigr)\subset V_{\mathbb{R}}(A)\bigr).$$

Thus the number A is called the *limit* of the function $f:E\to\mathbb{R}$ as x tends to a while remaining in the set E (a must be a limit point of E) if for every neighborhood of A there is a deleted neighborhood of a in E whose image under the mapping $f:E\to\mathbb{R}$ is contained in the given neighborhood of A.

We have given several statements of the definition of the limit of a function. For numerical functions, when a and A belong to $\mathbb{R}$, as we have seen, these statements are equivalent. In this connection, we note that one or another of these statements may be more convenient in different situations. For example, the original form is convenient in numerical computations, since it shows the allowable magnitude of the deviation of x from a needed to ensure that the deviation of $f(x)$ from A will not exceed a specified value. But from the point of view of extending the concept of a limit to more general functions the last statement the definition is most convenient. It shows that we can define the concept of a limit of a mapping $f:X\to Y$ provided we have been told what is meant by a neighborhood of a point in X and Y, that is, as we say, a *topology* is given on X and Y.

Let us consider a few more examples that are illustrative of the main definition.

Example 2 The function

$$\operatorname{sgn} x=\begin{cases}1 & \text{if } x>0,\\ 0 & \text{if } x=0,\\ -1 & \text{if } x<0\end{cases}$$

(read "signum x"[8]) is defined on the whole real line. We shall show that it has no limit as x tends to 0. The nonexistence of this limit is expressed by

$$\forall A\in\mathbb{R}\ \exists V(A)\ \forall \overset{\circ}{U}(0)\ \exists x\in \overset{\circ}{U}(0)\ \bigl(f(x)\notin V(A)\bigr),$$

that is, no matter what A we take (claiming to be the limit of $\operatorname{sgn} x$ as $x\to 0$), there is a neighborhood $V(A)$ of A such that no matter how small a deleted neighborhood $\overset{\circ}{U}(0)$ of 0 we take, that deleted neighborhood contains at least one point x at which the value of the function does not lie in $V(A)$.

[8]The Latin word for *sign*.

Since $\operatorname{sgn} x$ assumes only the values -1, 0, and 1, it is clear that no number distinct from them can be the limit of the function. For any such number has a neighborhood that does not contain any of these three numbers.

But if $A \in \{-1, 0, 1\}$ we choose as $V(A)$ the ε-neighborhood of A with $\varepsilon = \frac{1}{2}$. The points -1 and 1 certainly cannot both lie in this neighborhood. But, no matter what deleted neighborhood $\overset{\circ}{U}(0)$ of 0 we may take, that neighborhood contains both positive and negative numbers, that is, points x where $f(x) = 1$ and points where $f(x) = -1$.

Hence there is a point $x \in \overset{\circ}{U}(0)$ such that $f(x) \notin V(A)$.

If the function $f : E \to \mathbb{R}$ is defined on a whole deleted neighborhood of a point $a \in \mathbb{R}$, that is, when $\overset{\circ}{U}_E(a) = \overset{\circ}{U}_{\mathbb{R}}(a) = \overset{\circ}{U}(a)$, we shall agree to write more briefly $x \to a$ instead of $E \ni x \to a$.

Example 3 Let us show that $\lim_{x\to 0} |\operatorname{sgn} x| = 1$.

Indeed, for $x \in \mathbb{R}\backslash 0$ we have $|\operatorname{sgn} x| = 1$, that is, the function is constant and equal to 1 in any deleted neighborhood $\overset{\circ}{U}(0)$ of 0. Hence for any neighborhood $V(1)$ we obtain $f(\overset{\circ}{U}(0)) = 1 \in V(1)$.

Note carefully that although the function $|\operatorname{sgn} x|$ is defined at the point 0 itself and $|\operatorname{sgn} 0| = 0$, this value has no influence on the value of the limit in question. Thus one must not confuse the value $f(a)$ of the function at the point a with the limit $\lim_{x\to a} f(x)$ that the function has as $x \to a$.

Let $\mathbb{R}_-$ and $\mathbb{R}_+$ be the sets of negative and positive numbers respectively.

Example 4 We saw in Example 2 that the limit $\lim_{\mathbb{R}\ni x\to 0} \operatorname{sgn} x$ does not exist. Remarking, however, that the restriction $\operatorname{sgn}|_{\mathbb{R}_-}$ of sgn to $\mathbb{R}_-$ is a constant function equal to -1 and $\operatorname{sgn}|_{\mathbb{R}_+}$ is a constant function equal to 1, we can show, as in Example 3, that

$$\lim_{\mathbb{R}_-\ni x\to 0} \operatorname{sgn} x = -1, \quad \text{and} \quad \lim_{\mathbb{R}_+\ni x\to 0} \operatorname{sgn} x = 1,$$

that is, the restrictions of the same function to different sets may have different limits at the same point, or even fail to have a limit, as shown in Example 2.

Example 5 Developing the idea of Example 2, one can show similarly that $\sin \frac{1}{x}$ has no limit as $x \to 0$.

Indeed, in any deleted neighborhood $\overset{\circ}{U}(0)$ of 0 there are always points of the form $\frac{1}{-\pi/2+2\pi n}$ and $\frac{1}{\pi/2+2\pi n}$, where $n \in \mathbb{N}$. At these points the function assumes the values -1 and 1 respectively. But these two numbers cannot both lie in the ε-neighborhood $V(A)$ of a point $A \in \mathbb{R}$ if $\varepsilon < 1$. Hence no number $A \in \mathbb{R}$ can be the limit of this function as $x \to 0$.

Example 6 If

$$E_- = \left\{x \in \mathbb{R} \mid x = \frac{1}{-\pi/2 + 2\pi n}, n \in \mathbb{N}\right\}$$

and

$$E_+ = \left\{ x \in \mathbb{R} \mid x = \frac{1}{\pi/2 + 2\pi n}, n \in \mathbb{N} \right\},$$

then, as shown in Example 4, we find that

$$\lim_{E_- \ni x \to 0} \sin \frac{1}{x} = -1 \quad \text{and} \quad \lim_{E_+ \ni x \to 0} \sin \frac{1}{x} = 1.$$

There is a close connection between the concept of the limit of a sequence studied in the preceding section and the limit of an arbitrary numerical-valued function introduced in the present section, expressed by the following proposition.

Proposition 1 [9] *The relation* $\lim_{E \ni x \to a} f(x) = A$ *holds if and only if for every sequence* $\{x_n\}$ *of points* $x_n \in E \setminus a$ *converging to* a, *the sequence* $\{f(x_n)\}$ *converges to* A.

Proof The fact that $(\lim_{E \ni x \to a} f(x) = A) \Rightarrow (\lim_{n \to \infty} f(x_n) = A)$ follows immediately from the definitions. Indeed, if $\lim_{E \ni x \to a} f(x) = A$, then for any neighborhood $V(A)$ of A there exists a deleted neighborhood $\overset{\circ}{U}_E(a)$ of the point a in E such that for $x \in \overset{\circ}{U}_E(a)$ we have $f(x) \in V(A)$. If the sequence $\{x_n\}$ of points in $E \setminus a$ converges to a, there exists an index N such that $x_n \in \overset{\circ}{U}_E(a)$ for $n > N$, and then $f(x_n) \in V(A)$. By definition of the limit of a sequence, we then conclude that $\lim_{n \to \infty} f(x_n) = A$.

We now prove the converse. If A is not the limit of $f(x)$ as $E \ni x \to a$, then there exists a neighborhood $V(A)$ such that for any $n \in \mathbb{N}$, there is a point x_n in the deleted $\frac{1}{n}$-neighborhood of a in E such that $f(x_n) \notin V(A)$. But this means that the sequence $\{f(x_n)\}$ does not converge to A, even though $\{x_n\}$ converges to a. □

3.2.2 Properties of the Limit of a Function

We now establish a number of properties of the limit of a function that are constantly being used. Many of them are analogous to the properties of the limit of a sequence that we have already established, and for that reason are essentially already known to us. Moreover, by Proposition 1 just proved, many properties of the limit of a function follow obviously and immediately from the corresponding properties of the limit of a sequence: the uniqueness of the limit, the arithmetic properties of the limit, and passage to the limit in inequalities. Nevertheless, we shall carry out all the proofs again. As will be seen, there is some value in doing so.

[9] This proposition is sometimes called the statement of the equivalence of the Cauchy definition of a limit (in terms of neighborhoods) and the Heine definition (in terms of sequences).

E. Heine (1821–1881) – German mathematician.

We call the reader's attention to the fact that, in order to establish the properties of the limit of a function, we need only two properties of deleted neighborhoods of a limit point of a set:

B$_1$) $\overset{\circ}{U}_E(a) \neq \varnothing$, that is, the deleted neighborhood of the point in E is nonempty;
B$_2$) $\forall \overset{\circ}{U}{}'_E(a)\ \forall \overset{\circ}{U}{}''_E(a)\ \exists \overset{\circ}{U}_E(a)\ (\overset{\circ}{U}_E(a) \subset \overset{\circ}{U}{}'_E(a) \cap \overset{\circ}{U}{}''_E(a))$, that is, the intersection of any pair of deleted neighborhoods contains a deleted neighborhood. This observation leads us to a general concept of a limit of a function and the possibility of using the theory of limits in the future not only for functions defined on sets of numbers. To keep the discussion from becoming a mere repetition of what was said in Sect. 3.1, we shall employ some useful new devices and concepts that were not proved in that section.

a. General Properties of the Limit of a Function

We begin with some definitions.

Definition 4 As before, a function $f : E \to \mathbb{R}$ assuming only one value is called *constant*. A function $f : E \to \mathbb{R}$ is called *ultimately constant* as $E \ni x \to a$ if it is constant in some deleted neighborhood $\overset{\circ}{U}_E(a)$, where a is a limit point of E.

Definition 5 A function $f : E \to \mathbb{R}$ is *bounded*, *bounded above*, or *bounded below* respectively if there is a number $C \in \mathbb{R}$ such that $|f(x)| < C$, $f(x) < C$, or $C < f(x)$ for all $x \in E$.

If one of these three relations holds only in some deleted neighborhood $\overset{\circ}{U}_E(a)$, the function is said to be *ultimately bounded*, *ultimately bounded above*, or *ultimately bounded below* as $E \ni x \to a$ respectively.

Example 7 The function $f(x) = \sin \frac{1}{x} + x \cos \frac{1}{x}$ defined by this formula for $x \neq 0$ is not bounded on its domain of definition, but it is ultimately bounded as $x \to 0$.

Example 8 The same is true for the function $f(x) = x$ on $\mathbb{R}$.

Theorem 1 a) $(f : E \to \mathbb{R}$ *is ultimately the constant* A *as* $E \ni x \to a) \Rightarrow (\lim_{E \ni x \to a} f(x) = A)$.

b) $(\exists \lim_{E \ni x \to a} f(x)) \Rightarrow (f : E \to \mathbb{R}$ *is ultimately bounded as* $E \ni x \to a)$.

c) $(\lim_{E \ni x \to a} f(x) = A_1) \wedge (\lim_{E \ni x \to a} f(x) = A_2) \Rightarrow (A_1 = A_2)$.

Proof The assertion a) that an ultimately constant function has a limit, and assertion b) that a function having a limit is ultimately bounded, follow immediately from the corresponding definitions. We now turn to the proof of the uniqueness of the limit.

Suppose $A_1 \neq A_2$. Choose neighborhoods $V(A_1)$ and $V(A_2)$ having no points in common, that is, $V(A_1) \cap V(A_2) = \varnothing$. By definition of a limit, we have

$$\lim_{E\ni x\to a} f(x) = A_1 \Rightarrow \exists \overset{\circ}{U}{}'_E(a) \left(f\left(\overset{\circ}{U}{}'_E(a)\right) \subset V(A_1)\right),$$

$$\lim_{E\ni x\to a} f(x) = A_2 \Rightarrow \exists \overset{\circ}{U}{}''_E(a) \left(f\left(\overset{\circ}{U}{}''_E(a)\right) \subset V(A_2)\right).$$

We now take a deleted neighborhood $\overset{\circ}{U}_E(a)$ of a (which is a limit point of E) such that $\overset{\circ}{U}_E(a) \subset \overset{\circ}{U}{}'_E(a) \cap \overset{\circ}{U}{}''_E(a)$. (For example, we could take $\overset{\circ}{U}_E(a) = \overset{\circ}{U}{}'_E(a) \cap \overset{\circ}{U}{}''_E(a)$, since this intersection is also a deleted neighborhood.)

Since $\overset{\circ}{U}_E(a) \neq \varnothing$, we take $x \in \overset{\circ}{U}_E(a)$. We then have $f(x) \in V(A_1) \cap V(A_2)$, which is impossible since the neighborhoods $V(A_1)$ and $V(A_2)$ have no points in common. □

b. Passage to the Limit and Arithmetic Operations

Definition 6 If two numerical-valued functions $f : E \to \mathbb{R}$ and $g : E \to \mathbb{R}$ have a common domain of definition E, their *sum*, *product*, and *quotient* are respectively the functions defined on the same set by the following formulas:

$$(f + g)(x) := f(x) + g(x),$$

$$(f \cdot g)(x) := f(x) \cdot g(x),$$

$$\left(\frac{f}{g}\right)(x) := \frac{f(x)}{g(x)}, \quad \text{if } g(x) \neq 0 \text{ for } x \in E.$$

Theorem 2 *Let $f : E \to \mathbb{R}$ and $g : E \to \mathbb{R}$ be two functions with a common domain of definition.*

If $\lim_{E\ni x\to a} f(x) = A$ and $\lim_{E\ni x\to a} g(x) = B$, then

a) $\lim_{E\ni x\to a}(f + g)(x) = A + B$;
b) $\lim_{E\ni x\to a}(f \cdot g)(x) = A \cdot B$;
c) $\lim_{E\ni x\to a}(\frac{f}{g}) = \frac{A}{B}$, *if $B \neq 0$ and $g(x) \neq 0$ for $x \in E$.*

As already noted at the beginning of Sect. 3.2.2, this theorem is an immediate consequence of the corresponding theorem on limits of sequences, given Proposition 1. The theorem can also be obtained by repeating the proof of the theorem on the algebraic properties of the limit of a sequence. The changes needed in the proof in order to do this reduce to referring to some deleted neighborhood $\overset{\circ}{U}_E(a)$ of a in E, where previously we had referred to statements holding "from some $N \in \mathbb{N}$ on". We advise the reader to verify this.

Here we shall obtain the theorem from its simplest special case when $A = B = 0$. Of course assertion c) will then be excluded from consideration.

A function $f : E \to \mathbb{R}$ is said to be *infinitesimal* as $E \ni x \to a$ if $\lim_{E\ni x\to a} f(x) = 0$.

Proposition 2 a) *If $\alpha : E \to \mathbb{R}$ and $\beta : E \to \mathbb{R}$ are infinitesimal functions as $E \ni x \to a$, then their sum $\alpha + \beta : E \to \mathbb{R}$ is also infinitesimal as $E \ni x \to a$.*

b) *If $\alpha : E \to \mathbb{R}$ and $\beta : E \to \mathbb{R}$ are infinitesimal functions as $E \ni x \to a$, then their product $\alpha \cdot \beta : E \to \mathbb{R}$ is also infinitesimal as $E \ni x \to a$.*

c) *If $\alpha : E \to \mathbb{R}$ is infinitesimal as $E \ni x \to a$ and $\beta : E \to \mathbb{R}$ is ultimately bounded as $E \ni x \to a$, then the product $\alpha \cdot \beta : E \to \mathbb{R}$ is infinitesimal as $E \ni x \to a$.*

Proof a) We shall verify that

$$\Big(\lim_{E\ni x\to a} \alpha(x) = 0\Big) \wedge \Big(\lim_{E\ni x\to a} \beta(x) = 0\Big) \Rightarrow \Big(\lim_{E\ni x\to a} (\alpha+\beta)(x) = 0\Big).$$

Let $\varepsilon > 0$ be given. By definition of the limit, we have

$$\Big(\lim_{E\ni x\to a} \alpha(x) = 0\Big) \Rightarrow \Big(\exists \mathring{U}'_E(a)\ \forall x \in \mathring{U}'_E(a)\ \Big(\big|\alpha(x)\big| < \frac{\varepsilon}{2}\Big)\Big),$$

$$\Big(\lim_{E\ni x\to a} \beta(x) = 0\Big) \Rightarrow \Big(\exists \mathring{U}''_E(a)\ \forall x \in \mathring{U}''_E(a)\ \Big(\big|\beta(x)\big| < \frac{\varepsilon}{2}\Big)\Big).$$

Then for the deleted neighborhood $\mathring{U}_E(a) \subset \mathring{U}'_E(a) \cap \mathring{U}''_E(a)$ we obtain

$$\forall x \in \mathring{U}_E(a)\ \big|(\alpha+\beta)(x)\big| = \big|\alpha(x) + \beta(x)\big| \le \big|\alpha(x)\big| + \big|\beta(x)\big| < \varepsilon.$$

That is, we have verified that $\lim_{E\ni x\to a}(\alpha+\beta)(x) = 0$.

b) This assertion is a special case of assertion c), since every function that has a limit is ultimately bounded.

c) We shall verify that

$$\Big(\lim_{E\ni x\to a} \alpha(x) = 0\Big) \wedge \big(\exists M \in \mathbb{R}\ \exists \mathring{U}_E(a)\ \forall x \in \mathring{U}_E(a)\ \big(\big|\beta(x)\big| < M\big)\big) \Rightarrow$$
$$\Rightarrow \Big(\lim_{E\ni x\to a} \alpha(x)\beta(x) = 0\Big).$$

Let $\varepsilon > 0$ be given. By definition of limit we have

$$\Big(\lim_{E\ni x\to a} \alpha(x) = 0\Big) \Rightarrow \Big(\exists \mathring{U}'_E(a)\ \forall x \in \mathring{U}'_E(a)\ \Big(\big|\alpha(x)\big| < \frac{\varepsilon}{M}\Big)\Big).$$

Then for the deleted neighborhood $\mathring{U}''_E(a) \subset \mathring{U}'_E(a) \cap \mathring{U}_E(a)$, we obtain

$$\forall x \in \mathring{U}''_E(a)\ \big|(\alpha \cdot \beta)(x)\big| = \big|\alpha(x)\beta(x)\big| = \big|\alpha(x)\big|\big|\beta(x)\big| < \frac{\varepsilon}{M} \cdot M = \varepsilon.$$

Thus we have verified that $\lim_{E\ni x\to a} \alpha(x)\beta(x) = 0$. □

The following remark is very useful:

Remark 1

$$\boxed{\Big(\lim_{E\ni x\to a} f(x)=A\Big) \Leftrightarrow \Big(f(x)=A+\alpha(x) \wedge \lim_{E\ni x\to a} \alpha(x)=0\Big).}$$

In other words, the function $f : E \to \mathbb{R}$ tends to A if and only if it can be represented as a sum $A+\alpha(x)$, where $\alpha(x)$ is infinitesimal as $E \ni x \to a$. (The function $\alpha(x)$ is the deviation of $f(x)$ from A.)[10]

This remark follows immediately from the definition of limit, by virtue of which

$$\lim_{E\ni x\to a} f(x)=A \Leftrightarrow \lim_{E\ni x\to a} \big(f(x)-A\big)=0.$$

We now give the proof of the theorem on the arithmetic properties of the limit of a function, based on this remark and the properties of infinitesimal functions that we have established.

Proof a) If $\lim_{E\ni x\to \alpha} f(x) = A$ and $\lim_{E\ni x\to a} g(x) = B$, then $f(x) = A + \alpha(x)$ and $g(x) = B + \beta(x)$, where $\alpha(x)$ and $\beta(x)$ are infinitesimal as $E \ni x \to a$. Then $(f+g)(x) = f(x)+g(x) = A+\alpha(x)+B+\beta(x) = (A+B)+\gamma(x)$, where $\gamma(x) = \alpha(x) + \beta(x)$, being the sum of two infinitesimals, is infinitesimal as $E \ni x \to a$.

Thus $\lim_{E\ni x\to \alpha}(f+g)(x) = A+B$.

b) Again representing $f(x)$ and $g(x)$ in the form $f(x) = A + \alpha(x)$, $g(x) = B + \beta(x)$, we have

$$(f\cdot g)(x) = f(x)g(x) = \big(A+\alpha(x)\big)\big(B+\beta(x)\big) = A\cdot B + \gamma(x),$$

where $\gamma(x) = A\beta(x) + B\alpha(x) + \alpha(x)\beta(x)$ is infinitesimal as $E \ni x \to a$ because of the properties just proved for such functions.

Thus, $\lim_{E\ni x\to a}(f\cdot g)(x) = A\cdot B$.

c) We once again write $f(x) = A + \alpha(x)$ and $g(x) = B + \beta(x)$, where $\lim_{E\ni x\to a} \alpha(x) = 0$ and $\lim_{E\ni x\to a} \beta(x) = 0$.

Since $B \neq 0$, there exists a deleted neighborhood $\overset{\circ}{U}_E(a)$, at all points of which $|\beta(x)| < \frac{|B|}{2}$, and hence $|g(x)| = |B+\beta(x)| \geq |B| - |\beta(x)| > \frac{|B|}{2}$. Then in $\overset{\circ}{U}_E(a)$ we shall also have $\frac{1}{|g(x)|} < \frac{2}{|B|}$, that is, the function $\frac{1}{g(x)}$ is ultimately bounded as $E \ni x \to a$. We then write

$$\Big(\frac{f}{g}\Big)(x) - \frac{A}{B} = \frac{f(x)}{g(x)} - \frac{A}{B} = \frac{A+\alpha(x)}{B+\beta(x)} - \frac{A}{B} =$$
$$= \frac{1}{g(x)}\cdot\frac{1}{B}\big(B\alpha(x) + A\beta(x)\big) = \gamma(x).$$

---

[10]Here is a curious detail. This very obvious representation, which is nevertheless very useful on the computational level, was specially noted by the French mathematician and specialist in mechanics Lazare Carnot (1753–1823), a revolutionary general and academician, the father of Sadi Carnot (1796–1832), who in turn was the creator of thermodynamics.

By the properties of infinitesimals (taking account of the ultimate boundedness of $\frac{1}{g(x)}$) we find that the function $\gamma(x)$ is infinitesimal as $E \ni x \to a$. Thus we have proved that $\lim_{E\ni x\to a}(\frac{f}{g})(x) = \frac{A}{B}$. □

c. Passage to the Limit and Inequalities

Theorem 3 a) *If the functions $f : E \to \mathbb{R}$ and $g : E \to \mathbb{R}$ are such that $\lim_{E\ni x\to a} f(x) = A$, and $\lim_{E\ni x\to a} g(x) = B$ and $A < B$, then there exists a deleted neighborhood $\overset{\circ}{U}_E(a)$ of a in E at each point of which $f(x) < g(x)$.*

b) *If the relations $f(x) \le g(x) \le h(x)$ hold for the functions $f : E \to \mathbb{R}$, $g : E \to \mathbb{R}$, and $h : E \to \mathbb{R}$, and if $\lim_{E\ni x\to a} f(x) = \lim_{E\ni x\to a} h(x) = C$, then the limit of $g(x)$ exists as $E \ni x \to a$, and $\lim_{E\ni x\to a} g(x) = C$.*

Proof a) Choose a number C such that $A < C < B$. By definition of limit, we find deleted neighborhoods $\overset{\circ}{U}{}'_E(a)$ and $\overset{\circ}{U}{}''_E(a)$ of a in E such that $|f(x) - A| < C - A$ for $x \in \overset{\circ}{U}{}'_E(a)$ and $|g(x) - B| < B - C$ for $x \in \overset{\circ}{U}{}''_E(a)$. Then at any point of a deleted neighborhood $\overset{\circ}{U}_E(a)$ contained in $\overset{\circ}{U}{}'_E(a) \cap \overset{\circ}{U}{}''_E(a)$, we find

$$f(x) < A + (C - A) = C = B - (B - C) < g(x).$$

b) If $\lim_{E\ni x\to a} f(x) = \lim_{E\ni x\to a} h(x) = C$, then for any fixed $\varepsilon > 0$ there exist deleted neighborhoods $\overset{\circ}{U}{}'_E(a)$ and $\overset{\circ}{U}{}''_E(a)$ of a in E such that $C - \varepsilon < f(x)$ for $x \in \overset{\circ}{U}{}'_E(a)$ and $h(x) < C + \varepsilon$ for $x \in \overset{\circ}{U}{}''_E(a)$. Then at any point of a deleted neighborhood $\overset{\circ}{U}_E(a)$ contained in $\overset{\circ}{U}{}'_E(a) \cap \overset{\circ}{U}{}''_E(a)$, we have $C - \varepsilon < f(x) \le g(x) \le h(x) < C + \varepsilon$, that is, $|g(x) - C| < \varepsilon$, and consequently $\lim_{E\ni x\to a} g(x) = C$. □

Corollary *Suppose $\lim_{E\ni x\to a} f(x) = A$ and $\lim_{E\ni x\to a} g(x) = B$. Let $\overset{\circ}{U}_E(a)$ be a deleted neighborhood of a in E.*

a) *If $f(x) > g(x)$ for all $x \in \overset{\circ}{U}_E(a)$, then $A \ge B$;*
b) *$f(x) \ge g(x)$ for all $x \in \overset{\circ}{U}_E(a)$, then $A \ge B$;*
c) *$f(x) > B$ for all $x \in \overset{\circ}{U}_E(a)$, then $A \ge B$;*
d) *$f(x) \ge B$ for all $x \in \overset{\circ}{U}_E(a)$, then $A \ge B$.*

Proof Using proof by contradiction, we immediately obtain assertions a) and b) of the corollary from assertion a) of Theorem 3. Assertions c) and d) follow from a) and b) by taking $g(x) \equiv B$. □

d. Two Important Examples

Before developing the theory of the limit of a function further, we shall illustrate the use of the theorems just proved by two important examples.

Fig. 3.1

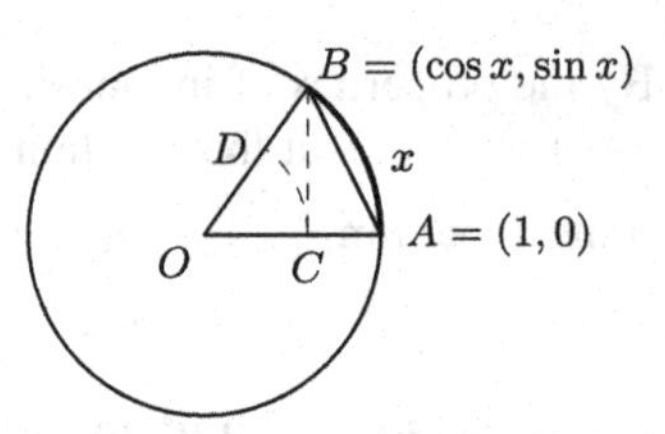

Example 9

$$\boxed{\lim_{x\to 0} \frac{\sin x}{x} = 1.}$$

Here we shall appeal to the definition of $\sin x$ given in high school, that is, $\sin x$ is the ordinate of the point to which the point $(1, 0)$ moves under a rotation of x radians about the origin. The completeness of such a definition is entirely a matter of the care with which the connection between rotations and real numbers is established. Since the system of real numbers itself was not described in sufficient detail in high school, one may consider that we need to sharpen the definition of $\sin x$ (and the same is true of $\cos x$).

We shall do so at the appropriate time and justify the reasoning that for now will rely on intuition.

a) We shall show that

$$\cos^2 x < \frac{\sin x}{x} < 1 \quad \text{for } 0 < |x| < \frac{\pi}{2}.$$

Proof Since $\cos^2 x$ and $\frac{\sin x}{x}$ are even functions, it suffices to consider the case $0 < x < \pi/2$. By Fig. 3.1 and the definition of $\cos x$ and $\sin x$, comparing the area of the sector $\sphericalangle OCD$, the triangle $\triangle OAB$, and the sector $\sphericalangle OAB$, we have

$$\begin{aligned}
S_{\sphericalangle OCD} &= \frac{1}{2}|OC| \cdot |\overset{\frown}{CD}| = \frac{1}{2}(\cos x)(x \cos x) = \frac{1}{2} x \cos^2 x < \\
&< S_{\triangle OAB} = \frac{1}{2}|OA| \cdot |BC| = \frac{1}{2} \cdot 1 \cdot \sin x = \frac{1}{2} \sin x < \\
&< S_{\sphericalangle OAB} = \frac{1}{2}|OA| \cdot |\overset{\frown}{AB}| = \frac{1}{2} \cdot 1 \cdot x = \frac{1}{2} x.
\end{aligned}$$

Dividing these inequalities by $\frac{1}{2}x$, we find that the result is what was asserted. □

b) It follows from a) that

$$|\sin x| \le |x|$$

for any $x \in \mathbb{R}$, equality holding only at $x = 0$.

Proof For $0 < |x| < \pi/2$, as shown in a), we have

$$|\sin x| < |x|.$$

But $|\sin x| \leq 1$, so that this last inequality also holds for $|x| \geq \pi/2 > 1$. Only for $x = 0$ do we find $\sin x = x = 0$. □

c) It follows from b) that

$$\lim_{x\to 0} \sin x = 0.$$

Proof Since $0 \leq |\sin x| \leq |x|$ and $\lim_{x\to 0} |x| = 0$, we find by the theorem on the limit of a function and inequalities (Theorem 3) that $\lim_{x\to 0} |\sin x| = 0$, so that $\lim_{x\to 0} \sin x = 0$. □

d) We shall now prove that $\lim_{x\to 0} \frac{\sin x}{x} = 1$.

Proof Assuming that $|x| < \pi/2$, from the inequality in a) we have

$$1 - \sin^2 x < \frac{\sin x}{x} < 1.$$

But $\lim_{x\to 0}(1 - \sin^2 x) = 1 - \lim_{x\to 0} \sin x \cdot \lim_{x\to 0} \sin x = 1 - 0 = 1$, so that by the theorem on passage to the limit and inequalities, we conclude that $\lim_{x\to 0} \frac{\sin x}{x} = 1$. □

Example 10 (Definition of the exponential, logarithmic, and power functions using limits) We shall now illustrate how the high-school definition of the exponential and logarithmic functions can be completed by means of the theory of real numbers and limits.

For convenience in reference and to give a complete picture, we shall start from the beginning.

a) *The exponential function.* Let $a > 1$.

1^0 For $n \in \mathbb{N}$ we define inductively $a^1 := a$, $a^{n+1} := a^n \cdot a$.

In this way we obtain a function a^n defined on $\mathbb{N}$, which, as can be seen from the definition, has the property

$$\frac{a^m}{a^n} = a^{m-n}$$

if $m, n \in \mathbb{N}$ and $m > n$.

2^0 This property leads to the natural definitions

$$a^0 := 1, \qquad a^{-n} := \frac{1}{a^n} \quad \text{for } n \in \mathbb{N},$$

which, when carried out, extend the function a^n to the set $\mathbb{Z}$ of all integers, and then

$$a^m \cdot a^n = a^{m+n}$$

for any $m, n \in \mathbb{Z}$.

3^0 In the theory of real numbers we have observed that for $a > 0$ and $n \in N$ there exists a unique nth root of a, that is, a number $x > 0$ such that $x^n = a$. For that number we use the notation $a^{1/n}$. It is convenient, since it allows us to retain the law of addition for exponents:

$$a = a^1 = \left(a^{1/n}\right)^n = a^{1/n} \cdots a^{1/n} = a^{1/n+\cdots+1/n}.$$

For the same reason it is natural to set $a^{m/n} := (a^{1/n})^m$ and $a^{-1/n} := (a^{1/n})^{-1}$ for $n \in \mathbb{N}$ and $m \in \mathbb{Z}$. If it turns out that $a^{(mk)/(nk)} = a^{m/n}$ for $k \in \mathbb{Z}$, we can consider that we have defined a^r for $r \in \mathbb{Q}$.

4^0 For numbers $0 < x$, $0 < y$, we verify by induction that for $n \in \mathbb{N}$

$$(x < y) \Leftrightarrow \left(x^n < y^n\right),$$

so that, in particular,

$$(x = y) \Leftrightarrow \left(x^n = y^n\right).$$

5^0 This makes it possible to prove the rules for operating with rational exponents, in particular, that

$$a^{(mk)/(nk)} = a^{m/n} \quad \text{for } k \in \mathbb{Z}$$

and

$$a^{m_1/n_1} \cdot a^{m_2/n_2} = a^{m_1/n_1+m_2/n_2}.$$

Proof Indeed, $a^{(mk)/(nk)} > 0$ and $a^{m/n} > 0$. Further, since

$$\begin{aligned}\left(a^{(mk)/(nk)}\right)^{nk} &= \left(\left(a^{1/(nk)}\right)^{mk}\right)^{nk} = \\ &= \left(a^{1/(nk)}\right)^{mk\cdot nk} = \left(\left(a^{1/(nk)}\right)^{nk}\right)^{mk} = a^{mk}\end{aligned}$$

and

$$\left(a^{m/n}\right)^{nk} = \left(\left(a^{1/n}\right)^n\right)^{mk} = a^{mk},$$

it follows that the first of the inequalities that needed to be verified in connection with point 4^0 is now established.

Similarly, since

$$\begin{aligned}\left(a^{m_1/n_1} \cdot a^{m_2/n_2}\right)^{n_1 n_2} &= \left(a^{m_1/n_1}\right)^{n_1 n_2} \cdot \left(a^{m_2/n_2}\right)^{n_1 n_2} = \\ &= \left(\left(a^{1/n_1}\right)^{n_1}\right)^{m_1 n_2} \cdot \left(\left(a^{1/n_2}\right)^{n_2}\right)^{m_2 n_1} = a^{m_1 n_2} \cdot a^{m_2 n_1} \\ &= a^{m_1 n_2 + m_2 n_1}\end{aligned}$$

and

$$\left(a^{m_1/n_1+m_2/n_2}\right)^{n_1n_2} = \left(a^{(m_1n_2+m_2n_1)/(n_1n_2)}\right)^{n_1n_2} =$$
$$= \left(\left(a^{1/(n_1n_2)}\right)^{n_1n_2}\right)^{m_1n_2+m_2n_1} = a^{m_1n_2+m_2n_1},$$

the second equality is also proved. □

Thus we have defined a^r for $r \in \mathbb{Q}$ and $a^r > 0$; and for any $r_1, r_2 \in \mathbb{Q}$,

$$a^{r_1} \,.\, a^{r_2} = a^{r_1+r_2}.$$

6^0 It follows from 4^0 that for $r_1, r_2 \in \mathbb{Q}$

$$(r_1 < r_2) \Rightarrow \left(a^{r_1} < a^{r_2}\right).$$

Proof Since $(1 < a) \Leftrightarrow (1 < a^{1/n})$ for $n \in \mathbb{N}$, which follows immediately from 4^0, we have $(a^{1/n})^m = a^{m/n} > 1$ for $n, m \in \mathbb{N}$, as again follows from 4^0. Thus for $1 < a$ and $r > 0$, $r \in \mathbb{Q}$, we have $a^r > 1$.

Then for $r_1 < r_2$ we obtain by 5^0

$$a^{r_2} = a^{r_1} \cdot a^{r_2-r_1} > a^{r_1} \cdot 1 = a^{r_1}.$$ □

7^0 We shall show that for $r_0 \in \mathbb{Q}$

$$\lim_{\mathbb{Q} \ni r \to r_0} a^r = a^{r_0}.$$

Proof We shall verify that $a^p \to 1$ as $\mathbb{Q} \ni p \to 0$. This follows from the fact that for $|p| < \frac{1}{n}$ we have by 6^0

$$a^{-1/n} < a^p < a^{1/n}.$$

We know that $a^{1/n} \to 1$ (and $a^{-1/n} \to 1$) as $n \to \infty$. Then by standard reasoning we verify that for $\varepsilon > 0$ there exists $\delta > 0$ such that for $|p| < \delta$ we have

$$1 - \varepsilon < a^p < 1 + \varepsilon.$$

We can take $\frac{1}{n}$ as δ here if $1 - \varepsilon < a^{-1/n}$ and $a^{1/n} < 1 + \varepsilon$.

We now prove the main assertion.

Given $\varepsilon > 0$, we choose δ so that

$$1 - \varepsilon a^{-r_0} < a^p < 1 + \varepsilon a^{-r_0}$$

for $|p| < \delta$. If now $|r - r_0| < \delta$, we have

$$a^{r_0}\left(1 - \varepsilon a^{-r_0}\right) < a^r = a^{r_0} \cdot a^{r-r_0} < a^{r_0}\left(1 + \varepsilon a^{-r_0}\right),$$

which says

$$a^{r_0} - \varepsilon < a^r < a^{r_0} + \varepsilon.$$ □

Thus we have defined a function a^r on $\mathbb{Q}$ having the following properties:

$$\begin{aligned} a^1 &= a > 1; \\ a^{r_1} \cdot a^{r_2} &= a^{r_1+r_2}; \\ a^{r_1} &< a^{r_2} \quad \text{for } r_1 < r_2; \\ a^{r_1} &\to a^{r_2} \quad \text{as } \mathbb{Q} \ni r_1 \to r_2. \end{aligned}$$

We now extend this function to the entire real line as follows.

8^0 Let $x \in \mathbb{R}$, $s = \sup_{\mathbb{Q}\ni r<x} a^r$, and $i = \inf_{\mathbb{Q}\ni r>x} a^r$. It is clear that $s, i \in \mathbb{R}$, since for $r_1 < x < r_2$ we have $a^{r_1} < a^{r_2}$.

We shall show that actually $s = i$ (and then we shall denote this common value by a^x).

Proof By definition of s and i we have

$$a^{r_1} \le s \le i \le a^{r_2}$$

for $r_1 < x < r_2$. Then $0 \le i - s \le a^{r_2} - a^{r_1} = a^{r_1}(a^{r_2-r_1} - 1) < s(a^{r_2-r_1} - 1)$. But $a^p \to 1$ as $\mathbb{Q} \ni p \to 0$, so that for any $\varepsilon > 0$ there exists $\delta > 0$ such that $a^{r_2-r_1} - 1 < \varepsilon/s$ for $0 < r_2 - r_1 < \delta$. We then find that $0 \le i - s \le \varepsilon$, and since $\varepsilon > 0$ is arbitrary, we conclude that $i = s$. □

We now define $a^x := s = i$.

9^0 Let us show that $a^x = \lim_{\mathbb{Q}\ni r\to x} a^r$.

Proof Taking 8^0 into account, for $\varepsilon > 0$ we find $r' < x$ such that $s - \varepsilon < a^{r'} \le s = a^x$ and r'' such that $a^x = i \le a^{r''} < i + \varepsilon$. Since $r' < r < r''$ implies $a^{r'} < a^r < a^{r''}$, we then have, for all $r \in \mathbb{Q}$ in the open interval $]r', r''[$,

$$a^x - \varepsilon < a^r < a^x + \varepsilon.$$ □

We now study the properties of the function a^x so defined on $\mathbb{R}$.

10^0 For $x_1, x_2 \in \mathbb{R}$ and $a > 1$, $(x_1 < x_2) \Rightarrow (a^{x_1} < a^{x_2})$.

Proof On the open interval $]x_1, x_2[$ there exist two rational numbers $r_1 < r_2$. If $x_1 \le r_1 < r_2 \le x_2$, by the definition of a^x given in 8^0 and the properties of the function a^x on $\mathbb{Q}$, we have

$$a^{x_1} \le a^{r_1} < a^{r_2} \le a^{x_2}.$$ □

11^0 For any $x_1, x_2 \in \mathbb{R}$, $a^{x_1} \cdot a^{x_2} = a^{x_1+x_2}$.

Proof By the estimates that we know for the absolute error in the product and by property 9^0, we can assert that for any $\varepsilon > 0$ there exists $\delta' > 0$ such that

$$a^{x_1} \cdot a^{x_2} - \frac{\varepsilon}{2} < a^{r_1} \cdot a^{r_2} < a^{x_1} \cdot a^{x_2} + \frac{\varepsilon}{2}$$

for $|x_1 - r_1| < \delta'$, and $|x_2 - r_2| < \delta'$. Making δ' smaller if necessary, we can choose $\delta < \delta'$ such that we also have

$$a^{r_1+r_2} - \frac{\varepsilon}{2} < a^{x_1+x_2} < a^{r_1+r_2} + \frac{\varepsilon}{2}$$

for $|x_1 - r_1| < \delta$ and $|x_2 - r_2| < \delta$, that is, $|(x_1 + x_2) - (r_1 + r_2)| < 2\delta$.

But $a^{r_1} \cdot a^{r_2} = a^{r_1+r_2}$, for $r_1, r_2 \in \mathbb{Q}$, so that these inequalities imply

$$a^{x_1} \cdot a^{x_2} - \varepsilon < a^{x_1+x_2} < a^{x_1} \cdot a^{x_2} + \varepsilon.$$

Since $\varepsilon > 0$ is arbitrary, we conclude that

$$a^{x_1} \cdot a^{x_2} = a^{x_1+x_2}.$$

□

12^0 $\lim_{x\to x_0} a^x = a^{x_0}$. (We recall that "$x \to x_0$" is an abbreviation of "$\mathbb{R} \ni x \to x_0$".)

Proof We first verify that $\lim_{x\to 0} a^x = 1$. Given $\varepsilon > 0$, we find $n \in \mathbb{N}$ such that

$$1 - \varepsilon < a^{-1/n} < a^{1/n} < 1 + \varepsilon.$$

Then by 10^0, for $|x| < 1/n$ we have

$$1 - \varepsilon < a^{-1/n} < a^x < a^{1/n} < 1 + \varepsilon,$$

that is, we have verified that $\lim_{x\to 0} a^x = 1$.

If we now take $\delta > 0$ so that $|a^{x-x_0} - 1| < \varepsilon a^{-x_0}$ for $|x - x_0| < \delta$, we find

$$a^{x_0} - \varepsilon < a^x = a^{x_0}\left(a^{x-x_0} - 1\right) < a^{x_0} + \varepsilon,$$

which verifies that $\lim_{x\to x_0} a^x = a^{x_0}$.

□

13^0 We shall show that the range of values of the function $x \mapsto a^x$ is the set $\mathbb{R}_+$ of positive real numbers.

Proof Let $y_0 \in \mathbb{R}_+$. If $a > 1$, then as we know, there exists $n \in \mathbb{N}$ such that $a^{-n} < y_0 < a^n$.

By virtue of this fact, the two sets

$$A = \left\{x \in \mathbb{R} \mid a^x < y_0\right\} \quad \text{and} \quad B = \left\{x \in \mathbb{R} \mid y_0 < a^x\right\}$$

are both nonempty. But since $(x_1 < x_2) \Leftrightarrow (a^{x_1} < a^{x_2})$ (when $a > 1$), for any numbers $x_1, x_2 \in \mathbb{R}$ such that $x_1 \in A$ and $x_2 \in B$ we have $x_1 < x_2$. Consequently, the axiom of completeness is applicable to the sets A and B, and it follows that there exists x_0 such that $x_1 \le x_0 \le x_2$ for all $x_1 \in A$ and $x_2 \in B$. We shall show that $a^{x_0} = y_0$.

If a^{x_0} were less than y_0, then, since $a^{x_0+1/n} \to a^{x_0}$ as $n \to \infty$, there would be a number $n \in \mathbb{N}$ such that $a^{x_0+1/n} < y_0$. Then we would have $(x_0 + \frac{1}{n}) \in A$, while the point x_0 separates A and B. Hence the assumption $a^{x_0} < y_0$ is untenable. Similarly we can verify that the inequality $a^{x_0} > y_0$ is also impossible. By the properties of real numbers, we conclude from this that $a^{x_0} = y_0$. □

14^0 We have assumed up to now that $a > 1$. But all the constructions could be repeated for $0 < a < 1$. Under this condition $0 < a^r < 1$ if $r > 0$, so that in 6^0 and 10^0 we now find that $(x_1 < x_2) \Rightarrow (a^{x_1} > a^{x_2})$ where $0 < a < 1$.

Thus for $a > 0$, $a \neq 1$, we have constructed a real-valued function $x \mapsto a^x$ on the set $\mathbb{R}$ of real numbers with the following properties:

1) $a^1 = a$;
2) $a^{x_1} \cdot a^{x_2} = a^{x_1+x_2}$;
3) $a^x \to a^{x_0}$ as $x \to x_0$;
4) $(a^{x_1} < a^{x_2}) \Leftrightarrow (x_1 < x_2)$ if $a > 1$, and $(a^{x_1} > a^{x_2}) \Leftrightarrow (x_1 < x_2)$ if $0 < a < 1$;
5) the range of values of the mapping $x \mapsto a^x$ is $\mathbb{R}_+ = \{y \in \mathbb{R} \mid 0 < y\}$, the set of positive numbers.

Definition 7 The mapping $x \mapsto a^x$ is called the *exponential* function with base a.

The mapping $x \mapsto \mathrm{e}^x$, which is the case $a = \mathrm{e}$, is encountered particularly often and is frequently denoted $\exp x$. In this connection, to denote the mapping $x \mapsto a^x$, we sometimes also use the notation $\exp_a x$.

b) *The logarithmic function.* The properties of the exponential function show that it is a bijective mapping $\exp_a : \mathbb{R} \to \mathbb{R}_+$. Hence it has an inverse.

Definition 8 The mapping inverse to $\exp_a : \mathbb{R} \to \mathbb{R}_+$ is called the *logarithm to base* a $(0 < a, a \neq 1)$, and is denoted

$$\log_a : \mathbb{R}_+ \to \mathbb{R}.$$

Definition 9 For base $a = \mathrm{e}$, the logarithm is called the *natural logarithm* and is denoted $\ln : \mathbb{R}_+ \to \mathbb{R}$.

The reason for the terminology becomes clear under a different approach to logarithms, one that is in many ways more natural and transparent, which we shall explain after constructing the fundamentals of differential and integral calculus.

By definition of the logarithm as the function inverse to the exponential function, we have

$$\forall x \in \mathbb{R} \left(\log_a\left(a^x\right) = x\right),$$
$$\forall y \in \mathbb{R}_+ \left(a^{\log_a y} = y\right).$$

It follows from this definition and the properties of the exponential function in particular that in its domain of definition $\mathbb{R}_+$ the logarithm has the following properties:

1′) $\log_a a = 1$;
2′) $\log_a(y_1 \cdot y_2) = \log_a y_1 + \log_a y_2$;
3′) $\log_a y \to \log_a y_0$ as $\mathbb{R}_+ \ni y \to y_0 \in \mathbb{R}_+$;
4′) $(\log_a y_1 < \log_a y_2) \Leftrightarrow (y_1 < y_2)$ if $a > 1$ and $(\log_a y_1 > \log_a y_2) \Leftrightarrow (y_1 < y_2)$ if $0 < a < 1$;
5′) the range of values of the function $\log_a : \mathbb{R}_+ \to \mathbb{R}$ is the set $\mathbb{R}$ of all real numbers.

Proof We obtain 1′) from property 1) of the exponential function and the definition of the logarithm.

We obtain property 2′) from property 2) of the exponential function. Indeed, let $x_1 = \log_a y_1$ and $x_2 = \log_a y_2$. Then $y_1 = a^{x_1}$ and $y_2 = a^{x_2}$, and so by 2), $y_1 \cdot y_2 = a^{x_1} \cdot a^{x_2} = a^{x_1+x_2}$, from which it follows that $\log_a(y_1 \cdot y_2) = x_1 + x_2$.

Similarly, property 4) of the exponential function implies property 4′) of the logarithm.

It is obvious that 5) ⇒ 5′).

Property 3′) remains to be proved.

By property 2′) of the logarithm we have

$$\log_a y - \log_a y_0 = \log_a\left(\frac{y}{y_0}\right),$$

and therefore the inequalities

$$-\varepsilon < \log_a y - \log_a y_0 < \varepsilon$$

are equivalent to the relation

$$\log_a\left(a^{-\varepsilon}\right) = -\varepsilon < \log_a\left(\frac{y}{y_0}\right) < \varepsilon = \log_a\left(a^{\varepsilon}\right),$$

which by property 4′) of the logarithm is equivalent to

$$-a^{\varepsilon} < \frac{y}{y_0} < a^{\varepsilon} \quad \text{for } a > 1,$$

$$a^{\varepsilon} < \frac{y}{y_0} < a^{-\varepsilon} \quad \text{for } 0 < a < 1.$$

In any case we find that if

$$y_0 a^{-\varepsilon} < y < y_0 a^{\varepsilon} \quad \text{when } a > 1$$

or

$$y_0 a^{\varepsilon} < y < y_0 a^{-\varepsilon} \quad \text{when } 0 < a < 1,$$

we have

$$-\varepsilon < \log_a y - \log_a y_0 < \varepsilon.$$

Fig. 3.2

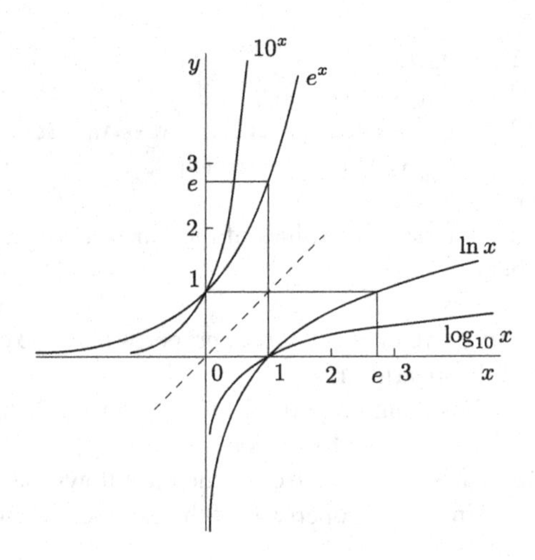

Thus we have proved that

$$\lim_{\mathbb{R}_+ \ni y \to y_0 \in \mathbb{R}_+} \log_a y = \log_a y_0. \qquad \square$$

Figure 3.2 shows the graphs of the functions e^x, 10^x, $\ln x$, and $\log_{10} x =: \log x$; Fig. 3.3 gives the graphs of $(\frac{1}{e})^x$, 0.1^x, $\log_{1/e} x$, and $\log_{0.1} x$.

We now give a more detailed discussion of one property of the logarithm that we shall have frequent occasion to use.

We shall show that the equality

6′)

$$\log_a\left(b^\alpha\right) = \alpha \log_a b$$

holds for any $b > 0$ and any $\alpha \in \mathbb{R}$.

Proof 1^0 The equality is true for $\alpha = n \in \mathbb{N}$. For by property 2′) of the logarithm and induction we find $\log_a(y_1 \cdots y_n) = \log_a y_1 + \cdots + \log_a y_n$, so that

$$\log_a\left(b^n\right) = \log_a b + \cdots + \log_a b = n \log_a b.$$

2^0 $\log_a(b^{-1}) = -\log_a b$, for if $\beta = \log_a b$, then

$$b = a^\beta, \qquad b^{-1} = a^{-\beta} \quad \text{and} \quad \log_a\left(b^{-1}\right) = -\beta.$$

3^0 From 1^0 and 2^0 we now conclude that the equality $\log_a(b^\alpha) = \alpha \log_a b$ holds for $\alpha \in \mathbb{Z}$.

4^0 $\log_a(b^{1/n}) = \frac{1}{n} \log_a b$ for $n \in \mathbb{Z}$. Indeed,

$$\log_a b = \log_a\left(b^{1/n}\right)^n = n \log_a\left(b^{1/n}\right).$$

Fig. 3.3

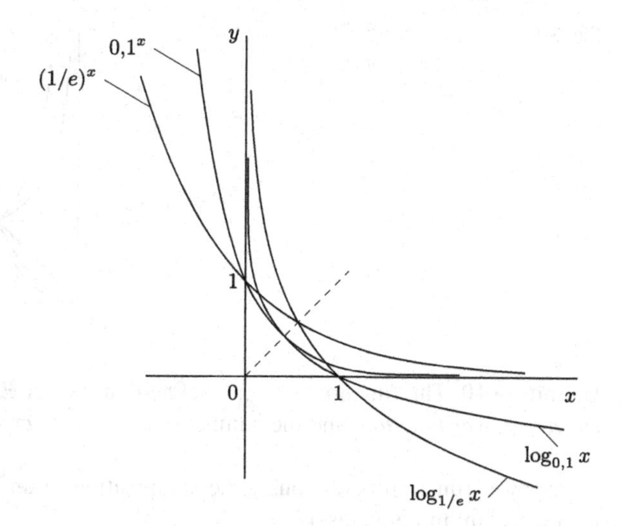

5^0 We can now verify that the assertion holds for any rational number $\alpha = \frac{m}{n} \in \mathbb{Q}$. In fact,

$$\frac{m}{n} \log_a b = m \log_a \left(b^{1/n}\right) = \log_a \left(b^{1/n}\right)^m = \log_a \left(b^{m/n}\right).$$

6^0 But if the equality $\log_a b^r = r \log_a b$ holds for all $r \in \mathbb{Q}$, then letting r in $\mathbb{Q}$ tend to α, we find by property 3) for the exponential function and 3′) for the logarithm that if r is sufficiently close to α, then b^r is close to b^α and $\log_a b^r$ is close to $\log_a b^\alpha$. This means that

$$\lim_{\mathbb{Q} \ni r \to \alpha} \log_a b^r = \log_a b^\alpha.$$

But $\log_a b^r = r \log_a b$, and therefore

$$\log_a b^\alpha = \lim_{\mathbb{Q} \ni r \to \alpha} \log_a b^r = \lim_{\mathbb{Q} \ni r \to \alpha} r \log_a b = \alpha \log_a b.$$

□

From the property of the logarithm just proved, one can conclude that the following equality holds for any $\alpha, \beta \in \mathbb{R}$ and $a > 0$:

6) $(a^\alpha)^\beta = a^{\alpha\beta}$.

Proof For $a = 1$ we have $1^\alpha = 1$ by definition for all $\alpha \in \mathbb{R}$. Thus the equality is trivial in this case.

If $a \neq 1$, then by what has just been proved we have

$$\log_a \left((a^\alpha)^\beta\right) = \beta \log_a \left(a^\alpha\right) = \beta \cdot \alpha \log_a a = \beta \cdot \alpha = \log_a \left(a^{\alpha\beta}\right),$$

which by property 4′) of the logarithm is equivalent to this equality. □

c) *The power function.* If we take $1^\alpha = 1$, then for all $x > 0$ and $\alpha \in \mathbb{R}$ we have defined the quantity x^α (read "x to power α").

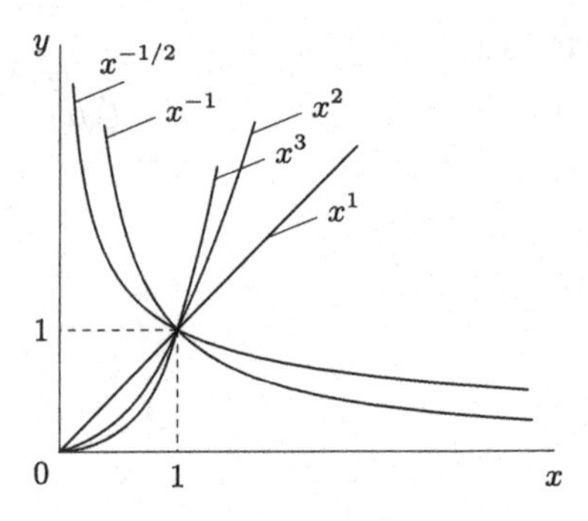

Fig. 3.4

Definition 10 The function $x \mapsto x^\alpha$ defined on the set $\mathbb{R}_+$ of positive numbers is called a *power function*, and the number α is called its *exponent*.

A power function is obviously the composition of an exponential function and the logarithm; more precisely

$$x^\alpha = a^{\log_a(x^\alpha)} = a^{\alpha \log_a x}.$$

Figure 3.4 shows the graphs of the function $y = x^\alpha$ for different values of the exponent.

3.2.3 The General Definition of the Limit of a Function (Limit over a Base)

When proving the properties of the limit of a function, we verified that the only requirements imposed on the deleted neighborhoods in which our functions were defined and which arose in the course of the proofs were the properties B_1) and B_2), mentioned in the introduction to the previous subsection. This fact justifies the definition of the following mathematical object.

a. Bases; Definition and Elementary Properties

Definition 11 A set $\mathcal{B}$ of subsets $B \subset X$ of a set X is called a *base* in X if the following conditions hold:

B_1) $\forall B \in \mathcal{B}\ (B \neq \varnothing)$;
B_2) $\forall B_1 \in \mathcal{B}\ \forall B_2 \in \mathcal{B}\ \exists B \in \mathcal{B}\ (B \subset B_1 \cap B_2)$.

In other words, the elements of the collection $\mathcal{B}$ are nonempty subsets of X and the intersection of any two of them always contains an element of the same collection.

In Table 3.1 we list some of the more useful bases in analysis.

Table 3.1

Notation for the base	Read	Sets (elements) of the base	Definition of and notation for elements
$x \to a$	x tends to a	Deleted neighborhoods of $a \in \mathbb{R}$	$\overset{\circ}{U}(a) := \{x \in \mathbb{R} \mid a - \delta_1 < < x < a + \delta_2 \wedge x \neq a\}$, where $\delta_1 > 0$, $\delta_2 > 0$
$x \to \infty$	x tends to infinity	Neighborhoods of infinity	$U(\infty) := \{x \in \mathbb{R} \mid \delta < \|x\|\}$, where $\delta \in \mathbb{R}$
$x \to a, x \in E$ or $E \ni x \to a$ or $x \underset{\in E}{\longrightarrow} a$	x tends to a in E	Deleted neighborhoods* of a in E	$\overset{\circ}{U}_E(a) := E \cap \overset{\circ}{U}(a)$
$x \to \infty, x \in E$ or $E \ni x \to \infty$ or $x \underset{\in E}{\longrightarrow} \infty$	x tends to infinity in E	Neighborhoods** of infinity in E	$U_E(\infty) := E \cap U(\infty)$

*It is assumed that a is a limit point of E

**It is assumed that E is not bounded

If $E = E_a^+ = \{x \in \mathbb{R} \mid x > a\}$ (resp. $E = E_a^- = \{x \in \mathbb{R} \mid x < a\}$) we write $x \to a + 0$ (resp. $x \to a - 0$) instead of $x \to a$, $x \in E$, and we say that x *tends to a from the right* (resp. x *tends to a from the left*) or *through larger values* (resp. *through smaller values*). When $a = 0$ it is customary to write $x \to +0$ (resp. $x \to -0$) instead of $x \to 0 + 0$ (resp. $x \to 0 - 0$).

The notation $E \ni x \to a + 0$ (resp. $E \ni x \to a - 0$) will be used instead of $x \to a$, $x \in E \cap E_a^+$ (resp. $x \to a$, $x \in E \cap E_a^-$). It means that x tends to a in E while remaining larger (resp. smaller) than a.

If

$$E = E_\infty^+ = \{x \in \mathbb{R} \mid c < x\} \quad \big(\text{resp. } E = E_\infty^- = \{x \in \mathbb{R} \mid x < c\}\big),$$

we write $x \to +\infty$ (resp. $x \to -\infty$) instead of $x \to \infty$, $x \in E$ and say that x *tends to positive infinity* (resp. x *tends to negative infinity*).

The notation $E \ni x \to +\infty$ (resp. $E \ni x \to -\infty$) will be used instead of $x \to \infty$, $x \in E \cap E_\infty^+$ (resp. $x \to \infty$, $x \in E \cap E_\infty^-$).

When $E = \mathbb{N}$, we shall write (when no confusion can arise), as is customary in the theory of limits of sequences, $n \to \infty$ instead of $x \to \infty$, $x \in \mathbb{N}$.

We remark that all the bases just listed have the property that the intersection of two elements of the base is itself an element of the base, not merely a set containing an element of the base. We shall meet with other bases in the study of functions defined on sets different from the real line.[11]

---

[11]For example, the set of open disks (not containing their boundary circles) containing a given point of the plane is a base. The intersection of two elements of the base is not always a disk, but always contains a disk from the collection.

We note also that the term "base" used here is an abbreviation for what is called a "filter base", and the limit over a base that we introduce below is, as far as analysis is concerned, the most important part of the concept of a limit over a filter,[12] created by the modern French mathematician H. Cartan.

b. The Limit of a Function over a Base

Definition 12 Let $f : X \to \mathbb{R}$ be a function defined on a set X and $\mathcal{B}$ a base in X. A number $A \in \mathbb{R}$ is called the *limit of the function* f *over the base* $\mathcal{B}$ if for every neighborhood $V(A)$ of A there is an element $B \in \mathcal{B}$ whose image $f(B)$ is contained in $V(A)$.

If A is the limit of $f : X \to \mathbb{R}$ over the base $\mathcal{B}$, we write

$$\lim_{\mathcal{B}} f(x) = A.$$

We now repeat the definition of the limit over a base in logical symbols:

$$\boxed{\left(\lim_{\mathcal{B}} f(x) = A\right) := \forall V(A)\ \exists B \in \mathcal{B}\ \big(f(B) \subset V(A)\big).}$$

Since we are considering numerical-valued functions at the moment, it is useful to keep in mind the following form of this fundamental definition:

$$\left(\lim_{\mathcal{B}} f(x) = A\right) := \forall \varepsilon > 0\ \exists B \in \mathcal{B}\ \forall x \in B\ \big(|f(x) - A| < \varepsilon\big).$$

In this form we take an ε-neighborhood (symmetric with respect to A) instead of an arbitrary neighborhood $V(A)$. The equivalence of these definitions for real-valued functions follows from the fact mentioned earlier that every neighborhood of a point contains a symmetric neighborhood of the same point (carry out the proof in full!).

We have now given the general definition of the limit of a function over a base. Earlier we considered examples of the bases most often used in analysis. In a specific problem in which one or another of these bases arises, one must know how to decode the general definition and write it in the form specific to that base.

Thus,

$$\left(\lim_{x \to a-0} f(x) = A\right) := \forall \varepsilon > 0\ \exists \delta > 0\ \forall x \in \,]a - \delta, a[\ \big(|f(x) - A| < \varepsilon\big),$$

$$\left(\lim_{x \to -\infty} f(x) = A\right) := \forall \varepsilon > 0\ \exists \delta \in \mathbb{R}\ \forall x < \delta\ \big(|f(x) - A| < \varepsilon\big).$$

---

[12]For more details, see Bourbaki's, *General topology*, Addison-Wesley, 1966.

In our study of examples of bases we have in particular introduced the concept of a neighborhood of infinity. If we use that concept, then it makes sense to adopt the following conventions in accordance with the general definition of limit:

$$\left(\lim_{\mathcal{B}} f(x) = \infty\right) := \forall V(\infty)\, \exists B \in \mathcal{B}\, \left(f(B) \subset V(\infty)\right),$$

or, what is the same,

$$\left(\lim_{\mathcal{B}} f(x) = \infty\right) := \forall \varepsilon > 0\, \exists B \in \mathcal{B}\, \forall x \in B\, \left(\varepsilon < |f(x)|\right),$$

$$\left(\lim_{\mathcal{B}} f(x) = +\infty\right) := \forall \varepsilon \in \mathbb{R}\, \exists B \in \mathcal{B}\, \forall x \in B\, \left(\varepsilon < f(x)\right),$$

$$\left(\lim_{\mathcal{B}} f(x) = -\infty\right) := \forall \varepsilon \in \mathbb{R}\, \exists B \in \mathcal{B}\, \forall x \in B\, \left(f(x) < \varepsilon\right).$$

The letter ε is usually assumed to represent a small number. Such is not the case in the definitions just given, of course. In accordance with the usual conventions, for example, we could write

$$\left(\lim_{x \to +\infty} f(x) = -\infty\right) := \forall \varepsilon \in \mathbb{R}\, \exists \delta \in \mathbb{R}\, \forall x > \delta\, \left(f(x) < \varepsilon\right).$$

We advise the reader to write out independently the full definition of limit for different bases in the cases of both finite (numerical) and infinite limits.

In order to regard the theorems on limits that we proved for the special base $E \ni x \to a$ in Sect. 3.2.2 as having been proved in the general case of a limit over an arbitrary base, we need to make suitable definitions of what it means for a function to be *ultimately constant*, *ultimately bounded*, and *infinitesimal* over a given base.

Definition 13 A function $f : X \to \mathbb{R}$ is *ultimately constant over the base* $\mathcal{B}$ if there exists a number $A \in \mathbb{R}$ and an element $B \in \mathcal{B}$ such that $f(x) = A$ for all $x \in B$.

Definition 14 A function $f : X \to \mathbb{R}$ is *ultimately bounded over the base* $\mathcal{B}$ if there exists a number $c > 0$ and an element $B \in \mathcal{B}$ such that $|f(x)| < c$ for all $x \in B$.

Definition 15 A function $f : X \to \mathbb{R}$ is *infinitesimal over the base* $\mathcal{B}$ if $\lim_{\mathcal{B}} f(x) = 0$.

After these definitions and the fundamental remark that all proofs of the theorems on limits used only the properties B_1) and B_2), we may regard all the properties of limits established in Sect. 3.2.2 as valid for limits over any base.

In particular, we can now speak of the limit of a function as $x \to \infty$ or as $x \to -\infty$ or as $x \to +\infty$.

In addition, we have now assured ourselves that we can also apply the theory of limits in the case when the functions are defined on sets that are not necessarily sets of numbers; this will turn out to be especially valuable later on. For example, the length of a curve is a numerical-valued function defined on a class of curves. If we know this function on broken lines, we can define it for more complicated curves, for example, for a circle, by passing to the limit.

At present the main use we have for this observation and the concept of a base introduced in connection with it is that they free us from the verifications and formal proofs of theorems on limits for each specific type of limiting passage, or, in our current terminology, for each specific type of base.

In order to master completely the concept of a limit over an arbitrary base, we shall carry out the proofs of the following properties of the limit of a function in general form.

3.2.4 Existence of the Limit of a Function

a. The Cauchy Criterion

Before stating the Cauchy criterion, we give the following useful definition.

Definition 16 The *oscillation* of a function $f : X \to \mathbb{R}$ on a set $E \subset X$ is

$$\omega(f, E) := \sup_{x_1, x_2 \in E} \left|f(x_1) - f(x_2)\right|,$$

that is, the least upper bound of the absolute value of the difference of the values of the function at two arbitrary points $x_1, x_2 \in E$.

Example 11 $\omega(x^2, [-1, 2]) = 4$;

Example 12 $\omega(x, [-1, 2]) = 3$,

Example 13 $\omega(x,]-1, 2[) = 3$;

Example 14 $\omega(\operatorname{sgn} x, [-1, 2]) = 2$;

Example 15 $\omega(\operatorname{sgn} x, [0, 2]) = 1$;

Example 16 $\omega(\operatorname{sgn} x,]0, 2]) = 0$.

Theorem 4 (The Cauchy criterion for the existence of a limit of a function) *Let X be a set and $\mathcal{B}$ a base in X.*

A function $f : X \to \mathbb{R}$ has a limit over the base $\mathcal{B}$ if and only if for every $\varepsilon > 0$ there exits $B \in \mathcal{B}$ such that the oscillation of f on B is less than ε.

Thus,

$$\exists \lim_{\mathcal{B}} f(x) \Leftrightarrow \forall \varepsilon > 0\, \exists B \in \mathcal{B}\, \big(\omega(f, B) < \varepsilon\big).$$

Proof Necessity. If $\lim_{\mathcal{B}} f(x) = A \in \mathbb{R}$, then, for all $\varepsilon > 0$, there exists an element $B \in \mathcal{B}$ such that $|f(x) - A| < \varepsilon/3$ for all $x \in B$. But then, for any $x_1, x_2 \in B$ we have

$$\left|f(x_1) - f(x_2)\right| \le \left|f(x_1) - A\right| + \left|f(x_2) - A\right| < \frac{2\varepsilon}{3},$$

and therefore $\omega(f; B) < \varepsilon$.

Sufficiency. We now prove the main part of the criterion, which asserts that if for every $\varepsilon > 0$ there exists $B \in \mathcal{B}$ for which $\omega(f, B) < \varepsilon$, then the function has a limit over $\mathcal{B}$.

Taking ε successively equal to $1, 1/2, \dots, 1/n, \dots$, we construct a sequence $B_1, B_2, \dots, B_n \dots$ of elements of $\mathcal{B}$ such that $\omega(f, B_n) < 1/n$, $n \in \mathbb{N}$. Since $B_n \ne \varnothing$, we can choose a point x_n in each B_n. The sequence $f(x_1), f(x_2), \dots, f(x_n), \dots$ is a Cauchy sequence. Indeed, $B_n \cap B_m \ne \varnothing$, and, taking an auxiliary point $x \in B_n \cap B_m$, we find that $|f(x_n) - f(x_m)| < |f(x_n) - f(x)| + |f(x) - f(x_m)| < 1/n + 1/m$. By the Cauchy criterion for convergence of a sequence, the sequence $\{f(x_n), n \in \mathbb{N}\}$ has a limit A. It follows from the inequality established above, if we let $m \to \infty$, that $|f(x_n) - A| \le 1/n$. We now conclude, taking account of the inequality $\omega(f; B_n) < 1/n$, that $|f(x) - A| < \varepsilon$ at every point $x \in B_n$ if $n > N = [2/\varepsilon] + 1$. □

Remark This proof, as we shall see below, remains valid for functions with values in any so-called *complete* space Y. If $Y = \mathbb{R}$, which is the case we are most interested in just now, we can if we wish use the same idea as in the proof of the sufficiency of the Cauchy criterion for sequences.

Proof Setting $m_B = \inf_{x \in B} f(x)$ and $M_B = \sup_{x \in B} f(x)$, and remarking that $m_{B_1} \le m_{B_1 \cap B_2} \le M_{B_1 \cap B_2} \le M_{B_2}$ for any elements B_1 and B_2 of the base $\mathcal{B}$, we find by the axiom of completeness that there exists a number $A \in \mathbb{R}$ separating the numerical sets $\{m_B\}$ and $\{M_B\}$, where $B \in \mathcal{B}$. Since $\omega(f; B) = M_B - m_B$, we can now conclude that, since $\omega(f; B) < \varepsilon$, we have $|f(x) - A| < \varepsilon$ at every point $x \in B$. □

Example 17 We shall show that when $X = \mathbb{N}$ and $\mathcal{B}$ is the base $n \to \infty$, $n \in \mathbb{N}$, the general Cauchy criterion just proved for the existence of the limit of a function coincides with the Cauchy criterion already studied for the existence of a limit of a sequence.

Indeed, an element of the base $n \to \infty$, $n \in \mathbb{N}$, is a set $B = \mathbb{N} \cap U(\infty) = \{n \in \mathbb{N} \mid N < n\}$ consisting of the natural numbers $n \in \mathbb{N}$ larger than some number $N \in \mathbb{R}$. Without loss of generality we may assume $N \in \mathbb{N}$. The relation $\omega(f; B) \le \varepsilon$ now means that $|f(n_1) - f(n_2)| \le \varepsilon$ for all $n_1, n_2 > N$.

Thus, for a function $f : \mathbb{N} \to \mathbb{R}$, the condition that for any $\varepsilon > 0$ there exists $B \in \mathcal{B}$ such that $\omega(f; B) < \varepsilon$ is equivalent to the condition that the sequence $\{f(n)\}$ be a Cauchy sequence.

b. The Limit of a Composite Function

Theorem 5 (The limit of a composite function) *Let Y be a set, $\mathcal{B}_Y$ a base in Y, and $g : Y \to \mathbb{R}$ a mapping having a limit over the base $\mathcal{B}_Y$. Let X be a set, $\mathcal{B}_X$ a base in X and $f : X \to Y$ a mapping of X into Y such that for every element $B_Y \in \mathcal{B}_Y$ there exists $B_X \in \mathcal{B}_X$ whose image $f(B_X)$ is contained in B_Y.*

Under these hypotheses, the composition $g \circ f : X \to \mathbb{R}$ of the mappings f and g is defined and has a limit over the base $\mathcal{B}_X$ and $\lim_{\mathcal{B}_X}(g \circ f)(x) = \lim_{\mathcal{B}_Y} g(y)$.

Proof The composite function $g \circ f : X \to \mathbb{R}$ is defined, since $f(X) \subset Y$. Suppose $\lim_{\mathcal{B}_Y} g(y) = A$. We shall show that $\lim_{\mathcal{B}_X}(g \circ f)(x) = A$. Given a neighborhood $V(A)$ of A, we find $B_Y \in \mathcal{B}_Y$ such that $g(B_Y) \subset V(A)$. By hypothesis, there exists $B_X \in \mathcal{B}_X$ such that $f(B_X) \subset B_Y$. But then $(g \circ f)(B_X) = g(f(B_X)) \subset g(B_Y) \subset V(A)$. We have thus verified that A is the limit of the function $(g \circ f) : X \to \mathbb{R}$ over the base $\mathcal{B}_X$. □

Example 18 Let us find the following limit:

$$\lim_{x\to 0} \frac{\sin 7x}{7x} = ?$$

If we set $g(y) = \frac{\sin y}{y}$ and $f(x) = 7x$, then $(g \circ f)(x) = \frac{\sin 7x}{7x}$. In this case $Y = \mathbb{R}\backslash 0$ and $X = \mathbb{R}$. Since $\lim_{y\to 0} g(y) = \lim_{y\to 0} \frac{\sin y}{y} = 1$, we can apply the theorem if we verify that for any element of the base $y \to 0$ there is an element of the base $x \to 0$ whose image under the mapping $f(x) = 7x$ is contained in the given element of the base $y \to 0$.

The elements of the base $y \to 0$ are the deleted neighborhoods $\mathring{U}_Y(0)$ of the point $0 \in \mathbb{R}$.

The elements of the base $x \to 0$ are also deleted neighborhoods $\mathring{U}_X(0)$ of the point $0 \in \mathbb{R}$. Let $\mathring{U}_Y(0) = \{y \in \mathbb{R} \mid \alpha < y < \beta, y \neq 0\}$ (where $\alpha, \beta \in \mathbb{R}$ and $\alpha < 0$, $\beta > 0$) be an arbitrary deleted neighborhood of 0 in Y. If we take $\mathring{U}_X(0) = \{x \in \mathbb{R} \mid \frac{\alpha}{7} < x < \frac{\beta}{7}, x \neq 0\}$, this deleted neighborhood of 0 in X has the property that $f(\mathring{U}_X(0)) = \mathring{U}_Y(0) \subset \mathring{U}_Y(0)$.

The hypotheses of the theorem are therefore satisfied, and we can now assert that

$$\lim_{x\to 0} \frac{\sin 7x}{7x} = \lim_{y\to 0} \frac{\sin y}{y} = 1.$$

Example 19 The function $g(y) = |\operatorname{sgn} y|$, as we have seen (see Example 3), has the limit $\lim_{y\to 0} |\operatorname{sgn} y| = 1$.

The function $y = f(x) = x \sin \frac{1}{x}$, which is defined for $x \neq 0$, also has the limit $\lim_{x\to 0} x \sin \frac{1}{x} = 0$ (see Example 1).

However, the function $(g \circ f)(x) = |\operatorname{sgn}(x \sin \frac{1}{x})|$ has no limit as $x \to 0$.

Indeed, in any deleted neighborhood of $x = 0$ there are zeros of the function $\sin \frac{1}{x}$, so that the function $|\operatorname{sgn}(x \sin \frac{1}{x})|$ assumes both the value 1 and the value 0 in any such neighborhood. By the Cauchy criterion, this function cannot have a limit as $x \to 0$.

But does this example not contradict Theorem 5?

Check, as we did in the preceding example, to see whether the hypotheses of the theorem are satisfied.

Example 20 Let us show that

$$\boxed{\lim_{x\to\infty}\left(1+\frac{1}{x}\right)^x = \mathrm{e}.}$$

Proof Let us make the following assumptions:

$$Y = \mathbb{N}, \quad \mathcal{B}_Y \text{ is the base } n \to \infty,\ n \in \mathbb{N};$$

$$X = \mathbb{R}_+ = \{x \in \mathbb{R} \mid x > 0\}, \quad \mathcal{B}_X \text{ is the base } x \to +\infty;$$

$$f : X \to Y \quad \text{is the mapping } x \overset{f}{\longmapsto} [x],$$

where $[x]$ is the integer part of x (that is, the largest integer not larger than x).

Then for any $B_Y = \{n \in \mathbb{N} \mid n > N\}$ in the base $n \to \infty$, $n \in \mathbb{N}$ there obviously exists an element $B_X = \{x \in R \mid x > N + 1\}$ of the base $x \to +\infty$ whose image under the mapping $x \mapsto [x]$ is contained in B_Y.

The functions $g(n) = (1 + \frac{1}{n})^n$, $g_1(n) = (1 + \frac{1}{n+1})^n$, and $g_2(n) = (1 + \frac{1}{n})^{n+1}$, as we know, have the number e as their limit in the base $n \to \infty$, $n \in \mathbb{N}$.

By Theorem 4 on the limit of a composite function, we can now assert that the functions

$$(g \circ f)(x) = \left(1 + \frac{1}{[x]}\right)^{[x]}, \qquad (g_1 \circ f) = \left(1 + \frac{1}{[x]+1}\right)^{[x]},$$

$$(g_2 \circ f) = \left(1 + \frac{1}{[x]}\right)^{[x]+1}$$

also have e as their limit over the base $x \to +\infty$.

It now remains for us only to remark that

$$\left(1 + \frac{1}{[x]+1}\right)^{[x]} < \left(1 + \frac{1}{x}\right)^x < \left(1 + \frac{1}{[x]}\right)^{[x]+1}$$

for $x \geq 1$. Since the extreme terms here tend to e as $x \to +\infty$, it follows from Theorem 3 on the properties of a limit that $\lim_{x\to+\infty}(1 + \frac{1}{x})^x = \mathrm{e}$. □

Using Theorem 5 on the limit of a composite function, we now show that $\lim_{x\to-\infty}(1 + \frac{1}{x})^x = \mathrm{e}$.

Proof We write

$$\begin{aligned}\lim_{x\to-\infty}\left(1+\frac{1}{x}\right)^x &= \lim_{(-t)\to-\infty}\left(1+\frac{1}{(-t)}\right)^{(-t)} = \lim_{t\to+\infty}\left(1-\frac{1}{t}\right)^{-t} = \\ &= \lim_{t\to+\infty}\left(1+\frac{1}{t-1}\right)^t = \\ &= \lim_{t\to+\infty}\left(1+\frac{1}{t-1}\right)^{t-1}\lim_{t\to+\infty}\left(1+\frac{1}{t-1}\right) = \\ &= \lim_{t\to+\infty}\left(1+\frac{1}{t-1}\right)^{t-1} = \lim_{u\to+\infty}\left(1+\frac{1}{u}\right)^u = \mathrm{e}.\end{aligned}$$

When we take account of the substitutions $u = t - 1$ and $t = -x$, these equalities can be justified in reverse order (!) using Theorem 5. Indeed, only after we have arrived at the limit $\lim_{u\to+\infty}(1+\frac{1}{u})^u$, whose existence has already been proved, does the theorem allow us to assert that the preceding limit also exists and has the same value. Then the limit before that one also exists, and by a finite number of such transitions we finally arrive at the original limit. This is a very typical example of the procedure for using the theorem on the limit of a composite function in computing limits.

Thus, we have

$$\lim_{x\to-\infty}\left(1+\frac{1}{x}\right)^x = \mathrm{e} = \lim_{x\to+\infty}\left(1+\frac{1}{x}\right)^x.$$

It follows that $\lim_{x\to\infty}(1+\frac{1}{x})^x = \mathrm{e}$. Indeed, let $\varepsilon > 0$ be given.

Since $\lim_{x\to-\infty}(1+\frac{1}{x})^x = \mathrm{e}$, there exists $c_1 \in \mathbb{R}$ such that $|(1+\frac{1}{x})^x - \mathrm{e}| < \varepsilon$ for $x < c_1$.

Since $\lim_{x\to+\infty}(1+\frac{1}{x})^x = \mathrm{e}$, there exists $c_2 \in \mathbb{R}$ such that $|(1+\frac{1}{x})^x - \mathrm{e}| < \varepsilon$ for $c_2 < x$.

Then for $|x| > c = \max\{|c_1|, |c_2|\}$ we have $|(1+\frac{1}{x})^x - \mathrm{e}| < \varepsilon$, which verifies that $\lim_{x\to\infty}(1+\frac{1}{x})^x = \mathrm{e}$. □

Example 21 We shall show that

$$\lim_{t\to 0}(1+t)^{1/t} = \mathrm{e}.$$

Proof After the substitution $x = 1/t$, we return to the limit considered in the preceding example. □

Example 22

$$\lim_{x\to+\infty}\frac{x}{q^x} = 0, \quad \text{if } q > 1.$$

Proof We know (see Example 11 in Sect. 3.1) that $\lim_{n\to\infty} \frac{n}{q^n} = 0$ if $q > 1$.

Now, as in Example 20, we can consider the auxiliary mapping $f : \mathbb{R}_+ \to \mathbb{N}$ given by the function $[x]$ (the integer part of x). Using the inequalities

$$\frac{1}{q} \cdot \frac{[x]}{q^{[x]}} < \frac{x}{q^x} < \frac{[x]+1}{q^{[x]+1}} \cdot q$$

and taking account of the theorem on the limit of a composite function, we find that the extreme terms here tend to 0 as $x \to +\infty$. We conclude that $\lim_{x\to+\infty} \frac{x}{q^x} = 0$. □

Example 23

$$\lim_{x\to+\infty} \frac{\log_a x}{x} = 0.$$

Proof Let $a > 1$. Set $t = \log_a x$, so that $x = a^t$. From the properties of the exponential function and the logarithm (taking account of the unboundedness of a^n for $n \in \mathbb{N}$) we have $(x \to +\infty) \Leftrightarrow (t \to +\infty)$. Using the theorem on the limit of a composite function and the result of Example 22, we obtain

$$\lim_{x\to+\infty} \frac{\log_a x}{x} = \lim_{t\to+\infty} \frac{t}{a^t} = 0.$$

If $0 < a < 1$ we set $-t = \log_a x, x = a^{-t}$. Then $(x \to +\infty) \Leftrightarrow (t \to +\infty)$, and since $1/a > 1$, we again have

$$\lim_{x\to+\infty} \frac{\log_a x}{x} = \lim_{t\to+\infty} \frac{-t}{a^{-t}} = - \lim_{t\to+\infty} \frac{t}{(1/a)^t} = 0.$$ □

c. The Limit of a Monotonic Function

We now consider a special class of numerical-valued functions, but one that is very useful, namely the monotonic functions.

Definition 17 A function $f : E \to \mathbb{R}$ defined on a set $E \subset \mathbb{R}$ is said to be

increasing on E if

$$\forall x_1, x_2 \in E \left(x_1 < x_2 \Rightarrow f(x_1) < f(x_2)\right);$$

nondecreasing on E if

$$\forall x_1, x_2 \in E \left(x_1 < x_2 \Rightarrow f(x_1) \le f(x_2)\right);$$

nonincreasing on E if

$$\forall x_1, x_2 \in E \left(x_1 < x_2 \Rightarrow f(x_1) \ge f(x_2)\right);$$

decreasing on E if

$$\forall x_1, x_2 \in E \left(x_1 < x_2 \Rightarrow f(x_1) > f(x_2)\right).$$

Functions of the types just listed are said to be *monotonic on the set E*.

Assume that the numbers (or symbols $-\infty$ or $+\infty$) $i = \inf E$ and $s = \sup E$ are limit points of the set E, and let $f : E \to \mathbb{R}$ be a monotonic function on E. Then the following theorem holds.

Theorem 6 (Criterion for the existence of a limit of a monotonic function) *A necessary and sufficient condition for a function $f : E \to \mathbb{R}$ that is nondecreasing on the set E to have a limit as $x \to s$, $x \in E$, is that it be bounded above. For this function to have a limit as $x \to i$, $x \in E$, it is necessary and sufficient that it be bounded below.*

Proof We shall prove this theorem for the limit $\lim_{E \ni x \to s} f(x)$.

If this limit exists, then, like any function having a limit, the function f is ultimately bounded over the base $E \ni x \to s$.

Since f is nondecreasing on E, it follows that f is bounded above. In fact, we can even assert that $f(x) \leq \lim_{E \ni x \to s} f(x)$. That will be clear from what follows.

Let us pass to the proof of the existence of the limit $\lim_{E \ni x \to s} f(x)$ when f is bounded above.

Given that f is bounded above, we see that there is a least upper bound of the values that the function assumes on $E \setminus \{s\}$. Let $A = \sup_{x \in E \setminus \{s\}} f(x)$. We shall show that $\lim_{E \ni x \to s} f(x) = A$. Given $\varepsilon > 0$, we use the definition of the least upper bound to find a point $x_0 \in E \setminus \{s\}$ for which $A - \varepsilon < f(x_0) \leq A$. Then, since f is nondecreasing on E, we have $A - \varepsilon < f(x) \leq A$ for $x_0 < x \in E \setminus \{s\}$. But the set $\{x \in E \mid x_0 < x < s\}$ is obviously an element of the base $x \to s, x \in E$ (since $s = \sup E$). Thus we have proved that $\lim_{E \ni x \to s} f(x) = A$.

For the limit $\lim_{E \ni x \to i} f(x)$ the reasoning is analogous. In this case we have $\lim_{E \ni x \to i} f(x) = \inf_{x \in E \setminus \{i\}} f(x)$. □

d. Comparison of the Asymptotic Behavior of Functions

We begin this discussion with some examples to clarify the subject.

Let $\pi(x)$ be the number of primes not larger than a given number $x \in \mathbb{R}$. Although for any fixed x we can find (if only by explicit enumeration) the value of $\pi(x)$, we are nevertheless not in a position to say, for example, how the function $\pi(x)$ behaves as $x \to +\infty$, or, what is the same, what the asymptotic law of distribution of prime numbers is. We have known since the time of Euclid that $\pi(x) \to +\infty$ as $x \to +\infty$, but the proof that $\pi(x)$ grows approximately like $\frac{x}{\ln x}$ was achieved only in the nineteenth century by P.L. Chebyshev.[13]

[13]P.L. Chebyshev (1821–1894) – outstanding Russian mathematician and specialist in theoretical mechanics, the founder of a large mathematical school in Russia.

When it becomes necessary to describe the behavior of a function near some point (or near infinity) at which, as a rule, the function itself is not defined, we say that we are interested in the *asymptotics* or *asymptotic behavior* of the function in a neighborhood of the point.

The asymptotic behavior of a function is usually characterized using a second function that is simpler or better studied and which reproduces the values of the function being studied in a neighborhood of the point in question with small relative error.

Thus, as $x \to +\infty$, the function $\pi(x)$ behaves like $\frac{x}{\ln x}$; as $x \to 0$, the function $\frac{\sin x}{x}$ behaves like the constant function 1. When we speak of the behavior of the function $x^2 + x + \sin\frac{1}{x}$ as $x \to \infty$, we shall obviously say that it behaves basically like x^2, while in speaking of its behavior as $x \to 0$, we shall say it behaves like $\sin\frac{1}{x}$.

We now give precise definitions of some elementary concepts involving the asymptotic behavior of functions. We shall make systematic use of these concepts at the very first stage of our study of analysis.

Definition 18 We shall say that a certain property of functions or a certain relation between functions holds *ultimately over a given base* $\mathcal{B}$ if there exists $B \in \mathcal{B}$ on which it holds.

We have already interpreted the notion of a function that is ultimately constant or ultimately bounded in a given base in this sense. In the same sense we shall say from now on that the relation $f(x) = g(x)h(x)$ holds ultimately between functions f, g, and h. These functions may have at the outset different domains of definition, but if we are interested in their asymptotic behavior over the base $\mathcal{B}$, all that matters to us is that they are all defined on some element of $\mathcal{B}$.

Definition 19 The function f is said to be *infinitesimal* compared with the function g over the base $\mathcal{B}$, and we write $f =_{\mathcal{B}} o(g)$ or $f = o(g)$ over $\mathcal{B}$ if the relation $f(x) = \alpha(x)g(x)$ holds ultimately over the $\mathcal{B}$, where $\alpha(x)$ is a function that is infinitesimal over $\mathcal{B}$.

Example 24 $x^2 = o(x)$ as $x \to 0$, since $x^2 = x \cdot x$.

Example 25 $x = o(x^2)$ as $x \to \infty$, since ultimately (as long as $x \neq 0$), $x = \frac{1}{x} \cdot x^2$.

From these examples one must conclude that it is absolutely necessary to indicate the base over which $f = o(g)$.

The notation $f = o(g)$ is read "f is little-oh of g".

It follows from the definition, in particular, that the notation $f =_{\mathcal{B}} o(1)$, which results when $g(x) \equiv 1$, means simply that f is infinitesimal over $\mathcal{B}$.

Definition 20 If $f =_{\mathcal{B}} o(g)$ and g is itself infinitesimal over $\mathcal{B}$, we say that f is an *infinitesimal of higher order than* g *over* B.

Example 26 $x^{-2} = \frac{1}{x^2}$ is an infinitesimal of higher order than $x^{-1} = \frac{1}{x}$ as $x \to \infty$.

Definition 21 A function that tends to infinity over a given base is said to be an *infinite function* or simply an *infinity* over the given base.

Definition 22 If f and g are infinite functions over $\mathcal{B}$ and $f =_{\mathcal{B}} o(g)$, we say that g is a *higher order infinity* than f over $\mathcal{B}$.

Example 27 $\frac{1}{x} \to \infty$ as $x \to 0$, $\frac{1}{x^2} \to \infty$ as $x \to 0$ and $\frac{1}{x} = o(\frac{1}{x^2})$. Therefore $\frac{1}{x^2}$ is a higher order infinity than $\frac{1}{x}$ as $x \to 0$.

At the same time, as $x \to \infty$, x^2 is a higher order infinity than x.

It should not be thought that we can characterize the order of every infinity or infinitesimal by choosing some power x^n and saying that it is of order n.

Example 28 We shall show that for $a > 1$ and any $n \in \mathbb{Z}$

$$\lim_{x\to+\infty} \frac{x^n}{a^x} = 0,$$

that is, $x^n = o(a^x)$ as $x \to +\infty$.

Proof If $n \le 0$ the assertion is obvious. If $n \in \mathbb{N}$, then, setting $q = \sqrt[n]{a}$, we have $q > 1$ and $\frac{x^n}{a^x} = (\frac{x}{q^x})^n$, and therefore

$$\lim_{x\to+\infty} \frac{x^n}{a^x} = \lim_{x\to+\infty} \left(\frac{x}{q^x}\right)^n = \underbrace{\lim_{x\to+\infty} \frac{x}{q^x} \cdot \ldots \cdot \lim_{x\to+\infty} \frac{x}{q^x}}_{n \text{ factors}} = 0.$$

We have used (with induction) the theorem on the limit of a product and the result of Example 22. □

Thus, for any $n \in \mathbb{Z}$ we obtain $x^n = o(a^x)$ as $x \to +\infty$ if $a > 1$.

Example 29 Extending the preceding example, let us show that

$$\lim_{x\to+\infty} \frac{x^\alpha}{a^x} = 0$$

for $a > 1$ and any $\alpha \in \mathbb{R}$, that is, $x^\alpha = o(a^x)$ as $x \to +\infty$.

Proof Indeed, let us choose $n \in \mathbb{N}$ such that $n > \alpha$. Then for $x > 1$ we obtain

$$0 < \frac{x^\alpha}{a^x} < \frac{x^n}{a^x}.$$

Using properties of the limit and the result of the preceding example, we find that $\lim_{x\to+\infty} \frac{x^\alpha}{a^x} = 0$. □

Example 30 Let us show that

$$\lim_{\mathbb{R}_+ \ni x \to 0} \frac{a^{-1/x}}{x^\alpha} = 0$$

for $a > 1$ and any $\alpha \in \mathbb{R}$, that is, $a^{-1/x} = o(x^\alpha)$ as $x \to 0$, $x \in \mathbb{R}_+$.

Proof Setting $x = -1/t$ in this case and using the theorem on the limit of a composite function and the result of the preceding example, we find

$$\lim_{\mathbb{R}_+ \ni x \to 0} \frac{a^{-1/x}}{x^\alpha} = \lim_{t \to +\infty} \frac{t^\alpha}{a^t} = 0. \qquad \square$$

Example 31 Let us show that

$$\lim_{x \to +\infty} \frac{\log_a x}{x^\alpha} = 0$$

for $\alpha > 0$, that is, for any positive exponent α we have $\log_a x = o(x^\alpha)$ as $x \to +\infty$.

Proof If $a > 1$, we set $x = a^{t/\alpha}$. Then by the properties of power functions and the logarithm, the theorem on the limit of a composite function, and the result of Example 29, we find

$$\lim_{x \to +\infty} \frac{\log_a x}{x^\alpha} = \lim_{t \to +\infty} \frac{(t/\alpha)}{a^t} = \frac{1}{\alpha} \lim_{t \to +\infty} \frac{t}{a^t} = 0.$$

If $0 < a < 1$, then $1/a > 1$, and after the substitution $x = a^{-t/\alpha}$, we obtain

$$\lim_{x \to +\infty} \frac{\log_a x}{x^\alpha} = \lim_{t \to +\infty} \frac{(-t/\alpha)}{a^{-t}} = -\frac{1}{\alpha} \lim_{t \to +\infty} \frac{t}{(1/a)^t} = 0. \qquad \square$$

Example 32 Let us show further that

$$x^\alpha \log_a x = o(1) \quad \text{as } x \to 0,\ x \in \mathbb{R}_+$$

for any $\alpha > 0$.

Proof We need to show that $\lim_{\mathbb{R}_+ \ni x \to 0} x^\alpha \log_a x = 0$ for $\alpha > 0$. Setting $x = 1/t$ and applying the theorem on the limit of a composite function and the result of the preceding example, we find

$$\lim_{\mathbb{R}_+ \ni x \to 0} x^\alpha \log_a x = \lim_{t \to +\infty} \frac{\log_a (1/t)}{t^\alpha} = -\lim_{t \to +\infty} \frac{\log_a t}{t^\alpha} = 0. \qquad \square$$

Definition 23 Let us agree that the notation $f =_{\mathcal{B}} O(g)$ or $f = O(g)$ over the base $\mathcal{B}$ (read "f is big-oh of g over $\mathcal{B}$") means that the relation $f(x) = \beta(x) g(x)$ holds ultimately over $\mathcal{B}$ where $\beta(x)$ is ultimately bounded over $\mathcal{B}$.

In particular $f =_{\mathcal{B}} O(1)$ means that the function f is ultimately bounded over $\mathcal{B}$.

Example 33 $(\frac{1}{x} + \sin x)x = O(x)$ as $x \to \infty$.

Definition 24 The functions f and g are *of the same order over* $\mathcal{B}$, and we write $f \asymp g$ over $\mathcal{B}$, if $f =_{\mathcal{B}} O(g)$ and $g =_{\mathcal{B}} O(f)$ simultaneously.

Example 34 The functions $(2 + \sin x)x$ and x are of the same order as $x \to \infty$, but $(1 + \sin x)x$ and x are not of the same order as $x \to \infty$.

The condition that f and g be of the same order over the base $\mathcal{B}$ is obviously equivalent to the condition that there exist $c_1 > 0$ and $c_2 > 0$ and an element $B \in \mathcal{B}$ such that the relations

$$c_1|g(x)| \le |f(x)| \le c_2|g(x)|$$

hold on B, or, what is the same,

$$\frac{1}{c_2}|f(x)| \le |g(x)| \le \frac{1}{c_1}|f(x)|.$$

Definition 25 If the relation $f(x) = \gamma(x)g(x)$ holds ultimately over $\mathcal{B}$ where $\lim_{\mathcal{B}} \gamma(x) = 1$, we say that *the function f behaves asymptotically like g over $\mathcal{B}$*, or, more briefly, that *f is equivalent to g over $\mathcal{B}$*.

In this case we shall write $f \sim_{\mathcal{B}} g$ or $f \sim g$ over $\mathcal{B}$.

The use of the word *equivalent* is justified by the relations

$$\left(f \underset{\mathcal{B}}{\sim} f\right),$$

$$\left(f \underset{\mathcal{B}}{\sim} g\right) \Rightarrow \left(g \underset{\mathcal{B}}{\sim} f\right),$$

$$\left(f \underset{\mathcal{B}}{\sim} g\right) \wedge \left(g \underset{\mathcal{B}}{\sim} h\right) \Rightarrow \left(f \underset{\mathcal{B}}{\sim} h\right).$$

Indeed, the relation $f \sim_{\mathcal{B}} f$ is obvious, since in this case $\gamma(x) \equiv 1$. Next, if $\lim_{\mathcal{B}} \gamma(x) = 1$, then $\lim_{\mathcal{B}} \frac{1}{\gamma(x)} = 1$ and $g(x) = \frac{1}{\gamma(x)} f(x)$. Here all we that need to explain is why it is permissible to assume that $\gamma(x) \ne 0$. If the relation $f(x) = \gamma(x)g(x)$ holds on $B_1 \in \mathcal{B}$, and $\frac{1}{2} < |\gamma(x)| < \frac{3}{2}$ on $B_2 \in \mathcal{B}$, then we can take $B \in \mathcal{B}$ with $B \subset B_1 \cap B_2$, on which both relations hold. Outside of B, if convenient, we may assume that $\gamma(x) \equiv 1$. Thus we do indeed have $(f \sim g) \Rightarrow (g \sim f)$.

Finally, if $f(x) = \gamma_1(x)g(x)$ on $B_1 \in \mathcal{B}$ and $g(x) = \gamma_2(x)h(x)$ on $B_2 \in \mathcal{B}$, then on an element $B \in \mathcal{B}$ such that $B \subset B_1 \cap B_2$, both of these relations hold simultaneously, and so $f(x) = \gamma_1(x)\gamma_2(x)h(x)$ on B. But $\lim_{\mathcal{B}} \gamma_1(x)\gamma_2(x) = \lim_{\mathcal{B}} \gamma_1(x) \cdot \lim_{\mathcal{B}} \gamma_2(x) = 1$, and hence we have verified that $f \sim_{\mathcal{B}} h$.

It is useful to note that since the relation $\lim_{\mathcal{B}} \gamma(x) = 1$ is equivalent to $\gamma(x) = 1 + \alpha(x)$, where $\lim_{\mathcal{B}} \alpha(x) = 0$, the relation $f \sim_{\mathcal{B}} g$ is equivalent to $f(x) = g(x) + \alpha(x)g(x) = g(x) + o(g(x))$ over $\mathcal{B}$.

We see that the relative error $|\alpha(x)| = |\frac{f(x)-g(x)}{g(x)}|$ in approximating $f(x)$ by a function $g(x)$ that is equivalent to $f(x)$ over $\mathcal{B}$ is infinitesimal over $\mathcal{B}$.

Let us now consider some examples.

Example 35 $x^2 + x = (1 + \frac{1}{x})x^2 \sim x^2$ as $x \to \infty$.

The absolute value of the difference of these functions

$$\left|(x^2 + x) - x^2\right| = |x|$$

tends to infinity. However, the relative error $\frac{|x|}{x^2} = \frac{1}{|x|}$ that results from replacing $x^2 + x$ by the equivalent function x^2 tends to zero as $x \to \infty$.

Example 36 At the beginning of this discussion we spoke of the famous asymptotic law of distribution of the prime numbers. We can now give a precise statement of this law:

$$\pi(x) = \frac{x}{\ln x} + o\left(\frac{x}{\ln x}\right) \quad \text{as } x \to +\infty.$$

Example 37 Since $\lim_{x\to 0} \frac{\sin x}{x} = 1$, we have $\sin x \sim x$ as $x \to 0$, which can also be written as $\sin x = x + o(x)$ as $x \to 0$.

Example 38 Let us show that $\ln(1 + x) \sim x$ as $x \to 0$.

Proof

$$\lim_{x\to 0} \frac{\ln(1+x)}{x} = \lim_{x\to 0} \ln(1+x)^{1/x} = \ln\left(\lim_{x\to 0} (1+x)^{1/x}\right) = \ln \mathrm{e} = 1.$$

Here we have used the relation $\log_a(b^\alpha) = \alpha \log_a b$ in the first equality and the relation $\lim_{t\to b} \log_a t = \log_a b = \log_a(\lim_{t\to b} t)$ in the second. □

Thus, $\ln(1 + x) = x + o(x)$ as $x \to 0$.

Example 39 Let us show that $\mathrm{e}^x = 1 + x + o(x)$ as $x \to 0$.

Proof

$$\lim_{x\to 0} \frac{\mathrm{e}^x - 1}{x} = \lim_{t\to 0} \frac{t}{\ln(1+t)} = 1.$$

Here we have made the substitution $x = \ln(1 + t)$, $\mathrm{e}^x - 1 = t$ and used the relations $\mathrm{e}^x \to \mathrm{e}^0 = 1$ as $x \to 0$ and $\mathrm{e}^x \neq 1$ for $x \neq 0$. Thus, using the theorem on the limit of a composite function and the result of the preceding example, we have proved the assertion. □

Thus, $e^x - 1 \sim x$ as $x \to 0$.

Example 40 Let us show that $(1+x)^\alpha = 1 + \alpha x + o(x)$ as $x \to 0$.

Proof

$$\lim_{x\to 0} \frac{(1+x)^\alpha - 1}{x} = \lim_{x\to 0} \frac{e^{\alpha \ln(1+x)} - 1}{\alpha \ln(1+x)} \cdot \frac{\alpha \ln(1+x)}{x} =$$

$$= \alpha \lim_{t\to 0} \frac{e^t - 1}{t} \cdot \lim_{x\to 0} \frac{\ln(1+x)}{x} = \alpha.$$

In this computation, assuming $\alpha \neq 0$, we made the substitution $\alpha \ln(1+x) = t$ and used the results of the two preceding examples.

If $\alpha = 0$, the assertion is obvious. □

Thus, $(1+x)^\alpha - 1 \sim \alpha x$ as $x \to 0$.

The following simple fact is sometimes useful in computing limits.

Proposition 3 *If* $f \sim_{\mathcal{B}} \tilde{f}$, *then* $\lim_{\mathcal{B}} f(x)g(x) = \lim_{\mathcal{B}} \tilde{f}(x)g(x)$, *provided one of these limits exists.*

Proof Indeed, given that $f(x) = \gamma(x)\tilde{f}(x)$ and $\lim_{\mathcal{B}} \gamma(x) = 1$, we have

$$\lim_{\mathcal{B}} f(x)g(x) = \lim_{\mathcal{B}} \gamma(x)\tilde{f}(x)g(x) = \lim_{\mathcal{B}} \gamma(x) \cdot \lim_{\mathcal{B}} \tilde{f}(x)g(x) = \lim_{\mathcal{B}} \tilde{f}(x)g(x).$$ □

Example 41

$$\lim_{x\to 0} \frac{\ln \cos x}{\sin(x^2)} = \frac{1}{2} \lim_{x\to 0} \frac{\ln \cos^2 x}{x^2} = \frac{1}{2} \lim_{x\to 0} \frac{\ln(1-\sin^2 x)}{x^2} =$$

$$= \frac{1}{2} \lim_{x\to 0} \frac{-\sin^2 x}{x^2} = -\frac{1}{2} \lim_{x\to 0} \frac{x^2}{x^2} = -\frac{1}{2}.$$

Here we have used the relations $\ln(1+\alpha) \sim \alpha$ as $\alpha \to 0$, $\sin x \sim x$ as $x \to 0$, $\frac{1}{\sin \beta} \sim \frac{1}{\beta}$ as $\beta \to 0$, and $\sin^2 x \sim x^2$ as $x \to 0$.

We have proved that one may replace functions by other functions equivalent to them in a given base when computing limits of monomials. This rule should not be extended to sums and differences of functions.

Example 42 $\sqrt{x^2+x} \sim x$ as $x \to +\infty$, but

$$\lim_{x\to +\infty} \left(\sqrt{x^2+x} - x\right) \neq \lim_{x\to +\infty} (x - x) = 0.$$

In fact,

$$\lim_{x\to+\infty}\left(\sqrt{x^2+x}-x\right)=\lim_{x\to+\infty}\frac{x}{\sqrt{x^2+x}+x}=\lim_{x\to+\infty}\frac{1}{\sqrt{1+\frac{1}{x}}+1}=\frac{1}{2}.$$

We note one more widely used rule for handling the symbols $o(\cdot)$ and $O(\cdot)$ in analysis.

Proposition 4 *For a given base*

a) $o(f)+o(f)=o(f)$;
b) $o(f)$ *is also* $O(f)$;
c) $o(f)+O(f)=O(f)$;
d) $O(f)+O(f)=O(f)$;
e) *if* $g(x)\neq 0$, *then* $\frac{o(f(x))}{g(x)}=o(\frac{f(x)}{g(x)})$ *and* $\frac{O(f(x))}{g(x)}=O(\frac{f(x)}{g(x)})$.

Notice some peculiarities of operations with the symbols $o(\cdot)$ and $O(\cdot)$ that follow from the meaning of these symbols. For example $2o(f)=o(f)$ and $o(f)+O(f)=O(f)$ (even though in general $o(f)\neq 0$); also, $o(f)=O(f)$, but $O(f)\neq o(f)$. Here the equality sign is used in the sense of "is". The symbols $o(\cdot)$ and $O(\cdot)$ do not really denote a function, but rather indicate its asymptotic behavior, a behavior that many functions may have simultaneously, for example, f and $2f$, and the like.

Proof a) After the clarification just given, this assertion ceases to appear strange. The first symbol $o(f)$ in it denotes a function of the form $\alpha_1(x)f(x)$, where $\lim_{\mathcal{B}}\alpha_1(x)=0$. The second symbol $o(f)$, which one can (or should) equip with some mark to distinguish it from the first, denotes a function of the form $\alpha_2(x)f(x)$, where $\lim_{\mathcal{B}}\alpha_2(x)=0$. Then $\alpha_1(x)f(x)+\alpha_2(x)f(x)=(\alpha_1(x)+\alpha_2(x))f(x)=\alpha_3(x)f(x)$, where $\lim_{\mathcal{B}}\alpha_3(x)=0$.

Assertion b) follows from the fact that any function having a limit is ultimately bounded.

Assertion c) follows from b) and d).

Assertion d) follows from the fact that the sum of ultimately bounded functions is ultimately bounded.

As for e), we have $\frac{o(f(x))}{g(x)}=\frac{\alpha(x)f(x)}{g(x)}=\alpha(x)\frac{f(x)}{g(x)}=o(\frac{f(x)}{g(x)})$.
The second part of assertion e) is verified similarly. □

Using these rules and the equivalences obtained in Example 40, we can now find the limit in Example 42 by the following direct method:

$$\lim_{x\to+\infty}\left(\sqrt{x^2+x}-x\right)=\lim_{x\to+\infty}x\left(\sqrt{1+\frac{1}{x}}-1\right)=$$

$$=\lim_{x\to+\infty}x\left(1+\frac{1}{2}\cdot\frac{1}{x}+o\left(\frac{1}{x}\right)-1\right)=$$

$$= \lim_{x\to+\infty}\left(\frac{1}{2} + x\cdot o\left(\frac{1}{x}\right)\right) =$$

$$= \lim_{x\to+\infty}\left(\frac{1}{2} + o(1)\right) = \frac{1}{2}.$$

We shall soon prove the following important relations, which should be memorized at this point like the multiplication table:

$$e^x = 1 + \frac{1}{1!}x + \frac{1}{2!}x^2 + \cdots + \frac{1}{n!}x^n + \cdots \quad \text{for } x \in \mathbb{R},$$

$$\cos x = 1 - \frac{1}{2!}x^2 + \frac{1}{4!}x^4 + \cdots + \frac{(-1)^k}{(2k)!}x^{2k} + \cdots \quad \text{for } x \in \mathbb{R},$$

$$\sin x = \frac{1}{1!}x - \frac{1}{3!}x^3 + \cdots + \frac{(-1)^k}{(2k+1)!}x^{2k+1} + \cdots \quad \text{for } x \in \mathbb{R},$$

$$\ln(1+x) = x - \frac{1}{2}x^2 + \frac{1}{3}x^3 + \cdots + \frac{(-1)^{n-1}}{n}x^n + \cdots \quad \text{for } |x| < 1,$$

$$(1+x)^\alpha = 1 + \frac{\alpha}{1!}x + \frac{\alpha(\alpha-1)}{2!}x^2 + \cdots + \frac{\alpha(\alpha-1)\cdots(\alpha-n+1)}{n!}x^n + \cdots \quad \text{for } |x| < 1.$$

On the one hand, these relations can already be used as computational formulas, and on the other hand they contain the following asymptotic formulas, which generalize the formulas contained in Examples 37–40:

$$e^x = 1 + \frac{1}{1!}x + \frac{1}{2!}x^2 + \cdots + \frac{1}{n!}x^n + O\left(x^{n+1}\right) \quad \text{as } x \to 0,$$

$$\cos x = 1 - \frac{1}{2!}x^2 + \frac{1}{4!}x^4 + \cdots + \frac{(-1)^k}{(2k)!}x^{2k} + O\left(x^{2k+2}\right) \quad \text{as } x \to 0,$$

$$\sin x = \frac{1}{1!}x - \frac{1}{3!}x^3 + \cdots + \frac{(-1)^k}{(2k+1)!}x^{2k+1} + O\left(x^{2k+3}\right) \quad \text{as } x \to 0,$$

$$\ln(1+x) = x - \frac{1}{2}x^2 + \frac{1}{3}x^3 + \cdots + \frac{(-1)^{n-1}}{n}x^n + O\left(x^{n+1}\right) \quad \text{as } x \to 0,$$

$$(1+x)^\alpha = 1 + \frac{\alpha}{1!}x + \frac{\alpha(\alpha-1)}{2!}x^2 + \cdots + \frac{\alpha(\alpha-1)\cdots(\alpha-n+1)}{n!}x^n + O\left(x^{n+1}\right) \quad \text{as } x \to 0.$$

These formulas are usually the most efficient method of finding the limits of the elementary functions. When doing so, it is useful to keep in mind that $O(x^{m+1}) = x^{m+1}\cdot O(1) = x^m \cdot xO(1) = x^m o(1) = o(x^m)$ as $x \to 0$.

In conclusion, let us consider a few examples showing these formulas in action.

Example 43

$$\lim_{x\to 0}\frac{x-\sin x}{x^3}=\lim_{x\to 0}\frac{x-(x-\frac{1}{3!}x^3+O(x^5))}{x^3}=\lim_{x\to 0}\left(\frac{1}{3!}+O(x^2)\right)=\frac{1}{3!}.$$

Example 44 Let us find

$$\lim_{x\to\infty} x^2\left(\sqrt[7]{\frac{x^3+x}{1+x^3}}-\cos\frac{1}{x}\right).$$

As $x\to\infty$ we have:

$$\frac{x^3+x}{1+x^3}=\frac{1+x^{-2}}{1+x^{-3}}=\left(1+\frac{1}{x^2}\right)\left(1+\frac{1}{x^3}\right)^{-1}=$$
$$=\left(1+\frac{1}{x^2}\right)\left(1-\frac{1}{x^3}+O\left(\frac{1}{x^6}\right)\right)=1+\frac{1}{x^2}+O\left(\frac{1}{x^3}\right),$$
$$\sqrt[7]{\frac{x^3+x}{1+x^3}}=\left(1+\frac{1}{x^2}+O\left(\frac{1}{x^3}\right)\right)^{1/7}=1+\frac{1}{7}\cdot\frac{1}{x^2}+O\left(\frac{1}{x^3}\right),$$
$$\cos\frac{1}{x}=1-\frac{1}{2!}\cdot\frac{1}{x^2}+O\left(\frac{1}{x^4}\right),$$

from which we obtain

$$\sqrt[7]{\frac{x^3+x}{1+x^3}}-\cos\frac{1}{x}=\frac{9}{14}\cdot\frac{1}{x^2}+O\left(\frac{1}{x^3}\right)\quad\text{as } x\to\infty.$$

Hence the required limit is

$$\lim_{x\to\infty} x^2\left(\frac{9}{14x^2}+O\left(\frac{1}{x^3}\right)\right)=\frac{9}{14}.$$

Example 45

$$\lim_{x\to\infty}\left[\frac{1}{\mathrm{e}}\left(1+\frac{1}{x}\right)^x\right]^x=\lim_{x\to\infty}\exp\left\{x\left(\ln\left(1+\frac{1}{x}\right)^x-1\right)\right\}=$$
$$=\lim_{x\to\infty}\exp\left\{x^2\ln\left(1+\frac{1}{x}\right)-x\right\}=$$
$$=\lim_{x\to\infty}\exp\left\{x^2\left(\frac{1}{x}-\frac{1}{2x^2}+O\left(\frac{1}{x^3}\right)\right)-x\right\}=$$
$$=\lim_{x\to\infty}\exp\left\{-\frac{1}{2}+O\left(\frac{1}{x}\right)\right\}=\mathrm{e}^{-1/2}.$$

3.2.5 Problems and Exercises

1. a) Prove that there exists a unique function defined on $\mathbb{R}$ and satisfying the following conditions:

$$f(1)=a \quad (a>0, a\neq 1),$$
$$f(x_1)\cdot f(x_2)=f(x_1+x_2),$$
$$f(x)\to f(x_0) \quad \text{as } x\to x_0.$$

b) Prove that there exists a unique function defined on $\mathbb{R}_+$ and satisfying the following conditions:

$$f(a)=1 \quad (a>0, a\neq 1),$$
$$f(x_1)+f(x_2)=f(x_1\cdot x_2),$$
$$f(x)\to f(x_0) \quad \text{for } x_0\in\mathbb{R}_+ \text{ and } \mathbb{R}_+\ni x\to x_0.$$

Hint: Look again at the construction of the exponential function and logarithm discussed in Example 10.

2. a) Establish a one-to-one correspondence $\varphi:\mathbb{R}\to\mathbb{R}_+$ such that $\varphi(x+y)=\varphi(x)\cdot\varphi(y)$ for any $x,y\in\mathbb{R}$, that is, so that the operation of multiplication in the image ($\mathbb{R}_+$) corresponds to the operation of addition in the pre-image ($\mathbb{R}$). The existence of such a mapping means that the groups $(\mathbb{R},+)$ and $(\mathbb{R}_+,\cdot)$ are identical as algebraic objects, or, as we say, they are *isomorphic*.

b) Prove that the groups $(\mathbb{R},+)$ and $(\mathbb{R}\backslash 0,\cdot)$ are not isomorphic.

3. Find the following limits.

a) $\lim_{x\to+0} x^x$;
b) $\lim_{x\to+\infty} x^{1/x}$;
c) $\lim_{x\to 0}\frac{\log_a(1+x)}{x}$;
d) $\lim_{x\to 0}\frac{a^x-1}{x}$.

4. Show that

$$1+\frac{1}{2}+\cdots+\frac{1}{n}=\ln n+c+o(1) \quad \text{as } n\to\infty,$$

where c is a constant. (The number $c=0.57721\ldots$ is called *Euler's constant.*)
Hint: One can use the relation

$$\ln\frac{n+1}{n}=\ln\left(1+\frac{1}{n}\right)=\frac{1}{n}+O\left(\frac{1}{n^2}\right) \quad \text{as } n\to\infty.$$

5. Show that

a) if two series $\sum_{n=1}^{\infty}a_n$ and $\sum_{n=1}^{\infty}b_n$ with positive terms are such that $a_n\sim b_n$ as $n\to\infty$, then the two series either both converge or both diverge;

b) the series $\sum_{n=1}^{\infty}\sin\frac{1}{n^p}$ converges only for $p>1$.

6. Show that

a) if $a_n \geq a_{n+1} > 0$ for all $n \in \mathbb{N}$ and the series $\sum_{n=1}^{\infty} a_n$ converges, then $a_n = o(\frac{1}{n})$ as $n \to \infty$;

b) if $b_n = o(\frac{1}{n})$, one can always construct a convergent series $\sum_{n=1}^{\infty} a_n$ such that $b_n = o(a_n)$ as $n \to \infty$;

c) if a series $\sum_{n=1}^{\infty} a_n$ with positive terms converges, then the series $\sum_{n=1}^{\infty} A_n$, where $A_n = \sqrt{\sum_{k=n}^{\infty} \alpha_k} - \sqrt{\sum_{k=n+1}^{\infty} a_k}$ also converges, and $a_n = o(A_n)$ as $n \to \infty$;

d) if a series $\sum_{n=1}^{\infty} a_n$ with positive terms diverges, then the series $\sum_{n=2}^{\infty} A_n$, where $A_n = \sqrt{\sum_{k=1}^{n} a_k} - \sqrt{\sum_{k=1}^{n-1} a_k}$ also diverges, and $A_n = o(a_n)$ as $n \to \infty$.

It follows from c) and d) that no convergent (resp. divergent) series can serve as a universal standard of comparison to establish the convergence (resp. divergence) of other series.

7. Show that

a) the series $\sum_{n=1}^{\infty} \ln a_n$, where $a_n > 0$, $n \in \mathbb{N}$, converges if and only if the sequence $\{\Pi_n = a_1 \cdots a_n\}$ has a finite nonzero limit.

b) the series $\sum_{n=1}^{\infty} \ln(1 + \alpha_n)$, where $|\alpha_n| < 1$, converges absolutely if and only if the series $\sum_{n=1}^{\infty} \alpha_n$ converges absolutely.

Hint: See part a) of Exercise 5.

8. An infinite product $\prod_{k=1}^{\infty} e_k$ is said to converge if the sequence of numbers $\Pi_n = \prod_{k=1}^{n} e_k$ has a finite nonzero limit Π. We then set $\Pi = \prod_{k=1}^{\infty} e_k$.
Show that

a) if an infinite product $\prod_{n=1}^{\infty} e_n$ converges, then $e_n \to 1$ as $n \to \infty$;

b) if $\forall n \in \mathbb{N}$ $(e_n > 0)$, then the infinite product $\prod_{n=1}^{\infty} e_n$ converges if and only if the series $\sum_{n=1}^{\infty} \ln e_n$ converges;

c) if $e_n = 1 + \alpha_n$ and the α_n are all of the same sign, then the infinite product $\prod_{n=1}^{\infty}(1 + \alpha_n)$ converges if and only if the series $\sum_{n=1}^{\infty} \alpha_n$ converges.

9. a) Find the product $\prod_{n=1}^{\infty}(1 + x^{2n-1})$.

b) Find $\prod_{n=1}^{\infty} \cos \frac{x}{2^n}$ and prove the following theorem of Viète[14]

$$\frac{\pi}{2} = \frac{1}{\sqrt{\frac{1}{2}} \cdot \sqrt{\frac{1}{2}+\frac{1}{2}\sqrt{\frac{1}{2}}} \cdot \sqrt{\frac{1}{2}+\frac{1}{2}\sqrt{\frac{1}{2}+\frac{1}{2}\sqrt{\frac{1}{2}}}} \cdots}.$$

c) Find the function $f(x)$ if

$$f(0) = 1,$$

[14] F. Viète (1540–1603) – French mathematician, one of the creators of modern symbolic algebra.

$$f(2x) = \cos^2 x \cdot f(x),$$

$$f(x) \to f(0) \quad \text{as } x \to 0.$$

Hint: $x = 2 \cdot \frac{x}{2}$.

10. Show that

a) if $\frac{b_n}{b_{n+1}} = 1 + \beta_n, n = 1, 2, \ldots$, and the series $\sum_{n=1}^{\infty} \beta_n$ converges absolutely, then the limit $\lim_{n\to\infty} b_n = b \in \mathbb{R}$ exists;

b) if $\frac{a_n}{a_{n+1}} = 1 + \frac{p}{n} + \alpha_n, n = 1, 2, \ldots$, and the series $\sum_{n=1}^{\infty} \alpha_n$ converges absolutely, then $a_n \sim \frac{c}{n^p}$ as $n \to \infty$;

c) if the series $\sum_{n=1}^{\infty} a_n$ is such that $\frac{a_n}{a_{n+1}} = 1 + \frac{p}{n} + \alpha_n$ and the series $\sum_{n=1}^{\infty} \alpha_n$ converges absolutely, then $\sum_{n=1}^{\infty} a_n$ converges absolutely for $p > 1$ and diverges for $p \leq 1$ (Gauss' test for absolute convergence of a series).

11. Show that

$$\overline{\lim_{n\to\infty}} \left(\frac{1 + a_{n+1}}{a_n} \right)^n \geq \mathrm{e}$$

for any sequence $\{a_n\}$ with positive terms, and that this estimate cannot be improved.

Chapter 4
Continuous Functions

4.1 Basic Definitions and Examples

4.1.1 Continuity of a Function at a Point

Let f be a real-valued function defined in a neighborhood of a point $a \in \mathbb{R}$. In intuitive terms the function f is *continuous* at a if its value $f(x)$ approaches the value $f(a)$ that it assumes at the point a itself as x gets nearer to a.

We shall now make this description of the concept of continuity of a function at a point precise.

Definition 0 A function f is *continuous at the point* a if for any neighborhood $V(f(a))$ of its value $f(a)$ at a there is a neighborhood $U(a)$ of a whose image under the mapping f is contained in $V(f(a))$.

We now give the expression of this concept in logical symbolism, along with two other versions of it that are frequently used in analysis.

$$(f \text{ is continuous at } a) := \big(\forall V\big(f(a)\big)\, \exists U(a)\, \big(f\big(U(a)\big) \subset V\big(f(a)\big)\big)\big),$$

$$\forall \varepsilon > 0\, \exists U(a)\, \forall x \in U(a)\, \big(\big|f(x) - f(a)\big| < \varepsilon\big),$$

$$\forall \varepsilon > 0\, \exists \delta > 0\, \forall x \in \mathbb{R}\, \big(|x - a| < \delta \Rightarrow \big|f(x) - f(a)\big| < \varepsilon\big).$$

The equivalence of these statements for real-valued functions follows from the fact (already noted several times) that any neighborhood of a point contains a symmetric neighborhood of the point.

For example, if for any ε-neighborhood $V^{\varepsilon}(f(a))$ of $f(a)$ one can choose a neighborhood $U(a)$ of a such that $\forall x \in U(a)$ $(|f(x) - f(a)| < \varepsilon)$, that is, $f(U(a)) \subset V^{\varepsilon}(f(a))$, then for any neighborhood $V(f(a))$ one can also choose a corresponding neighborhood of a. Indeed, it suffices first to take an ε-neighborhood of $f(a)$ with $V^{\varepsilon}(f(a)) \subset V(f(a))$, and then find $U(a)$ corresponding to $V^{\varepsilon}(f(a))$. Then $f(U(a)) \subseteq V^{\varepsilon}(f(a)) \subset V(f(a))$.

V.A. Zorich, *Mathematical Analysis I*, Universitext,
DOI 10.1007/978-3-662-48792-1_4

Thus, if a function is continuous at a in the sense of the second of these definitions, it is also continuous at a in the sense of the original definition. The converse is obvious, so that the equivalence of the two statements is established.

We leave the rest of the verification to the reader.

To avoid being distracted from the basic concept being defined, that of continuity at a point, we assumed for simplicity to begin with that the function f was defined in a whole neighborhood of a. We now consider the general case.

Let $f : E \to \mathbb{R}$ be a real-valued function defined on some set $E \subset \mathbb{R}$ and a a point of the domain of definition of the function.

Definition 1 A function $f : E \to \mathbb{R}$ is *continuous at the point* $a \in E$ if for every neighborhood $V(f(a))$ of the value $f(a)$ that the function assumes at a there exists a neighborhood $U_E(a)$ of a in E[1] whose image $f(U_E(a))$ is contained in $V(f(a))$.

Thus

$$(f : E \to \mathbb{R} \text{ is continuous at } a \in E) := \\ = \big(\forall V\,\big(f(a)\big)\,\exists U_E(a)\,\big(f\big(U_E(a)\big) \subset V\big(f(a)\big)\big)\big).$$

Of course, Definition 1 can also be written in the ε–δ-form discussed above. Where numerical estimates are needed, this will be useful, and even necessary.

We now write these versions of Definition 1.

$$(f : E \to \mathbb{R} \text{ is continuous at } a \in E) := \\ = \big(\forall \varepsilon > 0\, \exists U_E(a)\, \forall x \in U_E(a)\, \big(\big|f(x) - f(a)\big| < \varepsilon\big)\big),$$

or

$$(f : E \to \mathbb{R} \text{ is continuous at } a \in E) := \\ = \big(\forall \varepsilon > 0\, \exists \delta > 0\, \forall x \in E\, \big(|x - a| < \delta \Rightarrow \big|f(x) - f(a)\big| < \varepsilon\big)\big).$$

We now discuss in detail the concept of continuity of a function at a point.

1^0 If a is an isolated point, that is, not a limit point of E, there is a neighborhood $U(a)$ of a containing no points of E except a itself. In this case $U_E(a) = a$, and therefore $f(U_E(a)) = f(a) \subset V(f(a))$ for any neighborhood $V(f(a))$. Thus a function is obviously continuous at any isolated point of its domain of definition. This, however, is a degenerate case.

2^0 The substantive part of the concept of continuity thus involves the case when $a \in E$ and a is a limit point of E. It is clear from Definition 1 that

$$(f : E \to \mathbb{R} \text{ is continuous at } a \in E,\ \text{where } a \text{ is a limit point of } E) \Leftrightarrow \\ \Leftrightarrow \Big(\lim_{E \ni x \to a} f(x) = f(a)\Big).$$

[1] We recall that $U_E(a) = E \cap U(a)$.

Proof In fact, if a is a limit point of E, then the base $E \ni x \to a$ of deleted neighborhoods $\overset{\circ}{U}_E(a) = U_E(a) \setminus a$ of a is defined.

If f is continuous at a, then, by finding a neighborhood $U_E(a)$ for the neighborhood $V(f(a))$ such that $f(U_E(a)) \subset V(f(a))$, we will simultaneously have $f(\overset{\circ}{U}_E(a)) \subset V(f(a))$. By definition of limit, therefore, $\lim_{E \ni x \to a} f(x) = f(a)$.

Conversely, if we know that $\lim_{E \ni x \to a} f(x) = f(a)$, then, given a neighborhood $V(f(a))$, we find a deleted neighborhood $\overset{\circ}{U}_E(a)$ such that $f(\overset{\circ}{U}_E(a)) \subset V(f(a))$. But since $f(a) \in V(f(a))$, we then have also $f(U_E(a)) \subset V(f(a))$. By Definition 1 this means that f is continuous at $a \in E$. □

3^0 Since the relation $\lim_{E \ni x \to a} f(x) = f(a)$ can be rewritten as

$$\boxed{\lim_{E \ni x \to a} f(x) = f\Big(\lim_{E \ni x \to a} x\Big),}$$

we now arrive at the useful conclusion that the continuous functions (operations) and only the continuous ones commute with the operation of passing to the limit at a point. This means that the number $f(a)$ obtained by carrying out the operation f on the number a can be approximated as closely as desired by the values obtained by carrying out the operation f on values of x that approximate a with suitable accuracy.

4^0 If we remark that for $a \in E$ the neighborhoods $U_E(a)$ of a form a base $\mathcal{B}_a$ (whether a is a limit point or an isolated point of E), we see that Definition 1 of continuity of a function at the point a is the same as the definition of the statement that the number $f(a)$ – the value of the function at a – is the limit of the function over this base, that is

$$(f : E \to \mathbb{R} \text{ is continuous at } a \in E) \Leftrightarrow \Big(\lim_{\mathcal{B}_a} f(x) = f(a)\Big).$$

5^0 We remark, however, that if $\lim_{\mathcal{B}_a} f(x)$ exists, since $a \in U_E(a)$ for every neighborhood $U_E(a)$, it follows that this limit must necessarily be $f(a)$.

Thus, continuity of a function $f : E \to \mathbb{R}$ at a point $a \in E$ is equivalent to the existence of the limit of this function over the base $\mathcal{B}_a$ of neighborhoods (not deleted neighborhoods) $U_E(a)$ of $a \in E$.

Thus

$$(f : E \to \mathbb{R} \text{ is continuous at } a \in E) \Leftrightarrow \Big(\exists \lim_{\mathcal{B}_a} f(x)\Big).$$

6^0 By the Cauchy criterion for the existence of a limit, we can now say that a function is continuous at a point $a \in E$ if and only if for every $\varepsilon > 0$ there exists a neighborhood $U_E(a)$ of a in E on which the oscillation $\omega(f; U_E(a))$ of the function is less than ε.

Definition 2 The quantity $\omega(f; a) = \lim_{\delta \to +0} \omega(f; U_E^\delta(a))$ (where $U_E^\delta(a)$ is the δ-neighborhood of a in E) is called the *oscillation of* $f : E \to \mathbb{R}$ *at* a.

Formally the symbol $\omega(f; X)$ has already been taken; it denotes the oscillation of the function on the set X. However, we shall never consider the oscillation of a function on a set consisting of a single point (it would obviously be zero); therefore the symbol $\omega(f; a)$, where a is a point, will always denote the concept of oscillation at a point just defined in Definition 2.

The oscillation of a function on a subset of a set does not exceed its oscillation on the set itself, so that $\omega(f; U_E^\delta(a))$ is a nondecreasing function of δ. Since it is nonnegative, either it has a finite limit as $\delta \to +0$, or else $\omega(f; U_E^\delta(a)) = +\infty$ for every $\delta > 0$. In the latter case we naturally set $\omega(f; a) = +\infty$.

7^0 Using Definition 2 we can summarize what was said in 6^0 as follows: a function is continuous at a point if and only if its oscillation at that point is zero. Let us make this explicit:

$$(f : E \to \mathbb{R} \text{ is continuous at } a \in E) \Leftrightarrow \big(\omega(f; a) = 0\big).$$

Definition 3 A function $f : E \to \mathbb{R}$ is *continuous on the set* E if it is continuous at each point of E.

The set of all continuous real-valued functions defined on a set E will be denoted $C(E; \mathbb{R})$ or, more, $C(E)$.

We have now discussed the concept of continuity of a function. Let us consider some examples.

Example 1 If $f : E \to \mathbb{R}$ is a constant function, then $f \in C(E)$. This is obvious, since $f(E) = c \subset V(c)$, for any neighborhood $V(c)$ of $c \in \mathbb{R}$.

Example 2 The function $f(x) = x$ is continuous on $\mathbb{R}$. Indeed, for any point $x_0 \in \mathbb{R}$ we have $|f(x) - f(x_0)| = |x - x_0| < \varepsilon$ provided $|x - x_0| < \delta = \varepsilon$.

Example 3 The function $f(x) = \sin x$ is continuous on $\mathbb{R}$.

In fact, for any point $x_0 \in \mathbb{R}$ we have

$$|\sin x - \sin x_0| = \left|2\cos\frac{x + x_0}{2}\sin\frac{x - x_0}{2}\right| \le$$
$$\le 2\left|\sin\frac{x - x_0}{2}\right| \le 2\left|\frac{x - x_0}{2}\right| = |x - x_0| < \varepsilon,$$

provided $|x - x_0| < \delta = \varepsilon$.

Here we have used the inequality $|\sin x| \le |x|$ proved in Example 9 of Paragraph d) of Sect. 3.2.2.

Example 4 The function $f(x) = \cos x$ is continuous on $\mathbb{R}$.

Indeed, as in the preceding example, for any point $x_0 \in \mathbb{R}$ we have

$$|\cos x - \cos x_0| = \left|-2\sin\frac{x+x_0}{2}\sin\frac{x-x_0}{2}\right| \le$$
$$\le 2\left|\sin\frac{x-x_0}{2}\right| \le |x-x_0| < \varepsilon,$$

provided $|x-x_0| < \delta = \varepsilon$.

Example 5 The function $f(x) = a^x$ is continuous on $\mathbb{R}$.

Indeed by property 3) of the exponential function (see Paragraph d in Sect. 3.2.2, Example 10a), at any point $x_0 \in \mathbb{R}$ we have

$$\lim_{x\to x_0} a^x = a^{x_0},$$

which, as we now know, is equivalent to the continuity of the function a^x at the point x_0.

Example 6 The function $f(x) = \log_a x$ is continuous at any point x_0 in its domain of definition $\mathbb{R}_+ = \{x \in \mathbb{R} \mid x > 0\}$.

In fact, by property 3) of the logarithm (see Paragraph d in Sect. 3.2.2, Example 10b), at each point $x_0 \in \mathbb{R}_+$ we have

$$\lim_{\mathbb{R}_+ \ni x \to x_0} \log_a x = \log_a x_0,$$

which is equivalent to the continuity of the function $\log_a x$ at the point x_0.

Now, given $\varepsilon > 0$, let us try to find a neighborhood $U_{\mathbb{R}_+}(x_0)$ of the point x_0 so as to have

$$|\log_a x - \log_a x_0| < \varepsilon$$

at each point $x \in U_{\mathbb{R}_+}(x_0)$.

This inequality is equivalent to the relations

$$-\varepsilon < \log_a \frac{x}{x_0} < \varepsilon.$$

For definiteness assume $a > 1$; then these last relations are equivalent to

$$x_0 a^{-\varepsilon} < x < x_0 a^{\varepsilon}.$$

The open interval $]x_0 a^{-\varepsilon}, x_0 a^{\varepsilon}[$ is the neighborhood of the point x_0 that we are seeking. It is useful to note that this neighborhood depends on both ε and the point x_0, a phenomenon that did not occur in Examples 1–4.

Example 7 Any sequence $f : \mathbb{N} \to \mathbb{R}$ is a function that is continuous on the set $\mathbb{N}$ of natural numbers, since each point of $\mathbb{N}$ is isolated.

4.1.2 Points of Discontinuity

To improve our mastery of the concept of continuity, we shall explain what happens to a function in a neighborhood of a point where it is not continuous.

Definition 4 If the function $f : E \to \mathbb{R}$ is not continuous at a point of E, this point is called a *point of discontinuity* or simply a *discontinuity* of f.

By constructing the negation of the statement "the function $f : E \to \mathbb{R}$ is continuous at the point $a \in E$", we obtain the following expression of the definition of the statement that a is a point of discontinuity of f:

$$(a \in E \text{ is a point of discontinuity of } f) := \\ = \big(\exists V\big(f(a)\big)\ \forall U_E(a)\ \exists x \in U_E(a)\ \big(f(x) \notin V\big(f(a)\big)\big)\big).$$

In other words, $a \in E$ is a point of discontinuity of the function $f : E \to \mathbb{R}$ if there is a neighborhood $V(f(a))$ of the value $f(a)$ that the function assumes at a such that in any neighborhood $U_E(a)$ of a in E there is a point x whose image is not in $V(f(a))$.

In ε–δ-form, this definition has the following appearance:

$$\exists \varepsilon > 0\ \forall \delta > 0\ \exists x \in E\ \big(|x - a| < \delta \wedge \big|f(x) - f(a)\big| \geq \varepsilon\big).$$

Let us consider some examples.

Example 8 The function $f(x) = \operatorname{sgn} x$ is constant and hence continuous in the neighborhood of any point $a \in \mathbb{R}$ that is different from 0. But in any neighborhood of 0 its oscillation equals 2. Hence 0 is a point of discontinuity for $\operatorname{sgn} x$. We remark that this function has a left-hand limit $\lim_{x\to-0} \operatorname{sgn} x = -1$ and a right-hand limit $\lim_{x\to+0} \operatorname{sgn} x = 1$. However, in the first place, these limits are not the same; and in the second place, neither of them is equal to the value of $\operatorname{sgn} x$ at the point 0, namely $\operatorname{sgn} 0 = 0$. This is a direct verification that 0 is a point of discontinuity for this function.

Example 9 The function $f(x) = |\operatorname{sgn} x|$ has the limit $\lim_{x\to 0} |\operatorname{sgn} x| = 1$ as $x \to 0$, but $f(0) = |\operatorname{sgn} 0| = 0$, so that $\lim_{x\to 0} f(x) \neq f(0)$, and 0 is therefore a point of discontinuity of the function.

We remark, however, that in this case, if we were to change the value of the function at the point 0 and set it equal to 1 there, we would obtain a function that is continuous at 0, that is, we would remove the discontinuity.

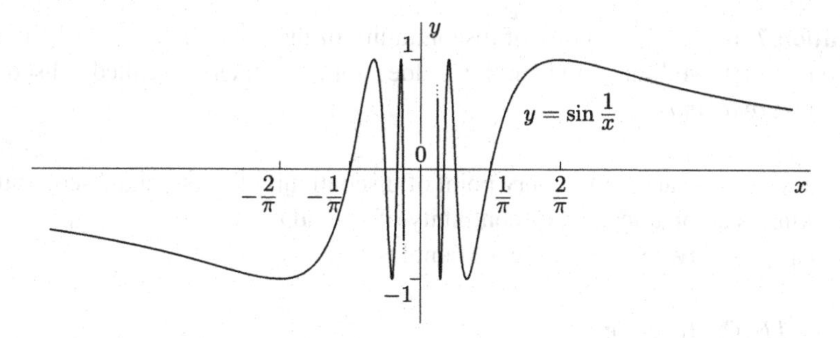

Fig. 4.1

Definition 5 If a point of discontinuity $a \in E$ of the function $f : E \to \mathbb{R}$ is such that there exists a continuous function $\tilde{f} : E \to \mathbb{R}$ such that $f|_{E\setminus a} = \tilde{f}|_{E\setminus a}$, then a is called a *removable discontinuity* of the function f.

Thus a removable discontinuity is characterized by the fact that the limit $\lim_{E\ni x\to a} f(x) = A$ exists, but $A \neq f(a)$, and it suffices to set

$$\tilde{f}(x) = \begin{cases} f(x) & \text{for } x \in E, x \neq a, \\ A & \text{for } x = a, \end{cases}$$

in order to obtain a function $\tilde{f} : E \to \mathbb{R}$ that is continuous at a.

Example 10 The function

$$f(x) = \begin{cases} \sin\frac{1}{x}, & \text{for } x \neq 0, \\ 0, & \text{for } x = 0, \end{cases}$$

is discontinuous at 0. Moreover, it does not even have a limit as $x \to 0$, since, as was shown Example 5 in Sect. 3.2.1, $\lim_{x\to 0} \sin\frac{1}{x}$ does not exist. The graph of the function $\sin\frac{1}{x}$ is shown in Fig. 4.1.

Examples 8, 9 and 10 explain the following terminology.

Definition 6 The point $a \in E$ is called a discontinuity *of first kind* for the function $f : E \to \mathbb{R}$ if the following limits[2] exist:

$$\lim_{E\ni x\to a-0} f(x) =: f(a-0), \qquad \lim_{E\ni x\to a+0} f(x) =: f(a+0),$$

but at least one of them is not equal to the value $f(a)$ that the function assumes at a.

[2]If a is a discontinuity, then a must be a limit point of the set E. It may happen, however, that all the points of E in some neighborhood of a lie on one side of a. In that case, only one of the limits in this definition is considered.

Definition 7 If $a \in E$ is a point of discontinuity of the function $f : E \to \mathbb{R}$ and at least one of the two limits in Definition 6 does not exist, then a is called a discontinuity *of second kind.*

Thus what is meant is that every point of discontinuity that is not a discontinuity of first kind is automatically a discontinuity of second kind.

Let us present two more classical examples.

Example 11 The function

$$\mathcal{D}(x) = \begin{cases} 1, & \text{if } x \in \mathbb{Q}, \\ 0, & \text{if } x \in \mathbb{R}\setminus\mathbb{Q}, \end{cases}$$

is called the *Dirichlet function.*[3]

This function is discontinuous at every point, and obviously all of its discontinuities are of second kind, since in every interval there are both rational and irrational numbers.

Example 12 Consider the *Riemann function*[4]

$$\mathcal{R}(x) = \begin{cases} \frac{1}{n}, & \text{if } x = \frac{m}{n} \in \mathbb{Q}, \text{ where } \frac{m}{n} \text{ is in lowest terms, } n \in \mathbb{N}, \\ 0, & \text{if } x \in \mathbb{R}\setminus\mathbb{Q}. \end{cases}$$

We remark that for any point $a \in \mathbb{R}$, any bounded neighborhood $U(a)$ of it, and any number $N \in \mathbb{N}$, the neighborhood $U(a)$ contains only a finite number of rational numbers $\frac{m}{n}$, $m \in \mathbb{Z}$, $n \in \mathbb{N}$, with $n < N$.

By shrinking the neighborhood, one can then assume that the denominators of all rational numbers in the neighborhood (except possibly for the point a itself if $a \in \mathbb{Q}$) are larger than N. Thus at any point $x \in \overset{\circ}{U}(a)$ we have $|\mathcal{R}(x)| < 1/N$.

We have thereby shown that

$$\lim_{x \to a} \mathcal{R}(x) = 0$$

at any point $a \in \mathbb{R}\setminus\mathbb{Q}$. Hence the Riemann function is continuous at any irrational number. At the remaining points, that is, at points $x \in \mathbb{Q}$, the function is discontinuous, except at the point $x = 0$, and all of these discontinuities are discontinuities of first kind.

---

[3]P.G. Dirichlet (1805–1859) – great German mathematician, an analyst who occupied the post of professor ordinarius at Göttingen University after the death of Gauss in 1855.

[4]B.F. Riemann (1826–1866) – outstanding German mathematician whose ground-breaking works laid the foundations of whole areas of modern geometry and analysis.

4.2 Properties of Continuous Functions

4.2.1 Local Properties

The *local* properties of functions are those that are determined by the behavior of the function in an arbitrarily small neighborhood of the point in its domain of definition.

Thus, the local properties themselves characterize the behavior of a function in any limiting relation when the argument of the function tends to the point in question. For example, the continuity of a function at a point of its domain of definition is obviously a local property.

We shall now exhibit the main local properties of continuous functions.

Theorem 1 *Let $f : E \to \mathbb{R}$ be a function that is continuous at the point $a \in E$. Then the following statements hold.*

1^0 *The function $f : E \to \mathbb{R}$ is bounded in some neighborhood $U_E(a)$ of a.*

2^0 *If $f(a) \neq 0$, then in some neighborhood $U_E(a)$ all the values of the function have the same sign as $f(a)$.*

3^0 *If the function $g : U_E(a) \to \mathbb{R}$ is defined in some neighborhood of a and, like f, is continuous at a, then the following functions are defined in some neighborhood of a and continuous at a:*

a) $(f + g)(x) := f(x) + g(x)$,
b) $(f \cdot g)(x) := f(x) \cdot g(x)$,
c) $(\frac{f}{g})(x) := \frac{f(x)}{g(x)}$ *(provided $g(a) \neq 0$).*

4^0 *If the function $g : Y \to \mathbb{R}$ is continuous at a point $b \in Y$ and f is such that $f : E \to Y$, $f(a) = b$, and f is continuous at a, then the composite function $(g \circ f)$ is defined on E and continuous at a.*

Proof To prove this theorem it suffices to recall (see Sect. 4.1) that the continuity of the function f or g at a point a of its domain of definition is equivalent to the condition that the limit of this function exists over the base $\mathcal{B}_a$ of neighborhoods of a and is equal to the value of the function at a: $\lim_{\mathcal{B}_a} f(x) = f(a)$, $\lim_{\mathcal{B}_a} g(x) = g(a)$.

Thus assertions 1^0, 2^0, and 3^0 of Theorem 1 follow immediately from the definition of continuity of a function at a point and the corresponding properties of the limit of a function.

The only explanation required is to verify that the ratio $\frac{f(x)}{g(x)}$ is actually defined in some neighborhood $\tilde{U}_E(a)$ of a. But by hypothesis $g(a) \neq 0$, and by assertion 2^0 of the theorem there exists a neighborhood $\tilde{U}_E(a)$ at every point of which $g(x) \neq 0$, that is, $\frac{f(x)}{g(x)}$ is defined in $\tilde{U}_E(a)$.

Assertion 4^0 of Theorem 1 is a consequence of the theorem on the limit of a composite function, by virtue of which

$$\lim_{\mathcal{B}_a}(g \circ f)(x) = \lim_{\mathcal{B}_b} g(y) = g(b) = g\big(f(a)\big) = (g \circ f)(a),$$

which is equivalent to the continuity of $(g \circ f)$ at a.

However, to apply the theorem on the limit of a composite function, we must verify that for any element $U_Y(b)$ of the base $\mathcal{B}_b$ there exists an element $U_E(a)$ of the base $\mathcal{B}_a$ such that $f(U_E(a)) \subset U_Y(b)$. But in fact, if $U_Y(b) = Y \cap U(b)$, then by definition of the continuity of $f : E \to Y$ at the point a, given a neighborhood $U(b) = U(f(a))$, there is a neighborhood $U_E(a)$ of a in E such that $f(U_E(a)) \subset U(f(a))$. Since the range of f is contained in Y, we have $f(U_E(a)) \subset Y \cap U(f(a)) = U_Y(b)$, and we have justified the application of the theorem on the limit of a composite function. □

Example 1 An algebraic polynomial $P(x) = a_0x^n + a_1x^{n-1} + \cdots + a_n$ is a continuous function on $\mathbb{R}$.

Indeed, it follows by induction from 3^0 of Theorem 1 that the sum and product of any finite number of functions that are continuous at a point are themselves continuous at that point. We have verified in Examples 1 and 2 of Sect. 4.1 that the constant function and the function $f(x) = x$ are continuous on $\mathbb{R}$. It then follows that the functions $ax^m = a \cdot \underbrace{x \cdot \ldots \cdot x}_{m \text{ factors}}$ are continuous, and consequently the polynomial $P(x)$ is also.

Example 2 A rational function $R(x) = \frac{P(x)}{Q(x)}$ – a quotient of polynomials – is continuous wherever it is defined, that is, where $Q(x) \neq 0$. This follows from Example 1 and assertion 3^0 of Theorem 1.

Example 3 The composition of a finite number of continuous functions is continuous at each point of its domain of definition. This follows by induction from assertion 4^0 of Theorem 1. For example, the function $e^{\sin^2(\ln|\cos x|)}$ is continuous on all of $\mathbb{R}$, except at the points $\frac{\pi}{2}(2k+1)$, $k \in \mathbb{Z}$, where it is not defined.

4.2.2 Global Properties of Continuous Functions

A *global* property of a function, intuitively speaking, is a property involving the entire domain of definition of the function.

Theorem 2 (The Bolzano–Cauchy intermediate-value theorem) *If a function that is continuous on a closed interval assumes values with different signs at the endpoints of the interval, then there is a point in the interval where it assumes the value* 0.

In logical symbols, this theorem has the following expression.[5]

$$\big(f \in C[a,b] \wedge f(a)\cdot f(b) < 0\big) \Rightarrow \exists c \in [a,b]\ \big(f(c)=0\big).$$

Proof Let us divide the interval $[a,b]$ in half. If the function does not assume the value 0 at the point of division, then it must assume opposite values at the endpoints of one of the two subintervals. In that interval we proceed as we did with the original interval, that is, we bisect it and continue the process.

Then either at some step we hit a point $c \in [a,b]$ where $f(c)=0$, or we obtain a sequence $\{I_n\}$ of nested closed intervals whose lengths tend to zero and at whose endpoints f assumes values with opposite signs. In the second case, by the nested interval lemma, there exists a unique point $c \in [a,b]$ common to all the intervals. By construction there are two sequences of endpoints $\{x'_n\}$ and $\{x''_n\}$ of the intervals I_n such that $f(x'_n) < 0$ and $f(x''_n) > 0$, while $\lim_{n\to\infty} x'_n = \lim_{n\to\infty} x''_n = c$. By the properties of a limit and the definition of continuity, we then find that $\lim_{n\to\infty} f(x'_n) = f(c) \le 0$ and $\lim_{n\to\infty} f(x''_n) = f(c) \ge 0$. Thus $f(c)=0$. □

Remarks to Theorem 2 1^0 The proof of the theorem provides a very simple algorithm for finding a root of the equation $f(x)=0$ on an interval at whose endpoints a continuous function $f(x)$ has values with opposite signs.

2^0 Theorem 2 thus asserts that it is impossible to pass continuously from positive to negative values without assuming the value zero along the way.

3^0 One should be wary of intuitive remarks like Remark 2^0, since they usually assume more than they state. Consider, for example, the function equal to -1 on the closed interval [0, 1] and equal to 1 on the closed interval [2, 3]. It is clear that this function is continuous on its domain of definition and assumes values with opposite signs, yet never assumes the value 0. This remark shows that the property of a continuous function expressed by Theorem 2 is actually the result of a certain property of the domain of definition (which, as will be made clear below, is the property of being *connected*).

Corollary to Theorem 2 *If the function φ is continuous on an open interval and assumes values $\varphi(a)=A$ and $\varphi(b)=B$ at points a and b, then for any number C between A and B, there is a point c between a and b at which $\varphi(c)=C$.*

Proof The closed interval I with endpoints a and b lies inside the open interval on which φ is defined. Therefore the function $f(x)=\varphi(x)-C$ is defined and continuous on I. Since $f(a)\cdot f(b)=(A-C)(B-C)<0$, Theorem 2 implies that there is a point c between a and b at which $f(c)=\varphi(c)-C=0$. □

Theorem 3 (The Weierstrass maximum-value theorem) *A function that is continuous on a closed interval is bounded on that interval. Moreover there is a point in*

---

[5] We recall that $C(E)$ denotes the set of all continuous functions on the set E. In the case $E=[a,b]$ we often write, more briefly, $C[a,b]$ instead of $C([a,b])$.

the interval where the function assumes its maximum value and a point where it assumes its minimal value.

Proof Let $f : E \to \mathbb{R}$ be a continuous function on the closed interval $E = [a, b]$. By the local properties of a continuous function (see Theorem 1) for any point $x \in E$ there exists a neighborhood $U(x)$ such that the function is bounded on the set $U_E(x) = E \cap U(x)$. The set of such neighborhoods $U(x)$ constructed for all $x \in E$ forms a covering of the closed interval $[a, b]$ by open intervals. By the finite covering lemma, one can extract a finite system $U(x_1), \dots, U(x_n)$ of open intervals that together cover the closed interval $[a, b]$. Since the function is bounded on each set $E \cap U(x_k) = U_E(x_k)$, that is, $m_k \le f(x) \le M_k$, where m_k and M_k are real numbers and $x \in U_E(x_k)$, we have

$$\min\{m_1, \dots, m_n\} \le f(x) \le \max\{M_1, \dots, M_N\}$$

at any point $x \in E = [a, b]$. It is now established that $f(x)$ is bounded on $[a, b]$.

Now let $M = \sup_{x \in E} f(x)$. Assume that $f(x) < M$ at every point $x \in E$. Then the continuous function $M - f(x)$ on E is nowhere zero, although (by the definition of M) it assumes values arbitrarily close to 0. It then follows that the function $\frac{1}{M - f(x)}$ is, on the one hand, continuous on E because of the local properties of continuous functions, but on the other hand not bounded on E, which contradicts what has just been proved about a function continuous on a closed interval.

Thus there must be a point $x_M \in [a, b]$ at which $f(x_M) = M$.

Similarly, by considering $m = \inf_{x \in E} f(x)$ and the auxiliary function $\frac{1}{f(x) - m}$, we prove that there exists a point $x_m \in [a, b]$ at which $f(x_m) = m$. □

We remark that, for example, the functions $f_1(x) = x$ and $f_2(x) = \frac{1}{x}$ are continuous on the open interval $E =]0, 1[$, but f_1 has neither a maximal nor a minimal value on E, and f_2 is unbounded on E. Thus, the properties of a continuous function expressed in Theorem 3 involve some property of the domain of definition, namely the property that from every covering of E by open intervals one can extract a finite subcovering. From now on we shall call such sets *compact*.

Before passing to the next theorem, we give a definition.

Definition 1 A function $f : E \to \mathbb{R}$ is *uniformly continuous* on a set $E \subset \mathbb{R}$ if for every $\varepsilon > 0$ there exists $\delta > 0$ such that $|f(x_1) - f(x_2)| < \varepsilon$ for all points $x_1, x_2 \in E$ such that $|x_1 - x_2| < \delta$.

More briefly,

$$(f : E \to \mathbb{R} \text{ is uniformly continuous}) :=$$
$$= \big(\forall \varepsilon > 0\ \exists \delta > 0\ \forall x_1 \in E\ \forall x_2 \in E\ \big(|x_1 - x_2| < \delta \Rightarrow \big|f(x_1) - f(x_2)\big| < \varepsilon\big)\big).$$

Let us now discuss the concept of uniform continuity.

1^0 If a function is uniformly continuous on a set, it is continuous at each point of that set. Indeed, in the definition just given it suffices to set $x_1 = x$ and $x_2 = a$, and we see that the definition of continuity of a function $f : E \to \mathbb{R}$ at a point $a \in E$ is satisfied.

2^0 Generally speaking, the continuity of a function does not imply its uniform continuity.

Example 4 The function $f(x) = \sin \frac{1}{x}$, which we have encountered many times, is continuous on the open interval $]0, 1[= E$. However, in every neighborhood of 0 in the set E the function assumes both values -1 and 1. Therefore, for $\varepsilon < 2$, the condition $|f(x_1) - f(x_2)| < \varepsilon$ does not hold.

In this connection it is useful to write out explicitly the negation of the property of uniform continuity for a function:

$$(f : E \to \mathbb{R} \text{ is not uniformly continuous}) :=$$
$$= \big(\exists \varepsilon > 0 \, \forall \delta > 0 \, \exists x_1 \in E \, \exists x_2 \in E \, \big(|x_1 - x_2| < \delta \wedge \big|f(x_1) - f(x_2)\big| \ge \varepsilon\big)\big).$$

This example makes the difference between continuity and uniform continuity of a function on a set intuitive. To point out the place in the definition of uniform continuity from which this difference proceeds, we give a detailed expression of what it means for a function $f : E \to \mathbb{R}$ to be continuous on E:

$$(f : E \to \mathbb{R} \text{ is continuous on } E) :=$$
$$= \big(\forall a \in E \, \forall \varepsilon > 0 \, \exists \delta > 0 \, \forall x \in E \, \big(|x - a| < \delta \Rightarrow \big|f(x) - f(a)\big| < \varepsilon\big)\big).$$

Thus the number δ is chosen knowing the point $a \in E$ and the number ε, and so for a fixed ε the number δ may vary from one point to another, as happens in the case of the function $\sin \frac{1}{x}$ considered in Example 1, or in the case of the function $\log_a x$ or a^x studied over their full domain of definition.

In the case of uniform continuity we are guaranteed the possibility of choosing δ knowing only $\varepsilon > 0$ so that $|x - a| < \delta$ implies $|f(x) - f(a)| < \varepsilon$ for all $x \in E$ and $a \in E$.

Example 5 If the function $f : E \to \mathbb{R}$ is unbounded in every neighborhood of a fixed point $x_0 \in E$, then it is not uniformly continuous.

Indeed, in that case for any $\delta > 0$ there are points x_1 and x_2 in every $\frac{\delta}{2}$-neighborhood of x_0 such that $|f(x_1) - f(x_2)| > 1$ although $|x_1 - x_2| < \delta$.

Such is the situation with the function $f(x) = \frac{1}{x}$ on the set $\mathbb{R} \backslash 0$. In this case $x_0 = 0$. The same situation holds in regard to $\log_a x$, which is defined on the set of positive numbers and unbounded in a neighborhood of $x_0 = 0$.

Example 6 The function $f(x) = x^2$, which is continuous on $\mathbb{R}$, is not uniformly continuous on $\mathbb{R}$.

In fact, at the points $x_n' = \sqrt{n+1}$ and $x_n'' = \sqrt{n}$, where $n \in \mathbb{N}$, we have $f(x_n') = n+1$ and $f(x_n'') = n$, so that $f(x_n') - f(x_n'') = 1$. But

$$\lim_{n\to\infty}(\sqrt{n+1}-\sqrt{n}) = \lim_{n\to\infty}\frac{1}{\sqrt{n+1}+\sqrt{n}} = 0,$$

so that for any $\delta > 0$ there are points x_n' and x_n'' such that $|x_n' - x_n''| < \delta$, yet $f(x_n') - f(x_n'') = 1$.

Example 7 The function $f(x) = \sin(x^2)$, which is continuous and bounded on $\mathbb{R}$, is not uniformly continuous on $\mathbb{R}$. Indeed, at the points $x_n' = \sqrt{\frac{\pi}{2}(n+1)}$ and $x_n'' = \sqrt{\frac{\pi}{2}n}$, where $n \in \mathbb{N}$, we have $|f(x_n') - f(x_n'')| = 1$, while $\lim_{n\to\infty}|x_n' - x_n''| = 0$.

After this discussion of the concept of uniform continuity of a function and comparison of continuity and uniform continuity, we can now appreciate the following theorem.

Theorem 4 (The Cantor–Heine theorem on uniform continuity) *A function that is continuous on a closed interval is uniformly continuous on that interval.*

We note that this theorem is usually called Cantor's theorem in the literature. To avoid unconventional terminology we shall preserve this common name in subsequent references.

Proof Let $f : E \to \mathbb{R}$ be a given function, $E = [a, b]$, and $f \in C(E)$. Since f is continuous at every point $x \in E$, it follows (see 6^0 in Sect. 4.1.1) that, knowing $\varepsilon > 0$ we can find a δ-neighborhood $U^\delta(x)$ of x such that the oscillation $\omega(f; U_E^\delta(x))$ of f on the set $U_E^\delta(x) = E \cap U^\delta(x)$, consisting of the points in the domain of definition E lying in $U^\delta(x)$, is less than ε. For each point $x \in E$ we construct a neighborhood $U^\delta(x)$ having this property. The quantity δ may vary from one point to another, so that it would be more accurate, if more cumbersome, to denote the neighborhood by the symbol $U^{\delta(x)}(x)$, but since the whole symbol is determined by the point x, we can agree on the following abbreviated notation: $U(x) = U^{\delta(x)}(x)$ and $V(x) = U^{\delta(x)/2}(x)$.

The open intervals $V(x)$, $x \in E$, taken together, cover the closed interval $[a, b]$, and so by the finite covering lemma one can select a finite covering $V(x_1), \dots, V(x_n)$. Let $\delta = \min\{\frac{1}{2}\delta(x_1), \dots, \frac{1}{2}\delta(x_n)\}$. We shall show that $|f(x') - f(x'')| < \varepsilon$ for any points $x', x'' \in E$ such that $|x' - x''| < \delta$. Indeed, since the system of open intervals $V(x_1), \dots, V(x_n)$ covers E, there exists an interval $V(x_i)$ of this system that contains x', that is $|x' - x_i| < \frac{1}{2}\delta(x_i)$. But in that case

$$|x'' - x_i| \le |x' - x''| + |x' - x_i| < \delta + \frac{1}{2}\delta(x_i) \le \frac{1}{2}\delta(x_i) + \frac{1}{2}\delta(x_i) = \delta(x_i).$$

Consequently $x', x'' \in U_E^{\delta(x_i)}(x_i) = E \cap U^{\delta(x_i)}(x_i)$ and so $|f(x') - f(x'')| \leq \omega(f; U_E^{\delta(x_i)}(x_i)) < \varepsilon$. □

The examples given above show that Cantor's theorem makes essential use of a certain property of the domain of definition of the function. It is clear from the proof that, as in Theorem 3, this property is that from every covering of E by neighborhoods of its points one can extract a finite covering.

Now that Theorem 4 has been proved, it is useful to return once again to the examples studied earlier of functions that are continuous but not uniformly continuous, in order to clarify how it happens that $\sin(x^2)$ for example, which is uniformly continuous on each closed interval of the real line by Cantor's theorem, is nevertheless not uniformly continuous on $\mathbb{R}$. The reason is completely analogous to the reason why a continuous function in general fails to be uniformly continuous. This time we invite our readers to investigate this question on their own.

We now pass to the last theorem of this section, the inverse function theorem. We need to determine the conditions under which a real-valued function on a closed interval has an inverse and the conditions under which the inverse is continuous.

Proposition 1 *A continuous mapping $f : E \to \mathbb{R}$ of a closed interval $E = [a, b]$ into $\mathbb{R}$ is injective if and only if the function f is strictly monotonic on $[a, b]$.*

Proof If f is increasing or decreasing on any set $E \subset \mathbb{R}$ whatsoever, the mapping $f : E \to \mathbb{R}$ is obviously injective: at different points of E the function assumes different values.

Thus the more substantive part of Proposition 1 consists of the assertion that every continuous injective mapping $f : [a, b] \to \mathbb{R}$ is realized by a strictly monotonic function.

Assuming that such is not the case, we find three points $x_1 < x_2 < x_3$ in $[a, b]$ such that $f(x_2)$ does not lie between $f(x_1)$ and $f(x_3)$. In that case, either $f(x_3)$ lies between $f(x_1)$ and $f(x_2)$ or $f(x_1)$ lies between $f(x_2)$ and $f(x_3)$. For definiteness assume that the latter is the case. By hypothesis f is continuous on $[x_2, x_3]$. Therefore, by Theorem 2, there is a point x_1' in this interval such that $f(x_1') = f(x_1)$. We then have $x_1 < x_1'$, but $f(x_1) = f(x_1')$, which is inconsistent with the injectivity of the mapping. The case when $f(x_3)$ lies between $f(x_1)$ and $f(x_2)$ is handled similarly. □

Proposition 2 *Each strictly monotonic function $f : X \to \mathbb{R}$ defined on a numerical set $X \subset \mathbb{R}$ has an inverse $f^{-1} : Y \to \mathbb{R}$ defined on the set $Y = f(X)$ of values of f, and has the same kind of monotonicity on Y that f has on X.*

Proof The mapping $f : X \to Y = f(X)$ is surjective, that is, it is a mapping of X *onto* Y. For definiteness assume that $f : X \to Y$ is increasing on X. In that case

$$\forall x_1 \in X\ \forall x_2 \in X\ \big(x_1 < x_2 \Leftrightarrow f(x_1) < f(x_2)\big). \tag{4.1}$$

Thus the mapping $f : X \to Y$ assumes different values at different points, and so is injective. Consequently $f : X \to Y$ is bijective, that is, it is a one-to-one correspondence between X and Y. Therefore the inverse mapping $f^{-1} : Y \to X$ is defined by the formula $x = f^{-1}(y)$ when $y = f(x)$.

Comparing the definition of the mapping $f^{-1} : Y \to X$ with relation (4.1), we arrive at the relation

$$\forall y_1 \in Y \ \forall y_2 \in Y \ \big(f^{-1}(y_1) < f^{-1}(y_2) \Leftrightarrow y_1 < y_2\big), \tag{4.2}$$

which means that the function f^{-1} is also increasing on its domain of definition.

The case when $f : X \to Y$ is decreasing on X is obviously handled similarly. □

In accordance with Proposition 2 just proved, if we are interested in the continuity of the function inverse to a real-valued function, it is useful to investigate the continuity of monotonic functions.

Proposition 3 *The discontinuities of a function $f : E \to \mathbb{R}$ that is monotonic on the set $E \subset \mathbb{R}$ can be only discontinuities of first kind.*

Proof For definiteness let f be nondecreasing. Assume that $a \in E$ is a point of discontinuity of f. Since a cannot be an isolated point of E, a must be a limit point of at least one of the two sets $E_a^- = \{x \in E \mid x < a\}$ and $E_a^+ = \{x \in E \mid x > a\}$. Since f is nondecreasing, for any point $x \in E_a^-$ we have $f(x) \le f(a)$, and the restriction $f|_{E_a^-}$ of f to E_a^- is a nondecreasing function that is bounded from above. It then follows that the limit

$$\lim_{E_a^- \ni x \to a} (f|_{E_a^-})(x) = \lim_{E \ni x \to a-0} f(x) = f(a-0)$$

exists.

The proof that the limit $\lim_{E \ni x \to a+0} f(x) = f(a+0)$ exists when a is a limit point of E_a^+ is analogous.

The case when f is a nonincreasing function can be handled either by repeating the reasoning just given or passing to the function $-f$, so as to reduce the question to the case already considered. □

Corollary 1 *If a is a point of discontinuity of a monotonic function $f : E \to \mathbb{R}$, then at least one of the limits*

$$\lim_{E \ni x \to a-0} f(x) = f(a-0), \qquad \lim_{E \ni x \to a+0} f(x) = f(a+0)$$

exists, and strict inequality holds in at least one of the inequalities $f(a-0) \le f(a) \le f(a+0)$ when f is nondecreasing and $f(a-0) \ge f(a) \ge f(a+0)$ when f is nonincreasing. The function assumes no values in the open interval defined by the strict inequality. Open intervals of this kind determined by different points of discontinuity have no points in common.

Proof Indeed, if a is a point of discontinuity, it must be a limit point of the set E, and by Proposition 3 is a discontinuity of first kind. Thus at least one of the bases $E \ni x \to a - 0$ and $E \ni x \to a + 0$ is defined, and the limit of the function over that base exists. (When both bases are defined, the limits over both bases exist.) For definiteness assume that f is nondecreasing. Since a is a point of discontinuity, strict inequality must actually hold in at least one of the inequalities $f(a-0) \le f(a) \le f(a+0)$. Since $f(x) \le \lim_{E\ni x\to a-0} f(x) = f(a-0)$, if $x \in E$ and $x < a$, the open interval $(f(a-0), f(a))$ defined by the strict inequality $f(a-0) < f(a)$ is indeed devoid of values of the function. Analogously, since $f(a+0) \le f(x)$ if $x \in E$ and $a < x$, the open interval $(f(a), f(a+0))$ defined by the strict inequality $f(a) < f(a+0)$ contains no values of f.

Let a_1 and a_2 be two different points of discontinuity of f, and assume $a_1 < a_2$. Then, since the function is nondecreasing,

$$f(a_1 - 0) \le f(a_1) \le f(a_1 + 0) \le f(a_2 - 0) \le f(a_2) \le f(a_2 + 0).$$

It follows from this that the intervals containing no values of f and corresponding to different points of discontinuity are disjoint. □

Corollary 2 *The set of points of discontinuity of a monotonic function is at most countable.*

Proof With each point of discontinuity of a monotonic function we associate the corresponding open interval in Corollary 1 containing no values of f. These intervals are pairwise disjoint. But on the line there cannot be more than a countable number of pairwise disjoint open intervals. In fact, one can choose a rational number in each of these intervals, so that the collection of intervals is equipollent with a subset of the set $\mathbb{Q}$ of rational numbers. Hence it is at most countable. Therefore, the set of points of discontinuity, which is in one-to-one correspondence with a set of such intervals, is also at most countable. □

Proposition 4 (A criterion for continuity of a monotonic function) *A monotonic function $f : E \to \mathbb{R}$ defined on a closed interval $E = [a, b]$ is continuous if and only if its set of values $f(E)$ is the closed interval with endpoints $f(a)$ and $f(b)$.*[6]

Proof If f is a continuous monotonic function, the monotonicity implies that all the values that f assumes on the closed interval $[a, b]$ lie between the values $f(a)$ and $f(b)$ that it assumes at the endpoints. By continuity, the function must assume all the values intermediate between $f(a)$ and $f(b)$. Hence the set of values of a function that is monotonic and continuous on a closed interval $[a, b]$ is indeed the closed interval with endpoints $f(a)$ and $f(b)$.

Let us now prove the converse. Let f be monotonic on the closed interval $[a, b]$. If f has a discontinuity at some point $c \in [a, b]$, by Corollary 1 one of the open intervals $]f(c-0), f(c)[$ and $]f(c), f(c+0)[$ is defined and nonempty and contains

[6]Here $f(a) \le f(b)$ if f is nondecreasing, and $f(b) \le f(a)$ if f is nonincreasing.

no values of f. But, since f is monotonic, that interval is contained in the interval with endpoints $f(a)$ and $f(b)$. Hence if a monotonic function has a point of discontinuity on the closed interval $[a, b]$, then the closed interval with endpoints $f(a)$ and $f(b)$ cannot be contained in the range of values of the function. □

Theorem 5 (The inverse function theorem) *A function $f : X \to \mathbb{R}$ that is strictly monotonic on a set $X \subset \mathbb{R}$ has an inverse $f^{-1} : Y \to \mathbb{R}$ defined on the set $Y = f(X)$ of values of f. The function $f^{-1} : Y \to \mathbb{R}$ is monotonic and has the same type of monotonicity on Y that f has on X.*

If in addition X is a closed interval $[a, b]$ and f is continuous on X, then the set $Y = f(X)$ is the closed interval with endpoints $f(a)$ and $f(b)$ and the function $f^{-1} : Y \to \mathbb{R}$ is continuous on it.

Proof The assertion that the set $Y = f(X)$ is the closed interval with endpoints $f(a)$ and $f(b)$ when $X = [a, b]$ and f is continuous follows from Proposition 4 proved above. It remains to be verified that $f^{-1} : Y \to \mathbb{R}$ is continuous. But f^{-1} is monotonic on Y, Y is a closed interval, and $f^{-1}(Y) = X = [a, b]$ is also a closed interval. We conclude by Proposition 4 that f^{-1} is continuous on the interval Y with endpoints $f(a)$ and $f(b)$. □

Example 8 The function $y = f(x) = \sin x$ is increasing and continuous on the closed interval $[-\frac{\pi}{2}, \frac{\pi}{2}]$. Hence the restriction of the function to the closed interval $[-\frac{\pi}{2}, \frac{\pi}{2}]$ has an inverse $x = f^{-1}(y)$, which we denote $x = \arcsin y$; this function is defined on the closed interval $[\sin(-\frac{\pi}{2}), \sin(\frac{\pi}{2})] = [-1, 1]$, increases from $-\frac{\pi}{2}$ to $\frac{\pi}{2}$, and is continuous on this closed interval.

Example 9 Similarly, the restriction of the function $y = \cos x$ to the closed interval $[0, \pi]$ is a decreasing continuous function, which by Theorem 5 has an inverse denoted $x = \arccos y$, defined on the closed interval $[-1, 1]$ and decreasing from π to 0 on that interval.

Example 10 The restriction of the function $y = \tan x$ to the open interval $X =]-\frac{\pi}{2}, \frac{\pi}{2}[$ is a continuous function that increases from $-\infty$ to $+\infty$. By the first part of Theorem 5 it has an inverse denoted $x = \arctan y$, defined for all $y \in \mathbb{R}$, and increasing within the open interval $]-\frac{\pi}{2}, \frac{\pi}{2}[$ of its values. To prove that the function $x = \arctan y$ is continuous at each point y_0 of its domain of definition, we take the point $x_0 = \arctan y_0$ and a closed interval $[x_0 - \varepsilon, x_0 + \varepsilon]$ containing x_0 and contained in the open interval $]-\frac{\pi}{2}, \frac{\pi}{2}[$. If $x_0 - \varepsilon = \arctan(y_0 - \delta_1)$ and $x_0 + \varepsilon = \arctan(y_0 + \delta_2)$, then for every $y \in \mathbb{R}$ such that $y_0 - \delta_1 < y < y_0 + \delta_2$ we shall have $x_0 - \varepsilon < \arctan y < x_0 + \varepsilon$. Hence $|\arctan y - \arctan y_0| < \varepsilon$ for $-\delta_1 < y - y_0 < \delta_2$. The former inequality holds in particular if $|y - y_0| < \delta = \min\{\delta_1, \delta_2\}$, which verifies that the function $x = \arctan y$ is continuous at the point $y_0 \in \mathbb{R}$.

Example 11 By reasoning analogous to that of the preceding example, we establish that since the restriction of the function $y = \cot x$ to the open interval $]0, \pi[$ is a continuous function that decreases from $+\infty$ to $-\infty$, it has an inverse denoted $x = \operatorname{arccot} y$, defined, continuous, and decreasing on the entire real line $\mathbb{R}$ from π to 0 and assuming values in the range $]0, \pi[$.

Remark In constructing the graphs of mutually inverse functions $y = f(x)$ and $x = f^{-1}(y)$ it is useful to keep in mind that in a given coordinate system the points with coordinates $(x, f(x)) = (x, y)$ and $(y, f^{-1}(y)) = (y, x)$ are symmetric with respect to the bisector of the angle in the first quadrant.

Thus the graphs of mutually inverse functions, when drawn in the same coordinate system, are symmetric with respect to this angle bisector.

4.2.3 Problems and Exercises

1. Show that

a) if $f \in C(A)$ and $B \subset A$, then $f|_B \in C(B)$;

b) if a function $f : E_1 \cup E_2 \to \mathbb{R}$ is such that $f|_{E_i} \in C(E_i)$, $i = 1, 2$, it is not always the case that $f \in C(E_1 \cup E_2)$.

c) the Riemann function $\mathcal{R}$, and its restriction $\mathcal{R}|_{\mathbb{Q}}$ to the set of rational numbers are both discontinuous at each point of $\mathbb{Q}$ except 0, and all the points of discontinuity are removable (see Example 12 of Sect. 4.1).

2. Show that for a function $f \in C[a, b]$ the functions

$$m(x) = \min_{a \le t \le x} f(t) \quad \text{and} \quad M(x) = \max_{a \le t \le x} f(t)$$

are also continuous on the closed interval $[a, b]$.

3. a) Prove that the function inverse to a function that *is* monotonic on an open interval is continuous on its domain of definition.

b) Construct a monotonic function with a countable set of discontinuities.

c) Show that if functions $f : X \to Y$ and $f^{-1} : Y \to X$ are mutually inverse (here X and Y are subsets of $\mathbb{R}$), and f is continuous at a point $x_0 \in X$, the function f^{-1} need not be continuous at $y_0 = f(x_0)$ in Y.

4. Show that

a) if $f \in C[a, b]$ and $g \in C[a, b]$, and, in addition, $f(a) < g(a)$ and $f(b) > g(b)$, then there exists a point $c \in [a, b]$ at which $f(c) = g(c)$;

b) any continuous mapping $f : [0, 1] \to [0, 1]$ of a closed interval into itself has a fixed point, that is, a point $x \in [0, 1]$ such that $f(x) = x$;

c) if two continuous mappings f and g of an interval into itself commute, that is, $f \circ g = g \circ f$, then they do not always have a common fixed point;

d) a continuous mapping $f : \mathbb{R} \to \mathbb{R}$ may fail to have a fixed point;

e) a continuous mapping $f :]0, 1[\to]0, 1[$ may fail to have a fixed point;

f) if a mapping $f : [0, 1] \to [0, 1]$ is continuous, $f(0) = 0$, $f(1) = 1$, and $(f \circ f)(x) \equiv x$ on $[0, 1]$, then $f(x) \equiv x$.

5. Show that the set of values of any function that is continuous on a closed interval is a closed interval.

6. Prove the following statements.

a) If a mapping $f : [0, 1] \to [0, 1]$ is continuous, $f(0) = 0$, $f(1) = 1$, and $f^n(x) := \underbrace{f \circ \cdots \circ f}_{n \text{ factors}}(x) \equiv x$ on $[0, 1]$, then $f(x) \equiv x$.

b) If a function $f : [0, 1] \to [0, 1]$ is continuous and nondecreasing, then for any point $x \in [0, 1]$ at least one of the following situations must occur: either x is a fixed point, or $f^n(x)$ tends to a fixed point. (Here $f^n(x) = f \circ \cdots \circ f(x)$ is the nth iteration of f.)

7. Let $f : [0, 1] \to \mathbb{R}$ be a continuous function such that $f(0) = f(1)$. Show that

a) for any $n \in \mathbb{N}$ there exists a horizontal closed interval of length $\frac{1}{n}$ with endpoints on the graph of this function;

b) if the number l is not of the form $\frac{1}{n}$ there exists a function of this form on whose graph one cannot inscribe a horizontal chord of length l.

8. The *modulus of continuity* of a function $f : E \to \mathbb{R}$ is the function $\omega(\delta)$ defined for $\delta > 0$ as follows:

$$\omega(\delta) = \sup_{\substack{|x_1 - x_2| < \delta \\ x_1, x_2 \in E}} \left|f(x_1) - f(x_2)\right|.$$

Thus, the least upper bound is taken over all pairs of points x_1, x_2 of E whose distance apart is less than δ.

Show that

a) the modulus of continuity is a nondecreasing nonnegative function having the limit[7] $\omega(+0) = \lim_{\delta \to +0} \omega(\delta)$;

b) for every $\varepsilon > 0$ there exists $\delta > 0$ such that for any points $x_1, x_2 \in E$ the relation $|x_1 - x_2| < \delta$ implies $|f(x_1) - f(x_2)| < \omega(+0) + \varepsilon$;

c) if E is a closed interval, an open interval, or a half-open interval, the relation

$$\omega(\delta_1 + \delta_2) \le \omega(\delta_1) + \omega(\delta_2)$$

holds for the modulus of continuity of a function $f : E \to \mathbb{R}$;

d) the moduli of continuity of the functions x and $\sin(x^2)$ on the whole real axis are respectively $\omega(\delta) = \delta$ and the constant $\omega(\delta) = 2$ in the domain $\delta > 0$;

e) a function f is uniformly continuous on E if and only if $\omega(+0) = 0$.

9. Let f and g be bounded functions defined on the same set X. The quantity $\Delta = \sup_{x \in X} |f(x) - g(x)|$ is called the *distance* between f and g. It shows how well one function approximates the other on the given set X. Let X be a closed interval $[a, b]$. Show that if $f, g \in C[a, b]$, then $\exists x_0 \in [a, b]$, where $\Delta = |f(x_0) - g(x_0)|$, and that such is not the case in general for arbitrary bounded functions.

---

[7] For this reason the modulus of continuity is usually considered for $\delta \ge 0$, setting $\omega(0) = \omega(+0)$.

10. Let $P_n(x)$ be a polynomial of degree n. We are going to approximate a bounded function $f : [a, b] \to \mathbb{R}$ by polynomials. Let

$$\Delta(P_n) = \sup_{x \in [a,b]} \left|f(x) - P_n(x)\right| \quad \text{and} \quad E_n(f) = \inf_{P_n} \Delta(P_n),$$

where the infimum is taken over all polynomials of degree n. A polynomial P_n is called a *polynomial of best approximation* of f if $\Delta(P_n) = E_n(f)$.
Show that

a) there exists a polynomial $P_0(x) \equiv a_0$ of best approximation of degree zero;

b) among the polynomials $Q_\lambda(x)$ of the form $\lambda P_n(x)$, where P_n is a fixed polynomial, there is a polynomial Q_{λ_0} such that

$$\Delta(Q_{\lambda_0}) = \min_{\lambda \in \mathbb{R}} \Delta(Q_\lambda);$$

c) if there exists a polynomial of best approximation of degree n, there also exists a polynomial of best approximation of degree $n + 1$;

d) for any bounded function on a closed interval and any $n = 0, 1, 2, \ldots$ there exists a polynomial of best approximation of degree n.

11. Prove the following statements.

a) A polynomial of odd degree with real coefficients has at least one real root.

b) If P_n is a polynomial of degree n, the function $\operatorname{sgn} P_n(x)$ has at most n points of discontinuity.

c) If there are $n + 2$ points $x_0 < x_1 < \cdots < x_{n+1}$ in the closed interval $[a, b]$ such that the quantity

$$\operatorname{sgn}\big[\big(f(x_i) - P_n(x_i)\big)(-1)^i\big]$$

assumes the same value for $i = 0, \ldots, n + 1$, then $E_n(f) \ge \min_{0 \le i \le n+1} |f(x_i) - P_n(x_i)|$. (This result is known as *Vallée Poussin's theorem.*[8] For the definition of $E_n(f)$ see Problem 10.)

12. a) Show that for any $n \in \mathbb{N}$ the function $T_n(x) = \cos(n \arccos x)$ defined on the closed interval $[-1, 1]$ is an algebraic polynomial of degree n. (These are the *Chebyshev polynomials.*)

b) Find an explicit algebraic expression for the polynomials T_1, T_2, T_3, and T_4 and draw their graphs.

c) Find the roots of the polynomial $T_n(x)$ on the closed interval $[-1, 1]$ and the points of the interval where $|T_n(x)|$ assumes its maximum value.

d) Show that among all polynomials $P_n(x)$ of degree n whose leading coefficient is 1 the polynomial $T_n(x)$ is the unique polynomial closest to zero, that is, $E_n(0) = \max_{|x| \le 1} |T_n(x)|$. (For the definition of $E_n(f)$ see Problem 10.)

---

[8]Ch.J. de la Vallée Poussin (1866–1962) – Belgian mathematician and specialist in theoretical mechanics.

13. Let $f \in C[a, b]$.

a) Show that if the polynomial $P_n(x)$ of degree n is such that there are $n+2$ points $x_0 < \cdots < x_{n+1}$ (called *Chebyshev alternant points*) for which $f(x_i) - P_n(x_i) = (-1)^i \Delta(P_n) \cdot \alpha$, where $\Delta(P_n) = \max_{x \in [a,b]} |f(x) - P_n(x)|$ and α is a constant equal to 1 or -1, then $P_n(x)$ is the unique polynomial of best approximation of degree n to f (see Problem 10).

b) Prove *Chebyshev's theorem: A polynomial* $P_n(x)$ *of degree n is a polynomial of best approximation to the function* $f \in C[a, b]$ *if and only if there are at least* $n+2$ *Chebyshev alternant points on the closed interval* $[a, b]$.

c) Show that for discontinuous functions the preceding statement is in general not true.

d) Find the polynomials of best approximation of degrees zero and one for the function $|x|$ on the interval $[-1, 2]$.

14. In Sect. 4.2 we discussed the local properties of continuous functions. The present problem makes the concept of a local property more precise.

Two functions f and g are considered equivalent if there is a neighborhood $U(a)$ of a given point $a \in \mathbb{R}$ such that $f(x) = g(x)$ for all $x \in U(a)$. This relation between functions is obviously reflexive, symmetric, and transitive, that is, it really is an equivalence relation.

A class of functions that are all equivalent to one another at a point a is called a *germ of functions* at a. If we consider only continuous functions, we speak of a germ of continuous functions at a.

The *local properties* of functions are properties of the germs of functions.

a) Define the arithmetic operations on germs of numerical-valued functions defined at a given point.

b) Show that the arithmetic operations on germs of continuous functions do not lead outside this class of germs.

c) Taking account of a) and b), show that the germs of continuous functions form a ring – the ring of germs of continuous functions.

d) A subring I of a ring K is called an *ideal* of K if the product of every element of the ring K with an element of the subring I belongs to I. Find an ideal in the ring of germs of continuous functions at a.

15. An ideal in a ring is *maximal* if it is not contained in any larger ideal except the ring itself. The set $C[a, b]$ of functions continuous on a closed interval forms a ring under the usual operations of addition and multiplication of numerical-valued functions. Find the maximal ideals of this ring.

Chapter 5
Differential Calculus

5.1 Differentiable Functions

5.1.1 Statement of the Problem and Introductory Considerations

Suppose, following Newton,[1] we wish to solve the Kepler problem[2] of two bodies, that is, we wish to explain the law of motion of one celestial body m (a planet) relative to another body M (a star). We take a Cartesian coordinate system in the plane of motion with origin at M (Fig. 5.1). Then the position of m at time t can be characterized numerically by the coordinates $(x(t), y(t))$ of the point in that coordinate system. We wish to find the functions $x(t)$ and $y(t)$.

The motion of m relative to M is governed by Newton's two famous laws: *the general law of motion*

$$m\mathbf{a} = \mathbf{F}, \tag{5.1}$$

connecting the force vector with the acceleration vector that it produces via the coefficient of proportionality m – the inertial mass of the body,[3] and *the law of universal gravitation*, which makes it possible to find the gravitational action of the bodies m and M on each other according to the formula

$$\mathbf{F} = G\frac{mM}{|\mathbf{r}|^3}\mathbf{r}, \tag{5.2}$$

[1] I. Newton (1642–1727) – British physicist, astronomer, and mathematician, an outstanding scholar, who stated the basic laws of classical mechanics, discovered the law of universal gravitation, and developed (along with Leibniz) the foundations of differential and integral calculus. He was appreciated even by his contemporaries, who inscribed on his tombstone: "Hic depositum est, quod mortale fuit Isaaci Newtoni" (here lies what was mortal of Isaac Newton).

[2] J. Kepler (1571–1630) – famous German astronomer who discovered the laws of motion of the planets (Kepler's laws).

[3] We have denoted the mass by the same symbol we used for the body itself, but this will not lead to any confusion. We remark also that if $m \ll M$, the coordinate system chosen can be considered inertial.

V.A. Zorich, *Mathematical Analysis I*, Universitext,
DOI 10.1007/978-3-662-48792-1_5

Fig. 5.1

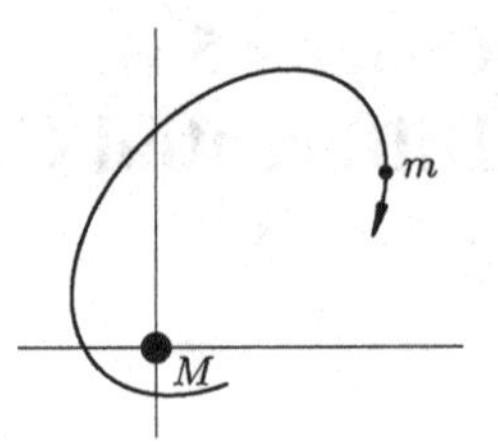

where $\mathbf{r}$ is a vector with its initial point in the body to which the force is applied and its terminal point in the other body and $|\mathbf{r}|$ is the length of the vector $\mathbf{r}$, that is, the distance between m and M.

Knowing the masses m and M, we can easily use Eq. (5.2) to express the right-hand side of Eq. (5.1) in terms of the coordinates $x(t)$ and $y(t)$ of the body m at time t, and thereby take account of all the data for the given motion.

To obtain the relations on $x(t)$ and $y(t)$ contained in Eq. (5.1), we must learn how to express the left-hand side of Eq. (5.1) in terms of $x(t)$ and $y(t)$.

Acceleration is a characteristic of a change in velocity $\mathbf{v}(t)$. More precisely, it is simply the rate at which the velocity changes. Therefore, to solve the problem we must first of all learn how to compute the velocity $\mathbf{v}(t)$ at time t possessed by a body whose motion is described by the radius-vector $\mathbf{r}(t) = (x(t), y(t))$.

Thus we wish to define and learn how to compute the instantaneous velocity of a body that is implicit in the law of motion (5.1).

To measure a thing is to compare it to a standard. In the present case, what can serve as a standard for determining the instantaneous velocity of motion?

The simplest kind of motion is that of a free body moving under inertia. This is a motion under which equal displacements of the body in space (as vectors) occur in equal intervals of time. It is the so-called uniform (rectilinear) motion. If a point is moving uniformly, and $\mathbf{r}(0)$ and $\mathbf{r}(1)$ are its radius-vectors relative to an inertial coordinate system at times $t = 0$ and $t = 1$ respectively, then at any time t we shall have

$$\mathbf{r}(t) - \mathbf{r}(0) = \mathbf{v} \cdot t, \tag{5.3}$$

where $\mathbf{v} = \mathbf{r}(1) - \mathbf{r}(0)$. Thus the displacement $\mathbf{r}(t) - \mathbf{r}(0)$ turns out to be a *linear function of time* in this simplest case, where the role of the constant of proportionality between the displacement $\mathbf{r}(t) - \mathbf{r}(0)$ and the time t is played by the vector $\mathbf{v}$ that is the displacement in unit time. It is this vector that we call the velocity of uniform motion. The fact that the motion is rectilinear can be seen from the parametric representation of the trajectory: $\mathbf{r}(t) = \mathbf{r}(0) + \mathbf{v} \cdot t$, which is the equation of a straight line, as you will recall from analytic geometry.

We thus know the velocity $\mathbf{v}$ of uniform rectilinear motion given by Eq. (5.3). By the law of inertia, if no external forces are acting on a body, it moves uniformly in a straight line. Hence if the action of M on m were to cease at time t, the latter would continue its motion, in a straight line at a certain velocity from that time on. It is natural to regard that velocity as the instantaneous velocity of the body at time t.

Fig. 5.2

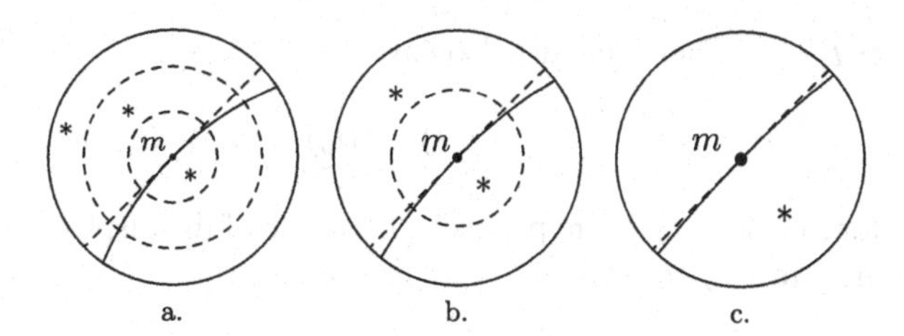

However, such a definition of instantaneous velocity would remain a pure abstraction, giving us no guidance for explicit computation of the quantity, if not for the circumstance of primary importance that we are about to discuss.

While remaining within the circle we have entered (logicians would call it a "vicious" circle) when we wrote down the equation of motion (5.1) and then undertook to determine what is meant by instantaneous velocity and acceleration, we nevertheless remark that, even with the most general ideas about these concepts, one can draw the following heuristic conclusions from Eq. (5.1). If there is no force, that is, $\mathbf{F} \equiv \mathbf{0}$, then the acceleration is also zero. But if the rate of change $\mathbf{a}(t)$ of the velocity $\mathbf{v}(t)$ is zero, then the velocity $\mathbf{v}(t)$ itself must not vary over time. In that way, we arrive at the law of inertia, according to which the body indeed moves in space with a velocity that is constant in time.

From this same Eq. (5.1) we can see that forces of bounded magnitude are capable of creating only accelerations of bounded magnitude. But if the absolute magnitude of the rate of change of a quantity $P(t)$ over a time interval $[0,t]$ does not exceed some constant c, then, in our picture of the situation, the change $|P(t) - P(0)|$ in the quantity P over time t cannot exceed $c \cdot t$, that is, in this situation, the quantity changes by very little in a small interval of time. (In any case, the function $P(t)$ turns out to be continuous.) Thus, in a real mechanical system the parameters change by small amounts over a small time interval.

In particular, *at all times t close to some time t_0 the velocity $\mathbf{v}(t)$ of the body m must be close to the value* $\mathbf{v}(t_0)$ that we wish to determine. But in that case, *in a small neighborhood of the time t_0 the motion itself must differ by only a small amount from uniform motion at velocity* $\mathbf{v}(t_0)$, and the closer to t_0, the less it differs.

If we photographed the trajectory of the body m through a telescope, depending on the power of the telescope, we would see approximately what is shown in Fig. 5.2.

The portion of the trajectory shown in Fig. 5.2c corresponds to a time interval so small that it is difficult to distinguish the actual trajectory from a straight line, since this portion of the trajectory really does resemble a straight line, and the motion resembles uniform rectilinear motion. From this observation, as it happens, we can conclude that by solving the problem of determining the instantaneous velocity (velocity being a vector quantity) we will at the same time solve the purely geometric problem of defining and finding the tangent to a curve (in the present case the curve is the trajectory of motion).

Thus we have observed that in this problem we must have $\mathbf{v}(t) \approx \mathbf{v}(t_0)$ for t close to t_0, that is, $\mathbf{v}(t) \to \mathbf{v}(t_0)$ as $t \to t_0$, or, what is the same, $\mathbf{v}(t) = \mathbf{v}(t_0) + o(1)$

as $t \to t_0$. Then we must also have

$$\mathbf{r}(t) - \mathbf{r}(t_0) \approx \mathbf{v}(t_0) \cdot (t - t_0)$$

for t close to t_0. More precisely, the value of the displacement $\mathbf{r}(t) - \mathbf{r}(t_0)$ is equivalent to $\mathbf{v}(t_0)(t - t_0)$ as $t \to t_0$, or

$$\mathbf{r}(t) - \mathbf{r}(t_0) = \mathbf{v}(t_0)(t - t_0) + o\big(\mathbf{v}(t_0)(t - t_0)\big), \tag{5.4}$$

where $o(\mathbf{v}(t_0)(t - t_0))$ is a correction vector whose magnitude tends to zero faster than the magnitude of the vector $\mathbf{v}(t_0)(t - t_0)$ as $t \to t_0$. Here, naturally, we must except the case when $\mathbf{v}(t_0) = \mathbf{0}$. So as not to exclude this case from consideration in general, it is useful to observe that[4] $|\mathbf{v}(t_0)(t - t_0)| = |\mathbf{v}(t_0)||t - t_0|$. Thus, if $|\mathbf{v}(t_0)| \neq 0$, then the quantity $|\mathbf{v}(t_0)(t - t_0)|$ is of the same order as $|t - t_0|$, and therefore $o(\mathbf{v}(t_0)(t - t_0)) = o(t - t_0)$. Hence, instead of (5.4) we can write the relation

$$\mathbf{r}(t) - \mathbf{r}(t_0) = \mathbf{v}(t_0)(t - t_0) + o(t - t_0), \tag{5.5}$$

which does not exclude the case $\mathbf{v}(t_0) = \mathbf{0}$.

Thus, starting from the most general, and perhaps vague ideas about velocity, we have arrived at Eq. (5.5), which the velocity must satisfy. But the quantity $\mathbf{v}(t_0)$ can be found unambiguously from Eq. (5.5):

$$\mathbf{v}(t_0) = \lim_{t \to t_0} \frac{\mathbf{r}(t) - \mathbf{r}(t_0)}{t - t_0}. \tag{5.6}$$

Therefore both the fundamental relation (5.5) and the relation (5.6) equivalent to it can now be taken as the definition of the quantity $\mathbf{v}(t_0)$, the instantaneous velocity of the body at time t_0.

At this point we shall not allow ourselves to be distracted into a detailed discussion of the problem of the limit of a vector-valued function. Instead, we shall confine ourselves to reducing it to the case of the limit of a real-valued function, which has already been discussed in complete detail. Since the vector $\mathbf{r}(t) - \mathbf{r}(t_0)$ has coordinates $(x(t) - x(t_0), y(t) - y(t_0))$, we have $\frac{\mathbf{r}(t) - \mathbf{r}(t_0)}{t - t_0} = (\frac{x(t) - x(t_0)}{t - t_0}, \frac{y(t) - y(t_0)}{t - t_0})$ and hence, if we regard vectors as being close together if their coordinates are close together, the limit in (5.6) should be interpreted as follows:

$$\mathbf{v}(t_0) = \lim_{t \to t_0} \frac{\mathbf{r}(t) - \mathbf{r}(t_0)}{t - t_0} = \left(\lim_{t \to t_0} \frac{x(t) - x(t_0)}{t - t_0}, \lim_{t \to t_0} \frac{y(t) - y(t_0)}{t - t_0} \right),$$

and the term $o(t - t_0)$ in (5.5) should be interpreted as a vector depending on t such that the vector $\frac{o(t - t_0)}{t - t_0}$ tends (coordinatewise) to zero as $t \to t_0$.

[4]Here $|t - t_0|$ is the absolute value of the number $t - t_0$ while $|\mathbf{v}|$ is the absolute value, or length of the vector $\mathbf{v}$.

Finally, we remark that if $\mathbf{v}(t_0) \neq \mathbf{0}$, then the equation

$$\mathbf{r} - \mathbf{r}(t_0) = \mathbf{v}(t_0) \cdot (t - t_0) \tag{5.7}$$

defines a line, which by the circumstances indicated above should be regarded as the *tangent* to the trajectory at the point $(x(t_0), y(t_0))$.

Thus, *the standard for defining the velocity of a motion is the velocity of uniform rectilinear motion defined by the linear relation* (5.7). *The standard motion* (5.7) *is connected with the motion being studied as shown by relation* (5.5). *The value* $\mathbf{v}(t_0)$ *at which* (5.5) *holds can be found by passing to the limit in* (5.6) *and is called the velocity of motion at time* t_0. The motions studied in classical mechanics, which are described by the law (5.1), must admit comparison with this standard, that is, they must admit of the linear approximation indicated in (5.5).

If $\mathbf{r}(t) = (x(t), y(t))$ is the radius-vector of a moving point m at time t, then $\dot{\mathbf{r}}(t) = (\dot{x}(t), \dot{y}(t)) = \mathbf{v}(t)$ is the vector that gives the rate of change of $\mathbf{r}(t)$ at time t, and $\ddot{\mathbf{r}}(t) = (\ddot{x}(t), \ddot{y}(t)) = \mathbf{a}(t)$ is the vector that gives the rate of change of $\mathbf{v}(t)$ (acceleration) at time t, then Eq. (5.1) can be written in the form

$$m \cdot \ddot{\mathbf{r}}(t) = \mathbf{F}(t),$$

from which we obtain in coordinate form for motion in a gravitational field

$$\begin{cases} \ddot{x}(t) = -GM\frac{x(t)}{[x^2(t)+y^2(t)]^{3/2}}, \\ \ddot{y}(t) = -GM\frac{y(t)}{[x^2(t)+y^2(t)]^{3/2}}. \end{cases} \tag{5.8}$$

This is a precise mathematical expression of our original problem. Since we know how to find $\dot{\mathbf{r}}(t)$ from $\mathbf{r}(t)$ and then how to find $\ddot{\mathbf{r}}(t)$, we are already in a position to answer the question whether a pair of functions $(x(t), y(t))$ can describe the motion of the body m about the body M. To answer this question, one must find $\ddot{x}(t)$ and $\ddot{y}(t)$ and check whether Eqs. (5.8) hold. The system (5.8) is an example of a system of so-called *differential equations*. At this point we can only check whether a set of functions is a solution of the system. How to find the solution or, better expressed, how to investigate the properties of solutions of differential equations, is studied in a special and, as one can now appreciate, critical area of analysis – the theory of differential equations.

The operation of finding the rate of change of a vector quantity, as has been shown, reduces to finding the rates of change of several numerical-valued functions – the coordinates of the vector. Thus we must first of all learn how to carry out this operation easily in the simplest case of real-valued functions of a real-valued argument, which we now take up.

5.1.2 Functions Differentiable at a Point

We begin with two preliminary definitions that we shall shortly make precise.

Definition 0_1 A function $f : E \to \mathbb{R}$ defined on a set $E \subset \mathbb{R}$ is *differentiable* at a point $a \in E$ that is a limit point of E if there exists a linear function $A \cdot (x - a)$ of the increment $x - a$ of the argument such that $f(x) - f(a)$ can be represented as

$$f(x) - f(a) = A \cdot (x - a) + o(x - a) \quad \text{as } x \to a, x \in E. \tag{5.9}$$

In other words, a function is differentiable at a point a if the change in its values in a neighborhood of the point in question is linear up to a correction that is infinitesimal compared with the magnitude of the displacement $x - a$ from the point a.

Remark As a rule we have to deal with functions defined in an entire neighborhood of the point in question, not merely on a subset of the neighborhood.

Definition 0_2 The linear function $A \cdot (x - a)$ in Eq. (5.9) is called the *differential* of the function f at a.

The differential of a function at a point is uniquely determined; for it follows from (5.9) that

$$\lim_{E \ni x \to a} \frac{f(x) - f(a)}{x - a} = \lim_{E \ni x \to a} \left(A + \frac{o(x - a)}{x - a} \right) = A,$$

so that the number A is unambiguously determined due to the uniqueness of the limit.

Definition 1 The number

$$f'(a) = \lim_{E \ni x \to a} \frac{f(x) - f(a)}{x - a} \tag{5.10}$$

is called the *derivative* of the function f at a.

Relation (5.10) can be rewritten in the equivalent form

$$\frac{f(x) - f(a)}{x - a} = f'(a) + \alpha(x),$$

where $\alpha(x) \to 0$ as $x \to a, x \in E$, which in turn is equivalent to

$$f(x) - f(a) = f'(a)(x - a) + o(x - a) \quad \text{as } x \to a, x \in E. \tag{5.11}$$

Thus, differentiability of a function at a point is equivalent to the existence of its derivative at the same point.

If we compare these definitions with what was said in Sect. 5.1.1, we can conclude that the derivative characterizes the rate of change of a function at the point under consideration, while the differential provides the best linear approximation to the increment of the function in a neighborhood of the same point.

If a function $f : E \to \mathbb{R}$ is differentiable at different points of the set E, then in passing from one point to another both the quantity A and the function $o(x - a)$ in

Eq. (5.9) may change (a result at which we have already arrived explicitly in (5.11)). This circumstance should be noted in the very definition of a differentiable function, and we now write out this fundamental definition in full.

Definition 2 A function $f : E \to \mathbb{R}$ defined on a set $E \subset \mathbb{R}$ is *differentiable* at a point $x \in E$ that is a limit point of E if

$$\boxed{f(x+h) - f(x) = A(x)h + \alpha(x; h),} \tag{5.12}$$

where $h \mapsto A(x)h$ is a linear function in h and $\alpha(x; h) = o(h)$ as $h \to 0$, $x + h \in E$.

The quantities

$$\Delta x(h) := (x+h) - x = h$$

and

$$\Delta f(x; h) := f(x+h) - f(x)$$

are called respectively the *increment of the argument* and the *increment of the function* (corresponding to this increment in the argument).

They are often denoted (not quite legitimately, to be sure) by the symbols Δx and $\Delta f(x)$ representing functions of h.

Thus, a function is differentiable at a point if its increment at that point, regarded as a function of the increment h in its argument, is linear up to a correction that is infinitesimal compared to h as $h \to 0$.

Definition 3 The function $h \mapsto A(x)h$ of Definition 2, which is linear in h, is called the *differential of the function* $f : E \to \mathbb{R}$ at the point $x \in E$ and is denoted $\mathrm{d}f(x)$ or $Df(x)$.

Thus, $\mathrm{d}f(x)(h) = A(x)h$.

From Definitions 2 and 3 we have

$$\Delta f(x; h) - \mathrm{d}f(x)(h) = \alpha(x; h),$$

and $\alpha(x; h) = o(h)$ as $h \to 0$, $x + h \in E$; that is, the difference between the increment of the function due to the increment h in its argument and the value of the function $\mathrm{d}f(x)$, which is linear in h, at the same h, is an infinitesimal of higher order than the first in h.

For that reason, we say that the differential is the (*principal*) *linear part of the increment of the function*.

As follows from relation (5.12) and Definition 1,

$$A(x) = f'(x) = \lim_{\substack{h \to 0 \\ x+h, x \in E}} \frac{f(x+h) - f(x)}{h},$$

and so the differential can be written as

$$df(x)(h) = f'(x)h. \tag{5.13}$$

In particular, if $f(x) \equiv x$, we obviously have $f'(x) \equiv 1$ and

$$dx(h) = 1 \cdot h = h,$$

so that it is sometimes said that "the differential of an independent variable equals its increment".

Taking this equality into account, we deduce from (5.13) that

$$df(x)(h) = f'(x)\,dx(h), \tag{5.14}$$

that is,

$$df(x) = f'(x)\,dx. \tag{5.15}$$

The equality (5.15) should be understood as the equality of two functions of h.

From (5.14) we obtain

$$\frac{df(x)(h)}{dx(h)} = f'(x), \tag{5.16}$$

that is, the function $\frac{df(x)}{dx}$ (the ratio of the functions $df(x)$ and dx) is constant and equals $f'(x)$. For this reason, following Leibniz, we frequently denote the derivative by the symbol $\frac{df(x)}{dx}$, alongside the notation $f'(x)$ proposed by Lagrange.[5]

In mechanics, in addition to these symbols, the symbol $\dot{\varphi}(t)$ (read "phi-dot of t") is also used to denote the derivative of the function $\varphi(t)$ with respect to time t.

5.1.3 The Tangent Line; Geometric Meaning of the Derivative and Differential

Let $f : E \to \mathbb{R}$ be a function defined on a set $E \subset \mathbb{R}$ and x_0 a given limit point of E. We wish to choose the constant c_0 so as to give the best possible description of the behavior of the function in a neighborhood of the point x_0 among constant functions. More precisely, we want the difference $f(x) - c_0$ to be infinitesimal compared with any nonzero constant as $x \to x_0$, $x \in E$, that is

$$f(x) = c_0 + o(1) \quad \text{as } x \to x_0, x \in E. \tag{5.17}$$

This last relation is equivalent to saying $\lim_{E \ni x \to x_0} f(x) = c_0$. If, in particular, the function is continuous at x_0, then $\lim_{E \ni x \to x_0} f(x) = f(x_0)$, and naturally $c_0 = f(x_0)$.

---

[5] J.L. Lagrange (1736–1831) – famous French mathematician and specialist in theoretical mechanics.

Now let us try to choose the function $c_0 + c_1(x - x_0)$ so as to have

$$f(x) = c_0 + c_1(x - x_0) + o(x - x_0) \quad \text{as } x \to x_0, x \in E. \tag{5.18}$$

This is obviously a generalization of the preceding problem, since the formula (5.17) can be rewritten as

$$f(x) = c_0 + o\big((x - x_0)^0\big) \quad \text{as } x \to x_0, x \in E.$$

It follows immediately from (5.18) that $c_0 = \lim_{E \ni x \to x_0} f(x)$, and if the function is continuous at this point, then $c_0 = f(x_0)$.

If c_0 has been found, it then follows from (5.18) that

$$c_1 = \lim_{E \ni x \to x_0} \frac{f(x) - c_0}{x - x_0}.$$

And, in general, if we were seeking a polynomial $P_n(x_0; x) = c_0 + c_1(x - x_0) + \cdots + c_n(x - x_0)^n$ such that

$$f(x) = c_0 + c_1(x - x_0) + \cdots + c_n(x - x_0)^n + o\big((x - x_0)^n\big) \quad \text{as } x \to x_0, x \in E, \tag{5.19}$$

we would find successively, with no ambiguity, that

$$\begin{aligned}
c_0 &= \lim_{E \ni x \to x_0} f(x), \\
c_1 &= \lim_{E \ni x \to x_0} \frac{f(x) - c_0}{x - x_0}, \\
&\vdots \\
c_n &= \lim_{E \ni x \to x_0} \frac{f(x) - [c_0 + \cdots + c_{n-1}(x - x_0)^{n-1}]}{(x - x_0)^n},
\end{aligned}$$

assuming that all these limits exist. Otherwise condition (5.19) cannot be fulfilled, and the problem has no solution.

If the function f is continuous at x_0, it follows from (5.18), as already pointed out, that $c_0 = f(x_0)$, and we then arrive at the relation

$$f(x) - f(x_0) = c_1(x - x_0) + o(x - x_0) \quad \text{as } x \to x_0, x \in E,$$

which is equivalent to the condition that $f(x)$ be differentiable at x_0.

From this we find

$$c_1 = \lim_{E \ni x \to x_0} \frac{f(x) - f(x_0)}{x - x_0} = f'(x_0).$$

We have thus proved the following proposition.

Proposition 1 *A function $f : E \to \mathbb{R}$ that is continuous at a point $x_0 \in E$ that is a limit point of $E \subset \mathbb{R}$ admits a linear approximation* (5.18) *if and only if it is differentiable at the point.*

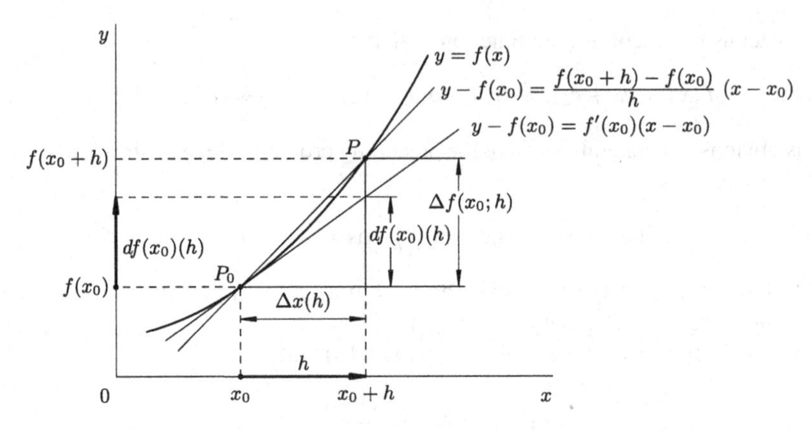

Fig. 5.3

The function

$$\varphi(x) = c_0 + c_1(x - x_0) \tag{5.20}$$

with $c_0 = f(x_0)$ and $c_1 = f'(x_0)$ is the only function of the form (5.20) that satisfies (5.18).

Thus the function

$$\varphi(x) = f(x_0) + f'(x_0)(x - x_0) \tag{5.21}$$

provides the best linear approximation to the function f in a neighborhood of x_0 in the sense that for any other function $\varphi(x)$ of the form (5.20) we have $f(x) - \varphi(x) \neq o(x - x_0)$ as $x \to x_0$, $x \in E$.

The graph of the function (5.21) is the straight line

$$y - f(x_0) = f'(x_0)(x - x_0), \tag{5.22}$$

passing through the point $(x_0, f(x_0))$ and having slope $f'(x_0)$.

Since the line (5.22) provides the optimal linear approximation of the graph of the function $y = f(x)$ in a neighborhood of the point $(x_0, f(x_0))$, it is natural to make the following definition.

Definition 4 If a function $f : E \to \mathbb{R}$ is defined on a set $E \subset \mathbb{R}$ and differentiable at a point $x_0 \in E$, the line defined by Eq. (5.22) is called the *tangent* to the graph of this function at the point $(x_0, f(x_0))$.

Figure 5.3 illustrates all the basic concepts we have so far introduced in connection with differentiability of a function at a point: the increment of the argument, the increment of the function corresponding to it, and the value of the differential. The figure shows the graph of the function, the tangent to the graph at the point $P_0 = (x_0, f(x_0))$, and for comparison, an arbitrary line (usually called a *secant*) passing through P_0 and some point $P \neq P_0$ of the graph of the function.

The following definition extends Definition 4.

Definition 5 If the mappings $f : E \to \mathbb{R}$ and $g : E \to \mathbb{R}$ are continuous at a point $x_0 \in E$ that is a limit point of E and $f(x) - g(x) = o((x - x_0)^n)$ as $x \to x_0, x \in E$, we say that f and g have *nth order contact* at x_0 (more precisely, contact *of order at least n*).

For $n = 1$ we say that the mappings f and g are *tangent* to each other at x_0.

According to Definition 5 the mapping (5.21) is tangent at x_0 to a mapping $f : E \to \mathbb{R}$ that is differentiable at that point.

We can now also say that the polynomial $P_n(x_0; x) = c_0 + c_1(x - x_0) + \cdots + c_n(x - x_0)^n$ of relation (5.19) has contact of order at least n with the function f.

The number $h = x - x_0$, that is, the increment of the argument, can be regarded as a vector attached to the point x_0 and defining the transition from x_0 to $x = x_0 + h$. We denote the set of all such vectors by $T\mathbb{R}(x_0)$ or $T\mathbb{R}_{x_0}$.[6] Similarly, we denote by $T\mathbb{R}(y_0)$ or $T\mathbb{R}_{y_0}$ the set of all displacement vectors from the point y_0 along the y-axis (see Fig. 5.3). It can then be seen from the definition of the differential that the mapping

$$\mathrm{d}f(x_0) : T\mathbb{R}(x_0) \to T\mathbb{R}\big(f(x_0)\big), \tag{5.23}$$

defined by the differential $h \mapsto f'(x_0)h = \mathrm{d}f(x_0)(h)$ is tangent to the mapping

$$h \mapsto f(x_0 + h) - f(x_0) = \Delta f(x_0; h), \tag{5.24}$$

defined by the increment of a differentiable function.

We remark (see Fig. 5.3) that if the mapping (5.24) is the increment of the ordinate of the graph of the function $y = f(x)$ as the argument passes from x_0 to $x_0 + h$, then the differential (5.23) gives the increment in the ordinate of the tangent to the graph of the function for the same increment h in the argument.

5.1.4 The Role of the Coordinate System

The analytic definition of a tangent (Definition 4) may be the cause of some vague uneasiness. We shall try to state what it is exactly that makes one uneasy. However, we shall first point out a more geometric construction of the tangent to a curve at one of its points P_0 (see Fig. 5.3).

Take an arbitrary point P of the curve different from P_0. The line determined by the pair of points P_0 and P, as already noted, is called a secant in relation to the curve. We now force the point P to approach P_0 along the curve. If the secant tends to some limiting position as we do so, that limiting position of the secant is the tangent to the curve at P_0.

Despite its intuitive nature, such a definition of the tangent is not available to us at the moment, since we do not know what a curve is, what it means to say that

---

[6]This is a slight deviation from the more common notation $T_{x_0}\mathbb{R}$ or $T_{x_0}(\mathbb{R})$.

"a point tends to another point along a curve", and finally, in what sense we are to interpret the phrase "limiting position of the secant".

Rather than make all these concepts precise, we point out a fundamental difference between the two definitions of tangent that we have introduced. The second was purely geometric, unconnected (at least until it is made more precise) with any coordinate system. In the first case, however, we have defined the tangent to a curve that is the graph of a differentiable function in some coordinate system. The question naturally arises whether, if the curve is written in a different coordinate system, it might not cease to be differentiable, or might be differentiable but yield a different line as tangent when the computations are carried out in the new coordinates.

This question of invariance, that is, independence of the coordinate system, always arises when a concept is introduced using a coordinate system. The question applies in equal measure to the concept of velocity, which we discussed in Sect. 5.1.1 and which, as we have mentioned already, includes the concept of a tangent.

Points, vectors, lines, and so forth have different numerical characteristics in different coordinate systems (coordinates of a point, coordinates of a vector, equation of a line). However, knowing the formulas that connect two coordinate systems, one can always determine from two numerical representations of the same type whether or not they are expressions for the same geometric object in different coordinate systems. Intuition suggests that the procedure for defining velocity described in Sect. 5.1.1 leads to the same vector independently of the coordinate system in which the computations are carried out. At the appropriate time in the study of functions of several variables we shall give a detailed discussion of questions of this sort. The invariance of the definition of velocity with respect to different coordinate systems will be verified in the next section.

Before passing to the study of specific examples, we now summarize some of the results.

We have encountered the problem of the describing mathematically the instantaneous velocity of a moving body.

This problem led us to the problem of approximating a given function in the neighborhood of a given point by a linear function, which on the geometric level led to the concept of the *tangent*. Functions describing the motion of a real mechanical system are assumed to admit such a linear approximation.

In this way we have distinguished the class of *differentiable functions* in the class of all functions.

The concept of the *differential* of a function at a point has been introduced. The differential is a linear mapping defined on displacements from the point under consideration that describes the behavior of the increment of a differentiable function in a neighborhood of the point, up to a quantity that is infinitesimal in comparison with the displacement.

The differential $\mathrm{d}f(x_0)h = f'(x_0)h$ is completely determined by the number $f'(x_0)$, the *derivative* of the function f at x_0, which can be found by taking the limit

$$f'(x_0) = \lim_{E \ni x \to x_0} \frac{f(x) - f(x_0)}{x - x_0}.$$

The physical meaning of the derivative is the *rate* of change of the quantity $f(x)$ at time x_0; its geometrical meaning is the *slope of the tangent* to the graph of the function $y = f(x)$ at the point $(x_0, f(x_0))$.

5.1.5 Some Examples

Example 1 Let $f(x) = \sin x$. We shall show that $f'(x) = \cos x$.

Proof

$$\lim_{h \to 0} \frac{\sin(x+h) - \sin x}{h} = \lim_{h \to 0} \frac{2\sin(\frac{h}{2})\cos(x + \frac{h}{2})}{h} =$$
$$= \lim_{h \to 0} \cos\left(x + \frac{h}{2}\right) \cdot \lim_{h \to 0} \frac{\sin(\frac{h}{2})}{(\frac{h}{2})} = \cos x.$$

Here we have used the theorem on the limit of a product, the continuity of the function $\cos x$, the equivalence $\sin t \sim t$ as $t \to 0$, and the theorem on the limit of a composite function. □

Example 2 We shall show that $\cos' x = -\sin x$.

Proof

$$\lim_{h \to 0} \frac{\cos(x+h) - \cos x}{h} = \lim_{h \to 0} \frac{-2\sin(\frac{h}{2})\sin(x + \frac{h}{2})}{h} =$$
$$= -\lim_{h \to 0} \sin\left(x + \frac{h}{2}\right) \cdot \lim_{h \to 0} \frac{\sin(\frac{h}{2})}{(\frac{h}{2})} = -\sin x.$$

□

Example 3 We shall show that if $f(t) = r\cos\omega t$, then $f'(t) = -r\omega\sin\omega t$.

Proof

$$\lim_{h \to 0} \frac{r\cos\omega(t+h) - r\cos\omega t}{h} = r\lim_{h \to 0} \frac{-2\sin(\frac{\omega h}{2})\sin\omega(t + \frac{h}{2})}{h} =$$
$$= -r\omega \lim_{h \to 0} \sin\omega(t + \frac{h}{2}) \cdot \lim_{h \to 0} \frac{\sin(\frac{\omega h}{2})}{(\frac{\omega h}{2})} =$$
$$= -r\omega\sin\omega t.$$

□

Example 4 If $f(t) = r\sin\omega t$, then $f'(t) = r\omega\cos\omega t$.

Proof The proof is analogous to that of Examples 1 and 3. □

Example 5 (The instantaneous velocity and instantaneous acceleration of a point mass) Suppose a point mass is moving in a plane and that in some given coordinate system its motion is described by differentiable functions of time

$$x = x(t), \qquad y = y(t)$$

or, what is the same, by a vector

$$\mathbf{r}(t) = \big(x(t), y(t)\big).$$

As we have explained in Sect. 5.1.1, the velocity of the point at time t is the vector

$$\mathbf{v}(t) = \dot{\mathbf{r}}(t) = \big(\dot{x}(t), \dot{y}(t)\big),$$

where $\dot{x}(t)$ and $\dot{y}(t)$ are the derivatives of $x(t)$ and $y(t)$ with respect to time t.

The acceleration $\mathbf{a}(t)$ is the rate of change of the vector $\mathbf{v}(t)$, so that

$$\mathbf{a}(t) = \dot{\mathbf{v}}(t) = \ddot{\mathbf{r}}(t) = \big(\ddot{x}(t), \ddot{y}(t)\big),$$

where $\ddot{x}(t)$ and $\ddot{y}(t)$ are the derivatives of the functions $\dot{x}(t)$ and $\dot{y}(t)$ with respect to time, the so-called second derivatives of $x(t)$ and $y(t)$.

Thus, in the sense of the physical problem, functions $x(t)$ and $y(t)$ that describe the motion of a point mass must have both first and second derivatives.

In particular, let us consider the uniform motion of a point along a circle of radius r. Let ω be the angular velocity of the point, that is, the magnitude of the central angle over which the point moves in unit time.

In Cartesian coordinates (by the definitions of the functions $\cos x$ and $\sin x$) this motion is written in the form

$$\mathbf{r}(t) = \big(r\cos(\omega t + \alpha), r\sin(\omega t + \alpha)\big),$$

and if $\mathbf{r}(0) = (r, 0)$, it assumes the form

$$\mathbf{r}(t) = (r\cos\omega t, r\sin\omega t).$$

Without loss of generality in our subsequent deductions, for the sake of brevity, we shall assume that $\mathbf{r}(0) = (r, 0)$.

Then by the results of Examples 3 and 4 we have

$$\mathbf{v}(t) = \dot{\mathbf{r}}(t) = (-r\omega\sin\omega t, r\omega\cos\omega t).$$

From the computation of the inner product

$$\big\langle \mathbf{v}(t),\ \mathbf{r}(t)\big\rangle = -r^2\omega\sin\omega t\cos\omega t + r^2\omega\cos\omega t\sin\omega t = 0,$$

as one should expect in this case, we find that the velocity vector $\mathbf{v}(t)$ is orthogonal to the radius-vector $\mathbf{r}(t)$ and is therefore directed along the tangent to the circle.

Fig. 5.4

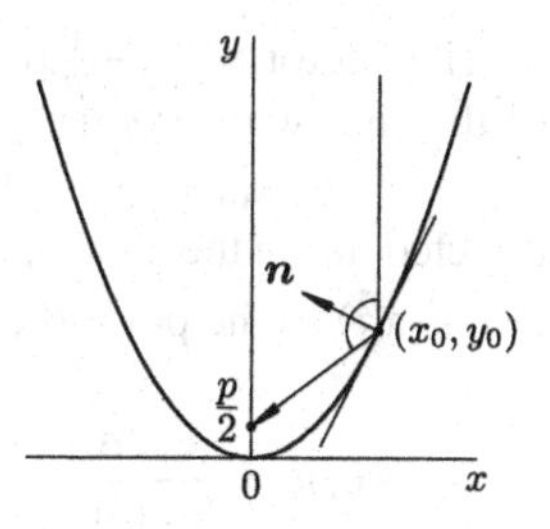

Next, for the acceleration, we have

$$\mathbf{a}(t) = \dot{\mathbf{v}}(t) = \ddot{\mathbf{r}}(t) = \big(-r\omega^2 \cos\omega t, -r\omega^2 \sin\omega t\big),$$

that is, $\mathbf{a}(t) = -\omega^2 \mathbf{r}(t)$, and the acceleration is thus indeed centripetal, since it has the direction opposite to that of the radius-vector $\mathbf{r}(t)$.

Moreover,

$$\big|\mathbf{a}(t)\big| = \omega^2 \big|\mathbf{r}(t)\big| = \omega^2 r = \frac{|\mathbf{v}(t)|^2}{r} = \frac{v^2}{r},$$

where $v = |\mathbf{v}(t)|$.

Starting from these formulas, let us compute, for example, the speed of a low-altitude satellite of the Earth. In this case r equals the radius of the earth, that is, $r = 6400$ km, while $|\mathbf{a}(t)| = g$, where $g \approx 10\ \mathrm{m/s}^2$ is the acceleration of free fall at the surface of the earth.

Thus, $v^2 = |\mathbf{a}(t)|r \approx 10\ \mathrm{m/s}^2 \times 64 \times 10^5\ \mathrm{m} = 64 \times 10^6\ (\mathrm{m/s})^2$, and so $v \approx 8 \times 10^3\ \mathrm{m/s}$.

Example 6 (The optic property of a parabolic mirror) Let us consider the parabola $y = \frac{1}{2p}x^2$ ($p > 0$, see Fig. 5.4), and construct the tangent to it at the point $(x_0, y_0) = (x_0, \frac{1}{2p}x_0^2)$.

Since $f(x) = \frac{1}{2p}x^2$, we have

$$f'(x_0) = \lim_{x \to x_0} \frac{\frac{1}{2p}x^2 - \frac{1}{2p}x_0^2}{x - x_0} = \frac{1}{2p} \lim_{x \to x_0}(x + x_0) = \frac{1}{p}x_0.$$

Hence the required tangent has the equation

$$y - \frac{1}{2p}x_0^2 = \frac{1}{p}x_0(x - x_0)$$

or

$$\frac{1}{p}x_0(x - x_0) - (y - y_0) = 0, \tag{5.25}$$

where $y_0 = \frac{1}{2p}x_0^2$.

The vector $\mathbf{n} = (-\frac{1}{p}x_0, 1)$, as can be seen from this last equation, is orthogonal to the line whose equation is (5.25). We shall show that the vectors $\mathbf{e}_y = (0, 1)$ and $\mathbf{e}_f = (-x_0, \frac{p}{2} - y_0)$ form equal angles with $\mathbf{n}$. The vector $\mathbf{e}_y$ is a unit vector directed along the y-axis, while $\mathbf{e}_f$ is directed from the point of tangency $(x_0, y_0) = (x_0, \frac{1}{2p}x_0^2)$ to the point $(0, \frac{p}{2})$, which is the focus of the parabola. Thus

$$\cos\widehat{\mathbf{e}_y\mathbf{n}} = \frac{\langle \mathbf{e}_y, \mathbf{n}\rangle}{|\mathbf{e}_f||\mathbf{n}|} = \frac{1}{|\mathbf{n}|},$$

$$\cos\widehat{\mathbf{e}_f\mathbf{n}} = \frac{\langle \mathbf{e}_f, \mathbf{n}\rangle}{|\mathbf{e}_y||\mathbf{n}|} = \frac{\frac{1}{p}x_0^2 + \frac{p}{2} - \frac{1}{2p}x_0^2}{|\mathbf{n}|\sqrt{x_0^2 + (\frac{p}{2} - \frac{1}{2p}x_0^2)^2}} = \frac{\frac{p}{2} + \frac{1}{2p}x_0^2}{|\mathbf{n}|\sqrt{(\frac{p}{2} + \frac{1}{2p}x_0^2)^2}} = \frac{1}{|\mathbf{n}|}.$$

Thus we have shown that a wave source located at the point $(0, \frac{p}{2})$, the focus of the parabola, will emit a ray parallel to the axis of the mirror (the y-axis), and that a wave arriving parallel to the axis of the mirror will pass through the focus (see Fig. 5.4).

Example 7 With this example we shall show that the tangent is merely the best linear approximation to the graph of a function in a neighborhood of the point of tangency and does not necessarily have only one point in common with the curve, as was the case with a circle, or in general, with convex curves. (For convex curves we shall give a separate discussion.)

Let the function be given by

$$f(x) = \begin{cases} x^2 \sin\frac{1}{x}, & \text{if } x \neq 0, \\ 0 & \text{if } x = 0. \end{cases}$$

The graph of this function is shown by the thick line in Fig. 5.5.

Let us find the tangent to the graph at the point $(0, 0)$. Since

$$f'(0) = \lim_{x\to 0} \frac{x^2 \sin\frac{1}{x} - 0}{x - 0} = \lim_{x\to 0} x \sin\frac{1}{x} = 0,$$

the tangent has the equation $y - 0 = 0 \cdot (x - 0)$, or simply $y = 0$.

Thus, in this example the tangent is the x-axis, which the graph intersects infinitely many times in any neighborhood of the point of tangency.

By the definition of differentiability of a function $f : E \to \mathbb{R}$ at a point $x_0 \in E$, we have

$$f(x) - f(x_0) = A(x_0)(x - x_0) + o(x - x_0) \quad \text{as } x \to x_0, x \in E.$$

Since the right-hand side of this equality tends to zero as $x \to x_0$, $x \in E$, it follows that $\lim_{E\ni x\to x_0} f(x) = f(x_0)$, so that a function that is differentiable at a point is necessarily continuous at that point.

We shall show that the converse, of course, is not always true.

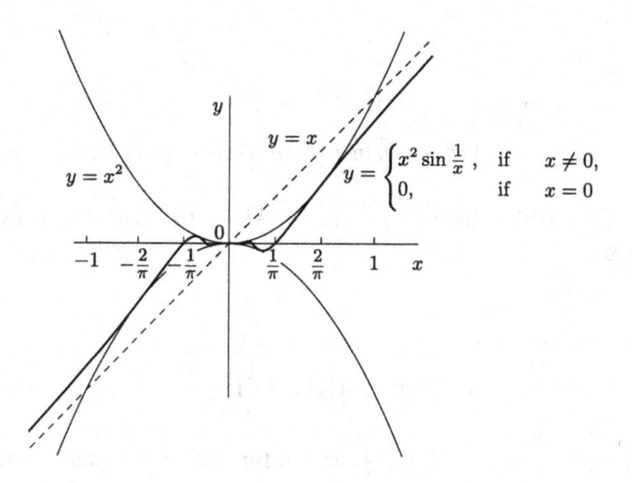

Fig. 5.5

Fig. 5.6

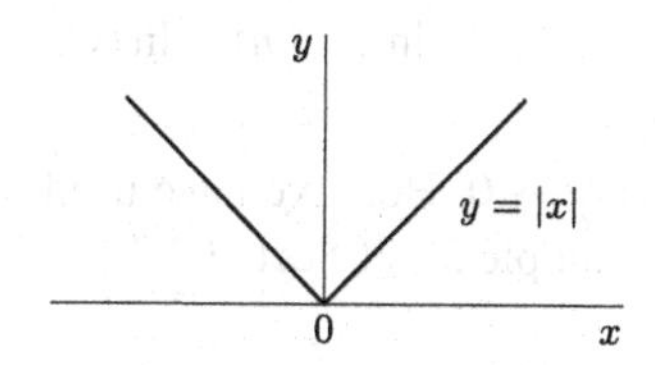

Example 8 Let $f(x) = |x|$, (Fig. 5.6). Then at the point $x_0 = 0$ we have

$$\lim_{x \to x_0 - 0} \frac{f(x) - f(x_0)}{x - x_0} = \lim_{x \to -0} \frac{|x| - 0}{x - 0} = \lim_{x \to -0} \frac{-x}{x} = -1,$$

$$\lim_{x \to x_0 + 0} \frac{f(x) - f(x_0)}{x - x_0} = \lim_{x \to +0} \frac{|x| - 0}{x - 0} = \lim_{x \to +0} \frac{x}{x} = 1.$$

Consequently, at this point the function has no derivative and hence is not differentiable at the point.

Example 9 We shall show that $e^{x+h} - e^x = e^x h + o(h)$ as $h \to 0$.

Thus, the function $\exp(x) = e^x$ is differentiable and $d\exp(x)h = \exp(x)h$, or $de^x = e^x\,dx$, and therefore $\exp' x = \exp x$, or $\frac{de^x}{dx} = e^x$.

Proof

$$e^{x+h} - e^x = e^x(e^h - 1) = e^x(h + o(h)) = e^x h + o(h).$$

Here we have used the formula $e^h - 1 = h + o(h)$ obtained in Example 39 of Sect. 3.2.4. □

Example 10 If $a > 0$, then $a^{x+h} - a^x = a^h(\ln a)h + o(h)$ as $h \to 0$. Thus $da^x = a^x(\ln a)\,dx$ and $\frac{da^x}{dx} = a^x \ln a$.

Proof

$$a^{x+h} - a^x = a^x\left(a^h - 1\right) = a^x\left(e^{h\ln a} - 1\right) =$$
$$= a^x\left(h\ln a + o(h\ln a)\right) = a^x(\ln a)h + o(h) \quad \text{as } h \to 0. \qquad \square$$

Example 11 If $x \neq 0$, then $\ln|x+h| - \ln|x| = \frac{1}{x}h + o(h)$ as $h \to 0$. Thus $\mathrm{d}\ln|x| = \frac{1}{x}\,\mathrm{d}x$ and $\frac{\mathrm{d}\ln|x|}{\mathrm{d}x} = \frac{1}{x}$.

Proof

$$\ln|x+h| - \ln|x| = \ln\left|1 + \frac{h}{x}\right|.$$

For $|h| < |x|$ we have $|1 + \frac{h}{x}| = 1 + \frac{h}{x}$, and so for sufficiently small values of h we can write

$$\ln|x+h| - \ln|x| = \ln\left(1 + \frac{h}{x}\right) = \frac{h}{x} + o\left(\frac{h}{x}\right) = \frac{1}{x}h + o(h)$$

as $h \to 0$. Here we have used the relation $\ln(1+t) = t + o(t)$ as $t \to 0$, shown in Example 38 of Sect. 3.2.4. $\square$

Example 12 If $x \neq 0$ and $0 < a \neq 1$, then $\log_a|x+h| - \log_a|x| = \frac{1}{x\ln a}h + o(h)$ as $h \to 0$. Thus, $\mathrm{d}\log_a|x| = \frac{1}{x\ln a}\,\mathrm{d}x$ and $\frac{\mathrm{d}\log_a|x|}{\mathrm{d}x} = \frac{1}{x\ln a}$.

Proof

$$\log_a|x+h| - \log_a|x| = \log_a\left|1 + \frac{h}{x}\right| = \log_a\left(1 + \frac{h}{x}\right) =$$
$$= \frac{1}{\ln a}\ln\left(1 + \frac{h}{x}\right) = \frac{1}{\ln a}\left(\frac{h}{x} + o\left(\frac{h}{x}\right)\right) = \frac{1}{x\ln a}h + o(h).$$

Here we have used the formula for transition from one base of logarithms to another and the considerations explained in Example 11. $\square$

5.1.6 Problems and Exercises

1. Show that

a) the tangent to the ellipse

$$\frac{x^2}{a^2} + \frac{y^2}{b^2} = 1$$

at the point (x_0, y_0) has the equation

$$\frac{xx_0}{a^2} + \frac{yy_0}{b^2} = 1;$$

b) light rays from a source located at a focus $F_1 = (-\sqrt{a^2 - b^2}, 0)$ or $F_2 = (\sqrt{a^2 - b^2}, 0)$ of an ellipse with semiaxes $a > b > 0$ are gathered at the other focus by an elliptical mirror.

2. Write the formulas for approximate computation of the following values:

a) $\sin(\frac{\pi}{6} + \alpha)$ for values of α near 0;
b) $\sin(30^\circ + \alpha^\circ)$ for values of α° near 0;
c) $\cos(\frac{\pi}{4} + \alpha)$ for values of α near 0;
d) $\cos(45^\circ + \alpha^\circ)$ for values of α° near 0.

3. A glass of water is rotating about its axis at constant angular velocity ω. Let $y = f(x)$ denote the equation of the curve obtained by cutting the surface of the liquid with a plane passing through its axis of rotation.

a) Show that $f'(x) = \frac{\omega^2}{g} x$, where g is the acceleration of free fall. (See Example 5.)

b) Choose a function $f(x)$ that satisfies the condition given in part a). (See Example 6.)

c) Does the condition on the function $f(x)$ given in part a) change if its axis of rotation does not coincide with the axis of the glass?

4. A body that can be regarded as a point mass is sliding down a smooth hill under the influence of gravity. The hill is the graph of a differentiable function $y = f(x)$.

a) Find the horizontal and vertical components of the acceleration vector that the body has at the point (x_0, y_0).

b) For the case $f(x) = x^2$ when the body slides from a great height, find the point of the parabola $y = x^2$ at which the horizontal component of the acceleration is maximal.

5. Set

$$\Psi_0(x) = \begin{cases} x, & \text{if } 0 \le x \le \frac{1}{2}, \\ 1 - x, & \text{if } \frac{1}{2} \le x \le 1, \end{cases}$$

and extend this function to the entire real line so as to have period 1. We denote the extended function by φ_0. Further, let

$$\varphi_n(x) = \frac{1}{4^n} \varphi_0(4^n x).$$

The function φ_n has period 4^{-n} and a derivative equal to $+1$ or -1 everywhere except at the points $x = \frac{k}{2^{2n+1}}$, $k \in \mathbb{Z}$. Let

$$f(x) = \sum_{n=1}^{\infty} \varphi_n(x).$$

Show that the function f is defined and continuous on $\mathbb{R}$, but does not have a derivative at any point. (This example is due to the well-known Dutch mathematician B.L. van der Waerden (1903–1996). The first examples of continuous functions having no derivatives were constructed by Bolzano (1830) and Weierstrass (1860).)

5.2 The Basic Rules of Differentiation

Constructing the differential of a given function or, equivalently, the process of finding its derivative, is called *differentiation*.[7]

5.2.1 Differentiation and the Arithmetic Operations

Theorem 1 *If functions $f : X \to \mathbb{R}$ and $g : X \to \mathbb{R}$ are differentiable at a point $x \in X$, then*

a) *their sum is differentiable at x, and*

$$(f + g)'(x) = (f' + g')(x);$$

b) *their product is differentiable at x, and*

$$(f \cdot g)'(x) = f'(x) \cdot g(x) + f(x) \cdot g'(x);$$

c) *their quotient is differentiable at x if $g(x) \neq 0$, and*

$$\left(\frac{f}{g}\right)'(x) = \frac{f'(x)g(x) - f(x)g'(x)}{g^2(x)}.$$

Proof In the proof we shall rely on the definition of a differentiable function and the properties of the symbol $o(\cdot)$ proved in Sect. 3.2.4.

[7] Although the problems of finding the differential and finding the derivative are mathematically equivalent, the derivative and the differential are nevertheless not the same thing. For that reason, for example, there are two terms in French – *dérivation*, for finding the derivative, and *différentiation*, for finding the differential.

a) $(f+g)(x+h)-(f+g)(x)=$

$$=\big(f(x+h)+g(x+h)\big)-\big(f(x)+g(x)\big)=$$
$$=\big(f(x+h)-f(x)\big)+\big(g(x+h)-g(x)\big)=$$
$$=\big(f'(x)h+o(h)\big)+\big(g'(x)h+o(h)\big)=\big(f'(x)+g'(x)\big)h+o(h)=$$
$$=\big(f'+g'\big)(x)h+o(h).$$

b) $(f\cdot g)(x+h)-(f\cdot g)(x)=$

$$=f(x+h)g(x+h)-f(x)g(x)=$$
$$=\big(f(x)+f'(x)h+o(h)\big)\big(g(x)+g'(x)h+o(h)\big)-f(x)g(x)=$$
$$=\big(f'(x)g(x)+f(x)g'(x)\big)h+o(h).$$

c) Since a function that is differentiable at a point $x \in X$ is continuous at that point, taking account of the relation $g(x) \neq 0$ and the properties of continuous functions, we can guarantee that $g(x+h) \neq 0$ for sufficiently small values of h. In the following computations it is assumed that h is small:

$$\left(\frac{f}{g}\right)(x+h)-\left(\frac{f}{g}\right)(x)=$$
$$=\frac{f(x+h)}{g(x+h)}-\frac{f(x)}{g(x)}=\frac{1}{g(x)g(x+h)}\big(f(x+h)g(x)-f(x)g(x+h)\big)=$$
$$=\left(\frac{1}{g^2(x)}+o(1)\right)\Big(\big(f(x)+f'(x)h+o(h)\big)g(x)-$$
$$-f(x)\big(g(x)+g'(x)h+o(h)\big)\Big)=$$
$$=\left(\frac{1}{g^2(x)}+o(1)\right)\Big(\big(f'(x)g(x)-f(x)g'(x)\big)h+o(h)\Big)=$$
$$=\frac{f'(x)g(x)-f(x)g'(x)}{g^2(x)}h+o(h).$$

Here we have used the continuity of g at the point x and the relation $g(x) \neq 0$ to deduce that

$$\lim_{h\to 0}\frac{1}{g(x)g(x+h)}=\frac{1}{g^2(x)},$$

that is,

$$\frac{1}{g(x)g(x+h)}=\frac{1}{g^2(x)}+o(1),$$

where $o(1)$ is infinitesimal as $h \to 0$, $x+h \in X$. □

Corollary 1 *The derivative of a linear combination of differentiable functions equals the same linear combination of the derivatives of these functions.*

Proof Since a constant function is obviously differentiable and has a derivative equal to 0 at every point, taking $f \equiv \text{const} = c$ in statement b) of Theorem 1, we find $(cg)'(x) = cg'(x)$.

Now, using statement a) of Theorem 1, we can write

$$(c_1 f + c_2 g)'(x) = (c_1 f)'(x) + (c_2 g)'(x) = c_1 f'(x) + c_2 g'(x).$$

Taking account of what has just been proved, we verify by induction that

$$(c_1 f_1 + \cdots + c_n f_n)'(x) = c_1 f_1'(x) + \cdots + c_n f_n'(x).$$ □

Corollary 2 *If the functions $f_1, \dots, f_n$ are differentiable at x, then*

$$(f_1 \cdots f_n)'(x) = f_1'(x) f_2(x) \cdots f_n(x) + \\ + f_1(x) f_2'(x) f_3(x) \cdots f_n(x) + \cdots + f_1(x) \cdots f_{n-1}(x) f_n'(x).$$

Proof For $n = 1$ the statement is obvious.

If it holds for some $n \in \mathbb{N}$, then by statement b) of Theorem 1 it also holds for $(n+1) \in \mathbb{N}$. By the principle of induction, we conclude that the formula is valid for any $n \in \mathbb{N}$. □

Corollary 3 *It follows from the relation between the derivative and the differential that Theorem 1 can also be written in terms of differentials. To be specific:*

a) $\mathrm{d}(f + g)(x) = \mathrm{d}f(x) + \mathrm{d}g(x)$;
b) $\mathrm{d}(f \cdot g)(x) = g(x)\,\mathrm{d}f(x) + f(x)\,\mathrm{d}g(x)$;
c) $\mathrm{d}(\frac{f}{g})(x) = \frac{g(x)\,\mathrm{d}f(x) - f(x)\,\mathrm{d}g(x)}{g^2(x)}$ *if* $g(x) \neq 0$.

Proof Let us verify, for example, statement a).

$$\begin{aligned}\mathrm{d}(f+g)(x)h &= (f+g)'(x)h = \big(f' + g'\big)(x)h = \\ &= \big(f'(x) + g'(x)\big)h = f'(x)h + g'(x)h = \\ &= \mathrm{d}f(x)h + \mathrm{d}g(x)h = \big(\mathrm{d}f(x) + \mathrm{d}g(x)\big)h,\end{aligned}$$

and we have verified that $\mathrm{d}(f + g)(x)$ and $\mathrm{d}f(x) + \mathrm{d}g(x)$ are the same function. □

Example 1 (Invariance of the definition of velocity) We are now in a position to verify that the instantaneous velocity vector of a point mass defined in Sect. 5.1.1 is independent of the Cartesian coordinate system used to define it. In fact we shall verify this for all affine coordinate systems.

Let (x^1, x^2) and $(\tilde{x}^1, \tilde{x}^2)$ be the coordinates of the same point of the plane in two different coordinate systems connected by the relations

$$\begin{aligned}\tilde{x}^1 &= a_1^1 x^1 + a_2^1 x^2 + b^1, \\ \tilde{x}^2 &= a_1^2 x^1 + a_2^2 x^2 + b^2.\end{aligned} \tag{5.26}$$

Since any vector (in affine space) is determined by a pair of points and its coordinates are the differences of the coordinates of the terminal and initial points of the vector, it follows that the coordinates of a given vector in these two coordinate systems must be connected by the relations

$$\begin{aligned}\tilde{v}^1 &= a_1^1 v^1 + a_2^1 v^2,\\ \tilde{v}^2 &= a_1^2 v^1 + a_2^2 v^2.\end{aligned} \tag{5.27}$$

If the law of motion of the point is given by functions $x^1(t)$ and $x^2(t)$ in one system of coordinates, it is given in the other system by functions $\tilde{x}^1(t)$ and $\tilde{x}^2(t)$ connected with the first set by relations (5.26).

Differentiating relations (5.26) with respect to t, we find by the rules for differentiation

$$\begin{aligned}\dot{\tilde{x}}^1 &= a_1^1 \dot{x}^1 + a_2^1 \dot{x}^2,\\ \dot{\tilde{x}}^2 &= a_1^2 \dot{x}^1 + a_2^2 \dot{x}^2.\end{aligned} \tag{5.28}$$

Thus the coordinates $(v^1, v^2) = (\dot{x}^1, \dot{x}^2)$ of the velocity vector in the first system and the coordinates $(\tilde{v}^1, \tilde{v}^2) = (\dot{\tilde{x}}^1, \dot{\tilde{x}}^2)$ of the velocity vector in the second system are connected by relations (5.27), telling us that we are dealing with two different expressions for the same vector.

Example 2 Let $f(x) = \tan x$. We shall show that $f'(x) = \frac{1}{\cos^2 x}$ at every point where $\cos x \neq 0$, that is, in the domain of definition of the function $\tan x = \frac{\sin x}{\cos x}$.

It was shown in Examples 1 and 2 of Sect. 5.1 that $\sin'(x) = \cos x$ and $\cos' x = -\sin x$, so that by statement c) of Theorem 1 we find, when $\cos x \neq 0$,

$$\begin{aligned}\tan' x &= \left(\frac{\sin}{\cos}\right)'(x) = \frac{\sin' x \cos x - \sin x \cos' x}{\cos^2 x} =\\ &= \frac{\cos x \cos x + \sin x \sin x}{\cos^2 x} = \frac{1}{\cos^2 x}.\end{aligned}$$

Example 3 $\cot' x = -\frac{1}{\sin^2 x}$ wherever $\sin x \neq 0$, that is, in the domain of definition of $\cot x = \frac{\cos x}{\sin x}$.

Indeed,

$$\begin{aligned}\cot' x &= \left(\frac{\cos}{\sin}\right)'(x) = \frac{\cos' x \sin x - \cos x \sin' x}{\sin^2 x} =\\ &= \frac{-\sin x \sin x - \cos x \cos x}{\sin^2 x} = -\frac{1}{\sin^2 x}.\end{aligned}$$

Example 4 If $P(x) = c_0 + c_1 x + \cdots + c_n x^n$ is a polynomial, then $P'(x) = c_1 + 2c_2 x + \cdots + nc_n x^{n-1}$.

Indeed, since $\frac{\mathrm{d}x}{\mathrm{d}x} = 1$, by Corollary 2 we have $\frac{\mathrm{d}x^n}{\mathrm{d}x} = nx^{n-1}$, and the statement now follows from Corollary 1.

5.2.2 Differentiation of a Composite Function (Chain Rule)

Theorem 2 (Differentiation of a composite function) *If the function $f : X \to Y \subset \mathbb{R}$ is differentiable at a point $x \in X$ and the function $g : Y \to \mathbb{R}$ is differentiable at the point $y = f(x) \in Y$, then the composite function $g \circ f : X \to \mathbb{R}$ is differentiable at x, and the differential $\mathrm{d}(g \circ f)(x) : T\mathbb{R}(x) \to T\mathbb{R}(g(f(x)))$ of their composition equals the composition $\mathrm{d}f(y) \circ \mathrm{d}f(x)$ of their differentials*

$$\mathrm{d}f(x) : T\mathbb{R}(x) \to T\mathbb{R}\big(y = f(x)\big) \quad \textit{and} \quad \mathrm{d}g\big(y = f(x)\big) : T\mathbb{R}(y) \to T\mathbb{R}\big(g(y)\big).$$

Proof The conditions for differentiability of the functions f and g have the form

$$f(x+h) - f(x) = f'(x)h + o(h) \quad \text{as } h \to 0,\ x+h \in X,$$

$$g(y+t) - g(y) = g'(y)t + o(t) \quad \text{as } t \to 0,\ y+t \in Y.$$

We remark that in the second equality here the function $o(t)$ can be considered to be defined for $t = 0$, and in the representation $o(t) = \gamma(t)t$, where $\gamma(t) \to 0$ as $t \to 0$, $y+t \in Y$, we may assume $\gamma(0) = 0$. Setting $f(x) = y$ and $f(x+h) = y+t$, by the differentiability (and hence continuity) of f at the point x we conclude that $t \to 0$ as $h \to 0$, and if $x + h \in X$, then $y + t \in Y$. By the theorem on the limit of a composite function, we now have

$$\gamma\big(f(x+h) - f(x)\big) = \alpha(h) \to 0 \quad \text{as } h \to 0,\ x+h \in X,$$

and thus if $t = f(x+h) - f(x)$ then

$$\begin{aligned} o(t) &= \gamma\big(f(x+h) - f(x)\big)\big(f(x+h) - f(x)\big) = \\ &= \alpha(h)\big(f'(x)h + o(h)\big) = \alpha(h)f'(x)h + \alpha(h)o(h) = \\ &= o(h) + o(h) = o(h) \quad \text{as } h \to 0, x+h \in X, \end{aligned}$$

$$\begin{aligned} &(g \circ f)(x+h) - (g \circ f)(x) = g\big(f(x+h)\big) - g\big(f(x)\big) = \\ &= g(y+t) - g(y) = g'(y)t + o(t) = \\ &= g'\big(f(x)\big)\big(f(x+h) - f(x)\big) + o\big(f(x+h) - f(x)\big) = \\ &= g'\big(f(x)\big)\big(f'(x)h + o(h)\big) + o\big(f(x+h) - f(x)\big) = \\ &= g'\big(f(x)\big)\big(f'(x)h\big) + g'\big(f(x)\big)\big(o(h)\big) + o\big(f(x+h) - f(x)\big). \end{aligned}$$

Since we can interpret the quantity $g'(f(x))(f'(x)h)$ as the value $\mathrm{d}g(f(x)) \circ \mathrm{d}f(x)h$ of the composition $h \overset{\mathrm{d}g(y)\circ \mathrm{d}f(x)}{\longmapsto} g'(f(x)) \cdot f'(x)h$ of the mappings $h \overset{\mathrm{d}f(x)}{\mapsto}$

$f'(x)h, \tau \overset{dg(y)}{\mapsto} g'(y)\tau$ at the displacement h, to complete the proof it remains only for us to remark that the sum

$$g'(f(x))(o(h)) + o(f(x+h) - f(x))$$

is infinitesimal compared with h as $h \to 0$, $x + h \in X$, or, as we have already established,

$$o(f(x+h) - f(x)) = o(h) \quad \text{as } h \to 0,\ x + h \in X.$$

Thus we have proved that

$$(g \circ f)(x+h) - (g \circ f)(x) =$$
$$= g'(f(x)) \cdot f'(x)h + o(h) \quad \text{as } h \to 0,\ x + h \in X.$$

□

Corollary 4 *The derivative $(g \circ f)'(x)$ of the composition of differentiable real-valued functions equals the product $g'(f(x)) \cdot f'(x)$ of the derivatives of these functions computed at the corresponding points.*

There is a strong temptation to give a short proof of this last statement in Leibniz' notation for the derivative, in which if $z = z(y)$ and $y = y(x)$, we have

$$\frac{dz}{dx} = \frac{dz}{dy} \cdot \frac{dy}{dx},$$

which appears to be completely natural, if one regards the symbol $\frac{dz}{dy}$ or $\frac{dy}{dx}$ not as a unit, but as the ratio of dz to dy or dy to dx.

The idea for a proof that thereby arises is to consider the difference quotient

$$\frac{\Delta z}{\Delta x} = \frac{\Delta z}{\Delta y} \cdot \frac{\Delta y}{\Delta x}$$

and then pass to the limit as $\Delta x \to 0$. The difficulty that arises here (which we also have had to deal with in part!) is that Δy may be 0 even if $\Delta x \neq 0$.

Corollary 5 *If the composition $(f_n \circ \cdots \circ f_1)(x)$ of differentiable functions $y_1 = f_1(x), \ldots, y_n = f_n(y_{n-1})$ exists, then*

$$(f_n \circ \cdots \circ f_1)'(x) = f_n'(y_{n-1}) f_{n-1}'(y_{n-2}) \cdots f_1'(x).$$

Proof The statement is obvious if $n = 1$.

If it holds for some $n \in \mathbb{N}$, then by Theorem 2 it also holds for $n + 1$, so that by the principle of induction, it holds for any $n \in \mathbb{N}$. □

Example 5 Let us show that for $\alpha \in \mathbb{R}$ we have $\frac{dx^\alpha}{dx} = \alpha x^{\alpha-1}$ in the domain $x > 0$, that is, $dx^\alpha = \alpha x^{\alpha-1}\,dx$ and

$$(x+h)^\alpha - x^\alpha = \alpha x^{\alpha-1} h + o(h) \quad \text{as } h \to 0.$$

Proof We write $x^\alpha = \mathrm{e}^{\alpha \ln x}$ and apply the theorem, taking account of the results of Examples 9 and 11 from Sect. 5.1 and statement b) of Theorem 1.

Let $g(y) = \mathrm{e}^y$ and $y = f(x) = \alpha \ln(x)$. Then $x^\alpha = (g \circ f)(x)$ and

$$(g \circ f)'(x) = g'(y) \cdot f'(x) = \mathrm{e}^y \cdot \frac{\alpha}{x} = \mathrm{e}^{\alpha \ln x} \cdot \frac{\alpha}{x} = \alpha x^{\alpha - 1}. \qquad \square$$

Example 6 The derivative of the logarithm of the absolute value of a differentiable function is often called its *logarithmic derivative.*

Since $F(x) = \ln|f(x)| = (\ln \circ |\ | \circ f)(x)$, by Example 11 of Sect. 5.1, we have $F'(x) = (\ln|f|)'(x) = \frac{f'(x)}{f(x)}$.

Thus

$$\mathrm{d}\big(\ln|f|\big)(x) = \frac{f'(x)}{f(x)}\,\mathrm{d}x = \frac{\mathrm{d}f(x)}{f(x)}.$$

Example 7 (The absolute and relative errors in the value of a differentiable function caused by errors in the data for the argument) If the function f is differentiable at x, then

$$f(x+h) - f(x) = f'(x)h + \alpha(x; h),$$

where $\alpha(x; h) = o(h)$ as $h \to 0$.

Thus, if in computing the value $f(x)$ of a function, the argument x is determined with absolute error h, the absolute error $|f(x+h) - f(x)|$ in the value of the function due to this error in the argument can be replaced for small values of h by the absolute value of the differential $|\mathrm{d}f(x)h| = |f'(x)h|$ at displacement h.

The relative error can then be computed as the ratio $\frac{|f'(x)h|}{f(x)|} = \frac{|\mathrm{d}f(x)h|}{|f(x)|}$ or as the absolute value of the product $|\frac{f'(x)}{f(x)}||h|$ of the logarithmic derivative of the function and the magnitude of the absolute error in the argument.

We remark by the way that if $f(x) = \ln x$, then $\mathrm{d}\ln x = \frac{\mathrm{d}x}{x}$, and the absolute error in determining the value of a logarithm equals the relative error in the argument. This circumstance can be beautifully exploited for example, in the slide rule (and many other devices with nonuniform scales). To be specific, let us imagine that with each point of the real line lying right of zero we connect its coordinate y and write it down above the point, while below the point we write the number $x = \mathrm{e}^y$. Then $y = \ln x$. The same real half-line has now been endowed with a uniform scale y and a nonuniform scale x (called logarithmic). To find $\ln x$, one need only set the cursor on the number x and read the corresponding number y written above it. Since the precision in setting the cursor on a particular point is independent of the number x or y corresponding to it and is measured by some quantity Δy (the length of the interval of possible deviation) on the uniform scale, we shall have approximately the same absolute error in determining both a number x and its logarithm y; and in determining a number from its logarithm we shall have approximately the same relative error in all parts of the scale.

Example 8 Let us differentiate a function $u(x)^{v(x)}$, where $u(x)$ and $v(x)$ are differentiable functions and $u(x) > 0$. We write $u(x)^{v(x)} = \mathrm{e}^{v(x)\ln u(x)}$ and use Corollary 5. Then

$$\frac{\mathrm{d}\mathrm{e}^{v(x)\ln u(x)}}{\mathrm{d}x} = \mathrm{e}^{v(x)\ln u(x)}\left(v'(x)\ln u(x) + v(x)\frac{u'(x)}{u(x)}\right) =$$
$$= u(x)^{v(x)} \cdot v'(x)\ln u(x) + v(x)u(x)^{v(x)-1} \cdot u'(x).$$

5.2.3 Differentiation of an Inverse Function

Theorem 3 (The derivative of an inverse function) *Let the functions $f : X \to Y$ and $f^{-1} : Y \to X$ be mutually inverse and continuous at points $x_0 \in X$ and $f(x_0) = y_0 \in Y$ respectively. If f is differentiable at x_0 and $f'(x_0) \neq 0$, then f^{-1} is also differentiable at the point y_0, and*

$$\left(f^{-1}\right)'(y_0) = \left(f'(x_0)\right)^{-1}.$$

Proof Since the functions $f : X \to Y$ and $f^{-1} : Y \to X$ are mutually inverse, the quantities $f(x) - f(x_0)$ and $f^{-1}(y) - f^{-1}(y_0)$, where $y = f(x)$, are both nonzero if $x \neq x_0$. In addition, we conclude from the continuity of f at x_0 and f^{-1} at y_0 that $(X \ni x \to x_0) \Leftrightarrow (Y \ni y \to y_0)$. Now using the theorem on the limit of a composite function and the arithmetic properties of the limit, we find

$$\lim_{Y \ni y \to y_0} \frac{f^{-1}(y) - f^{-1}(y_0)}{y - y_0} = \lim_{X \ni x \to x_0} \frac{x - x_0}{f(x) - f(x_0)} =$$
$$= \lim_{X \ni x \to x_0} \frac{1}{\left(\frac{f(x)-f(x_0)}{x-x_0}\right)} = \frac{1}{f'(x_0)}.$$

Thus we have shown that the function $f^{-1} : Y \to X$ has a derivative at y_0 and that

$$\left(f^{-1}\right)'(y_0) = \left(f'(x_0)\right)^{-1}. \qquad \square$$

Remark 1 If we knew in advance that the function f^{-1} was differentiable at y_0, we would find immediately by the identity $(f^{-1} \circ f)(x) = x$ and the theorem on differentiation of a composite function that $(f^{-1})'(y_0) \cdot f'(x_0) = 1$.

Remark 2 The condition $f'(x_0) \neq 0$ is obviously equivalent to the statement that the mapping $h \mapsto f'(x_0)h$ realized by the differential $\mathrm{d}f(x_0) : T\mathbb{R}(x_0) \to T\mathbb{R}(y_0)$ has the inverse mapping $[\mathrm{d}f(x_0)]^{-1} : T\mathbb{R}(y_0) \to T\mathbb{R}(x_0)$ given by the formula $\tau \mapsto (f'(x_0))^{-1}\tau$.

Hence, in terms of differentials we can write the second statement in Theorem 3 as follows:

If a function f is differentiable at a point x_0 and its differential $\mathrm{d}f(x_0): T\mathbb{R}(x_0) \to T\mathbb{R}(y_0)$ is invertible at that point, then the differential of the function f^{-1} inverse to f exists at the point $y_0 = f(x_0)$ and is the mapping

$$\mathrm{d}f^{-1}(y_0) = \left[\mathrm{d}f(x_0)\right]^{-1} : T\mathbb{R}(y_0) \to T\mathbb{R}(x_0),$$

inverse to $\mathrm{d}f(x_0): T\mathbb{R}(x_0) \to T\mathbb{R}(y_0)$.

Example 9 We shall show that $\arcsin' y = \frac{1}{\sqrt{1-y^2}}$ for $|y| < 1$. The functions $\sin : [-\pi/2, \pi/2] \to [-1, 1]$ and $\arcsin : [-1, 1] \to [-\pi/2, \pi/2]$ are mutually inverse and continuous (see Example 8 of Sect. 4.2) and $\sin'(x) = \cos x \neq 0$ if $|x| < \pi/2$. For $|x| < \pi/2$ we have $|y| < 1$ for the values $y = \sin x$. Therefore, by Theorem 3

$$\arcsin' y = \frac{1}{\sin' x} = \frac{1}{\cos x} = \frac{1}{\sqrt{1 - \sin^2 x}} = \frac{1}{\sqrt{1 - y^2}}.$$

The sign in front of the radical is chosen taking account of the inequality $\cos x > 0$ for $|x| < \pi/2$.

Example 10 Reasoning as in the preceding example, one can show (taking account of Example 9 of Sect. 4.2) that

$$\arccos' y = -\frac{1}{\sqrt{1 - y^2}} \quad \text{for } |y| < 1.$$

Indeed,

$$\arccos' y = \frac{1}{\cos' x} = -\frac{1}{\sin x} = -\frac{1}{\sqrt{1 - \cos^2 x}} = -\frac{1}{\sqrt{1 - y^2}}.$$

The sign in front of the radical is chosen taking account of the inequality $\sin x > 0$ if $0 < x < \pi$.

Example 11 $\arctan' y = \frac{1}{1+y^2}$, $y \in \mathbb{R}$.

Indeed,

$$\arctan' y = \frac{1}{\tan' x} = \frac{1}{\left(\frac{1}{\cos^2 x}\right)} = \cos^2 x = \frac{1}{1 + \tan^2 x} = \frac{1}{1 + y^2}.$$

Example 12 $\operatorname{arccot}' y = -\frac{1}{1+y^2}$, $y \in \mathbb{R}$.

Indeed

$$\operatorname{arccot}' y = \frac{1}{\cot' x} = \frac{1}{\left(-\frac{1}{\sin^2 x}\right)} = -\sin^2 x = -\frac{1}{1 + \cot^2 x} = -\frac{1}{1 + y^2}.$$

Example 13 We already know (see Examples 10 and 12 of Sect. 5.1) that the functions $y = f(x) = a^x$ and $x = f^{-1}(y) = \log_a y$ have the derivatives $f'(x) = a^x \ln a$ and $(f^{-1})'(y) = \frac{1}{y \ln a}$.

Let us see how this is consistent with Theorem 3:

$$(f^{-1})'(y) = \frac{1}{f'(x)} = \frac{1}{a^x \ln a} = \frac{1}{y \ln a},$$

$$f'(x) = \frac{1}{(f^{-1})'(y)} = \frac{1}{(\frac{1}{y \ln a})} = y \ln a = a^x \ln a.$$

Example 14 The hyperbolic and inverse hyperbolic functions and their derivatives. The functions

$$\sinh x = \frac{1}{2}(\mathrm{e}^x - \mathrm{e}^{-x}),$$

$$\cosh x = \frac{1}{2}(\mathrm{e}^x + \mathrm{e}^{-x})$$

are called respectively the *hyperbolic sine* and *hyperbolic cosine*[8] of x.

These functions, which for the time being have been introduced purely formally, arise just as naturally in many problems as the circular functions $\sin x$ and $\cos x$.

We remark that

$$\sinh(-x) = -\sinh x,$$

$$\cosh(-x) = \cosh x,$$

that is, the hyperbolic sine is an odd function and the hyperbolic cosine is an even function.

Moreover, the following basic identity is obvious:

$$\cosh^2 x - \sinh^2 x = 1.$$

The graphs of the functions $y = \sinh x$ and $y = \cosh x$ are shown in Fig. 5.7.

It follows from the definition of $\sinh x$ and the properties of the function e^x that $\sinh x$ is a continuous strictly increasing function mapping $\mathbb{R}$ in a one-to-one manner onto itself. The inverse function to $\sinh x$ thus exists, is defined on $\mathbb{R}$, is continuous, and is strictly increasing.

This inverse is denoted $\operatorname{arsinh} y$ (read "area-sine of y").[9] This function is easily expressed in terms of known functions. In solving the equation

$$\frac{1}{2}(\mathrm{e}^x - \mathrm{e}^{-x}) = y$$

---

[8]From the Latin phrases *sinus hyperbolici* and *cosinus hyperbolici*.

[9]The full name is *area sinus hyperbolici* (Lat.); the reason for using the term *area* here instead of *arc*, as in the case of the circular functions, will be explained later.

Fig. 5.7

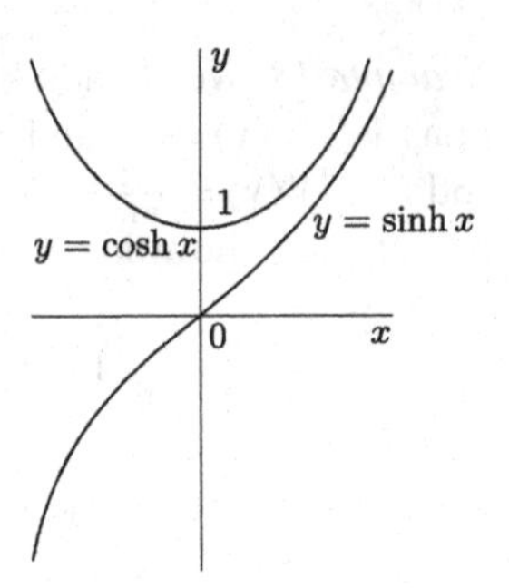

for x, we find successively

$$e^x = y + \sqrt{1+y^2}$$

($e^x > 0$, and so $e^x \neq y - \sqrt{1+y^2}$) and

$$x = \ln\left(y + \sqrt{1+y^2}\right).$$

Thus,

$$\operatorname{arsinh} y = \ln\left(y + \sqrt{1+y^2}\right), \quad y \in \mathbb{R}.$$

Similarly, using the monotonicity of the function $y = \cosh x$ on the two intervals $\mathbb{R}_- = \{x \in \mathbb{R} \mid x \leq 0\}$ and $\mathbb{R}_+ = \{x \in \mathbb{R} \mid x \geq 0\}$, we can construct functions $\operatorname{arcosh}_- y$ and $\operatorname{arcosh}_+ y$, defined for $y \geq 1$ and inverse to the function $\cosh x$ on $\mathbb{R}_-$ and $\mathbb{R}_+$ respectively.

They are given by the formulas

$$\operatorname{arcosh}_- y = \ln\left(y - \sqrt{y^2-1}\right),$$

$$\operatorname{arcosh}_+ y = \ln\left(y + \sqrt{y^2-1}\right).$$

From the definitions given above, we find

$$\sinh' x = \frac{1}{2}\left(e^x + e^{-x}\right) = \cosh x,$$

$$\cosh' x = \frac{1}{2}\left(e^x - e^{-x}\right) = \sinh x,$$

and by the theorem on the derivative of an inverse function, we find

$$\operatorname{arsinh}' y = \frac{1}{\sinh' x} = \frac{1}{\cosh x} = \frac{1}{\sqrt{1+\sinh^2 x}} = \frac{1}{\sqrt{1+y^2}},$$

$$\operatorname{arcosh}'_- y = \frac{1}{\cosh' x} = \frac{1}{\sinh x} = \frac{1}{-\sqrt{\cosh^2 x - 1}} = -\frac{1}{\sqrt{y^2-1}}, \quad y > 1,$$

$$\operatorname{arcosh}'_+ y = \frac{1}{\cosh' x} = \frac{1}{\sinh x} = \frac{1}{\sqrt{\cosh^2 x - 1}} = \frac{1}{\sqrt{y^2 - 1}}, \quad y > 1.$$

These last three relations can be verified by using the explicit expressions for the inverse hyperbolic functions $\operatorname{arsinh} y$ and $\operatorname{arcosh} y$.

For example,

$$\operatorname{arsinh}' y = \frac{1}{y + \sqrt{1 + y^2}} \left(1 + \frac{1}{2} (1 + y^2)^{-1/2} \cdot 2y \right) =$$

$$= \frac{1}{y + \sqrt{1 + y^2}} \cdot \frac{\sqrt{1 + y^2} + y}{\sqrt{1 + y^2}} = \frac{1}{\sqrt{1 + y^2}}.$$

Like $\tan x$ and $\cot x$ one can consider the functions

$$\tanh x = \frac{\sinh x}{\cosh x} \quad \text{and} \quad \coth x = \frac{\cosh x}{\sinh x},$$

called the *hyperbolic tangent* and *hyperbolic cotangent* respectively, and also the functions inverse to them, the *area tangent*

$$\operatorname{artanh} y = \frac{1}{2} \ln \frac{1 + y}{1 - y}, \quad |y| < 1,$$

and the *area cotangent*

$$\operatorname{arcoth} y = \frac{1}{2} \ln \frac{y + 1}{y - 1}, \quad |y| > 1.$$

We omit the solutions of the elementary equations that lead to these formulas.

By the rules for differentiation we have

$$\tanh' x = \frac{\sinh' x \cosh x - \sinh x \cosh' x}{\cosh^2 x} =$$

$$= \frac{\cosh x \cosh x - \sinh x \sinh x}{\cosh^2 x} = \frac{1}{\cosh^2 x},$$

$$\coth' x = \frac{\cosh' x \sinh x - \cosh x \sinh' x}{\sinh^2 x} =$$

$$= \frac{\sinh x \sinh x - \cosh x \cosh x}{\sinh^2 x} = -\frac{1}{\sinh^2 x}.$$

By the theorem on the derivative of an inverse function

$$\operatorname{artanh}' y = \frac{1}{\tanh' x} = \frac{1}{\left(\frac{1}{\cosh^2 x}\right)} = \cosh^2 x =$$

$$= \frac{1}{1 - \tanh^2 x} = \frac{1}{1 - y^2}, \quad |y| < 1,$$

$$\operatorname{arcoth}' y = \frac{1}{\coth' x} = \frac{1}{\left(-\frac{1}{\sinh^2 x}\right)} = -\sinh^2 x =$$

$$= -\frac{1}{\coth^2 x - 1} = -\frac{1}{y^2 - 1}, \quad |y| > 1.$$

The last two formulas can also be verified by direct differentiation of the explicit formulas for the functions artanh y and arcoth y.

5.2.4 Table of Derivatives of the Basic Elementary Functions

We now write out (see Table 5.1) the derivatives of the basic elementary functions computed in Sects. 5.1 and 5.2.

5.2.5 Differentiation of a Very Simple Implicit Function

Let $y = y(t)$ and $x = x(t)$ be differentiable functions defined in a neighborhood $U(t_0)$ of a point $t_0 \in \mathbb{R}$. Assume that the function $x = x(t)$ has an inverse $t = t(x)$ defined in a neighborhood $V(x_0)$ of $x_0 = x(t_0)$. Then the quantity $y = y(t)$, which depends on t, can also be regarded as an implicit function of x, since $y(t) = y(t(x))$. Let us find the derivative of this function with respect to x at the point x_0, assuming that $x'(t_0) \neq 0$. Using the theorem on the differentiation of a composite function and the theorem on differentiation of an inverse function, we obtain

$$y'_x\big|_{x=x_0} = \frac{\mathrm{d}y(t(x))}{\mathrm{d}x}\bigg|_{x=x_0} = \frac{\mathrm{d}y(t)}{\mathrm{d}t}\bigg|_{t=t_0} \cdot \frac{\mathrm{d}t(x)}{\mathrm{d}x}\bigg|_{x=x_0} = \frac{\frac{\mathrm{d}y(t)}{\mathrm{d}t}\big|_{t=t_0}}{\frac{\mathrm{d}x(t)}{\mathrm{d}t}\big|_{t=t_0}} = \frac{y'_t(t_0)}{x'_t(t_0)}.$$

(Here we have used the standard notation $f(x)|_{x=x_0} := f(x_0)$.)

If the same quantity is regarded as a function of different arguments, in order to avoid misunderstandings in differentiation, we indicate explicitly the variable with respect to which the differentiation is carried out, as we have done here.

Example 15 (The law of addition of velocities) The motion of a point along a line is completely determined if we know the coordinate x of the point in our chosen coordinate system (the real line) at each instant t in a system we have chosen for measuring time. Thus the pair of numbers (x, t) determines the position of the point in space and time. The law of motion is written in the form of a function $x = x(t)$.

Suppose we wish to express the motion of this point in terms of a different coordinate system $(\tilde{x}, \tilde{t})$. For example, the new real line may be moving uniformly with speed $-v$ relative to the first system. (The velocity vector in this case may be identified with the single number that defines it.) For simplicity we shall assume that the coordinates (0,0) refer to the same point in both systems; more precisely,

Table 5.1

Function $f(x)$	Derivative $f'(x)$	Restrictions on domain of $x \in \mathbb{R}$
1. C (const)	0	
2. x^α	$\alpha x^{\alpha-1}$	$x > 0$ for $\alpha \in \mathbb{R}$
		$x \in \mathbb{R}$ for $\alpha \in \mathbb{N}$
3. a^x	$a^x \ln a$	$x \in \mathbb{R}$ $(a > 0, a \neq 1)$
4. $\log_a \lvert x\rvert$	$\frac{1}{x \ln a}$	$x \in \mathbb{R}\backslash 0$ $(a > 0, a \neq 1)$
5. $\sin x$	$\cos x$	
6. $\cos x$	$-\sin x$	
7. $\tan x$	$\frac{1}{\cos^2 x}$	$x \neq \frac{\pi}{2} + \pi k, k \in \mathbb{Z}$
8. $\cot x$	$-\frac{1}{\sin^2 x}$	$x \neq \pi k, k \in \mathbb{Z}$
9. $\arcsin x$	$\frac{1}{\sqrt{1-x^2}}$	$\lvert x\rvert < 1$
10. $\arccos x$	$-\frac{1}{\sqrt{1-x^2}}$	$\lvert x\rvert < 1$
11. $\arctan x$	$\frac{1}{1+x2}$	
12. $\operatorname{arccot} x$	$-\frac{1}{1+x^2}$	
13. $\sinh x$	$\cosh x$	
14. $\cosh x$	$\sinh x$	
15. $\tanh x$	$\frac{1}{\cosh^2 x}$	
16. $\coth x$	$-\frac{1}{\sinh^2 x}$	$x \neq 0$
17. $\operatorname{arsinh} x = \ln(x + \sqrt{1+x^2})$	$\frac{1}{\sqrt{1+x^2}}$	
18. $\operatorname{arcosh} x = \ln(x \pm \sqrt{x^2-1})$	$\pm\frac{1}{\sqrt{x^2-1}}$	$\lvert x\rvert > 1$
19. $\operatorname{artanh} x = \frac{1}{2} \ln \frac{1+x}{1-x}$	$\frac{1}{1-x^2}$	$\lvert x\rvert < 1$
20. $\operatorname{arcoth} x = \frac{1}{2} \ln \frac{x+1}{x-1}$	$\frac{1}{x^2-1}$	$\lvert x\rvert > 1$

that at time $\tilde{t} = 0$ the point $\tilde{x} = 0$ coincided with the point $x = 0$ at which the clock showed $t = 0$.

Then one of the possible connections between the coordinate systems (x, t) and $(\tilde{x}, \tilde{t})$ describing the motion of the same point observed from different coordinate systems is provided by the classical Galilean transformations:

$$\begin{aligned} \tilde{x} &= x + vt, \\ \tilde{t} &= t. \end{aligned} \tag{5.29}$$

Let us consider a somewhat more general linear connection

$$\begin{aligned} \tilde{x} &= \alpha x + \beta t, \\ \tilde{t} &= \gamma x + \delta t, \end{aligned} \tag{5.30}$$

assuming, of course, that this connection is invertible, that is, the determinant of the matrix $\left(\begin{smallmatrix} \alpha & \beta \\ \gamma & \delta \end{smallmatrix}\right)$ is not zero.

Let $x = x(t)$ and $\tilde{x} = \tilde{x}(\tilde{t})$ be the law of motion for the point under observation, written in these coordinate systems.

We remark that, knowing the relation $x = x(t)$, we find by formula (5.30) that

$$\begin{aligned} \tilde{x}(t) &= \alpha x(t) + \beta t, \\ \tilde{t}(t) &= \gamma x(t) + \delta t, \end{aligned} \tag{5.31}$$

and since the transformation (5.30) is invertible, after writing

$$\begin{aligned} x &= \tilde{\alpha}\tilde{x} + \tilde{\beta}\tilde{t}, \\ t &= \tilde{\gamma}\tilde{x} + \tilde{\delta}\tilde{t}, \end{aligned} \tag{5.32}$$

knowing $\tilde{x} = \tilde{x}(\tilde{t})$, we find

$$\begin{aligned} x(\tilde{t}) &= \tilde{\alpha}\tilde{x}(\tilde{t}) + \tilde{\beta}\tilde{t}, \\ t(\tilde{t}) &= \tilde{\gamma}\tilde{x}(\tilde{t}) + \tilde{\delta}\tilde{t}. \end{aligned} \tag{5.33}$$

It is clear from relations (5.31) and (5.33) that for the given point there exist mutually inverse functions $\tilde{t} = \tilde{t}(t)$ and $t = t(\tilde{t})$.

We now consider the problem of the connection between the velocities

$$V(t) = \frac{\mathrm{d}x(t)}{\mathrm{d}t} = \dot{x}_t(t) \quad \text{and} \quad \tilde{V}(t) = \frac{\mathrm{d}\tilde{x}(\tilde{t})}{\mathrm{d}\tilde{t}} = \dot{\tilde{x}}_{\tilde{t}}(\tilde{t})$$

of the point computed in the coordinate systems (x, t) and $(\tilde{x}, \tilde{t})$ respectively.

Using the rule for differentiating an implicit function and formula (5.31), we have

$$\frac{\mathrm{d}\tilde{x}}{\mathrm{d}\tilde{t}} = \frac{\frac{\mathrm{d}\tilde{x}}{\mathrm{d}t}}{\frac{\mathrm{d}\tilde{t}}{\mathrm{d}t}} = \frac{\alpha\frac{\mathrm{d}x}{\mathrm{d}t} + \beta}{\gamma\frac{\mathrm{d}x}{\mathrm{d}t} + \delta}$$

or

$$\tilde{V}(\tilde{t}) = \frac{\alpha V(t) + \beta}{\gamma V(t) + \delta}, \tag{5.34}$$

where $\tilde{t}$ and t are the coordinates of the same instant of time in the systems (x, t) and $(\tilde{x}, \tilde{t})$. This is always to be kept in mind in the abbreviated notation

$$\tilde{V} = \frac{\alpha V + \beta}{\gamma V + \delta} \tag{5.35}$$

for formula (5.34).

In the case of the Galilean transformations (5.29) we obtain the classical law of addition of velocities from formula (5.35)

$$\tilde{V} = V + v. \tag{5.36}$$

It has been established experimentally with a high degree of precision (and this became one of the postulates of the special theory of relativity) that in a vacuum light propagates with a certain velocity c that is independent of the state of motion of the radiating body. This means that if an explosion occurs at time $t = \tilde{t} = 0$ at the point $x = \tilde{x} = 0$, the light will reach the points x with coordinates such that $x^2 = (ct)^2$ after time t in the coordinate system (x, t), while in the system $(\tilde{x}, \tilde{t})$ this event will correspond to time $\tilde{t}$ and coordinates $\tilde{x}$, where again $\tilde{x}^2 = (c\tilde{t})^2$.

Thus, if $x^2 - c^2t^2 = 0$, then $\tilde{x}^2 - c\tilde{t}^2 = 0$ also, and conversely. By virtue of certain additional physical considerations, one must consider that, in general

$$x^2 - c^2t^2 = \tilde{x}^2 - c^2\tilde{t}^2, \tag{5.37}$$

if (x, t) and $(\tilde{x}, \tilde{t})$ correspond to the same event in the different coordinate systems connected by relation (5.30). Conditions (5.37) give the following relations on the coefficients α, β, γ, and δ of the transformation (5.30):

$$\begin{aligned} \alpha^2 - c^2\gamma^2 &= 1, \\ \alpha\beta - c^2\gamma\delta &= 0, \\ \beta^2 - c^2\delta^2 &= -c^2. \end{aligned} \tag{5.38}$$

If $c = 1$, we would have, instead of (5.38),

$$\begin{aligned} \alpha^2 - \gamma^2 &= 1, \\ \frac{\beta}{\delta} &= \frac{\gamma}{\alpha}, \\ \beta^2 - \delta^2 &= -1, \end{aligned} \tag{5.39}$$

from which it follows easily that the general solution of (5.39) (up to a change of sign in the pairs (α, β) and (γ, δ)) can be given as

$$\alpha = \cosh\varphi, \qquad \gamma = \sinh\varphi, \qquad \beta = \sinh\varphi, \qquad \delta = \cosh\varphi,$$

where φ is a parameter.

The general solution of the system (5.38) then has the form

$$\begin{pmatrix} \alpha & \beta \\ \gamma & \delta \end{pmatrix} = \begin{pmatrix} \cosh\varphi & c\sinh\varphi \\ \frac{1}{c}\sinh\varphi & \cosh\varphi \end{pmatrix}$$

and the transformation (5.30) can be made specific:

$$\begin{aligned} \tilde{x} &= \cosh\varphi x + c\sinh\varphi t, \\ \tilde{t} &= \frac{1}{c}\sinh\varphi x + \cosh\varphi t. \end{aligned} \tag{5.40}$$

This is the *Lorentz transformation*.

In order to clarify the way in which the free parameter φ is determined, we recall that the $\tilde{x}$-axis is moving with speed $-v$ relative to the x-axis, that is, the point $\tilde{x}=0$ of this axis, when observed in the system (x, t) has velocity $-v$. Setting $\tilde{x}=0$ in (5.40), we find its law of motion in the system (x, t):

$$x = -c \tanh \varphi t.$$

Therefore,

$$\tanh \varphi = \frac{v}{c}. \tag{5.41}$$

Comparing the general law (5.35) of transformation of velocities with the Lorentz transformation (5.40), we obtain

$$\widetilde{V} = \frac{\cosh \varphi V + c \sinh \varphi}{\frac{1}{c} \sinh \varphi V + \cosh \varphi},$$

or, taking account of (5.41),

$$\widetilde{V} = \frac{V + v}{1 + \frac{vV}{c^2}}. \tag{5.42}$$

Formula (5.42) is the relativistic law of addition of velocities, which for $|vV| \ll c^2$, that is, as $c \to \infty$, becomes the classical law expressed by formula (5.36).

The Lorentz transformation (5.40) itself can be rewritten taking account of relation (5.41) in the following more natural form:

$$\begin{aligned} \tilde{x} &= \frac{x + vt}{\sqrt{1 - (\frac{v}{c})^2}}, \\ \tilde{t} &= \frac{t + \frac{v}{c^2} x}{\sqrt{1 - (\frac{v}{c})^2}}, \end{aligned} \tag{5.43}$$

from which one can see that for $|v| \ll c$, that is, as $c \to \infty$, they become the classical Galilean transformations (5.29).

5.2.6 Higher-Order Derivatives

If a function $f : E \to \mathbb{R}$ is differentiable at every point $x \in E$, then a new function $f' : E \to \mathbb{R}$ arises, whose value at a point $x \in E$ equals the derivative $f'(x)$ of the function f at that point.

The function $f' : E \to \mathbb{R}$ may itself have a derivative $(f')' : E \to \mathbb{R}$ on E, called the *second derivative* of the original function f and denoted by one of the following two symbols:

$$f''(x), \quad \frac{\mathrm{d}^2 f(x)}{\mathrm{d}x^2},$$

and if we wish to indicate explicitly the variable of differentiation in the first case, we also write, for example, $f''_{xx}(x)$.

Definition By induction, if the derivative $f^{(n-1)}(x)$ of order $n-1$ of f has been defined, then the *derivative of order n* is defined by the formula

$$f^{(n)}(x) := \left(f^{(n-1)}\right)'(x).$$

The following notations are conventional for the derivative of order n:

$$f^{(n)}(x), \quad \frac{\mathrm{d}^n f(x)}{\mathrm{d}x^n}.$$

Also by convention, $f^{(0)}(x) := f(x)$.

The set of functions $f : E \to \mathbb{R}$ having continuous derivatives up to order n inclusive will be denoted $C^{(n)}(E, \mathbb{R})$, and by the simpler symbol $C^{(n)}(E)$, or $C^n(E, \mathbb{R})$ and $C^n(E)$ respectively wherever no confusion can arise.

In particular $C^{(0)}(E) = C(E)$ by our convention that $f^{(0)}(x) = f(x)$.

Let us now consider some examples of the computation of higher-order derivatives.

Examples

	$f(x)$	$f'(x)$	$f''(x)$	$\cdots$	$f^{(n)}(x)$
16)	a^x	$a^x \ln a$	$a^x \ln^2 a$	$\cdots$	$a^x \ln^n a$
17)	e^x	e^x	e^x	$\cdots$	e^x
18)	$\sin x$	$\cos x$	$-\sin x$	$\cdots$	$\sin(x + n\pi/2)$
19)	$\cos x$	$-\sin x$	$-\cos x$	$\cdots$	$\cos(x + n\pi/2)$
20)	$(1+x)^\alpha$	$\alpha(1+x)^{\alpha-1}$	$\alpha(\alpha-1)(1+x)^{\alpha-2}$	$\cdots$	$\alpha(\alpha-1)\cdots(\alpha-n+1)(1+x)^{\alpha-n}$
21)	x^α	$\alpha x^{\alpha-1}$	$\alpha(\alpha-1)x^{\alpha-2}$	$\cdots$	$\alpha(\alpha-1)\cdots(\alpha-n+1)x^{\alpha-n}$
22)	$\log_a \lvert x\rvert$	$\frac{1}{\ln a}x^{-1}$	$\frac{-1}{\ln a}x^{-2}$	$\cdots$	$\frac{(-1)^{n-1}(n-1)!}{\ln a}x^{-n}$
23)	$\ln \lvert x\rvert$	x^{-1}	$(-1)x^{-2}$	$\cdots$	$(-1)^{n-1}(n-1)!x^{-n}$

Example 24 (Leibniz' formula) Let $u(x)$ and $v(x)$ be functions having derivatives up to order n inclusive on a common set E. The following formula of Leibniz holds for the nth derivative of their product:

$$(uv)^{(n)} = \sum_{m=0}^{n} \binom{n}{m} u^{(n-m)} v^{(m)}. \tag{5.44}$$

Leibniz' formula bears a strong resemblance to Newton's binomial formula, and in fact the two are directly connected.

Proof For $n = 1$ formula (5.44) agrees with the rule already established for the derivative of a product.

If the functions u and v have derivatives up to order $n+1$ inclusive, then, assuming that formula (5.44) holds for order n, after differentiating the left- and right-hand sides, we find

$$(uv)^{(n+1)} = \sum_{m=0}^{n} \binom{n}{m} u^{(n-m+1)} v^{(m)} + \sum_{m=0}^{n} \binom{n}{m} u^{(n-m)} v^{(m+1)} =$$

$$= u^{(n+1)} v^{(0)} + \sum_{k=1}^{n} \left(\binom{n}{k} + \binom{n}{k-1} \right) u^{((n+1)-k)} v^{(k)} + u^{(0)} v^{(n+1)} =$$

$$= \sum_{k=0}^{n+1} \binom{n+1}{k} u^{((n+1)-k)} v^{(k)}.$$

Here we have combined the terms containing like products of derivatives of the functions u and v and used the binomial relation $\binom{n}{k} + \binom{n}{k-1} = \binom{n+1}{k}$.

Thus by induction we have established the validity of Leibniz' formula. □

Example 25 If $P_n(x) = c_0 + c_1 x + \cdots + c_n x^n$, then

$$P_n(0) = c_0,$$

$$P_n'(x) = c_1 + 2c_2 x + \cdots + nc_n x^{n-1} \text{ and } P_n'(0) = c_1,$$

$$P_n''(x) = 2c_2 + 3 \cdot 2c_3 x + \cdots + n(n-1)c_n x^{n-2} \text{ and } P_n''(0) = 2!c_2,$$

$$P_n^{(3)}(x) = 3 \cdot 2c_3 + \cdots n(n-1)(n-2)c_n x^{n-3} \text{ and } P_n^{(3)}(0) = 3!c_3,$$

$$\vdots$$

$$P_n^{(n)}(x) = n(n-1)(n-2) \cdots 2c_n \text{ and } P_n^{(n)}(0) = n!c_n,$$

$$P_n^{(k)}(x) = 0 \quad \text{for } k > n.$$

Thus, the polynomial $P_n(x)$ can be written as

$$P_n(x) = P_n^{(0)}(0) + \frac{1}{1!} P_n^{(1)}(0)x + \frac{1}{2!} P_n^{(2)}(0)x^2 + \cdots + \frac{1}{n!} P_n^{(n)}(0)x^n.$$

Example 26 Using Leibniz' formula and the fact that all the derivatives of a polynomial of order higher than the degree of the polynomial are zero, we can find the nth derivative of $f(x) = x^2 \sin x$:

$$f^{(n)}(x) = \sin^{(n)}(x) \cdot x^2 + \binom{n}{1} \sin^{(n-1)} x \cdot 2x + \binom{n}{2} \sin^{(n-2)} x \cdot 2 =$$

$$= x^2 \sin\left(x + n\frac{\pi}{2}\right) + 2nx \sin\left(x + (n-1)\frac{\pi}{2}\right) +$$

$$+\left(-n(n-1)\sin\left(x+n\frac{\pi}{2}\right)\right)=$$

$$=\left(x^2-n(n-1)\right)\sin\left(x+n\frac{\pi}{2}\right)-2nx\cos\left(x+n\frac{\pi}{2}\right).$$

Example 27 Let $f(x)=\arctan x$. Let us find the values $f^{(n)}(0)$ $(n=1,2,\ldots)$.

Since $f'(x)=\dfrac{1}{1+x^2}$, it follows that $(1+x^2)f'(x)=1$.

Applying Leibniz' formula to this last equality, we find the recursion relation

$$\left(1+x^2\right)f^{(n+1)}(x)+2nxf^{(n)}(x)+n(n-1)f^{(n-1)}(x)=0,$$

from which one can successively find all the derivatives of $f(x)$.

Setting $x=0$, we obtain

$$f^{(n+1)}(0)=-n(n-1)f^{(n-1)}(0).$$

For $n=1$ we find $f^{(2)}(0)=0$, and therefore $f^{(2n)}(0)=0$. For derivatives of odd order we have

$$f^{(2m+1)}(0)=-2m(2m-1)f^{(2m-1)}(0)$$

and since $f'(0)=1$, we obtain

$$f^{(2m+1)}(0)=(-1)^m(2m)!.$$

Example 28 (Acceleration) If $x=x(t)$ denotes the time dependence of a point mass moving along the real line, then $\frac{\mathrm{d}x(t)}{\mathrm{d}t}=\dot{x}(t)$ is the velocity of the point, and then $\frac{\mathrm{d}\dot{x}(t)}{\mathrm{d}t}=\frac{\mathrm{d}^2x(t)}{\mathrm{d}t^2}=\ddot{x}(t)$ is its acceleration at time t.

If $x(t)=\alpha t+\beta$, then $\dot{x}(t)=\alpha$ and $\ddot{x}(t)\equiv 0$, that is, the acceleration in a uniform motion is zero. We shall soon verify that if the second derivative equals zero, then the function itself has the form $\alpha t+\beta$. Thus, in uniform motions, and only in uniform motions, is the acceleration equal to zero.

But if we wish for a body moving under inertia in empty space to move uniformly in a straight line when observed in two different coordinate systems, it is necessary for the transition formulas from one inertial system to the other to be linear. That is the reason why, in Example 15, the linear formulas (5.30) were chosen for the coordinate transformations.

Example 29 (The second derivative of a simple implicit function) Let $y=y(t)$ and $x=x(t)$ be twice-differentiable functions. Assume that the function $x=x(t)$ has a differentiable inverse function $t=t(x)$. Then the quantity $y(t)$ can be regarded as an implicit function of x, since $y=y(t)=y(t(x))$. Let us find the second derivative y''_{xx} assuming that $x'(t)\neq 0$.

By the rule for differentiating such a function, studied in Sect. 5.2.5, we have

$$y'_x=\frac{y'_t}{x'_t},$$

so that

$$y''_{xx} = \left(y'_x\right)'_x = \frac{(y'_x)'_t}{x'_t} = \frac{\left(\frac{y'_t}{x'_t}\right)'_t}{x'_t} = \frac{\frac{y''_{tt}x'_t - y'_t x''_{tt}}{(x'_t)^2}}{x'_t} = \frac{x'_t y''_{tt} - x''_{tt} y'_t}{(x'_t)^3}.$$

We remark that the explicit expressions for all the functions that occur here, including y''_{xx}, depend on t, but they make it possible to obtain the value of y''_{xx} at the particular point x after substituting for t the value $t = t(x)$ corresponding to the value x.

For example, if $y = \mathrm{e}^t$ and $x = \ln t$, then

$$y'_x = \frac{y'_t}{x'_t} = \frac{\mathrm{e}^t}{1/t} = t\mathrm{e}^t, \qquad y''_{xx} = \frac{(y'_x)'_t}{x'_t} = \frac{\mathrm{e}^t + t\mathrm{e}^t}{1/t} = t(t+1)\mathrm{e}^t.$$

We have deliberately chosen this simple example, in which one can explicitly express t in terms of x as $t = \mathrm{e}^x$ and, by substituting $t = \mathrm{e}^x$ into $y(t) = \mathrm{e}^t$, find the explicit dependence of $y = \mathrm{e}^{\mathrm{e}^x}$ on x. Differentiating this last function, one can justify the results obtained above.

It is clear that in this way one can find the derivatives of any order by successively applying the formula

$$y^{(n)}_{x^n} = \frac{(y^{(n-1)}_{x^{n-1}})'_t}{x'_t}.$$

5.2.7 Problems and Exercises

1. Let $\alpha_0, \alpha_1, \dots, \alpha_n$ be given real numbers. Exhibit a polynomial $P_n(x)$ of degree n having the derivatives $P_n^{(k)}(x_0) = \alpha_k$, $k = 0, 1, \dots, n$, at a given point $x_0 \in \mathbb{R}$.

2. Compute $f'(x)$ if

a) $f(x) = \begin{cases} \exp(-\frac{1}{x^2}) & \text{for } x \neq 0, \\ 0 & \text{for } x = 0; \end{cases}$

b) $f(x) = \begin{cases} x^2 \sin\frac{1}{x} & \text{for } x \neq 0, \\ 0 & \text{for } x = 0. \end{cases}$

c) Verify that the function in part a) is infinitely differentiable on $\mathbb{R}$, and that $f^{(n)}(0) = 0$.

d) Show that the derivative in part b) is defined on $\mathbb{R}$ but is not a continuous function on $\mathbb{R}$.

e) Show that the function

$$f(x) = \begin{cases} \exp(-\frac{1}{(1+x)^2} - \frac{1}{(1-x)^2}) & \text{for } -1 < x < 1, \\ 0 & \text{for } 1 \le |x| \end{cases}$$

is infinitely differentiable on $\mathbb{R}$.

3. Let $f \in C^{(\infty)}(\mathbb{R})$. Show that for $x \neq 0$

$$\frac{1}{x^{n+1}} f^{(n)}\left(\frac{1}{x}\right) = (-1)^n \frac{\mathrm{d}^n}{\mathrm{d}x^n}\left(x^{n-1} f\left(\frac{1}{x}\right)\right).$$

4. Let f be a differentiable function on $\mathbb{R}$. Show that

a) if f is an even function, then f' is an odd function;
b) if f is an odd function, then f' is an even function;
c) (f' is odd) $\Leftrightarrow$ (f is even).

5. Show that

a) the function $f(x)$ is differentiable at the point x_0 if and only if $f(x) - f(x_0) = \varphi(x)(x - x_0)$, where $\varphi(x)$ is a function that is continuous at x_0 (and in that case $\varphi(x_0) = f'(x_0)$);

b) if $f(x) - f(x_0) = \varphi(x)(x - x_0)$ and $\varphi \in C^{(n-1)}(U(x_0))$, where $U(x_0)$ is a neighborhood of x_0, then $f(x)$ has a derivative ($f^{(n)}(x_0)$) of order n at x_0.

6. Give an example showing that the assumption that f^{-1} be continuous at the point y_0 cannot be omitted from Theorem 3.

7. a) Two bodies with masses m_1 and m_2 respectively are moving in space under the action of their mutual gravitation alone. Using Newton's laws (formulas (5.1) and (5.2) of Sect. 5.1), verify that the quantity

$$E = \left(\frac{1}{2}m_1 v_1^2 + \frac{1}{2}m_2 v_2^2\right) + \left(-G\frac{m_1 m_2}{r}\right) =: K + U,$$

where v_1 and v_2 are the velocities of the bodies and r the distance between them, does not vary during this motion.

b) Give a physical interpretation of the quantity $E = K + U$ and its components.
c) Extend this result to the case of the motion of n bodies.

5.3 The Basic Theorems of Differential Calculus

5.3.1 Fermat's Lemma and Rolle's Theorem

Definition 1 A point $x_0 \in E \subset \mathbb{R}$ is called a *local maximum* (resp. *local minimum*) and the value of a function $f : E \to \mathbb{R}$ at that point a *local maximum value* (resp. *local minimum value*) if there exists a neighborhood $U_E(x_0)$ of x_0 in E such that at any point $x \in U_E(x_0)$ we have $f(x) \leq f(x_0)$ (resp. $f(x) \geq f(x_0)$).

Definition 2 If the strict inequality $f(x) < f(x_0)$ (resp. $f(x) > f(x_0)$) holds at every point $x \in U_E(x_0) \setminus x_0 = \overset{\circ}{U}_E(x_0)$, the point x_0 is called *strict local maximum* (resp. *strict local minimum*) and the value of the function $f : E \to \mathbb{R}$ a *strict local maximum value* (resp. *strict local minimum value*).

Fig. 5.8

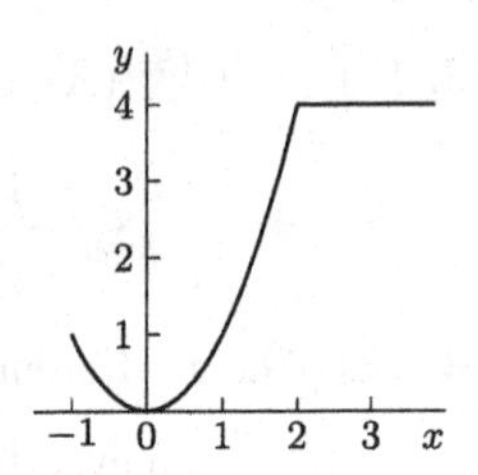

Definition 3 The local maxima and minima are called *local extrema* and the values of the function at these points *local extreme values of the function.*

Example 1 Let

$$f(x) = \begin{cases} x^2, & \text{if } -1 \le x < 2, \\ 4, & \text{if } 2 \le x \end{cases}$$

(see Fig. 5.8). For this function

$x = -1$ is a strict local maximum;
$x = 0$ is a strict local minimum;
$x = 2$ is a local maximum;

the points $x > 2$ are all local extrema, being simultaneously maxima and minima, since the function is locally constant at these points.

Example 2 Let $f(x) = \sin\frac{1}{x}$ on the set $E = \mathbb{R}\backslash 0$.

The points $x = (\frac{\pi}{2} + 2k\pi)^{-1}$, $k \in \mathbb{Z}$, are strict local maxima, and the points $x = (-\frac{\pi}{2} + 2k\pi)^{-1}$, $k \in \mathbb{Z}$, are strict local minima for $f(x)$ (see Fig. 4.1).

Definition 4 An extremum $x_0 \in E$ of the function $f : E \to \mathbb{R}$ is called an *interior* extremum if x_0 is a limit point of both sets $E_- = \{x \in E \mid x < x_0\}$ and $E_+ = \{x \in E \mid x > x_0\}$.

In Example 2, all the extrema are interior extrema, while in Example 1 the point $x = -1$ is not an interior extremum.

Lemma 1 (Fermat) *If a function* $f : E \to \mathbb{R}$ *is differentiable at an interior extremum,* $x_0 \in E$*, then its derivative at* x_0 *is* $0 : f'(x_0) = 0$.

Proof By definition of differentiability at x_0 we have

$$f(x_0 + h) - f(x_0) = f'(x_0)h + \alpha(x_0; h)h,$$

where $\alpha(x_0; h) \to 0$ as $h \to x$, $x_0 + h \in E$.

Let us rewrite this relation as follows:

$$f(x_0 + h) - f(x_0) = \big[f'(x_0) + \alpha(x_0; h)\big]h. \tag{5.45}$$

Since x_0 is an extremum, the left-hand side of Eq. (5.45) is either non-negative or nonpositive for all values of h sufficiently close to 0 and for which $x_0 + h \in E$.

If $f'(x_0) \neq 0$, then for h sufficiently close to 0 the quantity $f'(x_0) + \alpha(x_0; h)$ would have the same sign as $f'(x_0)$, since $\alpha(x_0; h) \to 0$ as $h \to 0$, $x_0 + h \in E$.

But the value of h can be both positive or negative, given that x_0 is an interior extremum.

Thus, assuming that $f'(x_0) \neq 0$, we find that the right-hand side of (5.45) changes sign when h does (for h sufficiently close to 0), while the left-hand side cannot change sign when h is sufficiently close to 0. This contradiction completes the proof. □

Remarks on Fermat's Lemma 1^0. Fermat's lemma thus gives a necessary condition for an interior extremum of a differentiable function. For noninterior extrema (such as the point $x = -1$ in Example 1) it is generally not true that $f'(x_0) = 0$.

2^0. Geometrically this lemma is obvious, since it asserts that at an extremum of a differentiable function the tangent to its graph is horizontal. (After all, $f'(x_0)$ is the tangent of the angle the tangent line makes with the x-axis.)

3^0. Physically this lemma means that in motion along a line the velocity must be zero at the instant when the direction reverses (which is an extremum!).

This lemma and the theorem on the maximum (or minimum) of a continuous function on a closed interval together imply the following proposition.

Proposition 1 (Rolle's[10] theorem) *If a function* $f : [a, b] \to \mathbb{R}$ *is continuous on a closed interval* $[a, b]$ *and differentiable on the open interval* $]a, b[$ *and* $f(a) = f(b)$, *then there exists a point* $\xi \in]a, b[$ *such that* $f'(\xi) = 0$.

Proof Since the function f is continuous on $[a, b]$, there exist points $x_m, x_M \in [a, b]$ at which it assumes its minimal and maximal values respectively. If $f(x_m) = f(x_M)$, then the function is constant on $[a, b]$; and since in that case $f'(x) \equiv 0$, the assertion is obviously true. If $f(x_m) < f(x_M)$, then, since $f(a) = f(b)$, one of the points x_m and x_M must lie in the open interval $]a, b[$. We denote it by ξ. Fermat's lemma now implies that $f'(\xi) = 0$. □

5.3.2 The Theorems of Lagrange and Cauchy on Finite Increments

The following proposition is one of the most frequently used and important methods of studying numerical-valued functions.

Theorem 1 (Lagrange's finite-increment theorem) *If a function* $f : [a, b] \to \mathbb{R}$ *is continuous on a closed interval* $[a, b]$ *and differentiable on the open interval,* $]a, b[$,

[10] M. Rolle (1652–1719) – French mathematician.

Fig. 5.9

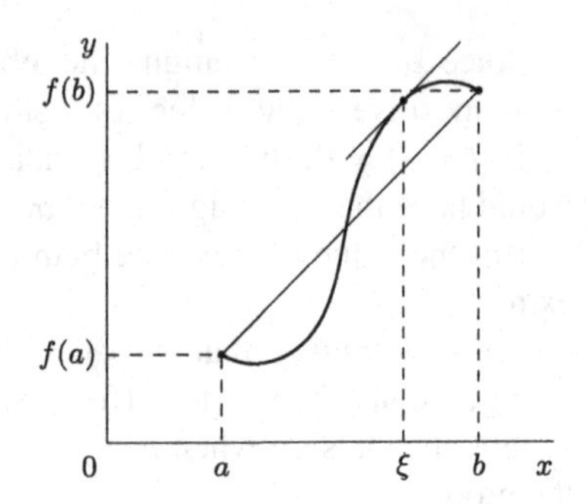

there exists a point $\xi \in]a, b[$ such that

$$\boxed{f(b) - f(a) = f'(\xi)(b - a).} \tag{5.46}$$

Proof Consider the auxiliary function

$$F(x) = f(x) - \frac{f(b) - f(a)}{b - a}(x - a),$$

which is obviously continuous on the closed interval $[a, b]$ and differentiable on the open interval $]a, b[$ and has equal values at the endpoints: $F(a) = F(b) = f(a)$. Applying Rolle's theorem to $F(x)$, we find a point $\xi \in]a, b[$ at which

$$F'(\xi) = f'(\xi) - \frac{f(b) - f(a)}{b - a} = 0.$$ □

Remarks on Lagrange's Theorem 1^0. In geometric language Lagrange's theorem means (see Fig. 5.9) that at some point $(\xi, f(\xi))$, where $\xi \in]a, b[$, the tangent to the graph of the function is parallel to the chord joining the points $(a, f(a))$ and $(b, f(b))$, since the slope of the chord equals $\frac{f(b)-f(a)}{b-a}$.

2^0. If x is interpreted as time and $f(b) - f(a)$ as the amount of displacement over the time $b - a$ of a particle moving along a line, Lagrange's theorem says that the velocity $f'(x)$ of the particle at some time $\xi \in]a, b[$ is such that if the particle had moved with the constant velocity $f'(\xi)$ over the whole time interval, it would have been displaced by the same amount $f(b) - f(a)$. It is natural to call $f'(\xi)$ the *average velocity* over the time interval $[a, b]$.

3^0. We note nevertheless that for motion that is not along a straight line there may be no average speed in the sense of Remark 2^0. Indeed, suppose the particle is moving around a circle of unit radius at constant angular velocity $\omega = 1$. Its law of motion, as we know, can be written as

$$\mathbf{r}(t) = (\cos t, \sin t).$$

Then

$$\dot{\mathbf{r}}(t) = \mathbf{v}(t) = (-\sin t, \cos t)$$

and $|\mathbf{v}| = \sqrt{\sin^2 t + \cos^2 t} = 1$.

The particle is at the same point $\mathbf{r}(0) = \mathbf{r}(2\pi) = (1, 0)$ at times $t = 0$ and $t = 2\pi$, and the equality

$$\mathbf{r}(2\pi) - \mathbf{r}(0) = \mathbf{v}(\xi)(2\pi - 0)$$

would mean that $\mathbf{v}(\xi) = \mathbf{0}$. But this is impossible.

Even so, we shall learn that there is still a relation between the displacement over a time interval and the velocity. It consists of the fact that the full length L of the path traversed cannot exceed the maximal absolute value of the velocity multiplied by the time interval of the displacement. What has just been said can be written in the following more precise form:

$$\boxed{\left|\mathbf{r}(b) - \mathbf{r}(a)\right| \le \sup_{t \in]a,b[} \left|\dot{\mathbf{r}}(t)\right| |b - a|.} \tag{5.47}$$

As will be shown later, this natural inequality does indeed always hold. It is also called Lagrange's finite-increment theorem, while relation (5.46), which is valid only for numerical-valued functions, is often called the *Lagrange mean-value theorem* (the role of the mean in this case is played by both the value $f'(\xi)$ of the velocity and by the point ξ between a and b).

4^0. Lagrange's theorem is important in that it connects the increment of a function over a finite interval with the derivative of the function on that interval. Up to now we have not had such a theorem on finite increments and have characterized only the local (infinitesimal) increment of a function in terms of the derivative or differential at a given point.

Corollaries of Lagrange's Theorem

Corollary 1 (Criterion for monotonicity of a function) *If the derivative of a function is nonnegative* (*resp. positive*) *at every point of an open interval, then the function is nondecreasing* (*resp. increasing*) *on that interval.*

Proof Indeed, if x_1 and x_2 are two points of the interval and $x_1 < x_2$, that is, $x_2 - x_1 > 0$, then by formula (5.46)

$$f(x_2) - f(x_1) = f'(\xi)(x_2 - x_1), \quad \text{where } x_1 < \xi < x_2,$$

and therefore, the sign of the difference on the left-hand side of this equality is the same as the sign of $f'(\xi)$. □

Naturally an analogous assertion can be made about the nonincreasing (resp. decreasing) nature of a function with a nonpositive (resp. negative) derivative.

Remark By the inverse function theorem and Corollary 1 we can conclude, in particular, that if a numerical-valued function $f(x)$ on some interval I has a derivative that is always positive or always negative, then the function is continuous and monotonic on I and has an inverse function f^{-1} that is defined on the interval $I' = f(I)$ and is differentiable on it.

Corollary 2 (Criterion for a function to be constant) *A function that is continuous on a closed interval* $[a, b]$ *is constant on it if and only if its derivative equals zero at every point of the interval* $[a, b]$ *(or only the open interval* $]a, b[$*).*

Proof Only the fact that $f'(x) \equiv 0$ on $]a, b[$ implies that $f(x_1) = f(x_2)$ for all $x_1, x_2, \in [a, b]$ is of interest. But this follows from Lagrange's formula, according to which

$$f(x_2) - f(x_1) = f'(\xi)(x_2 - x_1) = 0,$$

since ξ lies between x_1 and x_2, that is, $\xi \in]a, b[$, and so $f'(\xi) = 0$. □

Remark From this we can draw the following conclusion (which as we shall see, is very important for integral calculus): *If the derivatives* $F_1'(x)$ *and* $F_2'(x)$ *of two functions* $F_1(x)$ *and* $F_2(x)$ *are equal on some interval, that is,* $F_1'(x) = F_2'(x)$ *on the interval, then the difference* $F_1(x) - F_2(x)$ *is constant.*

The following proposition is a useful generalization of Lagrange's theorem, and is also based on Rolle's theorem.

Proposition 2 (Cauchy's finite-increment theorem) *Let* $x = x(t)$ *and* $y = y(t)$ *be functions that are continuous on a closed interval* $[\alpha, \beta]$ *and differentiable on the open interval* $]\alpha, \beta[$.

Then there exists a point $\tau \in [\alpha, \beta]$ *such that*

$$x'(\tau)\big(y(\beta) - y(\alpha)\big) = y'(\tau)\big(x(\beta) - x(\alpha)\big).$$

If in addition $x'(t) \neq 0$ *for each* $t \in]\alpha, \beta[$, *then* $x(\alpha) \neq x(\beta)$ *and we have the equality*

$$\frac{y(\beta) - y(\alpha)}{x(\beta) - x(\alpha)} = \frac{y'(\tau)}{x'(\tau)}. \tag{5.48}$$

Proof The function $F(t) = x(t)(y(\beta) - y(\alpha)) - y(t)(x(\beta) - x(\alpha))$ satisfies the hypotheses of Rolle's theorem on the closed interval $[\alpha, \beta]$. Therefore there exists a point $\tau \in]\alpha, \beta[$ at which $F'(\tau) = 0$, which is equivalent to the equality to be proved. To obtain relation (5.48) from it, it remains only to observe that if $x'(t) \neq 0$ on $]\alpha, \beta[$, then $x(\alpha) \neq x(\beta)$, again by Rolle's theorem. □

Remarks on Cauchy's Theorem 1^0. If we regard the pair $x(t), y(t)$ as the law of motion of a particle, then $(x'(t), y'(t))$ is its velocity vector at time t, and $(x(\beta) - x(\alpha), y(\beta) - y(\alpha))$ is its displacement vector over the time interval $[\alpha, \beta]$. The theorem then asserts that at some instant of time $\tau \in [\alpha, \beta]$ these two vectors are collinear. However, this fact, which applies to motion in a plane, is the same kind of pleasant exception as the mean-velocity theorem in the case of motion along a line. Indeed, imagine a particle moving at uniform speed along a helix. Its velocity makes a constant nonzero angle with the vertical, while the displacement vector can be purely vertical (after one complete turn).

2^0. Lagrange's formula can be obtained from Cauchy's by setting $x = x(t) = t$, $y(t) = y(x) = f(x)$, $\alpha = a$, $\beta = b$.

5.3.3 Taylor's Formula

From the amount of differential calculus that has been explained up to this point one may obtain the correct impression that the more derivatives of two functions coincide (including the derivative of zeroth order) at a point, the better these functions approximate each other in a neighborhood of that point. We have mostly been interested in approximations of a function in the neighborhood of a point by a polynomial $P_n(x) = P_n(x_0; x) = c_0 + c_1(x - x_0) + \cdots + c_n(x - x_0)^n$, and that will continue to be our main interest. We know (see Example 25 in Sect. 5.2.6) that an algebraic polynomial can be represented as

$$P_n(x) = P_n(x_0) + \frac{P_n'(x_0)}{1!}(x - x_0) + \cdots + \frac{P_n^{(n)}(x_0)}{n!}(x - x_0)^n,$$

that is, $c_k = \frac{P_n^{(k)}(x_0)}{k!}$ $(k = 0, 1, \dots, n)$. This can easily be verified directly.

Thus, if we are given a function $f(x)$ having derivatives up to order n inclusive at x_0, we can immediately write the polynomial

$$P_n(x_0; x) = P_n(x) = f(x_0) + \frac{f'(x_0)}{1!}(x - x_0) + \cdots + \frac{f^{(n)}(x_0)}{n!}(x - x_0)^n, \quad (5.49)$$

whose derivatives up to order n inclusive at the point x_0 are the same as the corresponding derivatives of $f(x)$ at that point.

Definition 5 The algebraic polynomial given by (5.49) is the *Taylor*[11] *polynomial of order n of $f(x)$ at x_0.*

We shall be interested in the value of

$$f(x) - P_n(x_0; x) = r_n(x_0; x) \quad (5.50)$$

of the discrepancy between the polynomial $P_n(x)$ and the function $f(x)$, which is often called the *remainder*, more precisely, the *nth remainder* or the *nth remainder*

[11]B. Taylor (1685–1731) – British mathematician.

term in Taylor's formula:

$$f(x) = f(x_0) + \frac{f'(x_0)}{1!}(x - x_0) + \cdots + \frac{f^{(n)}(x_0)}{n!}(x - x_0)^n + r_n(x_0; x). \quad (5.51)$$

The equality (5.51) itself is of course of no interest if we know nothing more about the function $r_n(x_0; x)$ than its definition (5.50).

We shall now use a highly artificial device to obtain information on the remainder term. A more natural route to this information will come from the integral calculus.

Theorem 2 *If the function f is continuous on the closed interval with end-points x_0 and x along with its first n derivatives, and it has a derivative of order $n+1$ at the interior points of this interval, then for any function φ that is continuous on this closed interval and has a nonzero derivative at its interior points, there exists a point ξ between x_0 and x such that*

$$r_n(x_0; x) = \frac{\varphi(x) - \varphi(x_0)}{\varphi'(\xi) n!} f^{(n+1)}(\xi)(x - \xi)^n. \quad (5.52)$$

Proof On the closed interval I with endpoints x_0 and x we consider the auxiliary function

$$F(t) = f(x) - P_n(t; x) \quad (5.53)$$

of the argument t. We now write out the definition of the function $F(t)$ in more detail:

$$F(t) = f(x) - \left[f(t) + \frac{f'(t)}{1!}(x - t) + \cdots + \frac{f^{(n)}(t)}{n!}(x - t)^n \right]. \quad (5.54)$$

We see from the definition of the function $F(t)$ and the hypotheses of the theorem that F is continuous on the closed interval I and differentiable at its interior points, and that

$$F'(t) = -\left[f'(t) - \frac{f'(t)}{1!} + \frac{f''(t)}{1!}(x - t) - \frac{f''(t)}{1!}(x - t) + \right.$$

$$\left. + \frac{f'''(t)}{2!}(x - t)^2 - \cdots + \frac{f^{(n+1)}(t)}{n!}(x - t)^n \right] = -\frac{f^{(n+1)}(t)}{n!}(x - t)^n.$$

Applying Cauchy's theorem to the pair of functions $F(t), \varphi(t)$ on the closed interval I (see relation (5.48)), we find a point ξ between x_0 and x at which

$$\frac{F(x) - F(x_0)}{\varphi(x) - \varphi(x_0)} = \frac{F'(\xi)}{\varphi'(\xi)}.$$

Substituting the expression for $F'(\xi)$ here and observing from comparison of formulas (5.50), (5.53) and (5.54) that $F(x) - F(x_0) = 0 - F(x_0) = -r_n(x_0; x)$, we obtain formula (5.52). □

Setting $\varphi(t) = x - t$ in (5.52), we obtain the following corollary.

Corollary 1 (Cauchy's formula for the remainder term)

$$\boxed{r_n(x_0; x) = \frac{1}{n!} f^{(n+1)}(\xi)(x-\xi)^n(x-x_0).} \tag{5.55}$$

A particularly elegant formula results if we set $\varphi(t) = (x-t)^{n+1}$ in (5.52):

Corollary 2 (The Lagrange form of the remainder)

$$\boxed{r_n(x_0; x) = \frac{1}{(n+1)!} f^{(n+1)}(\xi)(x-x_0)^{n+1}.} \tag{5.56}$$

We remark that when $x_0 = 0$ Taylor's formula (5.51) is often called *MacLaurin's formula.*[12]

Let us consider some examples.

Example 3 For the function $f(x) = \mathrm{e}^x$ with $x_0 = 0$ Taylor's formula has the form

$$\mathrm{e}^x = 1 + \frac{1}{1!}x + \frac{1}{2!}x^2 + \cdots + \frac{1}{n!}x^n + r_n(0; x), \tag{5.57}$$

and by (5.56) we can assume that

$$r_n(0; x) = \frac{1}{(n+1)!}\mathrm{e}^{\xi} \cdot x^{n+1},$$

where $|\xi| < |x|$.

Thus

$$\bigl|r_n(0; x)\bigr| = \frac{1}{(n+1)!}\mathrm{e}^{\xi} \cdot |x|^{n+1} < \frac{|x|^{n+1}}{(n+1)!}\mathrm{e}^{|x|}. \tag{5.58}$$

But for each fixed $x \in \mathbb{R}$, if $n \to \infty$, the quantity $\frac{|x|^{n+1}}{(n+1)!}$, as we know (see Example 12 of Sect. 3.1.3), tends to zero. Hence it follows from the estimate (5.58) and the definition of the sum of a series that

$$\mathrm{e}^x = 1 + \frac{1}{1!}x + \frac{1}{2!}x^2 + \cdots + \frac{1}{n!}x^n + \cdots \tag{5.59}$$

for all $x \in \mathbb{R}$.

---

[12] C. MacLaurin (1698–1746) – British mathematician.

Example 4 We obtain the expansion of the function a^x for any a, $0 < a$, $a \neq 1$, similarly:

$$a^x = 1 + \frac{\ln a}{1!}x + \frac{\ln^2 a}{2!}x^2 + \cdots + \frac{\ln^n a}{n!}x^n + \cdots.$$

Example 5 Let $f(x) = \sin x$. We know (see Example 18 of Sect. 5.2.6) that $f^{(n)}(x) = \sin(x + \frac{\pi}{2}n)$, $n \in \mathbb{N}$, and so by Lagrange's formula (5.56) with $x_0 = 0$ and any $x \in \mathbb{R}$ we find

$$r_n(0; x) = \frac{1}{(n+1)!} \sin\left(\xi + \frac{\pi}{2}(n+1)\right)x^{n+1}, \tag{5.60}$$

from which it follows that $r_n(0; x)$ tends to zero for any $x \in \mathbb{R}$ as $n \to \infty$. Thus we have the expansion

$$\sin x = x - \frac{1}{3!}x^3 + \frac{1}{5!}x^5 - \cdots + \frac{(-1)^n}{(2n+1)!}x^{2n+1} + \cdots \tag{5.61}$$

for every $x \in \mathbb{R}$.

Example 6 Similarly, for the function $f(x) = \cos x$, we obtain

$$r_n(0; x) = \frac{1}{(n+1)!} \cos\left(\xi + \frac{\pi}{2}(n+1)\right)x^{n+1} \tag{5.62}$$

and

$$\cos x = 1 - \frac{1}{2!}x^2 + \frac{1}{4!}x^4 - \cdots + \frac{(-1)^n}{(2n)!}x^{2n} + \cdots \tag{5.63}$$

for $x \in \mathbb{R}$.

Example 7 Since $\sinh' x = \cosh x$ and $\cosh' x = \sinh x$, formula (5.56) yields the following expression for the remainder in the Taylor series of $f(x) = \sinh x$:

$$r_n(0; x) = \frac{1}{(n+1)!} f^{(n+1)}(\xi)x^{n+1},$$

where $f^{(n+1)}(\xi) = \sinh \xi$ if n is even and $f^{(n+1)}(\xi) = \cosh \xi$ if n is odd. In any case $|f^{(n+1)}(\xi)| \leq \max\{|\sinh x|, |\cosh x|\}$, since $|\xi| < |x|$. Hence for any given value $x \in \mathbb{R}$ we have $r_n(0, x) \to 0$ as $n \to \infty$, and we obtain the expansion

$$\sinh x = x + \frac{1}{3!}x^3 + \frac{1}{5!}x^5 + \cdots + \frac{1}{(2n+1)!}x^{2n+1} + \cdots, \tag{5.64}$$

valid for all $x \in \mathbb{R}$.

Example 8 Similarly we obtain the expansion

$$\cosh x = 1 + \frac{1}{2!}x^2 + \frac{1}{4!}x^4 + \cdots + \frac{1}{(2n)!}x^{2n} + \cdots, \tag{5.65}$$

valid for any $x \in \mathbb{R}$.

Example 9 For the function $f(x) = \ln(1+x)$ we have $f^{(n)}(x) = \frac{(-1)^{n-1}(n-1)!}{(1+x)^n}$, so that the Taylor series of this function at $x_0 = 0$ is

$$\ln(1+x) = x - \frac{1}{2}x^2 + \frac{1}{3}x^3 - \cdots + \frac{(-1)^{n-1}}{n}x^n + r_n(0;x). \tag{5.66}$$

This time we represent $r_n(0;x)$ using Cauchy's formula (5.55):

$$r_n(0;x) = \frac{1}{n!}\frac{(-1)^n n!}{(1+\xi)^n}(x-\xi)^n x,$$

or

$$r_n(0;x) = (-1)^n x\left(\frac{x-\xi}{1+\xi}\right)^n, \tag{5.67}$$

where ξ lies between 0 and x.

If $|x| < 1$, it follows from the condition that ξ lies between 0 and x that

$$\left|\frac{x-\xi}{1+\xi}\right| = \frac{|x| - |\xi|}{|1+\xi|} \le \frac{|x| - |\xi|}{1 - |\xi|} = 1 - \frac{1-|x|}{1-|\xi|} \le 1 - \frac{1-|x|}{1-|0|} = |x|. \tag{5.68}$$

Thus for $|x| < 1$

$$\left|r_n(0;x)\right| \le |x|^{n+1}, \tag{5.69}$$

and consequently the following expansion is valid for $|x| < 1$:

$$\ln(1+x) = x - \frac{1}{2}x^2 + \frac{1}{3}x^3 - \cdots + \frac{(-1)^{n-1}}{n}x^n + \cdots. \tag{5.70}$$

We remark that outside the closed interval $|x| \le 1$ the series on the right-hand side of (5.70) diverges at every point, since its general term does not tend to zero if $|x| > 1$.

Example 10 For the function $(1+x)^\alpha$, where $\alpha \in \mathbb{R}$, we have $f^{(n)}(x) = \alpha(\alpha-1)\cdots(\alpha-n+1)(1+x)^{\alpha-n}$, so that Taylor's formula at $x_0 = 0$ for this function has the form

$$(1+x)^\alpha = 1 + \frac{\alpha}{1!}x + \frac{\alpha(\alpha-1)}{2!}x^2 + \cdots + \\ + \frac{\alpha(\alpha-1)\cdots(\alpha-n+1)}{n!}x^n + r_n(0;x). \tag{5.71}$$

Using Cauchy's formula (5.55), we find

$$r_n(0;x) = \frac{\alpha(\alpha-1)\cdots(\alpha-n)}{n!}(1+\xi)^{\alpha-n-1}(x-\xi)^n x, \tag{5.72}$$

where ξ lies between 0 and x.

If $|x| < 1$, then, using the estimate (5.68), we have

$$\left|r_n(0;x)\right| \le \left|\alpha\left(1-\frac{\alpha}{1}\right)\cdots\left(1-\frac{\alpha}{n}\right)\right|(1+\xi)^{\alpha-1}|x|^{n+1}. \tag{5.73}$$

When n is increased by 1, the right-hand side of Eq. (5.73) is multiplied by $|(1-\frac{\alpha}{n+1})x|$. But since $|x| < 1$, we shall have $|(1-\frac{\alpha}{n+1})x| < q < 1$, independently of the value of α, provided $|x| < q < 1$ and n is sufficiently large.

It follows from this that $r_n(0;x) \to 0$ as $n \to \infty$ for any $\alpha \in \mathbb{R}$ and any x in the open interval $|x| < 1$. Therefore the expansion obtained by Newton (*Newton's binomial theorem*) is valid on the open interval $|x| < 1$:

$$(1+x)^\alpha = 1 + \frac{\alpha}{1!}x + \frac{\alpha(\alpha-1)}{2!}x^2 + \cdots + \frac{\alpha(\alpha-1)\cdots(\alpha-n+1)}{n!}x^n + \cdots. \tag{5.74}$$

We remark that d'Alembert's test (see Paragraph b of Sect. 3.1.4) implies that for $|x| > 1$ the series (5.74) generally diverges if $\alpha \notin \mathbb{N}$. Let us now consider separately the case when $\alpha = n \in \mathbb{N}$.

In this case $f(x) = (1+x)^\alpha = (1+x)^n$ is a polynomial of degree n and hence all of its derivatives of order higher than n are equal to 0. Therefore Taylor's formula, together with, for example, the Lagrange form of the remainder, enables us to write the following equality:

$$(1+x)^n = 1 + \frac{n}{1!}x + \frac{n(n-1)}{2!}x^2 + \cdots + \frac{n(n-1)\cdots 1}{n!}x^n, \tag{5.75}$$

which is the Newton binomial theorem known from high school for a natural-number exponent:

$$(1+x)^n = 1 + \binom{n}{1}x + \binom{n}{2}x^2 + \cdots + \binom{n}{n}x^n.$$

Thus we have defined Taylor's formula (5.51) and obtained the forms (5.52), (5.55), and (5.56) for the remainder term in the formula. We have obtained the relations (5.58), (5.60), (5.62), (5.69), and (5.73), which enable us to estimate the error in computing the important elementary functions using Taylor's formula. Finally, we have obtained the power-series expansions of these functions.

Definition 6 If the function $f(x)$ has derivatives of all orders $n \in \mathbb{N}$ at a point x_0, the series

$$f(x_0) + \frac{1}{1!}f'(x_0)(x-x_0) + \cdots + \frac{1}{n!}f^{(n)}(x_0)(x-x_0)^n + \cdots$$

is called the *Taylor series* of f at the point x_0.

It should not be thought that the Taylor series of an infinitely differentiable function converges in some neighborhood of x_0, for given any sequence $c_0, c_1, \dots, c_n, \dots$ of numbers, one can construct (although this is not simple to do) a function $f(x)$ such that $f^{(n)}(x_0) = c_n$, $n \in \mathbb{N}$.

It should also not be thought that if the Taylor series converges, it necessarily converges to the function that generated it. A Taylor series converges to the function that generated it only when the generating function belongs to the class of so-called *analytic functions*.

Here is Cauchy's example of a nonanalytic function:

$$f(x) = \begin{cases} e^{-1/x^2}, & \text{if } x \neq 0, \\ 0, & \text{if } x = 0. \end{cases}$$

Starting from the definition of the derivative and the fact that $x^k e^{-1/x^2} \to 0$ as $x \to 0$, independently of the value of k (see Example 30 in Sect. 3.2), one can verify that $f^{(n)}(0) = 0$ for $n = 0, 1, 2, \dots$. Thus, the Taylor series in this case has all its terms equal to 0 and hence its sum is identically equal to 0, while $f(x) \neq 0$ if $x \neq 0$.

In conclusion, we discuss a local version of Taylor's formula.

We return once again to the problem of the local representation of a function $f : E \to \mathbb{R}$ by a polynomial, which we began to discuss in Sect. 5.1.3. We wish to choose the polynomial $P_n(x_0; x) = x_0 + c_1(x - x_0) + \dots + c_n(x - x_0)^n$ so as to have

$$f(x) = P_n(x) + o\big((x - x_0)^n\big) \quad \text{as } x \to x_0, x \in E,$$

or, in more detail,

$$f(x) = c_0 + c_1(x - x_0) + \dots + c_n(x - x_0)^n + o\big((x - x_0)^n\big) \quad \text{as } x \to x_0, x \in E. \tag{5.76}$$

We now state explicitly a proposition that has already been proved in all its essentials.

Proposition 3 *If there exists a polynomial* $P_n(x_0; x) = c_0 + c_1(x - x_0) + \dots + c_n(x - x_0)^n$ *satisfying condition* (5.76), *that polynomial is unique.*

Proof Indeed, from relation (5.76) we obtain the coefficients of the polynomial successively and completely unambiguously

$$\begin{aligned} c_0 &= \lim_{E \ni x \to x_0} f(x), \\ c_1 &= \lim_{E \ni x \to x_0} \frac{f(x) - c_0}{x - x_0}, \\ &\vdots \\ c_n &= \lim_{E \ni x \to x_0} \frac{f(x) - [c_0 + \dots + c_{n-1}(x - x_0)^{n-1}]}{(x - x_0)^n}. \end{aligned}$$

□

We now prove the local version of Taylor's theorem.

Proposition 4 (The local Taylor formula) *Let E be a closed interval having $x_0 \in \mathbb{R}$ as an endpoint. If the function $f : E \to \mathbb{R}$ has derivatives $f'(x_0), \dots, f^{(n)}(x_0)$ up to order n inclusive at the point x_0, then the following representation holds*:

$$f(x) = f(x_0) + \frac{f'(x_0)}{1!}(x - x_0) + \cdots + \frac{f^{(n)}(x_0)}{n!}(x - x_0)^n + \\ + o\bigl((x - x_0)^n\bigr) \quad \text{as } x \to x_0,\ x \in E. \tag{5.77}$$

Thus the problem of the local approximation of a differentiable function is solved by the Taylor polynomial of the appropriate order.

Since the Taylor polynomial $P_n(x_0; x)$ is constructed from the requirement that its derivatives up to order n inclusive must coincide with the corresponding derivatives of the function f at x_0, it follows that $f^{(k)}(x_0) - P_n^{(k)}(x_0; x_0) = 0$ $(k = 0, 1, \dots, n)$ and the validity of formula (5.77) is established by the following lemma.

Lemma 2 *If a function $\varphi : E \to \mathbb{R}$, defined on a closed interval E with endpoint x_0, is such that it has derivatives up to order n inclusive at x_0 and $\varphi(x_0) = \varphi'(x_0) = \cdots = \varphi^{(n)}(x_0) = 0$, then $\varphi(x) = o((x - x_0)^n)$ as $x \to x_0$, $x \in E$.*

Proof For $n = 1$ the assertion follows from the definition of differentiability of the function φ at x_0, by virtue of which

$$\varphi(x) = \varphi(x_0) + \varphi'(x_0)(x - x_0) + o(x - x_0) \quad \text{as } x \to x_0,\ x \in E,$$

and, since $\varphi(x_0) = \varphi'(x_0) = 0$, we have

$$\varphi(x) = o(x - x_0) \quad \text{as } x \to x_0,\ x \in E.$$

Suppose the assertion has been proved for order $n = k - 1 \geq 1$. We shall show that it is then valid for order $n = k \geq 2$.

We make the preliminary remark that since

$$\varphi^{(k)}(x_0) = \bigl(\varphi^{(k-1)}\bigr)'(x_0) = \lim_{E \ni x \to x_0} \frac{\varphi^{(k-1)}(x) - \varphi^{(k-1)}(x_0)}{x - x_0},$$

the existence of $\varphi^{(k)}(x_0)$ presumes that the function $\varphi^{(k-1)}(x)$ is defined on E, at least near the point x_0. Shrinking the closed interval E if necessary, we can assume from the outset that the functions $\varphi(x), \varphi'(x), \dots, \varphi^{(k-1)}(x)$, where $k \geq 2$, are all defined on the whole closed interval E with endpoint x_0. Since $k \geq 2$, the function $\varphi(x)$ has a derivative $\varphi'(x)$ on E, and by hypothesis

$$\bigl(\varphi'\bigr)'(x_0) = \cdots = \bigl(\varphi'\bigr)^{(k-1)}(x_0) = 0.$$

Therefore, by the induction assumption,

$$\varphi'(x) = o\bigl((x - x_0)^{k-1}\bigr) \quad \text{as } x \to x_0, x \in E.$$

Then, using Lagrange's theorem, we obtain

$$\varphi(x)=\varphi(x)-\varphi(x_0)=\varphi'(\xi)(x-x_0)=\alpha(\xi)(\xi-x_0)^{(k-1)}(x-x_0),$$

where ξ lies between x_0 and x, that is, $|\xi-x_0|<|x-x_0|$, and $\alpha(\xi)\to 0$ as $\xi\to x_0$, $\xi\in E$. Hence as $x\to x_0$, $x\in E$, we have simultaneously $\xi\to x_0$, $\xi\in E$, and $\alpha(\xi)\to 0$. Since

$$\bigl|\varphi(x)\bigr|\le\bigl|\alpha(\xi)\bigr||x-x_0|^{k-1}|x-x_0|,$$

we have verified that

$$\varphi(x)=o\bigl((x-x_0)^k\bigr)\quad\text{as } x\to x_0, x\in E.$$

Thus, the assertion of Lemma 2 has been verified by mathematical induction. □

Relation (5.77) is called the *local Taylor formula* since the form of the remainder term given in it (the so-called *Peano form*)

$$r_n(x_0;x)=o\bigl((x-x_0)^n\bigr), \tag{5.78}$$

makes it possible to draw inferences only about the asymptotic nature of the connection between the Taylor polynomial and the function as $x\to x_0$, $x\in E$.

Formula (5.77) is therefore convenient in computing limits and describing the asymptotic behavior of a function as $x\to x_0$, $x\in E$, but it cannot help with the approximate computation of the values of the function until some actual estimate of the quantity $r_n(x_0;x)=o((x-x_0)^n)$ is available.

Let us now summarize our results. We have defined the Taylor polynomial

$$P_n(x_0;x)=f(x_0)+\frac{f'(x_0)}{1!}(x-x_0)+\cdots+\frac{f^{(n)}(x_0)}{n!}(x-x_0)^n,$$

written the Taylor formula

$$f(x)=f(x_0)+\frac{f'(x_0)}{1!}(x-x_0)+\cdots+\frac{f^{(n)}(x_0)}{n!}(x-x_0)^n+r_n(x_0;x),$$

and obtained the following very important specific form of it:

If f has a derivative of order $n+1$ on the open interval with endpoints x_0 and x, then

$$f(x)=f(x_0)+\frac{f'(x_0)}{1!}(x-x_0)+\cdots+\frac{f^{(n)}(x_0)}{n!}(x-x_0)^n+$$
$$+\frac{f^{(n+1)}(\xi)}{(n+1)!}(x-x_0)^{n+1}, \tag{5.79}$$

where ξ is a point between x_0 and x.

If f has derivatives of orders up to $n \geq 1$ inclusive at the point x_0, then

$$f(x) = f(x_0) + \frac{f'(x_0)}{1!}(x - x_0) + \cdots + \frac{f^{(n)}(x_0)}{n!}(x - x_0)^n + o\big((x - x_0)^n\big). \quad (5.80)$$

Relation (5.79), called *Taylor's formula with the Lagrange form of the remainder*, is obviously a generalization of Lagrange's mean-value theorem, to which it reduces when $n = 0$.

Relation (5.80), called *Taylor's formula with the Peano form of the remainder*, is obviously a generalization of the definition of differentiability of a function at a point, to which it reduces when $n = 1$.

We remark that formula (5.79) is nearly always the more substantive of the two. For, on the one hand, as we have seen, it enables us to estimate the absolute magnitude of the remainder term. On the other hand, when, for example, $f^{(n+1)}(x)$ is bounded in a neighborhood of x_0, it also implies the asymptotic formula

$$f(x) = f(x_0) + \frac{f'(x_0)}{1!}(x - x_0) + \cdots + \frac{f^{(n)}(x_0)}{n!}(x - x_0)^n + O\big((x - x_0)^{n+1}\big). \quad (5.81)$$

Thus for infinitely differentiable functions, with which classical analysis deals in the overwhelming majority of cases, formula (5.79) contains the local formula (5.80).

In particular, on the basis of (5.81) and Examples 3–10 just studied, we can now write the following table of asymptotic formulas as $x \to 0$:

$$\mathrm{e}^x = 1 + \frac{1}{1!}x + \frac{1}{2!}x^2 + \cdots + \frac{1}{n!}x^n + O\big(x^{n+1}\big),$$

$$\cos x = 1 - \frac{1}{2!}x^2 + \frac{1}{4!}x^4 - \cdots + \frac{(-1)^n}{(2n)!}x^{2n} + O\big(x^{2n+2}\big),$$

$$\sin x = x - \frac{1}{3!}x^3 + \frac{1}{5!}x^5 - \cdots + \frac{(-1)^n}{(2n+1)!}x^{2n+1} + O\big(x^{2n+3}\big),$$

$$\cosh x = 1 + \frac{1}{2!}x^2 + \frac{1}{4!}x^4 + \cdots + \frac{1}{(2n)!}x^{2n} + O\big(x^{2n+2}\big),$$

$$\sinh x = x + \frac{1}{3!}x^3 + \frac{1}{5!}x^5 + \cdots + \frac{1}{(2n+1)!}x^{2n+1} + O\big(x^{2n+3}\big),$$

$$\ln(1 + x) = x - \frac{1}{2}x^2 + \frac{1}{3}x^3 - \cdots + \frac{(-1)^n}{n}x^n + O\big(x^{n+1}\big),$$

$$(1 + x)^\alpha = 1 + \frac{\alpha}{1!}x + \frac{\alpha(\alpha - 1)}{2!}x^2 + \cdots + \frac{\alpha(\alpha - 1)\cdots(\alpha - n + 1)}{n!}x^n + O\big(x^{n+1}\big).$$

Let us now consider a few more examples of the use of Taylor's formula.

Example 11 We shall write a polynomial that makes it possible to compute the values of $\sin x$ on the interval $-1 \leq x \leq 1$ with absolute error at most 10^{-3}.

One can take this polynomial to be a Taylor polynomial of suitable degree obtained from the expansion of $\sin x$ in a neighborhood of $x_0 = 0$. Since

$$\sin x = x - \frac{1}{3!}x^3 + \frac{1}{5!}x^5 - \cdots + \frac{(-1)^n}{(2n+1)!}x^{2n+1} + 0 \cdot x^{2n+2} + r_{2n+2}(0; x),$$

where by Lagrange's formula

$$r_{2n+2}(0; x) = \frac{\sin(\xi + \frac{\pi}{2}(2n+3))}{(2n+3)!}x^{2n+3},$$

we have, for $|x| \le 1$,

$$|r_{2n+2}(0; x)| \le \frac{1}{(2n+3)!},$$

but $\frac{1}{(2n+3)!} < 10^{-3}$ for $n \ge 2$. Thus the approximation $\sin x \approx x - \frac{1}{3!} + \frac{1}{5!}x^5$ has the required precision on the closed interval $|x| \le 1$.

Example 12 We shall show that $\tan x = x + \frac{1}{3}x^3 + o(x^3)$ as $x \to 0$. We have

$$\tan' x = \cos^{-2} x,$$
$$\tan'' x = 2\cos^{-3} x \sin x,$$
$$\tan''' x = 6\cos^{-4} x \sin^2 x + 2\cos^{-2} x.$$

Thus, $\tan 0 = 0$, $\tan' 0 = 1$, $\tan'' 0 = 0$, $\tan''' 0 = 2$, and the relation now follows from the local Taylor formula.

Example 13 Let $\alpha > 0$. Let us study the convergence of the series $\sum_{n=1}^{\infty} \ln \cos \frac{1}{n^\alpha}$. For $\alpha > 0$ we have $\frac{1}{n^\alpha} \to 0$ as $n \to \infty$. Let us estimate the order of a term of the series:

$$\ln \cos \frac{1}{n^\alpha} = \ln\left(1 - \frac{1}{2!} \cdot \frac{1}{n^{2\alpha}} + o\left(\frac{1}{n^{2\alpha}}\right)\right) = -\frac{1}{2} \cdot \frac{1}{n^{2\alpha}} + o\left(\frac{1}{n^{2\alpha}}\right).$$

Thus we have a series of terms of constant sign whose terms are equivalent to those of the series $\sum_{n=1}^{\infty} \frac{-1}{2n^{2\alpha}}$. Since this last series converges only for $\alpha > \frac{1}{2}$, when $\alpha > 0$ the original series converges only for $\alpha > \frac{1}{2}$ (see Problem 15b) below).

Example 14 Let us show that $\ln \cos x = -\frac{1}{2}x^2 - \frac{1}{12}x^4 - \frac{1}{45}x^6 + O(x^8)$ as $x \to 0$.

This time, instead of computing six successive derivatives, we shall use the already-known expansions of $\cos x$ as $x \to 0$ and $\ln(1+u)$ as $u \to 0$:

$$\ln\cos x = \ln\left(1 - \frac{1}{2!}x^2 + \frac{1}{4!}x^4 - \frac{1}{6!}x^6 + O(x^8)\right) = \ln(1+u) =$$
$$= u - \frac{1}{2}u^2 + \frac{1}{3}u^3 + O(u^4) =$$
$$= \left(-\frac{1}{2!}x^2 + \frac{1}{4!}x^4 - \frac{1}{6!}x^6 + O(x^8)\right) -$$
$$-\frac{1}{2}\left(\frac{1}{(2!)^2}x^4 - 2\cdot\frac{1}{2!4!}x^6 + O(x^8)\right) + \frac{1}{3}\left(-\frac{1}{(2!)^3}x^6 + O(x^8)\right) =$$
$$= -\frac{1}{2}x^2 - \frac{1}{12}x^4 - \frac{1}{45}x^6 + O(x^8).$$

Example 15 Let us find the values of the first six derivatives of the function $\ln\cos x$ at $x = 0$.

We have $(\ln\cos)'x = \frac{-\sin x}{\cos x}$, and it is therefore clear that the function has derivatives of all orders at 0, since $\cos 0 \neq 0$. We shall not try to find functional expressions for these derivatives, but rather we shall make use of the uniqueness of the Taylor polynomial and the result of the preceding example.

If

$$f(x) = c_0 + c_1 x + \cdots + c_n x^n + o(x^n) \quad \text{as } x \to 0,$$

then

$$c_k = \frac{f^{(k)}(0)}{k!} \quad \text{and} \quad f^{(k)}(0) = k!c_k.$$

Thus, in the present case we obtain

$$(\ln\cos)(0) = 0, \qquad (\ln\cos)'(0) = 0, \qquad (\ln\cos)''(0) = -\frac{1}{2}\cdot 2!,$$
$$(\ln\cos)^{(3)}(0) = 0, \qquad (\ln\cos)^{(4)}(0) = -\frac{1}{12}\cdot 4!,$$
$$(\ln\cos)^{(5)}(0) = 0, \qquad (\ln\cos)^{(6)}(0) = -\frac{1}{45}\cdot 6!.$$

Example 16 Let $f(x)$ be an infinitely differentiable function at the point x_0, and suppose we know the expansion

$$f'(x) = c'_0 + c'_1 x + \cdots + c'_n x^n + O(x^{n+1})$$

of its derivative in a neighborhood of zero. Then, from the uniqueness of the Taylor expansion we have

$$(f')^{(k)}(0) = k!c'_k,$$

and so $f^{(k+1)}(0) = k!c'_k$. Thus for the function $f(x)$ itself we have the expansion

$$f(x) = f(0) + \frac{c'_0}{1!}x + \frac{1!c'_1}{2!}x^2 + \cdots + \frac{n!c'_n}{(n+1)!}x^{n+1} + O\big(x^{n+2}\big),$$

or, after simplification,

$$f(x) = f(0) + \frac{c'_0}{1}x + \frac{c'_1}{2}x^2 + \cdots + \frac{c'_n}{n+1}x^{n+1} + O\big(x^{n+2}\big).$$

Example 17 Let us find the Taylor expansion of the function $f(x) = \arctan x$ at 0.

Since $f'(x) = \frac{1}{1+x^2} = (1+x^2)^{-1} = 1 - x^2 + x^4 - \cdots + (-1)^n x^{2n} + O(x^{2n+2})$, by the considerations explained in the preceding example,

$$f(x) = f(0) + \frac{1}{1}x - \frac{1}{3}x^3 + \frac{1}{5}x^5 - \cdots + \frac{(-1)^n}{2n+1}x^{2n+1} + O\big(x^{2n+3}\big),$$

that is,

$$\arctan x = x - \frac{1}{3}x^3 + \frac{1}{5}x^5 - \cdots + \frac{(-1)^n}{2n+1}x^{2n+1} + O\big(x^{2n+3}\big).$$

Example 18 Similarly, by expanding the function $\arcsin' x = (1-x^2)^{-1/2}$ by Taylor's formula in a neighborhood of zero, we find successively,

$$(1+u)^{-1/2} = 1 + \frac{-\frac{1}{2}}{1!}u + \frac{-\frac{1}{2}(-\frac{1}{2}-1)}{2!}u^2 + \cdots + \\ + \frac{-\frac{1}{2}(-\frac{1}{2}-1)\cdots(-\frac{1}{2}-n+1)}{n!}u^n + O\big(u^{n+1}\big),$$

$$(1-x)^{-1/2} = 1 + \frac{1}{2}x^2 + \frac{1\cdot 3}{2^2\cdot 2!}x^4 + \cdots + \\ + \frac{1\cdot 3\cdots(2n-1)}{2^n\cdot n!}x^{2n} + O\big(x^{2n+2}\big),$$

$$\arcsin x = x + \frac{1}{2\cdot 3}x^3 + \frac{1\cdot 3}{2^2\cdot 2!\cdot 5}x^5 + \cdots + \\ + \frac{(2n-1)!!}{(2n)!!(2n+1)}x^{2n+1} + O\big(x^{2n+3}\big),$$

or, after elementary transformations,

$$\arcsin x = x + \frac{1}{3!}x^3 + \frac{[3!!]^2}{5!}x^5 + \cdots + \frac{[(2n-1)!!]^2}{(2n+1)!}x^{2n+1} + O\big(x^{2n+3}\big).$$

Here $(2n-1)!! := 1\cdot 3\cdots(2n-1)$ and $(2n)!! := 2\cdot 4\cdots(2n)$.

Example 19 We use the results of Examples 5, 12, 17, and 18 and find

$$\lim_{x\to 0}\frac{\arctan x-\sin x}{\tan x-\arcsin x}=\lim_{x\to 0}\frac{[x-\frac{1}{3}x^3+O(x^5)]-[x-\frac{1}{3!}x^3+O(x^5)]}{[x+\frac{1}{3}x^3+O(x^5)]-[x+\frac{1}{3!}x^3+O(x^5)]}=$$

$$=\lim_{x\to 0}\frac{-\frac{1}{6}x^3+O(x^5)}{\frac{1}{6}x^3+O(x^5)}=-1.$$

5.3.4 Problems and Exercises

1. Choose numbers a and b so that the function $f(x)=\cos x-\frac{1+ax^2}{1+bx^2}$ is an infinitesimal of highest possible order as $x\to 0$.
2. Find $\lim_{x\to\infty} x[\frac{1}{\mathrm{e}}-(\frac{x}{x+1})^x]$.
3. Write a Taylor polynomial of e^x at zero that makes it possible to compute the values of e^x on the closed interval $-1\le x\le 2$ within 10^{-3}.
4. Let f be a function that is infinitely differentiable at 0. Show that

a) if f is even, then its Taylor series at 0 contains only even powers of x;
b) if f is odd, then its Taylor series at 0 contains only odd powers of x.

5. Show that if $f\in C^{(\infty)}[-1,1]$ and $f^{(n)}(0)=0$ for $n=0,1,2,\dots$, and there exists a number C such that $\sup_{-1\le x\le 1}|f^{(n)}(x)|\le n!C$ for $n\in\mathbb{N}$, then $f\equiv 0$ on $[-1,1]$.
6. Let $f\in C^{(n)}(]-1,1[)$ and $\sup_{-1<x<1}|f(x)|\le 1$. Let $m_k(I)=\inf_{x\in I}|f^{(k)}(x)|$, where I is an interval contained in $]-1,1[$. Show that

a) if I is partitioned into three successive intervals I_1, I_2, and I_3 and μ is the length of I_2, then

$$m_k(I)\le\frac{1}{\mu}\big(m_{k-1}(I_1)+m_{k-1}(I_3)\big);$$

b) if I has length λ, then

$$m_k(I)\le\frac{2^{k(k+1)/2}k^k}{\lambda^k};$$

c) there exists a number α_n depending only on n such that if $|f'(0)|\ge\alpha_n$, then the equation $f^{(n)}(x)=0$ has at least $n-1$ distinct roots in $]-1,1[$.

Hint: In part b) use part a) and mathematical induction; in c) use a) and prove by induction that there exists a sequence $x_{k_1}<x_{k_2}<\dots<x_{k_k}$ of points of the open interval $]-1,1[$ such that $f^{(k)}(x_{k_i})\cdot f^{(k)}(x_{k_{i+1}})<0$ for $1\le i\le k-1$.
7. Show that if a function f is defined and differentiable on an open interval I and $[a,b]\subset I$, then

a) the function $f'(x)$ (even if it is not continuous!) assumes on $[a,b]$ all the values between $f'(a)$ and $f'(b)$ (*the theorem of Darboux*);[13]

b) if $f''(x)$ also exists in $]a,b[$, then there is a point $\xi \in]a,b[$ such that $f'(b) - f'(a) = f''(\xi)(b-a)$.

8. A function $f(x)$ may be differentiable on the entire real line, without having a continuous derivative $f'(x)$ (see Example 7 in Sect. 5.1.5).

a) Show that $f'(x)$ can have only discontinuities of second kind.

b) Find the flaw in the following "proof" that $f'(x)$ is continuous.

Proof Let x_0 be an arbitrary point on $\mathbb{R}$ and $f'(x_0)$ the derivative of f at the point x_0. By definition of the derivative and Lagrange's theorem

$$f'(x_0) = \lim_{x \to x_0} \frac{f(x) - f(x_0)}{x - x_0} = \lim_{x \to x_0} f'(\xi) = \lim_{\xi \to x_0} f'(\xi),$$

where ξ is a point between x_0 and x and therefore tends to x_0 as $x \to x_0$. □

9. Let f be twice differentiable on an interval I. Let $M_0 = \sup_{x \in I} |f(x)|$, $M_1 = \sup_{x \in I} |f'(x)|$ and $M_2 = \sup_{x \in I} |f''(x)|$. Show that

a) if $I = [-a, a]$, then

$$|f'(x)| \le \frac{M_0}{a} + \frac{x^2 + a^2}{2a} M_2;$$

b) $\begin{cases} M_1 \le 2\sqrt{M_0 M_2}, & \text{if the length of } I \text{ is not less than } 2\sqrt{M_0/M_2}, \\ M_1 \le \sqrt{2 M_0 M_2}, & \text{if } I = \mathbb{R}; \end{cases}$

c) the numbers 2 and $\sqrt{2}$ in part b) cannot be replaced by smaller numbers;

d) if f is differentiable p times on $\mathbb{R}$ and the quantities M_0 and $M_p = \sup_{x \in \mathbb{R}} |f^{(p)}(x)|$ are finite, then the quantities $M_k = \sup_{x \in \mathbb{R}} |f^{(k)}(x)|$, $1 \le k < p$, are also finite and

$$M_k \le 2^{k(p-k)/2} M_0^{1-k/p} M_p^{k/p}.$$

Hint: Use Exercises 6b) and 9b) and mathematical induction.

10. Show that if a function f has derivatives up to order $n+1$ inclusive at a point x_0 and $f^{(n+1)}(x_0) \ne 0$, then in the Lagrange form of the remainder in Taylor's formula

$$r_n(x_0; x) = \frac{1}{n!} f^{(n)}\big(x_0 + \theta(x - x_0)\big)(x - x_0)^n,$$

where $0 < \theta < 1$ and the quantity $\theta = \theta(x)$ tends to $\frac{1}{n+1}$ as $x \to x_0$.

11. Let f be a function that is differentiable n times on an interval I. Prove the following statements.

---

[13] G. Darboux (1842–1917) – French mathematician.

a) If f vanishes at $(n+1)$ points of I, there exists a point $\xi \in I$ such that $f^{(n)}(\xi) = 0$.

b) If $x_1, x_2, \dots, x_p$ are points of the interval I, there exists a unique polynomial $L(x)$ (the *Lagrange interpolation polynomial*) of degree at most $(n-1)$ such that $f(x_i) = L(x_i)$, $i = 1, \dots, n$. In addition, for $x \in I$ there exists a point $\xi \in I$ such that

$$f(x) - L(x) = \frac{(x-x_1)\cdots(x-x_n)}{n!} f^{(n)}(\xi).$$

c) If $x_1 < x_2 < \dots < x_p$ are points of I and n_i, $1 \le i \le p$, are natural numbers such that $n_1 + n_2 + \dots + n_p = n$ and $f^{(k)}(x_i) = 0$ for $0 \le k \le n_i - 1$, then there exists a point ξ in the closed interval $[x_1, x_p]$ at which $f^{(n-1)}(\xi) = 0$.

d) There exists a unique polynomial $H(x)$ (the *Hermite interpolating polynomial*)[14] of degree $(n-1)$ such that $f^{(k)}(x_i) = H^{(k)}(x_i)$ for $0 \le k \le n_i - 1$. Moreover, inside the smallest interval containing the points x and x_i, $i = 1, \dots, p$, there is a point ξ such that

$$f(x) = H(x) + \frac{(x-x_1)^{n_1}\cdots(x-x_p)^{n_p}}{n!} f^{(n)}(\xi).$$

This formula is called the *Hermite interpolation formula*. The points $x_i, i = 1, \dots, p$, are called *interpolation nodes of multiplicity* n_i respectively. Special cases of the Hermite interpolation formula are the Lagrange interpolation formula, which is part b) of this exercise, and Taylor's formula with the Lagrange form of the remainder, which results when $p = 1$, that is, for interpolation with a single node of multiplicity n.

12. Show that

a) between two real roots of a polynomial $P(x)$ with real coefficients there is a root of its derivative $P'(x)$;

b) if the polynomial $P(x)$ has a multiple root, the polynomial $P'(x)$ has the same root, but its multiplicity as a root of $P'(x)$ is one less than its multiplicity as a root of $P(x)$;

c) if $Q(x)$ is the greatest common divisor of the polynomials $P(x)$ and $P'(x)$, where $P'(x)$ is the derivative of $P(x)$, then the polynomial $\frac{P(x)}{Q(x)}$ has the roots of $P(x)$ as its roots, all of them being roots of multiplicity 1.

13. Show that

a) any polynomial $P(x)$ admits a representation in the form $c_0 + c_1(x - x_0) + \dots + c_n(x - x_0)^n$;

b) there exists a unique polynomial of degree n for which $f(x) - P(x) = o((x - x_0)^n)$ as $E \ni x \to x_0$. Here f is a function defined on a set E and x_0 is a limit point of E.

---

[14] Ch. Hermite (1822–1901) – French mathematician who studied problems of analysis; in particular, he proved that e is transcendental.

14. Using induction on k, $1 \le k$, we define the *finite differences of order* k of the function f at x_0:

$$\begin{aligned}
\Delta^1 f(x_0; h_1) &:= \Delta f(x_0; h_1) = f(x_0 + h_1) - f(x_0), \\
\Delta^2 f(x_0; h_1, h_2) &:= \Delta\Delta f(x_0; h_1, h_2) = \\
&= \big(f(x_0 + h_1 + h_2) - f(x_0 + h_2)\big) - \\
&\quad - \big(f(x_0 + h_1) - f(x_0)\big) = \\
&= f(x_0 + h_1 + h_2) - f(x_0 + h_1) - f(x_0 + h_2) + f(x_0), \\
&\ \vdots \\
\Delta^k f(x_0; h_1, \dots, h_k) &:= \Delta^{k-1} g_k(x_0; h_1, \dots, h_{k-1}),
\end{aligned}$$

where $g_k(x) = \Delta^1 f(x; h_k) = f(x + h_k) - f(x)$.

a) Let $f \in C^{(n-1)}[a, b]$ and suppose that $f^{(n)}(x)$ exists at least in the open interval $]a, b[$. Show that if all the points $x_0, x_0 + h_1, x_0 + h_2, x_0 + h_1 + h_2, \dots, x_0 + h_1 + \dots + h_n$ lie in $[a, b]$, then inside the smallest closed interval containing all of them there is a point ξ such that

$$\Delta^n f(x_0; h_1, \dots, h_n) = f^{(n)}(\xi) h_1 \cdots h_n.$$

b) (Continuation.) If $f^{(n)}(x_0)$ exists, then the following estimate holds:

$$\begin{aligned}
&\big|\Delta^n f(x_0; h_1, \dots, h_n) - f^{(n)}(x_0) h_1 \cdots h_n\big| \le \\
&\le \sup_{x \in]a,b[} \big|f^{(n)}(x) - f^{(n)}(x_0)\big| \cdot |h_1| \cdots |h_n|.
\end{aligned}$$

c) (Continuation.) Set $\Delta^n f(x_0; h, \dots, h) =: \Delta^n f(x_0; h^n)$. Show that if $f^{(n)}(x_0)$ exists, then

$$f^{(n)}(x_0) = \lim_{h \to 0} \frac{\Delta^n f(x_0; h^n)}{h^n}.$$

d) Show by example that the preceding limit may exist even when $f^{(n)}(x_0)$ does not exist.

Hint: Consider, for example, $\Delta^2 f(0; h^2)$ for the function

$$f(x) = \begin{cases} x^3 \sin \frac{1}{x}, & x \ne 0, \\ 0, & x = 0, \end{cases}$$

and show that

$$\lim_{h \to 0} \frac{\Delta^2 f(0; h^2)}{h^2} = 0.$$

15. a) Applying Lagrange's theorem to the function $\frac{1}{x^\alpha}$, where $\alpha > 0$, show that the inequality

$$\frac{1}{n^{1+\alpha}} < \frac{1}{\alpha}\left(\frac{1}{(n-1)^\alpha} - \frac{1}{n^\alpha}\right)$$

holds for $n \in \mathbb{N}$ and $\alpha > 0$.

b) Use the result of a) to show that the series $\sum_{n=1}^{\infty} \frac{1}{n^\sigma}$ converges for $\sigma > 1$.

5.4 The Study of Functions Using the Methods of Differential Calculus

5.4.1 Conditions for a Function to be Monotonic

Proposition 1 *The following relations hold between the monotonicity properties of a function $f : E \to \mathbb{R}$ that is differentiable on an open interval $]a, b[= E$ and the sign (positivity) of its derivative f' on that interval:*

$$\begin{aligned}
f'(x) > 0 \Rightarrow &\quad f \text{ is increasing} &&\Rightarrow f'(x) \geq 0,\\
f'(x) \geq 0 \Rightarrow &\quad f \text{ is nondecreasing} &&\Rightarrow f'(x) \geq 0,\\
f'(x) \equiv 0 \Rightarrow &\quad f \equiv \text{const.} &&\Rightarrow f'(x) \equiv 0,\\
f'(x) \leq 0 \Rightarrow &\quad f \text{ is nonincreasing} &&\Rightarrow f'(x) \leq 0,\\
f'(x) < 0 \Rightarrow &\quad f \text{ is decreasing} &&\Rightarrow f'(x) \leq 0.
\end{aligned}$$

Proof The left-hand column of implications is already known to us from Lagrange's theorem, by virtue of which $f(x_2) - f(x_1) = f'(\xi)(x_2 - x_1)$, where $x_1, x_2 \in]a, b[$ and ξ is a point between x_1 and x_2. It can be seen from this formula that for $x_1 < x_2$ the difference $f(x_2) - f(x_1)$ is positive if and only if $f'(\xi)$ is positive.

The right-hand column of implications can be obtained immediately from the definition of the derivative. Let us show, for example, that if a function f that is differentiable on $]a, b[$ is increasing, then $f'(x) \geq 0$ on $]a, b[$. Indeed,

$$f'(x) = \lim_{h \to 0} \frac{f(x+h) - f(x)}{h}.$$

If $h > 0$, then $f(x+h) - f(x) > 0$; and if $h < 0$, then $f(x+h) - f(x) < 0$. Therefore the fraction after the limit sign is positive.

Consequently, its limit $f'(x)$ is nonnegative, as asserted. □

Remark 1 It is clear from the example of the function $f(x) = x^3$ that a strictly increasing function has a nonnegative derivative, not necessarily one that is always positive. In this example, $f'(0) = 3x^2|_{x=0} = 0$.

Remark 2 In the expression $A \Rightarrow B$, as we noted at the appropriate point, A is a sufficient condition for B and B a necessary condition for A. Hence, one can make the following inferences from Proposition 1.

A function is constant on an open interval if and only if its derivative is identically zero on that interval.

A sufficient condition for a function that is differentiable on an open interval to be decreasing on that interval is that its derivative be negative at every point of the interval.

A necessary condition for a function that is differentiable on an open interval to be nonincreasing on that interval is that its derivative be nonpositive on the interval.

Example 1 Let $f(x) = x^3 - 3x + 2$ on $\mathbb{R}$. Then $f'(x) = 3x^2 - 3 = 3(x^2 - 1)$, and since $f'(x) < 0$ for $|x| < 1$ and $f'(x) > 0$ for $|x| > 1$, we can say that the function is increasing on the open interval $]-\infty, -1[$, decreasing on $]-1, 1[$, and increasing again on $]1, +\infty[$.

5.4.2 Conditions for an Interior Extremum of a Function

Taking account of Fermat's lemma (Lemma 1 of Sect. 5.3), we can state the following proposition.

Proposition 2 (Necessary conditions for an interior extremum) *In order for a point x_0 to be an extremum of a function $f : U(x_0) \to \mathbb{R}$ defined on a neighborhood $U(x_0)$ of that point, a necessary condition is that one of the following two conditions hold: either the function is not differentiable at x_0 or $f'(x_0) = 0$.*

Simple examples show that these necessary conditions are not sufficient.

Example 2 Let $f(x) = x^3$ on $\mathbb{R}$. Then $f'(0) = 0$, but there is no extremum at $x_0 = 0$.

Example 3 Let

$$f(x) = \begin{cases} x & \text{for } x > 0, \\ 2x & \text{for } x < 0. \end{cases}$$

This function has a bend at 0 and obviously has neither a derivative nor an extremum at 0.

Example 4 Let us find the maximum of $f(x) = x^2$ on the closed interval $[-2, 1]$. It is obvious in this case that the maximum will be attained at the endpoint -2, but here is a systematic procedure for finding the maximum. We find $f'(x) = 2x$, then we find all points of the open interval $]-2, 1[$ at which $f'(x) = 0$. In this case, the only such point is $x = 0$. The maximum of $f(x)$ must be either among the points where $f'(x) = 0$, or at one of the endpoints, about which Proposition 2 is silent. Thus we need to compare $f(-2) = 4$, $f(0) = 0$, and $f(1) = 1$, from which we conclude that the maximal value of $f(x) = x^2$ on the closed interval $[-2, 1]$ equals 4 and is assumed at -2, which is an endpoint of the interval.

Using the connection established in Sect. 5.4.1 between the sign of the derivative and the nature of the monotonicity of the function, we arrive at the following sufficient conditions for the presence or absence of a local extremum at a point.

Proposition 3 (Sufficient conditions for an extremum in terms of the first derivative) *Let $f : U(x_0) \to \mathbb{R}$ be a function defined on a neighborhood $U(x_0)$ of the point x_0, which is continuous at the point itself and differentiable in a deleted neighborhood $\overset{\circ}{U}(x_0)$. Let $\overset{\circ}{U}{}^-(x_0) = \{x \in U(x_0) \mid x < x_0\}$ and $\overset{\circ}{U}{}^+(x_0) = \{x \in U(x_0) \mid x > x_0\}$.*

Then the following conclusions are valid:

a) $(\forall x \in \overset{\circ}{U}{}^-(x_0)\ (f'(x) < 0)) \wedge (\forall x \in \overset{\circ}{U}{}^+(x_0)\ (f'(x) < 0)) \Rightarrow (f$ *has no extremum at* $x_0)$;

b) $(\forall x \in \overset{\circ}{U}{}^-(x_0)\ (f'(x) < 0)) \wedge (\forall x \in \overset{\circ}{U}{}^+(x_0)\ (f'(x) > 0)) \Rightarrow (x_0$ *is a strict local minimum of* $f)$;

c) $(\forall x \in \overset{\circ}{U}{}^-(x_0)\ (f'(x) > 0)) \wedge (\forall x \in \overset{\circ}{U}{}^+(x_0)\ (f'(x) < 0)) \Rightarrow (x_0$ *is a strict local maximum of* $f)$;

d) $(\forall x \in \overset{\circ}{U}{}^-(x_0)\ (f'(x) > 0)) \wedge (\forall x \in \overset{\circ}{U}{}^+(x_0)\ (f'(x) > 0)) \Rightarrow (f$ *has no extremum at* $x_0)$.

Briefly, but less precisely, one can say that if the derivative changes sign in passing through the point, then the point is an extremum, while if the derivative does not change sign, the point is not an extremum.

We remark immediately, however, that these sufficient conditions are not necessary for an extremum, as one can verify using the following example.

Example 5 Let

$$f(x) = \begin{cases} 2x^2 + x^2 \sin \frac{1}{x} & \text{for } x \neq 0, \\ 0 & \text{for } x = 0. \end{cases}$$

Since $x^2 \leq f(x) \leq 3x^2$, it is clear that the function has a strict local minimum at $x_0 = 0$, but the derivative $f'(x) = 4x + 2x \sin \frac{1}{x} - \cos \frac{1}{x}$ is not of constant sign in any deleted one-sided neighborhood of this point. This same example shows the misunderstandings that can arise in connection with the abbreviated statement of Proposition 3 just given.

We now turn to the proof of Proposition 3.

Proof a) It follows from Proposition 2 that f is strictly decreasing on $\overset{\circ}{U}{}^-(x_0)$. Since it is continuous at x_0, we have $\lim_{\overset{\circ}{U}{}^-(x_0) \ni x \to x_0} f(x) = f(x_0)$, and consequently $f(x) > f(x_0)$ for $x \in \overset{\circ}{U}{}^-(x_0)$. By the same considerations we have $f(x_0) > f(x)$

for $x \in \overset{\circ}{U}^{+}(x_0)$. Thus the function is strictly decreasing in the whole neighborhood $U(x_0)$ and x_0 is not an extremum.

b) We conclude to begin with, as in a), that since $f(x)$ is decreasing on $\overset{\circ}{U}^{-}(x_0)$ and continuous at x_0, we have $f(x) > f(x_0)$ for $x \in \overset{\circ}{U}^{-}(x_0)$. We conclude from the increasing nature of f on $\overset{\circ}{U}^{+}(x_0)$ that $f(x_0) < f(x)$ for $x \in \overset{\circ}{U}^{+}(x_0)$. Thus f has a strict local minimum at x_0.

Statements c) and d) are proved similarly. □

Proposition 4 (Sufficient conditions for an extremum in terms of higher-order derivatives) *Suppose a function $f : U(x_0) \to \mathbb{R}$ defined on a neighborhood $U(x_0)$ of x_0 has derivatives of order up to n inclusive at x_0 $(n \geq 1)$.*

If $f'(x_0) = \cdots = f^{(n-1)}(x_0) = 0$ and $f^{(n)}(x_0) \neq 0$, then there is no extremum at x_0 if n is odd. If n is even, the point x_0 is a local extremum, in fact a strict local minimum if $f^{(n)}(x_0) > 0$ and a strict local maximum if $f^{(n)}(x_0) < 0$.

Proof Using the local Taylor formula

$$f(x) - f(x_0) = \frac{1}{n!} f^{(n)}(x_0)(x - x_0)^n + \alpha(x)(x - x_0)^n, \tag{5.82}$$

where $\alpha(x) \to 0$ as $x \to x_0$, we shall reason as in the proof of Fermat's lemma. We rewrite Eq. (5.82) as

$$f(x) - f(x_0) = \left(\frac{1}{n!} f^{(n)}(x_0) + \alpha(x)\right)(x - x_0)^n. \tag{5.83}$$

Since $f^{(n)}(x_0) \neq 0$ and $\alpha(x) \to 0$ as $x \to x_0$, the sum $f^{(n)}(x_0) + \alpha(x)$ has the sign of $f^{(n)}(x_0)$ when x is sufficiently close to x_0. If n is odd, the factor $(x - x_0)^n$ changes sign when x passes through x_0, and then the sign of the right-hand side of Eq. (5.83) also changes sign. Consequently, the left-hand side changes sign as well, and so for $n = 2k + 1$ there is no extremum.

If n is even, then $(x - x_0)^n > 0$ for $x \neq x_0$ and hence in some small neighborhood of x_0 the sign of the difference $f(x) - f(x_0)$ is the same as the sign of $f^{(n)}(x_0)$, as is clear from Eq. (5.83). □

Let us now consider some examples.

Example 6 (The law of refraction in geometric optics (Snell's law)[15]) According to *Fermat's principle*, the actual trajectory of a light ray between two points is such that the ray requires minimum time to pass from one point to the other compared with all paths joining the two points.

It follows from Fermat's principle and the fact that the shortest path between two points is a straight line segment having the points as endpoints that in a homogeneous and isotropic medium (having identical structure at each point and in each direction) light propagates in straight lines.

[15] W. Snell (1580–1626) – Dutch astronomer and mathematician.

Fig. 5.10

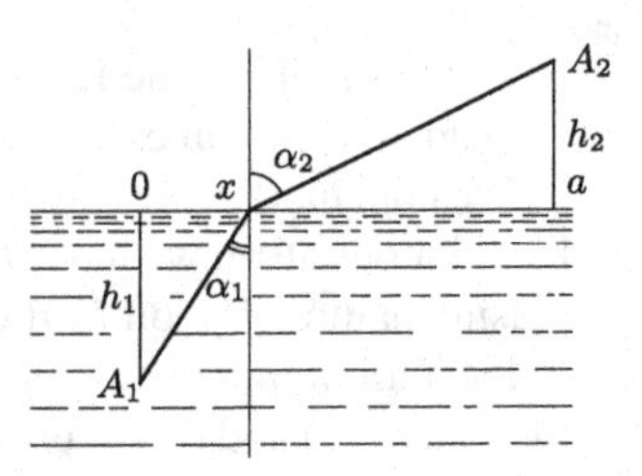

Now consider two such media, and suppose that light propagates from point A_1 to A_2, as shown in Fig. 5.10.

If c_1 and c_2 are the velocities of light in these media, the time required to traverse the path is

$$t(x) = \frac{1}{c_1}\sqrt{h_1^2 + x^2} + \frac{1}{c_2}\sqrt{h_2^2 + (a-x)^2}.$$

We now find the extremum of the function $t(x)$:

$$t'(x) = \frac{1}{c_1}\frac{x}{\sqrt{h_1^2 + x^2}} - \frac{1}{c_2}\frac{a-x}{\sqrt{h_2^2 + (a-x)^2}} = 0,$$

which in accordance with the notation of the figure, yields $c_1^{-1}\sin\alpha_1 = c_2^{-1}\sin\alpha_2$.

It is clear from physical considerations, or directly from the form of the function $t(x)$, which increases without bound as $x \to \infty$, that the point where $t'(x) = 0$ is an absolute minimum of the continuous function $t(x)$. Thus Fermat's principle implies the law of refraction $\frac{\sin\alpha_1}{\sin a_2} = \frac{c_1}{c_2}$.

Example 7 We shall show that for $x > 0$

$$x^\alpha - \alpha x + \alpha - 1 \le 0, \quad \text{when } 0 < \alpha < 1, \tag{5.84}$$

$$x^\alpha - \alpha x + \alpha - 1 \ge 0, \quad \text{when } \alpha < 0 \text{ or } 1 < \alpha. \tag{5.85}$$

Proof Differentiating the function $f(x) = x^\alpha - \alpha x + \alpha - 1$, we find $f'(x) = \alpha(x^{\alpha-1} - 1)$ and $f'(x) = 0$ when $x = 1$. In passing through the point 1 the derivative passes from positive to negative values if $0 < \alpha < 1$ and from negative to positive values if $\alpha < 0$ or $\alpha > 1$. In the first case the point 1 is a strict maximum, and in the second case a strict minimum (and, as follows from the monotonicity of f on the intervals $0 < x < 1$ and $1 < x$, not merely a local minimum). But $f(1) = 0$ and hence both inequalities (5.84) and (5.85) are established. In doing so, we have even shown that both inequalities are strict if $x \ne 1$. □

We remark that if x is replaced by $1 + x$, we find that (5.84) and (5.85) are extensions of Bernoulli's inequality (Sect. 2.2; see also Problem 2 below), which we already know for a natural-number exponent α.

By elementary algebraic transformations one can obtain a number of classical inequalities of great importance for analysis from the inequalities just proved. We shall now derive these inequalities.

a. Young's Inequalities[16]

If $a > 0$ and $b > 0$, and the numbers p and q such that $p \neq 0, 1$, $q \neq 0, 1$ and $\frac{1}{p} + \frac{1}{q} = 1$, then

$$a^{1/p}b^{1/q} \le \frac{1}{p}a + \frac{1}{q}b, \quad \text{if } p > 1, \tag{5.86}$$

$$a^{1/p}b^{1/q} \ge \frac{1}{p}a + \frac{1}{q}b, \quad \text{if } p < 1, \tag{5.87}$$

and equality holds in (5.86) *and* (5.87) *only when $a = b$.*

Proof It suffices to set $x = \frac{a}{b}$ and $\alpha = \frac{1}{p}$ in (5.84) and (5.85), and then introduce the notation $\frac{1}{q} = 1 - \frac{1}{p}$. □

b. Hölder's Inequalities[17]

Let $x_i \ge 0$, $y_i \ge 0$, $i = 1, \dots, n$, and $\frac{1}{p} + \frac{1}{q} = 1$. Then

$$\sum_{i=1}^{n} x_i y_i \le \left(\sum_{i=1}^{n} x_i^p\right)^{1/p} \left(\sum_{i=1}^{n} y_i^q\right)^{1/q} \quad \text{for } p > 1, \tag{5.88}$$

and

$$\sum_{i=1}^{n} x_i y_i \ge \left(\sum_{i=1}^{n} x_i^p\right)^{1/p} \left(\sum_{i=1}^{n} y_i^q\right)^{1/q} \quad \text{for } p < 1, p \neq 0. \tag{5.89}$$

In the case $p < 0$ it is assumed in (5.89) *that $x_i > 0$ ($i = 1, \dots, n$). Equality is possible in* (5.88) *and* (5.89) *only when the vectors $(x_1^p, \dots, x_n^p)$ and $(y_1^q, \dots, y_n^q)$ are proportional.*

Proof Let us verify the inequality (5.88). Let $X = \sum_{i=1}^{n} x_i^p > 0$ and $Y = \sum_{i=1}^{n} y_i^q > 0$. Setting $a = \frac{x_i^p}{X}$ and $b = \frac{y_i^q}{Y}$ in (5.86), we obtain

$$\frac{x_i y_i}{X^{1/p} Y^{1/q}} \le \frac{1}{p}\frac{x_i^p}{X} + \frac{1}{q}\frac{y_i^q}{Y}.$$

[16] W.H. Young (1882–1946) – British mathematician.

[17] O. Hölder (1859–1937) – German mathematician.

Summing these inequalities over i from 1 to n, we obtain

$$\frac{\sum_{i=1}^{n} x_i y_i}{X^{1/p} Y^{1/q}} \leq 1,$$

which is equivalent to relation (5.88).

We obtain (5.89) similarly from (5.87). Since equality occurs in (5.86) and (5.87) only when $a = b$, we conclude that it is possible in (5.88) and (5.89) only when a proportionality $x_i^p = \lambda y_i^q$ or $y_i^q = \lambda x_i^p$ holds. □

c. Minkowski's Inequalities[18]

Let $x_i \geq 0$, $y_i \geq 0$, $i = 1, \ldots, n$. *Then*

$$\left(\sum_{i=1}^{n} (x_i + y_i)^p\right)^{1/p} \leq \left(\sum_{i=1}^{n} x_i^p\right)^{1/p} + \left(\sum_{i=1}^{n} y_i^p\right)^{1/p} \quad \text{when } p > 1, \tag{5.90}$$

and

$$\left(\sum_{i=1}^{n} (x_i + y_i)^p\right)^{1/p} \geq \left(\sum_{i=1}^{n} x_i^p\right)^{1/p} + \left(\sum_{i=1}^{n} y_i^p\right)^{1/p} \quad \text{when } p < 1, p \neq 0. \tag{5.91}$$

Proof We apply Hölder's inequality to the terms on the right-hand side of the identity

$$\sum_{i=1}^{n} (x_i + y_i)^p = \sum_{i=1}^{n} x_i (x_i + y_i)^{p-1} + \sum_{i=1}^{n} y_i (x_i + y_i)^{p-1}.$$

The left-hand side is then bounded from above (for $p > 1$) or below (for $p < 1$) in accordance with inequalities (5.88) and (5.89) by the quantity

$$\left(\sum_{i=1}^{n} x_i^p\right)^{1/p} \left(\sum_{i=1}^{n} (x_i + y_i)^p\right)^{1/q} + \left(\sum_{i=1}^{n} y_i^p\right)^{1/p} \left(\sum_{i=1}^{n} (x_i + y_i)^p\right)^{1/q}.$$

After dividing these inequalities by $(\sum_{i=1}^{n} (x_i + y_i)^p)^{1/q}$, we arrive at (5.90) and (5.91).

Knowing the conditions for equality in Hölder's inequalities, we verify that equality is possible in Minkowski's inequalities only when the vectors $(x_1, \ldots, x_n)$ and $(y_1, \ldots, y_n)$ are collinear. □

For $n = 3$ and $p = 2$, Minkowski's inequality (5.90) is obviously the triangle inequality in three-dimensional Euclidean space.

[18] H. Minkowski (1864–1909) – German mathematician who proposed a mathematical model adapted to the special theory of relativity (a space with a sign-indefinite metric).

Fig. 5.11

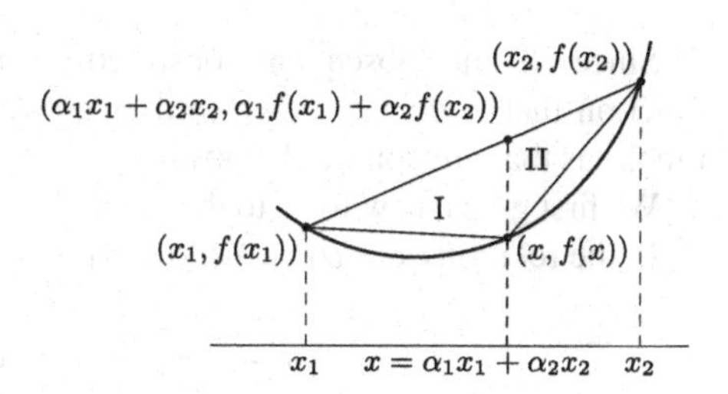

Example 8 Let us consider another elementary example of the use of higher-order derivatives to find local extrema. Let $f(x) = \sin x$. Since $f'(x) = \cos x$ and $f''(x) = -\sin x$, all the points where $f'(x) = \cos x = 0$ are local extrema of $\sin x$, since $f''(x) = -\sin x \neq 0$ at these points. Here $f''(x) < 0$ if $\sin x > 0$ and $f''(x) > 0$ if $\sin x < 0$. Thus the points where $\cos x = 0$ and $\sin x > 0$ are local maxima and those where $\cos x = 0$ and $\sin x < 0$ are local minima for $\sin x$ (which, of course, was already well-known).

5.4.3 Conditions for a Function to be Convex

Definition 1 A function $f :]a, b[\to \mathbb{R}$ defined on an open interval $]a, b[\subset \mathbb{R}$ is *convex* if the inequalities

$$f(\alpha_1 x_1 + \alpha_2 x_2) \leq \alpha_1 f(x_1) + \alpha_2 f(x_2) \tag{5.92}$$

hold for any points $x_1, x_2 \in]a, b[$ and any numbers $\alpha_1 \geq 0$, $\alpha_2 \geq 0$ such that $\alpha_1 + \alpha_2 = 1$. If this inequality is strict whenever $x_1 \neq x_2$ and $\alpha_1 \alpha_2 \neq 0$, the function is *strictly convex* on $]a, b[$.

Geometrically, condition (5.92) for convexity of a function $f :]a, b[\to \mathbb{R}$ means that the points of any arc of the graph of the function lie below the chord subtended by the arc (see Fig. 5.11).

In fact, the left-hand side of (5.92) contains the value $f(x)$ of the function at the point $x = \alpha_1 x_1 + \alpha_2 x_2 \in [x_1, x_2]$ and the right-hand side contains the value at the same point of the linear function whose (straight-line) graph passes through the points $(x_1, f(x_1))$ and $(x_2, f(x_2))$.

Relation (5.92) means that the set $E = \{(x, y) \in \mathbb{R}^2 \mid x \in]a, b[, f(x) < y\}$ of the points of the plane lying above the graph of the function is convex; hence the term "convex", as applied to the function itself.

Definition 2 If the opposite inequality holds for a function $f :]a, b[\to \mathbb{R}$, that function is said to be *concave* on the interval $]a, b[$, or, more often, *convex upward* in the interval, as opposed to a convex function, which is then said to be *convex downward* on $]a, b[$.

Since all our subsequent constructions are carried out in the same way for a function that is convex downward or convex upward, we shall limit ourselves to functions that are convex downward.

We first give a new form to the inequality (5.92), better adapted for our purposes. In the relations $x = \alpha_1 x_1 + \alpha_2 x_2$, $\alpha_1 + \alpha_2 = 1$, we have

$$\alpha_1 = \frac{x_2 - x}{x_2 - x_1}, \qquad \alpha_2 = \frac{x - x_1}{x_2 - x_1},$$

so that (5.92) can be rewritten as

$$f(x) \le \frac{x_2 - x}{x_2 - x_1} f(x_1) + \frac{x - x_1}{x_2 - x_1} f(x_2).$$

Taking account of the inequalities $x_1 \le x \le x_2$ and $x_1 < x_2$, we multiply by $x_2 - x_1$ and obtain

$$(x_2 - x) f(x_1) + (x_1 - x_2) f(x) + (x - x_1) f(x_2) \ge 0.$$

Remarking that $x_2 - x_1 = (x_2 - x) + (x - x_1)$ we obtain from the last inequality, after elementary transformations,

$$\frac{f(x) - f(x_1)}{x - x_1} \le \frac{f(x_2) - f(x)}{x_2 - x} \tag{5.93}$$

for $x_1 < x < x_2$ and any $x_1, x_2 \in]a, b[$.

Inequality (5.93) is another way of writing the definition of convexity of the function $f(x)$ on an open interval $]a, b[$. Geometrically, (5.93) means (see Fig. 5.11) that the slope of the chord I joining $(x_1, f(x_1))$ to $(x, f(x))$ is not larger than (and in the case of strict convexity is less than) the slope of the chord II joining $(x, f(x))$ to $(x_2, f(x_2))$.

Now let us assume that the function $f :]a, b[\to \mathbb{R}$ is differentiable on $]a, b[$. Then, letting x in (5.93) tend first to x_1, then to x_2, we obtain

$$f'(x_1) \le \frac{f(x_2) - f(x_1)}{x_2 - x_1} \le f'(x_2),$$

which establishes that the derivative of f is monotonic.

Taking this fact into account, for a strictly convex function we find, using Lagrange's theorem, that

$$f'(x_1) < f'(\xi_1) = \frac{f(x) - f(x_1)}{x - x_1} < \frac{f(x_2) - f(x)}{x_2 - x} = f'(\xi_2) \le f'(x_2)$$

for $x_1 < \xi_1 < x < \xi_2 < x_2$, that is, strict convexity implies that the derivative is strictly monotonic.

Thus, if a differentiable function f is convex on an open interval $]a, b[$, then f' is nondecreasing on $]a, b[$; and in the case when f is strictly convex, its derivative f' is increasing on $]a, b[$.

These conditions turn out to be not only necessary, but also sufficient for convexity of a differentiable function.

In fact, for $a < x_1 < x < x_2 < b$, by Lagrange's theorem

$$\frac{f(x) - f(x_1)}{x - x_1} = f'(\xi_1), \qquad \frac{f(x_2) - f(x)}{x_2 - x} = f'(\xi_2),$$

where $x_1 < \xi_1 < x < \xi_2 < x_2$; and if $f'(\xi_1) \le f'(\xi_2)$, then condition (5.93) for convexity holds (with strict convexity if $f'(\xi_1) < f'(\xi_2)$).

We have thus proved the following proposition.

Proposition 5 *A necessary and sufficient condition for a function $f :]a, b[\to \mathbb{R}$ that is differentiable on the open interval $]a, b[$ to be convex (downward) on that interval is that its derivative f' be nondecreasing on $]a, b[$. A strictly increasing f' corresponds to a strictly convex function.*

Comparing Proposition 5 with Proposition 3, we obtain the following corollary.

Corollary *A necessary and sufficient condition for a function $f :]a, b[\to \mathbb{R}$ having a second derivative on the open interval $]a, b[$ to be convex (downward) on $]a, b[$ is that $f''(x) \ge 0$ on that interval. The condition $f''(x) > 0$ on $]a, b[$ is sufficient to guarantee that f is strictly convex.*

We are now in a position to explain, for example, why the graphs of the simplest elementary functions are drawn with one form of convexity or another.

Example 9 Let us study the convexity of $f(x) = x^\alpha$ on the set $x > 0$. Since $f''(x) = \alpha(\alpha - 1)x^{\alpha-2}$, we have $f''(x) > 0$ for $\alpha < 0$ or $\alpha > 1$, that is, for these values of the exponent α the power function x^α is strictly convex (downward). For $0 < \alpha < 1$ we have $f''(x) < 0$, so that for these exponents it is strictly convex upward. For example, we always draw the parabola $f(x) = x^2$ as convex downward. The other cases $\alpha = 0$ and $\alpha = 1$ are trivial: $x^0 \equiv 1$ and $x^1 = x$. In both of these cases the graph of the function is a ray (see Fig. 5.18 on p. 251).

Example 10 Let $f(x) = a^x$, $0 < a$, $a \ne 1$. Since $f''(x) = a^x \ln^2 a > 0$, the exponential function a^x is strictly convex (downward) on $\mathbb{R}$ for any allowable value of the base a (see Fig. 5.12).

Example 11 For the function $f(x) = \log_a x$ we have $f''(x) = -\frac{1}{x^2 \ln a}$, so that the function is strictly convex (downward) if $0 < a < 1$, and strictly convex upward if $1 < a$ (see Fig. 5.13).

Example 12 Let us study the convexity of $f(x) = \sin x$ (see Fig. 5.14).

Since $f''(x) = -\sin x$, we have $f''(x) < 0$ on the intervals $\pi \cdot 2k < x < \pi(2k + 1)$ and $f''(x) > 0$ on $\pi(2k - 1) < x < \pi \cdot 2k$, where $k \in \mathbb{Z}$. It follows

Fig. 5.12

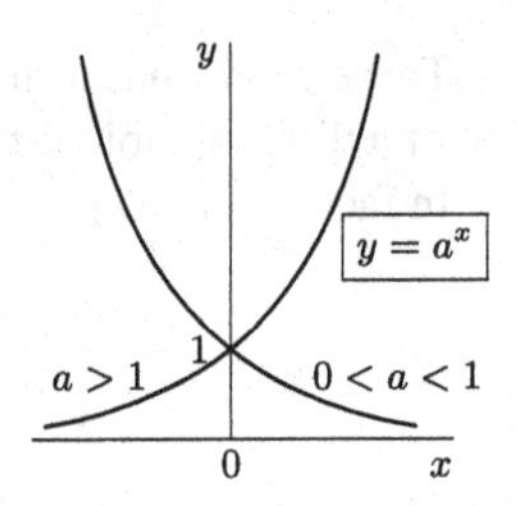

Fig. 5.13

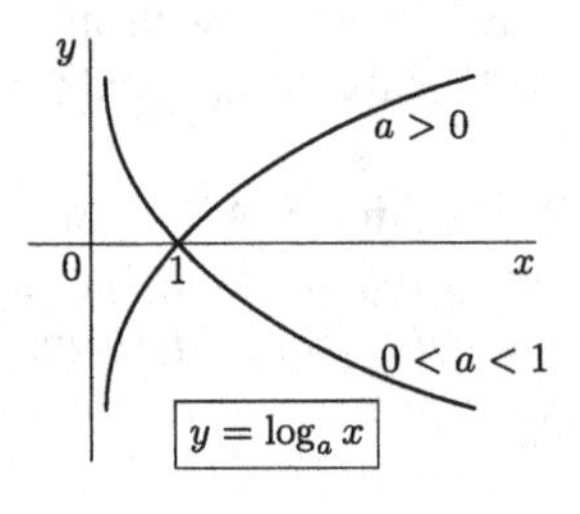

Fig. 5.14

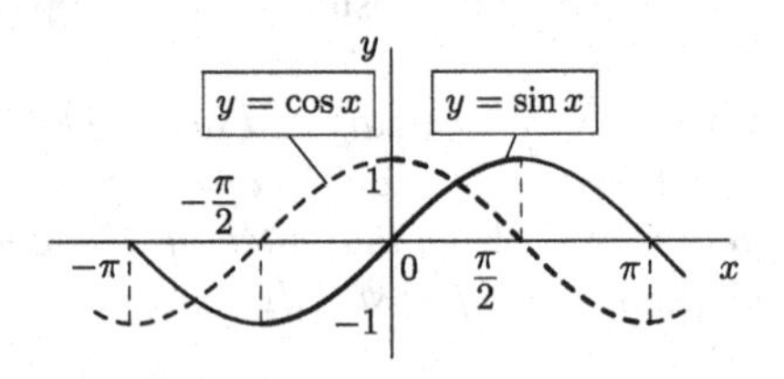

from this, for example, that the arc of the graph of $\sin x$ on the closed interval $0 \le x \le \frac{\pi}{2}$ lies above the chord it subtends everywhere except at the endpoints; therefore $\sin x > \frac{2}{\pi}x$ for $0 < x < \frac{\pi}{2}$.

We now point out another characteristic of a convex function, geometrically equivalent to the statement that a convex region of the plane lies entirely on one side of a tangent line to its boundary.

Proposition 6 *A function $f :]a, b[\to \mathbb{R}$ that is differentiable on the open interval $]a, b[$ is convex (downward) on $]a, b[$ if and only if its graph contains no points below any tangent drawn to it. In that case, a necessary and sufficient condition for strict convexity is that all points of the graph except the point of tangency lie strictly above the tangent line.*

Proof Necessity. Let $x_0 \in]a, b[$. The equation of the tangent line to the graph at $(x_0, f(x_0))$ has the form

$$y = f(x_0) + f'(x_0)(x - x_0),$$

so that

$$f(x) - y(x) = f(x) - f(x_0) - f'(x_0)(x - x_0) = \big(f'(\xi) - f'(x_0)\big)(x - x_0),$$

where ξ is a point between x and x_0. Since f is convex, the function $f'(x)$ is nondecreasing on $]a, b[$ and so the sign of the difference $f(\xi) - f'(x_0)$ is the same as the sign of the difference $x - x_0$. Therefore $f(x) - y(x) \geq 0$ at each point $x \in]a, b[$. If f is strictly convex, then f' is strictly increasing on $]a, b[$ and so $f(x) - y(x) > 0$ for $x \in]a, b[$ and $x \neq x_0$.

Sufficiency. If the inequality

$$f(x) - y(x) = f(x) - f(x_0) - f'(x_0)(x - x_0) \geq 0 \tag{5.94}$$

holds for any points $x, x_0 \in]a, b[$, then

$$\frac{f(x) - f(x_0)}{x - x_0} \leq f'(x_0) \quad \text{for } x < x_0,$$

$$\frac{f(x) - f(x_0)}{x - x_0} \geq f'(x_0) \quad \text{for } x_0 < x.$$

Thus, for any triple of points $x_1, x, x_2 \in]a, b[$ such that $x_1 < x < x_2$ we obtain

$$\frac{f(x) - f(x_1)}{x - x_1} \leq \frac{f(x_2) - f(x)}{x_2 - x},$$

and strict inequality in (5.94) implies strict inequality in this last relation, which, as we see, is the same as the definition (5.93) for convexity of a function. □

Let us now consider some examples.

Example 13 The function $f(x) = \mathrm{e}^x$ is strictly convex. The straight line $y = x + 1$ is tangent to the graph of this function at $(0, 1)$, since $f(0) = \mathrm{e}^0 = 1$ and $f'(0) = \mathrm{e}^x|_{x=0} = 1$. By Proposition 6 we conclude that for any $x \in \mathbb{R}$

$$\mathrm{e}^x \geq 1 + x,$$

and this inequality is strict for $x \neq 0$.

Example 14 Similarly, using the strict upward convexity of $\ln x$, one can verify that the inequality

$$\ln x \leq x - 1$$

holds for $x > 0$, the inequality being strict for $x \neq 1$.

In constructing the graphs of functions, it is useful to distinguish the points of inflection of a graph.

Definition 3 Let $f : U(x_0) \to \mathbb{R}$ be a function defined and differentiable on a neighborhood $U(x_0)$ of $x_0 \in \mathbb{R}$. If the function is convex downward (resp. upward) on the set $\overset{\circ}{U}{}^-(x_0) = \{x \in U(x_0) \mid x < x_0\}$ and convex upward (resp. downward) on $\overset{\circ}{U}{}^+(x_0) = \{x \in U(x_0) \mid x > x_0\}$, then $(x_0, f(x_0))$ is called a *point of inflection of the graph*.

Thus when we pass through a point of inflection, the direction of convexity of the graph changes. This means, in particular, that at the point $(x_0, f(x_0))$ the graph of the function passes from one side of the tangent line to the other.

An analytic criterion for the abscissa x_0 of a point of inflection is easy to surmise, if we compare Proposition 5 with Proposition 3. To be specific, one can say that if f is twice differentiable at x_0, then since $f'(x)$ has either a maximum or a minimum at x_0, we must have $f''(x_0) = 0$.

Now if the second derivative $f''(x)$ is defined on $U(x_0)$ and has one sign everywhere on $\mathring{U}^-(x_0)$ and the opposite sign everywhere on $\mathring{U}^+(x_0)$, this is sufficient for $f'(x)$ to be monotonic in $\mathring{U}^-(x_0)$ and monotonic in $\mathring{U}^+(x_0)$ but with the opposite monotonicity. By Proposition 5, a change in the direction of convexity occurs at $(x_0, f(x_0))$, and so that point is a point of inflection.

Example 15 When considering the function $f(x) = \sin x$ in Example 12 we found the regions of convexity and concavity for its graph. We shall now show that the points of the graph with abscissas $x = \pi k$, $k \in \mathbb{Z}$, are points of inflection.

Indeed, $f''(x) = -\sin x$, so that $f''(x) = 0$ at $x = \pi k$, $k \in \mathbb{Z}$. Moreover, $f''(x)$ changes sign as we pass through these points, which is a sufficient condition for a point of inflection (see Fig. 5.14 on p. 244).

Example 16 It should not be thought that the passing of a curve from one side of its tangent line to the other at a point is a sufficient condition for the point to be a point of inflection. It may, after all, happen that the curve does not have any constant convexity on either a left- or a right-hand neighborhood of the point. An example is easy to construct, by improving Example 5, which was given for just this purpose.

Let

$$f(x) = \begin{cases} 2x^3 + x^3 \sin \frac{1}{x^2} & \text{for } x \neq 0, \\ 0 & \text{for } x = 0. \end{cases}$$

Then $x^3 \le f(x) \le 3x^3$ for $0 \le x$ and $3x^3 \le f(x) \le x^3$ for $x \le 0$, so that the graph of this function is tangent to the x-axis at $x = 0$ and passes from the lower half-plane to the upper at that point. At the same time, the derivative of $f(x)$

$$f'(x) = \begin{cases} 6x^2 + 3x^2 \sin \frac{1}{x^2} - 2\cos \frac{1}{x^2} & \text{for } x \neq 0, \\ 0 & \text{for } x = 0 \end{cases}$$

is not monotonic in any one-sided neighborhood of $x = 0$.

In conclusion, we return again to the definition (5.92) of a convex function and prove the following proposition.

Proposition 7 (Jensen's inequality)[19] *If* $f :]a, b[\to \mathbb{R}$ *is a convex function,* $x_1, \dots, x_n$ *are points of* $]a, b[$*, and* $\alpha_1, \dots, \alpha_n$ *are nonnegative numbers such that*

[19] J.L. Jensen (1859–1925) – Danish mathematician.

$\alpha_1 + \cdots + \alpha_n = 1$, *then*

$$f(\alpha_1 x_1 + \cdots + \alpha_n x_n) \leq \alpha_1 f(x_1) + \cdots + \alpha_n f(x_n). \tag{5.95}$$

Proof For $n = 2$, condition (5.95) is the same as the definition (5.92) of a convex function.

We shall now show that if (5.95) is valid for $n = m - 1$, it is also valid for $n = m$.

For the sake of definiteness, assume that $\alpha_n \neq 0$ in the set $\alpha_1, \ldots, \alpha_n$. Then $\beta = \alpha_2 + \cdots + \alpha_n > 0$ and $\frac{\alpha_2}{\beta} + \cdots + \frac{\alpha_n}{\beta} = 1$. Using the convexity of the function, we find

$$f(\alpha_1 x_1 + \cdots + \alpha_n x_n) = f\left(\alpha_1 x_1 + \beta\left(\frac{\alpha_2}{\beta} x_2 + \cdots + \frac{\alpha_n}{\beta} x_n\right)\right) \leq$$

$$\leq \alpha_1 f(x_1) + \beta f\left(\frac{\alpha_2}{\beta} x_2 + \cdots + \frac{\alpha_n}{\beta} x_n\right),$$

since $\alpha_1 + \beta = 1$ and $(\frac{\alpha_2}{\beta} x_1 + \cdots + \frac{\alpha_n}{\beta} x_n) \in]a, b[$.

By the induction hypothesis, we now have

$$f\left(\frac{\alpha_2}{\beta} x_2 + \cdots + \frac{\alpha_n}{\beta} x_n\right) \leq \frac{\alpha_2}{\beta} f(x_2) + \cdots + \frac{\alpha_n}{\beta} f(x_n).$$

Consequently

$$f(\alpha_1 x_1 + \cdots + \alpha_n x_n) \leq \alpha_1 f(x_1) + \beta f\left(\frac{\alpha_2}{\beta} x_2 + \cdots + \frac{\alpha_n}{\beta} x_n\right) \leq$$

$$\leq \alpha_1 f(x_1) + \alpha_2 f(x_2) + \cdots + \alpha_n f(x_n).$$

By induction we now conclude that (5.95) holds for any $n \in \mathbb{N}$. (For $n = 1$, relation (5.95) is trivial.) □

We remark that, as the proof shows, a strict Jensen's inequality corresponds to strict convexity, that is, if the numbers $\alpha_1, \ldots, \alpha_n$ are nonzero, then equality holds in (5.95) if and only if $x_1 = \cdots = x_n$.

For a function that is convex upward, of course, the opposite relation to inequality (5.95) is obtained:

$$f(\alpha_1 x_1 + \cdots + \alpha_n x_n) \geq \alpha_1 f(x_1) + \cdots + \alpha_n f(x_n). \tag{5.96}$$

Example 17 The function $f(x) = \ln x$ is strictly convex upward on the set of positive numbers, and so by (5.96)

$$\alpha_1 \ln x_1 + \cdots + \alpha_n \ln x_n \leq \ln(\alpha_1 x_1 + \cdots + \alpha_n x_n)$$

or,

$$x_1^{\alpha_1} \cdots x_n^{\alpha_n} \le \alpha_1 x_1 + \cdots + \alpha_n x_n \tag{5.97}$$

for $x_i \ge 0$, $\alpha_i \ge 0$, $i = 1, \dots, n$, and $\sum_{i=1}^n \alpha_i = 1$.

In particular, if $\alpha_1 = \cdots = \alpha_n = \frac{1}{n}$, we obtain the classical inequality

$$\sqrt[n]{x_1 \cdots x_n} \le \frac{x_1 + \cdots + x_n}{n} \tag{5.98}$$

between the geometric and arithmetic means of n nonnegative numbers. Equality holds in (5.98), as noted above, only when $x_1 = x_2 = \cdots = x_n$. If we set $n = 2$, $\alpha_1 = \frac{1}{p}$, $\alpha_2 = \frac{1}{q}$, $x_1 = a$, $x_2 = b$ in (5.97), we again obtain the known equality (5.86).

Example 18 Let $f(x) = x^p$, $x \ge 0$, $p > 1$. Since such a function is convex, we have

$$\left(\sum_{i=1}^n \alpha_i x_i\right)^p \le \sum_{i=1}^n \alpha_i x_i^p.$$

Setting $q = \frac{p}{p-1}$, $\alpha_i = b_i^q (\sum_{i=1}^n b_i^q)^{-1}$, and $x_i = a_i b_i^{-1/(p-1)} \sum_{i=1}^n b_i^q$ here, we obtain Hölder's inequality (5.88):

$$\sum_{i=1}^n a_i b_i \le \left(\sum_{i=1}^n a_i^p\right)^{1/p} \left(\sum_{i=1}^n b_i^q\right)^{1/q},$$

where $\frac{1}{p} + \frac{1}{q} = 1$ and $p > 1$.

For $p < 1$ the function $f(x) = x^p$ is convex upward, and so analogous reasoning can be carried out in Hölder's other inequality (5.89).

5.4.4 L'Hôpital's Rule

We now pause to discuss a special, but very useful device for finding the limit of a ratio of functions, known as l'Hôpital's rule.[20]

Proposition 8 (l'Hôpital's rule) *Suppose the functions $f :]a, b[\to \mathbb{R}$ and $g :]a, b[\to \mathbb{R}$ are differentiable on the open interval $]a, b[$ $(-\infty \le a < b \le +\infty)$ with $g'(x) \ne 0$ on $]a, b[$ and*

$$\frac{f'(x)}{g'(x)} \to A \quad \text{as } x \to a + 0 \quad (-\infty \le A \le +\infty).$$

[20]G.F. de l'Hôpital's (1661–1704) – French mathematician, a capable student of Johann Bernoulli, a marquis for whom the latter wrote the first textbook of analysis in the years 1691–1692. The portion of this textbook devoted to differential calculus was published in slightly altered form by l'Hôpital's under his own name. Thus "l'Hôpital's rule" is really due to Johann Bernoulli.

Then

$$\frac{f(x)}{g(x)} \to A \quad \textit{as } x \to a+0$$

in each of the following two cases:

1^0 $(f(x) \to 0) \wedge (g(x) \to 0)$ *as* $x \to a+0$,

or

2^0 $g(x) \to \infty$ *as* $x \to a+0$.

A similar assertion holds as $x \to b-0$.

L'Hôpital's rule can be stated succinctly, but not quite accurately, as follows. *The limit of a ratio of functions equals the limit of the ratio of their derivatives if the latter exists.*

Proof If $g'(x) \neq 0$, we conclude on the basis of Rolle's theorem that $g(x)$ is strictly monotonic on $]a, b[$. Hence, shrinking the interval $]a, b[$ if necessary by shifting toward the endpoint a, we can assume that $g(x) \neq 0$ on $]a, b[$. By Cauchy's theorem, for $x, y \in]a, b[$ there exists a point $\xi \in]a, b[$ such that

$$\frac{f(x)-f(y)}{g(x)-g(y)} = \frac{f'(\xi)}{g'(\xi)}.$$

Let us rewrite this equality in a form convenient for us at this point:

$$\frac{f(x)}{g(x)} = \frac{f(y)}{g(x)} + \frac{f'(\xi)}{g'(\xi)}\left(1 - \frac{g(y)}{g(x)}\right).$$

As $x \to a+0$, we shall make y tend to $a+0$ in such a way that

$$\frac{f(y)}{g(x)} \to 0 \quad \text{and} \quad \frac{g(y)}{g(x)} \to 0.$$

This is obviously possible under each of the two hypotheses 1^0 and 2^0 that we are considering. Since ξ lies between x and y, we also have $\xi \to a+0$. Hence the right-hand side of the last inequality (and therefore the left-hand side also) tends to A. □

Example 19 $\lim_{x\to 0} \frac{\sin x}{x} = \lim_{x\to 0} \frac{\cos x}{1} = 1$.

This example should not be looked on as a new, independent proof of the relation $\frac{\sin x}{x} \to 1$ as $x \to 0$. The fact is that in deriving the relation $\sin' x = \cos x$ we already made use of the limit just calculated.

We always verify the legitimacy of applying l'Hôpital's rule after we find the limit of the ratio of the derivatives. In doing so, one must not forget to verify condition 1^0 or 2^0. The importance of these conditions can be seen in the following example.

Example 20 Let $f(x) = \cos x$, $g(x) = \sin x$. Then $f'(x) = -\sin x$, $g'(x) = \cos x$, and $\frac{f(x)}{g(x)} \to +\infty$ as $x \to +0$, while $\frac{f'(x)}{g'(x)} \to 0$ as $x \to +0$.

Example 21

$$\lim_{x\to+\infty} \frac{\ln x}{x^\alpha} = \lim_{x\to+\infty} \frac{(\frac{1}{x})}{\alpha x^{\alpha-1}} = \lim_{x\to+\infty} \frac{1}{\alpha x^\alpha} = 0 \quad \text{for } \alpha > 0.$$

Example 22

$$\lim_{x\to+\infty} \frac{x^\alpha}{a^x} = \lim_{x\to+\infty} \frac{\alpha x^{\alpha-1}}{a^x \ln a} = \cdots = \lim_{x\to+\infty} \frac{\alpha(\alpha-1)\cdots(\alpha-n+1)x^{a-n}}{a^x (\ln a)^n} = 0$$

for $a > 1$, since for $n > \alpha$ and $a > 1$ it is obvious that $\frac{x^{\alpha-n}}{a^x} \to 0$ if $x \to +\infty$.

We remark that this entire chain of equalities was hypothetical until we arrived at an expression whose limit we could find.

5.4.5 Constructing the Graph of a Function

A graphical representation is often used to gain a visualizable description of a function. As a rule, such a representation is useful in discussing qualitative questions about the behavior of the function being studied.

For precise computations graphs are used more rarely. In this connection what is important is not so much a scrupulous reproduction of the function in the form of a graph as the construction of a sketch of the graph of the function that correctly reflects the main elements of its behavior. In this subsection we shall study some general devices that are encountered in constructing a sketch of the graph of a function.

a. Graphs of the Elementary Functions

We recall first of all what the graphs of the main elementary functions look like. A complete mastery of these is needed for what follows (Figs. 5.12–5.18).

b. Examples of Sketches of Graphs of Functions (Without Application of the Differential Calculus)

Let us now consider some examples in which a sketch of the graph of a function can be easily constructed if we know the graphs and properties of the simplest elementary functions.

Fig. 5.15

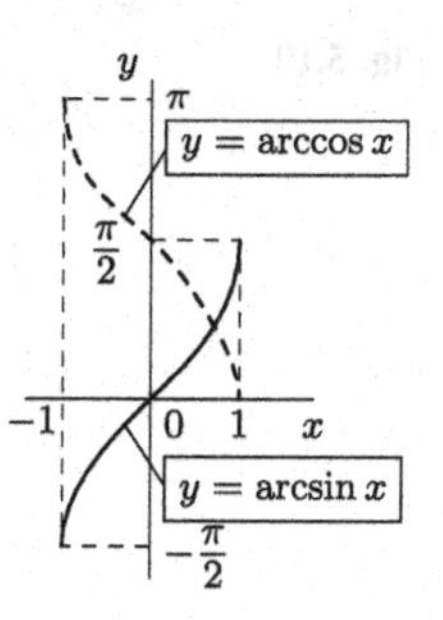

Fig. 5.16

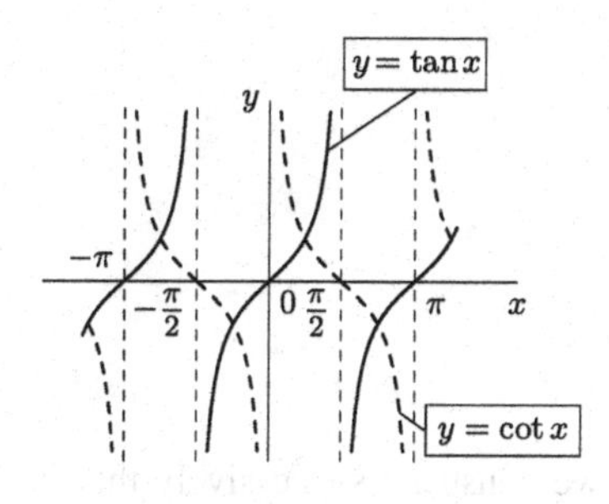

Fig. 5.17

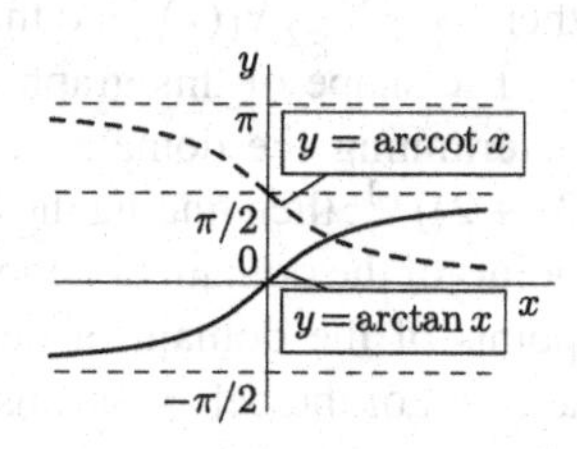

Fig. 5.18

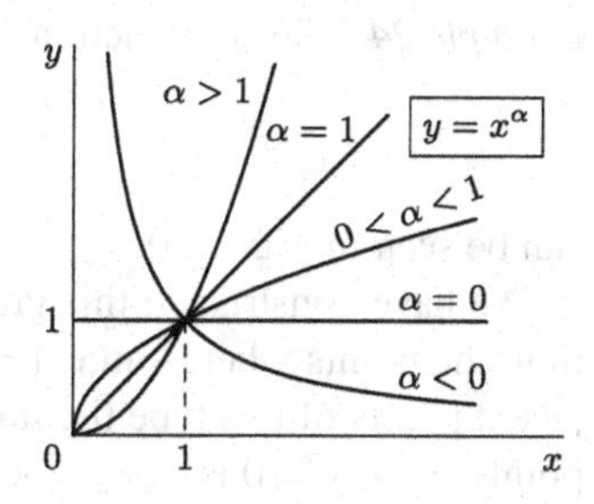

Example 23 Let us construct a sketch of the graph of the function

$$h = \log_{x^2-3x-2} 2.$$

Taking account of the relation

$$y = \log_{x^2-3x+2} 2 = \frac{1}{\log_2(x^2-3x+2)} = \frac{1}{\log_2(x-1)(x-2)},$$

Fig. 5.19

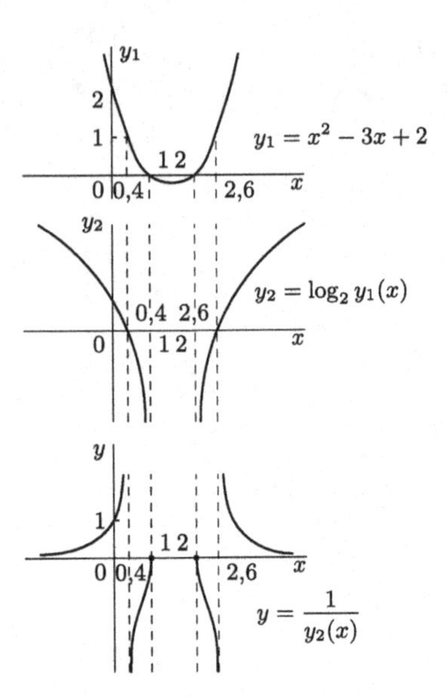

we construct successively the graph of the quadratic trinomial $y_1 = x^2 - 3x + 2$, then $y_2 = \log_2 y_1(x)$, and then $y = \frac{1}{y_2(x)}$ (Fig. 5.19).

The shape of this graph could have been "guessed" in a different way: by first determining the domain of definition of the function $\log_{x^2-3x+2} 2 = (\log_2(x^2 - 3x + 2))^{-1}$, then finding the behavior of the function under approach to the boundary points of the domain of definition and on intervals whose endpoints are the boundary points of the domain of definition, and finally drawing a "smooth curve" taking account of the behavior thus determined at the ends of the interval.

Example 24 The construction of a sketch of the graph of the function

$$y = \sin(x^2)$$

can be seen in Fig. 5.20.

We have constructed this graph using certain characteristic points for this function, the points where $\sin(x^2) = -1$, $\sin(x^2) = 0$, or $\sin(x^2) = 1$. Between two adjacent points of this type the function is monotonic. The form of the graph near the point $x = 0$, $y = 0$ is determined by the fact that $\sin(x^2) \sim x^2$ as $x \to 0$. Moreover, it is useful to note that this function is even.

Since we will be speaking only of sketches rather than a precise construction of the graph of a function, let us agree for the sake of brevity to understand that "constructing the graph of a function" is equivalent to "constructing a sketch of the graph of the function".

Example 25 Let us construct the graph of the function

$$y = x + \arctan(x^3 - 1)$$

Fig. 5.20

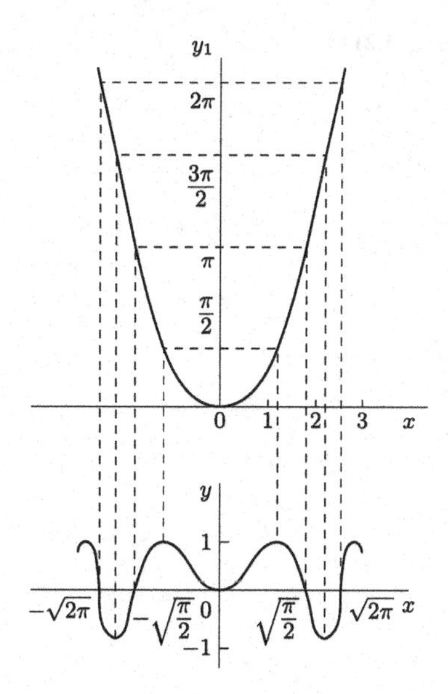

(Fig. 5.21). As $x \to -\infty$ the graph is well approximated by the line $y = x - \frac{\pi}{2}$, while for $x \to +\infty$ it is approximated by $y = x + \frac{\pi}{2}$.

We now introduce a useful concept.

Definition 4 The line $c_0 + c_1 x$ is called an *asymptote* of the graph of the function $y = f(x)$ as $x \to -\infty$ (or $x \to +\infty$) if $f(x) - (c_0 + c_1 x) = o(1)$ as $x \to -\infty$ (or $x \to +\infty$).

Thus in the present example the graph has the asymptote $y = x - \frac{\pi}{2}$ as $x \to -\infty$ and $y = x + \frac{\pi}{2}$ as $x \to +\infty$.

If $|f(x)| \to \infty$ as $x \to a - 0$ (or as $x \to a + 0$) it is clear that the graph of the function will move ever closer to the vertical line $x = a$ as x approaches a. We call this line a *vertical asymptote* of the graph, in contrast to the asymptotes introduced in Definition 4, which are always oblique.

Thus, the graph in Example 23 (see Fig. 5.19) has two vertical asymptotes and one horizontal asymptote (the same asymptote as $x \to -\infty$ and as $x \to +\infty$).

It obviously follows from Definition 4 that

$$c_1 = \lim_{x \to -\infty} \frac{f(x)}{x},$$

$$c_0 = \lim_{x \to -\infty} \big(f(x) - c_1 x\big).$$

In general, if $f(x) - (c_0 + c_1 x + \cdots + c_n x^n) = o(1)$ as $x \to -\infty$, then

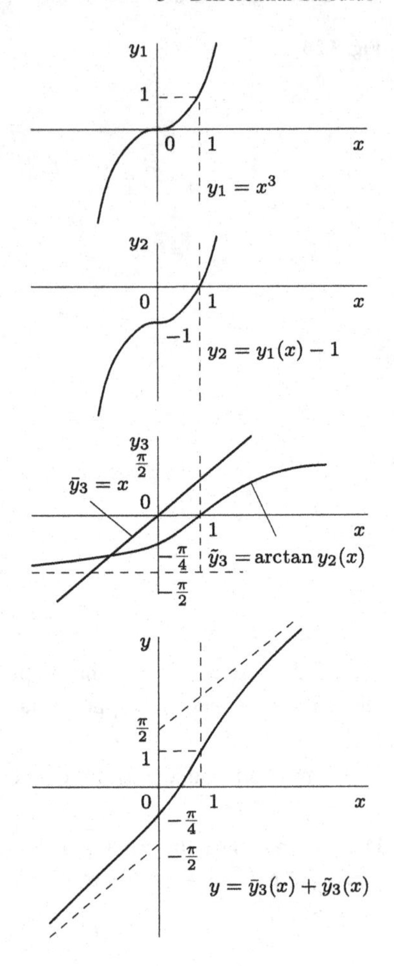

Fig. 5.21

$$c_n = \lim_{x\to-\infty} \frac{f(x)}{x^n},$$

$$c_{n-1} = \lim_{x\to-\infty} \frac{f(x) - c_n x^n}{x^{n-1}},$$

$$\vdots$$

$$c_0 = \lim_{x\to-\infty} \big(f(x) - (c_1 x + \cdots + c_n x^n)\big).$$

These relations, written out here for the case $x \to -\infty$, are of course also valid in the case $x \to +\infty$ and can be used to describe the asymptotic behavior of the graph of a function $f(x)$ using the graph of the corresponding algebraic polynomial $c_1 + c_1 x + \cdots + c_n x^n$.

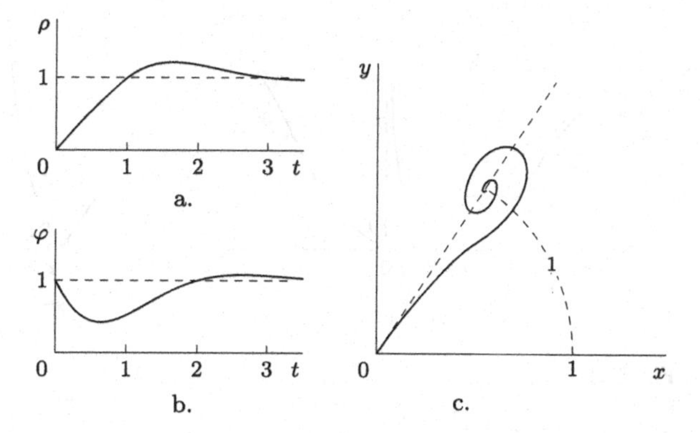

Fig. 5.22

Example 26 Let (ρ,φ) be polar coordinates in the plane and suppose a point is moving in the plane in such a way that

$$\rho = \rho(t) = 1 - \mathrm{e}^{-t}\cos\frac{\pi}{2}t,$$

$$\varphi = \varphi(t) = 1 - \mathrm{e}^{-t}\sin\frac{\pi}{2}t$$

at time t $(t \geq 0)$. Draw the trajectory of the point.

In order to do this, we first draw the graphs of $\rho(t)$ and $\varphi(t)$ (Figs. 5.22a and 5.22b).

Then, looking simultaneously at both of the graphs just constructed, we can describe the general form of the trajectory of the point (Fig. 5.22c).

c. The Use of Differential Calculus in Constructing the Graph of a Function

As we have seen, the graphs of many functions can be drawn in their general features without going beyond the most elementary considerations. However, if we want to make the sketch more precise, we can use the machinery of differential calculus in cases where the derivative of the function being studied is not too complicated. We shall illustrate this using examples.

Example 27 Construct the graph of the function $y = f(x)$ when

$$f(x) = |x+2|\mathrm{e}^{-1/x}.$$

The function $f(x)$ is defined for $x \in \mathbb{R}\backslash 0$. Since $\mathrm{e}^{-1/x} \to 1$ as $x \to \infty$, it follows that

$$|x+2|\mathrm{e}^{-1/x} \sim \begin{cases} -(x+2) & \text{as } x \to -\infty, \\ (x+2) & \text{as } x \to +\infty. \end{cases}$$

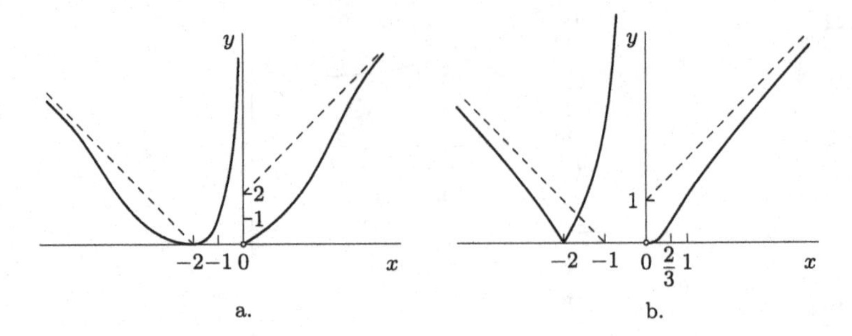

Fig. 5.23

Table 5.2

Interval	$]-\infty,-2[$	$]-2,0[$	$]0,+\infty[$
Sign of $f'(x)$	$-$	$+$	$+$
Behavior of $f(x)$	$+\infty \searrow 0$	$0 \nearrow +\infty$	$0 \nearrow +\infty$

Next, it is obvious that $|x+2|e^{-1/x} \to +\infty$ as $x \to -0$, and $|x+2|e^{-1/x} \to 0$ as $x \to +0$. Finally, it is clear that $f(x) \geq 0$ and $f(-2) = 0$. On the basis of these observations, one can already make a first draft of the graph (Fig. 5.23a).

Let us now see for certain whether this function is monotonic on the intervals $]-\infty,-2[$, $[-2,0[$, and $]0,+\infty[$, whether it really does have these asymptotics, and whether the convexity of the graph is correctly shown.

Since

$$f'(x) = \begin{cases} -\frac{x^2+x+2}{x^2} e^{-1/x}, & \text{if } x < -2, \\ \frac{x^2+x+2}{x^2} e^{-1/x}, & \text{if } -2 < x \text{ and } x \neq 0, \end{cases}$$

and $f'(x) \neq 0$, we can form Table 5.2.

On the regions of constant sign of the derivative, as we know, the function exhibits the corresponding monotonicity. In the bottom row of the table the symbol $+\infty \searrow 0$ denotes a monotonic decrease in the values of the function from $+\infty$ to 0, and $0 \nearrow +\infty$ denotes monotonic increase from 0 to $+\infty$.

We observe that $f'(x) \to -4e^{-1/2}$ as $x \to -2-0$ and $f'(x) \to 4e^{-1/2}$ as $x \to -2+0$, so that the point $(-2,0)$ must be a cusp in the graph (a bend of the same type as in the graph of the function $|x|$), and not a regular point, as depicted in Fig. 5.23a). Next, $f'(x) \to 0$ as $x \to +0$, so that the graph should emanate from the origin tangentially to the x-axis (remember the geometric meaning of $f'(x)$!).

We now make the asymptotics of the function as $x \to -\infty$ and $x \to +\infty$ more precise.

Since $e^{-1/x} = 1 - x^{-1} + o(x^{-1})$ as $x \to \infty$, it follows that

$$|x+2|e^{-1/x} \sim \begin{cases} -x-1+o(1) & \text{as } x \to -\infty, \\ x+1+o(1) & \text{as } x \to +\infty, \end{cases}$$

Table 5.3

Interval	$]-\infty, -2[$	$]-2, 0[$	$]0, 2/3[$	$]2/3, +\infty[$
Sign of $f''(x)$	$-$	$+$	$+$	$-$
Convexity of $f(x)$	Upward	Downward	Downward	Upward

so that in fact the oblique asymptotes of the graph are $y = -x - 1$ as $x \to -\infty$ and $y = x + 1$ as $x \to +\infty$.

From these data we can already construct a quite reliable sketch of the graph, but we shall go further and find the regions of convexity of the graph by computing the second derivative:

$$f''(x) = \begin{cases} -\frac{2-3x}{x^4}\mathrm{e}^{-1/x}, & \text{if } x < -2, \\ \frac{2-3x}{x^4}\mathrm{e}^{-1/x}, & \text{if } -2 < x \text{ and } x \neq 0. \end{cases}$$

Since $f''(x) = 0$ only at $x = 2/3$, we have Table 5.3.

Since the function is differentiable at $x = 2/3$ and $f''(x)$ changes sign as x passes through that point, the point $(2/3, f(2/3))$ is a point of inflection of the graph.

Incidentally, if the derivative $f'(x)$ had had a zero, it would have been possible to judge using the table of values of $f'(x)$ whether the corresponding point was an extremum. In this case, however, $f'(x)$ has no zeros, even though the function has a local minimum at $x = -2$. It is continuous at that point and $f'(x)$ changes from negative to positive as x passes through that point. Still, the fact that the function has a minimum at $x = -2$ can be seen just from the description of the variation of values of $f(x)$ on the corresponding intervals, taking into account, of course, the relation $f(-2) = 0$.

We can now draw a more precise sketch of the graph of this function (Fig. 5.23b).

We conclude with one more example.

Example 28 Let (x, y) be Cartesian coordinates in the plane and suppose a moving point has coordinates

$$x = \frac{t}{1 - t^2}, \qquad y = \frac{t - 2t^3}{1 - t^2}$$

at time t $(t \geq 0)$. Describe the trajectory of the point.

We begin by sketching the graphs of each of the two coordinate functions $x = x(t)$ and $y = y(t)$ (Figs. 5.24a and 5.24b).

The second of these graphs is somewhat more interesting than the first, and so we shall describe how to construct it.

We can see the behavior of the function $y = y(t)$ as $t \to +0$, $t \to 1 - 0$, $t \to 1 + 0$, and the asymptote $y(t) = 2t + o(1)$ as $t \to +\infty$ immediately from the form of the analytic expression for $y(t)$.

After computing the derivative

$$\dot{y}(t) = \frac{1 - 5t^2 + 2t^4}{(1 - t^2)^2},$$

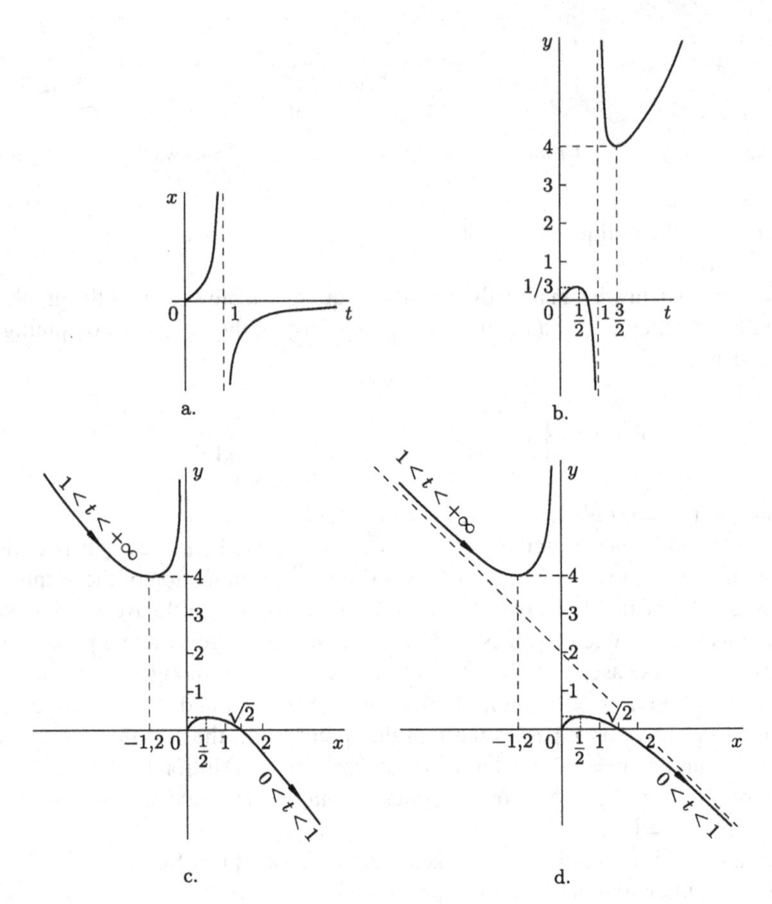

Fig. 5.24

Table 5.4

Interval	$]0,t_1[$	$]t_1,1[$	$]1,t_2[$	$]t_2,+\infty[$
Sign of $\dot{y}(t)$	+	−	−	+
Behavior of $y(t)$	$0 \nearrow y(t_1)$	$y(t_1) \searrow -\infty$	$+\infty \searrow y(t_2)$	$y(t_2) \nearrow +\infty$

we find its zeros: $t_1 \approx 0.5$ and $t_2 \approx 1.5$ in the region $t \geq 0$.

Then, by compiling Table 5.4 we find the regions of monotonicity and the local extreme values $y(t_1) \approx \frac{1}{3}$ (a maximum) and $y(t_2) \approx 4$ (a minimum).

Now, by studying both graphs $x = x(t)$ and $y = y(t)$ simultaneously, we make a sketch of the trajectory of the point in the plane (Fig. 5.24c).

This sketch can be made more precise. For example, one can determine the asymptotics of the trajectory.

Since $\lim_{t\to 1} \frac{y(t)}{x(t)} = -1$ and $\lim_{t\to 1}(y(t)+x(t)) = 2$, the line $y = -x+2$ is an asymptote for both ends of the trajectory, corresponding to t approaching 1. It is also clear that the line $x = 0$ is a vertical asymptote for the portion of the trajectory corresponding to $t \to +\infty$.

We find next

$$y'_x = \frac{\dot{y}_t}{\dot{x}_t} = \frac{1 - 5t^2 + 2t^4}{1 + t^2}.$$

As one can easily see, the function $\frac{1-5u+2u^2}{1+u}$ decreases monotonically from 1 to -1 as u increases from 0 to 1 and increases from -1 to $+\infty$ as u increases from 1 to $+\infty$.

From the monotonic nature of y'_x, one can draw conclusions about the convexity of the trajectory on the corresponding regions. Taking account of what has just been said, one can construct the following, more precise sketch of the trajectory of the point (Fig. 5.24d).

If we had considered the trajectory for $t < 0$ as well, the fact that $x(t)$ and $y(t)$ are odd functions would have added to the curves already drawn in the xy-plane the curves obtained from them by reflection in the origin.

We now summarize some of these results as very general recommendations for the order in which to proceed when constructing the graph of a function given analytically. Here they are:

1^0 Give the domain of definition of the function.
2^0 Note the specific properties of the function if they are obvious (for example, evenness or oddness, periodicity, identity to some well-known functions up to simple coordinate changes).
3^0 Determine the asymptotic behavior of the function under approach to boundary points of the domain of definition and, in particular, find asymptotes if they exist.
4^0 Find the intervals of monotonicity of the function and exhibit its local extreme values.
5^0 Determine the convexity properties of the graph and indicate the points of inflection.
6^0 Note any characteristic points of the graph, in particular points of intersection with the coordinate axes, provided there are such and they are amenable to computation.

5.4.6 Problems and Exercises

1. Let $x = (x_1, \dots, x_n)$ and $\alpha = (\alpha_1, \dots, \alpha_n)$, where $x_i \geq 0, \alpha_i > 0$ for $i = 1, \dots, n$ and $\sum_{i=1}^{n} \alpha_i = 1$. For any number $t \neq 0$ we consider the *mean of order t of the numbers $x_1, \dots, x_n$ with weights α_i*:

$$M_t(x, \alpha) = \left(\sum_{i=1}^{n} \alpha_x x_i^t \right)^{1/t}.$$

In particular, when $\alpha_1 = \cdots = \alpha_n = \frac{1}{n}$, we obtain the harmonic, arithmetic, and quadratic means for $t = -1, 1, 2$ respectively.
Show that

a) $\lim_{t\to 0} M_t(x,\alpha) = x_1^{\alpha_1} \cdots x_n^{\alpha_n}$, that is, in the limit one can obtain the geometric mean;
b) $\lim_{t\to +\infty} M_t(x,\alpha) = \max_{1\le i\le n} x_i$;
c) $\lim_{t\to -\infty} M_t(x,\alpha) = \min_{1\le i\le n} x_i$;
d) $M_t(x,\alpha)$ is a nondecreasing function of t on $\mathbb{R}$ and is strictly increasing if $n > 1$ and the numbers x_i are all nonzero.

2. Show that $|1+x|^p \ge 1 + px + c_p\varphi_p(x)$, where c_p is a constant depending only on p,

$$\varphi_p(x) = \begin{cases} |x|^2 & \text{for } |x| \le 1, \\ |x|^p & \text{for } |x| > 1, \end{cases} \quad \text{if } 1 < p \le 2,$$

and $\varphi_p(x) = |x|^\rho$ on $\mathbb{R}$ if $2 < p$.
3. Verify that $\cos x < (\frac{\sin x}{x})^3$ for $0 < |x| < \frac{\pi}{2}$.
4. Study the function $f(x)$ and construct its graph if

a) $f(x) = \arctan \log_2 \cos(\pi x + \frac{\pi}{4})$;
b) $f(x) = \arccos(\frac{3}{2} - \sin x)$;
c) $f(x) = \sqrt[3]{x(x+3)^2}$;
d) Construct the curve defined in polar coordinates by the equation $\varphi = \frac{\rho}{\rho^2+1}$, $\rho \ge 0$, and exhibit its asymptotics;
e) Show how, knowing the graph of the function $y = f(x)$, one can obtain the graph of the following functions $f(x) + B$, $Af(x)$, $f(x+b)$, $f(ax)$, and, in particular $-f(x)$ and $f(-x)$.

5. Show that if $f \in C(]a,b[)$ and the inequality

$$f\left(\frac{x_1+x_2}{2}\right) \le \frac{f(x_1)+f(x_2)}{2}$$

holds for any points $x_1, x_2 \in]a,b[$, then the function f is convex on $]a,b[$.
6. Show that

a) if a convex function $f : \mathbb{R} \to \mathbb{R}$ is bounded, it is constant;
b) if

$$\lim_{x\to -\infty} \frac{f(x)}{x} = \lim_{x\to +\infty} \frac{f(x)}{x} = 0,$$

for a convex function $f : \mathbb{R} \to \mathbb{R}$, then f is constant.
c) for any convex function f defined on an open interval $a < x < +\infty$ (or $-\infty < x < a$), the ratio $\frac{f(x)}{x}$ tends to a finite limit or to infinity as x tends to infinity in the domain of definition of the function.

7. Show that if $f :]a,b[\to \mathbb{R}$ is a convex function, then

a) at any point $x \in]a, b[$ it has a left-hand derivative f'_- and a right-hand derivative f'_+, defined as

$$f'_-(x) = \lim_{h \to -0} \frac{f(x+h) - f(x)}{h},$$

$$f'_+(x) = \lim_{h \to +0} \frac{f(x+h) - f(x)}{h},$$

and $f'_-(x) \le f'_+(x)$;

b) the inequality $f'_+(x_1) \le f'_-(x_2)$ holds for $x_1, x_2 \in]a, b[$ and $x_1 < x_2$;

c) the set of cusps of the graph of $f(x)$ (for which $f'_-(x) \ne f'_+(x)$) is at most countable.

8. *The Legendre transform*[21] of a function $f : I \to \mathbb{R}$ defined on an interval $I \subset \mathbb{R}$ is the function

$$f^*(t) = \sup_{x \in I} \bigl(tx - f(x)\bigr).$$

Show that

a) The set I^* of values of $t \in \mathbb{R}$ for which $f^*(t) \in \mathbb{R}$ (that is, $f^*(t) \ne \infty$) is either empty or consists of a single point, or is an interval of the line, and in this last case the function $f^*(t)$ is convex on I^*.

b) If f is a convex function, then $I^* \ne \varnothing$, and for $f^* \in C(I^*)$

$$\bigl(f^*\bigr)^* = \sup_{t \in I^*} \bigl(xt - f^*(t)\bigr) = f(x)$$

for any $x \in I$. Thus the Legendre transform of a convex function is *involutive*, (its square is the identity transform).

c) The following inequality holds:

$$xt \le f(x) + f^*(t) \quad \text{for } x \in I \text{ and } t \in I^*.$$

d) When f is a convex differentiable function, $f^*(t) = tx_t - f(x_t)$, where x_t is determined from the equation $t = f'(x)$. Use this relation to obtain a geometric interpretation of the Legendre transform f^* and its argument t, showing that the Legendre transform is a function defined on the set of tangents to the graph of f.

e) The Legendre transform of the function $f(x) = \frac{1}{\alpha}x^\alpha$ for $\alpha > 1$ and $x \ge 0$ is the function $f^*(t) = \frac{1}{\beta}t^\beta$, where $t \ge 0$ and $\frac{1}{\alpha} + \frac{1}{\beta} = 1$. Taking account of c), use this fact to obtain Young's inequality, which we already know:

$$xt \le \frac{1}{\alpha}x^\alpha + \frac{1}{\beta}t^\beta.$$

---

[21] A.M. Legendre (1752–1833) – famous French mathematician.

f) The Legendre transform of the function $f(x) = \mathrm{e}^x$ is the function $f^*(t) = t \ln \frac{t}{e}$, $t > 0$, and the inequality

$$xt \le \mathrm{e}^x + t \ln \frac{t}{\mathrm{e}}$$

holds for $x \in \mathbb{R}$ and $t > 0$.

9. *Curvature and the radius and center of curvature of a curve at a point.* Suppose a point is moving in the plane according to a law given by a pair of twice-differentiable coordinate functions of time: $x = x(t)$, $y = y(t)$. In doing so, it describes a certain curve, which is said to be given in the parametric form $x = x(t)$, $y = y(t)$. A special case of such a definition is that of the graph of a function $y = f(x)$, where one may take $x = t$, $y = f(t)$. We wish to find a number that characterizes the curvature of the curve at a point, as the reciprocal of the radius of a circle serves as an indication of the amount of bending of the circle. We shall make use of this comparison.

a) Find the tangential and normal components $\mathbf{a}_t$ and $\mathbf{a}_n$ respectively of the acceleration $\mathbf{a} = (\ddot{x}(t), \ddot{y}(t))$ of the point, that is, write $\mathbf{a}$ as the sum $\mathbf{a}_t + \mathbf{a}_n$, where $\mathbf{a}_t$ is collinear with the velocity vector $\mathbf{v}(t) = (\dot{x}(t), \dot{y}(t))$, so that $\mathbf{a}_t$ points along the tangent to the trajectory and $\mathbf{a}_n$ is directed along the normal to the trajectory.

b) Show that the relation

$$\mathbf{r} = \frac{|\mathbf{v}(t)|}{|\mathbf{a}_n(t)|}$$

holds for motion along a circle of radius r.

c) For motion along any curve, taking account of b), it is natural to call the quantity

$$r(t) = \frac{|\mathbf{v}(t)|}{|\mathbf{a}_n(t)|}$$

the *radius of curvature* of the curve at the point $(x(t), y(t))$.

Show that the radius of curvature can be computed from the formula

$$r(t) = \frac{(\dot{x}^2 + \dot{y}^2)^{3/2}}{|\dot{x}\ddot{y} - \ddot{x}\dot{y}|}.$$

d) The reciprocal of the radius of curvature is called the *absolute curvature* of a plane curve at the point $(x(t), y(t))$. Along with the absolute curvature we consider the quantity

$$k(t) = \frac{\dot{x}\ddot{y} - \ddot{x}\dot{y}}{(\dot{x}^2 + \dot{y}^2)^{3/2}},$$

called the *curvature*.

Show that the sign of the curvature characterizes the direction of turning of the curve relative to its tangent. Determine the physical dimension of the curvature.

e) Show that the curvature of the graph of a function $y = f(x)$ at a point $(x, f(x))$ can be computed from the formula

$$k(x) = \frac{y''(x)}{[1 + (y')^2]^{3/2}}.$$

Compare the signs of $k(x)$ and $y''(x)$ with the direction of convexity of the graph.

f) Choose the constants a, b, and R so that the circle $(x - a)^2 + (y - b)^2 = R^2$ has the highest possible order of contact with the given parametrically defined curve $x = x(t)$, $y = y(t)$. It is assumed that $x(t)$ and $y(t)$ are twice differentiable and that $(\dot{x}(t_0), \dot{y}(t_0)) \neq (0, 0)$.

This circle is called the *osculating circle* of the curve at the point (x_0, y_0). Its center is called the *center of curvature* of the curve at the point (x_0, y_0). Verify that its radius equals the radius of curvature of the curve at that point, as defined in b).

g) Under the influence of gravity a particle begins to slide without any preliminary impetus from the tip of an iceberg of parabolic cross-section. The equation of the cross-section is $x + y^2 = 1$, where $x \geq 0$, $y \geq 0$. Compute the trajectory of motion of the particle until it reaches the ground.

5.5 Complex Numbers and the Connections Among the Elementary Functions

5.5.1 Complex Numbers

Just as the equation $x^2 = 2$ has no solutions in the domain $\mathbb{Q}$ of rational numbers, the equation $x^2 = -1$ has no solutions in the domain $\mathbb{R}$ of real numbers. And, just as we adjoin the symbol $\sqrt{2}$ as a solution of $x^2 = 2$ and connect it with rational numbers to get new numbers of the form $r_1 + \sqrt{2}r_2$, where $r_1, r_2 \in \mathbb{Q}$, we introduce the symbol i as a solution of $x^2 = -1$ and attach this number, which lies outside the real numbers, to real numbers and arithmetic operations in $\mathbb{R}$.

One remarkable feature of this enlargement of the field $\mathbb{R}$ of real numbers, among many others, is that in the resulting field $\mathbb{C}$ of complex numbers, every algebraic equation with real or complex coefficients now has a solution.

Let us now carry out this program.

a. Algebraic Extension of the Field $\mathbb{R}$

Thus, following Euler, we introduce a number i, the *imaginary unit*, such that $i^2 = -1$. The interaction between i and the real numbers is to consist of the following. One may multiply i by numbers $y \in \mathbb{R}$, that is, numbers of the form iy necessarily arise, and one may add such numbers to real numbers, that is, numbers of the form $x + iy$ occur, where $x, y \in \mathbb{R}$.

If we wish to have the usual operations of a commutative addition and a commutative multiplication that is distributive with respect to addition defined on the set of objects of the form $x + iy$ (which, following Gauss, we shall call the *complex numbers*), then we must make the following definitions:

$$(x_1 + iy_1) + (x_2 + iy_2) := (x_1 + x_2) + i(y_1 + y_2) \tag{5.99}$$

and

$$(x_1 + iy_1) \cdot (x_2 + iy_2) := (x_1x_2 - y_1y_2) + i(x_1y_2 + x_2y_1). \tag{5.100}$$

Two complex numbers $x_1 + iy_1$ and $x_2 + iy_2$ are considered equal if and only if $x_1 = x_2$ and $y_1 = y_2$.

We identify the real numbers $x \in \mathbb{R}$ with the numbers of the form $x + i \cdot 0$, and i with the number $0 + i \cdot 1$. The role of 0 in the complex numbers, as can be seen from Eq. (5.99), is played by the number $0 + i \cdot 0 = 0 \in \mathbb{R}$; the role of 1, as can be seen from Eq. (5.100), is played by $1 + i \cdot 0 = 1 \in \mathbb{R}$.

It follows from properties of the real numbers and definitions (5.99) and (5.100) that the set of complex numbers is a field containing $\mathbb{R}$ as a subfield.

We shall denote the field of complex numbers by $\mathbb{C}$ and typical elements of it usually by z and w.

The only nonobvious point in the verification that $\mathbb{C}$ is a field is the assertion that every non-zero complex number $z = x + iy$ has an inverse z^{-1} with respect to multiplication (a reciprocal), that is $z \cdot z^{-1} = 1$. Let us verify this.

We call the number $x - iy$ the *conjugate* of $z = x + iy$, and we denote it $\bar{z}$.

We observe that $z \cdot \bar{z} = (x^2 + y^2) + i \cdot 0 = x^2 + y^2 \neq 0$ if $z \neq 0$. Thus z^{-1} should be taken as $\frac{1}{x^2+y^2} \cdot \bar{z} = \frac{x}{x^2+y^2} - i\frac{y}{x^2+y^2}$.

b. Geometric Interpretation of the Field $\mathbb{C}$

We remark that once the algebraic operations (5.99) and (5.100) on complex numbers have been introduced, the symbol i, which led us to these definitions, is no longer needed. We can identify the complex number $z = x + iy$ with the ordered pair (x, y) of real numbers, called respectively the *real part* and the *imaginary part* of the complex number z. (The notation for this is $x = \operatorname{Re} z$, $y = \operatorname{Im} z$.)

But then, regarding the pair (x, y) as the Cartesian coordinates of a point of the plane $\mathbb{R}^2 = \mathbb{R} \times \mathbb{R}$, one can identify complex numbers with the points of this plane or with two-dimensional vectors having coordinates (x, y).

In such a vector interpretation the coordinatewise addition (5.99) of complex numbers corresponds to vector addition. Moreover such an interpretation naturally leads to the idea of the *absolute value* or *modulus* $|z|$ of a complex number as the absolute value or length of the vector (x, y) corresponding to it, that is

$$|z| = \sqrt{x^2 + y^2}, \quad \text{if } z = x + iy, \tag{5.101}$$

and also to a way of measuring the distance between complex numbers z_1 and z_2 as the distance between the points of the plane corresponding to them, that is, as

$$|z_1 - z_2| = \sqrt{(x_1 - x_2)^2 + (y_1 - y_2)^2}. \tag{5.102}$$

The set of complex numbers, interpreted as the set of points of the plane, is called the *complex plane* and also denoted by $\mathbb{C}$, just as the set of real numbers and the real line are both denoted by $\mathbb{R}$.

Since a point of the plane can also be defined in polar coordinates (r, φ) connected with Cartesian coordinates by the relations

$$\begin{aligned} x &= r\cos\varphi, \\ y &= r\sin\varphi, \end{aligned} \tag{5.103}$$

the complex number

$$z = x + iy \tag{5.104}$$

can be represented in the form

$$z = r(\cos\varphi + i\sin\varphi). \tag{5.105}$$

The expressions (5.104) and (5.105) are called respectively the *algebraic* and *trigonometric* (*polar*) forms of the complex number.

In the expression (5.105) the number $r \geq 0$ is called the *modulus* or *absolute value* of the complex number z (since, as one can see from (5.103), $r = |z|$), and φ the *argument* of z. The argument has meaning only for $z \neq 0$. Since the functions $\cos\varphi$ and $\sin\varphi$ are periodic, the argument of a complex number is determined only up to a multiple of 2π, and the symbol $\operatorname{Arg} z$ denotes the set of angles of the form $\varphi + 2\pi k$, $k \in \mathbb{Z}$, where φ is any angle satisfying (5.105). When it is desirable for every complex number to determine uniquely some angle $\varphi \in \operatorname{Arg} z$, one must agree in advance on the range from which the argument is to be chosen. This range is usually either $0 \leq \varphi < 2\pi$ or $-\pi < \varphi \leq \pi$. If such a choice has been made, we say that a *branch* (or the *principal branch*) of the argument has been chosen. The values of the argument within the chosen range are usually denoted $\arg z$.

The trigonometric form (5.105) for writing complex numbers is convenient in carrying out the operation of multiplication of complex numbers. In fact, if

$$\begin{aligned} z_1 &= r_1(\cos\varphi_1 + i\ \sin\varphi_1), \\ z_2 &= r_2(\cos\varphi_2 + i\ \sin\varphi_2), \end{aligned}$$

then

$$\begin{aligned} z_1 \cdot z_2 &= (r_1\cos\varphi_1 + ir_1\sin\varphi_1)(r_2\cos\varphi_2 + ir_2\sin\varphi_2) = \\ &= (r_1r_2\cos\varphi_1\cos\varphi_2 - r_1r_2\sin\varphi_1\sin\varphi_2) + \\ &\quad + i(r_1r_2\sin\varphi_1\cos\varphi_2 + r_1r_2\cos\varphi_2\sin\varphi_2), \end{aligned}$$

or

$$z_1 \cdot z_2 = r_1 r_2 \bigl(\cos(\varphi_2 + \varphi_2) + i\ \sin(\varphi_1 + \varphi_2)\bigr). \tag{5.106}$$

Thus, when two complex numbers are multiplied, their moduli are multiplied and their arguments are added.

We remark that what we have actually shown is that if $\varphi_1 \in \operatorname{Arg} z_1$ and $\varphi_2 \in \operatorname{Arg} z_2$, then $\varphi_1 + \varphi_2 \in \operatorname{Arg}(z_1 \cdot z_2)$. But since the argument is defined only up to a multiple of 2π, we can write that

$$\operatorname{Arg}(z_1 \cdot z_2) = \operatorname{Arg} z_1 + \operatorname{Arg} z_2, \tag{5.107}$$

interpreting this equality as set equality, the set on the right-hand side being the set of all numbers of the form $\varphi_1 + \varphi_2$, where $\varphi_1 \in \operatorname{Arg} z_1$ and $\varphi_2 \in \operatorname{Arg} z_2$. Thus it is useful to interpret the sum of the arguments in the sense of the set equality (5.107).

With this understanding of equality of arguments, one can assert, for example, that two complex numbers are equal if and only if their moduli and arguments are equal.

The following formula of de Moivre[22] follows by induction from formula (5.106):

$$\text{if } z = r(\cos\varphi + i \sin\varphi), \quad \text{then } z^n = r^n(\cos n\varphi + i \sin n\varphi). \tag{5.108}$$

Taking account of the explanations given in connection with the argument of a complex number, one can use de Moivre's formula to write out explicitly all the complex solutions of the equation $z^n = a$.

Indeed, if

$$a = \rho(\cos\psi + i \sin\psi)$$

and, by formula (5.108)

$$z^n = r^n(\cos n\varphi + i \sin n\varphi),$$

we have $r = \sqrt[n]{\rho}$ and $n\varphi = \psi + 2\pi k$, $k \in \mathbb{Z}$, from which we have $\varphi_k = \frac{\psi}{n} + \frac{2\pi}{n} k$. Different complex numbers are obviously obtained only for $k = 0, 1, \dots, n-1$. Thus we find n distinct roots of a:

$$z_k = \sqrt[n]{\rho}\left(\cos\left(\frac{\psi}{n} + \frac{2\pi}{n} k\right) + i \sin\left(\frac{\psi}{n} + \frac{2\pi}{n} k\right)\right) \quad (k = 0, 1, \dots, n-1).$$

In particular, if $a = 1$, that is, $\rho = 1$ and $\psi = 0$, we have

$$z_k = \sqrt[n]{{}_k 1} = \cos\left(\frac{2\pi}{n} k\right) + i \sin\left(\frac{2\pi}{n} k\right) \quad (k = 0, 1, \dots, n-1).$$

These points are located on the unit circle at the vertices of a regular n-gon.

[22] A. de Moivre (1667–1754) – British mathematician.

In connection with the geometric interpretation of the complex numbers themselves, it is useful to recall the geometric interpretation of the arithmetic operations on them.

For a fixed $b \in \mathbb{C}$, the sum $z+b$ can be interpreted as the mapping of $\mathbb{C}$ into itself given by the formula $z \mapsto z+b$. This mapping is a translation of the plane by the vector b.

For a fixed $a = |a|(\cos\varphi + i\sin\varphi) \neq 0$, the product az can be interpreted as the mapping $z \to az$ of $\mathbb{C}$ into itself, which is the composition of a dilation by a factor of $|a|$ and a rotation through the angle $\varphi \in \operatorname{Arg} a$. This is clear from formula (5.106).

5.5.2 Convergence in $\mathbb{C}$ and Series with Complex Terms

The distance (5.102) between complex numbers enables us to define the ε-neighborhood of a number $z_0 \in \mathbb{C}$ as the set $\{z \in \mathbb{C} \mid |z - z_0| < \varepsilon\}$. This set is a disk (without the boundary circle) of radius ε centered at the point (x_0, y_0) if $z_0 = x_i + iy_0$.

We shall say that a sequence $\{z_n\}$ of complex numbers *converges to* $z_0 \in \mathbb{C}$ if $\lim_{n\to\infty} |z_n - z_0| = 0$.

It is clear from the inequalities

$$\max\{|x_n - x_0|, |y_n - y_0|\} \le |z_n - z_0| \le |x_n - x_0| + |y_n - y_0| \tag{5.109}$$

that a sequence of complex numbers converges if and only if the sequences of real and imaginary parts of the terms of the sequence both converge.

By analogy with sequences of real numbers, a sequence of complex numbers $\{z_n\}$ is called a *fundamental* or *Cauchy* sequence if for every $\varepsilon > 0$ there exists an index $N \in \mathbb{N}$ such that $|z_n - z_m| < \varepsilon$ for all $n, m > N$.

It is clear from inequalities (5.109) that a sequence of complex numbers is a Cauchy sequence if and only if the sequences of real and imaginary parts of its terms are both Cauchy sequences.

Taking the Cauchy convergence criterion for sequences of real numbers into account, we conclude on the basis of (5.109) that the following proposition holds.

Proposition 1 (The Cauchy criterion) *A sequence of complex numbers converges if and only if it is a Cauchy sequence.*

If we interpret the sum of a series of complex numbers

$$z_1 + z_2 + \cdots + z_n + \cdots \tag{5.110}$$

as the limit of its partial sums $s_n = z_1 + \cdots + z_n$ as $n \to \infty$, we also obtain the Cauchy criterion for convergence of the series (5.110).

Proposition 2 *The series* (5.110) *converges if and only if for every $\varepsilon > 0$ there exists $N \in \mathbb{N}$ such that*

$$|z_m + \cdots + z_n| < \varepsilon \tag{5.111}$$

for any natural numbers $n \geq m > N$.

From this one can see that a necessary condition for convergence of the series (5.110) is that $z_n \to 0$ as $n \to \infty$. (This, however, is also clear from the very definition of convergence.)

As in the real case, the series (5.110) is *absolutely convergent* if the series

$$|z_1| + |z_2| + \cdots + |z_n| + \cdots \tag{5.112}$$

converges.

It follows from the Cauchy criterion and the inequality

$$|z_m + \cdots + z_n| \leq |z_m| + \cdots + |z_n|$$

that if the series (5.110) converges absolutely, then it converges.

Examples The series

1) $1 + \frac{1}{1!}z + \frac{1}{2!}z^2 + \cdots + \frac{1}{n!}z^n + \cdots$,
2) $z - \frac{1}{3!}z^3 + \frac{1}{5!}z^5 - \cdots$, and
3) $1 - \frac{1}{2!}z^2 + \frac{1}{4!}z^4 - \cdots$

converge absolutely for all $z \in \mathbb{C}$, since the series

1′) $1 + \frac{1}{1!}|z| + \frac{1}{2!}|z|^2 + \cdots$,
2′) $|z| + \frac{1}{3!}|z|^3 + \frac{1}{5!}|z|^5 + \cdots$,
3′) $1 + \frac{1}{2!}|z|^2 + \frac{1}{4!}|z|^4 + \cdots$,

all converge for any value of $|z| \in \mathbb{R}$. We remark that we have used the equality $|z^n| = |z|^n$ here.

Example 4 The series $1 + z + z^2 + \cdots$ converges absolutely for $|z| < 1$ and its sum is $s = \frac{1}{1-z}$. For $|z| \geq 1$ it does not converge, since in that case the general term does not tend to zero.

Series of the form

$$c_0 + c_1(z - z_0) + \cdots + c_n(z - z_0)^n + \cdots \tag{5.113}$$

are called *power series*.

By applying the Cauchy criterion (Sect. 3.1.4) to the series

$$|c_0| + \left|c_1(z - z_0)\right| + \cdots + \left|c_n(z - z_0)^n\right| + \cdots, \tag{5.114}$$

we conclude that this series converges if

$$|z - z_0| < \left(\overline{\lim_{n\to\infty}} \sqrt[n]{|c_n|}\right)^{-1},$$

and that the general term does not tend to zero if $|z - z_0| \geq (\overline{\lim}_{n\to\infty} \sqrt[n]{|c_n|})^{-1}$. From this we obtain the following proposition.

Proposition 3 (The Cauchy–Hadamard[23] formula) *The power series* (5.113) *converges inside the disk $|z - z_0| < R$ with center at z_0 and radius given by the Cauchy–Hadamard formula*

$$R = \frac{1}{\overline{\lim}_{n\to\infty} \sqrt[n]{|c_n|}}. \tag{5.115}$$

At any point exterior to this disk the power series diverges.
At any point interior to the disk, the power series converges absolutely.

Remark In regard to convergence on the boundary circle $|z - z_0| = R$ Proposition 3 is silent, since all the logically admissible possibilities really can occur.

Examples The series

5) $\sum_{n=1}^{\infty} z^n$,
6) $\sum_{n=1}^{\infty} \frac{1}{n} z^n$,

and

7) $\sum_{n=1}^{\infty} \frac{1}{n^2} z^n$

converge in the unit disk $|z| < 1$, but the series 5) diverges at every point z where $|z| = 1$. The series 6) diverges for $z = 1$ and (as one can show) converges for $z = -1$. The series 7) converges absolutely for $|z| = 1$, since $|\frac{1}{n^2} z^n| = \frac{1}{n^2}$.

One must keep in mind the possible degenerate case when $R = 0$ in (5.115), which was not taken account of in Proposition 3. In this case, of course, the entire *disk of convergence* degenerates to the single point z_0 of convergence of the series (5.113).

The following result is an obvious corollary of Proposition 3.

Corollary (Abel's first theorem on power series) *If the power series* (5.113) *converges at some value z^*, then it converges, and indeed even absolutely, for any value of z satisfying the inequality $|z - z_0| < |z^* - z_0|$.*

The propositions obtained up to this point can be regarded as simple extensions of facts already known to us. We shall now prove two general propositions about

[23] J. Hadamard (1865–1963) – well-known French mathematician.

series that we have not proved up to now in any form, although we have partly discussed some of the questions they address.

Proposition 4 *If a series* $z_1 + z_2 + \cdots + z_n + \cdots$ *of complex numbers converges absolutely, then a series* $z_{n_1} + z_{n_2} + \cdots + z_{n_k} + \cdots$ *obtained by rearranging*[24] *its terms also converges absolutely and has the same sum.*

Proof Using the convergence of the series $\sum_{n=1}^{\infty} |z_n|$, given a number $\varepsilon > 0$, we choose $N \in \mathbb{N}$ such that $\sum_{n=N+1}^{\infty} |z_n| < \varepsilon$.

We then find an index $K \in \mathbb{N}$ such that all the terms in the sum $S_N = z_1 + \cdots + z_N$ are among the terms of the sum $\tilde{s}_k = z_{n_1} + \cdots + z_{n_k}$ for $k > K$. If $s = \sum_{n=1}^{\infty} z_n$, we find that for $k > K$

$$|s - \tilde{s}_k| \leq |s - s_N| + |s_N - \tilde{s}_k| \leq \sum_{n=N+1}^{\infty} |z_n| + \sum_{n=N+1}^{\infty} |z_n| < 2\varepsilon.$$

Thus we have shown that $\tilde{s}_k \to s$ as $k \to \infty$. If we apply what has just been proved to the series $|z_1| + |z_2| + \cdots + |z_n| + \cdots$ and $|z_{n_1}| + |z_{n_2}| + \cdots + |z_{n_k}| + \cdots$, we find that the latter series converges. Thus Proposition 4 is now completely proved.

Our next proposition will involve the product of two series

$$(a_1 + a_2 + \cdots + a_n + \cdots) \cdot (b_1 + b_2 + \cdots + b_n + \cdots).$$

The problem is that if we remove the parentheses and form all possible pair-wise products $a_i b_j$, there is no natural order for summing these products, since we have two indices of summation. The set of pairs (i, j), where $i, j \in \mathbb{N}$, is countable, as we know. Therefore we could write down a series having the products $a_i b_j$ as terms in some order. The sum of such a series might depend on the order in which these terms are taken. But, as we have just seen, in absolutely convergent series the sum is independent of any rearrangement of the terms. Thus, it is desirable to determine when the series with terms $a_i b_j$ converges absolutely. □

Proposition 5 *The product of absolutely convergent series is an absolutely convergent series whose sum equals the product of the sums of the factor series.*

Proof We begin by remarking that whatever finite sum $\sum a_i b_j$ of terms of the form $a_i b_j$ we take, we can always find N such that the product of the sums $A_N = a_1 + \cdots + a_N$ and $B_N = b_1 + \cdots + b_N$ contains all the terms in that sum. Therefore

$$\left|\sum a_i b_j\right| \leq \sum |a_i b_j| \leq \sum_{i,j=1}^{N} |a_i b_j| = \sum_{i=1}^{N} |a_i| \cdot \sum_{j=1}^{N} |b_j| \leq \sum_{i=1}^{\infty} |a_i| \cdot |\sum_{j=1}^{\infty} |b_j|,$$

[24]The term with index k in this series is the term z_{n_k} with index n_k in the original series. Here the mapping $\mathbb{N} \ni k \mapsto n_k \in \mathbb{N}$ is assumed to be a bijective mapping on the set $\mathbb{N}$.

from which it follows that the series $\sum_{i,j=1}^{\infty} a_i b_j$ converges absolutely and that its sum is uniquely determined independently of the order of the factors. In that case the sum can be obtained, for example, as the limit of the products of the sums $A_n = a_1 + \cdots + a_n$ and $B_n = b_1 + \cdots + b_n$. But $A_n B_n \to AB$ as $n \to \infty$, where $A = \sum_{n=1}^{\infty} a_n$ and $B = \sum_{n=1}^{\infty} b_n$, which completes the proof of Proposition 5. □

The following example is very important.

Example 8 The series $\sum_{n=0}^{\infty} \frac{1}{n!} a^n$ and $\sum_{m=0}^{\infty} \frac{1}{m!} b^m$ converge absolutely. In the product of these series let us group together all monomials of the form $a^n b^m$ having the same total degree $n + m = k$. We then obtain the series

$$\sum_{k=0}^{\infty} \Bigl(\sum_{n+m=k} \frac{1}{n!} a^n \frac{1}{m!} b^m \Bigr).$$

But

$$\sum_{m+n=k} \frac{1}{n!m!} a^n b^m = \frac{1}{k!} \sum_{n=0}^{k} \frac{k!}{n!(k-n)!} a^n b^{k-n} = \frac{1}{k!}(a+b)^k,$$

and therefore we find that

$$\sum_{n=0}^{\infty} \frac{1}{n!} a^n \cdot \sum_{m=0}^{\infty} \frac{1}{m!} b^m = \sum_{k=0}^{\infty} \frac{1}{k!} (a+b)^k. \tag{5.116}$$

5.5.3 Euler's Formula and the Connections Among the Elementary Functions

In Examples 1)–3) we established the absolute convergence in $\mathbb{C}$ of the series obtained by extending into the complex domain the Taylor series of the functions e^x, $\sin x$, and $\cos x$, which are defined on $\mathbb{R}$. For that reason, the following definitions are natural ones to make for the functions e^z, $\cos z$, and $\sin z$ in $\mathbb{C}$:

$$e^z = \exp z := 1 + \frac{1}{1!} z + \frac{1}{2!} z^2 + \frac{1}{3!} z^3 + \cdots, \tag{5.117}$$

$$\cos z := 1 - \frac{1}{2!} z^2 + \frac{1}{4!} z^4 - \cdots, \tag{5.118}$$

$$\sin z := z - \frac{1}{3!} z^3 + \frac{1}{5!} z^5 - \cdots. \tag{5.119}$$

Following Euler,[25] let us make the substitution $z = iy$ in Eq. (5.117). By suitably grouping the terms of the partial sums of the resulting series, we find that

$$1 + \frac{1}{1!}(iy) + \frac{1}{2!}(iy)^2 + \frac{1}{3!}(iy)^3 + \frac{1}{4!}(iy)^4 + \frac{1}{5!}(iy)^5 + \cdots =$$
$$= \left(1 - \frac{1}{2!}y^2 + \frac{1}{4!}y^4 - \cdots\right) + i\left(\frac{1}{1!}y - \frac{1}{3!}y^3 + \frac{1}{5!}y^5 - \cdots\right),$$

that is,

$$\boxed{e^{iy} = \cos y + i \sin y.} \tag{5.120}$$

This is the famous *Euler formula*.

In deriving it we used the fact that $i^2 = -1$, $i^3 = -i$, $i^4 = 1$, $i^5 = i$, and so forth. The number y in formula (5.120) may be either a real number or an arbitrary complex number.

It follows from the definitions (5.118) and (5.119) that

$$\cos(-z) = \cos z,$$
$$\sin(-z) = -\sin z,$$

that is, $\cos z$ is an even function and $\sin z$ is an odd function. Thus

$$e^{-iy} = \cos y - i \sin y.$$

Comparing this last equality with formula (5.120), we obtain

$$\cos y = \frac{1}{2}\left(e^{iy} + e^{-iy}\right),$$
$$\sin y = \frac{1}{2i}\left(e^{iy} - e^{-iy}\right).$$

Since y is any complex number, it would be better to rewrite these equalities using notation that leaves no doubt of this fact:

$$\cos z = \frac{1}{2}\left(e^{iz} + e^{-iz}\right),$$
$$\sin z = \frac{1}{2i}\left(e^{iz} - e^{-iz}\right). \tag{5.121}$$

Thus, if we assume that $\exp z$ is defined by relation (5.117), then formulas (5.121), which are equivalent to the expansions (5.118) and (5.119), like the for-

[25] L. Euler (1707–1783) – eminent mathematician and specialist in theoretical mechanics, of Swiss extraction, who lived the majority of his life in St. Petersburg. In the words of Laplace, "Euler is the common teacher of all mathematicians of the second half of the eighteenth century."

mulas

$$\cosh y = \frac{1}{2}(e^z + e^{-z}),$$
$$\sinh z = \frac{1}{2}(e^z - e^{-z}), \tag{5.122}$$

can be taken as the definitions of the corresponding circular and hyperbolic functions. Disregarding all the considerations about trigonometric functions that led us to this step, which have not been rigorously justified (even though they did lead us to Euler's formula), we can now perform a typical mathematical trick and take formulas (5.121) and (5.122) as definitions and obtain from them in a completely formal manner all the properties of the circular and trigonometric functions.

For example, the fundamental identities

$$\cos^2 z + \sin^2 z = 1,$$
$$\cosh^2 z - \sinh^2 z = 1,$$

like the parity properties, can be verified immediately.

The deeper properties, such as, for example, the formula for the cosine and sine of a sum follow from the characteristic property of the exponential function:

$$\exp(z_1 + z_2) = \exp(z_1) \cdot \exp(z_2), \tag{5.123}$$

which obviously follows from the definition (5.117) and formula (5.116). Let us derive the formulas for the cosine and sine of a sum:

On the one hand, by Euler's formula

$$e^{i(z_1+z_2)} = \cos(z_1 + z_2) + i \sin(z_1 + z_2). \tag{5.124}$$

On the other hand, by the property of the exponential function and Euler's formula

$$e^{i(z_1+z_2)} = e^{iz_1} e^{iz_2} = (\cos z_1 + i \sin z_1)(\cos z_2 + i \sin z_2) =$$
$$= (\cos z_1 \cos z_2 - \sin z_1 \sin z_2) + i(\sin z_1 \cos z_2 + \cos z_2 \sin z_2). \tag{5.125}$$

If z_1 and z_2 were real numbers, then, equating the real and imaginary parts of the numbers in formulas (5.124) and (5.125), we would now have obtained the required formulas. Since we are trying to prove them for any $z_1, z_2 \in \mathbb{C}$, we use the fact that $\cos z$ is even and $\sin z$ is odd to obtain yet another equality:

$$e^{-i(z_1+z_2)} = (\cos z_1 \cos z_2 - \sin z_1 \sin z_2) - i(\sin z_1 \cos z_2 + \cos z_1 \sin z_2). \tag{5.126}$$

Comparing (5.125) and (5.126), we find

$$\cos(z_1 + z_2) = \frac{1}{2}\left(e^{i(z_1+z_2)} + e^{-i(z_1+z_2)}\right) = \cos z_1 \cos z_2 - \sin z_1 \sin z_2,$$

$$\sin(z_1 + z_2) = \frac{1}{2i}\left(e^{i(z_1+z_2)} - e^{-i(z_1+z_2)}\right) = \sin z_1 \cos z_2 + \cos z_1 \sin z_2.$$

The corresponding formulas for the hyperbolic functions $\cosh z$ and $\sinh z$ could be obtained in a completely analogous manner. Incidentally, as can be seen from formulas (5.121) and (5.122), these functions are connected with $\cos z$ and $\sin z$ by the relations

$$\cosh z = \cos iz,$$

$$\sinh z = -i \sin iz.$$

However, to obtain even such geometrically obvious facts as the equality $\sin \pi = 0$ or $\cos(z+2\pi) = \cos z$ from the definitions (5.121) and (5.122) is very difficult. Hence, while striving for precision, one must not forget the problems where these functions naturally arise. For that reason, we shall not attempt at this point to overcome the potential difficulties connected with the definitions (5.121) and (5.122) when describing the properties of the trigonometric functions. We shall return to these functions after presenting the theory of integration. Our purpose at present was only to demonstrate the remarkable unity of seemingly completely different functions, which would have been impossible to detect without going into the domain of complex numbers.

If we take as known that for $x \in \mathbb{R}$

$$\cos(x + 2\pi) = \cos x, \qquad \sin(x + 2\pi) = \sin x,$$

$$\cos 0 = 1, \qquad \sin 0 = 0,$$

then from Euler's formula (5.120) we obtain the relation

$$\boxed{e^{i\pi} + 1 = 0,} \tag{5.127}$$

in which all the most important constants of the different areas of mathematics are represented: 1 (arithmetic), π (geometry), e (analysis), and i (algebra).

From (5.123) and (5.127), as well as from (5.120), one can see that

$$\exp(z + i2\pi) = \exp z,$$

that is, the exponential function is a periodic function on $\mathbb{C}$ with the purely imaginary period $T = i2\pi$.

Taking account of Euler's formula, we can now represent the trigonometric notation (5.105) for a complex number in the form

$$z = re^{i\varphi},$$

where r is the modulus of z and φ its argument.

The formula of de Moivre now becomes very simple:

$$z^n = r^n e^{in\varphi}. \tag{5.128}$$

Fig. 5.25

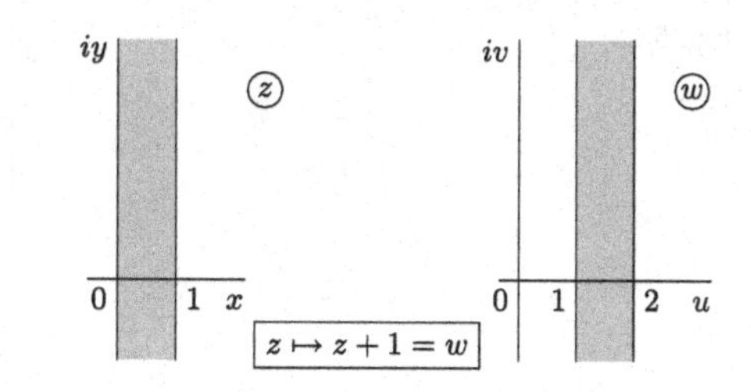

Fig. 5.26

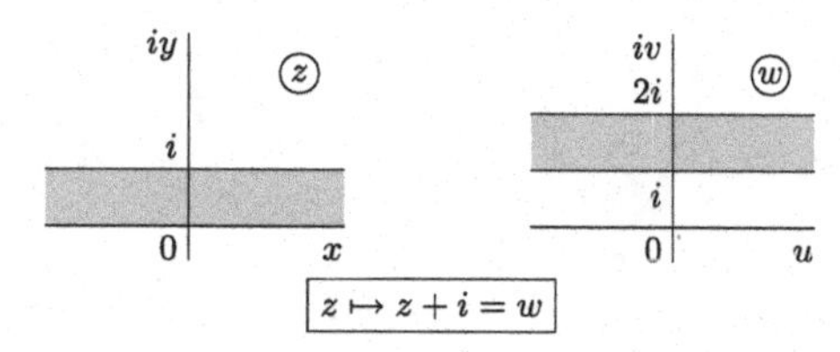

Fig. 5.27

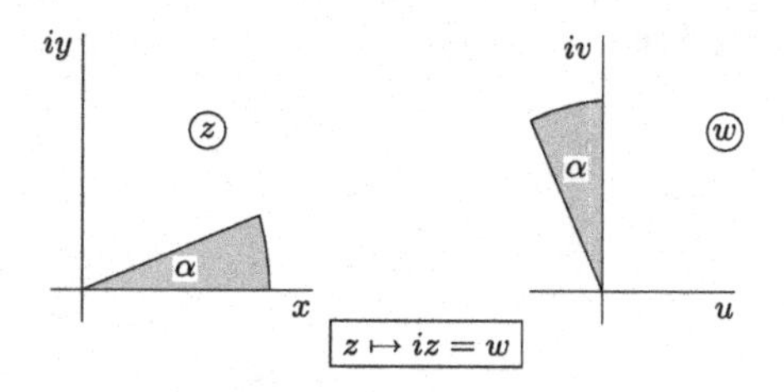

5.5.4 Power Series Representation of a Function. Analyticity

A function $w = f(z)$ of a complex variable z with complex values w, defined on a set $E \subset \mathbb{C}$, is a mapping $f : E \to \mathbb{C}$. The graph of such a function is a subset of $\mathbb{C} \times \mathbb{C} = \mathbb{R}^2 \times \mathbb{R}^2 = \mathbb{R}^4$, and therefore is not visualizable in the traditional way. To compensate for this loss to some extent, one usually keeps two copies of the complex plane $\mathbb{C}$, indicating points of the domain of definition in one and points of the range of values in the other.

In the examples below the domain E and its image under the corresponding mapping are indicated.

Example 9 See Fig. 5.25.

Example 10 See Fig. 5.26.

Example 11 See Fig. 5.27.

These correspondences follow from the equalities $i = e^{i\pi/2}$, $z = re^{i\varphi}$, and $iz = re^{i(r+\pi/2)}$, that is, a rotation through angle $\frac{\pi}{2}$ has occurred.

Example 12 See Fig. 5.28.

Fig. 5.28

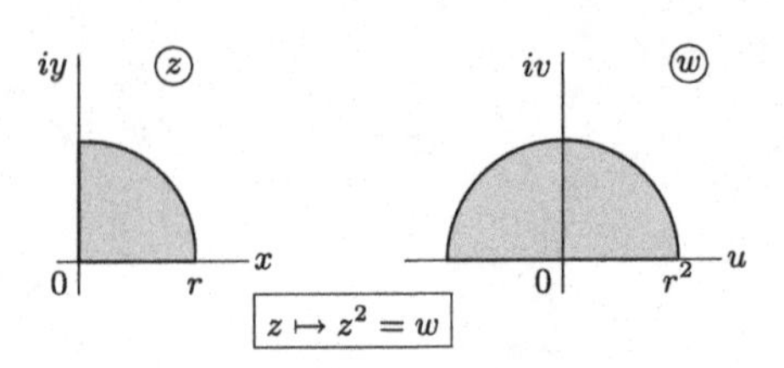

Fig. 5.29

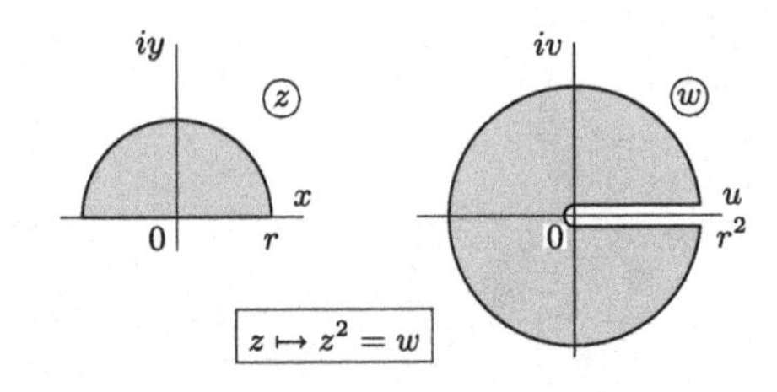

Fig. 5.30

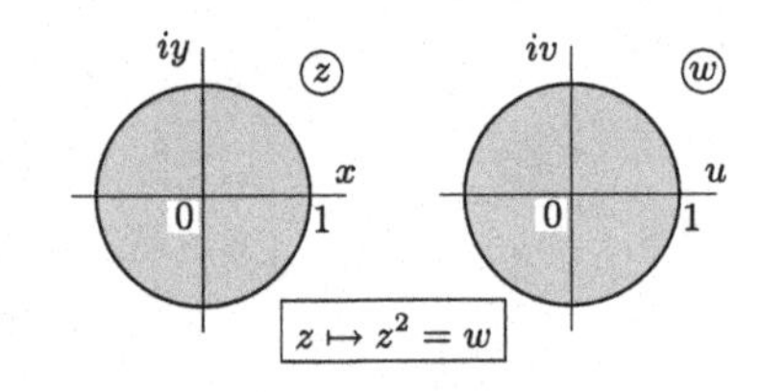

Fig. 5.31

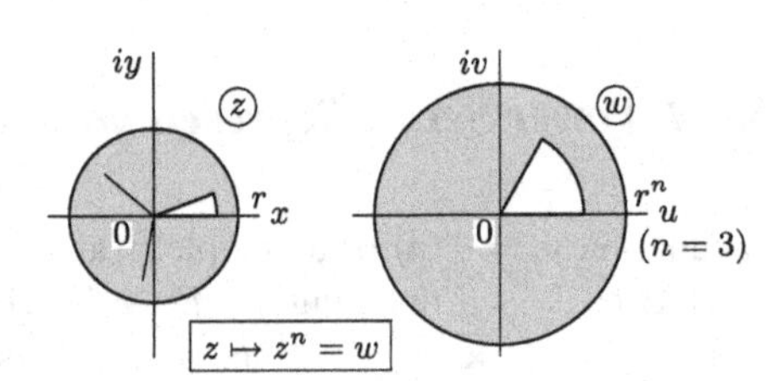

For, if $z = re^{i\varphi}$, then $z^2 = r^2 e^{i2\varphi}$.

Example 13 See Fig. 5.29.

Example 14 See Fig. 5.30.

It is clear from Examples 12 and 13 that under this function the unit disk maps into itself, but is covered twice.

Example 15 See Fig. 5.31.

If $z = re^{i\varphi}$, then by (5.128), we have $z^n = r^n e^{in\varphi}$, so that in this case the image of the disk of radius r is the disk of radius r^n, each point of which is the image of n points in the original disk (located, as it happens, at the vertices of a regular n-gon).

The only exception is the point $w = 0$, whose pre-image is the point $z = 0$. However, as $z \to 0$, the function z^n is an infinitesimal of order n, and so we say that at $z = 0$ the function has a zero of order n. Taking account of this kind of multiplicity, one can now say that the number of pre-images of every point w under the mapping $z \mapsto z^n = w$ is n. In particular, the equation $z^n = 0$ has the n coincident roots $z_1 = \cdots = z_n = 0$.

In accordance with the general definition of continuity, a function $f(z)$ of a complex variable is called *continuous* at a point $z_0 \in \mathbb{C}$ if for any neighborhood V $(f(z_0))$ of its value $f(z_0)$ there exists a neighborhood $U(z_0)$ such that $f(z) \in V(f(z_0))$ for all $z \in U(z_0)$. In short,

$$\lim_{z \to z_0} f(z) = f(z_0).$$

The *derivative* of a function $f(z)$ at a point z_0, as for the real-valued case, is defined as

$$f'(z_0) = \lim_{z \to z_0} \frac{f(z) - f(z_0)}{z - z_0}, \tag{5.129}$$

if this limit exists.

The equality (5.129) is equivalent to

$$f(z) - f(z_0) = f'(z_0)(z - z_0) + o(z - z_0) \tag{5.130}$$

as $z \to z_0$, corresponding to the definition of *differentiability* of a function at the point z_0.

Since the definition of differentiability in the complex-valued case is the same as the corresponding definition for real-valued functions and the arithmetic properties of the fields $\mathbb{C}$ and $\mathbb{R}$ are the same, one may say that all the general rules for differentiation hold also in the complex-valued case.

Example 16

$$(f + g)'(z) = f'(z) + g'(z),$$
$$(f \cdot g)'(z) = f'(z)g(z) + f(z)g'(z),$$
$$(g \circ f)'(z) = g'\big(f(z)\big) \cdot f'(z),$$

so that if $f(z) = z^2$, then $f'(z) = 1 \cdot z + z \cdot 1 = 2z$, or if $f(z) = z^n$, then $f'(z) = nz^{n-1}$, and if

$$P_n(z) = c_0 + c_1(z - z_0) + \cdots + c_n(z - z_0)^n,$$

then

$$P_n'(z) = c_1 + 2c_2(z - z_0) + \cdots + nc_n(z - z_0)^{n-1}. .$$

Theorem 1 *The sum* $f(z)=\sum_{n=0}^{\infty}c_n(z-z_0)^n$ *of a power series is an infinitely differentiable function inside the entire disk in which convergence occurs. Moreover,*

$$f^{(k)}(z)=\sum_{n=0}^{\infty}\frac{\mathrm{d}^k}{\mathrm{d}z^k}\big(c_n(z-z_0)^n\big),\quad k=0,1,\ldots,$$

and

$$c_n=\frac{1}{n!}f^{(n)}(z_0),\quad n=0,1,\ldots.$$

Proof The expressions for the coefficient follows in an obvious way from the expressions for $f^{(k)}(z)$ for $k=n$ and $z=z_0$.

As for the formula for $f^{(k)}(z)$, it suffices to verify this formula for $k=1$, since the function $f'(z)$ will then be the sum of a power series.

Thus, let us verify that the function $\varphi(z)=\sum_{n=1}^{\infty}nc_n(z-z_0)^{n-1}$ is indeed the derivative of $f(z)$.

We begin by remarking that by the Cauchy–Hadamard formula (5.115) the radius of convergence of the derived series is the same as the radius of convergence R of the original power series for $f(z)$.

For simplicity of notation from now on we shall assume that $z_0=0$, that is, $f(z)=\sum_{n=0}^{\infty}c_nz^n$, $\varphi(z)=\sum_{n=1}^{\infty}nc_nz^{n-1}$ and that these series converge for $|z|<R$.

Since a power series converges absolutely on the interior of its disk of convergence, we note (and this is crucial) that the estimate $|nc_nz^{n-1}|=n|c_n||z|^{n-1}\le n|c_n|r^{n-1}$ holds for $|z|\le r<R$, and that series $\sum_{n=1}^{\infty}n|c_n|r^{n-1}$ converges. Hence, for any $\varepsilon>0$ there exists an index N such that

$$\left|\sum_{n=N+1}^{\infty}nc_nz^{n-1}\right|\le\sum_{n=N+1}^{\infty}nc_nr^{n-1}\le\frac{\varepsilon}{3}$$

for $|z|\le r$.

Thus at any point of the disk $|z|<r$ the function $\varphi(z)$ is within $\frac{\varepsilon}{3}$ of the Nth partial sum of the series that defines it.

Now let ζ and z be arbitrary points of this disk. The transformation

$$\frac{f(\zeta)-f(z)}{\zeta-z}=\sum_{n=1}^{\infty}c_n\frac{\zeta^n-z^n}{\zeta-z}=$$

$$=\sum_{n=1}^{\infty}c_n\big(\zeta^{n-1}+\zeta^{n-2}z+\cdots+\zeta z^{n-2}+z^{n-1}\big)$$

and the estimate $|c_n(\zeta^{n-1}+\cdots+z^{n-1})|\le|c_n|nr^{n-1}$ enable us to conclude, as above, that the difference quotient we are interested in is equal within $\frac{\varepsilon}{3}$ to the partial sum of the series that defines it, provided $|\zeta| < r$ and $|z| < r$. Hence, for $|\zeta| < r$ and $|z| < r$ we have

$$\left|\frac{f(\zeta)-f(z)}{\zeta - z} - \varphi(z)\right| \leq \left|\sum_{n=1}^{N} c_n \frac{\zeta^n - z^n}{\zeta - z} - \sum_{n=1}^{N} n c_n z^{n-1}\right| + 2\frac{\varepsilon}{3}.$$

If we now fix z and let ζ tend to z, passing to the limit in the finite sum, we see that the right-hand side of this last inequality will be less than ε for ζ sufficiently close to z, and hence the left-hand side will be also.

Thus, for any point z in the disk $|z| < r < R$, we have verified that $f'(z) = \varphi(z)$. Since r is arbitrary, this relation holds for any point of the disk $|z| < R$. □

This theorem enables us to specify the class of functions whose Taylor series converge to them.

A function is *analytic* at a point $z_0 \in \mathbb{C}$ if it can be represented in a neighborhood of the point in the following ("analytic") form:

$$f(z) = \sum_{n=0}^{\infty} c_n (z - z_0)^n,$$

that is, as the sum of a power series in $z - z_0$.

It is not difficult to verify (see Problem 7 below) that the sum of a power series is analytic at any interior point of the disk of convergence of the series.

Taking account of the definition of analyticity, we deduce the following corollary from the definition of analyticity.

Corollary a) *If a function is analytic at a point, then it is infinitely differentiable at that point, and its Taylor series converges to it in a neighborhood of the point.*

b) *The Taylor series of a function defined in a neighborhood of a point and infinitely differentiable at that point converges to the function in some neighborhood of the point if and only if the function is analytic.*

In the theory of functions of a complex variable one can prove a remarkable fact that has no analogue in the theory of functions of a real variable. It turns out that if a function $f(z)$ is differentiable in a neighborhood of a point $z_0 \in \mathbb{C}$, then it is analytic at that point. This is certainly an amazing fact, since it then follows from the theorem just proved that if a function $f(z)$ has one derivative $f'(z)$ in a neighborhood of a point, it also has derivatives of all orders in that neighborhood.

At first sight this result is just as surprising as the fact that by adjoining to $\mathbb{R}$ a root i of the one particular equation $z^2 = -1$ we obtain a field $\mathbb{C}$ in which every algebraic polynomial $P(z)$ has a root. We intend to make use of the fact that an algebraic equation $P(z) = 0$ has a solution in $\mathbb{C}$, and for that reason we shall prove it as a good illustration of the elementary concepts of complex numbers and functions of a complex variable introduced in this section.

Fig. 5.32

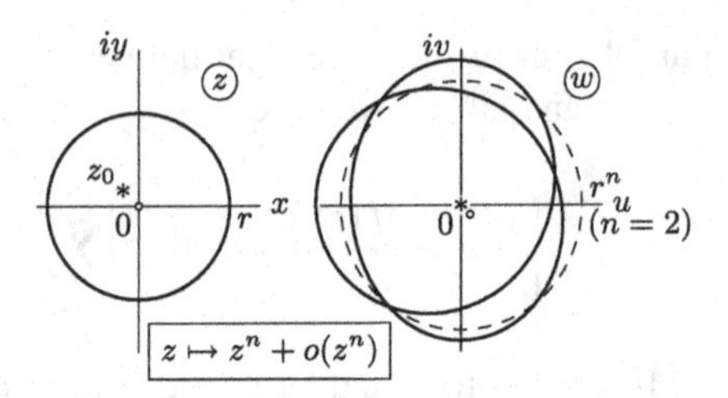

5.5.5 Algebraic Closedness of the Field $\mathbb{C}$ of Complex Numbers

If we prove that every polynomial $P(z) = c_0 + c_1 z + \cdots + c_n z^n$, $n \geq 1$, with complex coefficients has a root in $\mathbb{C}$, then there will be no need to enlarge the field $\mathbb{C}$ because some algebraic equation is not solvable in $\mathbb{C}$. In this sense the assertion that every polynomial $P(z)$ has a root establishes that *the field $\mathbb{C}$ is algebraically closed.*

To obtain a clear idea of the reason why every polynomial has a root in $\mathbb{C}$ while there can fail to be a root in $\mathbb{R}$, we use the geometric interpretation of complex numbers and functions of a complex variable.

We remark that

$$P(z) = z^n \left(\frac{c_0}{z^n} + \frac{c_1}{z^{n-1}} + \cdots + \frac{c_{n-1}}{z} + c_n \right),$$

so that $P(z) = c_n z^n + o(z^n)$ as $|z| \to \infty$. Since we are interested in finding a root of the equation $P(z) = 0$, dividing both sides of the equation by c_n, we may assume that the leading coefficient c_n of $P(z)$ equals 1, and hence

$$P(z) = z^n + o(z^n) \quad \text{as } |z| \to \infty. \tag{5.131}$$

If we recall (Example 15) that the circle of radius r maps to the circle of radius r^n with center at 0 under the mapping $z \mapsto z^n$, we see that for sufficiently large values of r the image of the circle $|z| = r$ under the mapping $w = P(z)$ will be, with small relative error, the circle $|w| = r^n$ in the w-plane (Fig. 5.32). What is important is that, in any case, it will be a curve that encloses the point $w = 0$.

If the disk $|z| \leq r$ is regarded as a film stretched over the circle $|z| = r$, this film is mapped into a film stretched over the image of that disk under the mapping $w = P(z)$. But, since the latter encloses the point $w = 0$, some point of that film must coincide with $w = 0$, and hence there is a point z_0 in the disk $|z| < r$ that maps to $w = 0$ under the mapping $w = P(z)$, that is, $P(z_0) = 0$.

This intuitive reasoning leads to a number of important and useful concepts of topology (the *index* of a path with respect to a point, and the *degree* of a mapping), by means of which it can be made into a complete proof that is valid not only for polynomials, as one can see. However, these considerations would unfortunately distract us from the main subject we are now studying. For that reason, we shall give another proof that is more in the mainstream of the ideas we have already mastered.

Theorem 2 *Every polynomial*

$$P(z)=c_0+c_1z+\cdots+c_nz^n$$

of degree $n\geq 1$ with complex coefficients has a root in $\mathbb{C}$.

Proof Without loss of generality, we may obviously assume that $c_n=1$.

Let $\mu=\inf_{z\in\mathbb{C}}|P(z)|$. Since $P(z)=z^n(1+\frac{c_{n-1}}{z}+\cdots+\frac{c_0}{z^n})$, we have

$$\big|P(z)\big|\geq |z|^n\biggl(1-\frac{|c_{n-1}|}{|z|}-\cdots-\frac{|c_0|}{|z|^n}\biggr),$$

and obviously $|P(z)|>\max\{1,2\mu\}$ for $|z|>R$ if R is sufficiently large. Consequently, the points of a sequence $\{z_k\}$ at which $0<|P(z_k)|-\mu<\frac{1}{k}$ lie inside the disk $|z|\leq R$.

We shall verify that there is a point z_0 in $\mathbb{C}$ (in fact, in this disk) at which $|P(z_0)|=\mu$. To do this, we remark that if $z_k=x_k+iy_k$, then $\max\{|x_k|,\ |y_k|\}\leq |z_k|\leq R$ and hence the sequences of real numbers $\{x_k\}$ and $\{y_k\}$ are bounded. Choosing first a convergent subsequence $\{x_{k_l}\}$ from $\{x_k\}$ and then a convergent subsequence $\{y_{k_{l_m}}\}$ from $\{y_{k_l}\}$, we obtain a subsequence $z_{k_{l_m}}=x_{k_{l_m}}+iy_{k_{l_m}}$ of the sequence $\{z_k\}$ that has a limit $\lim_{m\to\infty}z_{k_{l_m}}=\lim_{m\to\infty}x_{k_{l_m}}+i\lim_{m\to\infty}y_{k_{l_m}}=x_0+iy_0=z_0$, and since $|z_{k_{l_m}}|\to|z_0|$ as $m\to\infty$, it follows that $|z_0|\leq R$. So as to avoid cumbersome notation, and not have to pass to subsequences, we shall assume that the sequence $\{z_k\}$ itself converges. It follows from the continuity of $P(z)$ at $z_0\in\mathbb{C}$ that $\lim_{k\to\infty}P(z_k)=P(z_0)$. But then[26] $|P(z_0)|=\lim_{k\to\infty}|P(z_k)|=\mu$.

We shall now assume that $\mu>0$, and use this assumption to derive a contradiction. If $P(z_0)\neq 0$, consider the polynomial $Q(z)=\frac{P(z+z_0)}{P(z_0)}$. By construction $Q(0)=1$ and $|Q(z)|=\frac{|P(z+z_0)|}{|P(z_0)|}\geq 1$.

Since $Q(0)=1$, the polynomial $Q(z)$ has the form

$$Q(z)=1+q_kz^k+q_{k+1}z^{k+1}+\cdots+q_nz^n,$$

where $|q_k|\neq 0$ and $1\leq k\leq n$. If $q_k=\rho e^{i\psi}$, then for $\varphi=\frac{\pi-\psi}{k}$ we shall have $q_k\cdot(e^{i\varphi})^k=\rho e^{i\psi}e^{i(\pi-\psi)}=\rho e^{i\pi}=-\rho=-|q_k|$. Then, for $z=re^{i\varphi}$ we obtain

$$\begin{aligned}\big|Q\big(re^{i\varphi}\big)\big|&\leq\big|1+q_kz^k\big|+\big(\big|q_{k+1}z^{k+1}\big|+\cdots+\big|q_nz^n\big|\big)=\\&=\big|1-r^k|q_k|\big|+r^{k+1}\big(|q_{k+1}|+\cdots+|q_n|r^{n-k-1}\big)=\\&=1-r^k\big(|q_k|-r|q_{k+1}|-\cdots-r^{n-k}|q_n|\big)<1,\end{aligned}$$

[26] Observe that on the one hand we have shown that from every sequence of complex numbers whose moduli are bounded one can extract a convergent subsequence, while on the other hand we have given another possible proof of the theorem that a continuous function on a closed interval has a minimum, as was done here for the disk $|z|\leq R$.

if r is sufficiently close to 0. But $|Q(z)| \geq 1$ for $z \in \mathbb{C}$. This contradiction shows that $P(z_0) = 0$. □

Remark 1 The first proof of the theorem that every algebraic equation with complex coefficients has a solution in $\mathbb{C}$ (which is traditionally known as the fundamental theorem of algebra) was given by Gauss, who in general breathed real life into the so-called "imaginary" numbers by finding a variety of profound applications for them.

Remark 2 A polynomial with real coefficients $P(z) = a_0 + \cdots + a_n z^n$, as we know, does not always have real roots. However, compared with an arbitrary polynomial having complex coefficients, it does have the unusual property that if $P(z_0) = 0$, then $P(\overline{z}_0) = 0$ also. Indeed, it follows from the definition of the complex conjugate and the rules for adding complex numbers that $\overline{(z_1 + z_0)} = \overline{z}_1 + \overline{z}_2$. It follows from the trigonometric form of writing a complex number and the rules for multiplying complex numbers that

$$\overline{(z_1 \cdot z_2)} = \overline{\left(r_1 e^{i\varphi_1} \cdot r_2 e^{i\varphi_2}\right)} = \overline{r_1 r_2 e^{i(\varphi_1+\varphi_2)}} =$$
$$= r_1 r_2 e^{-i(\varphi_1+\varphi_2)} = r_1 e^{-i\varphi_1} \cdot r_2 e^{-i\varphi_2} = \overline{z}_1 \cdot \overline{z}_2.$$

Thus,

$$\overline{P(z_0)} = \overline{a_0 + \cdots + a_n z_0^n} = \overline{a}_0 + \cdots + \overline{a}_n \overline{z}_0^n = a_0 + \cdots + a_n \overline{z}_0^n = P(\overline{z}_0),$$

and if $P(z_0) = 0$, then $\overline{P(z_0)} = P(\overline{z}_0) = 0$.

Corollary 1 *Every polynomial $P(z) = c_0 + \cdots + c_n z^n$ of degree $n \geq 1$ with complex coefficients admits a representation in the form*

$$P(z) = c_n(z - z_1) \cdots (z - z_n), \tag{5.132}$$

where $z_1, \ldots, z_n \in \mathbb{C}$ (and the numbers $z_1, \ldots, z_n$ are not necessarily all distinct). This representation is unique up to the order of the factors.

Proof From the long division algorithm for dividing one polynomial $P(z)$ by another polynomial $Q(z)$ of lower degree, we find that $P(z) = q(z)Q(z) + r(z)$, where $q(z)$ and $r(z)$ are polynomials, the degree of $r(z)$ being less than the degree m of $Q(z)$. Thus if $m = 1$, then $r(z) = r$ is simply a constant.

Let z_1 be a root of the polynomial $P(z)$. Then $P(z) = q(z)(z - z_1) + r$, and since $P(z_1) = r$, it follows that $r = 0$. Hence if z_1 is a root of $P(z)$, we have the representation $P(z) = (z - z_1)q(z)$. The degree of the polynomial $q(z)$ is $n - 1$, and we can repeat the reasoning with $q(z)$ if $n - 1 \geq 1$. By induction we find that $P(z) = c(z - z_1) \cdots (z - z_n)$. Since we must have $cz^n = c_n z^n$, it follows that $c = c_n$. □

Corollary 2 *Every polynomial $P(z) = a_0 + \cdots + a_n z^n$ with real coefficients can be expanded as a product of linear and quadratic polynomials with real coefficients.*

Proof This follows from Corollary 1 and Remark 2, by virtue of which for any root z_k of $P(z)$ the number $\bar{z}_k$ is also a root. Then, carrying out the multiplication $(z - z_k)(z - \bar{z}_k)$ in the product (5.132), we obtain the quadratic polynomial $z^2 - (z_k + \bar{z}_k)z + |z_k|^2$ with real coefficients. The number c_n, which equals a_n, is a real number in this case and can be moved inside one of the sets of parentheses without changing the degree of that factor. □

By multiplying out all the identical factors in (5.132), we can rewrite that product:

$$P(z) = c_n(z - z_1)^{k_1} \cdots (z - z_p)^{k_p}. \tag{5.133}$$

The number k_j is called the *multiplicity* of the root z_j.

Since $P(z) = (z - z_j)^{k_j} Q(z)$, where $Q(z_j) \neq 0$, it follows that

$$P'(z) = k_j(z - z_j)^{k_j - 1} Q(z) + (z - z_j)^{k_j} Q'(z) = (z - z_j)^{k_j - 1} R(z),$$

where $R(z_j) = k_j Q(z_j) \neq 0$. We thus arrive at the following conclusion.

Corollary 3 *Every root z_j of multiplicity $k_j > 1$ of a polynomial $P(z)$ is a root of multiplicity $k_j - 1$ of the derivative $P'(z)$.*

Not yet being in a position to find the roots of the polynomial $P(z)$, we can use this last proposition and the representation (5.133) to find a polynomial $p(z) = (z - z_1) \cdots (z - z_p)$ whose roots are the same as those of $P(z)$ but are of multiplicity 1.

Indeed, by the Euclidean algorithm, we first find the greatest common divisor $q(z)$ of $P(z)$ and $P'(z)$. By Corollary 3, the expansion (5.133), and Theorem 2, the polynomial $q(z)$ is equal, apart from a constant factor, to $(z - z_1)^{k_1 - 1} \cdots (z - z_p)^{k_p - 1}$. Hence by dividing $P(z)$ by $q(z)$ we obtain, apart from a constant factor that can be removed by dividing out the coefficient of z^p, a polynomial $p(z) = (z - z_1) \cdots (z - z_p)$.

Now consider the ratio $R(x) = \frac{P(x)}{Q(x)}$ of two polynomials, where $Q(x) \not\equiv \text{const}$. If the degree of $P(x)$ is larger or equal than that of $Q(x)$, we apply the division algorithm and represent $P(x)$ as $p(x)Q(x) + r(x)$, where $p(x)$ and $r(x)$ are polynomials, the degree of $r(x)$ being less than that of $Q(x)$. Thus we obtain a representation of the form $R(x) = p(x) + \frac{r(x)}{Q(x)}$, where the fraction $\frac{r(x)}{Q(x)}$ is now a proper fraction in the sense that the degree of $r(x)$ is less than that of $Q(x)$.

The corollary we are about to state involves the representation of a proper fraction as a sum of fractions called *partial fractions*.

Corollary 4 *a) If $Q(z) = (z - z_1)^{k_1} \cdots (z - z_p)^{k_p}$ and $\frac{P(z)}{Q(z)}$ is a proper fraction, there exists a unique representation of the fraction $\frac{P(z)}{Q(z)}$ in the form*

$$\frac{P(z)}{Q(z)} = \sum_{j=1}^{p} \left(\sum_{k=1}^{k_j} \frac{a_{jk}}{(z - z_j)^k} \right). \tag{5.134}$$

b) If $P(x)$ and $Q(x)$ are polynomials with real coefficients and

$$Q(x) = (x - x_1)^{k_1} \cdots (x - x_l)^{k_l} \left(x^2 + p_1 x + q_1\right)^{m_1} \cdots \left(x^2 + p_n x + q_n\right)^{m_n},$$

there exists a unique representation of the proper fraction $\frac{P(x)}{Q(x)}$ in the form

$$\frac{P(x)}{Q(x)} = \sum_{j=1}^{l} \left(\sum_{k=1}^{k_j} \frac{a_{jk}}{(x - x_j)^k} \right) + \sum_{j=1}^{n} \left(\sum_{k=1}^{m_j} \frac{b_{jk} x + c_{jk}}{(x^2 + p_j x + q_j)^k} \right), \tag{5.135}$$

where a_{jk}, b_{jk}, and c_{jk} are real numbers.

We remark that there is a universal method of finding the expansions (5.134) and (5.135) known as the method of undetermined coefficients, although this method is not always the shortest way. It consists of putting all the terms on the right-hand side of (5.134) or (5.135) over a common denominator, then equating the coefficients of the resulting numerator to the corresponding coefficients of $P(x)$. The system of linear equations that results always has a unique solution because of Corollary 4.

Since we shall as a rule be interested in the expansion of a specific fraction, which we shall obtain by the method of undetermined coefficients, we require nothing more from Corollary 4 than the assurance that it is always possible to do so. For that reason, we shall not bother to go through the proof. It is usually couched in algebraic language in a course of modern algebra and in analytic language in a course in the theory of functions of a complex variable.

Let us consider a specially chosen example to illustrate what has just been explained.

Example 17 Let

$$P(x) = 2x^6 + 3x^5 + 6x^4 + 6x^3 + 10x^2 + 3x + 2,$$
$$Q(x) = x^7 + 3x^6 + 5x^5 + 7x^4 + 7x^3 + 5x^2 + 3x + 1.$$

Find the partial-fraction expansion (5.135) of the fraction $\frac{P(x)}{Q(x)}$.

First of all, the problem is complicated by the fact that we do not know the factors of the polynomial $Q(x)$. Let us try to simplify the situation by eliminating any multiple roots there may be of $Q(x)$. We find

$$Q'(x) = 7x^6 + 18x^5 + 25x^4 + 28x^3 + 21x^2 + 10x + 3.$$

By a rather fatiguing, but feasible computation using the Euclidean algorithm, we find the greatest common divisor

$$d(x) = x^4 + 2x^3 + 2x^2 + 2x + 1$$

of $Q(x)$ and $Q'(x)$. We have written the greatest common divisor with leading coefficient 1.

Dividing $Q(x)$ by $d(x)$, we obtain the polynomial

$$q(x) = x^3 + x^2 + x + 1,$$

which has the same roots as $Q(x)$, but each with multiplicity 1. The root -1 is easily guessed. After $q(x)$ is divided by $x + 1$, we find a quotient of $x^2 + 1$. Thus

$$q(x) = (x + 1)(x^2 + 1),$$

and then by successively dividing $d(x)$ by $x^2 + 1$ and $x + 1$, we find the factorization of $d(x)$;

$$d(x) = (x + 1)^2(x^2 + 1),$$

and then the factorization

$$Q(x) = (x + 1)^3(x^2 + 1)^2.$$

Thus, by Corollary 4b, we are seeking an expansion of the fraction $\frac{P(x)}{Q(x)}$ in the form

$$\frac{P(x)}{Q(x)} = \frac{a_{11}}{x+1} + \frac{a_{12}}{(x+1)^2} + \frac{a_{13}}{(x+1)^3} + \frac{b_{11}x + c_{11}}{x^2+1} + \frac{b_{12}x + c_{12}}{(x^2+1)^2}.$$

Putting the right-hand side over a common denominator and equating the coefficients of the resulting numerator to those of $P(x)$, we arrive at a system of seven equations in seven unknowns, solving which, we finally obtain

$$\frac{P(x)}{Q(x)} = \frac{1}{x+1} - \frac{2}{(x+1)^2} + \frac{1}{(x+1)^3} + \frac{x-1}{x^2+1} + \frac{x+1}{(x^2+1)^2}.$$

5.5.6 Problems and Exercises

1. Using the geometric interpretation of complex numbers

a) explain the inequalities $|z_1 + z_2| \le |z_1| + |z_2|$ and $|z_1| + \cdots + |z_n| \le |z_1| + \cdots + |z_n|$;

b) exhibit the locus of points in the plane $\mathbb{C}$ satisfying the relation $|z - 1| + |z + 1| \le 3$;

c) describe all the nth roots of unity and find their sum;

d) explain the action of the transformation of the plane $\mathbb{C}$ defined by the formula $z \mapsto \bar{z}$.

2. Find the following sums:

a) $1 + q + \cdots + q^n$;
b) $1 + q + \cdots + q^n + \cdots$ for $|q| < 1$;
c) $1 + e^{i\varphi} + \cdots + e^{in\varphi}$;

d) $1+re^{i\varphi}+\cdots+r^n e^{in\varphi}$;
e) $1+re^{i\varphi}+\cdots+r^n e^{in\varphi}+\cdots$ for $|r|<1$;
f) $1+r\cos\varphi+\cdots+r^n\cos n\varphi$;
g) $1+r\cos\varphi+\cdots+r^n\cos n\varphi+\cdots$ for $|r|<1$;
h) $1+r\sin\varphi+\cdots+r^n\sin n\varphi$;
i) $1+r\sin\varphi+\cdots+r^n\sin n\varphi+\cdots$ for $|r|<1$.

3. Find the modulus and argument of the complex number $\lim_{n\to\infty}(1+\frac{z}{n})^n$ and verify that this number is e^z.

4. a) Show that the equation $e^w=z$ in w has the solution $w=\ln|z|+i\operatorname{Arg} z$. It is natural to regard w as the *natural logarithm* of z. Thus $w=\operatorname{Ln} z$ is not a functional relation, since $\operatorname{Arg} z$ is multi-valued.

b) Find $\operatorname{Ln} 1$ and $\operatorname{Ln} i$.

c) Set $z^\alpha=e^{\alpha\operatorname{Ln} z}$. Find 1^π and i^i.

d) Using the representation $w=\sin z=\frac{1}{2i}(e^{iz}-e^{-iz})$, obtain an expression for $z=\arcsin w$.

e) Are there points in $\mathbb{C}$ where $|\sin z|=2$?

5. a) Investigate whether the function $f(z)=\frac{1}{1+z^2}$ is continuous at all points of the plane $\mathbb{C}$.

b) Expand the function $\frac{1}{1+z^2}$ in a power series around $z_0=0$ and find its radius of convergence.

c) Solve parts a) and b) for the function $\frac{1}{1+\lambda^2 z^2}$, where $\lambda\in\mathbb{R}$ is a parameter.

Can you make a conjecture as to how the radius of convergence is determined by the relative location of certain points in the plane $\mathbb{C}$? Could this relation have been understood on the basis of the real line alone, that is, by expanding the function $\frac{1}{1+\lambda^2 x^2}$, where $\lambda\in\mathbb{R}$ and $x\in\mathbb{R}$?

6. a) Investigate whether the Cauchy function

$$f(z)=\begin{cases} e^{-1/z^2}, & z\neq 0,\\ 0, & z=0\end{cases}$$

is continuous at $z=0$.

b) Is the restriction $f|_{\mathbb{R}}$ of the function f in a) to the real line continuous?

c) Does the Taylor series of the function f in a) exist at the point $z_0=0$?

d) Are there functions analytic at a point $z_0\in\mathbb{C}$ whose Taylor series converge only at the point z_0?

e) Invent a power series $\sum_{n=0}^{\infty}c_n(z-z_0)^n$ that converges only at the one point z_0.

7. a) Making the formal substitution $z-a=(z-z_0)+(z_0-a)$ in the power series $\sum_{n=0}^{\infty}A_n(z-a)^n$ and gathering like terms, obtain a series $\sum_{n=0}^{\infty}C_n(z-z_0)^n$ and expressions for its coefficients in terms of A_k and $(z_0-a)^k$, $k=0,1,\dots$.

b) Verify that if the original series converges in the disk $|z-a|<R$ and $|z_0-a|=r<R$, then the series defining C_n, $n=0,1,\dots$, converge absolutely and the series $\sum_{n=0}^{\infty}C_n(z-z_0)^n$ converges for $|z-z_0|<R-r$.

Fig. 5.33

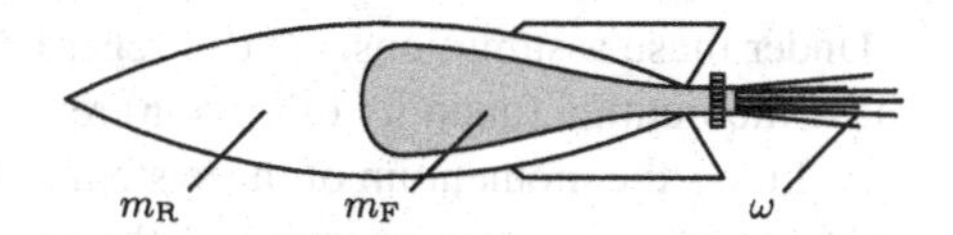

c) Show that if $f(z)=\sum_{n=0}^{\infty} A_n(z-a)^n$ in the disk $|z-a|<R$ and $|z_0-a|<R$, then in the disk $|z-z_0|<R-|z_0-a|$ the function f admits the representation $f(z)=\sum_{n=0}^{\infty} C_n(z-z_0)^n$.

8. Verify that

a) as the point $z\in\mathbb{C}$ traverses the circle $|z|=r>1$ the point $w=z+z^{-1}$ traverses an ellipse with center at zero and foci at ± 2;

b) when a complex number is squared (more precisely, under the mapping $w\mapsto w^2$), such an ellipse maps to an ellipse with a focus at 0, traversed twice;

c) under squaring of complex numbers, any ellipse with center at zero maps to an ellipse with a focus at 0.

5.6 Some Examples of the Application of Differential Calculus in Problems of Natural Science

In this section we shall study some problems from natural science that are very different from one another in their statement, but which, as will be seen, have closely related mathematical models. That model is none other than a very simple differential equation for the function we are interested in. From the study of one such example – the two-body problem – we really began the construction of differential calculus. The study of the system of equations we obtained for this problem was inaccessible at the time. Here we shall consider some problems that can be solved completely at our present level of knowledge. In addition to the pleasure of seeing mathematical machinery in action in a specific case, from the series of examples in this section we shall in particular acquire additional confidence in both the naturalness with which the exponential function $\exp x$ arises and in the usefulness of extending it to the complex domain.

5.6.1 Motion of a Body of Variable Mass

Consider a rocket moving in a straight line in outer space, far from gravitating bodies (Fig. 5.33).

Let $M(t)$ be the mass of the rocket (including fuel) at time t, $V(t)$ its velocity at time t, and ω the speed (relative to the rocket) with which fuel flows out of the nozzle of the rocket as it burns.

We wish to establish the connection among these quantities.

Under these assumptions, we can regard the rocket with fuel as a closed system whose momentum (quantity of motion) remains constant over time.

At time t the momentum of the system is $M(t)V(t)$.

At time $t+h$ the momentum of the rocket with the remaining fuel is $M(t+h)V(f+h)$ and the momentum ΔI of the mass of fuel ejected over that time $|\Delta M| = |M(t+h) - M(t)| = -(M(t+h) - M(t))$ lies between the bounds

$$\bigl(V(t) - \omega\bigr)|\Delta M| < \Delta I < \bigl(V(t+h) - \omega\bigr)|\Delta M|,$$

that is, $\Delta I = (V(t) - \omega)|\Delta M| + \alpha(h)|\Delta M$, and it follows from the continuity of $V(t)$ that $\alpha(h) \to 0$ as $h \to 0$.

Equating the momenta of the system at times t and $t+h$, we have

$$M(t)V(t) = M(t+h)V(t+h) + \bigl(V(t) - \omega\bigr)|\Delta M| + \alpha(h)|\Delta M|,$$

or, after substituting $|\Delta M| = -(M(t+h) - M(t))$ and simplifying,

$$\begin{aligned} M(t+h)\bigl(V(t+h) - V(t)\bigr) = \\ = -\omega\bigl(M(t+h) - M(t)\bigr) + \alpha(h)\bigl(M(t+h) - M(t)\bigr). \end{aligned}$$

Dividing this last equation by h and passing to the limit as $h \to 0$, we obtain

$$M(t)V'(t) = -\omega M'(t). \tag{5.136}$$

This is the relation we were seeking between the functions we were interested in, $V(t)$, $M(t)$, and their derivatives.

We now must find the relation between the functions $V(t)$ and $M(t)$ themselves, using the relation between their derivatives. In general a problem of this type is more difficult than the problem of finding the relations between the derivatives knowing the relation between the functions. However, in the present case, this problem has a completely elementary solution.

Indeed, after dividing Eq. (5.136) by $M(t)$, we can rewrite it in the form

$$V'(t) = (-\omega \ln M)'(t). \tag{5.137}$$

But if the derivatives of two functions are equal on an interval, then the functions themselves differ by at most a constant on that interval.

Thus it follows from (5.137) that

$$V(t) = -\omega \ln M(t) + c. \tag{5.138}$$

If it is known, for example, that $V(0) = V_0$, this initial condition determines the constant c completely. Indeed, from (5.138) we find

$$c = V_0 + \omega \ln M(0),$$

and we then find the formula we were seeking[27]

$$V(t) = V_0 + \omega \ln \frac{M(0)}{M(t)}. \tag{5.139}$$

It is useful to remark that if m_R is the mass of the body of the rocket and m_F is the mass of the fuel and V is the terminal velocity achieved by the rocket when all the fuel is expended, substituting $M(0) = m_R + m_F$ and $M(t) = m_R$, we find

$$V = V_0 + \omega \ln\left(1 + \frac{m_F}{m_R}\right).$$

This last formula shows very clearly that the terminal velocity is affected not so much by the ratio m_F/m_R inside the logarithm as by the outflow speed ω, which depends on the type of fuel used. It follows in particular from this formula that if $V_0 = 0$, then in order to impart a velocity V to a rocket whose own mass is m_R one must have the following initial supply of fuel:

$$m_F = m_R\left(e^{V/\omega} - 1\right).$$

5.6.2 The Barometric Formula

This is the name given to the formula that exhibits the dependence of atmospheric pressure on elevation above sea level.

Let $p(h)$ be the pressure at elevation h. Since $p(h)$ is the weight of the column of air above an area of 1 cm^2 at elevation h, it follows that $p(h + \Delta)$ differs from $p(h)$ by the weight of the portion of the gas lying in the parallelepiped whose base is the original area of 1 cm^2 and the same area at elevation $h + \Delta$. Let $\rho(h)$ be the density of air at elevation h. Since $\rho(h)$ depends continuously on h, one may assume that the mass of this portion of air is calculated from the formula

$$\rho(\xi)\ \mathrm{g/cm^3} \cdot 1\ \mathrm{cm^2} \cdot \Delta\ \mathrm{cm} = p(\xi)\Delta \mathrm{g},$$

where ξ is some height between h and $h + \Delta$. Hence the weight of that mass[28] is $g \cdot \rho(\xi)\Delta$.

Thus,

$$p(h + \Delta) - p(h) = -g\rho(\xi)\Delta.$$

---

[27] This formula is sometimes connected with the name of K.E. Tsiolkovskii (1857–1935), a Russian scientist and the founder of the theory of space flight. But it seems to have been first obtained by the Russian specialist in theoretical mechanics I.V. Meshcherskii (1859–1935) in an 1897 paper devoted to the dynamics of a point of variable mass.

[28] Within the region where the atmosphere is noticeable, g may be regarded as constant.

After dividing this equality by Δ and passing to the limit as $\Delta \to 0$, taking account of the relation $\xi \to h$, we obtain

$$p'(h) = -g\rho(h). \tag{5.140}$$

Thus the rate of variation in atmospheric pressure has turned out to be proportional to the density of the air at the corresponding elevation.

To obtain an equation for the function $p(h)$, we eliminate the function $\rho(h)$ from (5.140). By Clapeyron's law[29] (the ideal gas law) the pressure p, molar volume V, and temperature T of the gas (on the Kelvin[30] scale) are connected by the relation

$$\frac{pV}{T} = R, \tag{5.141}$$

where R is the so-called universal gas constant. If M is the mass of one mole of air and V its volume, then $\rho = \frac{M}{V}$, so that from (5.141) we find

$$p = \frac{1}{V} \cdot R \cdot T = \frac{M}{V} \cdot \frac{R}{M} \cdot T = \rho \cdot \frac{R}{M} T.$$

Setting $\lambda = \frac{R}{M} T$, we thus have

$$p = \lambda(T)\rho. \tag{5.142}$$

If we now assume that the temperature of the layer of air we are describing is constant, we finally obtain from (5.140) and (5.142)

$$p'(h) = -\frac{g}{\lambda} p(h). \tag{5.143}$$

This differential equation can be rewritten as

$$\frac{p'(h)}{p(h)} = -\frac{g}{\lambda}$$

or

$$(\ln p)'(h) = \left(-\frac{g}{\lambda} h\right)',$$

from which we derive

$$\ln p(h) = -\frac{g}{\lambda} h + c,$$

or

$$p(h) = \mathrm{e}^c \cdot \mathrm{e}^{-(g/\lambda)h}.$$

---

[29] B.P.E. Clapeyron (1799–1864) – French physicist who studied thermodynamics.

[30] W. Thomson (Lord Kelvin) (1824–1907) – famous British physicist.

The factor e^c can be determined from the known initial condition $p(0) = p_0$, from which it follows that $e^c = p_0$.

Thus, we have found the following dependence of pressure on elevation:

$$p = p_0 e^{-(g/\lambda)h}. \tag{5.144}$$

For the air at sea level (at zero altitude) and at zero temperature ($273\ \mathrm{K} = 0\ ^\circ\mathrm{C}$), by the above formula $\lambda = \frac{RT}{M}$, it would be possible to obtain for λ the value $7.8 \times 10^4\ (\mathrm{m/s})^2$. We set $g = 10\ \mathrm{m/s}^2$. The formula (5.144) acquires a completely finished form after these numerical values are substituted. In particular, for such values λ and g, the formula (5.144) shows that the pressure drops by a factor of e (≈ 3) at the height $h = \frac{\lambda}{g} = 7.8 \times 10^3\ \mathrm{m} = 7.8\ \mathrm{km}$. It will increase by the same amount, if one descends to a mine at a depth of order 7.8 km.[31]

5.6.3 Radioactive Decay, Chain Reactions, and Nuclear Reactors

It is known that the nuclei of heavy elements are subject to sporadic (spontaneous) decay. This phenomenon is called *natural radioactivity*.

The main statistical law of radioactivity (which is consequently valid for amounts and concentrations of a substance that are not too small) is that the number of decay events over a small interval of time h starting at time t is proportional to h and to the number $N(t)$ of atoms of the substance that have not decayed up to time t, that is,

$$N(t+h) - N(t) \approx -\lambda N(t)h,$$

where $\lambda > 0$ is a numerical coefficient that is characteristic of the chemical element.

Thus the function $N(t)$ satisfies the now familiar differential equation

$$N'(t) = -\lambda N(t), \tag{5.145}$$

from which it follows that

$$N(t) = N_0 e^{-\lambda t},$$

where $N_0 = N(0)$ is the initial number of atoms of the substance.

The time T required for half of the initial number of atoms to decay is called the *half-life* of the substance. The quantity T can thus be found from the equation $e^{-\lambda T} = \frac{1}{2}$, that is, $T = \frac{\ln 2}{\lambda} \approx \frac{0.69}{\lambda}$. For example, for polonium-210 (Po^{210}) the half-life T is approximately 138 days, for radium-226 (Ra^{226}), $T \approx 1600$ years, for uranium-235 (U^{235}), $T \approx 7.1 \times 10^8$ years, and for its isotope U^{238}, $T \approx 4.5 \times 10^9$ years.

[31] It should be noted that this barometric formula corresponds only approximately to the real distribution of pressure and density of the air in Earth's atmosphere. Some corrections and additions are given in Problem 2 at the end of this section.

A *nuclear reaction* is an interaction of nuclei or of a nucleus with elementary particles resulting in the appearance of a nucleus of a new type. This may be nuclear fusion, in which the coalescence of the nuclei of lighter elements leads to the formation of nuclei of a heavier element (for example, two nuclei of heavy hydrogen – deuterium – yield a helium nucleus along with a release of energy); or it may be the decay of a nucleus and the formation of one or more nuclei of lighter elements. In particular, such decay occurs in approximately half of the cases when a neutron collides with a U^{235} nucleus. The breakup of the uranium nucleus leads to the formation of 2 or 3 new neutrons, which may then participate in further interactions with nuclei, causing them to split and thereby leading to further multiplication of the number of neutrons. A nuclear reaction of this type is called a *chain reaction.*

We shall describe a theoretical mathematical model of a chain reaction in a radioactive element and obtain the law of variation in the number $N(t)$ of neutrons as a function of time.

We take the substance to have the shape of a sphere of radius r. If r is not too small, on the one hand new neutrons will be generated over the time interval h measured from some time t in a number proportional to h and $N(t)$, while on the other hand some of the neutrons will be lost, having moved outside the sphere.

If v is the velocity of a neutron, then the only ones that can leave the sphere in time h are those lying within vh of its boundary, and of those only the ones whose direction of motion is approximately along a radius. Assuming that those neutrons constitute a fixed proportion of the ones lying in this zone, and that neutrons are distributed approximately uniformly throughout the sphere, one can say that the number of neutrons lost over the time interval h is proportional to $N(t)$ and the ratio of the volume of this boundary layer to the volume of the sphere.

What has just been said leads to the equality

$$N(t+h) - N(t) \approx \alpha N(t)h - \frac{\beta}{r} N(t)h \tag{5.146}$$

(since the volume of the boundary layer is approximately $4\pi r^2 vh$, and the volume of the sphere is $\frac{4}{3}\pi r^3$). Here the coefficients α and β depend only on the particular radioactive substance.

After dividing by h and passing to the limit in (5.146) as $h \to 0$, we obtain

$$N'(t) = \left(\alpha - \frac{\beta}{r}\right) N(t), \tag{5.147}$$

from which

$$N(t) = N_0 \exp\left\{\left(\alpha - \frac{\beta}{r}\right)t\right\}.$$

It can be seen from this formula that when $(\alpha - \frac{\beta}{r}) > 0$, the number of neutrons will increase exponentially with time. The nature of this increase, independently of the initial condition N_0, is such that practically total decay of the substance occurs over a very short time interval, releasing a colossal amount of energy – that is an *explosion*.

If $(\alpha - \frac{\beta}{r}) < 0$, the reaction ceases very quickly since more neutrons are being lost than are being generated.

If the boundary condition between the two conditions just considered holds, that is, $\alpha - \frac{\beta}{r} = 0$, an equilibrium occurs between the generation of neutrons and their exit from the reaction, as a result of which the number of neutrons remains approximately constant.

The value of r at which $\alpha - \frac{\beta}{r} = 0$ is called the *critical radius*, and the mass of the substance in a sphere of that volume is called the *critical mass* of the substance.

For U^{235} the critical radius is approximately 8.5 cm, and the critical mass approximately 50 kg.

In nuclear reactors, where steam is produced by a chain reaction in a radioactive substance there is an artificial source of neutrons, providing the fissionable matter with a certain number n of neutrons per unit time. Thus for an atomic reactor Eq. (5.147) is slightly altered:

$$N'(t) = \left(\alpha - \frac{\beta}{r}\right)N(t) + n. \tag{5.148}$$

This equation can be solved by the same device as Eq. (5.147), since $\frac{N'(t)}{(\alpha-\beta/r)N(t)+n}$ is the derivative of the function $\frac{1}{\alpha-\beta/r}\ln[(\alpha - \frac{\beta}{r})N(t) + n]$ if $\alpha - \frac{\beta}{r} \neq 0$. Consequently the solution of Eq. (5.148) has the form

$$N(t) = \begin{cases} N_0 e^{(\alpha-\beta/r)t} - \frac{n}{\alpha-\beta/r}[1 - e^{(\alpha-\beta/r)t}] & \text{if } \alpha - \frac{\beta}{r} \neq 0, \\ N_0 + nt & \text{if } \alpha - \frac{\beta}{f} = 0. \end{cases}$$

It can be seen from this solution that if $\alpha - \frac{\beta}{r} > 0$ (supercritical mass), an explosion occurs. If the mass is pre-critical, however, that is, $\alpha - \frac{\beta}{r} < 0$, we shall very soon have

$$N(t) \approx \frac{n}{\frac{\beta}{r} - \alpha}.$$

Thus, if the mass of radioactive substance is maintained in a pre-critical state but close to critical, then independently of the power of the additional neutron source, that is, independently of n, one can obtain higher values of $N(t)$ and consequently greater power from the reactor. Keeping the process in the pre-critical zone is a delicate matter and is achieved by a rather complicated automatic control system.

5.6.4 Falling Bodies in the Atmosphere

We are now interested in the velocity $v(t)$ of a body falling to Earth under the influence of gravity.

If there were no air resistance, the relation

$$\dot{v}(t) = g, \tag{5.149}$$

would hold for fall from relatively low altitudes. This law follows from Newton's second law $ma = F$ and the law of universal gravitation, by virtue of which for $h \ll R$ (where R is the radius of the Earth)

$$F(t) = G\frac{Mm}{(R + h(t))^2} \approx G\frac{Mm}{R^2} = gm.$$

A body moving in the atmosphere experiences a resistance that depends on the velocity of the motion, and as a result, the velocity of free fall for a heavy body does not grow indefinitely, but stabilizes at a certain level. For instance, during skydiving or parachuting, if the parachute of a skydiver does not open for a long time, the speed of a midsized skydiver in the lower strata of the atmosphere stabilizes at 50–60 m/s.

We shall assume that under these conditions (that is, for this body and a speed range from 0 to 60), the resistance force is proportional to the velocity of the body moving in the air. By equating the acting forces on the body, we arrive at the following equation, which must satisfy the velocity of the free-falling body in the atmosphere:

$$m\dot{v}(t) = mg - \alpha v. \tag{5.150}$$

Dividing this equation by m and denoting $\frac{\alpha}{m}$ by β, we finally obtain

$$\dot{v}(t) = -\beta v + g. \tag{5.148'}$$

We have now arrived at an equation that differs from Eq. (5.148) only in notation. We remark that if we set $-\beta v(t) + g = f(t)$, then, since $f'(t) = -\beta v'(t)$, one can obtain from (5.148′) the equivalent equation

$$f'(t) = -\beta f(t),$$

which is the same as Eq. (5.143) or (5.145) except for notation. Thus we have once again arrived at an equation whose solution is the exponential function

$$f(t) = f(0)\mathrm{e}^{-\beta t}.$$

It follows from this that the solution of Eq. (5.148′) has the form

$$v(t) = \frac{1}{\beta}g + \left(v_0 - \frac{1}{\beta}\right)\mathrm{e}^{-\beta t},$$

and the solution of the basic equation (5.150) has the form

$$v(t) = \frac{m}{\alpha}g + \left(v_0 - \frac{m}{\alpha}g\right)\mathrm{e}^{-(\alpha/m)t}, \tag{5.151}$$

where $v_0 = v(0)$ is the initial vertical velocity of the body.

It can be seen from (5.151) that for $\alpha > 0$ a body falling in the atmosphere reaches a steady state at which $v(t) \approx \frac{m}{\alpha} g$. Thus, in contrast to fall in airless space, the velocity of descent in the atmosphere depends not only on the shape of the body, but also on its mass. As $\alpha \to 0$, the right-hand side of (5.151) tends to $v_0 + gt$, that is, to the solution of Eq. (5.149) obtained from (5.150) when $\alpha = 0$.

Using formula (5.151), one can get an idea of how quickly the limiting velocity of fall in the atmosphere is reached.

For example, if a parachute is designed to that a person of average size will fall with a velocity of the order 10 meters per second when the parachute is open, then, if the parachute opens after a free fall during which a velocity of approximately 50 meters per second has been attained, the person will have a velocity of about 12 meters per second three seconds after the parachute opens.

Indeed, from the data just given and relation (5.151) we find $\frac{m}{\alpha} g \approx 10$, $\frac{m}{\alpha} \approx 1$, $v_0 = 50$ m/s, so that relation (5.151) assumes the form

$$v(t) = 10 + 40\mathrm{e}^{-t}.$$

Since $\mathrm{e}^3 \approx 20$, for $t = 3$, we obtain $v \approx 12$ m/s.

Note that at the opening of the parachute, which is significantly bigger on the front surface compared with the body of the skydiver, the force of resistance is already proportional to the square of the speed. This means that after the opening of the parachute, the slowdown will be quicker than that obtained from the calculation with formula (5.151), corresponding to Eq. (5.150).

It is useful to rewrite and solve the equation of type (5.150) with this new assumption of the quadratic dependence of the resistance force on the speed of movement and to make the necessary correction to the result obtained in the first calculation.[32]

5.6.5 The Number e and the Function $\exp x$ Revisited

By examples we have verified (see also Problems 3 and 4 at the end of this section) that a number of natural phenomena can be described from the mathematical point of view by the same differential equation, namely

$$f'(x) = \alpha f(x), \tag{5.152}$$

[32]It should be noted that here, like in many others studied examples of problems of natural science, we illustrate the calculation method. In the calculation are included some original data (empirical constants, various assumptions) on which depends the level of concordance of the calculations with the reality. This data are not given by mathematics but by a corresponding science. For example, in order to obtain data on the aerodynamic properties of the bodies, in particular the nature of the change in the resistance force occurring during their movement in the atmosphere, it is necessary often to perform a long series of experiments in wind tunnels. Mathematics helps to calculate one or another mathematical model of the phenomenon, it often provides a ready calculation procedure, but the correctness of the initial data and the adequacy of the model is the work of another relevant science (biology, chemistry, physics, ...).

whose solution $f(x)$ is uniquely determined when the "initial condition" $f(0)$ is specified. Then

$$f(x) = f(0)e^{\alpha x}.$$

We introduced the number e and the function $e^x = \exp x$ in a rather formal way earlier, assuring the reader that e really was an important number and $\exp x$ really was an important function. It is now clear that even if we had not introduced this function earlier, it would certainly have been necessary to introduce it as the solution of the important, though very simple equation (5.152). More precisely, it would have sufficed to introduce the function that is the solution of Eq. (5.152) for some specific value of α, for example, $\alpha = 1$; for the general equation (5.152) can be reduced to this case by changing to a new variable t connected with x by the relation $x = \frac{t}{\alpha}$ $(\alpha \neq 0)$.

Indeed, we then have

$$f(x) = f\left(\frac{t}{\alpha}\right) = F(t), \qquad \frac{df(x)}{dx} = \frac{\frac{dF(t)}{dt}}{\frac{dx}{dt}} = \alpha F'(t)$$

and instead of the equation $f'(x) = \alpha f(x)$ we then have $\alpha F'(t) = \alpha F(t)$, or $F'(t) = F(t)$.

Thus, let us consider the equation

$$f'(x) = f(x) \tag{5.153}$$

and denote the solution of this equation satisfying $f(0) = 1$ by $\exp x$.

Let us check to see whether this definition agrees with our previous definition of $\exp x$.

Let us try to calculate the value of $f(x)$ starting from the relation $f(0) = 1$ and the assumption that f satisfies (5.153). Since f is differentiable, it is continuous. But then Eq. (5.153) implies that $f'(x)$ is also continuous. Moreover, it follows from (5.153) that f also has a second derivative $f''(x) = f'(x)$, and in general that f is infinitely differentiable. Since the rate of variation $f'(x)$ of the function $f(x)$ is continuous, the function f' changes very little over a small interval h of variation of its argument. Therefore $f(x_0 + h) = f(x_0) + f'(\xi)h \approx f(x_0) + f'(x_0)h$. Let us use this approximate formula and traverse the interval from 0 to x in small steps of size $h = \frac{x}{n}$, where $n \in \mathbb{N}$. If $x_0 = 0$ and $x_{k+1} = x_k + h$, we should have

$$f(x_{k+1}) \approx f(x_k) + f'(x_k)h.$$

Taking account of (5.153) and the condition $f(0) = 1$, we have

$$\begin{aligned} f(x) = f(x_n) &\approx f(x_{n-1}) + f'(x_{n-1})h = \\ &= f(x_{n-1})(1 + h) \approx \big(f(x_{n-2}) + f'(x_{n-2})h\big)(1 + h) = \end{aligned}$$

Fig. 5.34

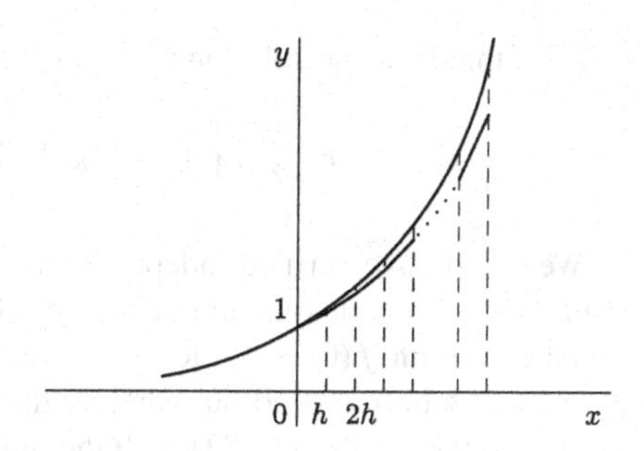

$$= f(x_{n-2})(1+h)^2 \approx \cdots \approx f(x_0)(1+h)^n =$$
$$= f(0)(1+h)^n = \left(1+\frac{x}{n}\right)^n.$$

It seems natural (and this can be proved) that the smaller the step $h = \frac{x}{n}$, the closer the approximation in the formula $f(x) \approx (1+\frac{x}{n})^n$.

Thus we arrive at the conclusion that

$$f(x) = \lim_{n\to\infty}\left(1+\frac{x}{n}\right)^n.$$

In particular, if we denote the quantity $f(1) = \lim_{n\to\infty}(1+\frac{1}{n})^n$ by e and show that $\mathrm{e} \neq 1$, we shall have obtained

$$f(x) = \lim_{n\to\infty}\left(1+\frac{x}{n}\right)^n = \lim_{t\to 0}(1+t)^{x/t} = \lim_{t\to 0}\big[(1+t)^{1/t}\big]^x = \mathrm{e}^x, \tag{5.154}$$

since we know that $u^\alpha \to v^\alpha$ if $u \to v$.

This method of solving Eq. (5.153) numerically, which enabled us to obtain formula (5.154), was proposed by Euler long ago, and is called *Euler's polygonal method*. This name is connected with the fact that the computations carried out in it have a geometric interpretation as the replacement of the solution $f(x)$ of the equation (or rather its graph) by an approximating graph consisting of a broken line whose links on the corresponding closed intervals $[x_k, x_{k+1}]$ $(k = 0, \ldots, n-1)$ are given by the equations $y = f(x_k) + f'(x_k)(x - x_k)$ (see Fig. 5.34).

We have also encountered the definition of the function $\exp x$ as the sum of the power series $\sum_{n=0}^{\infty}\frac{1}{n!}x^n$. This definition can also be reached from Eq. (5.153) by using the following frequently-applied device, called the *method of undetermined coefficients*. We seek a solution of Eq. (5.153) as the sum of a power series

$$f(x) = c_0 + c_1x + \cdots + c_nx^n + \cdots, \tag{5.155}$$

whose coefficients are to be determined.

As we have seen (Theorem 1 of Sect. 5.5) Eq. (5.155) implies that $c_n = \frac{f^{(n)}(0)}{n!}$. But, by (5.153), $f(0) = f'(0) = \cdots = f^{(n)}(0) = \cdots$, and since $f(0) = 1$, we have

$c_n = \frac{1}{n!}$, that is, if the solution has the form (5.155) and $f(0) = 1$, then necessarily

$$f(x) = 1 + \frac{1}{1!}x + \frac{1}{2!}x^2 + \cdots + \frac{1}{n!}x^n + \cdots .$$

We could have verified independently that the function defined by this series is indeed differentiable (and not only at $x = 0$) and that it satisfies Eq. (5.153) and the initial condition $f(0) = 1$. However, we shall not linger over this point, since our purpose was only to find out whether the introduction of the exponential function as the solution of Eq. (5.153) with the initial condition $f(0) = 1$ was in agreement with what we had previously meant by the function $\exp x$.

We remark that Eq. (5.153) could have been studied in the complex plane, that is, we could have regarded x as an arbitrary complex number. When this is done, the reasoning we have carried out remains valid, although some of the geometric intuitiveness of Euler's method may be lost.

Thus it is natural to expect that the function

$$\mathrm{e}^z = 1 + \frac{1}{1!}z + \frac{1}{2!}z^2 + \cdots + \frac{1}{n!}z^n + \cdots$$

is the unique solution of the equation

$$f'(z) = f(z)$$

satisfying the condition $f(0) = 1$.

5.6.6 Oscillations

If a body suspended from a spring is displaced from its equilibrium position, for example by lifting it and then dropping it, it will oscillate about its equilibrium position. Let us describe this process in its general form.

Suppose it is known that a force is acting on a point mass m that is free to move along the x-axis, and that the force $F = -kx$ is proportional[33] to the displacement of the point from the origin. Suppose also that we know the initial position $x_0 = x(0)$ of the point mass and its initial velocity $v_0 = \dot{x}(0)$. Let us find the dependence $x = x(t)$ of the position of the point on time.

By Newton's law, this problem can be rewritten in the following purely mathematical form: Solve the equation

$$m\ddot{x}(t) = -kx(t) \tag{5.156}$$

under the initial conditions $x_0 = x(0)$, $\dot{x}(0) = v_0$.

---

[33] In the case of a spring, the coefficient $k > 0$ characterizing its stiffness is called the *modulus*.

Let us rewrite Eq. (5.156) as

$$\ddot{x}(t) + \frac{k}{m}x(t) = 0 \tag{5.157}$$

and again try to make use of the exponential. Specifically, let us try to choose the number λ so that the function $x(t) = e^{\lambda t}$ satisfies Eq. (5.157).

Making the substitution $x(t) = e^{\lambda t}$ in (5.157), we obtain

$$\left(\lambda^2 + \frac{k}{m}\right)e^{\lambda t} = 0,$$

or

$$\lambda^2 + \frac{k}{m} = 0, \tag{5.158}$$

that is, $\lambda_1 = -\sqrt{-\frac{k}{m}}$, $\lambda_2 = \sqrt{-\frac{k}{m}}$. Since $m > 0$, we have the two imaginary numbers $\lambda_1 = -i\sqrt{\frac{k}{m}}$, $\lambda_2 = i\sqrt{\frac{k}{m}}$ when $k > 0$. We had not reckoned on this possibility; however, let us continue our study. By Euler's formula

$$e^{-i\sqrt{k/m}t} = \cos\sqrt{\frac{k}{m}}t - i\sin\sqrt{\frac{k}{m}}t,$$

$$e^{i\sqrt{k/m}t} = \cos\sqrt{\frac{k}{m}}t + i\sin\sqrt{\frac{k}{m}}t.$$

Since differentiating with respect to the real variable t amounts to differentiating the real and imaginary parts of the function $e^{\lambda t}$ separately, Eq. (5.157) must be satisfied by both functions $\cos\sqrt{\frac{k}{m}}t$ and $\sin\sqrt{\frac{k}{m}}t$. And this is indeed the case, as one can easily verify directly. Thus the complex exponential function has enabled us to guess two solutions of Eq. (5.157), any linear combination of which

$$x(t) = c_1\cos\sqrt{\frac{k}{m}}t + c_2\sin\sqrt{\frac{k}{m}}t, \tag{5.159}$$

is obviously also a solution of Eq. (5.157).

We choose the coefficients c_1 and c_2 in (5.159) from the condition

$$x_0 = x(0) = c_1,$$

$$v_0 = \dot{x}(0) = \left(-c_1\sqrt{\frac{k}{m}}\sin\sqrt{\frac{k}{m}}t + c_2\sqrt{\frac{k}{m}}\cos\sqrt{\frac{k}{m}}t\right)\Bigg|_{t=0} = c_2\sqrt{\frac{k}{m}}.$$

Thus the function

$$x(t) = x_0\cos\sqrt{\frac{k}{m}}t + v_0\sqrt{\frac{m}{k}}\sin\sqrt{\frac{k}{m}}t \tag{5.160}$$

is the required solution.

By making standard transformations we can rewrite (5.160) in the form

$$x(t) = \sqrt{x_0^2 + v_0^2 \frac{m}{k}} \sin\left(\sqrt{\frac{k}{m}} t + \alpha\right), \tag{5.161}$$

where

$$\alpha = \arcsin \frac{x_0}{\sqrt{x_0^2 + v_0^2 \frac{m}{k}}}.$$

Thus, for $k > 0$ the point will make periodic oscillations with period $T = 2\pi\sqrt{\frac{m}{k}}$, that is, with frequency $\frac{1}{T} = \frac{1}{2\pi}\sqrt{\frac{k}{m}}$, and amplitude $\sqrt{x_0^2 + v_0^2 \frac{m}{k}}$. We state this because it is clear from physical considerations that the solution (5.160) is unique. (See Problem 5 at the end of this section.)

The motion described by (5.161) is called a *simple harmonic oscillation*, and Eq. (5.157) the *equation of a simple harmonic oscillator.*

Let us now turn to the case when $k < 0$ in Eq. (5.158). Then the two functions $e^{\lambda_1 t} = \exp(-\sqrt{-\frac{k}{m}} t)$ and $e^{\lambda_2 t} = \exp(\sqrt{-\frac{k}{m}} t)$ are real-valued solutions of Eq. (5.157) and the function

$$x(t) = c_1 e^{\lambda_1 t} + c_2 e^{\lambda_2 t} \tag{5.162}$$

is also a solution. We choose the constants c_1 and c_2 from the conditions

$$\begin{cases} x_0 = x(0) = c_1 + c_2, \\ v_0 = \dot{x}(0) = c_1\lambda_1 + c_2\lambda_2. \end{cases}$$

This system of linear equations always has a unique solution, since its determinant $\lambda_2 - \lambda_1$ is not 0.

Since the numbers λ_1 and λ_2 are of opposite sign, it can be seen from (5.162) that for $k < 0$ the force $F = -kx$ not only has no tendency to restore the point to its equilibrium position at $x = 0$, but in fact as time goes on, carries it an unlimited distance away from this position if x_0 or v_0 is nonzero. That is, in this case $x = 0$ is a point of unstable equilibrium.

In conclusion let us consider a very natural modification of Eq. (5.156), in which the usefulness of the exponential function and Euler's formula connecting the basic elementary functions shows up even more clearly.

Let us assume that the particle we are considering moves in a medium (the air or a liquid) whose resistance cannot be neglected. Suppose the resisting force is proportional to the velocity of the point. Then, instead of Eq. (5.156) we must write

$$m\ddot{x}(t) = -\alpha\dot{x}(t) - kx(t),$$

which we rewrite as

$$\ddot{x}(t) + \frac{\alpha}{m}\dot{x}(t) + \frac{k}{m}x(t) = 0. \tag{5.163}$$

If once again we seek a solution of the form $x(t) = e^{\lambda t}$, we arrive at the quadratic equation

$$\lambda^2 + \frac{\alpha}{m}\lambda + \frac{k}{m} = 0,$$

whose roots are $\lambda_{1,2} = -\frac{\alpha}{2m} \pm \frac{\sqrt{\alpha^2-4mk}}{2m}$.

The case when $\alpha^2 - 4mk > 0$ leads to two real roots λ_1 and λ_2, and the solution can be found in the form (5.162).

We shall study in more detail the case in which we are more interested, when $\alpha^2 - 4mk < 0$. Then both roots λ_1 and λ_2 are complex, but not purely imaginary:

$$\lambda_1 = -\frac{\alpha}{2m} - i\frac{\sqrt{4mk-\alpha^2}}{2m},$$

$$\lambda_2 = -\frac{\alpha}{2m} + i\frac{\sqrt{4mk-\alpha^2}}{2m}.$$

In this case Euler s formula yields

$$e^{\lambda_1 t} = \exp\left(-\frac{\alpha}{2m}t\right)(\cos\omega t - \mathrm{i}\sin(\omega t)),$$

$$e^{\lambda_2 t} = \exp\left(-\frac{\alpha}{2m}t\right)(\cos\omega t + \mathrm{i}\sin\omega t),$$

where $\omega = \frac{\sqrt{4mk-\alpha^2}}{2m}$. Thus we find the two real-valued solutions $\exp(-\frac{\alpha}{2m})\cos\omega t$ and $\exp(-\frac{\alpha}{2m})\sin\omega t$ of Eq. (5.163), which would have been very difficult to guess. We then seek a solution of the original equation in the form of a linear combination of these two

$$x(t) = \exp\left(-\frac{\alpha}{2m}t\right)(c_1\cos\omega t + c_2\sin\omega t), \tag{5.164}$$

choosing c_1 and c_2 so that the initial conditions $x(0) = x_0$ and $\dot{x}(0) = v_0$ are satisfied.

The system of linear equations that results, as one can verify, always has a unique solution. Thus, after transformations, we obtain the solution of the problem from (5.164) in the form

$$x(t) = A\exp\left(-\frac{\alpha}{2m}t\right)\sin(\omega t + a), \tag{5.165}$$

where A and a are constants determined by the initial conditions.

It can be seen from this formula that, because of the factor $\exp(-\frac{\alpha}{2m}t)$, when $\alpha > 0$ and $m > 0$, the oscillations will be *damped* and the rate of damping of the amplitude depends on the ratio $\frac{\alpha}{m}$. The frequency of the oscillations $\frac{1}{2\pi}\omega = \frac{1}{2\pi}\sqrt{\frac{k}{m} - (\frac{\alpha}{2m})^2}$ will not vary over time. The quantity ω also depends only on the ratios $\frac{k}{m}$ and $\frac{\alpha}{m}$, which, however, could have been foreseen from the form (5.163) of the original equation. When $\alpha = 0$, we again return to undamped harmonic oscillations (5.161) and Eq. (5.157).

5.6.7 Problems and Exercises

1. *Efficiency in rocket propulsion.*

a) Let Q be the chemical energy of a unit mass of rocket fuel and ω the outflow speed of the fuel. Then $\frac{1}{2}\omega^2$ is the kinetic energy of a unit mass of fuel when ejected. The coefficient α in the equation $\frac{1}{2}\omega^2 = \alpha Q$ is the efficiency of the processes of burning and outflow of the fuel. For engines of solid fuel (smokeless powder) $\omega = 2$ km/s and $Q = 1000$ kcal/kg, and for engines of liquid fuel (gasoline with oxygen) $\omega = 3$ km/s and $Q = 2500$ kcal/kg. Determine the efficiency α for these cases.

b) The efficiency of a rocket is defined as the ratio of its final kinetic energy $m_{\mathrm{R}}\frac{v^2}{2}$ to the chemical energy of the fuel burned $m_{\mathrm{F}}Q$. Using formula (5.139), obtain a formula for the efficiency of a rocket in terms of m_{R}, m_{F}, Q, and α (see part a)).

c) Evaluate the efficiency of an automobile with a liquid-fuel jet engine, if the automobile is accelerated to the usual city speed limit of 60 km/h.

d) Evaluate the efficiency of a liquid-fuel rocket carrying a satellite into low orbit around the earth.

e) Determine the final speed for which rocket propulsion using liquid fuel is maximally efficient.

f) Which ratio of masses $m_{\mathrm{F}}/m_{\mathrm{R}}$ yields the highest possible efficiency for any kind of fuel?

2. *The barometric formula.*

a) Using the data from Sect. 5.6.2, obtain a formula for a correction term to take account of the dependence of pressure on the temperature of the air column, if the temperature is subject to variation (for example, seasonal) within the range ± 40 °C.

b) Use formula (5.144) to determine the dependence of pressure on elevation at temperatures of -40 °C, 0 °C, and 40 °C, and compare these results with the results given by your approximate formula from part a).

c) The change in temperature of the atmosphere at an altitude of 10–11 km is well described by the following empirical formula: $T(h) = T_0 - \alpha h$, where T_0 is the temperature at sea level (at $h = 0$ m), the coefficient $\alpha = 6.5 \times 10^{-3}$ K/m, and h is the height in meters. Deduce under these conditions the formula for the dependence of the pressure on the height. (T_0 it is often given the value 288 K, corresponding to 15 °C.)

d) Find the pressure in a mine shaft at depths of 1 km, 3 km and 9 km using formula (5.144) and the formula that you obtained in c).

e) Independently of altitude, air consists of approximately $1/5$ oxygen. The partial pressure of oxygen is also approximately $1/5$ of the air pressure. A certain species of fish can live under a partial pressure of oxygen not less than 0.15 atmospheres. Should one expect to find this species in a river at sea level? Could it be found in a river emptying into Lake Titicaca at an elevation of 3.81 km?

3. *Radioactive decay.*

a) By measuring the amount of a radioactive substance and its decay products in ore samples of the Earth, assuming that no decay products were originally present, one can estimate the age of the Earth (at least from the time when the substance appeared). Suppose that in a rock there are m grams of a radioactive substance and r grams of its decay product. Knowing the half-life T of the substance, find the time elapsed since the decay began and the amount of radioactive substance in a sample of the same volume at the initial time.

b) Atoms of radium in an ore constitute approximately 10^{-12} of the total number of atoms. What was the radium content 10^5, 10^6, and 5×10^9 years ago? (The age of the Earth is estimated at 5×10^9 years.)

c) In the diagnosis of kidney diseases one often measures the ability of the kidneys to remove from the blood various substances deliberately introduced into the body, for example creatin (the "clearance test"). An example of an opposite process of the same type is the restoration of the concentration of hemoglobin in the blood of a donor or of a patient who has suddenly lost a large amount of blood. In all these cases the decrease in the quantity of the substance introduced (or, conversely, the restoration of an insufficient quantity) is subject to the law $N = N_0 e^{-t/\tau}$, where N is the amount (in other words, the number of molecules) of the substance remaining in the body after time t has elapsed from the introduction of the amount N_0 and τ is the so-called *lifetime*: the time elapsed when $1/e$ of the quantity originally introduced remains in the body. The lifetime, as one can easily verify, is 1.44 times larger than the half-life, which is the time elapsed when half of the original quantity of the substance remains.

Suppose a radioactive substance leaves the body at a rate characterized by the lifetime τ_0, and at the same time decays spontaneously with lifetime τ_d. Show that in this case the lifetime τ characterizing the time the substance remains in the body is determined by the relation $\tau^{-1} = \tau_0^{-1} + \tau_d^{-1}$.

d) A certain quantity of blood containing 201 mg of iron has been taken from a donor. To make up for this loss of iron, the donor was ordered to take iron sulfate tablets three times a day for a week, each tablet containing 67 mg of iron. The amount of iron in the donor's blood returns to normal according to an exponential law with lifetime equal to approximately seven days. Assuming that the iron from the tablets enters the bloodstream most rapidly immediately after the blood is taken, determine approximately the portion of the iron in the tablets that will enter the blood over the time needed to restore the normal iron content in the blood.

e) A certain quantity of radioactive phosphorus P^{32} was administered to diagnose a patient with a malignant tumor, after which the radioactivity of the skin of the thigh was measured at regular time intervals. The decrease in radioactivity was subject to an exponential law. Since the half-life of phosphorus is known to be 14.3 days, it was possible to use the data thus obtained to determine the lifetime for the process of decreasing radioactivity as a result of biological causes. Find this constant if it has been established by observation that the lifetime for the overall decrease in radioactivity was 9.4 days (see part c) above).

4. *Absorption of radiation.* The passage of radiation through a medium is accompanied by partial absorption of the radiation. In many cases (the linear theory) one can assume that the absorption in passing through a layer two units thick is the same as the absorption in successively passing through two layers, each one unit thick.

a) Show that under this condition the absorption of radiation is subject to the law $I = I_0 e^{-kl}$, where I_0 is the intensity of the radiation falling on the absorbing substance, I is the intensity after passing through a layer of thickness l, and k is a coefficient having the physical dimension inverse to length.

b) In the case of absorption of light by water, the coefficient k depends on the wave length of the incident light, for example as follows: for ultraviolet $k = 1.4 \times 10^{-2}\ \text{cm}^{-1}$; for blue $k = 4.6 \times 10^{-4}\ \text{cm}^{-1}$; for green $k = 4.4 \times 10^{-4}\ \text{cm}^{-1}$; for red $k = 2.9 \times 10^{-3}\ \text{cm}^{-1}$. Sunlight is falling vertically on the surface of a pure lake 10 meters deep. Compare the intensities of these components of sunlight listed above the surface of the lake and at the bottom.

5. Show that if the law of motion of a point $x = x(t)$ satisfies the equation $m\ddot{x} + kx = 0$ for harmonic oscillations, then

a) the quantity $E = \frac{m\dot{x}^2(t)}{2} + \frac{mx^2(t)}{2}$ is constant ($E = K + U$ is the sum of the kinetic energy $K = \frac{m\dot{x}^2(t)}{2}$ of the point and its potential energy $U = \frac{kx^2(t)}{2}$ at time t);

b) if $x(0) = 0$ and $\dot{x}(0) = 0$, then $x(t) \equiv 0$;

c) there exists a unique motion $x = x(t)$ with initial conditions $x(0) = x_0$ and $\dot{x}(0) = v_0$;

d) Verify that if the point moves in a medium with friction and $x = x(t)$ satisfies the equation $m\ddot{x} + \alpha\dot{x} + kx = 0$, $\alpha > 0$, then the quantity E (see part a)) decreases. Find its rate of decrease and explain the physical meaning of the result, taking account of the physical meaning of E.

6. *Motion under the action of a Hooke*[34] *central force* (the plane oscillator). To develop Eq. (5.156) for a linear oscillator in Sect. 5.6.6 and in Problem 5 let us consider the equation $m\ddot{\mathbf{r}}(t) = -k\mathbf{r}(t)$ satisfied by the radius-vector $\mathbf{r}(t)$ of a point of mass m moving in space under the attraction of a centripetal force proportional to the distance $|\mathbf{r}(t)|$ from the center with constant of proportionality (modulus) $k > 0$. Such a force arises if the point is joined to the center by a Hooke elastic connection, for example, a spring with constant k.

[34] R. Hooke (1635–1703) – British scientist, a versatile scholar and experimenter. He discovered the cell structure of tissues and introduced the word *cell*. He was one of the founders of the mathematical theory of elasticity and the wave theory of light; he stated the hypothesis of gravitation and the inverse-square law for gravitational interaction.

a) By differentiating the vector product $\mathbf{r}(t) \times \dot{\mathbf{r}}(t)$, show that the motion takes place in the plane passing through the center and containing the initial position vector $\mathbf{r}_0 = \mathbf{r}(t_0)$ and the initial velocity vector $\dot{\mathbf{r}}_0 = \dot{\mathbf{r}}(t_0)$ (a plane oscillator). If the vectors $\mathbf{r}_0 = \mathbf{r}(t_0)$ and $\dot{\mathbf{r}}_0 = \dot{\mathbf{r}}(t_0)$ are collinear, the motion takes place along the line containing the center and the vector $\mathbf{r}_0$ (the linear oscillator considered in Sect. 5.6.6).

b) Verify that the orbit of a plane oscillator is an ellipse and that the motion is periodic. Find the period of revolution.

c) Show that the quantity $E = m\dot{\mathbf{r}}^2(t) + k\mathbf{r}^2(t)$ is conserved (constant in time).

d) Show that the initial data $\mathbf{r}_0 = \mathbf{r}(t_0)$ and $\dot{\mathbf{r}}_0 = \dot{\mathbf{r}}(t_0)$ completely determine the subsequent motion of the point.

7. *Ellipticity of planetary orbits.* The preceding problem makes it possible to regard the motion of a point under the action of a central Hooke force as taking place in a plane. Suppose this plane is the plane of the complex variable $z = x + iy$. The motion is determined by two real-valued functions $x = x(t)$, $y = y(t)$ or, what is the same, by one complex-valued function $z = z(t)$ of time t. Assuming for simplicity in Problem 6 that $m = 1$ and $k = 1$, consider the simplest form of the equation of such motion $\ddot{z}(t) = -z(t)$.

a) Knowing from Problem 6 that the solution of this equation corresponding to the specific initial data $z_0 = z(t_0)$, $\dot{z}_0 = \dot{z}(t_0)$ is unique, find it in the form $z(t) = c_1 e^{it} + c_2 e^{-it}$ and, using Euler's formula, verify once again that the trajectory of motion is an ellipse with center at zero. (In certain cases it may become a circle or degenerate into a line segment – determine when.)

b) Taking account of the invariance of the quantity $|\dot{z}(t)|^2 + |z(t)|^2$ during the motion of a point $z(t)$ subject to the equation $\ddot{z}(t) = -z(t)$, verify that, in terms of a new (time) parameter τ connected with t by a relation $\tau = \tau(t)$ such that $\frac{d\tau}{dt} = |z(t)|^2$, the point $w(t) = z^2(t)$ moves subject to the equation $\frac{d^2w}{d\tau^2} = -c\frac{w}{|w|^3}$, where c is a constant and $w = w(t(\tau))$. Thus motion in a central Hooke force field and motion in a Newtonian gravitational field turn out to be connected.

c) Compare this with the result of Problem 8 of Sect. 5.5 and prove that planetary orbits are ellipses.

d) If you have access to a computer, looking again at Euler's method, explained in Sect. 5.6.5, first compute several values of e^x using this method. (Observe that this method uses nothing except the definition of the differential, more precisely the formula $f(x_n) \approx f(x_{n-1}) + f'(x_{n-1})h$, where $h = x_n - x_{n-1}$.)

Now let $\mathbf{r}(t) = (x(t), y(t))$, $\mathbf{r}_0 = \mathbf{r}(0) = (1, 0)$, $\dot{\mathbf{r}}_0 = \dot{\mathbf{r}}(0) = (0, 1)$ and $\ddot{\mathbf{r}}(t) = -\frac{\mathbf{r}(t)}{|\mathbf{r}(t)|^3}$. Using the formulas

$$\mathbf{r}(t_n) \approx \mathbf{r}(t_{n-1}) + \mathbf{v}(t_{n-1})h,$$

$$\mathbf{v}(t_n) \approx \mathbf{v}(t_{n-1}) + \mathbf{a}(t_{n-1})h,$$

where $\mathbf{v}(t) = \dot{\mathbf{r}}(t)$, $\mathbf{a}(t) = \dot{\mathbf{v}}(t) = \ddot{\mathbf{r}}(t)$, use Euler's method to compute the trajectory of the point. Observe its shape and how it is traversed by a point as time passes.

5.7 Primitives

In differential calculus, as we have verified on the examples of the previous section, in addition to knowing how to differentiate functions and write relations between their derivatives, it is also very valuable to know how to find functions from relations satisfied by their derivatives. The simplest such problem, but, as will be seen below, a very important one, is the problem of finding a function $F(x)$ knowing its derivative $F'(x) = f(x)$. The present section is devoted to an introductory discussion of that problem.

5.7.1 The Primitive and the Indefinite Integral

Definition 1 A function $F(x)$ is a *primitive* of a function $f(x)$ on an interval if F is differentiable on the interval and satisfies the equation $F'(x) = f(x)$, or, what is the same, $\mathrm{d}F(x) = f(x)\,\mathrm{d}x$.

Example 1 The function $F(x) = \arctan x$ is a primitive of $f(x) = \frac{1}{1+x^2}$ on the entire real line, since $\arctan' x = \frac{1}{1+x^2}$.

Example 2 The function $F(x) = \operatorname{arccot} \frac{1}{x}$ is a primitive of $f(x) = \frac{1}{1+x^2}$ on the set of positive real numbers and on the set of negative real numbers, since for $x \neq 0$

$$F'(x) = -\frac{1}{1+(\frac{1}{x})^2} \cdot \left(-\frac{1}{x^2}\right) = \frac{1}{1+x^2} = f(x).$$

What is the situation in regard to the existence of a primitive, and what is the set of primitives of a given function?

In the integral calculus we shall prove the fundamental fact that every function that is continuous on an interval has a primitive on that interval.

We present this fact for the reader's information, but in the present section we shall essentially use only the following characteristic of the set of primitives of a given function on an interval, already known to us (see Sect. 5.3.1) from Lagrange's theorem.

Proposition 1 *If $F_1(x)$ and $F_2(x)$ are two primitives of $f(x)$ on the same interval, then the difference $F_1(x) - F_2(x) is$ constant on that interval.*

The hypothesis that F_1 and F_2 are being compared on a connected interval is essential, as was pointed out in the proof of this proposition. One can also see this by comparing Examples 1 and 2, in which the derivatives of $F_1(x) = \arctan x$ and $F_2(x) = \operatorname{arccot} \frac{1}{x}$ agree on the entire domain $\mathbb{R}\backslash 0$ that they have in common. However,

$$F_1(x) - F_2(x) = \arctan x - \operatorname{arccot} \frac{1}{x} = \arctan x - \arctan x = 0,$$

for $x > 0$ while $F_1(x) - F_2(x) \equiv -\pi$ for $x < 0$. For if $x < 0$, we have $\operatorname{arccot}\frac{1}{x} = \pi + \arctan x$.

Like the operation of taking the differential, which has the name "differentiation" and the mathematical notation $\mathrm{d}F(x) = F'(x)\,\mathrm{d}x$, the operation of finding a primitive has the name "indefinite integration" and the mathematical notation

$$\int f(x)\,\mathrm{d}x, \tag{5.166}$$

called the *indefinite integral* of $f(x)$ on the given interval.

Thus we shall interpret the expression (5.166) as a notation for any of the primitives of f on the interval in question.

In the notation (5.166) the sign $\int$ is called the *indefinite integral* sign, f is called the *integrand*, and $f(x)\,\mathrm{d}x$ is called a *differential form*.

It follows from Proposition 1 that if $F(x)$ is any particular primitive of $f(x)$ on the interval, then on that interval

$$\int f(x)\,\mathrm{d}x = F(x) + C, \tag{5.167}$$

that is, any other primitive can be obtained from the particular primitive $F(x)$ by adding a constant.

If $F'(x) = f(x)$, that is, F is a primitive of f on some interval, then by (5.167) we have

$$\boxed{\mathrm{d}\int f(x)\,\mathrm{d}x = \mathrm{d}F(x) = F'(x)\,\mathrm{d}x = f(x)\,\mathrm{d}x.} \tag{5.168}$$

Moreover, in accordance with the concept of an indefinite integral as any primitive, it also follows from (5.167) that

$$\boxed{\int \mathrm{d}F(x) = \int F'(x)\,\mathrm{d}x = F(x) + C.} \tag{5.169}$$

Formulas (5.168) and (5.169) establish a reciprocity between the operations of differentiation and indefinite integration. These operations are mutually inverse up to the undetermined constant C that appears in (5.169).

Up to this point we have discussed only the mathematical nature of the constant C in (5.167). We now give its physical meaning using a simple example. Suppose a point is moving along a line in such a way that its velocity $v(t)$ is known as a function of time (for example, $v(t) \equiv v$). If $x(t)$ is the coordinate of the point at time t, the function $x(t)$ satisfies the equation $\dot{x}(t) = v(t)$, that is, $x(t)$ is a primitive of $v(t)$. Can the position of a point on a line be recovered knowing its velocity over a certain time interval? Clearly not. From the velocity and the time interval one can determine the length s of the path traversed during this time, but not the position on the line. However, the position will also be completely determined if it is given at even one instant, for example, $t = 0$, that is, we give the initial condition

$x(0) = x_0$. Until the initial condition is given, the law of motion could be any law of the form $x(t) = \tilde{x}(t) + c$, where $\tilde{x}(t)$ is any particular primitive of $v(t)$ and c is an arbitrary constant. But once the initial condition $x(0) = \tilde{x}(0) + c = x_0$ is given, all the indeterminacy disappears; for we must have $x(0) = \tilde{x}(0) + c = x_0$, that is, $c = x_0 - \tilde{x}(0)$ and $x(t) = x_0 + [\tilde{x}(t) - \tilde{x}(0)]$. This last formula is entirely physical, since the arbitrary primitive $\tilde{x}$ appears in it only as the difference that determines the path traversed or the magnitude of the displacement from the known initial point $x(0) = x_0$.

5.7.2 The Basic General Methods of Finding a Primitive

In accordance with the definition of the expression (5.166) for the indefinite integral, this expression denotes a function whose derivative is the integrand. From this definition, taking account of (5.167) and the laws of differentiation, one can assert that the following relations hold:

a. $$\int \big(\alpha u(x) + \beta v(x)\big)\,\mathrm{d}x = \alpha \int u(x)\,\mathrm{d}x + \beta \int v(x)\,\mathrm{d}x + c. \tag{5.170}$$

b. $$\int (uv)'\,\mathrm{d}x = \int u'(x)v(x)\,\mathrm{d}x + \int u(x)v'(x)\,\mathrm{d}x + c. \tag{5.171}$$

c. *If*

$$\int f(x)\,\mathrm{d}x = F(x) + c$$

on an interval I_x and $\varphi : I_t \to I_x$ is a smooth (continuously differentiable) mapping of the interval I_t into I_x, then

$$\int (f \circ \varphi)(t)\varphi'(t)\,\mathrm{d}t = (F \circ \varphi)(t) + c. \tag{5.172}$$

The equalities (5.170), (5.171), and (5.172) can be verified by differentiating the left- and right-hand sides using the linearity of differentiation in (5.170), the rule for differentiating a product in (5.171), and the rule for differentiating a composite function in (5.172).

Just like the rules for differentiation, which make it possible to differentiate linear combinations, products, and compositions of known functions, relations (5.170), (5.171), and (5.172), as we shall see, make it possible in many cases to reduce the search for a primitive of a function either to the construction of primitives for simpler functions or to primitives that are already known. A set of such known primitives can be provided, for example, by the following short table of indefinite integrals, obtained by rewriting the table of derivatives of the basic elementary functions (see Sect. 5.2.3):

$$\int x^\alpha \, dx = \frac{1}{\alpha+1} x^{\alpha+1} + c \quad (\alpha \neq -1),$$

$$\int \frac{1}{x} \, dx = \ln|x| + c,$$

$$\int a^x \, dx = \frac{1}{\ln a} a^x + c \quad (0 < a \neq 1),$$

$$\int e^x \, dx = e^x + c,$$

$$\int \sin x \, dx = -\cos x + c,$$

$$\int \cos x \, dx = \sin x + c,$$

$$\int \frac{1}{\cos^2 x} \, dx = \tan x + c,$$

$$\int \frac{1}{\sin^2 x} \, dx = -\cot x + c,$$

$$\int \frac{1}{\sqrt{1-x^2}} \, dx = \begin{cases} \arcsin x + c, \\ -\arccos x + \tilde{c}, \end{cases}$$

$$\int \frac{1}{1+x^2} \, dx = \begin{cases} \arctan x + c, \\ -\operatorname{arccot} x + \tilde{c}, \end{cases}$$

$$\int \sinh x \, dx = \cosh x + c,$$

$$\int \cosh x \, dx = \sinh x + c,$$

$$\int \frac{1}{\cosh^2 x} \, dx = \tanh x + c,$$

$$\int \frac{1}{\sinh^2 x} \, dx = -\coth x + c,$$

$$\int \frac{1}{\sqrt{x^2 \pm 1}} \, dx = \ln\left|x + \sqrt{x^2 \pm 1}\right| + c,$$

$$\int \frac{1}{1-x^2} \, dx = \frac{1}{2} \ln\left|\frac{1+x}{1-x}\right| + c.$$

Each of these formulas is used on the intervals of the real line $\mathbb{R}$ on which the corresponding integrand is defined. If more than one such interval exists, the constant c on the right-hand side may change from one interval to another.

Let us now consider some examples that show relations (5.170), (5.171) and (5.172) in action. We begin with a preliminary remark.

Given that, once a primitive has been found for a given function on an interval the other primitives can be found by adding constants, we shall agree to save writing below by adding the arbitrary constant only to the final result, which is a particular primitive of the given function.

a. Linearity of the Indefinite Integral

This heading means that by relation (5.170) the primitive of a linear combination of functions can be found as the same linear combination of the primitives of the functions.

Example 3

$$\begin{aligned}&\int \left(a_0 + a_1 x + \cdots + a_n x^n\right) \mathrm{d}x = \\ &= a_0 \int 1 \,\mathrm{d}x + a_1 \int x \,\mathrm{d}x + \cdots + a_n \int x^n \,\mathrm{d}x = \\ &= c + a_0 x + \frac{1}{2} a_1 x^2 + \cdots + \frac{1}{n+1} a_n x^{n+1}.\end{aligned}$$

Example 4

$$\begin{aligned}\int \left(x + \frac{1}{\sqrt{x}}\right)^2 \mathrm{d}x &= \int \left(x^2 + 2\sqrt{x} + \frac{1}{x}\right) \mathrm{d}x = \\ &= \int x^2 \,\mathrm{d}x + 2 \int x^{1/2} \,\mathrm{d}x + \int \frac{1}{x} \,\mathrm{d}x = \\ &= \frac{1}{3} x^3 + \frac{4}{3} x^{3/2} + \ln|x| + c.\end{aligned}$$

Example 5

$$\begin{aligned}\int \cos^2 \frac{x}{2} \,\mathrm{d}x &= \int \frac{1}{2}(1 + \cos x) \,\mathrm{d}x = \frac{1}{2} \int (1 + \cos x) \,\mathrm{d}x = \\ &= \frac{1}{2} \int 1 \,\mathrm{d}x + \frac{1}{2} \int \cos x \,\mathrm{d}x = \frac{1}{2} x + \frac{1}{2} \sin x + c.\end{aligned}$$

b. Integration by Parts

Formula (5.171) can be rewritten as

$$u(x)v(x) = \int u(x) \,\mathrm{d}v(x) + \int v(x) \,\mathrm{d}u(x) + c$$

or, what is the same, as

$$\int u(x)\,\mathrm{d}v(x) = u(x)v(x) - \int v(x)\,\mathrm{d}u(x) + c. \tag{5.171$'$}$$

This means that in seeking a primitive for the function $u(x)v'(x)$ one can reduce the problem to finding a primitive for $v(x)u'(x)$, throwing the differentiation onto the other factor and partially integrating the function, as shown in (5.171′), separating the term $u(x)v(x)$ when doing so. Formula (5.171′) is called the formula for *integration by parts*.

Example 6

$$\int \ln x\,\mathrm{d}x = x\ln x - \int x\,\mathrm{d}\ln x = x\ln x - \int x\cdot\frac{1}{x}\,\mathrm{d}x =$$
$$= x\ln x - \int 1\,\mathrm{d}x = x\ln x - x + c.$$

Example 7

$$\int x^2\mathrm{e}^x\,\mathrm{d}x = \int x^2\,\mathrm{d}\mathrm{e}^x = x^2\mathrm{e}^x - \int \mathrm{e}^x\,\mathrm{d}x^2 = x^2\mathrm{e}^x - 2\int x\mathrm{e}^x\,\mathrm{d}x =$$
$$= x^2\mathrm{e}^x - 2\int x\,\mathrm{d}\mathrm{e}^x = x^2\mathrm{e}^x - 2\Big(x\mathrm{e}^x - \int \mathrm{e}^x\,\mathrm{d}x\Big) =$$
$$= x^2\mathrm{e}^x - 2x\mathrm{e}^x + 2\mathrm{e}^x + c = (x^2 - 2x + 2)\mathrm{e}^x + c.$$

c. Change of Variable in an Indefinite Integral

Formula (5.172) shows that in seeking a primitive for the function $(f\circ\varphi)(t)\cdot\varphi'(t)$ one may proceed as follows:

$$\int (f\circ\varphi)(t)\cdot\varphi'(t)\,\mathrm{d}t = \int f\big(\varphi(t)\big)\,\mathrm{d}\varphi(t) =$$
$$= \int f(x)\,\mathrm{d}x = F(x) + c = F\big(\varphi(t)\big) + c,$$

that is, first make the change of variable $\varphi(t) = x$ in the integrand and pass to the new variable x, then, after finding the primitive as a function of x, return to the old variable t by the substitution $x = \varphi(t)$.

Example 8

$$\int \frac{t\,\mathrm{d}t}{1+t^2} = \frac{1}{2}\int \frac{\mathrm{d}(t^2+1)}{1+t^2} = \frac{1}{2}\int \frac{\mathrm{d}x}{x} = \frac{1}{2}\ln|x| + c = \frac{1}{2}\ln(t^2+1) + c.$$

Example 9

$$\int \frac{\mathrm{d}x}{\sin x} = \int \frac{\mathrm{d}x}{2\sin\frac{x}{2}\cos\frac{x}{2}} = \int \frac{\mathrm{d}(\frac{x}{2})}{\tan\frac{x}{2}\cos^2\frac{x}{2}} =$$
$$= \int \frac{\mathrm{d}u}{\tan u \cos^2 u} = \int \frac{\mathrm{d}(\tan u)}{\tan u} = \int \frac{\mathrm{d}v}{v} =$$
$$= \ln|v| + c = \ln|\tan u| + c = \ln\left|\tan\frac{x}{2}\right| + c.$$

We have now considered several examples in which properties a, b, and c of the indefinite integral have been used individually. Actually, in the majority of cases, these properties are used together.

Example 10

$$\int \sin 2x \cos 3x\,\mathrm{d}x = \frac{1}{2}\int (\sin 5x - \sin x)\mathrm{d}x =$$
$$= \frac{1}{2}\left(\int \sin 5x\,\mathrm{d}x - \int \sin x\,\mathrm{d}x\right) =$$
$$= \frac{1}{2}\left(\frac{1}{5}\int \sin 5x\,\mathrm{d}(5x) + \cos x\right) =$$
$$= \frac{1}{10}\int \sin u\,\mathrm{d}u + \frac{1}{2}\cos x = -\frac{1}{10}\cos u + \frac{1}{2}\cos x + c =$$
$$= \frac{1}{2}\cos x - \frac{1}{10}\cos 5x + c.$$

Example 11

$$\int \arcsin x\,\mathrm{d}x = x\arcsin x - \int x\,\mathrm{d}\arcsin x =$$
$$= x\arcsin x - \int \frac{x}{\sqrt{1-x^2}}\mathrm{d}x = x\arcsin x + \frac{1}{2}\int \frac{\mathrm{d}(1-x^2)}{\sqrt{1-x^2}} =$$
$$= x\arcsin x + \frac{1}{2}\int u^{-1/2}\,\mathrm{d}u = x\arcsin x + u^{1/2} + c =$$
$$= x\arcsin x + \sqrt{1-x^2} + c.$$

Example 12

$$\int \mathrm{e}^{ax}\cos bx\,\mathrm{d}x = \frac{1}{a}\int \cos bx\,\mathrm{d}\mathrm{e}^{ax} =$$

$$= \frac{1}{a}\mathrm{e}^{ax}\cos bx - \frac{1}{a}\int \mathrm{e}^{ax}\,\mathrm{d}\cos bx =$$

$$= \frac{1}{a}\mathrm{e}^{ax}\cos bx + \frac{b}{a}\int \mathrm{e}^{ax}\sin bx\,\mathrm{d}x =$$

$$= \frac{1}{a}\mathrm{e}^{ax}\cos bx + \frac{b}{a^2}\int \sin bx\,\mathrm{d}\mathrm{e}^{ax} =$$

$$= \frac{1}{a}\mathrm{e}^{ax}\cos bx + \frac{b}{a^2}\mathrm{e}^{ax}\sin bx - \frac{b}{a^2}\int \mathrm{e}^{ax}\,\mathrm{d}\sin bx =$$

$$= \frac{a\cos bx + b\sin bx}{a^2} - \frac{b^2}{a^2}\int \mathrm{e}^{ax}\cos bx\,\mathrm{d}x.$$

From this result we conclude that

$$\int \mathrm{e}^{ax}\cos bx\,\mathrm{d}x = \frac{a\cos bx + b\sin bx}{a^2+b^2}\mathrm{e}^{ax} + c.$$

We could have arrived at this result by using Euler's formula and the fact that the primitive of the function $\mathrm{e}^{(a+ib)x} = \mathrm{e}^{ax}\cos bx + i\mathrm{e}^{ax}\sin bx$ is

$$\frac{1}{a+ib}\mathrm{e}^{(a+ib)x} = \frac{a-ib}{a^2+b^2}\mathrm{e}^{(a+ib)x} =$$

$$= \frac{a\cos bx + b\sin bx}{a^2+b^2}\mathrm{e}^{ax} + i\frac{a\sin x - b\cos bx}{a^2+b^2}\mathrm{e}^{ax}.$$

It will be useful to keep this in mind in the future. For real values of x this can easily be verified directly by differentiating the real and imaginary parts of the function $\frac{1}{a+ib}\mathrm{e}^{(a+ib)x}$.

In particular, we also find from this result that

$$\int \mathrm{e}^{ax}\sin bx\,\mathrm{d}x = \frac{a\sin bx - b\cos bx}{a^2+b^2}\mathrm{e}^{ax} + c.$$

Even the small set of examples we have considered suffices to show that in seeking primitives for even the elementary functions one is often obliged to resort to auxiliary transformations and clever devices, which was not at all the case in finding the derivatives of compositions of the functions whose derivatives we knew. It turns out that this difficulty is not accidental. For example, in contrast to differentiation, finding the primitive of an elementary function may lead to a function that is no longer a composition of elementary functions. For that reason, one should not conflate the phrase "finding a primitive" with the sometimes impossible task of "expressing the primitive of a given elementary function in terms of elementary functions". In general, the class of elementary functions is a rather artificial object. There are very many special functions of importance in applications that have been studied and tabulated at least as well as, say $\sin x$ or e^x.

For example, the *sine integral* $\operatorname{Si} x$ is the primitive $\int \frac{\sin x}{x}\,\mathrm{d}x$ of the function $\frac{\sin x}{x}$ that tends to zero as $x \to 0$. There exists such a primitive, but, like all the other primitives of $\frac{\sin x}{x}$, it is not a composition of elementary functions.

Similarly, the function

$$\operatorname{Ci} x = \int \frac{\cos x}{x}\,\mathrm{d}x,$$

specified by the condition $\operatorname{Ci} x \to 0$ as $x \to \infty$ is not elementary. The function $\operatorname{Ci} x$ is called the *cosine integral*.

The primitive $\int \frac{\mathrm{d}x}{\ln x}$ of the function $\frac{1}{\ln x}$ is also not elementary. One of the primitives of this function is denoted $\operatorname{li} x$ and is called the *logarithmic integral*. It satisfies the condition $\operatorname{li} x \to 0$ as $x \to +0$. (More details about the functions $\operatorname{Si} x$, $\operatorname{Ci} x$, and $\operatorname{li} x$ will be given in Sect. 6.5.)

Because of these difficulties in finding primitives, rather extensive tables of indefinite integrals have been compiled. However, in order to use these tables successfully and avoid having to resort to them when the problem is very simple, one must acquire some skill in dealing with indefinite integrals.

The remainder of this section is devoted to integrating some special classes of functions whose primitives can be expressed as compositions of elementary functions.

5.7.3 Primitives of Rational Functions

Let us consider the problem of integrating $\int R(x)\,\mathrm{d}x$, where $R(x) = \frac{P(x)}{Q(x)}$ is a ratio of polynomials.

If we work in the domain of real numbers, then, without going outside this domain, we can express every such fraction, as we know from algebra (see formula (5.135) in Sect. 5.5.4) as a sum

$$\frac{P(x)}{Q(x)} = p(x) + \sum_{j=1}^{l}\left(\sum_{k=1}^{k_j} \frac{a_{jk}}{(x - x_j)^k}\right) + \sum_{j=1}^{n}\left(\sum_{k=1}^{m_j} \frac{b_{jk}x + c_{jk}}{(x^2 + p_j x + q_j)^k}\right), \tag{5.173}$$

where $p(x)$ is a polynomial (which arises when $P(x)$ is divided by $Q(x)$, but only when the degree of $P(x)$ is not less than the degree of $Q(x)$), a_{jk}, b_{jk}, and c_{jk} are uniquely determined real numbers, and $Q(x) = (x - x_1)^{k_1} \cdots (x - x_l)^{k_l}(x^2 + p_1 x + q_1)^{m_1} \cdots (x^2 + p_n x + q_n)^{m_n}$.

We have already discussed how to find the expansion (5.173) in Sect. 5.5. Once the expansion (5.173) has been constructed, integrating $R(x)$ reduces to integrating the individual terms.

We have already integrated a polynomial in Example 1, so that it remains only to consider the integration of fractions of the forms

$$\frac{1}{(x - a)^k} \quad \text{and} \quad \frac{bx + c}{(x^2 + px + q)^k}, \qquad \text{where } k \in \mathbb{N}.$$

The first of these problems can be solved immediately, since

$$\int \frac{1}{(x-a)^k}\,\mathrm{d}x = \begin{cases} \frac{1}{-k+1}(x-a)^{-k+1} + c & \text{for } k \neq 1, \\ \ln|x-a| + c & \text{for } k = 1. \end{cases} \tag{5.174}$$

With the integral

$$\int \frac{bx+c}{(x^2+px+q)^k}\,\mathrm{d}x$$

we proceed as follows. We represent the polynomial x^2+px+q as $(x+\frac{1}{2}p)^2 + (q-\frac{1}{4}p^2)$, where $q-\frac{1}{4}p^2 > 0$, since the polynomial x^2+px+q has no real roots. Setting $x+\frac{1}{2}p = u$ and $q-\frac{1}{4}p^2 = a^2$, we obtain

$$\int \frac{bx+c}{(x^2+px+q)^k}\,\mathrm{d}x = \int \frac{\alpha u+\beta}{(u^2+a^2)^k}\,\mathrm{d}u,$$

where $\alpha = b$ and $\beta = c - \frac{1}{2}bp$.

Next,

$$\int \frac{u}{(u^2+a^2)^k}\,\mathrm{d}u = \frac{1}{2}\int \frac{\mathrm{d}(u^2+a^2)}{(u^2+a^2)^k} = \\ = \begin{cases} \frac{1}{2(1-k)}(u^2+a^2)^{-k+1} & \text{for } k \neq 1, \\ \frac{1}{2}\ln(u^2+a^2) & \text{for } k = 1, \end{cases} \tag{5.175}$$

and it remains only to study the integral

$$I_k = \int \frac{\mathrm{d}u}{(u^2+a^2)^k}. \tag{5.176}$$

Integrating by parts and making elementary transformations, we have

$$I_k = \int \frac{\mathrm{d}u}{(u^2+a^2)^k} = \frac{u}{(u^2+a^2)^k} + 2k\int \frac{u^2\,\mathrm{d}u}{(u^2+a^2)^{k+1}} = \\ = \frac{u}{(u^2+a^2)^k} + 2k\int \frac{(u^2+a^2)-a^2}{(u^2+a^2)^{k+1}}\,\mathrm{d}u = \frac{u}{(u^2+a^2)^k} + 2kI_k - 2ka^2 I_{k+1},$$

from which we obtain the recursion relation

$$I_{k+1} = \frac{1}{2ka^2}\frac{u}{(u^2+a^2)^k} + \frac{2k-1}{2ka^2}I_k, \tag{5.177}$$

which makes it possible to lower the exponent k in the integral (5.176). But I_1 is easy to compute:

$$I_1 = \int \frac{\mathrm{d}u}{u^2+a^2} = \frac{1}{a}\int \frac{\mathrm{d}(\frac{u}{a})}{1+(\frac{u}{a})^2} = \frac{1}{a}\arctan\frac{u}{a} + c. \tag{5.178}$$

Thus, by using (5.177) and (5.178), one can also compute the primitive (5.176).

Thus we have proved the following proposition.

Proposition 2 *The primitive of any rational function $R(x) = \frac{P(x)}{Q(x)}$ can be expressed in terms of rational functions and the transcendental functions* ln *and* arctan. *The rational part of the primitive, when placed over a common denominator, will have a denominator containing all the factors of the polynomial $Q(x)$ with multiplicities one less than they have in $Q(x)$.*

Example 13 Let us calculate $\displaystyle\int \frac{2x^2+5x+5}{(x^2-1)(x+2)}\,\mathrm{d}x$.

Since the integrand is a proper fraction, and the factorization of the denominator into the product $(x-1)(x+1)(x+2)$ is also known, we immediately seek a partial fraction expansion

$$\frac{2x^2+5x+5}{(x-1)(x+1)(x+2)} = \frac{A}{x-1} + \frac{B}{x+1} + \frac{C}{x+2}. \tag{5.179}$$

Putting the right-hand side of Eq. (5.179) over a common denominator,we have

$$\frac{2x^2+5x+5}{(x-1)(x+1)(x+2)} = \frac{(A+B+C)x^2+(3A+B)x+(2A-2B-C)}{(x-1)(x+1)(x+2)}.$$

Equating the corresponding coefficients in the numerators, we obtain the system

$$\begin{cases} A+B+C=2, \\ 3A+B=5, \\ 2A-2B-C=5, \end{cases}$$

from which we find $(A, B, C) = (2, -1, 1)$.

We remark that in this case these numbers could have been found in one's head. Indeed, multiplying (5.179) by $x-1$ and then setting $x=1$ in the resulting equality, we would have A on the right-hand side, while the left-hand side would have been the value at $x=1$ of the fraction obtained by striking out the factor $x-1$ in the denominator, that is, $A = \frac{2+5+5}{2\cdot 3} = 2$. One could proceed similarly to find B and C.

Thus,

$$\begin{aligned} \int \frac{2x^2+5x+5}{(x^2-1)(x+2)}\,\mathrm{d}x &= 2\int \frac{\mathrm{d}x}{x-1} - \int \frac{\mathrm{d}x}{x+1} + \int \frac{\mathrm{d}x}{x+2} = \\ &= 2\ln|x-1| - \ln|x+1| + \ln|x+2| + c \\ &= \ln\left|\frac{(x-1)^2(x+2)}{x-1}\right| + c. \end{aligned}$$

Example 14 Let us compute a primitive of the function

$$R(x) = \frac{x^7 - 2x^6 + 4x^5 - 5x^4 + 4x^3 - 5x^2 - x}{(x-1)^2(x^2+1)^2}.$$

We begin by remarking that this is an improper fraction, so that, removing the parentheses and finding the denominator $Q(x) = x^6 - 2x^5 + 3x^4 - 4x^3 + 3x^2 - 2x + 1$, we divide the numerator by it, after which we obtain

$$R(x) = x + \frac{x^5 - x^4 + x^3 - 3x^2 - 2x}{(x-1)^2(x^2+1)^2},$$

and we then seek a partial-fraction expansion of the proper fraction

$$\frac{x^5 - x^4 + x^3 - 3x^2 - 2x}{(x-1)^2(x^2+1)^2} = \frac{A}{(x-1)^2} + \frac{B}{x-1} + \frac{Cx+D}{(x^2+1)^2} + \frac{Ex+F}{x^2+1}. \quad (5.180)$$

Of course the expansion could be obtained in the canonical way, by writing out a system of six equations in six unknowns. However, instead of doing that, we shall demonstrate some other technical possibilities that are sometimes used.

We find the coefficient A by multiplying Eq. (5.180) by $(x-1)^2$ and then setting $x = 1$. The result is $A = -1$. We then transpose the fraction $\frac{A}{(x-1)^2}$, in which A is now the known quantity -1, to the left-hand side of Eq. (5.180). We then have

$$\frac{x^4 + x^3 + 2x^2 + x - 1}{(x-1)(x^2+1)^2} = \frac{B}{x-1} + \frac{Cx+D}{(x^2+1)^2} + \frac{Ex+F}{x^2+1} \quad (5.181)$$

from which, multiplying (5.181) by $x - 1$ and then setting $x = 1$, we find $B = 1$.

Now, transposing the fraction $\frac{1}{x-1}$ to the left-hand side of (5.181), we obtain

$$\frac{x^2 + x + 2}{(x^2+1)^2} = \frac{Cx+D}{(x^2+1)^2} + \frac{Ex+F}{x^2+1}. \quad (5.182)$$

Now, after putting the right-hand side of (5.182) over a common denominator, we equate the numerators

$$x^2 + x + 2 = Ex^3 + Fx^2 + (C+E)x + (D+F),$$

from which it follows that

$$\begin{cases} E = 0, \\ F = 1, \\ C + E = 1, \\ D + F = 2, \end{cases}$$

or $(C, D, E, F) = (1, 1, 0, 1)$.

We now know all the coefficients in (5.180). Upon integration, the first two fractions yield respectively $\frac{1}{x-1}$ and $\ln|x-1|$. Then

$$\int \frac{Cx+D}{(x^2+1)^2}\,\mathrm{d}x = \int \frac{x+1}{(x^2+1)^2}\,\mathrm{d}x =$$

$$= \frac{1}{2}\int \frac{\mathrm{d}(x^2+1)}{(x^2+1)^2} + \int \frac{\mathrm{d}x}{(x^2+1)^2} = \frac{-1}{2(x^2+1)} + I_2,$$

where

$$I_2 = \int \frac{\mathrm{d}x}{(x^2+1)^2} = \frac{1}{2}\frac{x}{(x^2+1)^2} + \frac{1}{2}\arctan x,$$

which follows from (5.177) and (5.178).

Finally,

$$\int \frac{Ex+F}{x^2+1}\,\mathrm{d}x = \int \frac{1}{x^2+1}\,\mathrm{d}x = \arctan x.$$

Gathering all the integrals, we finally have

$$\int R(x)\,\mathrm{d}x = \frac{1}{2}x^2 + \frac{1}{x-1} + \frac{x}{2(x^2+1)^2} + \ln|x-1| + \frac{3}{2}\arctan x + c.$$

Let us now consider some frequently encountered indefinite integrals whose computation can be reduced to finding the primitive of a rational function.

5.7.4 Primitives of the Form $\int R(\cos x, \sin x)\,\mathrm{d}x$

Let $R(u, v)$ be a rational function in u and v, that is a quotient of polynomials $\frac{P(u,v)}{Q(u,v)}$, which are linear combinations of monomials $u^m v^n$, where $m = 0, 1, 2\ldots$ and $n = 0, 1, \ldots$.

Several methods exist for computing the integral $\int R(\cos x, \sin x)\,\mathrm{d}x$, one of which is completely general, although not always the most efficient.

a.

We make the change of variable $t = \tan\frac{x}{2}$. Since

$$\cos x = \frac{1-\tan^2\frac{x}{2}}{1+\tan^2\frac{x}{2}}, \qquad \sin x = \frac{2\tan\frac{x}{2}}{1+\tan^2\frac{x}{2}},$$

$$\mathrm{d}t = \frac{\mathrm{d}x}{2\cos^2\frac{x}{2}}, \quad \text{that is,} \quad \mathrm{d}x = \frac{2\,\mathrm{d}t}{1+\tan^2\frac{x}{2}},$$

it follows that

$$\int R(\cos x, \sin x)\,\mathrm{d}x = \int R\left(\frac{1-t^2}{1+t^2}, \frac{2t}{1+t^2}\right)\frac{2}{1+t^2}\,\mathrm{d}t,$$

and the problem has been reduced to integrating a rational function.

However, this way leads to a very cumbersome rational function; for that reason one should keep in mind that in many cases there are other possibilities for rationalizing the integral.

b.

In the case of integrals of the form $\int R(\cos^2 x, \sin^2 x)\,\mathrm{d}x$ or $\int r(\tan x)\,\mathrm{d}x$, where $r(u)$ is a rational function, a convenient substitution is $t = \tan x$, since

$$\cos^2 x = \frac{1}{1+\tan^2 x}, \qquad \sin^2 x = \frac{\tan^2 x}{1+\tan^2 x},$$

$$\mathrm{d}t = \frac{\mathrm{d}x}{\cos^2 x}, \quad \text{that is,} \quad \mathrm{d}x = \frac{\mathrm{d}t}{1+t^2}.$$

Carrying out this substitution, we obtain respectively

$$\int R\left(\cos^2 x, \sin^2 x\right)\mathrm{d}x = \int R\left(\frac{1}{1+t^2}, \frac{t^2}{1+t^2}\right)\frac{\mathrm{d}t}{1+t^2},$$

$$\int r(\tan x)\,\mathrm{d}x = \int r(t)\frac{\mathrm{d}t}{1+t^2}.$$

c.

In the case of integrals of the form

$$\int R\left(\cos x, \sin^2 x\right)\sin x\,\mathrm{d}x \quad \text{or} \quad \int R\left(\cos^2 x, \sin x\right)\cos x\,\mathrm{d}x.$$

One can move the functions $\sin x$ and $\cos x$ into the differential and make the substitution $t = \cos x$ or $t = \sin x$ respectively. After these substitutions, the integrals will have the form

$$-\int R\left(t, 1-t^2\right)\mathrm{d}t \quad \text{or} \quad \int R\left(1-t^2, t\right)\mathrm{d}t.$$

Example 15

$$\int \frac{\mathrm{d}x}{3+\sin x} = \int \frac{1}{3+\frac{2t}{1+t^2}} \cdot \frac{2\,\mathrm{d}t}{1+t^2} =$$

$$= 2\int \frac{\mathrm{d}t}{3t^2+2t+3} = \frac{2}{3}\int \frac{\mathrm{d}(t+\frac{1}{3})}{(t+\frac{1}{3})^2+\frac{8}{9}} = \frac{2}{3}\int \frac{\mathrm{d}u}{u^2+(\frac{2\sqrt{2}}{3})^2} =$$

$$= \frac{1}{\sqrt{2}}\arctan\frac{3u}{2\sqrt{2}} + c = \frac{1}{\sqrt{2}}\arctan\frac{3t+1}{2\sqrt{2}} + c =$$

$$= \frac{1}{\sqrt{2}}\arctan\frac{3\tan\frac{x}{2}+1}{2\sqrt{2}} + c.$$

Here we have used the universal change of variable $t = \tan\frac{x}{2}$.

Example 16

$$\int \frac{\mathrm{d}x}{(\sin x+\cos x)^2} = \int \frac{\mathrm{d}x}{\cos^2 x(\tan x+1)^2} =$$

$$= \int \frac{\mathrm{d}\tan x}{(\tan x+1)^2} = \int \frac{\mathrm{d}t}{(t+1)^2} = -\frac{1}{t+1} + c =$$

$$= c - \frac{1}{1+\tan x}.$$

Example 17

$$\int \frac{\mathrm{d}x}{2\sin^2 3x - 3\cos^2 3x + 1} = \int \frac{\mathrm{d}x}{\cos^2 3x(2\tan^2 3x - 3 + (1+\tan^2 3x))} =$$

$$= \frac{1}{3}\int \frac{\mathrm{d}\tan 3x}{3\tan^2 3x - 2} = \frac{1}{3}\int \frac{\mathrm{d}t}{3t^2-2} =$$

$$= \frac{1}{3\cdot 2}\sqrt{\frac{2}{3}}\int \frac{\mathrm{d}\sqrt{\frac{3}{2}}t}{\frac{3}{2}t^2-1} =$$

$$= \frac{1}{3\sqrt{6}}\int \frac{\mathrm{d}u}{u^2-1} = \frac{1}{6\sqrt{6}}\ln\left|\frac{u-1}{u+1}\right| + c =$$

$$= \frac{1}{6\sqrt{6}}\ln\left|\frac{\sqrt{\frac{3}{2}}t-1}{\sqrt{\frac{3}{2}}t+1}\right| + c =$$

$$= \frac{1}{6\sqrt{6}}\ln\left|\frac{\tan 3x - \sqrt{\frac{2}{3}}}{\tan 3x + \sqrt{\frac{2}{3}}}\right| + c.$$

Example 18

$$\int \frac{\cos^3 x}{\sin^7 x}\,\mathrm{d}x = \int \frac{\cos^2 x\,\mathrm{d}\sin x}{\sin^7 x} = \int \frac{(1-t^2)\,\mathrm{d}t}{t^7} =$$

$$= \int \left(t^{-7} - t^{-5}\right)\mathrm{d}t = -\frac{1}{6}t^{-6} + \frac{1}{4}t^{-4} + c = \frac{1}{4\sin^4 x} - \frac{1}{6\sin^6 x} + c.$$

5.7.5 Primitives of the Form $\int R(x, y(x))\,\mathrm{d}x$

Let $R(x, y)$ be, as in Sect. 5.7.4, a rational function. Let us consider some special integrals of the form

$$\int R\big(x, y(x)\big)\,\mathrm{d}x,$$

where $y = y(x)$ is a function of x.

First of all, it is clear that if one can make a change of variable $x = x(t)$ such that both functions $x = x(t)$ and $y = y(x(t))$ are rational functions of t, then $x'(t)$ is also a rational function and

$$\int R\big(x, y(x)\big)\,\mathrm{d}x = \int R\big(x(t), y\big(x(t)\big)\big)x'(t)\,\mathrm{d}t,$$

that is, the problem will have been reduced to integrating a rational function.

Consider the following special choices of the function $y = y(x)$.

a.

If $y = \sqrt[n]{\frac{ax+b}{cx+d}}$, where $n \in \mathbb{N}$, then, setting $t^n = \frac{ax+b}{cx+d}$, we obtain

$$x = \frac{d \cdot t^n - b}{a - c \cdot t^n}, \qquad y = t,$$

and the integrand rationalizes.

Example 19

$$\int \sqrt[3]{\frac{x-1}{x+1}}\,\mathrm{d}x = \int t\,\mathrm{d}\left(\frac{t^3+1}{1-t^3}\right) = t \cdot \frac{t^3+1}{1-t^3}\,\mathrm{d}t - \int \frac{t^3+1}{1-t^3}\,\mathrm{d}t =$$

$$= t \cdot \frac{t^3+1}{1-t^3} - \int \left(\frac{2}{1-t^3} - 1\right)\mathrm{d}t =$$

$$= t \cdot \frac{t^3+1}{1-t^3} + t - 2\int \frac{\mathrm{d}t}{(1-t)(1+t+t^2)} =$$

$$= \frac{2t}{1-t^3} - 2\int \left(\frac{1}{3(1-t)} + \frac{2+t}{3(1+t+t^2)} \right) \mathrm{d}t =$$

$$= \frac{2t}{1-t^3} + \frac{2}{3}\ln|1-t| - \frac{2}{3}\int \frac{(t+\frac{1}{2})+\frac{3}{2}}{(t+\frac{1}{2})^2+\frac{3}{4}} \mathrm{d}t =$$

$$= \frac{2t}{1-t^3} + \frac{2}{3}\ln|1-t| - \frac{1}{3}\ln\left[\left(t+\frac{1}{2}\right)+\frac{3}{4}\right] -$$

$$-\frac{2}{\sqrt{3}}\arctan\frac{2}{\sqrt{3}}\left(t+\frac{1}{2}\right)+c, \quad \text{where } t = \sqrt[3]{\frac{x-1}{x+1}}.$$

b.

Let us now consider the case when $y = \sqrt{ax^2+bx+c}$, that is, integrals of the form

$$\int R\big(x, \sqrt{ax^2+bx+c}\big)\,\mathrm{d}x.$$

By completing the square in the trinomial ax^2+bx+c and making a suitable linear substitution, we reduce the general case to one of the following three simple cases:

$$\int R\big(t, \sqrt{t^2+1}\big)\,\mathrm{d}t, \quad \int R\big(t, \sqrt{t^2-1}\big)\,\mathrm{d}t, \quad \int R\big(t, \sqrt{1-t^2}\big)\,\mathrm{d}t. \tag{5.183}$$

To rationalize these integrals it now suffices to make the following substitutions, respectively:

$$\sqrt{t^2+1} = tu+1, \quad \text{or} \quad \sqrt{t^2+1} = tu-1, \quad \text{or} \quad \sqrt{t^2+1} = t-u;$$

$$\sqrt{t^2-1} = u(t-1), \quad \text{or} \quad \sqrt{t^2-1} = u(t+1), \quad \text{or} \quad \sqrt{t^2-1} = t-u;$$

$$\sqrt{1-t^2} = u(1-t), \quad \text{or} \quad \sqrt{1-t^2} = u(1+t), \quad \text{or} \quad \sqrt{1-t^2} = tu \pm 1.$$

These substitutions were proposed long ago by Euler (see Problem 3 at the end of this section).

Let us verify, for example, that after the first substitution we will have reduced the first integral to the integral of a rational function.

In fact, if $\sqrt{t^2+1} = tu+1$, then $t^2+1 = t^2u^2+2tu+1$, from which we find

$$t = \frac{2u}{1-u^2}$$

and then

$$\sqrt{t^2+1} = \frac{1+u^2}{1-u^2}.$$

Thus t and $\sqrt{t^2+1}$ have been expressed rationally in terms of u, and consequently the integral has been reduced to the integral of a rational function.

The integrals (5.183) can also be reduced, by means of the substitutions $t = \sinh\varphi$, $t = \cosh\varphi$, and $t = \sin\varphi$ (or $t = \cos\varphi$) respectively, to the following forms:

$$\int R(\sinh\varphi, \cosh\varphi)\cosh\varphi\, \mathrm{d}\varphi, \qquad \int R(\cosh\varphi, \sinh\varphi)\sinh\varphi\, \mathrm{d}\varphi$$

and

$$\int R(\sin\varphi, \cos\varphi)\cos\varphi\, \mathrm{d}\varphi \quad \text{or} \quad -\int R(\cos\varphi, \sin\varphi)\sin\varphi\, \mathrm{d}\varphi.$$

Example 20

$$\int \frac{\mathrm{d}x}{x+\sqrt{x^2+2x+2}} = \int \frac{\mathrm{d}x}{x+\sqrt{(x+1)^2+1}} = \int \frac{\mathrm{d}t}{t-1+\sqrt{t^2+1}}.$$

Setting $\sqrt{t^2+1} = u - t$, we have $1 = u^2 - 2tu$, from which it follows that $t = \frac{u^2-1}{2u}$. Therefore

$$\begin{aligned}\int \frac{\mathrm{d}t}{t-1+\sqrt{t^2+1}} &= \frac{1}{2}\int \frac{1}{u-1}\left(1+\frac{1}{u^2}\right)\mathrm{d}u = \frac{1}{2}\int \frac{1}{u-1}\,\mathrm{d}u + \\ &+ \frac{1}{2}\int \frac{\mathrm{d}u}{u^2(u-1)} \\ &= \frac{1}{2}\ln|u-1| + \frac{1}{2}\int\left(\frac{1}{\mathrm{u}-1} - \frac{1}{u^2} - \frac{1}{u}\right)\mathrm{d}u = \\ &= \frac{1}{2}\ln|u-1| + \frac{1}{2}\ln\left|\frac{u-1}{u}\right| + \frac{1}{2u} + c.\end{aligned}$$

It now remains to retrace the path of substitutions: $u = t + \sqrt{t^2+1}$ and $t = x+1$.

c. Elliptic Integrals

Another important class of integrals consists of those of the form

$$\int R\big(x, \sqrt{P(x)}\big)\,\mathrm{d}x, \tag{5.184}$$

where $P(x)$ is a polynomial of degree $n > 2$. As Abel and Liouville showed, such an integral cannot in general be expressed in terms of elementary functions.

For $n = 3$ and $n = 4$ the integral (5.184) is called an *elliptic integral*, and for $n > 4$ it is called *hyperelliptic*.

It can be shown that by elementary substitutions the general elliptic integral can be reduced to the following three standard forms up to terms expressible in elementary functions:

$$\int \frac{dx}{\sqrt{(1-x^2)(1-k^2x^2)}}, \tag{5.185}$$

$$\int \frac{x^2\,dx}{\sqrt{(1-x^2)(1-k^2x^2)}}, \tag{5.186}$$

$$\int \frac{dx}{(1+hx^2)\sqrt{(1-x^2)(1-k^2x^2)}}, \tag{5.187}$$

where h and k are parameters, the parameter k lying in the interval $]0, 1[$ in all three cases.

By the substitution $x = \sin\varphi$ these integrals can be reduced to the following canonical integrals and combinations of them:

$$\int \frac{d\varphi}{\sqrt{1-k^2\sin^2\varphi}}, \tag{5.188}$$

$$\int \sqrt{1-k^2\sin^2\varphi}\,d\varphi, \tag{5.189}$$

$$\int \frac{d\varphi}{(1+h\sin^2\varphi)\sqrt{1-k^2\sin^2\varphi}}. \tag{5.190}$$

The integrals (5.188), (5.189) and (5.190) are called respectively the *elliptic integral of first kind, second kind*, and *third kind* (in the Legendre form).

The symbols $F(k,\varphi)$ and $E(k,\varphi)$ respectively denote the particular elliptic integrals (5.188) and (5.189) of first and second kind that satisfy $F(k, 0) = 0$ and $E(k,0) = 0$.

The functions $F(k,\varphi)$ and $E(k,\varphi)$ are frequently used, and for that reason very detailed tables of their values have been compiled for $0 < k < 1$ and $0 \le \varphi \le \pi/2$.

As Abel showed, it is natural to study elliptic integrals in the complex domain, in intimate connection with the so-called elliptic functions, which functions, which are related to the elliptic integrals exactly as the function $\sin x$, for example, is related to the integral $\int \frac{d\varphi}{\sqrt{1-\varphi^2}} = \arcsin\varphi$.

5.7.6 Problems and Exercises

1. *Ostrogradskii's*[35] *method of separating off the rational part of the integral of a proper rational fraction.*

---

[35]M.V. Ostrogradskii (1801–1861) – prominent Russian specialist in theoretical mechanics and mathematician, one of the founders of the applied area of research in the Petersburg mathematical school.

Let $\frac{P(x)}{Q(x)}$ be a proper rational fraction, let $q(x)$ be the polynomial having the same roots as $Q(x)$, but with multiplicity 1, and let $Q_1(x) = \frac{Q(x)}{q(x)}$.
Show that

a) the following formula of Ostrogradskii holds:

$$\int \frac{P(x)}{Q(x)}\,dx = \frac{P_1(x)}{Q_1(x)} + \int \frac{p(x)}{q(x)}\,dx, \tag{5.191}$$

where $\frac{P_1(x)}{Q_1(x)}$ and $\frac{p(x)}{q(x)}$ are proper rational fractions and $\int \frac{p(x)}{q(x)}\,dx$ is a transcendental function.

(Because of this result, the fraction $\frac{P_1(x)}{Q_1(x)}$ in (5.191) is called the *rational part* of the integral $\int \frac{P(x)}{Q(x)}\,dx$.)

b) In the formula

$$\frac{P(x)}{Q(x)} = \left(\frac{P_1(x)}{Q_1(x)}\right)' + \frac{p(x)}{q(x)}$$

obtained by differentiating Ostrogradskii's formula, the sum at the right hand side can be given the denominator $Q(x)$ after suitable cancellations.

c) The polynomials $q(x)$, $Q_1(x)$, and then also the polynomials $p(x)$, $P_1(x)$ can be found algebraically, without even knowing the roots of $Q(x)$. Thus the rational part of the integral (5.191) can be found completely without even computing the whole primitive.

d) Separate off the rational part of the integral (5.191) if

$$P(x) = 2x^6 + 3x^5 + 6x^4 + 6x^3 + 10x^2 + 3x + 2,$$
$$Q(x) = x^7 + 3x^6 + 5x^5 + 7x^4 + 7x^3 + 5x^2 + 3x + 1$$

(see Example 17 in Sect. 5.5).

2. Suppose we are seeking the primitive

$$\int R(\cos x, \sin x)\,dx, \tag{5.192}$$

where $R(u, v) = \frac{P(u,v)}{Q(u,v)}$ is a rational function.
Show that

a) if $R(-u, v) = R(u, v)$, then $R(u, v)$ has the form $R_1(u^2, v)$;

b) if $R(-u, v) = -R(u, v)$, then $R(u, v) = u \cdot R_2(u^2, v)$ and the substitution $t = \sin x$ rationalizes the integral (5.192);

c) If $R(-u, -v) = R(u, v)$, then $R(u, v) = R_3(\frac{u}{v}, v^2)$, and the substitution $t = \tan x$ rationalizes the integral (5.192).

3. *Integrals of the form*

$$\int R\big(x, \sqrt{ax^2 + bx + c}\big)\,dx. \tag{5.193}$$

a) Verify that the integral (5.193) can be reduced to the integral of a rational function by the following *Euler substitutions*:

$t = \sqrt{ax^2 + bx + c} \pm \sqrt{a}x$, if $a > 0$,

$t = \sqrt{\frac{x-x_1}{x-x_2}}$ if x_2 and x_2 are real roots of the trinomial $ax^2 + bx + c$.

b) Let (x_0, y_0) be a point of the curve $y^2 = ax^2 + bx + c$ and t the slope of the line passing through (x_0, y_0) and intersecting this curve in the point (x, y). Express the coordinates (x, y) in terms of (x_0, y_0) and t and connect these formulas with Euler's substitutions.

c) A curve defined by an algebraic equation $P(x, y) = 0$ is *unicursal* if it admits a parametric description $x = x(t)$, $y = y(t)$ in terms of rational functions $x(t)$ and $y(t)$. Show that the integral $\int R(x, y(x))\,\mathrm{d}x$, where $R(u, v)$ is a rational function and $y(x)$ is an algebraic function satisfying the equation $P(x, y) = 0$ that defines the unicursal curve, can be reduced to the integral of a rational function.

d) Show that the integral (5.193) can always be reduced to computing integrals of the following three types:

$$\int \frac{P(x)}{\sqrt{ax^2 + bx + c}}\,\mathrm{d}x \int \frac{\mathrm{d}x}{(x - x_0)^k \cdot \sqrt{ax^2 + bx + c}},$$

$$\int \frac{(Ax + B)\,\mathrm{d}x}{(x^2 + px + a)^m \cdot \sqrt{ax^2 + bx + c}}.$$

4. a) Show that the integral

$$\int x^m (a + bx^n)^p\,\mathrm{d}x$$

whose differential is a binomial, where m, n, and p are rational numbers, can be reduced to the integral

$$\int (a + bt)^p t^q\,\mathrm{d}t, \tag{5.194}$$

where p and q are rational numbers.

b) The integral (5.194) can be expressed in terms of elementary functions if one of the three numbers p, q, and $p + q$ is an integer. (Chebyshev showed that there were no other cases in which the integral (5.194) could be expressed in elementary functions.)

5. *Elliptic integrals.*

a) Any polynomial of degree three with real coefficients has a real root x_0, and can be reduced to a polynomial of the form $t^2(at^4 + bt^3 + ct^2 + dt + e)$, where $a \neq 0$, by the substitution $x - x_0 = t^2$.

b) If $R(u, v)$ is a rational function and P a polynomial of degree 3 or 4, the function $R(x, \sqrt{P(x)})$ can be reduced to the form $R_1(t, \sqrt{at^4 + bt^3 + \cdots + e})$, where $a \neq 0$.

c) A fourth-degree polynomial $ax^4 + bx^3 + \cdots + e$ can be represented as a product $a(x^2 + p_1 x + q_1)(x^2 + p_2 x + q_2)$ and can always be brought into the form $\frac{(M_1+N_1t^2)(M_2+N_2t^2)}{(\gamma t+1)^2}$ by a substitution $x = \frac{\alpha t+\beta}{\gamma t+1}$.

d) A function $R(x, \sqrt{ax^4 + bx^3 + \cdots + e})$ can be reduced to the form

$$R_1\big(t, \sqrt{A(1+m_1t^2)(1+m_2t^2)}\big)$$

by a substitution $x = \frac{\alpha t+\beta}{\gamma t+1}$.

e) A function $R(x, \sqrt{y})$ can be represented as a sum $R_1(x, y) + \frac{R_2(x,y)}{\sqrt{y}}$, where R_1 and R_2 are rational functions.

f) Any rational function can be represented as the sum of even and odd rational functions.

g) If the rational function $R(x)$ is even, it has the form $r(x^2)$; if odd, it has the form $xr(x^2)$, where $r(x)$ is a rational function.

h) Any function $R(x, \sqrt{y})$ can be reduced to the form

$$R_1(x, y) + \frac{R_2(x^2, y)}{\sqrt{y}} + \frac{R_3(x^2, y)}{\sqrt{y}} x.$$

i) Up to a sum of elementary terms, any integral of the form $\int R(x, \sqrt{P(x)})\,\mathrm{d}x$, where $P(x)$ is a polynomial of degree four, can be reduced to an integral

$$\int \frac{r(t^2)\,\mathrm{d}t}{\sqrt{A(1+m_1t^2)(1+m_2t^2)}},$$

where $r(t)$ is a rational function and $A = \pm 1$.

j) If $|m_1| > |m_2| > 0$, one of the substitutions $\sqrt{m_1}t = x$, $\sqrt{m_1}t = \sqrt{1-x^2}$, $\sqrt{m_1}t = \frac{x}{\sqrt{1-x^2}}$, and $\sqrt{m_1}t = \frac{1}{\sqrt{1-x^2}}$ will reduce the integral $\int \frac{r(t^2)\,\mathrm{d}t}{\sqrt{A(1+m_1t^2)(1+m_2t^2)}}$ to the form $\int \frac{\tilde{r}(x^2)\,\mathrm{d}x}{\sqrt{(1-x^2)(1-k^2x^2)}}$, where $0 < k < 1$ and $\tilde{r}$ is a rational function.

k) Derive a formula for lowering the exponents $2n$ and m for the integrals

$$\int \frac{x^{2n}\,\mathrm{d}x}{\sqrt{(1-x^2)(1-k^2x^2)}}, \quad \int \frac{\mathrm{d}x}{(x^2-a)^m \cdot \sqrt{(1-x^2)(1-k^2x^2)}}.$$

l) Any elliptic integral

$$\int R\big(x, \sqrt{P(x)}\big)\,\mathrm{d}x,$$

where P is a fourth-degree polynomial, can be reduced to one of the canonical forms (5.185), (5.186), (5.187), up to a sum of terms consisting of elementary functions.

m) Express the integral $\int \frac{\mathrm{d}x}{\sqrt{1+x^3}}$ in terms of canonical elliptic integrals.

n) Express the primitives of the functions $\frac{1}{\sqrt{\cos 2x}}$ and $\frac{1}{\sqrt{\cos\alpha - \cos x}}$ in terms of elliptic integrals.

6. Using the notation introduced below, find primitives of the following nonelementary special functions, up to a linear function $Ax+B$;

a) $\mathrm{Ei}(x) = \int \frac{\mathrm{e}^x}{x}\,\mathrm{d}x$ (the exponential integral);

b) $\mathrm{Si}(x) = \int \frac{\sin x}{x}\,\mathrm{d}x$ (the sine integral);

c) $\mathrm{Ci}(x) = \int \frac{\cos x}{x}\,\mathrm{d}x$ (the cosine integral);

d) $\mathrm{Shi}(x) = \int \frac{\sinh x}{x}\,\mathrm{d}x$ (the hyperbolic sine integral);

e) $\mathrm{Chi}(x) = \int \frac{\cosh x}{x}\,\mathrm{d}x$ (the hyperbolic cosine integral);

f) $S(x) = \int \sin x^2\,\mathrm{d}x$

g) $C(x) = \int \cos x^2\,\mathrm{d}x$ (the Fresnel integrals);

h) $\Phi(x) = \int \mathrm{e}^{-x^2}\,\mathrm{d}x$ (the Euler–Poisson integral);

i) $\mathrm{li}(x) = \int \frac{\mathrm{d}x}{\ln x}$ (the logarithmic integral).

7. Verify that the following equalities hold, up to a constant:

a) $\mathrm{Ei}(x) = \mathrm{li}(x)$;
b) $\mathrm{Chi}(x) = \frac{1}{2}[\mathrm{Ei}(x) + \mathrm{Ei}(-x)]$;
c) $\mathrm{Shi}(x) = \frac{1}{2}[\mathrm{Ei}(x) - \mathrm{Ei}(-x)]$;
d) $\mathrm{Ei}(ix) = \mathrm{Ci}(x) + i\,\mathrm{Si}(x)$;
e) $\mathrm{e}^{i\pi/4}\Phi(x\mathrm{e}^{-i\pi/4}) = C(x) + iS(x)$.

8. A differential equation of the form

$$\frac{\mathrm{d}y}{\mathrm{d}x} = \frac{f(x)}{g(y)}$$

is called an equation with *variables separable*, since it can be rewritten in the form

$$g(y)\,\mathrm{d}y = f(x)\,\mathrm{d}x,$$

in which the variables x and y are separated. Once this is done, the equation can be solved as

$$\int g(y)\,\mathrm{d}y = \int f(x)\,\mathrm{d}x + c,$$

by computing the corresponding integrals.
Solve the following equations:

a) $2x^3yy' + y^2 = 2$;
b) $xyy' = \sqrt{1+x^2}$;

c) $y' = \cos(y + x)$, setting $u(x) = y(x) + x$;
d) $x^2 y' - \cos 2y = 1$, and exhibit the solution satisfying the condition $y(x) \to 0$ as $x \to +\infty$.
e) $\frac{1}{x} y'(x) = \mathrm{Si}(x)$;
f) $\frac{y'(x)}{\cos x} = C(x)$.

9. A parachutist has jumped from an altitude of 1.5 km and opened the parachute at an altitude of 0.5 km. For how long a time did he fall before opening the parachute? Assume the limiting velocity of fall for a human being in air of normal density is 50 m/s. Solve this problem assuming that the air resistance is proportional to:

a) the velocity;
b) the square of the velocity.

Neglect the variation of pressure with altitude.

10. It is known that the velocity of outflow of water from a small aperture at the bottom of a vessel can be computed quite precisely from the formula $0.6\sqrt{2gH}$, where g is the acceleration of gravity and H the height of the surface of the water above the aperture.
A cylindrical vat is set upright and has an opening in its bottom. Half of the water from the full vat flows out in 5 minutes. How long will it take for all the water to flow out?

11. What shape should a vessel be, given that it is to be a solid of revolution, in order for the surface of the water flowing out of the bottom to fall at a constant rate as water flows out its bottom? (For the initial data, see Exercise 10.)

12. In a workshop with a capacity of 10^4 m^3 fans deliver 10^3 m^3 of fresh air per minute, containing 0.04 % CO_2, and the same amount of air is vented to the outside. At 9:00 AM the workers arrive and after half an hour, the content of CO_2 in the air rises to 0.12 %. Evaluate the carbon dioxide content of the air by 2:00 PM.

Chapter 6
Integration

6.1 Definition of the Integral and Description of the Set of Integrable Functions

6.1.1 The Problem and Introductory Considerations

Suppose a point is moving along the real line, with $s(t)$ being its coordinate at time t and $v(t) = s'(t)$ its velocity at the same instant t. Assume that we know the position $s(t_0)$ of the point at time t_0 and that we receive information on its velocity. Having this information, we wish to compute $s(t)$ for any given value of time $t > t_0$.

If we assume that the velocity $v(t)$ varies continuously, the displacement of the point over a small time interval can be computed approximately as the product $v(\tau)\Delta t$ of the velocity at an arbitrary instant τ belonging to that time interval and the magnitude Δt of the time interval itself. Taking this observation into account, we partition the interval $[t_0, t]$ by marking some times t_i ($i = 0, \dots, n$) so that $t_0 < t_1 < \cdots < t_n = t$ and so that the intervals $[t_{i-1}, t_i]$ are small. Let $\Delta t_i = t_i - t_{i-1}$ and $\tau_i \in [t_{i-1}, t_i]$. Then we have the approximate equality

$$s(t) - s(t_0) \approx \sum_{i=1}^{n} v(\tau_i)\Delta t_i .$$

According to our picture of the situation, this approximate equality will become more precise if we partition the closed interval $[t_0, t]$ into smaller and smaller intervals. Thus we must conclude that in the limit as the length λ of the largest of these intervals tends to zero we shall obtain an exact equality

$$\lim_{\lambda \to 0} \sum_{i=1}^{n} v(\tau_i)\Delta t_i = s(t) - s(t_0). \tag{6.1}$$

This equality is none other than the Newton–Leibniz formula (fundamental theorem of calculus), which is fundamental in all of analysis. It enables us on the one

V.A. Zorich, *Mathematical Analysis I*, Universitext,
DOI 10.1007/978-3-662-48792-1_6

Fig. 6.1

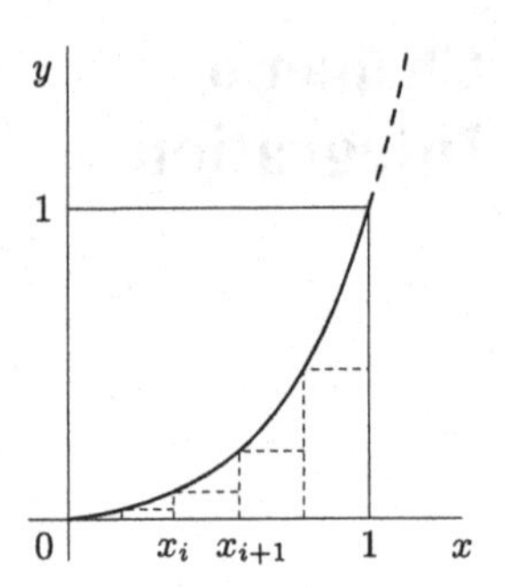

hand to find a primitive $s(t)$ numerically from its derivative $v(t)$, and on the other hand to find the limit of sums $\sum_{i=1}^{n} v(\tau_i)\Delta t_i$ on the left-hand side from a primitive $s(t)$ found by any means whatever.

Such sums, called *Riemann sums*, are encountered in a wide variety of situations.

Let us attempt, for example, following Archimedes, to find the area under the parabola $y = x^2$ above the closed interval $[0, 1]$ (see Fig. 6.1). Without going into detail here as to the meaning of the area of a figure, which we shall take up later, like Archimedes, we shall work by the method of exhausting the figure with simple figures – rectangles, whose areas we know how to compute. After partitioning the closed interval $[0, 1]$ by points $0 = x_0 < x_1 < \cdots < x_n = 1$ into tiny closed intervals $[x_{i-1}, x_i]$, we can obviously compute the required area σ as the sum of the areas of the rectangles shown in the figure:

$$\sigma \approx \sum_{i=1}^{n} x_{i-1}^2 \Delta x_i;$$

here $\Delta x_i = x_i - x_{i-1}$. Setting $f(x) = x^2$ and $\xi_i = x_{i-1}$, we rewrite the formula as

$$\sigma \approx \sum_{i=1}^{n} f(\xi_i)\Delta x_i.$$

In this notation we have, in the limit,

$$\lim_{\lambda\to 0} \sum_{i=1}^{n} f(\xi_i)\Delta x_i = \sigma \tag{6.2}$$

where, as above, λ is the length of the longest interval $[x_{i-1}, x_i]$ in the partition.

Formula (6.2) differs from (6.1) only in the notation. Forgetting for a moment the geometric meaning of $f(\xi_i)\Delta x_i$ and regarding x as time and $f(x)$ as velocity, we find a primitive $F(x)$ for the function $f(x)$ and then, by formula (6.1) we find that $\sigma = F(1) - F(0)$.

In our case $f(x) = x^2$, so that $F(x) = \frac{1}{3}x^3 + c$, and $\sigma = F(1) - F(0) = \frac{1}{3}$.

This is Archimedes' result, which he obtained by a direct computation of the limit in (6.2).

A limit of integral sums is called an *integral*. Thus the Newton–Leibniz formula (6.1) connects the integral with the primitive.

We now turn to precise formulations and the verification of what was obtained above on the heuristic level from general considerations.

6.1.2 Definition of the Riemann Integral

a. Partitions

Definition 1 A *partition* P of a closed interval $[a,b]$, $a<b$, is a finite system of points $x_0,\dots,x_n$ of the interval such that $a=x_0<x_1<\dots<x_n=b$.

The intervals $[x_{i-1},x_i]$, $(i=1,\dots,n)$ are called the *intervals* of the partition P.

The largest of the lengths of the intervals of the partition P, denoted $\lambda(P)$, is called the *mesh* of the partition.

Definition 2 We speak of a *partition with distinguished points* (P,ξ) on the closed interval $[a,b]$ if we have a partition P of $[a,b]$ and a point $\xi_i\in[x_{i-1},x_i]$ has been chosen in each of the intervals of the partition $[x_{i-1},x_i]$ $(i=1,\dots,n)$.

We denote the set of points $(\xi_1,\dots,\xi_n)$ by the single letter ξ.

b. A Base in the Set of Partitions

In the set $\mathcal{P}$ of partitions with distinguished points on a given interval $[a,b]$, we consider the following base $\mathcal{B}=\{B_d\}$. The element B_d, $d>0$, of the base $\mathcal{B}$ consists of all partitions with distinguished points (P,ξ) on $[a,b]$ for which $\lambda(P)<d$.

Let us verify that $\{B_d\}$, $d>0$ is actually a base in $\mathcal{P}$.

First $B_d\neq\varnothing$. In fact, for any number $d>0$, it is obvious that there exists a partition P of $[a,b]$ with mesh $\lambda(P)<d$ (for example, a partition into n congruent closed intervals). But then there also exists a partition (P,ξ) with distinguished points for which $\lambda(P)<d$.

Second, if $d_1>0$, $d_2>0$, and $d=\min\{d_1,d_2\}$, it is obvious that $B_{d_1}\cap B_{d_2}=B_d\in\mathcal{B}$.

Hence $\mathcal{B}=\{B_d\}$ is indeed a base in $\mathcal{P}$.

c. Riemann Sums

Definition 3 If a function f is defined on the closed interval $[a,b]$ and (P,ξ) is a partition with distinguished points on this closed interval, the sum

$$\sigma(f;P,\xi):=\sum_{i=1}^{n} f(\xi_i)\Delta x_i, \tag{6.3}$$

where $\Delta x_i = x_i - x_{i-1}$, is the *Riemann sum* of the function f corresponding to the partition (P, ξ) with distinguished points on $[a, b]$.

Thus, when the function f is fixed, the Riemann sum $\sigma(f; P, \xi)$ is a function $\Phi(p) = \sigma(f; p)$ on the set $\mathcal{P}$ of all partitions $p = (P, \xi)$ with distinguished points on the closed interval $[a, b]$.

Since there is a base $\mathcal{B}$ in $\mathcal{P}$, one can ask about the limit of the function $\Phi(p)$ over that base.

d. The Riemann Integral

Let f be a function defined on a closed interval $[a, b]$.

Definition 4 The number I is the *Riemann integral* of the function f on the closed interval $[a, b]$ if for every $\varepsilon > 0$ there exists $\delta > 0$ such that

$$\left| I - \sum_{i=1}^{n} f(\xi_i)\Delta x_i \right| < \varepsilon$$

for any partition (P, ξ) with distinguished points on $[a, b]$ whose mesh $\lambda(P)$ is less than δ.

Since the partitions $p = (P, \xi)$ for which $\lambda(P) < \delta$ form the element B_δ of the base $\mathcal{B}$ introduced above in the set $\mathcal{P}$ of partitions with distinguished points, Definition 4 is equivalent to the statement

$$I = \lim_{\mathcal{B}} \Phi(p),$$

that is, the integral I is the limit over $\mathcal{B}$ of the Riemann sums of the function f corresponding to partitions with distinguished points on $[a, b]$.

It is natural to denote the base $\mathcal{B}$ by $\lambda(P) \to 0$, and then the definition of the integral can be rewritten as

$$I = \lim_{\lambda(P)\to 0} \sum_{i=1}^{n} f(\xi_i)\Delta x_i. \tag{6.4}$$

The integral of $f(x)$ over $[a, b]$ is denoted

$$\int_a^b f(x)\,\mathrm{d}x,$$

in which the numbers a and b are called respectively the *lower* and *upper limits of integration*. The function f is called the *integrand*, $f(x)\,\mathrm{d}x$ is called the *differential form*, and x is the *variable of integration*. Thus

$$\boxed{\int_a^b f(x)\,\mathrm{d}x := \lim_{\lambda(P)\to 0} \sum_{i=1}^{n} f(\xi_i)\Delta x_i.} \tag{6.5}$$

Definition 5 A function f is *Riemann integrable* on the closed interval $[a, b]$ if the limit of the Riemann sums in (6.5) exists as $\lambda(P) \to 0$ (that is, the Riemann integral of f is defined).

The set of Riemann-integrable functions on a closed interval $[a, b]$ will be denoted $\mathcal{R}[a, b]$.

Since we shall not be considering any integrals except the Riemann integral for a while, we shall agree for the sake of brevity to say simply "integral" and "integrable function" instead of "Riemann integral" and "Riemann-integrable function".

6.1.3 The Set of Integrable Functions

By the definition of the integral (Definition 4) and its reformulation in the forms (6.4) and (6.5), an integral is the limit of a certain special function $\Phi(p) = \sigma(f; P, \xi)$, the Riemann sum, defined on the set $\mathcal{P}$ of partitions $p = (P, \xi)$ with distinguished points on $[a, b]$. This limit is taken with respect to the base $\mathcal{B}$ in $\mathcal{P}$ that we have denoted $\lambda(P) \to 0$.

Thus the integrability or nonintegrability of a function f on $[a, b]$ depends on the existence of this limit.

By the Cauchy criterion, this limit exists if and only if for every $\varepsilon > 0$ there exists an element $B_\delta \in \mathcal{B}$ in the base such that

$$\left|\Phi(p') - \Phi(p'')\right| < \varepsilon$$

for any two points p', p'' in B_δ.

In more detailed notation, what has just been said means that for any $\varepsilon > 0$ there exists $\delta > 0$ such that

$$\left|\sigma(f; P', \xi') - \sigma(f; P'', \xi'')\right| < \varepsilon$$

or, what is the same,

$$\left|\sum_{i=1}^{n'} f(\xi_i')\Delta x_i' - \sum_{i=1}^{n''} f(\xi_i'')\Delta x_i''\right| < \varepsilon \tag{6.6}$$

for any partitions (P', ξ') and (P'', ξ'') with distinguished points on the interval $[a, b]$ with $\lambda(P') < \delta$ and $\lambda(P'') < \delta$.

We shall use the Cauchy criterion just stated to find first a simple necessary condition, then a sufficient condition for Riemann integrability.

a. A Necessary Condition for Integrability

Proposition 1 *A necessary condition for a function f defined on a closed interval $[a, b]$ to be Riemann integrable on $[a, b]$ is that f be bounded on $[a, b]$.*

In short,

$$\big(f \in \mathcal{R}[a,b]\big) \Rightarrow \big(f \text{ is bounded on } [a,b]\big).$$

Proof If f is not bounded on $[a,b]$, then for any partition P of $[a,b]$ the function f is unbounded on at least one of the intervals $[x_{i-1},x_i]$ of P. This means that, by choosing the point $\xi_i \in [x_{i-1},x_i]$ in different ways, we can make the quantity $|f(\xi_i)\Delta x_i|$ as large as desired. But then the Riemann sum $\sigma(f;P,\xi) = \sum_{i=1}^{n} f(\xi_i)\Delta x_i$ can also be made as large as desired in absolute value by changing only the point ξ_i in this interval.

It is clear that there can be no possibility of a finite limit for the Riemann sums in such a case. That was in any case clear from the Cauchy criterion, since relation (6.6) cannot hold in that case, even for arbitrarily fine partitions. □

As we shall see, the necessary condition just obtained is far from being both necessary and sufficient for integrability. However, it does enable us to restrict consideration to bounded functions.

b. A Sufficient Condition for Integrability and the Most Important Classes of Integrable Functions

We begin with some notation and remarks that will be used in the explanation to follow.

We agree that when a partition P

$$a = x_0 < x_1 < \cdots < x_n = b$$

is given on the interval $[a,b]$, we shall use the symbol Δ_i to denote the interval $[x_{i-1},x_i]$ along with Δx_i as a notation for the difference $x_i - x_{i-1}$.

If a partition $\tilde{P}$ of the closed interval $[a,b]$ is obtained from the partition P by the adjunction of new points to P, we call $\tilde{P}$ a *refinement* of P.

When a refinement $\widetilde{P}$ of a partition P is constructed, some (perhaps all) of the closed intervals $\Delta_i = [x_{i-1},x_i]$ of the partition P themselves undergo partitioning: $x_{i-1} = x_{i0} < \cdots < x_{in_i} = x_i$. In that connection, it will be useful for us to label the points of $\tilde{P}$ by double indices. In the notation x_{ij} the first index means that $x_{ij} \in \Delta_i$, and the second index is the ordinal number of the point on the closed interval Δ_i. It is now natural to set $\Delta x_{ij} := x_{ij} - x_{ij-1}$ and $\Delta_{ij} := [x_{ij-1}, x_{ij}]$. Thus $\Delta x_i = \Delta x_{i1} + \cdots + \Delta x_{in_i}$.

As an example of a partition that is a refinement of both the partition P' and P'' one can take $\widetilde{P} = P' \cup P''$, obtained as the union of the points of the two partitions P' and P''.

We recall finally that, as before, $\omega(f;E)$ denotes the oscillation of the function f on the set E, that is

$$\omega(f;E) := \sup_{x',x'' \in E} \big|f(x') - f(x'')\big|.$$

In particular, $\omega(f; \Delta_i)$ is the oscillation of f on the closed interval Δ_i. This oscillation is necessarily finite if f is a bounded function.

We now state and prove a sufficient condition for integrability.

Proposition 2 *A sufficient condition for a bounded function f to be integrable on a closed interval $[a, b]$ is that for every $\varepsilon > 0$ there exists a number $\delta > 0$ such that*

$$\sum_{i=1}^{n} \omega(f; \Delta_i)\Delta x_i < \varepsilon$$

for any partition P of $[a, b]$ with mesh $\lambda(P) < \delta$.

Proof Let P be a partition of $[a, b]$ and $\widetilde{P}$ a refinement of P. Let us estimate the difference between the Riemann sums $\sigma(f; \widetilde{P}, \tilde{\xi}) - \sigma(f; P, \xi)$. Using the notation introduced above, we can write

$$\begin{aligned}
\left|\sigma(f; \tilde{P}, \tilde{\xi}) - \sigma(f; P, \xi)\right| &= \left|\sum_{i=1}^{n}\sum_{j=1}^{n_i} f(\xi_{ij})\Delta x_{ij} - \sum_{i=1}^{n} f(\xi_i)\Delta x_i\right| = \\
&= \left|\sum_{i=1}^{n}\sum_{j=1}^{n_i} f(\xi_{ij})\Delta x_{ij} - \sum_{i=1}^{n}\sum_{j=1}^{n_i} f(\xi_i)\Delta x_{ij}\right| = \\
&= \left|\sum_{i=1}^{n}\sum_{j=1}^{n_i}\bigl(f(\xi_{ij}) - f(\xi_i)\bigr)\Delta x_{ij}\right| \le \\
&\le \sum_{i=1}^{n}\sum_{j=1}^{n_i}\left|f(\xi_{ij}) - f(\xi_i)\right|\Delta x_{ij} \le \\
&\le \sum_{i=1}^{n}\sum_{j=1}^{n_i}\omega(f; \Delta_i)\Delta x_{ij} = \sum_{i=1}^{n}\omega(f; \Delta_i)\Delta x_i.
\end{aligned}$$

In this computation we have used the relation $\Delta x_i = \sum_{j=1}^{n_i} \Delta x_{ij}$ and the inequality $|f(\xi_{ij}) - f(\xi_i)| \le \omega(f; \Delta_i)$, which holds because $\xi_{ij} \in \Delta_{ij} \subset \Delta_i$ and $\xi_i \in \Delta_i$.

It follows from the estimate for the difference of the Riemann sums that if the function satisfies the sufficient condition given in the statement of Proposition 2, then for any $\varepsilon > 0$ we can find $\delta > 0$ such that

$$\left|\sigma(f; \tilde{P}, \tilde{\xi}) - \sigma(f; P, \xi)\right| < \frac{\varepsilon}{2}$$

for any partition P of $[a, b]$ with mesh $\lambda(P) < \delta$, any refinement $\tilde{P}$ of P, and any choice of the sets of distinguished points ξ and $\tilde{\xi}$.

Now if (P', ξ') and (P'', ξ'') are arbitrary partitions with distinguished points on $[a, b]$ whose meshes satisfy $\lambda(P') < \delta$ and $\lambda(P'') < \delta$, then, by what has just been proved, the partition $\widetilde{P} = P' \cup P''$, which is a refinement of both of them, must satisfy

$$\left|\sigma(f; \tilde{P}, \tilde{\xi}) - \sigma\left(f; P', \xi'\right)\right| < \frac{\varepsilon}{2},$$

$$\left|\sigma(f; \tilde{P}, \tilde{\xi}) - \sigma\left(f; P'', \xi''\right)\right| < \frac{\varepsilon}{2}.$$

It follows that

$$\left|\sigma\left(f; P', \xi'\right) - \sigma\left(f; P'', \xi''\right)\right| < \varepsilon,$$

provided $\lambda(P') < \delta$ and $\lambda(P'') < \delta$. Therefore, by the Cauchy criterion, the limit of the Riemann sums exists:

$$\lim_{\lambda(P)\to 0} \sum_{i=1}^{n} f(\xi_i)\Delta x_i,$$

that is $f \in \mathcal{R}[a, b]$. □

Corollary 1 $(f \in C[a, b]) \Rightarrow (f \in \mathcal{R}[a, b])$, *that is, every continuous function on a closed interval is integrable on that closed interval.*

Proof If a function is continuous on a closed interval, it is bounded there, so that the necessary condition for integrability is satisfied in this case. But a continuous function on a closed interval is uniformly continuous on that interval. Therefore, for every $\varepsilon > 0$ there exists $\delta > 0$ such that $\omega(f; \Delta) < \frac{\varepsilon}{b-a}$ on any closed interval $\Delta \subset [a, b]$ of length less than δ. Then for any partition P with mesh $\lambda(P) < \delta$ we have

$$\sum_{i=1}^{n} \omega(f; \Delta_i)\Delta x_i < \frac{\varepsilon}{b-a} \sum_{i=1}^{n} \Delta x_i = \frac{\varepsilon}{b-a}(b-a) = \varepsilon.$$

By Proposition 2, we can now conclude that $f \in \mathcal{R}[a, b]$. □

Corollary 2 *If a bounded function f on a closed interval $[a, b]$ is continuous everywhere except at a finite set of points, then $f \in \mathcal{R}[a, b]$.*

Proof Let $\omega(f; [a, b]) \leq C < \infty$, and suppose f has k points of discontinuity on $[a, b]$. We shall verify that the sufficient condition for integrability of the function f is satisfied.

For a given $\varepsilon > 0$ we choose the number $\delta_1 = \frac{\varepsilon}{8C\cdot k}$ and construct the δ_1-neighborhood of each of the k points of discontinuity of f on $[a, b]$. The complement of the union of these neighborhoods in $[a, b]$ consists of a finite number of closed intervals, on each of which f is continuous and hence uniformly continuous. Since the number of these intervals is finite, given $\varepsilon > 0$ there exists $\delta_2 > 0$ such that on each interval Δ whose length is less than δ_2 and which is entirely contained in one of the closed intervals just mentioned, on which f is continuous, we have $\omega(f; \Delta) < \frac{\varepsilon}{2(b-a)}$. We now choose $\delta = \min\{\delta_1, \delta_2\}$.

Let P be an arbitrary partition of $[a,b]$ for which $\lambda(P) < \delta$. We break the sum $\sum_{i=1}^{n} \omega(f;\Delta_i)\Delta x_i$ corresponding to the partition P into two parts:

$$\sum_{i=1}^{n} \omega(f;\Delta_i)\Delta x_i = \sum{}' \omega(f;\Delta_i)\Delta x_i + \sum{}'' \omega(f;\Delta_i)\Delta x_i .$$

The sum $\sum{}'$ contains the terms corresponding to intervals Δ_i of the partition having no points in common with any of the δ_1-neighborhoods of the points of discontinuity. For these intervals Δ_i we have $\omega(f;\Delta_i) < \frac{\varepsilon}{2(b-a)}$, and so

$$\sum{}' \omega(f;\Delta_i)\Delta x_i < \frac{\varepsilon}{2(b-a)} \sum{}' \Delta x_i \le \frac{\varepsilon}{2(b-a)}(b-a) = \frac{\varepsilon}{2}.$$

The sum of the lengths of the remaining intervals of the partition P, as one can easily see, is at most $(\delta + 2\delta_1 + \delta)k \le 4\frac{\varepsilon}{8C\cdot k}\cdot k = \frac{\varepsilon}{2C}$, and therefore

$$\sum{}'' \omega(f;\Delta_i)\Delta x_i \le C \sum{}'' \Delta x_i < C\cdot \frac{\varepsilon}{2C} = \frac{\varepsilon}{2}.$$

Thus we find that for $\lambda(P) < \delta$,

$$\sum_{i=1}^{n} \omega(f;\Delta_i)\Delta x_i < \varepsilon,$$

that is, the sufficient condition for integrability holds, and so $f \in \mathcal{R}[a,b]$. □

Corollary 3 *A monotonic function on a closed interval is integrable on that interval.*

Proof It follows from the monotonicity of f on $[a,b]$ that $\omega(f;[a,b]) = |f(b) - f(a)|$. Suppose $\varepsilon > 0$ is given. We set $\delta = \frac{\varepsilon}{|f(b)-f(a)|}$. We assume that $f(b) - f(a) \ne 0$, since otherwise f is constant, and there is no doubt as to its integrability. Let P be an arbitrary partition of $[a,b]$ with mesh $\lambda(P) < \delta$.

Then, taking account of the monotonicity of f, we have

$$\sum_{i=1}^{n} \omega(f;\Delta_i)\Delta x_i < \delta \sum_{i=1}^{n} \omega(f;\Delta_i) = \delta \sum_{i=1}^{n} \bigl|f(x_i) - f(x_{i-1})\bigr| =$$

$$= \delta \left| \sum_{i=1}^{n} \bigl(f(x_i) - f(x_{i-1})\bigr) \right| = \delta \bigl|f(b) - f(a)\bigr| = \varepsilon.$$

Thus f satisfies the sufficient condition for integrability, and therefore $f \in \mathcal{R}[a,b]$. □

A monotonic function may have a (countably) infinite set of discontinuities on a closed interval. For example, the function defined by the relations

$$f(x)=\begin{cases}1-\frac{1}{2^{n-1}} & \text{for } 1-\frac{1}{2^{n-1}}\le x<1-\frac{1}{2^n},\ n\in\mathbb{N},\\ 1 & \text{for } x=1\end{cases}$$

on $[0, 1]$ is nondecreasing and has a discontinuity at every point of the form $1-\frac{1}{2^n}$, $n \in \mathbb{N}$.

Remark We note that, although we are dealing at the moment with real-valued functions on an interval, we have made no use of the assumption that the functions are real-valued rather than complex-valued or even vector-valued functions of a point of the closed interval $[a, b]$, either in the definition of the integral or in the propositions proved above, except Corollary 3.

On the other hand, the concept of upper and lower Riemann sums, to which we now turn, is specific to real-valued functions.

Definition 6 Let $f : [a, b] \to \mathbb{R}$ be a real-valued function that is defined and bounded on the closed interval $[a, b]$, let P be a partition of $[a, b]$, and let Δ_i $(i = 1, \dots, n)$ be the intervals of the partition P. Let $m_i = \inf_{x\in\Delta_i} f(x)$ and $M_i = \sup_{x\in\Delta_i} f(x)$ $(i = 1, \dots, n)$.

The sums

$$s(f; P) := \sum_{i=1}^{n} m_i \Delta x_i$$

and

$$S(f; P) := \sum_{i=1}^{n} M_i \Delta x_i$$

are called respectively the *lower* and *upper Riemann sums* of the function f on the interval $[a, b]$ corresponding to the partition P of that interval.[1] The sums $s(f; P)$ and $S(f; P)$ are also called the lower and upper *Darboux* sums corresponding to the partition P of $[a, b]$.

If (P, ξ) is an arbitrary partition with distinguished points on $[a, b]$, then obviously

$$s(f; P) \le \sigma(f; P, \xi) \le S(f; P). \tag{6.7}$$

Lemma 1

$$s(f; P) = \inf_{\xi} \sigma(f; P, \xi),$$

$$S(f; P) = \sup_{\xi} \sigma(f; P, \xi).$$

[1] The term "Riemann sum" here is not quite accurate, since m_i and M_i are not always values of the function f at some point $\xi_i \in \Delta_i$.

Proof Let us verify, for example, that the upper Darboux sum corresponding to a partition P of the closed interval $[a,b]$ is the least upper bound of the Riemann sums corresponding to the partitions with distinguished points (P,ξ), the supremum being taken over all sets $\xi=(\xi_1,\dots,\xi_n)$ of distinguished points.

In view of (6.7), it suffices to prove that for any $\varepsilon>0$ there is a set $\overline{\xi}$ of distinguished points such that

$$S(f;P)<\sigma(f;P,\overline{\xi})+\varepsilon. \tag{6.8}$$

By definition of the numbers M_i, for each $i\in\{1,\dots,n\}$ there is a point $\overline{\xi}_i\in\Delta_i$ at which $M_i<f(\overline{\xi}_i)+\frac{\varepsilon}{b-a}$. Let $\overline{\xi}=(\overline{\xi}_1,\dots,\overline{\xi}_n)$. Then

$$\sum_{i=1}^{n}M_i\Delta x_i<\sum_{i=1}^{n}\left(f(\overline{\xi}_i)+\frac{\varepsilon}{b-a}\right)\Delta x_i=\sum_{i=1}^{n}f(\overline{\xi}_i)\Delta x_i+\varepsilon,$$

which completes the proof of the second assertion of the lemma. The first assertion is verified similarly. □

From this lemma and inequality (6.7), taking account of the definition of the Riemann integral, we deduce the following proposition.

Proposition 3 *A bounded real-valued function $f:[a,b]\to\mathbb{R}$ is Riemann-integrable on $[a,b]$ if and only if the following limits exist and are equal to each other:*

$$\underline{I}=\lim_{\lambda(P)\to 0}s(f;P),\qquad \overline{I}=\lim_{\lambda(P)\to 0}S(f;P). \tag{6.9}$$

When this happens, the common value $I=\underline{I}=\overline{I}$ is the integral

$$\int_a^b f(x)\,\mathrm{d}x.$$

Proof Indeed, if the limits (6.9) exist and are equal, we conclude by the properties of limits and by (6.7) that the Riemann sums have a limit and that

$$\underline{I}=\lim_{\lambda(P)\to 0}\sigma(f;P,\xi)=\overline{I}.$$

On the other hand, if $f\in\mathcal{R}[a,b]$, that is, the limit

$$\lim_{\lambda(P)\to 0}\sigma(f;P,\xi)=I$$

exists, we conclude from (6.7) and (6.8) that the limit $\lim_{\lambda(P)\to 0}S(f;P)=\overline{I}$ exists and $\overline{I}=I$.

Similarly one can verify that $\lim_{\lambda(P)\to 0}s(f;P)=\underline{I}=I$. □

As a corollary of Proposition 3, we obtain the following sharpening of Proposition 2.

Proposition 2′ *A necessary and sufficient condition for a function $f : [a, b] \to \mathbb{R}$ defined on a closed interval $[a, b]$ to be Riemann integrable on $[a, b]$ is the following relation*:

$$\lim_{\lambda(P)\to 0} \sum_{i=1}^{n} \omega(f; \Delta_i)\Delta x_i = 0. \tag{6.10}$$

Proof Taking account of Proposition 2, we have only to verify that condition (6.10) is necessary for f to be integrable.

We remark that $\omega(f; \Delta_i) = M_i - m_i$, and therefore

$$\sum_{i=1}^{n} \omega(f; \Delta_i)\Delta x_i = \sum_{i=1}^{n} (M_i - m_i)\Delta x_i = S(f; P) - s(f; P),$$

and (6.10) now follows from Proposition 3 if $f \in \mathcal{R}[a, b]$. □

c. The Vector Space $\mathcal{R}[a, b]$

Many operations can be performed on integrable functions without going outside the class of integrable functions $\mathcal{R}[a, b]$.

Proposition 4 *If* $f, g \in \mathcal{R}[a, b]$, *then*

a) $(f + g) \in \mathcal{R}[a, b]$;
b) $(\alpha f) \in \mathcal{R}[a, b]$, *where* α *is a numerical coefficient*;
c) $|f| \in \mathcal{R}[a, b]$;
d) $f|_{[c,d]} \in \mathcal{R}[c, d]$ *if* $[c, d] \subset [a, b]$;
e) $(f \cdot g) \in \mathcal{R}[a, b]$.

We are considering only real-valued functions at the moment, but it is worthwhile to note that properties a), b), c), and d) turn out to be valid for complex-valued and vector-valued functions. For vector-valued functions, in general, the product $f \cdot g$ is not defined, so that property e) is not considered for them. However, this property continues to hold for complex-valued functions.

We now turn to the proof of Proposition 4.

Proof a) This assertion is obvious since

$$\sum_{i=1}^{n} (f + g)(\xi_i)\Delta x_i = \sum_{i=1}^{n} f(\xi_i)\Delta x_i + \sum_{i=1}^{n} g(\xi_i)\Delta x_i.$$

b) This assertion is obvious, since

$$\sum_{i=1}^{n} (\alpha f)(\xi_i)\Delta x_i = \alpha \sum_{i=1}^{n} f(\xi_i)\Delta x_i.$$

c) Since $\omega(|f|; E) \le \omega(f; E)$, we can write

$$\sum_{i=1}^{n} \omega(|f|; \Delta_i)\Delta x_i \le \sum_{i=1}^{n} \omega(f; \Delta_i)\Delta x_i,$$

and conclude by Proposition 2 that $(f \in \mathcal{R}[a,b]) \Rightarrow (|f| \in \mathcal{R}[a,b])$.

d) We want to verify that the restriction $f|_{[c,d]}$ to $[c,d]$ of a function f that is integrable on the closed interval $[a,b]$ is integrable on $[c,d]$ if $[c,d] \subset [a,b]$. Let π be a partition of $[c,d]$. By adding points to π, we extend it to a partition P of the closed interval $[a,b]$ so as to have $\lambda(P) \le \lambda(\pi)$. It is clear that one can always do this.

We can then write

$$\sum\nolimits_{\pi} \omega(f|_{[c,d]}; \Delta_i)\Delta x_i \le \sum\nolimits_{P} \omega(f; \Delta_i)\Delta x_i,$$

where $\sum_{\pi}$ is the sum over all the intervals of the partition π and $\sum_P$ the sum over all the intervals of P.

By construction, as $\lambda(\pi) \to 0$ we have $\lambda(P) \to 0$ also, and so by Proposition 2′ we conclude from this inequality that $(f \in \mathcal{R}[a,b]) \Rightarrow (f \in \mathcal{R}[c,d])$ if $[c,d] \subset [a,b]$.

e) We first verify that if $f \in \mathcal{R}[a,b]$, then $f^2 \in \mathcal{R}[a,b]$.

If $f \in \mathcal{R}[a,b]$, then f is bounded on $[a,b]$. Let $|f(x)| \le C < \infty$ on $[a,b]$. Then

$$\left|f^2(x_1) - f^2(x_2)\right| = \left|\bigl(f(x_1) + f(x_2)\bigr) \cdot \bigl(f(x_1) - f(x_2)\bigr)\right| \le 2C\left|f(x_1) - f(x_2)\right|,$$

and therefore $\omega(f^2; E) \le 2C\omega(f; E)$ if $E \subset [a,b]$. Hence

$$\sum_{i=1}^{n} \omega(f^2; \Delta_i)\Delta x_i \le 2C\sum_{i=1}^{n} \omega(f; \Delta_i)\Delta x_i,$$

from which we conclude by Proposition 2′ that

$$\bigl(f \in \mathcal{R}[a,b]\bigr) \Rightarrow \bigl(f^2 \in \mathcal{R}[a,b]\bigr).$$

We now turn to the general case. We write the identity

$$(f \cdot g)(x) = \frac{1}{4}\bigl[(f+g)^2(x) - (f-g)^2(x)\bigr].$$

From this identity and the result just proved, together with a) and b), which have already been proved, we conclude that

$$\bigl(f \in \mathcal{R}[a,b]\bigr) \wedge \bigl(g \in \mathcal{R}[a,b]\bigr) \Rightarrow \bigl(f \cdot g \in \mathcal{R}[a,b]\bigr). \qquad \square$$

You already know what a vector space is from your study of algebra. The real-valued functions defined on a set can be added and multiplied by a real number, both operations being performed pointwise, and the result is another real-valued function on the same set. If functions are regarded as vectors, one can verify that all the axioms of a vector space over the field of real numbers hold, and the set of real-valued functions is a vector space with respect to the operations of pointwise addition and multiplication by real numbers.

In parts a) and b) of Proposition 4 it was asserted that addition of integrable functions and multiplication of an integrable function by a number do not lead outside the class $\mathcal{R}[a,b]$ of integrable functions. Thus $\mathcal{R}[a,b]$ is itself a vector space – a subspace of the vector space of real-valued functions defined on the closed interval $[a,b]$.

d. Lebesgue's Criterion for Riemann Integrability of a Function

In conclusion we present, without proof for the time being, a theorem of Lebesgue giving an intrinsic description of a Riemann-integrable function.

To do this, we introduce the following concept, which is useful in its own right.

Definition 7 A set $E \subset \mathbb{R}$ *has measure zero* or *is of measure zero* (in the sense of Lebesgue) if for every number $\varepsilon > 0$ there exists a covering of the set E by an at most countable system $\{I_k\}$ of intervals, the sum of whose lengths $\sum_{k=1}^{\infty} |I_k|$ is at most ε.

Since the series $\sum_{k=1}^{\infty} |I_k|$ converges absolutely, the order of summation of the lengths of the intervals of the covering does not affect the sum (see Proposition 4 of Sect. 5.5.2), so that this definition is unambiguous.

Lemma 2 a) *A single point and a finite number of points are sets of measure zero.*

b) *The union of a finite or countable number of sets of measure zero is a set of measure zero.*

c) *A subset of a set of measure zero is itself of measure zero.*

d) *A closed interval $[a,b]$ with $a < b$ is not a set of measure zero.*

Proof a) A point can be covered by one interval of length less than any preassigned number $\varepsilon > 0$; therefore a point is a set of measure zero. The rest of a) then follows from b).

b) Let $E = \bigcup_n E^n$ be an at most countable union of sets E^n of measure zero. Given $\varepsilon > 0$, for each E^n we construct a covering $\{I_k^n\}$ of E^n such that $\sum_k |I_k^n| < \frac{\varepsilon}{2^n}$.

Since the union of an at most countable collection of at most countably many sets is itself at most countable, the intervals I_k^n, $k, n \in \mathbb{N}$, form an at most countable covering of the set E, and

$$\sum_{n,k} |I_k^n| < \frac{\varepsilon}{2} + \frac{\varepsilon}{2^2} + \cdots + \frac{\varepsilon}{2^n} + \cdots = \varepsilon.$$

The order of summation $\sum_{n,k} |I_k^n|$ on the indices n and k is of no importance, since the series converges to the same sum for any order of summation if it converges in even one ordering. Such is the case here, since any partial sums of the series are bounded above by ε.

Thus E is a set of measure zero in the sense of Lebesgue.

c) This statement obviously follows immediately from the definition of a set of measure zero and the definition of a covering.

d) Since every covering of a closed interval by open intervals contains a finite covering, the sum of the lengths of which obviously does not exceed the sum of the lengths of the intervals in the original covering, it suffices to verify that the sum of the lengths of open intervals forming a finite covering of a closed interval $[a, b]$ is not less than the length $b - a$ of that closed interval.

We shall carry out an induction on the number of intervals in the covering.

For $n = 1$, that is, when the closed interval $[a, b]$ is contained in one open interval (α, β), it is obvious that $\alpha < a < b < \beta$ and $\beta - \alpha > b - a$.

Suppose the statement is proved up to index $k \in \mathbb{N}$ inclusive. Consider a covering consisting of $k + 1$ open intervals. We take an interval (α_1, α_2) containing the point a. If $\alpha_2 \geq b$, then $\alpha_2 - \alpha_1 > b - a$, and the result is proved. If $a < \alpha_2 < b$, the closed interval $[\alpha_2, b]$ is covered by a system of at most k intervals, the sum of whose lengths, by the induction assumption, must be at least $b - \alpha_2$. But

$$b - a = (b - \alpha_2) + \alpha_2 - a < (b - \alpha_2) + (\alpha_2 - \alpha_1),$$

and so the sum of the lengths of all the intervals of the original covering of the closed interval $[a, b]$ was greater than its length $b - a$. □

It is interesting to note that by a) and b) of Lemma 2 the set $\mathbb{Q}$ of rational points on the real line $\mathbb{R}$ is a set of measure zero, which seems rather surprising at first sight, upon comparison with part d) of the same lemma.

Definition 8 If a property holds at all points of a set X except possibly the points of a set of measure zero, we say that this property holds *almost everywhere* on X or *at almost every point* of X.

We now state Lebesgue's criterion for integrability.

Theorem *A function defined on a closed interval is Riemann integrable on that interval if and only if it is bounded and continuous at almost every point.*

Thus,

$$\big(f \in \mathcal{R}[a, b]\big) \Leftrightarrow \big(f \text{ is bounded on } [a, b]\big) \wedge$$
$$\wedge \big(f \text{ is continuous almost everywhere on } [a, b]\big).$$

It is obvious that one can easily derive Corollaries 1, 2, and 3 and Proposition 4 from the Lebesgue criterion and the properties of sets of measure zero proved in Lemma 2.

We shall not prove the Lebesgue criterion here, since we do not need it to work with the rather regular functions we shall be dealing with for the present. However, the essential ideas involved in the Lebesgue criterion can be explained immediately.

Proposition 2′ contained a criterion for integrability expressed by relation (6.10). The sum $\sum_{i=1}^{n} \omega(f;\Delta_i)\Delta x_i$ can be small on the one hand because of the factors $\omega(f;\Delta_i)$, which are small in small neighborhoods of points of continuity of the function. But if some of the closed intervals Δ_i contain points of discontinuity of the function,then $\omega(f;\Delta_i)$ does not tend to zero for these points, no matter how fine we make the partition P of the closed interval $[a,b]$. However, $\omega(f;\Delta_i) \le \omega(f;[a,b]) < \infty$ since f is bounded on $[a,b]$. Hence the sum of the terms containing points of discontinuity will also be small if the sum of the lengths of the intervals of the partition that cover the set of points of discontinuity is small; more precisely, if the increase in the oscillation of the function on some intervals of the partition is compensated for by the smallness of the total lengths of these intervals.

A precise realization and formulation of these observations amounts to the Lebesgue criterion.

We now give two classical examples to clarify the property of Riemann integrability for a function.

Example 1 The Dirichlet function

$$D(x) = \begin{cases} 1 & \text{for } x \in \mathbb{Q}, \\ 0 & \text{for } x \in \mathbb{R}\setminus\mathbb{Q}, \end{cases}$$

on the interval $[0,1]$ is not integrable on that interval, since for any partition P of $[0,1]$ one can find in each interval Δ_i of the partition both a rational point ξ_i' and an irrational point ξ_i''. Then

$$\sigma(f;P,\xi') = \sum_{i=1}^{n} 1 \cdot \Delta x_i = 1,$$

while

$$\sigma(f;P,\xi'') = \sum_{i=1}^{n} 0 \cdot \Delta x_i = 0.$$

Thus the Riemann sums of the function $\mathcal{D}(x)$ cannot have a limit as $\lambda(P) \to 0$.

From the point of view of the Lebesgue criterion the nonintegrability of the Dirichlet function is also obvious, since $\mathcal{D}(x)$ is discontinuous at every point of $[0,1]$ which, as was shown in Lemma 2, is not a set of measure zero.

Example 2 Consider the Riemann function

$$\mathcal{R}(x) = \begin{cases} \frac{1}{n}, & \text{if } x \in \mathbb{Q} \text{ and } x = \frac{m}{n} \text{ is in lowest terms, } n \in \mathbb{N}, \\ 0, & \text{if } x \in \mathbb{R}\setminus\mathbb{Q}. \end{cases}$$

We have already studied this function in Sect. 4.1.2, and we know that $\mathcal{R}(x)$ is continuous at all irrational points and discontinuous at all rational points except 0. Thus the set of points of discontinuity of $\mathcal{R}(x)$ is countable and hence has measure zero. By the Lebesgue criterion, $\mathcal{R}(x)$ is Riemann integrable on any interval $[a, b] \subset \mathbb{R}$, despite there being a discontinuity of this function in every interval of every partition of the interval of integration.

Example 3 Now let us consider a less classical problem and example.

Let $f : [a, b] \to \mathbb{R}$ be a function that is integrable on $[a, b]$, assuming values in the interval $[c, d]$ on which a continuous function $g : [c, d] \to \mathbb{R}$ is defined. Then the composition $g \circ f : [a, b] \to \mathbb{R}$ is obviously defined and continuous at all the points of $[a, b]$ where f is continuous. By the Lebesgue criterion, it follows that $(g \circ f) \in \mathcal{R}[a, b]$.

We shall now show that the composition of two arbitrary integrable functions is not always integrable.

Consider the function $g(x) = |\operatorname{sgn}|(x)$. This function equals 1 for $x \neq 0$ and 0 for $x = 0$. By inspection, we can verify that if we take, say the Riemann function f on the closed interval $[1, 2]$, then the composition $(g \circ f)(x)$ on that interval is precisely the Dirichlet function $\mathcal{D}(x)$. Thus the presence of even one discontinuity of the function $g(x)$ has led to nonintegrability of the composition $g \circ f$.

6.1.4 Problems and Exercises

1. *The theorem of Darboux.*

a) Let $s(f; P)$ and $S(f; P)$ be the lower and upper Darboux sums of a real-valued function f defined and bounded on the closed interval $[a, b]$ and corresponding to a partition P of that interval. Show that

$$s(f; P_1) \leq S(f; P_2)$$

for any two partitions P_1 and P_2 of $[a, b]$.

b) Suppose the partition $\tilde{P}$ is a refinement of the partition P of the interval $[a, b]$, and let $\Delta_{i_1}, \dots, \Delta_{i_k}$ be the intervals of the partition P that contain points of the partition $\widetilde{P}$ that do not belong to P. Show that the following estimates hold:

$$0 \leq S(f; P) - S(f; \widetilde{P}) \leq \omega\bigl(f; [a, b]\bigr) \cdot (\Delta x_{i_1} + \dots + \Delta x_{i_k}),$$
$$0 \leq s(f; \widetilde{P}) - s(f; P) \leq \omega\bigl(f; [a, b]\bigr) \cdot (\Delta x_{i_1} + \dots + \Delta x_{i_k}).$$

c) The quantities $\underline{I} = \sup_P s(f; P)$ and $\overline{I} = \inf_P S(f; P)$ are called respectively the *lower Darboux integral* and the *upper Darboux integral* of f on the closed interval $[a, b]$. Show that $\underline{I} \leq \overline{I}$.

d) Prove the theorem of Darboux:

$$\underline{I} = \lim_{\lambda(P)\to 0} s(f;P), \qquad \overline{I} = \lim_{\lambda(P)\to 0} S(f;P).$$

e) Show that $(f \in \mathcal{R}[a,b]) \Leftrightarrow (\underline{I} = \overline{I})$.

f) Show that $f \in \mathcal{R}[a,b]$ if and only if for every $\varepsilon > 0$ there exists a partition P of $[a,b]$ such that $S(f;P) - s(f;P) < \varepsilon$.

2. *The Cantor set of Lebesgue measure zero.*

a) The Cantor set described in Problem 7 of Sect. 2.4 is uncountable. Verify that it nevertheless is a set of measure 0 in the sense of Lebesgue. Show how to modify the construction of the Cantor set in order to obtain an analogous set "full of holes" that is not a set of measure zero. (Such a set is also called a Cantor set.)

b) Show that the function on $[0,1]$ defined to be zero outside a Cantor set and 1 on the Cantor set is Riemann integrable if and only if the Cantor set has measure zero.

c) Construct a nondecreasing continuous and nonconstant function on $[0,1]$ that has a derivative equal to zero everywhere except at the points of a Cantor set of measure zero.

3. *The Lebesgue criterion.*

a) Verify directly (without using the Lebesgue criterion) that the Riemann function of Example 2 is integrable.

b) Show that a bounded function f belongs to $\mathcal{R}[a,b]$ if and only if for any two numbers $\varepsilon > 0$ and $\delta > 0$ there is a partition P of $[a,b]$ such that the sum of the lengths of the intervals of the partition on which the oscillation of the function is larger than ε is at most δ.

c) Show that $f \in \mathcal{R}[a,b]$ if and only if f is bounded on $[a,b]$ and for any $\varepsilon > 0$ and $\delta > 0$ the set of points in $[a,b]$ at which f has oscillation larger than ε can be covered by a finite set of open intervals the sum of whose lengths is less than δ (the *du Bois-Reymond criterion*).[2]

d) Using the preceding problem, prove the Lebesgue criterion for Riemann integrability of a function.

4. Show that if $f, g \in \mathcal{R}[a,b]$ and f and g are real-valued, then $\max\{f,g\} \in \mathcal{R}[a,b]$ and $\min\{f,g\} \in \mathcal{R}[a,b]$.

5. Show that

a) if $f, g \in \mathcal{R}[a,b]$ and $f(x) = g(x)$ almost everywhere on $[a,b]$, then $\int_a^b f(x)\,dx = \int_a^b g(x)\,dx$;

b) if $f \in \mathcal{R}[a,b]$ and $f(x) = g(x)$ almost everywhere on $[a,b]$, then g can fail to be Riemann-integrable on $[a,b]$, even if g is defined and bounded on $[a,b]$.

---

[2] P. du Bois-Reymond (1831–1889) – German mathematician.

6. *Integration of vector-valued functions.*

a) Let $\mathbf{r}(t)$ be the radius-vector of a point moving in space, $\mathbf{r}_0 = \mathbf{r}(0)$ the initial position of the point, and $\mathbf{v}(t)$ the velocity vector as a function of time. Show how to recover $\mathbf{r}(t)$ from $\mathbf{r}_0$ and $\mathbf{v}(t)$.

b) Does the integration of vector-valued functions reduce to integrating real-valued functions?

c) Is the criterion for integrability stated in Proposition 2′ valid for vector-valued functions?

d) Is Lebesgue's criterion for integrability valid for vector-valued functions?

e) Which concepts and facts from this section extend to functions with complex values?

6.2 Linearity, Additivity and Monotonicity of the Integral

6.2.1 The Integral as a Linear Function on the Space $\mathcal{R}[a, b]$

Theorem 1 *If f and g are integrable functions on the closed interval $[a, b]$, a linear combination of them $\alpha f + \beta g$ is also integrable on $[a, b]$, and*

$$\int_a^b (\alpha f + \beta g)(x)\,\mathrm{d}x = \alpha \int_a^b f(x)\,\mathrm{d}x + \beta \int_a^b g(x)\,\mathrm{d}x. \tag{6.11}$$

Proof Consider a Riemann sum for the integral on the left-hand side of (6.11), and transform it as follows:

$$\sum_{i=1}^{n} (\alpha f + \beta g)(\xi_i)\Delta x_i = \alpha \sum_{i=1}^{n} f(\xi_i)\Delta x_i + \beta \sum_{i=1}^{n} g(\xi_i)\Delta x_i. \tag{6.12}$$

Since the right-hand side of this last equality tends to the linear combination of integrals that makes up the right-hand side of (6.11) if the mesh $\lambda(P)$ of the partition tends to 0, the left-hand side of (6.12) must also have a limit as $\lambda(P) \to 0$, and that limit must be the same as the limit on the right. Thus $(\alpha f + \beta g) \in \mathcal{R}[a, b]$ and Eq. (6.11) holds. □

If we regard $\mathcal{R}[a, b]$ as a vector space over the field of real numbers and the integral $\int_a^b f(x)\,\mathrm{d}x$ as a real-valued function defined on vectors of $\mathcal{R}[a, b]$, Theorem 1 asserts that the integral is a linear function on the vector space $\mathcal{R}[a, b]$.

To avoid any possible confusion, functions defined on functions are usually called *functionals*. Thus we have proved that the integral is a linear functional on the vector space of integrable functions.

6.2.2 The Integral as an Additive Function of the Interval of Integration

The value of the integral $\int_a^b f(x)\,\mathrm{d}x = I(f;[a,b])$ depends on both the integrand and the closed interval over which the integral is taken. For example, if $f \in \mathcal{R}[a,b]$, then, as we know, $f|_{[\alpha,\beta]} \in \mathcal{R}[\alpha,\beta]$ if $[\alpha,\beta] \subset [a,b]$, that is, the integral $\int_\alpha^\beta f(x)\,\mathrm{d}x$ is defined, and we can study its dependence on the closed interval $[\alpha,\beta]$ of integration.

Lemma 1 *If $a < b < c$ and $f \in \mathcal{R}[a,c]$, then $f|_{[a,b]} \in \mathcal{R}[a,b]$, $f|_{[b,c]} \in \mathcal{R}[b,c]$, and the following equality*[3] *holds;*

$$\int_a^c f(x)\,\mathrm{d}x = \int_a^b f(x)\,\mathrm{d}x + \int_b^c f(x)\,\mathrm{d}x. \tag{6.13}$$

Proof We first note that the integrability of the restrictions of f to the closed intervals $[a,b]$ and $[b,c]$ is guaranteed by Proposition 4 of Sect. 6.1.

Next, since $f \in \mathcal{R}[a,c]$, in computing the integral $\int_a^c f(x)\,\mathrm{d}x$ as the limit of Riemann sums we may choose any convenient partitions of $[a,c]$. We shall now consider only partitions P of $[a,c]$ that contain the point b. Obviously any such partition with distinguished points (P,ξ) generates partitions (P',ξ') and (P'',ξ'') of $[a,b]$ and $[b,c]$ respectively, and $P = P' \cup P''$ and $\xi = \xi' \cup \xi''$.

But then the following equality holds between the corresponding Riemann sums:

$$\sigma(f;P,\xi) = \sigma(f;P',\xi') + \sigma(f;P'',\xi'').$$

Since $\lambda(P') \le \lambda(P)$ and $\lambda(P'') \le \lambda(P)$, for $\lambda(P)$ sufficiently small, each of these Riemann sums is close to the corresponding integral in (6.13), which consequently must hold. □

To widen the application of this result slightly, we temporarily revert once again to the definition of the integral.

We defined the integral as the limit of Riemann sums

$$\sigma(f;P,\xi) = \sum_{i=1}^{n} f(\xi_i)\Delta x_i, \tag{6.14}$$

corresponding to partitions with distinguished points (P,ξ) of the closed interval of integration $[a,b]$. A partition P consisted of a finite monotonic sequence of points $x_0, x_1, \dots, x_n$, the point x_0 being the lower limit of integration a and x_n the upper limit of integration b. This construction was carried out assuming that $a < b$. If we

[3] We recall that $f|E$ denotes the restriction of the function f to a set E contained in the domain of definition of f. Formally we should have written the restriction of f to the intervals $[a,b]$ and $[b,c]$, rather than f, on the right-hand side of Eq. (6.13).

now take two arbitrary points a and b without requiring $a < b$ and, regarding a as the lower limit of integration and b as the upper, carry out this construction, we shall again obtain a sum of the form (6.14), in which now $\Delta x_i > 0$ $(i = 1, \dots, n)$ if $a < b$ and $\Delta x_i < 0$ $(i = 1, \dots, n)$ if $a > b$, since $\Delta x_i = x_i - x_{i-1}$. Thus for $a > b$ the sum (6.14) will differ from the Riemann sum of the corresponding partition of the closed interval $[b, a]$ $(b < a)$ only in sign.

From these considerations we adopt the following convention: if $a > b$, then

$$\int_a^b f(x)\,\mathrm{d}x := -\int_b^a f(x)\,\mathrm{d}x. \tag{6.15}$$

In this connection, it is also natural to set

$$\int_a^a f(x)\,\mathrm{d}x := 0. \tag{6.16}$$

After these conventions, taking account of Lemma 1, we arrive at the following important property of the integral.

Theorem 2 *Let $a, b, c \in \mathbb{R}$ and let f be a function integrable over the largest closed interval having two of these points as endpoints. Then the restriction of f to each of the other closed intervals is also integrable over those intervals and the following equality holds*:

$$\int_a^b f(x)\,\mathrm{d}x + \int_b^c f(x)\,\mathrm{d}x + \int_c^a f(x)\,\mathrm{d}x = 0. \tag{6.17}$$

Proof By the symmetry of Eq. (6.17) in a, b, and c, we may assume without loss of generality that $a = \min\{a, b, c\}$.

If $\max\{a, b, c\} = c$ and $a < b < c$, then by Lemma 1

$$\int_a^b f(x)\,\mathrm{d}x + \int_b^c f(x)\,\mathrm{d}x - \int_a^c f(x)\,\mathrm{d}x = 0,$$

which, when we take account of the convention (6.15) yields (6.17).

If $\max\{a, b, c\} = b$ and $a < c < b$, then by Lemma 1

$$\int_a^c f(x)\,\mathrm{d}x + \int_c^b f(x)\,\mathrm{d}x - \int_a^b f(x)\,\mathrm{d}x = 0,$$

which, when we take account of (6.15), again yields (6.17).

Finally, if two of the points a, b, and c are equal, then (6.17) follows from the conventions (6.15) and (6.16). □

Definition 1 Suppose that to each ordered pair (α, β) of points $\alpha, \beta \in [a, b]$ a number $I(\alpha, \beta)$ is assigned so that

$$I(\alpha, \gamma) = I(\alpha, \beta) + I(\beta, \gamma)$$

for any triple of points $\alpha, \beta, \gamma \in [a, b]$.

Then the function $I(\alpha, \beta)$ is called an *additive (oriented) interval function* defined on intervals contained in $[a, b]$.

If $f \in \mathcal{R}[A, B]$, and $a, b, c \in [A, B]$, then, setting $I(a, b) = \int_a^b f(x)\,dx$, we conclude from (6.17) that

$$\int_a^c f(x)\,dx = \int_a^b f(x)\,dx + \int_b^c f(x)\,dx, \tag{6.18}$$

that is, the integral is an additive interval function on the interval of integration. The orientation of the interval in this case amounts to the fact that we order the pair of endpoints of the interval by indicating which is to be first (the lower limit of integration) and which is to be second (the upper limit of integration).

6.2.3 Estimation of the Integral, Monotonicity of the Integral, and the Mean-Value Theorem

a. A General Estimate of the Integral

We begin with a general estimate of the integral, which, as will become clear later, holds for integrals of functions that are not necessarily real-valued.

Theorem 3 *If $a \le b$ and $f \in \mathcal{R}[a, b]$, then $|f| \in \mathcal{R}[a, b]$ and the following inequality holds:*

$$\left|\int_a^b f(x)\,dx\right| \le \int_a^b |f|(x)\,dx. \tag{6.19}$$

If $|f|(x) \le C$ on $[a, b]$ then

$$\int_a^b |f|(x)\,dx \le C(b-a). \tag{6.20}$$

Proof For $a = b$ the assertion is trivial, and so we shall assume that $a < b$.

To prove the theorem it now suffices to recall that $|f| \in \mathcal{R}[a, b]$ (see Proposition 4 of Sect. 6.1), and write the following estimate for the Riemann sum $\sigma(f; P, \xi)$:

$$\left|\sum_{i=1}^n f(\xi_i)\Delta x_i\right| \le \sum_{i=1}^n |f(\xi_i)||\Delta x_i| = \sum_{i=1}^n |f(\xi_i)|\Delta x_i \le C\sum_{i=1}^n \Delta x_i = C(b-a).$$

Passing to the limit as $\lambda(P) \to 0$, we obtain

$$\left|\int_a^b f(x)\,dx\right| \le \int_a^b |f|(x)\,dx \le C(b-a). \qquad \square$$

b. Monotonicity of the Integral and the First Mean-Value Theorem

The results that follow are specific to integrals of real-valued functions.

Theorem 4 *If $a \le b$, $f_1, f_2 \in \mathcal{R}[a,b]$, and $f_1(x) \le f_2(x)$ at each point $x \in [a,b]$, then*

$$\int_a^b f_1(x)\,dx \le \int_a^b f_2(x)\,dx. \tag{6.21}$$

Proof For $a = b$ the assertion is trivial. If $a < b$, it suffices to write the following inequality for the Riemann sums:

$$\sum_{i=1}^n f_1(\xi_i)\Delta x_i \le \sum_{i=1}^n f_2(\xi_i)\Delta x_i,$$

which is valid since $\Delta x_i > 0$ $(i = 1, \dots, n)$, and then pass to the limit as $\lambda(P) \to 0$.

Theorem 4 can be interpreted as asserting that the integral is monotonic as a function of the integrand.

Theorem 4 has a number of useful corollaries. □

Corollary 1 *If $a \le b$, $f \in \mathcal{R}[a,b]$; and $m \le f(x) \le M$ at each $x \in [a,b]$, then*

$$m \cdot (b-a) \le \int_a^b f(x)\,dx \le M \cdot (b-a), \tag{6.22}$$

and, in particular, if $0 \le f(x)$ on $[a,b]$, then

$$0 \le \int_a^b f(x)\,dx.$$

Proof Relation (6.22) is obtained by integrating each term in the inequality $m \le f(x) \le M$ and using Theorem 4. □

Corollary 2 *If $f \in \mathcal{R}[a,b]$, $m = \inf_{x\in[a,b]} f(x)$, and $M = \sup_{x\in[a,b]} f(x)$, then there exists a number $\mu \in [m, M]$ such that*

$$\int_a^b f(x)\,dx = \mu \cdot (b-a). \tag{6.23}$$

Proof If $a = b$, the assertion is trivial. If $a \ne b$, we set $\mu = \frac{1}{b-a}\int_a^b f(x)\,dx$. It then follows from (6.22) that $m \le \mu \le M$ if $a < b$. But both sides of (6.23) reverse sign if a and b are interchanged, and therefore (6.23) is also valid for $b < a$. □

Corollary 3 *If $f \in C[a,b]$, there is a point $\xi \in [a,b]$ such that*

$$\int_a^b f(x)\,dx = f(\xi)(b-a). \tag{6.24}$$

Proof By the intermediate-value theorem for a continuous function, there is a point ξ on $[a, b]$ at which $f(\xi) = \mu$ if

$$m = \min_{x\in[a,b]} f(x) \le \mu \le \max_{x\in[a,b]} f(x) = M.$$

Therefore (6.24) follows from (6.23). □

The equality (6.24) is often called the *first mean-value theorem* for the integral. We, however, reserve that name for the following somewhat more general proposition.

Theorem 5 (First mean-value theorem for the integral) *Let* $f, g \in \mathcal{R}[a, b]$, $m = \inf_{x\in[a,b]} f(x)$, *and* $M = \sup_{x\in[a,b]} f(x)$. *If* g *is nonnegative (or nonpositive) on* $[a, b]$, *then*

$$\int_a^b (f \cdot g)(x)\,\mathrm{d}x = \mu \int_a^b g(x)\,\mathrm{d}x, \tag{6.25}$$

where $\mu \in [m, M]$.

If, in addition, it is known that $f \in C[a, b]$, *then there exists a point* $\xi \in [a, b]$ *such that*

$$\int_a^b (f \cdot g)(x)\,\mathrm{d}x = f(\xi) \int_a^b g(x)\,\mathrm{d}x. \tag{6.26}$$

Proof Since interchanging the limits of integration leads to a simultaneous sign reversal on both sides of Eq. (6.25), it suffices to verify this equality for the case $a < b$. Reversing the sign of $g(x)$ also reverses the signs of both sides of (6.25), so that we may assume without loss of generality that $g(x) \ge 0$ on $[a, b]$.

Since $m = \inf_{x\in[a,b]} f(x)$ and $M = \sup_{x\in[a,b]} f(x)$, we have, for $g(x) \ge 0$,

$$mg(x) \le f(x)g(x) \le Mg(x).$$

Since $m \cdot g \in \mathcal{R}[a, b]$, $f \cdot g \in \mathcal{R}[a, b]$, and $M \cdot g \in \mathcal{R}[a, b]$, applying Theorem 4 and Theorem 1, we obtain

$$m \int_a^b g(x)\,\mathrm{d}x \le \int_a^b f(x)g(x)\,\mathrm{d}x \le M \int_a^b g(x)\,\mathrm{d}x. \tag{6.27}$$

If $\int_a^b g(x)\,\mathrm{d}x = 0$, it is obvious from these inequalities that (6.25) holds.

If $\int_a^b g(x)\,\mathrm{d}x \ne 0$, then, setting

$$\mu = \left(\int_a^b g(x)\,\mathrm{d}x\right)^{-1} \cdot \int_a^b (f \cdot g)(x)\,\mathrm{d}x,$$

we find by (6.27) that

$$m \le \mu \le M,$$

but this is equivalent to (6.25).

The equality (6.26) now follows from (6.25) and the intermediate-value theorem for a function $f \in C[a,b]$, if we take account of the fact that when $f \in C[a,b]$, we have

$$m = \min_{x\in[a,b]} f(x) \quad \text{and} \quad M = \max_{x\in[a,b]} f(x). \qquad \square$$

We remark that (6.23) results from (6.25) if $g(x) \equiv 1$ on $[a,b]$.

c. The Second Mean-Value Theorem for the Integral

The so-called *second mean-value theorem*[4] is significantly more special and delicate in the context of the Riemann integral.

So as not to complicate the proof of this theorem, we shall carry out a useful preparatory discussion that is of independent interest.

Abel's transformation. This is the name given to the following transformation of the sum $\sum_{i=1}^{n} a_i b_i$. Let $A_k = \sum_{i=1}^{k} a_i$; we also set $A_0 = 0$. Then

$$\sum_{i=l}^{m} a_i b_i = \sum_{i=1}^{n} (A_i - A_{i-1}) b_i = \sum_{i=1}^{n} A_i b_i - \sum_{i=1}^{n} A_{i-1} b_i =$$

$$= \sum_{i=1}^{n} A_i b_i - \sum_{i=0}^{n-1} A_i b_{i+1} = A_n b_n - A_0 b_1 + \sum_{i=1}^{n-1} A_i (b_i - b_{i+1}).$$

Thus

$$\sum_{i=1}^{n} a_i b_i = (A_n b_n - A_0 b_1) + \sum_{i=1}^{n-1} A_i (b_i - b_{i+1}), \tag{6.28}$$

or, since $A_0 = 0$,

$$\sum_{i=1}^{n} a_i b_i = A_n b_n + \sum_{i=1}^{n-1} A_i (b_i - b_{i+1}). \tag{6.29}$$

Abel's transformation provides an easy verification of the following lemma.

Lemma 2 *If the numbers $A_k = \sum_{i=1}^{k} a_i$ $(k = 1, \dots, n)$ satisfy the inequalities $m \le A_k \le M$ and the numbers b_i $(i = 1, \dots, n)$ are nonnegative and $b_i \ge b_{i+1}$ for $i = 1, \dots, n-1$, then*

$$m b_1 \le \sum_{i=1}^{n} a_i b_i \le M b_1. \tag{6.30}$$

[4]Under an additional hypothesis on the function, one that is often completely acceptable, Theorem 6 in this section could easily be obtained from the first mean-value theorem. On this point, see Problem 3 at the end of Sect. 6.3.

Proof Using the fact that $b_n \geq 0$ and $b_i - b_{i+1} \geq 0$ for $i = 1, \dots, n-1$, we obtain from (6.29),

$$\sum_{i=1}^{n} a_i b_i \leq M b_n + \sum_{i=1}^{n-1} M(b_i - b_{i+1}) = M b_n + M(b_1 - b_n) = M b_1.$$

The left-hand inequality of (6.30) is verified similarly. □

Lemma 3 *If $f \in \mathcal{R}[a,b]$, then for any $x \in [a,b]$ the function*

$$F(x) = \int_a^x f(t)\,\mathrm{d}t \tag{6.31}$$

is defined and $F(x) \in C[a,b]$.

Proof The existence of the integral in (6.31) for any $x \in [a,b]$ is already known from Proposition 4 of Sect. 6.1; therefore it remains only for us to verify that the function $F(x)$ is continuous. Since $f \in \mathcal{R}[a,b]$, we have $|f| \leq C < \infty$ on $[a,b]$. Let $x \in [a,b]$ and $x+h \in [a,b]$. Then, by the additivity of the integral and inequalities (6.19) and (6.20) we obtain

$$\begin{aligned}\left|F(x+h) - F(x)\right| &= \left|\int_a^{x+h} f(t)\,\mathrm{d}t - \int_a^x f(t)\,\mathrm{d}t\right| = \\ &= \left|\int_x^{x+h} f(t)\,\mathrm{d}t\right| \leq \left|\int_x^{x+h} \left|f(t)\right|\mathrm{d}t\right| \leq C|h|.\end{aligned}$$

Here we have used inequality (6.20) taking account of the fact that for $h < 0$ we have

$$\left|\int_x^{x+h} \left|f(t)\right|\mathrm{d}t\right| = \left|-\int_{x+h}^x \left|f(t)\right|\mathrm{d}t\right| = \int_{x+h}^x \left|f(t)\right|\mathrm{d}t.$$

Thus we have shown that if x and $x+h$ both belong to $[a,b]$, then

$$\left|F(x+h) - F(x)\right| \leq C|h| \tag{6.32}$$

from which it obviously follows that the function F is continuous at each point of $[a,b]$. □

We now prove a lemma that is a version of the second mean-value theorem.

Lemma 4 *If $f, g \in \mathcal{R}[a,b]$ and g is a nonnegative nonincreasing function on $[a,b]$ then there exists a point $\xi \in [a,b]$ such that*

$$\int_a^b (f \cdot g)(x)\,\mathrm{d}x = g(a) \int_a^\xi f(x)\,\mathrm{d}x. \tag{6.33}$$

Before turning to the proof, we note that, in contrast to relation (6.26) of the first mean-value theorem, it is the function $f(x)$ that remains under the integral sign in (6.33), not the monotonic function g.

Proof To prove (6.33), as in the cases considered above, we attempt to estimate the corresponding Riemann sum.

Let P be a partition of $[a,b]$. We first write the identity

$$\int_a^b (f\cdot g)\,\mathrm{d}x = \sum_{i=1}^n \int_{x_{i-1}}^{x_i} (f\cdot g)(x)\,\mathrm{d}x =$$
$$= \sum_{i=1}^n g(x_{i-1}) \int_{x_{i-1}}^{x_i} f(x)\,\mathrm{d}x + \sum_{i=1}^n \int_{x_{i-1}}^{x_i} \big[g(x)-g(x_{i-1})\big] f(x)\,\mathrm{d}x$$

and then show that the last sum tends to zero as $\lambda(P)\to 0$.

Since $f\in\mathcal{R}[a,b]$, it follows that $|f(x)|\le C<\infty$ on $[a,b]$. Then, using the properties of the integral already proved, we obtain

$$\left|\sum_{i=1}^n \int_{x_{i-1}}^{x_i} \big[g(x)-g(x_{i-1})\big] f(x)\,\mathrm{d}x\right| \le \sum_{i=1}^n \int_{x_{i-1}}^{x_i} \big|g(x)-g(x_i)\big|\big|f(x)\big|\,\mathrm{d}x \le$$
$$\le C\sum_{i=1}^n \int_{x_{i-1}}^{x_i} \big|g(x)-g(x_{i-1})\big|\,\mathrm{d}x \le$$
$$\le C\sum_{i=1}^n \omega(g;\Delta_i)\Delta x_i \to 0$$

as $\lambda(P)\to 0$, because $g\in\mathcal{R}[a,b]$ (see Proposition 2 of Sect. 6.1). Therefore

$$\int_a^b (f\cdot g)(x)\,\mathrm{d}x = \lim_{\lambda(P)\to 0} \sum_{i=1}^n g(x_{i-1}) \int_{x_{i-1}}^{x_i} f(x)\,\mathrm{d}x. \tag{6.34}$$

We now estimate the sum on the right-hand side of (6.34). Setting

$$F(x) = \int_a^x f(t)\,\mathrm{d}t,$$

by Lemma 3 we obtain a continuous function on $[a,b]$.

Let

$$m = \min_{x\in[a,b]} F(x) \quad\text{and}\quad M = \max_{x\in[a,b]} F(x).$$

Since $\int_{x_{i-1}}^{x_i} f(x)\,\mathrm{d}x = F(x_i) - F(x_{i-1})$, it follows that

$$\sum_{i=1}^{n} g(x_{i-1}) \int_{x_{i-1}}^{x_i} f(x)\,\mathrm{d}x = \sum_{i=1}^{n} \big(F(x_i) - F(x_{i-1})\big) g(x_{i-1}). \tag{6.35}$$

Taking account of the fact that g is nonnegative and nonincreasing on $[a, b]$, and setting

$$a_i = F(x_i) - F(x_{i-1}), \qquad b_i = g(x_{i-1}),$$

we find by Lemma 2 that

$$mg(a) \le \sum_{i=1}^{n} \big(F(x_i) - F(x_{i-1})\big) g(x_{i-1}) \le Mg(a), \tag{6.36}$$

since

$$A_k = \sum_{i=1}^{k} a_i = F(x_k) - F(x_0) = F(x_k) - F(a) = F(x_k).$$

Having now shown that the sums (6.35) satisfy the inequalities (6.36), and recalling relation (6.34), we have

$$mg(a) \le \int_a^b (f \cdot g)(x)\,\mathrm{d}x \le Mg(a). \tag{6.37}$$

If $g(a) = 0$, then, as inequalities (6.37) show, the relation to be proved (6.33) is obviously true.

If $g(a) > 0$, we set

$$\mu = \frac{1}{g(a)} \int_a^b (f \cdot g)(x)\,\mathrm{d}x.$$

It follows from (6.37) that $m \le \mu \le M$, and from the continuity of $F(x) = \int_a^x f(t)\,\mathrm{d}t$ on $[a, b]$ that there exists a point $\xi \in [a, b]$ at which $F(\xi) = \mu$. But that is precisely what formula (6.33) says. □

Theorem 6 (Second mean-value theorem for the integral) *If $f, g \in \mathcal{R}[a, b|$ and g is a monotonic function on $[a, b]$, then there exists a point $\xi \in [a, b|$ such that*

$$\int_a^b (f \cdot g)(x)\,\mathrm{d}x = g(a) \int_a^\xi f(x)\,\mathrm{d}x + g(b) \int_\xi^b f(x)\,\mathrm{d}x. \tag{6.38}$$

The equality (6.38) (like (6.33), as it happens) is often called *Bonnet's formula*.[5]

[5] P.O. Bonnet (1819–1892) – French mathematician and astronomer. His most important mathematical works are in differential geometry.

Proof Let g be a nondecreasing function on $[a, b]$. Then $G(x) = g(b) - g(x)$ is nonnegative, nondecreasing, and integrable on $[a, b]$. Applying formula (6.33), we find

$$\int_a^b (f \cdot G)(x)\,\mathrm{d}x = G(a) \int_a^\xi f(x)\,\mathrm{d}x. \tag{6.39}$$

But

$$\int_\alpha^b (f \cdot G)\,\mathrm{d}x = g(b) \int_a^b f(x)\,\mathrm{d}x - \int_a^b (f \cdot g)(x)\,\mathrm{d}x,$$

$$G(a) \int_a^\xi f(x)\,\mathrm{d}x = g(b) \int_a^\xi f(x)\,\mathrm{d}x - g(a) \int_a^\xi f(x)\,\mathrm{d}x.$$

Taking account of these relations and the additivity of the integral, we obtain the equality (6.38), which was to be proved, from (6.39).

If g is a nonincreasing function, setting $G(x) = g(x) - g(b)$, we find that $G(x)$ is a nonnegative, nonincreasing, integrable function on $[a, b]$. We then obtain (6.39) again, and then formula (6.38). □

6.2.4 Problems and Exercises

1. Show that if $f \in \mathcal{R}[a, b]$ and $f(x) \geq 0$ on $[a, b]$, then the following statements are true.

a) If the function $f(x)$ assumes a positive value $f(x_0) > 0$ at a point of continuity $x_0 \in [a, b]$, then the strict inequality

$$\int_a^b f(x)\,\mathrm{d}x > 0$$

holds.

b) The condition $\int_a^b f(x)\,\mathrm{d}x = 0$ implies that $f(x) = 0$ at almost all points of $[a, b]$.

2. Show that if $f \in \mathcal{R}[a, b]$, $m = \inf_{]a,b[} f(x)$, and $M = \sup_{]a,b[} f(x)$, then

a) $\int_a^b f(x)\,\mathrm{d}x = \mu(b - a)$, where $\mu \in [m, M]$ (see Problem 5a of Sect. 6.1);
b) if f is continuous on $[a, b]$, there exists a point $\xi \in\,]a, b[$ such that

$$\int_a^b f(x)\,\mathrm{d}x = f(\xi)(b - a).$$

3. Show that if $f \in C[a, b]$, $f(x) \geq 0$ on $[a, b]$, and $M = \max_{[a,b]} f(x)$, then

$$\lim_{n\to\infty} \left(\int_a^b f^n(x)\,\mathrm{d}x \right)^{1/n} = M.$$

4. a) Show that if $f \in \mathcal{R}[a,b]$, then $|f|^p \in \mathcal{R}[a,b]$ for $p \geq 0$.

b) Starting from Hölder's inequality for sums, obtain Hölder's inequality for integrals:[6]

$$\left|\int_a^b (f \cdot g)(x)\,\mathrm{d}x\right| \leq \left(\int_a^b |f|^p(x)\,\mathrm{d}x\right)^{1/p} \cdot \left(\int_a^b |g|^q(x)\,\mathrm{d}x\right)^{1/q},$$

if $f, g \in \mathcal{R}[a,b]$, $p > 1$, $q > 1$, and $\frac{1}{p} + \frac{1}{q} = 1$.

c) Starting from Minkowski's inequality for sums, obtain Minkowski's inequality for integrals:

$$\left(\int_a^b |f+g|^p(x)\,\mathrm{d}x\right)^{1/p} \leq \left(\int_a^b |f|^p(x)\,\mathrm{d}x\right)^{1/p} + \left(\int_a^b |g|^p(x)\,\mathrm{d}x\right)^{1/p},$$

if $f, g \in \mathcal{R}[a,b]$ and $p \geq 1$. Show that this inequality reverses direction if $0 < p < 1$.

d) Verify that if f is a continuous convex function on $\mathbb{R}$ and φ an arbitrary continuous function on $\mathbb{R}$, then Jensen's inequality

$$f\left(\frac{1}{c}\int_0^c \varphi(t)\,\mathrm{d}t\right) \leq \frac{1}{c}\int_0^c f\big(\varphi(t)\big)\,\mathrm{d}t$$

holds for $c \neq 0$.

6.3 The Integral and the Derivative

6.3.1 The Integral and the Primitive

Let f be a Riemann-integrable function on a closed interval $[a,b]$. On this interval let us consider the function

$$F(x) = \int_a^x f(t)\,\mathrm{d}t, \tag{6.40}$$

often called an *integral with variable upper limit.*

Since $f \in \mathcal{R}[a,b]$, it follows that $f|_{[a,x]} \in \mathcal{R}[a,x]$ if $[a,x] \subset [a,b]$; therefore the function $x \mapsto F(x)$ is unambiguously defined for $x \in [a,b]$.

---

[6]The algebraic Hölder inequality for $p = q = 2$ was first obtained in 1821 by Cauchy and bears his name. Hölder's inequality for integrals with $p = q = 2$ was first discovered in 1859 by the Russian mathematician B.Ya. Bunyakovskii (1804–1889). This important integral inequality (in the case $p = q = 2$) is called *Bunyakovskii's inequality* or the *Cauchy–Bunyakovskii inequality*. One also some-times sees the less accurate name "Schwarz inequality" after the German mathematician H.K.A. Schwarz (1843–1921), in whose work it appeared in 1884.

If $|f(t)| \le C < +\infty$ on $[a,b]$ (and f, being an integrable function, is bounded on $[a,b]$), it follows from the additivity of the integral and the elementary estimate of it that

$$\big|F(x+h) - F(x)\big| \le C|h|, \tag{6.41}$$

if $x, x+h \in [a,b]$.

Actually, we already discussed this while proving Lemma 3 in the preceding section.

It follows in particular from (6.41) that the function F is continuous on $[a,b]$, so that $F \in C[a,b]$.

We now investigate the function F more thoroughly.

The following lemma is fundamental for what follows.

Lemma 1 *If $f \in \mathcal{R}[a,b]$ and the function f is continuous at a point $x \in [a,b]$, then the function F defined on $[a,b]$ by (6.40) is differentiable at the point x, and the following equality holds*:

$$F'(x) = f(x).$$

Proof Let $x, x+h \in [a,b]$. Let us estimate the difference $F(x+h) - F(x)$. It follows from the continuity of f at x that $f(t) = f(x) + \Delta(t)$, where $\Delta(t) \to 0$ as $t \to x$, $t \in [a,b]$. If the point x is held fixed, the function $\Delta(t) = f(t) - f(x)$ is integrable on $[a,b]$, being the difference of the integrable function $t \mapsto f(t)$ and the constant $f(x)$. We denote by $M(h)$ the quantity $\sup_{t \in I(h)} |\Delta(t)|$, where $I(h)$ is the closed interval with endpoints $x, x+h \in [a,b]$. By hypothesis $M(h) \to 0$ as $h \to 0$.

We now write

$$\begin{aligned} F(x+h) - F(x) &= \int_a^{x+h} f(t)\,\mathrm{d}t - \int_a^x f(t)\,\mathrm{d}t = \int_x^{x+h} f(t)\,\mathrm{d}t = \\ &= \int_x^{x+h} \big(f(x) + \Delta(t)\big)\,\mathrm{d}t = \\ &= \int_x^{x+h} f(x)\,\mathrm{d}t + \int_x^{x+h} \Delta(t)\,\mathrm{d}t = f(x)h + \alpha(h)h, \end{aligned}$$

where we have set

$$\int_x^{x+h} \Delta(t)\,\mathrm{d}t = \alpha(h)h.$$

Since

$$\left|\int_x^{x+h} \Delta(t)\,\mathrm{d}t\right| \le \left|\int_x^{x+h} |\Delta(t)|\,\mathrm{d}t\right| \le \left|\int_x^{x+h} M(h)\,\mathrm{d}t\right| = M(h)|h|,$$

it follows that $|\alpha(h)| \le M(h)$, and so $\alpha(h) \to 0$ as $h \to 0$ (in such a way that $x+h \in [a,b]$).

Thus we have shown that if the function f is continuous at a point $x \in [a,b]$, then for displacements h from x such that $x+h \in [a,b]$ the following equality holds:

$$F(x+h) - F(x) = f(x)h + \alpha(h)h, \tag{6.42}$$

where $\alpha(h) \to 0$ as $h \to 0$.

But this means that the function $F(x)$ is differentiable on $[a,b]$ at the point $x \in [a,b]$ and that $F'(x) = f(x)$. □

A very important immediate corollary of Lemma 1 is the following.

Theorem 1 *Every continuous function $f : [a,b] \to \mathbb{R}$ on the closed interval $[a,b]$ has a primitive, and every primitive of f on $[a,b]$ has the form*

$$\mathcal{F}(x) = \int_a^x f(t)\,\mathrm{d}t + c, \tag{6.43}$$

where c is a constant.

Proof We have the implication $(f \in C[a,b]) \Rightarrow (f \in \mathcal{R}[a,b])$, so that by Lemma 1 the function (6.40) is a primitive for f on $[a,b]$. But two primitives $\mathcal{F}(x)$ and $F(x)$ of the same function on a closed interval can differ on that interval only by a constant; hence $\mathcal{F}(x) = F(x) + c$. □

For later applications it is convenient to broaden the concept of primitive slightly and adopt the following definition.

Definition 1 A continuous function $x \mapsto \mathcal{F}(x)$ on an interval of the real line is called a *primitive* (or *generalized primitive*) of the function $x \to f(x)$ defined on the same interval if the relation $\mathcal{F}'(x) = f(x)$ holds at all points of the interval, with only a finite number of exceptions.

Taking this definition into account, we can assert that the following theorem holds.

Theorem 1′ *A function $f : [a,b] \to \mathbb{R}$ that is defined and bounded on a closed interval $[a,b]$ and has only a finite number of points of discontinuity has a (generalized) primitive on that interval, and any primitive of f on $[a,b]$ has the form (6.43).*

Proof Since f has only a finite set of points of discontinuity, $f \in \mathcal{R}[a,b]$, and by Lemma 1 the function (6.40) is a generalized primitive for f on $[a,b]$. Here we have taken into account, as already pointed out, the fact that by (6.41) the function (6.40) is continuous on $[a,b]$. If $\mathcal{F}(x)$ is another primitive of f on $[a,b]$, then $\mathcal{F}(x) - F(x)$ is a continuous function and constant on each of the finite number of intervals into which the discontinuities of f divide the closed interval $[a,b]$. But it then follows from the continuity of $\mathcal{F}(x) - F(x)$ on all of $[a,b]$ that $\mathcal{F}(x) - F(x) \equiv \text{const}$ on $[a,b]$. □

6.3.2 The Newton–Leibniz Formula

Theorem 2 *If $f : [a, b] \to \mathbb{R}$ is a bounded function with a finite number of points of discontinuity, then $f \in \mathcal{R}[a, b]$ and*

$$\boxed{\int_a^b f(x)\,\mathrm{d}x = \mathcal{F}(b) - \mathcal{F}(a),} \tag{6.44}$$

where $\mathcal{F} : [a, b] \to \mathbb{R}$ is any primitive of f on $[a, b]$.

Proof We already know that a bounded function on a closed interval having only a finite number of discontinuities is integrable (see Corollary 2 after Proposition 2 in Sect. 6.1). The existence of a generalized primitive $\mathcal{F}(x)$ of the function f on $[a, b]$ is guaranteed by Theorem 1′, by virtue of which $\mathcal{F}(x)$ has the form (6.43). Setting $x = a$ in (6.43), we find that $\mathcal{F}(a) = c$, and so

$$\mathcal{F}(x) = \int_a^x f(t)\,\mathrm{d}t + \mathcal{F}(a).$$

In particular

$$\int_a^b f(t)\,\mathrm{d}t = \mathcal{F}(b) - \mathcal{F}(a),$$

which, up to the notation for the variable of integration, is exactly formula (6.44), which was to be proved. □

Relation (6.44), which is fundamental for all of analysis, is called the *Newton–Leibniz formula* (or the *fundamental theorem of calculus*).

The difference $\mathcal{F}(b) - \mathcal{F}(a)$ of values of any function is often written $\mathcal{F}(x)|_a^b$. In this notation, the Newton–Leibniz formula assumes the form

$$\boxed{\int_a^b f(x)\,\mathrm{d}x = \mathcal{F}(x)\big|_a^b.}$$

Since both sides of the formula reverse sign when a and b are interchanged, the formula is valid for any relation between the magnitudes of a and b, that is, both for $a \le b$ and for $a \ge b$.

In exercises of analysis the Newton–Leibniz formula is mostly used to compute the integral on the left-hand side, and that may lead to a some-what distorted idea of its use. The actual situation is that particular integrals are rarely found using a primitive; more often one resorts to direct computation on a computer using highly developed numerical methods. The Newton–Leibniz formula occupies a key position in the theory of mathematical analysis itself, since it links integration and differentiation. In analysis it has a very far-reaching extension in the form of the so-called *generalized Stokes' formula*.[7]

An example of the use of the Newton–Leibniz formula in analysis itself is provided by the material in the next subsection.

6.3.3 Integration by Parts in the Definite Integral and Taylor's Formula

Proposition 1 *If the functions $u(x)$ and $v(x)$ are continuously differentiable on a closed interval with endpoints a and b, then*

$$\int_a^b (u \cdot v')(x)\,\mathrm{d}x = (u \cdot v)\Big|_a^b - \int_a^b (v \cdot u')(x)\,\mathrm{d}x. \tag{6.45}$$

It is customary to write this formula in abbreviated form as

$$\int_a^b u\,\mathrm{d}v = u \cdot v\Big|_a^b - \int_a^b v\,\mathrm{d}u$$

and call it the formula for *integration by parts* in the definite integral.

Proof By the rule for differentiating a product of functions, we have

$$(u \cdot v)'(x) = \big(u' \cdot v\big)(x) + \big(u \cdot v'\big)(x).$$

By hypothesis, all the functions in this last equality are continuous, and hence integrable on the interval with endpoints a and b. Using the linearity of the integral and the Newton–Leibniz formula, we obtain

$$(u \cdot v)(x)\Big|_a^b = \int_a^b \big(u' \cdot v\big)(x)\,\mathrm{d}x + \int_a^b \big(u \cdot v'\big)(x)\,\mathrm{d}x.$$ □

As a corollary we now obtain the Taylor formula with integral form of the remainder.

Suppose on the closed interval with endpoints a and x the function $t \mapsto f(t)$ has n continuous derivatives. Using the Newton–Leibniz formula and formula (6.45), we carry out the following chain of transformations, in which all differentiations and substitutions are carried out on the variable t:

$$f(x) - f(a) = \int_a^x f'(t)\,\mathrm{d}t = -\int_a^x f'(t)(x-t)'\,\mathrm{d}t =$$

[7] G.G. Stokes (1819–1903) – British physicist and mathematician.

$$= -f'(t)(x-t)\big|_a^x + \int_a^x f''(t)(x-t)\,dt =$$

$$= f'(a)(x-a) - \frac{1}{2}\int_a^x f''(t)\big((x-t)^2\big)'\,dt =$$

$$= f'(a)(x-a) - \frac{1}{2}f''(t)(x-t)^2\big|_a^x + \frac{1}{2}\int_a^x f'''(t)(x-t)^2\,dt =$$

$$= f'(a)(x-a) + \frac{1}{2}f''(a)(x-a)^2 - \frac{1}{2\cdot 3}\int_a^x f'''(t)\big((x-t)^3\big)'\,dt =$$

$$= \cdots =$$

$$= f'(a)(x-a) + \frac{1}{2}f''(a)(x-a)^2 + \cdots +$$

$$+ \frac{1}{2\cdot 3\cdots(n-1)}f^{(n-1)}(a)(x-a)^{n-1} + r_{n-1}(a;x),$$

where

$$\boxed{r_{n-1}(a;x) = \frac{1}{(n-1)!}\int_a^x f^{(n)}(t)(x-t)^{n-1}\,dt.} \tag{6.46}$$

Thus we have proved the following proposition.

Proposition 2 *If the function $t \mapsto f(t)$ has continuous derivatives up to order n inclusive on the closed interval with endpoints a and x, then Taylor's formula holds:*

$$f(x) = f(a) + \frac{1}{1!}f'(a)(x-a) + \cdots + \frac{1}{(n-1)!}f^{(n-1)}(a)(x-a)^{n-1} + r_{n-1}(a;x)$$

with remainder term $r_{n-1}(a;x)$ represented in the integral form (6.46).

We note that the function $(x-t)^{n-1}$ does not change sign on the closed interval with endpoints a and x, and since $t \to f^{(n)}(t)$ is continuous on that interval, the first mean-value theorem implies that there exists a point ξ such that

$$r_{n-1}(a;x) = \frac{1}{(n-1)!}\int_a^x f^{(n)}(t)(x-t)^{n-1}\,dt =$$

$$= \frac{1}{(n-1)!}f^{(n)}(\xi)\int_a^x (x-t)^{n-1}\,dt =$$

$$= \frac{1}{(n-1)!}f^{(n)}(\xi)\left(-\frac{1}{n}(x-t)^n\right)\bigg|_a^x = \frac{1}{n!}f^{(n)}(\xi)(x-a)^n.$$

We have again obtained the familiar Lagrange form of the remainder in Taylor's theorem. By Problem 2d) of Sect. 6.2, we may assume that ξ lies in the open interval with endpoints a and x.

This reasoning can be repeated, taking the expression $f^{(n)}(\xi)(x-\xi)^{n-k}$, where $k \in [1, n]$, outside the integral in (6.46). The Cauchy and Lagrange forms of the remainder term that result correspond to the values $k = 1$ and $k = n$.

6.3.4 Change of Variable in an Integral

One of the basic formulas of integral calculus is the formula for change of variable in a definite integral. This formula is just as important in integration theory as the formula for differentiating a composite function is in differential calculus. Under certain conditions, the two formulas can be linked by the Newton–Leibniz formula.

Proposition 3 *If $\varphi : [\alpha, \beta] \to [a, b]$ is a continuously differentiable mapping of the closed interval $\alpha \le t \le \beta$ into the closed interval $a \le x \le b$ such that $\varphi(\alpha) = a$ and $\varphi(\beta) = b$, then for any continuous function $f(x)$ on $[a, b]$ the function $f(\varphi(t))\varphi'(t)$ is continuous on the closed interval $[\alpha, \beta]$, and*

$$\int_a^b f(x)\,\mathrm{d}x = \int_\alpha^\beta f\bigl(\varphi(t)\bigr)\varphi'(t)\,\mathrm{d}t. \tag{6.47}$$

Proof Let $\mathcal{F}(x)$ be a primitive of $f(x)$ on $[a, b]$. Then, by the theorem on differentiation of a composite function, the function $\mathcal{F}(\varphi(t))$ is a primitive of the function $f(\varphi(t))\varphi'(t)$, which is continuous, being the composition and product of continuous functions on the closed interval $[\alpha, \beta]$. By the Newton–Leibniz formula $\int_a^b f(x)\,\mathrm{d}x = \mathcal{F}(b) - \mathcal{F}(a)$ and $\int_\alpha^\beta f(\varphi(t))\varphi'(t)\mathrm{d}t = \mathcal{F}(\varphi(\beta)) - \mathcal{F}(\varphi(\alpha))$. But by hypothesis $\varphi(\alpha) = a$ and $\varphi(\beta) = b$, so that Eq. (6.47) does indeed hold. □

It is clear from formula (6.47) how convenient it is that we have not just the symbol for the function, but the entire differential $f(x)\,\mathrm{d}x$ in the symbol for integration, which makes it possible to obtain the correct integrand automatically when the new variable $x = \varphi(t)$ is substituted in the integral.

So as not to complicate matters with a cumbersome proof, in Proposition 3 we deliberately shrank the true range of applicability of (6.47) and obtained it by the Newton–Leibniz formula. We now turn to the basic theorem on change of variable, whose hypotheses differ somewhat from those of Proposition 3. The proof of this theorem will rely directly on the definition of the integral as the limit of Riemann sums.

Theorem 3 *Let $\varphi : [\alpha, \beta] \to [a, b]$ be a continuously differentiable strictly monotonic mapping of the closed interval $\alpha \le t \le \beta$ into the closed interval $a \le x \le b$ with the correspondence $\varphi(\alpha) = a$, $\varphi(\beta) = b$ or $\varphi(\alpha) = b$, $\varphi(\beta) = a$ at the endpoints. Then for any function $f(x)$ that is integrable on $[a, b]$ the function*

$f(\varphi(t))\varphi'(t)$ is integrable on $[\alpha, \beta]$ and

$$\boxed{\int_{\varphi(\alpha)}^{\varphi(\beta)} f(x)\,\mathrm{d}x = \int_{\alpha}^{\beta} f\big(\varphi(t)\big)\varphi'(t)\,\mathrm{d}t.} \tag{6.48}$$

Proof Since φ is a strictly monotonic mapping of $[\alpha, \beta]$ onto $[a, b]$ with endpoints corresponding to endpoints, every partition P_t $(\alpha = t_0 < \cdots < t_n = \beta)$ of the closed interval $[\alpha, \beta]$ generates a corresponding partition P_x of $[a, b]$ by means of the images $x_i = \varphi(t_i)$ $(i = 0, \dots, n)$; the partition P_x may be denoted $\varphi(P_t)$. Here $x_0 = a$ if $\varphi(\alpha) = a$ and $x_0 = b$ if $\varphi(\alpha) = b$. It follows from the uniform continuity of φ on $[\alpha, \beta]$ that if $\lambda(P_t) \to 0$, then $\lambda(P_x) = \lambda(\varphi(P_t))$ also tends to zero.

Using Lagrange's theorem, we transform the Riemann sum $\sigma(f; P_x, \xi)$ as follows:

$$\sum_{i=1}^{n} f(\xi_i)\Delta x_i = \sum_{i=1}^{n} f(\xi_i)(x_i - x_{i-1}) =$$
$$= \sum_{i=1}^{n} f\big(\varphi(\tau_i)\big)\varphi'(\tilde{\tau}_i)(t_i - t_{i-1}) = \sum_{i=1}^{n} f\big(\varphi(\tau_i)\big)\varphi'(\tilde{\tau}_i)\Delta t_i.$$

Here $x_i = \varphi(t_i)$, $\xi_i = \varphi(\tau_i)$, ξ_i lies in the closed interval with endpoints x_{i-1} and x_i, and the points τ_i and $\tilde{\tau}_i$ lie in the interval with endpoints t_{i-1} and t_i $(i = 1, \dots, n)$.

Next

$$\sum_{i=1}^{n} f\big(\varphi(\tau_i)\big)\varphi'(\tilde{\tau}_i)\Delta t_i = \sum_{i=1}^{n} f\big(\varphi(\tau_i)\big)\varphi'(\tau_i)\Delta t_i +$$
$$+ \sum_{i=1}^{n} f\big(\varphi(\tau_i)\big)\big(\varphi'(\tilde{\tau}_i) - \varphi'(\tau_i)\big)\Delta t_i.$$

Let us estimate this last sum. Since $f \in \mathcal{R}[a, b]$, the function f is bounded on $[a, b]$. Let $|f(x)| \le C$ on $[a, b]$. Then

$$\left|\sum_{i=1}^{n} f\big(\varphi(\tau)\big)\big(\varphi'(\tilde{\tau}_i) - \varphi'(\tau_i)\big)\Delta t_i\right| \le C \cdot \sum_{i=1}^{n} \omega(\varphi'; \Delta_i)\Delta t_i,$$

where Δ_i is the closed interval with endpoints t_{i-1} and t_i.

This last sum tends to zero as $\lambda(P_t) \to 0$, since φ' is continuous on $[\alpha, \beta]$.

Thus we have shown that

$$\sum_{i=1}^{n} f(\xi_i)\Delta x_i = \sum_{i=1}^{n} f\big(\varphi(\tau_i)\big)\varphi'(\tau_i)\Delta t_i + \alpha,$$

where $\alpha \to 0$ as $\lambda(P_t) \to 0$. As already pointed out, if $\lambda(P_t) \to 0$, then $\lambda(P_x) \to 0$ also. But $f \in \mathcal{R}[a,b]$, so that as $\lambda(P_x) \to 0$ the sum on the left-hand side of this last equality tends to the integral $\int_{\varphi(\alpha)}^{\varphi(\beta)} f(x)\,\mathrm{d}x$. Hence as $\lambda(P_t) \to 0$ the right-hand side of the equality also has the same limit.

But the sum $\sum_{i=1}^{n} f(\varphi(\tau_i))\varphi'(\tau_i)\Delta t_i$ can be regarded as a completely arbitrary Riemann sum for the function $f(\varphi(t))\varphi'(t)$ corresponding to the partition P_t with distinguished points $\tau = (\tau_1, \dots, \tau_n)$, since in view of the strict monotonicity of φ, any set of points τ can be obtained from some corresponding set $\xi = (\xi_1, \dots, \xi_n)$ of distinguished points in the partition $P_x = \varphi(P_t)$.

Thus, the limit of this sum is, by definition, the integral of the function $f(\varphi(t))\varphi'(t)$ over the closed interval $[\alpha, \beta]$, and we have simultaneously proved both the integrability of $f(\varphi(t))\varphi'(t)$ on $[\alpha, \beta]$ and formula (6.48). □

6.3.5 Some Examples

Let us now consider some examples of the use of these formulas and the theorems on properties of the integral proved in the last two sections.

Example 1

$$\int_{-1}^{1} \sqrt{1-x^2}\,\mathrm{d}x = \int_{-\pi/2}^{\pi/2} \sqrt{1-\sin^2 t}\cos t\,\mathrm{d}t = \int_{-\pi/2}^{\pi/2} \cos^2 t\,\mathrm{d}t =$$

$$= \frac{1}{2}\int_{-\pi/2}^{\pi/2} (1+\cos 2t)\,\mathrm{d}t = \frac{1}{2}\Big(t + \frac{1}{2}\sin 2t\Big)\Big|_{-\pi/2}^{\pi/2} = \frac{\pi}{2}.$$

In computing this integral we made the change of variable $x = \sin t$ and then, after finding a primitive for the integrand that resulted from this substitution, we applied the Newton–Leibniz formula.

Of course, we could have proceeded differently. We could have found the rather cumbersome primitive $\frac{1}{2}x\sqrt{1-x^2} + \frac{1}{2}\arcsin x$ for the function $\sqrt{1-x^2}$ and then used the Newton–Leibniz formula. This example shows that in computing a definite integral one can fortunately sometimes avoid having to find a primitive for the integrand.

Example 2 Let us show that

$$\text{a)} \int_{-\pi}^{\pi} \sin mx \cos nx\,\mathrm{d}x = 0, \qquad \text{b)} \int_{-\pi}^{\pi} \sin^2 mx\,\mathrm{d}x = \pi,$$

$$\text{c)} \int_{-\pi}^{\pi} \cos^2 nx\,\mathrm{d}x = \pi$$

for $m, n \in \mathbb{N}$.

a)

$$\int_{-\pi}^{\pi} \sin mx \cos nx \,\mathrm{d}x = \frac{1}{2}\int_{-\pi}^{\pi} \big(\sin(n+m)x - \sin(n-m)x\big)\,\mathrm{d}x =$$
$$= \frac{1}{2}\left(-\frac{1}{n+m}\cos(n+m)x + \frac{1}{n-m}\cos(n-m)x\right)\Bigg|_{-\pi}^{\pi} =$$
$$= 0,$$

if $n - m \neq 0$. The case when $n - m = 0$ can be considered separately, and in this case we obviously arrive at the same result.

b)

$$\int_{-\pi}^{\pi} \sin^2 mx \,\mathrm{d}x = \frac{1}{2}\int_{-\pi}^{\pi} (1 - \cos 2mx)\,\mathrm{d}x = \frac{1}{2}\left(x - \frac{1}{2m}\sin 2mx\right)\Bigg|_{-\pi}^{\pi} = \pi.$$

c)

$$\int_{-\pi}^{\pi} \cos^2 nx \,\mathrm{d}x = \frac{1}{2}\int_{-\pi}^{\pi} (1 + \cos 2nx)\,\mathrm{d}x = \frac{1}{2}\left(x + \frac{1}{2n}\sin 2nx\right)\Bigg|_{-\pi}^{\pi} = \pi.$$

Example 3 Let $f \in \mathcal{R}[-a, a]$. We shall show that

$$\int_{-a}^{a} f(x)\,\mathrm{d}x = \begin{cases} 2\int_0^a f(x)\,\mathrm{d}x, & \text{if } f \text{ is an even function,} \\ 0, & \text{if } f \text{ is an odd function.} \end{cases}$$

If $f(-x) = f(x)$, then

$$\int_{-a}^{a} f(x)\,\mathrm{d}x = \int_{-a}^{0} f(x)\,\mathrm{d}x + \int_{0}^{a} f(x)\,\mathrm{d}x = \int_{a}^{0} f(-t)(-1)\,\mathrm{d}t + \int_{0}^{a} f(x)\,\mathrm{d}x =$$
$$= \int_{0}^{a} f(-t)\,\mathrm{d}t + \int_{0}^{a} f(x)\,\mathrm{d}x = \int_{0}^{a} \big(f(-x) + f(x)\big)\,\mathrm{d}x =$$
$$= 2\int_{0}^{a} f(x)\,\mathrm{d}x.$$

If $f(-x) = -f(x)$, we obtain from the same computations that

$$\int_{-a}^{a} f(x)\,\mathrm{d}x = \int_{0}^{a} \big(f(-x) + f(x)\big)\,\mathrm{d}x = \int_{0}^{a} 0\,\mathrm{d}x = 0.$$

Example 4 Let f be a function defined on the entire real line $\mathbb{R}$ and having period T, that is $f(x + T) = f(x)$ for all $x \in \mathbb{R}$.

If f is integrable on each finite closed interval, then for any $a \in \mathbb{R}$ we have the equality

$$\int_{a}^{a+T} f(x)\,\mathrm{d}x = \int_{0}^{T} f(x)\,\mathrm{d}x,$$

that is, the integral of a periodic function over an interval whose length equals the period T of the function is independent of the location of the interval of integration on the real line:

$$\int_a^{a+T} f(x)\,\mathrm{d}x = \int_a^0 f(x)\,\mathrm{d}x + \int_0^T f(x)\,\mathrm{d}x + \int_T^{a+T} f(x)\,\mathrm{d}x =$$
$$= \int_0^T f(x)\,\mathrm{d}x + \int_a^0 f(x)\,\mathrm{d}x + \int_0^a f(t+T)\cdot 1\,\mathrm{d}t =$$
$$= \int_0^T f(x)\,\mathrm{d}x + \int_a^0 f(x)\,\mathrm{d}x + \int_0^a f(t)\,\mathrm{d}t = \int_0^T f(x)\,\mathrm{d}x.$$

Here we have made the change of variable $x = t + T$ and used the periodicity of the function $f(x)$.

Example 5 Suppose we need to compute the integral $\int_0^1 \sin(x^2)\,\mathrm{d}x$, for example within 10^{-2}.

We know that the primitive $\int \sin(x^2)\,\mathrm{d}x$ (the Fresnel integral) cannot be expressed in terms of elementary functions, so that it is impossible to use the Newton–Leibniz formula here in the traditional sense.

We take a different approach. When studying Taylor's formula in differential calculus, we found as an example (see Example 11 of Sect. 5.3) that on the interval [–1, 1] the equality

$$\sin x \approx x - \frac{1}{3!}x^3 + \frac{1}{5!}x^5 =: P(x)$$

holds within 10^{-3}.

But if $|\sin x - P(x)| < 10^{-3}$ on the interval [–1, 1], then $|\sin(x^2) - P(x^2)| < 10^{-3}$ also, for $0 \le x \le 1$.

Consequently,

$$\left|\int_0^1 \sin(x^2)\,\mathrm{d}x - \int_0^1 P(x^2)\,\mathrm{d}x\right| \le \int_0^1 \left|\sin(x^2) - P(x^2)\right|\mathrm{d}x < \int_0^1 10^{-3}\,\mathrm{d}x < 10^{-3}.$$

Thus, to compute the integral $\int_0^1 \sin(x^2)\,\mathrm{d}x$ with the required precision, it suffices to compute the integral $\int_0^1 P(x^2)\,\mathrm{d}x$. But

$$\int_0^1 P(x^2)\,\mathrm{d}x = \int_0^1 \left(x^2 - \frac{1}{3!}x^3 + \frac{1}{5!}x^{10}\right)\mathrm{d}x =$$
$$= \left(\frac{1}{3!}x^3 - \frac{1}{3!7}x^7 + \frac{1}{5!11}x^{11}\right)\Big|_0^1 = \frac{1}{3} - \frac{1}{3!7} + \frac{1}{5!11} =$$
$$= 0.310 \pm 10^{-3},$$

and therefore

$$\int_0^1 \sin(x^2)\,\mathrm{d}x = 0.310 \pm 2 \times 10^{-3} = 0.31 \pm 10^{-2}.$$

Example 6 The quantity $\mu = \frac{1}{b-a}\int_a^b f(x)\,\mathrm{d}x$ is called the *integral average value of the function on the closed interval* $[a, b]$.

Let f be a function that is defined on $\mathbb{R}$ and integrable on any closed interval. We use f to construct the new function

$$F_\delta(x) = \frac{1}{2\delta}\int_{x-\delta}^{x+\delta} f(t)\,\mathrm{d}t,$$

whose value at the point x is the integral average value of f in the δ-neighborhood of x.

We shall show that $F_\delta(x)$ (called the *average* of f) is, compared to f, more regular. More precisely, if f is integrable on any interval $[a, b]$, then $F_\delta(x)$ is continuous on $\mathbb{R}$, and if $f \in C(\mathbb{R})$, then $F_\delta(x) \in C^{(1)}(\mathbb{R})$.

We verify first that $F_\delta(x)$ is continuous:

$$\left|F_\delta(x+h) - F_\delta(x)\right| = \frac{1}{2\delta}\left|\int_{x+\delta}^{x+\delta+h} f(t)\,\mathrm{d}t + \int_{x-\delta+h}^{x-\delta} f(t)\,\mathrm{d}t\right| \le$$

$$\le \frac{1}{2\delta}\bigl(C|h| + C|h|\bigr) = \frac{C}{\delta}|h|,$$

if $|f(t)| \le C$, for example, in the 2δ-neighborhood of x and $|h| < \delta$. It is obvious that this estimate implies the continuity of $F_\delta(x)$.

Now if $f \in C(\mathbb{R})$, then by the rule for differentiating a composite function

$$\frac{\mathrm{d}}{\mathrm{d}x}\int_a^{\varphi(x)} f(t)\,\mathrm{d}t = \frac{\mathrm{d}}{\mathrm{d}\varphi}\int_a^{\varphi} f(t)\,\mathrm{d}t \cdot \frac{\mathrm{d}\varphi}{\mathrm{d}x} = f\bigl(\varphi(x)\bigr)\varphi'(x),$$

so that from the expression

$$F_\delta(x) = \frac{1}{2\delta}\int_a^{x+\delta} f(t)\,\mathrm{d}t - \frac{1}{2\delta}\int_a^{x-\delta} f(t)\,\mathrm{d}t$$

we find that

$$F'_\delta(x) = \frac{f(x+\delta) - f(x-\delta)}{2\delta}.$$

After the change of variable $t = x + u$ in the integral, the function $F_\delta(x)$ can be written as

$$F_\delta(x) = \frac{1}{2\delta}\int_{-\delta}^{\delta} f(x+u)\,\mathrm{d}u.$$

If $f \in C(\mathbb{R})$, then, applying the first mean-value theorem, we find that

$$F_\delta(x) = \frac{1}{2\delta} f(x+\tau) \cdot 2\delta = f(x+\tau),$$

where $|\tau| \le \delta$. It follows that

$$\lim_{\delta \to +0} F(\delta)(x) = f(x),$$

which is completely natural.

6.3.6 Problems and Exercises

1. Using the integral, find

a) $\lim_{n\to\infty}[\frac{n}{(n+1)^2} + \cdots + \frac{n}{(2n)^2}]$;
b) $\lim_{n\to\infty} \frac{1^\alpha + 2^\alpha + \cdots + n^\alpha}{n^{\alpha+1}}$, if $\alpha \ge 0$.

2. a) Show that any continuous function on an open interval has a primitive on that interval.

b) Show that if $f \in C^{(1)}[a,b]$, then f can be represented as the difference of two nondecreasing functions on $[a,b]$ (see Problem 4 of Sect. 6.1).

3. Show that if the function g is smooth, then the second mean-value theorem (Theorem 6 of Sect. 6.2) can be reduced to the first mean-value theorem through integration by parts.

4. Show that if $f \in C(\mathbb{R})$, then for any fixed closed interval $[a,b]$, given $\varepsilon > 0$ one can choose $\delta > 0$ so that the inequality $|F_\delta(x) - f(x)| < \varepsilon$ holds on $[a,b]$, where F_δ is the average of the function studied in Example 6.

5. Show that

$$\int_1^{x^2} \frac{\mathrm{e}^t}{t}\,\mathrm{d}t \sim \frac{1}{x^2}\mathrm{e}^{x^2} \quad \text{as } x \to +\infty.$$

6. a) Verify that the function $f(x) = \int_x^{x+1} \sin(t^2)\,\mathrm{d}t$ has the following representation as $x \to \infty$:

$$f(x) = \frac{\cos(x^2)}{2x} - \frac{\cos(x+1)^2}{2(x+1)} + O\left(\frac{1}{x^2}\right).$$

b) Find $\underline{\lim}_{x\to\infty} xf(x)$ and $\overline{\lim}_{x\to\infty} xf(x)$.

7. Show that if $f : \mathbb{R} \to \mathbb{R}$ is a periodic function that is integrable on every closed interval $[a,b] \subset \mathbb{R}$, then the function

$$F(x) = \int_a^x f(t)\,\mathrm{d}t$$

can be represented as the sum of a linear function and a periodic function.

8. a) Verify that for $x > 1$ and $n \in \mathbb{N}$ the function

$$P_n(x) = \frac{1}{\pi}\int_0^{\pi} \left(x + \sqrt{x^2-1}\cos\varphi\right)^n \mathrm{d}\varphi$$

is a polynomial of degree n (the nth *Legendre polynomial*).

b) Show that

$$P_n(x) = \frac{1}{\pi}\int_0^{\pi} \frac{\mathrm{d}\psi}{(x - \sqrt{x^2-1}\cos\psi)^n}.$$

9. Let f be a real-valued function defined on a closed interval $[a,b] \subset \mathbb{R}$ and $\xi_1, \dots, \xi_m$ distinct points of this interval. The values of the *Lagrange interpolating polynomial* of degree $m-1$

$$L_{m-1}(x) := \sum_{j=1}^{m} f(\xi_j) \prod_{i \neq j} \frac{x - \xi_i}{\xi_j - \xi_i}$$

are equal to the values of the function at the points $\xi_1, \dots, \xi_m$ (the nodes of the interpolation), and if $f \in C^{(m)}[a,b]$, then

$$f(x) - L_{m-1}(x) = \frac{1}{m!} f^{(m)}\big(\zeta(x)\big)\omega_m(x),$$

where $\omega_m(x) = \prod_{i=1}^{m}(x - \xi_i)$ and $\zeta(x) \in]a,b[$ (see Exercise 11 in Sect. 5.3).

Let $\xi_i = \frac{b+a}{2} + \frac{b-a}{2}\theta_i$; then $\theta_i \in [-1,1]$, $i = 1, \dots, m$.

a) Show that

$$\int_a^b L_{m-1}(x)\,\mathrm{d}x = \frac{b-a}{2}\sum_{i=1}^{m} c_i f(\xi_i),$$

where

$$c_i = \int_{-1}^{1}\left(\prod_{i \neq j}\frac{t - \theta_i}{\theta_j - \theta_i}\right)\mathrm{d}t.$$

In particular

α_1) $\int_a^b L_0(x)\,\mathrm{d}x = (b-a)f(\frac{a+b}{2})$, if $m = 1$, $\theta_1 = 0$;

α_2) $\int_a^b L_1(x)\,\mathrm{d}x = \frac{b-a}{2}[f(a) + f(b)]$, if $m = 2$, $\theta_1 = -1$, $\theta_2 = 1$;

α_3) $\int_a^b L_2(x)\,\mathrm{d}x = \frac{b-a}{6}[f(a) + 4f(\frac{a+b}{2}) + f(b)]$, if $m = 3$, $\theta_1 = -1$, $\theta_2 = 0$, $\theta_3 = 1$.

b) Assuming that $f \in C^{(m)}[a,b]$ and setting $M_m = \max_{x\in[a,b]}|f^{(m)}(x)|$, estimate the magnitude R_m of the absolute error in the formula

$$\int_a^b f(x)\,\mathrm{d}x = \int_a^b L_{m-1}(x)\,\mathrm{d}x + R_m \qquad (*)$$

and show that $|R_m| \le \frac{Mm}{m!}\int_a^b |\omega_m(x)|\,\mathrm{d}x$.

c) In cases α_1), α_2), and α_3) formula (*) is called respectively the *rectangular*, *trapezoidal*, and *parabolic* rule. In the last case it is also called *Simpson's rule*.[8] Show that the following formulas hold in cases α_1), α_2), and α_3):

$$R_1 = \frac{f'(\xi_1)}{4}(b-a)^2, \qquad R_2 = -\frac{f''(\xi_2)}{12}(b-a)^3,$$

$$R_3 = -\frac{f^{(4)}(\xi_3)}{2880}(b-a)^5,$$

where $\xi_1, \xi_2, \xi_3 \in [a,b]$ and the function f belongs to a suitable class $C^{(k)}[a,b]$.

d) Let f be a polynomial P. What is the highest degree of polynomials P for which the rectangular, trapezoidal, and parabolic rules respectively are exact?

Let $h = \frac{b-a}{n}$, $x_k = a + hk$ $(k = 0, 1, \dots, n)$, and $y_k = f(x_k)$.

e) Show that in the *rectangular rule*

$$\int_a^b f(x)\,\mathrm{d}x = h(y_0 + y_1 + \cdots + y_{n-1}) + R_1$$

the remainder has the form $R_1 = \frac{f'(\xi)}{2}(b-a)h$, where $\xi \in [a,b]$.

f) Show that in the *trapezoidal rule*

$$\int_a^b f(x)\,\mathrm{d}x = \frac{h}{2}\big[(y_0 + y_n) + 2(y_1 + y_2 + \cdots + y_{n-1})\big] + R_2$$

the remainder has the form $R_2 = -\frac{f''(\xi)}{12}(b-a)h^2$, where $\xi \in [a,b]$.

g) Show that in *Simpson's rule (the parabolic rule)*

$$\int_a^b f(x)\,\mathrm{d}x = \frac{h}{3}\big[(y_0 + y_n) + 4(y_1 + y_3 + \cdots + y_{n-1}) +$$
$$+ 2(y_2 + y_4 + \cdots + y_{n-2})\big] + R_3,$$

which can be written for even values of n, the remainder R_3 has the form

$$R_3 = -\frac{f^{(4)}(\xi)}{180}(b-a)h^4,$$

where $\xi \in [a,b]$.

h) Starting from the relation

$$\pi = 4\int_0^1 \frac{\mathrm{d}x}{1+x^2},$$

[8] T. Simpson (1710–1761) – British mathematician.

compute π within 10^{-3}, using the rectangular, trapezoidal, and parabolic rules. Note carefully the efficiency of Simpson's rule, which is, for that reason, the most widely used *quadrature* formula. (That is the name given to formulas for numerical integration in the one-dimensional case, in which the integral is identified with the area of the corresponding curvilinear trapezoid.)

10. By transforming formula (6.46), obtain the following forms for the remainder term in Taylor's formula, where we have set $h = x - a$:

a) $\frac{h^n}{(n-1)!} \int_0^1 f^{(n)}(a + \tau h)(1 - \tau)^{n-1}\, d\tau$;
b) $\frac{h^n}{n!} \int_0^1 f^{(n)}(x - h\sqrt[n]{t})\, dt$.

11. Show that the important formula (6.48) for change of variable in an integral remains valid without the assumption that the function in the substitution is monotonic.

6.4 Some Applications of Integration

There is a single pattern of ideas that often guides the use of integration in applications; for that reason it is useful to expound this pattern once in its pure form. The first subsection of this section is devoted to that purpose.

6.4.1 Additive Interval Functions and the Integral

In discussing the additivity of the integral over intervals in Sect. 6.2 we introduced the concept of an *additive* (*oriented*) *interval function*. We recall that this is a function $(\alpha, \beta) \mapsto I(\alpha, \beta)$ that assigns a number $I(\alpha, \beta)$ to each ordered pair of points (α, β) of a fixed closed interval $[a, b]$, in such a way that the following equality holds for any triple of points $\alpha, \beta, \gamma \in [a, b]$:

$$I(\alpha, \gamma) = I(\alpha, \beta) + I(\beta, \gamma). \tag{6.49}$$

It follows from (6.49) when $\alpha = \beta = \gamma$ that $I(\alpha, \alpha) = 0$, while for $\alpha = \gamma$ we find that $I(\alpha, \beta) + I(\beta, \alpha) = 0$, that is, $I(\alpha, \beta) = -I(\beta, \alpha)$. This relation shows the effect of the order of the points α, β.

Setting

$$\mathcal{F}(x) = I(a, x),$$

by the additivity of the function I we have

$$I(\alpha, \beta) = I(a, \beta) - I(a, \alpha) = \mathcal{F}(\beta) - \mathcal{F}(\alpha).$$

Thus, every additive oriented interval function has the form

$$I(\alpha, \beta) = \mathcal{F}(\beta) - \mathcal{F}(\alpha), \tag{6.50}$$

where $x \mapsto \mathcal{F}(x)$ is a function of points on the interval $[a, b]$.

It is easy to verify that the converse is also true, that is, any function $x \mapsto \mathcal{F}(x)$ defined on $[a, b]$ generates an additive (oriented) interval function by formula (6.50).

We now give two typical examples.

Example 1 If $f \in \mathcal{R}[a, b]$, the function $\mathcal{F}(x) = \int_a^x f(t)\,\mathrm{d}t$ generates via formula (6.50) the additive function

$$I(\alpha, \beta) = \int_\alpha^\beta f(t)\,\mathrm{d}t.$$

We remark that in this case the function $\mathcal{F}(x)$ is continuous on the closed interval $[a, b]$.

Example 2 Suppose the interval $[0, 1]$ is a weightless string with a bead of unit mass attached to the string at the point $x = 1/2$.

Let $\mathcal{F}(x)$ be the amount of mass located in the closed interval $[0, x]$ of the string. Then by hypothesis

$$\mathcal{F}(x) = \begin{cases} 0 & \text{for } x < 1/2, \\ 1 & \text{for } 1/2 \leq x \leq 1. \end{cases}$$

The physical meaning of the additive function

$$I(\alpha, \beta) = \mathcal{F}(\beta) - \mathcal{F}(\alpha)$$

for $\beta > \alpha$ is the amount of mass located in the half-open interval $]\alpha, \beta]$.

Since the function $\mathcal{F}$ is discontinuous, the additive function $I(\alpha, \beta)$ in this case cannot be represented as the Riemann integral of a function – a mass density. (This density, that is, the limit of the ratio of the mass in an interval to the length of the interval, would have to be zero at any point of the interval $[a, b]$ except the point $x = 1/2$, where it would have to be infinite.)

We shall now prove a sufficient condition for an additive interval function to be generated by an integral, one that will be useful in what follows.

Proposition 1 *Suppose the additive function $I(\alpha, \beta)$ defined for points α, β of a closed interval $[a, b]$ is such that there exists a function $f \in \mathcal{R}[a, b]$ connected with I as follows: the relation*

$$\inf_{x \in [\alpha, \beta]} f(x)(\beta - \alpha) \leq I(\alpha, \beta) \leq \sup_{x \in [\alpha, \beta]} f(x)(\beta - \alpha)$$

holds for any closed interval $[\alpha, \beta]$ such that $a \leq \alpha \leq \beta \leq b$. Then

$$I(a, b) = \int_a^b f(x)\,\mathrm{d}x.$$

Proof Let P be an arbitrary partition $a = x_0 < \cdots < x_n = b$ of the closed interval $[a, b]$, let $m_i = \inf_{x\in[x_{i-1},x_i]} f(x)$, and let $M_i = \sup_{x\in[x_{i-1},x_i]} f(x)$.

For each interval $[x_{i-1}, x_i]$ of the partition P we have by hypothesis

$$m_i \Delta x_i \le I(x_{i-1}, x_i) \le M_i \Delta x_i .$$

Summing these inequalities and using the additivity of the function $I(\alpha, \beta)$, we obtain

$$\sum_{i=1}^{n} m_i \Delta x_i \le I(a, b) \le \sum_{i=1}^{n} M_i \Delta x_i .$$

The extreme terms in this last relation are familiar to us, being the upper and lower Darboux sums of the function f corresponding to the partition P of the closed interval $[a, b]$. As $\lambda(P) \to 0$ they both have the integral of f over the closed interval $[a, b]$ as their limit. Thus, passing to the limit as $\lambda(P) \to 0$, we find that

$$I(a, b) = \int_a^b f(x)\, \mathrm{d}x .$$

□

Let us now illustrate Proposition 1 in action.

6.4.2 Arc Length

Suppose a particle is moving in space $\mathbb{R}^3$ and suppose its law of motion is known to be $\mathbf{r}(t) = (x(t), y(t), z(t))$, where $x(t)$, $y(t)$, and $z(t)$ are the rectangular Cartesian coordinates of the point at time t.

We wish to define the length $l[a, b]$ of the path traversed during the time interval $a \le t \le b$.

Let us make some concepts more precise.

Definition 1 A *path* in $\mathbb{R}^3$ is a mapping $t \mapsto (x(t), y(t), z(t))$ of an interval of the real line into $\mathbb{R}^3$ defined by functions $x(t)$, $y(t)$, $z(t)$ that are continuous on the interval.

Definition 2 If $t \mapsto (x(t), y(t), z(t))$ is a path for which the domain of the parameter t is the closed interval $[a, b]$ then the points

$$A = \big(x(a), y(a), z(a)\big) \quad \text{and} \quad B = \big(x(b), y(b), z(b)\big)$$

in $\mathbb{R}^3$ are called the *initial point* and *terminal point* of the path.

Definition 3 A path is *closed* if it has both an initial and terminal point, and these points coincide.

Definition 4 If $\Gamma : I \to \mathbb{R}^3$ is a path, the image $\Gamma(I)$ of the interval I in $\mathbb{R}^3$ is called the *support* of the path.

The support of an abstract path may turn out to be not at all what we would like to call a curve. There are examples of paths whose supports, for example, contain an entire three-dimensional cube (the so-called Peano "curves"). However, if the functions $x(t)$, $y(t)$, and $z(t)$ are sufficiently regular (as happens, for example, in the case of a mechanical motion, when they are differentiable), we are guaranteed that nothing contrary to our intuition will occur, as one can verify rigorously.

Definition 5 A path $\Gamma : I \to \mathbb{R}^3$ for which the mapping $I \to \Gamma(I)$ is one-to-one is called a *simple path* or *parametrized curve*, and its support is called *a curve* in $\mathbb{R}^3$.

Definition 6 A closed path $\Gamma : [a, b] \to \mathbb{R}^3$ is called a *simple closed path* or *simple closed curve* if the path $\Gamma : [a, b[\to \mathbb{R}^3$ is simple.

Thus a simple path differs from an arbitrary path in that when moving over its support we do not return to points reached earlier, that is, we do not intersect our trajectory anywhere except possibly at the terminal point, when the simple path is closed.

Definition 7 The path $\Gamma : I \to \mathbb{R}^3$ is called a path *of a given smoothness* if the functions $x(t)$, $y(t)$, and $z(t)$ have that smoothness.

(For example, the smoothness $C[a, b]$, $C^{(1)}[a, b]$, or $C^{(k)}[a, b]$.)

Definition 8 A path $\Gamma : [a, b] \to \mathbb{R}^3$ is *piecewise smooth* if the closed interval $[a, b]$ can be partitioned into a finite number of closed intervals on each of which the corresponding restriction of the mapping Γ is defined by continuously differentiable functions.

It is smooth paths, that is, paths of class $C^{(1)}$ and piecewise smooth paths that we intend to study just now.

Let us return to the original problem, which we can now state as the problem of defining the length of a smooth path $\Gamma : [a, b] \to \mathbb{R}^3$.

Our initial ideas about the length $l[a, b]$ of the path traversed during the time interval $\alpha \le t \le \beta$ are as follows: First, if $\alpha < \beta < \gamma$, then

$$l[\alpha, \gamma] = l[\alpha, \beta] + l[\beta, \gamma],$$

and second, if $\mathbf{v}(t) = (\dot{x}(t), \dot{y}(t), \dot{z}(t))$ is the velocity of the point at time t, then

$$\inf_{x\in[\alpha,\beta]} \left|\mathbf{v}(t)\right|(\beta - \alpha) \le l[\alpha, \beta] \le \sup_{x\in[\alpha,\beta]} \left|\mathbf{v}(t)\right|(\beta - \alpha).$$

Thus, if the functions $x(t)$, $y(t)$, and $z(t)$ are continuously differentiable on $[a, b]$, by Proposition 1 we arrive in a deterministic manner at the formula

$$l[a, b] = \int_a^b \left|\mathbf{v}(t)\right| \mathrm{d}t = \int_a^b \sqrt{\dot{x}^2(t) + \dot{y}^2(t) + \dot{z}^2(t)}\, \mathrm{d}t, \tag{6.51}$$

which we now take as the definition of the length of a smooth path $\Gamma : [a, b] \to \mathbb{R}^3$.

If $z(t) \equiv 0$, the support lies in a plane, and formula (6.51) assumes the form

$$l[a, b] = \int_a^b \sqrt{\dot{x}^2(t) + \dot{y}^2(t)}\, dt. \tag{6.52}$$

Example 3 Let us test formula (6.52) on a familiar object. Suppose the point moves according to the law

$$\begin{aligned} x &= R \cos 2\pi t, \\ y &= R \sin 2\pi t. \end{aligned} \tag{6.53}$$

Over the time interval [0, 1] the point will traverse a circle of radius R, that is, a path of length $2\pi R$ if the length of a circle can be computed from this formula.

Let us carry out the computation according to formula (6.52):

$$l[0, 1] = \int_0^1 \sqrt{(-2\pi R \sin 2\pi t)^2 + (2\pi R \cos 2\pi t)^2}\, dt = 2\pi R.$$

Despite the encouraging agreement of the results, the reasoning just carried out contains some logical gaps that are worth paying attention to.

The functions $\cos\alpha$ and $\sin\alpha$, if we use the high-school definition of them, are the Cartesian coordinates of the image p of the point $p_0 = (1, 0)$ under a rotation through angle α.

Up to sign, the quantity α is measured by the length of the arc of the circle $x^2 + y^2 = 1$ between p_0 and p. Thus, in this approach to trigonometric functions their definition relies on the concept of the length of an arc of a circle and hence, in computing the circumference of a circle above, we were in a certain sense completing a logical circle by giving the parametrization in the form (6.53).

However, this difficulty, as we shall now see, is not fundamental, since a parametrization of the circle can be given without resorting to trigonometric functions at all.

Let us consider the problem of computing the length of the graph of a function $y = f(x)$ defined on a closed interval $[a, b] \subset \mathbb{R}$. We have in mind the computation of the length of the path $\Gamma : [a, b] \to \mathbb{R}^2$ having the special parametrization

$$x \mapsto \bigl(x, f(x)\bigr),$$

from which one can conclude that the mapping $\Gamma : [a, b] \to \mathbb{R}^2$ is one-to-one. Hence, by Definition 5 the graph of a function is a curve in $\mathbb{R}^2$.

In this case formula (6.52) can be simplified, since by setting $x = t$ and $y = f(t)$ in it, we obtain

$$l[a, b] = \int_a^b \sqrt{1 + \bigl[f'(t)\bigr]^2}\, dt. \tag{6.54}$$

In particular, if we consider the semicircle

$$y = \sqrt{1-x^2}, \quad -1 \le x \le 1,$$

of the circle $x^2 + y^2 = 1$, we obtain for it

$$l = \int_{-1}^{+1} \sqrt{1 + \left[\frac{-x}{\sqrt{1-x^2}}\right]^2}\, \mathrm{d}x = \int_{-1}^{1} \frac{\mathrm{d}x}{\sqrt{1-x^2}}. \tag{6.55}$$

But the integrand in this last integral is an unbounded function, and hence does not exist in the traditional sense we have studied. Does this mean that a semicircle has no length? For the time being it means only that this parametrization of the semicircle does not satisfy the condition that the functions $\dot{x}$ and $\dot{y}$ be continuous, under which formula (6.52), and hence also formula (6.54), was written. For that reason we must either consider broadening the concept of integral or passing to a parametrization satisfying the conditions under which (6.54) can be applied.

We remark that if we consider this parametrization on any closed interval of the form $[-1+\delta, 1-\delta]$, where $-1 < -1+\delta < 1-\delta < 1$, then formula (6.54) applies on that interval, and we find the length

$$l[-1+\delta, 1-\delta] = \int_{-1+\delta}^{1-\delta} \frac{\mathrm{d}x}{\sqrt{1-x^2}}$$

for the arc of the circle lying above the closed interval $[-1+\delta, 1-\delta]$.

It is therefore natural to consider that the length l of the semicircle is the limit $\lim_{\delta \to +0} l[-1+\delta, 1-\delta]$. One can interpret the integral in (6.55) in the same sense. We shall study this naturally arising extension of the concept of a Riemann integral in the next section.

As for the particular problem we are studying, without even changing the parametrization one can find, for example, the length $l[-\frac{1}{2}, \frac{1}{2}]$ of an arc of the unit circle subtended by a chord congruent to the radius of the circle. Then (from geometric considerations alone) it must be that $l = 3 \cdot l[-\frac{1}{2}, \frac{1}{2}]$.

We remark also that

$$\int \frac{\mathrm{d}x}{\sqrt{1-x^2}} = \int \frac{(1-x^2+x^2)\,\mathrm{d}x}{\sqrt{1-x^2}} = \int \sqrt{1-x^2}\,\mathrm{d}x - \frac{1}{2}\int \frac{x\,\mathrm{d}(1-x^2)}{\sqrt{1-x^2}} =$$
$$= 2\int \sqrt{1-x^2}\,\mathrm{d}x - x\sqrt{1-x^2},$$

and therefore

$$l[-1+\delta, 1-\delta] = 2\int_{-1+\delta}^{1-\delta} \sqrt{1-x^2}\,\mathrm{d}x - \left(x\sqrt{1-x^2}\right)\Big|_{-1+\delta}^{1-\delta}.$$

Thus,

$$l = \lim_{\delta \to +0} l[-1+\delta, 1-\delta] = 2\int_{-1}^{1} \sqrt{1-x^2}\,\mathrm{d}x.$$

The length of a semicircle of unit radius is denoted π, and we thus arrive at the following formula

$$\pi = 2\int_{-1}^{1} \sqrt{1-x^2}\,dx.$$

This last integral is an ordinary (not generalized) Riemann integral and can be computed with any precision.

If for $x \in [-1, 1]$ we define $\arccos x$ as $l[x, 1]$, then by the computations carried out above

$$\arccos x = \int_x^1 \frac{dt}{\sqrt{1-t^2}},$$

or

$$\arccos x = x\sqrt{1-x^2} + 2\int_x^1 \sqrt{1-t^2}\,dt.$$

If we regard arc length as a primitive concept, then we must also regard the function $x \mapsto \arccos x$ introduced just now and the function $x \mapsto \arcsin x$, which can be introduced similarly, as primitive. But the functions $x \mapsto \cos x$ and $x \mapsto \sin x$ can then be obtained as the inverses of these functions on the corresponding intervals. In essence, this is what is done in elementary geometry.

The example of the length of a semicircle is instructive not only because while studying it we made a remark on the definition of the trigonometric functions that may be of use to someone, but also because it naturally raises the question whether the number defined by formula (6.51) depends on the coordinate system x, y, z and the parametrization of the curve when one is finding the length of a curve.

Leaving to the reader the analysis of the role played by three-dimensional Cartesian coordinates, we shall examine here the role of the parametrization.

We need to clarify that by a *parametrization* of a curve in $\mathbb{R}^3$, we mean a definition of a simple path $\Gamma : I \to \mathbb{R}^3$ whose support is that curve. The point or number $t \in I$ is called a *parameter* and the interval I the *domain* of the parameter.

If $\Gamma : I \to \mathcal{L}$ and $\widetilde{\Gamma} : \tilde{I} \to \mathcal{L}$ are two one-to-one mappings with the same set of values $\mathcal{L}$, there naturally arise one-to-one mappings $\widetilde{\Gamma}^{-1} \circ \Gamma : I \to \tilde{I}$ and $\Gamma^{-1} \circ \widetilde{\Gamma} : \tilde{I} \to I$ between the domains I and $\tilde{I}$ of these mappings.

In particular, if there are two parametrizations of the same curve, then there is a natural correspondence between the parameters $t \in I$ and $\tau \in \tilde{I}$, $t = t(\tau)$ or $\tau = \tau(t)$, making it possible, knowing the parameter of a point in one parametrization, to find its parameter in the other parametrization.

Let $\Gamma : [a, b] \to \mathcal{L}$ and $\widetilde{\Gamma} : [\alpha, \beta] \to \mathcal{L}$ be two parametrizations of the same curve with the correspondences $\Gamma(a) = \widetilde{\Gamma}(\alpha)$ and $\Gamma(b) = \widetilde{\Gamma}(\beta)$ between their initial and terminal points. Then the *transition functions* $t = t(\tau)$ and $\tau = \tau(t)$ from one parameter to another will be continuous, strictly monotonic mappings of the closed intervals $a \le t \le b$ and $\alpha \le \tau \le \beta$ onto each other with the initial points and terminal points corresponding: $a \leftrightarrow \alpha$, $b \leftrightarrow \beta$.

Here, if the curves Γ and $\widetilde{\Gamma}$ are defined by the triples $(x(t), y(t), z(t))$ and $(\tilde{x}(t), \tilde{y}(t), \tilde{z}(t))$ of smooth functions such that $|\mathbf{v}(t)|^2 = \dot{x}^2(t) + \dot{y}^2(t) + \dot{z}^2(t) \neq 0$

on $[a, b]$ and $\tilde{\mathbf{v}}(\tau)|^2 = \dot{\tilde{x}}^2(\tau) + \dot{\tilde{y}}^2(\tau) + \dot{\tilde{z}}^2(\tau) \neq 0$ on $[\alpha, \beta]$, then one can verify that in this case the transition functions $t = t(\tau)$ and $\tau = \tau(t)$ are smooth functions having positive derivatives on the intervals on which they are defined.

We shall not undertake to verify this assertion here; it will eventually be obtained as a corollary of the important implicit function theorem. At the moment this assertion is mostly intended as motivation for the following definition.

Definition 9 The path $\widetilde{\Gamma} : [\alpha, \beta] \to \mathbb{R}^3$ is obtained from $\Gamma : [a, b] \to \mathbb{R}^3$ by an *admissible change of parameter* if there exists a smooth mapping $T : [\alpha, \beta] \to [a, b]$ such that $T(\alpha) = a$, $T(\beta) = b$, $T'(\tau) > 0$ on $[\alpha, \beta]$ and $\widetilde{\Gamma} = \Gamma \circ T$.

We now prove a general proposition.

Proposition 2 *If a smooth path $\widetilde{\Gamma} : [\alpha, \beta] \to \mathbb{R}^3$ is obtained from a smooth path $\Gamma : [a, b] \to \mathbb{R}^3$ by an admissible change of parameter, then the lengths of the two paths are equal.*

Proof Let $\widetilde{\Gamma} : [\alpha, \beta] \to \mathbb{R}^3$ and $\Gamma : [a, b] \to \mathbb{R}^3$ be defined respectively by the triples of smooth functions $\tau \mapsto (\tilde{x}(\tau), \tilde{y}(\tau), \tilde{z}(\tau))$ and $t \mapsto (x(t), y(t), z(t))$, and let $t = t(\tau)$ be the admissible change of parameter under which

$$\tilde{x}(\tau) = x\big(t(\tau)\big), \qquad \tilde{y}(\tau) = y\big(t(\tau)\big), \qquad \tilde{z}(\tau) = z\big(t(\tau)\big).$$

Using the definition (6.51) of path length, the rule for differentiating a composite function, and the rule for change of variable in an integral, we have

$$\int_a^b \sqrt{\dot{x}^2(t) + \dot{y}^2(t) + \dot{z}^2(t)}\, \mathrm{d}t =$$

$$= \int_\alpha^\beta \sqrt{\dot{x}^2\big(t(\tau)\big) + \dot{y}^2\big(t(\tau)\big) + \dot{z}^2\big(t(\tau)\big)}\, t'(\tau)\, \mathrm{d}\tau =$$

$$= \int_\alpha^\beta \sqrt{\big[\dot{x}\big(t(\tau)\big)t'(\tau)\big]^2 + \big[\dot{y}\big(t(\tau)\big)t'(\tau)\big]^2 + \big[\dot{z}\big(t(\tau)\big)t'(\tau)\big]^2}\, \mathrm{d}\tau =$$

$$= \int_\alpha^\beta \sqrt{\dot{\tilde{x}}^2(\tau) + \dot{\tilde{y}}^2(\tau) + \dot{\tilde{z}}^2(\tau)}\, \mathrm{d}\tau.$$

□

Thus, in particular, we have shown that the length of a curve is independent of a smooth parametrization of it.

The length of a piecewise smooth path is defined as the sum of the lengths of the smooth paths into which it can be divided; for that reason it is easy to verify that the length of a piecewise smooth path also does not change under an admissible change of its parameter.

To conclude the discussion of the concept of the length of a path and the length of a curve (which, after Proposition 2, we now have the right to talk about), we consider another example.

Fig. 6.2

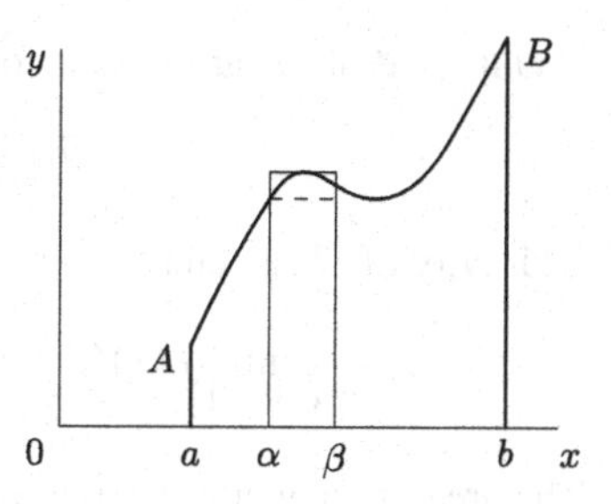

Example 4 Let us find the length of the ellipse defined by the canonical equation

$$\frac{x^2}{a^2} + \frac{y^2}{b^2} = 1 \quad (a \geq b > 0). \tag{6.56}$$

Taking the parametrization $x = a \sin \psi$, $y = b \cos \psi$, $0 \leq \psi \leq 2\pi$, we obtain

$$l = \int_0^{2\pi} \sqrt{(a \cos \psi)^2 + (-b \sin \psi)^2}\, d\psi = \int_0^{2\pi} \sqrt{a^2 - (a^2 - b^2) \sin^2 \psi}\, d\psi =$$

$$= 4a \int_0^{\pi/2} \sqrt{1 - \frac{a^2 - b^2}{a^2} \sin^2 \psi}\, d\psi = 4a \int_0^{\pi/2} \sqrt{1 - k^2 \sin^2 \psi}\, d\psi,$$

where $k^2 = 1 - \frac{b^2}{a^2}$ is the square of the eccentricity of the ellipse.

The integral

$$E(k, \varphi) = \int_0^{\varphi} \sqrt{1 - k^2 \sin^2 \psi}\, d\psi$$

cannot be expressed in elementary functions, and is called an elliptic integral because of the connection with the ellipse just discussed. More precisely, $E(k, \varphi)$ is the *elliptic integral of second kind* in the Legendre form. The value that it assumes for $\varphi = \pi/2$ depends only on k, is denoted $E(k)$, and is called the *complete elliptic integral of second kind*. Thus $E(k) = E(k, \pi/2)$, so that the length of an ellipse has the form $l = 4aE(k)$ in this notation.

6.4.3 The Area of a Curvilinear Trapezoid

Consider the figure $aABb$ of Fig. 6.2, which is called a *curvilinear trapezoid*. This figure is bounded by the vertical line segments aA and bB, the closed interval $[a, b]$ on the x-axis, and the curve $\overset{\frown}{AB}$, which is the graph of an integrable function $y = f(x)$ on $[a, b]$.

Let $[\alpha, \beta]$ be a closed interval contained in $[a, b]$. We denote by $S(\alpha, \beta)$ the area of the curvilinear trapezoid $\alpha f(\alpha) f(\beta) \beta$ corresponding to it.

Our ideas about area are as follows: if $a \le \alpha < \beta < \gamma \le b$, then

$$S(\alpha, \gamma) = S(\alpha, \beta) + S(\beta, \gamma)$$

(additivity of areas) and

$$\inf_{x \in [\alpha,\beta]} f(x)(\beta - \alpha) \le S(\alpha, \beta) \le \sup_{x \in [\alpha,\beta]} f(x)(\beta - \alpha).$$

(The area of an enclosing figure is not less than the area of the figure enclosed.)

Hence by Proposition 1, the area of this figure must be computed from the formula

$$S(a, b) = \int_a^b f(x)\,\mathrm{d}x. \tag{6.57}$$

Example 5 Let us use formula (6.57) to compute the area of the ellipse given by the canonical equation (6.56).

By the symmetry of the figure and the assumed additivity of areas, it suffices to find the area of just the part of the ellipse in the first quadrant, then quadruple the result. Here are the computations:

$$S = 4\int_0^a \sqrt{b^2\left(1 - \frac{x^2}{a^2}\right)}\,\mathrm{d}x = 4b\int_0^{\pi/2} \sqrt{1 - \sin^2 t}\, a\cos t\,\mathrm{d}t =$$

$$= 4ab\int_0^{\pi/2} \cos^2 t\,\mathrm{d}t = 2ab\int_0^{\pi/2} (1 - \cos 2t)\,\mathrm{d}t = \pi ab.$$

Along the way we have made the change of variable $x = a\sin t$, $0 \le t \le \pi/2$.

Thus $S = \pi ab$. In particular, when $a = b = R$, we obtain the formula πR^2 for the area of a disk of radius R.

Remark It should be noted that formula (6.57) gives the area of the curvilinear trapezoid under the condition that $f(x) \ge 0$ on $[a, b]$. If f is an arbitrary integrable function, then the integral (6.57) obviously gives the algebraic sum of the areas of corresponding curvilinear trapezoids lying above and below the x-axis. When this is done, the areas of trapezoids lying above the x-axis are summed with a positive sign and those below with a negative sign.

6.4.4 Volume of a Solid of Revolution

Now suppose the curvilinear trapezoid shown in Fig. 6.2 is revolved about the closed interval $[a, b]$. Let us determine the volume of the solid that results.

We denote by $V(\alpha, \beta)$ the volume of the solid obtained by revolving the curvilinear trapezoid $\alpha f(\alpha) f(\beta)\beta$ (see Fig. 6.2) corresponding to the closed interval $[\alpha, \beta] \subset [a, b]$.

According to our ideas about volume the following relations must hold: if $a \le \alpha < \beta < \gamma \le b$, then

$$V(\alpha, \gamma) = V(\alpha, \beta) + V(\beta, \gamma)$$

and

$$\pi\Big(\inf_{x\in[\alpha,\beta]} f(x)\Big)^2 (\beta - \alpha) \le V(\alpha, \beta) \le \pi\Big(\sup_{x\in[\alpha,\beta]} f(x)\Big)^2 (\beta - \alpha).$$

In this last relation we have estimated the volume $V(\alpha, \beta)$ by the volumes of inscribed and circumscribed cylinders and used the formula for the volume of a cylinder (which is not difficult to obtain, once the area of a disk has been found).

Then by Proposition 1

$$V(a, b) = \pi \int_a^b f^2(x)\,dx. \tag{6.58}$$

Example 6 By revolving about the x-axis the semicircle bounded by the closed interval $[-R, R]$ of the axis and the arc of the circle $y = \sqrt{R^2 - x^2}$, $-R \le x \le R$, one can obtain a three-dimensional ball of radius R whose volume is easily computed from (6.58):

$$V = \pi \int_{-R}^{R} (R^2 - x^2)\,dx = \frac{4}{3}\pi R^3.$$

More details on the measurement of lengths, areas, and volumes will be given in Part 2 of this course. At that time we shall solve the problem of the invariance of the definitions we have given.

6.4.5 Work and Energy

The energy expenditure connected with the movement of a body under the influence of a constant force in the direction in which the force acts is measured by the product $F \cdot S$ of the magnitude of the force and the magnitude of the displacement. This quantity is called the *work* done by the force in the displacement. In general the directions of the force and displacement may be noncollinear (for example, when we pull a sled by a rope), and then the work is defined as the inner product $\langle \mathbf{F}, \mathbf{S} \rangle$ of the force vector and the displacement vector.

Let us consider some examples of the computation of work and the use of the related concept of energy.

Example 7 The work that must be performed against the force of gravity to lift a body of mass m vertically from height h_1 above the surface of the Earth to height h_2 is, by the definition just given, $mg(h_2 - h_1)$. It is assumed that the entire operation occurs near the surface of the Earth, so that the variation of the gravitational force mg can be neglected. The general case is studied in Example 10.

Example 8 Suppose we have a perfectly elastic spring, one end of which is attached at the point 0 of the real line, while the other is at the point x. It is known that the force necessary to hold this end of the spring is kx, where k is the modulus of the spring.

Let us compute the work that must be done to move the free end of the spring from position $x = a$ to $x = b$.

Regarding the work $A(\alpha, \beta)$ as an additive function of the interval $[\alpha, \beta]$ and assuming valid the estimates

$$\inf_{x\in[\alpha,\beta]} (kx)(\beta - \alpha) \le A(\alpha, \beta) \le \sup_{x\in[\alpha,\beta]} (kx)(\beta - \alpha),$$

we arrive via Proposition 1 at the conclusion that

$$A(a, b) = \int_a^b kx\,\mathrm{d}x = \left.\frac{kx^2}{2}\right|_a^b.$$

This work is done against the force. The work done by the spring during the same displacement differs only in sign.

The function $U(x) = \frac{kx^2}{2}$ that we have found enables us to compute the work we do in changing the state of the spring, and hence the work that the spring must do in returning to its initial state. Such a function $U(x)$, which depends only on the configuration of the system, is called the *potential energy of the system.* It is clear from the construction that the derivative of the potential energy gives the force of the spring with the opposite sign.

If a point of mass m moves along the axis subject to this elastic force $F = -kx$, its coordinate $x(t)$ as a function of time satisfies the equation

$$m\ddot{x} = -kx. \tag{6.59}$$

We have already verified once (see Sect. 5.6.6) that the quantity

$$\frac{mv^2}{2} + \frac{kx^2}{2} = K(t) + U\bigl(x(t)\bigr) = E, \tag{6.60}$$

which is the sum of the kinetic and (as we now understand) potential energies of the system, remains constant during the motion.

Example 9 We now consider another example. In this example we shall encounter a number of concepts that we have introduced and become familiar with in differential and integral calculus.

We begin by remarking that by analogy with the function (6.60), which was written for a particular mechanical system satisfying Eq. (6.59), one can verify that for an arbitrary equation of the form

$$\ddot{s}(t) = f\bigl(s(t)\bigr), \tag{6.61}$$

Fig. 6.3

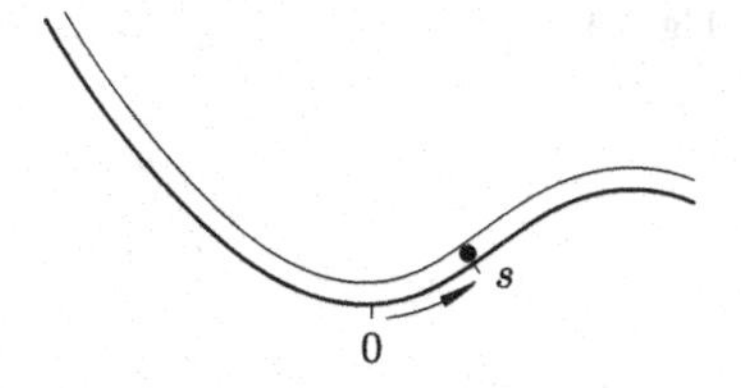

where $f(s)$ is a given function, the sum

$$\frac{\dot{s}^2}{2} + U(s) = E \tag{6.62}$$

does not vary over time if $U'(s) = -f(s)$.

Indeed,

$$\frac{\mathrm{d}E}{\mathrm{d}t} = \frac{1}{2}\frac{\mathrm{d}\dot{s}^2}{\mathrm{d}t} + \frac{\mathrm{d}U(s)}{\mathrm{d}t} = \dot{s}\ddot{s} + \frac{\mathrm{d}U}{\mathrm{d}s}\cdot\frac{\mathrm{d}s}{\mathrm{d}t} = \dot{s}\big(\ddot{s} - f(s)\big) = 0.$$

Thus by (6.62), regarding E as a constant, we obtain successively, first

$$\dot{s} = \pm\sqrt{2\big(E - U(s)\big)}$$

(where the sign must correspond to the sign of the derivative $\frac{\mathrm{d}s}{\mathrm{d}t}$), then

$$\frac{\mathrm{d}t}{\mathrm{d}s} = \pm\frac{1}{\sqrt{2(E - U(s))}},$$

and finally

$$t = c_1 \pm \int \frac{\mathrm{d}s}{\sqrt{2(E - U(s))}}.$$

Consequently, using the law of conservation of the "energy" (6.62) in Eq. (6.61), we have succeeded theoretically in solving this equation by finding not the function $s(t)$, but its inverse $t(s)$.

Equation (6.61) arises, for example, in describing the motion of a point along a given curve. Suppose a particle moves under the influence of the force of gravity along a narrow ideally smooth track (Fig. 6.3).

Let $s(t)$ be the distance along the track (that is, the length of the path) from a fixed point O – the origin of the measurement – to the point where the particle is at time t. It is clear that then $\dot{s}(t)$ is the magnitude of the velocity of the particle and $\ddot{s}(t)$ is the magnitude of the tangential component of its acceleration, which must equal the magnitude of the tangential component of the force of gravity at a given point of the track. It is also clear that the tangential component of the force of gravity depends only on the point of the track, that is, it depends only on s, since s can be

Fig. 6.4

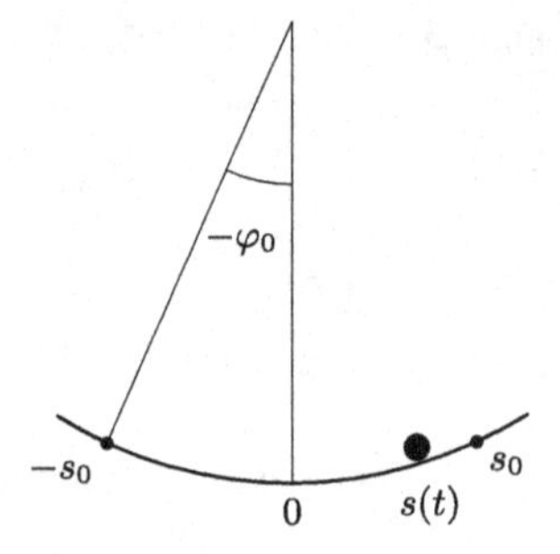

regarded as a parameter that parametrizes the curve[9] with which we are identifying the track. If we denote this component of the force of gravity by $f(s)$, we find that

$$m\ddot{s} = f(s).$$

For this equation the following quantity will be preserved:

$$\frac{1}{2}m\dot{s}^2 + U(s) = E,$$

where $U'(s) = -f(s)$.

Since the term $\frac{1}{2}m\dot{s}^2$ is the kinetic energy of the point and the motion along the track is frictionless, we can guess, avoiding calculations, that the function $U(s)$, up to a constant term, must have the form $mgh(s)$, where $mgh(s)$ is the potential energy of a point of height $h(s)$ in the gravitational field.

If the relations $\dot{s}(0) = 0$, $s(0) = s_0$, and $h(s_0) = h_0$ held at the initial time $t = 0$, then by the relations

$$\frac{2E}{m} = \dot{s}^2 + 2gh(s) = C$$

we find that $C = 2gh(s_0)$, and therefore $\dot{s}^2 = 2g(h_0 - h(s))$ and

$$t = \int_{s_0}^{s} \frac{\mathrm{d}s}{\sqrt{2g(h_0 - h(s))}}. \tag{6.63}$$

In particular if, as in the case of a pendulum, the point moves along a circle of radius R, the length s is measured from the lowest point O of the circle, and the initial conditions amount to the equality $\dot{s}(0) = 0$ at $t = 0$ and a given initial angle of displacement $-\varphi_0$ (see Fig. 6.4). Then, as one can verify, expressing s and $h(s)$ in terms of the angle of displacement φ, we obtain

$$t = \int_{s_0}^{s} \frac{\mathrm{d}s}{\sqrt{2g(h_0 - h(s))}} = \int_{-\varphi_0}^{\varphi} \frac{R\,\mathrm{d}\psi}{\sqrt{2gR(\cos\psi - \cos\varphi_0)}},$$

[9] The parametrization of a curve by its own arc length is called its *natural* parametrization, and s is called the *natural parameter*.

or

$$t = \frac{1}{2}\sqrt{\frac{R}{g}} \int_{-\varphi_0}^{\varphi} \frac{\mathrm{d}\psi}{\sqrt{(\sin^2 \frac{\varphi_0}{2} - \sin^2 \frac{\psi}{2})}}. \tag{6.64}$$

Thus for a half-period $\frac{1}{2}T$ of oscillation of the pendulum we obtain

$$\frac{1}{2}T = \frac{1}{2}\sqrt{\frac{R}{g}} \int_{-\varphi_0}^{\varphi_0} \frac{\mathrm{d}\psi}{\sqrt{\sin^2 \frac{\varphi_0}{2} - \sin^2 \frac{\psi}{2}}}, \tag{6.65}$$

from which, after the substitution $\frac{\sin(\psi/2)}{\sin(\varphi_0/2)} = \sin\theta$, we find

$$T = 4\sqrt{\frac{R}{g}} \int_0^{\pi/2} \frac{\mathrm{d}\theta}{\sqrt{1 - k^2 \sin^2 \theta}}, \tag{6.66}$$

where $k^2 = \sin^2 \frac{\varphi_0}{2}$.

We recall that the function

$$F(k, \varphi) = \int_0^{\varphi} \frac{\mathrm{d}\theta}{\sqrt{1 - k^2 \sin^2 \theta}}$$

is called an *elliptic integral of first kind* in the Legendre form. For $\varphi = \pi/2$ it depends only on k^2, is denoted $K(k)$, and is called the *complete elliptic integral of first kind*. Thus, the period of oscillation of the pendulum is

$$T = 4\sqrt{\frac{R}{g}} K(k). \tag{6.67}$$

If the initial displacement angle φ_0 is small, we can set $k = 0$, and then we obtain the approximate formula

$$T \approx 2\pi\sqrt{\frac{R}{g}}. \tag{6.68}$$

Now that formula (6.66) has been obtained it is still necessary to examine the whole chain of reasoning. When we do, we notice that the integrands in the integrals (6.63)–(6.65) are unbounded functions on the interval of integration. We encountered a similar difficulty in studying the length of a curve, so that we have an approximate idea of how to interpret the integrals (6.63)–(6.65).

However, given that this problem has arisen for the second time, we should study it in a precise mathematical formulation, as will be done in the next section.

Example 10 A body of mass m rises above the surface of the Earth along the trajectory $t \mapsto (x(t), y(t), z(t))$, where t is time, $a \le t \le b$, and x, y, z are the Cartesian coordinates of the point in space. It is required to compute the work of the body against the force of gravity during the time interval $[a, b]$.

The work $A(\alpha, \beta)$ is an additive function of the time interval $[\alpha, \beta] \subset [a, b]$.

A constant force $\mathbf{F}$ acting on a body moving with constant velocity $\mathbf{v}$ performs work $\langle \mathbf{F}, \mathbf{v}h \rangle = \langle \mathbf{F}, \mathbf{v} \rangle h$ in time h, and so the estimate

$$\inf_{t\in[\alpha,\beta]} \big\langle \mathbf{F}\big(p(t)\big), \mathbf{v}(t)\big\rangle(\beta - \alpha) \leq A(\alpha, \beta) \leq \sup_{t\in[\alpha,\beta]} \big\langle \mathbf{F}\big(p(t)\big), \mathbf{v}(t)\big\rangle(\beta - \alpha)$$

seems natural, where $\mathbf{v}(t)$ is the velocity of the body at time t, $p(t)$ is the point in space where the body is located at time t, and $\mathbf{F}(p(t))$ is the force acting on the body at the point $p = p(t)$.

If the function $\langle \mathbf{F}(p(t)), \mathbf{v}(t) \rangle$ happens to be integrable, then by Proposition 1 we must conclude that

$$A(a, b) = \int_a^b \big\langle \mathbf{F}\big(p(t)\big), \mathbf{v}(t)\big\rangle \, \mathrm{d}t.$$

In the present case $\mathbf{v}(t) = (\dot{x}(t), \dot{y}(t), \dot{z}(t))$, and if $\mathbf{r}(t) = (x(t), y(t), z(t))$, then by the law of universal gravitation, we find

$$\mathbf{F}(p) = \mathbf{F}(x, y, z) = G\frac{mM}{|\mathbf{r}|^3}\mathbf{r} = \frac{GmM}{(x^2 + y^2 + z^2)^{3/2}}(x, y, z),$$

where M is the mass of the Earth and its center is taken as the origin of the coordinate system.

Then,

$$\langle \mathbf{F}, \mathbf{v} \rangle(t) = GmM\frac{x(t)\dot{x}(t) + y(t)\dot{y}(t) + z(t)\dot{z}(t)}{(x^2(t) + y^2(t) + z^2(t))^{3/2}},$$

and therefore

$$\int_a^b \langle \mathbf{F}, \mathbf{v} \rangle(t) \, \mathrm{d}t = \frac{1}{2}GmM \int_a^b \frac{(x^2(t) + y^2(t) + z^2(t))'}{(x^2(t) + y^2(t) + z^2(t))^{3/2}} \, \mathrm{d}t =$$

$$= -\frac{GmM}{(x^2(t) + y^2(t) + z^2(t))^{1/2}}\bigg|_a^b = -\frac{GmM}{|\mathbf{r}(t)|}\bigg|_a^b.$$

Thus

$$A(a, b) = \frac{GmM}{|\mathbf{r}(b)|} - \frac{GmM}{|\mathbf{r}(a)|}.$$

We have discovered that the work we were seeking depends only on the magnitudes $|\mathbf{r}(a)|$ and $|\mathbf{r}(b)|$ of the distance of the body of mass m from the center of the Earth at the initial and final instants of time in the interval $[a, b]$.

Setting

$$U(r) = \frac{GM}{r},$$

Fig. 6.5

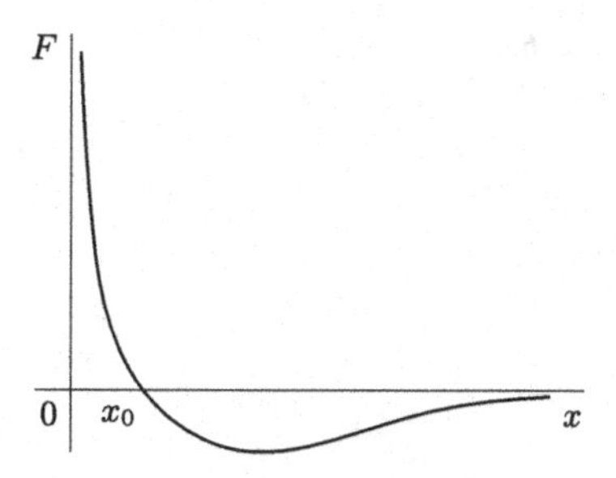

we find that the work done against gravity in displacing the mass m from any point of a sphere of radius r_0 to any point of a sphere of radius r_1 is computed by the formula

$$A_{r_0 r_1} = m\big(U(r_0) - U(r_1)\big).$$

The function $U(r)$ is called a *Newtonian potential.* If we denote the radius of the Earth by R, then, since $\frac{GM}{R^2} = g$, we can rewrite $U(r)$ as

$$U(r) = \frac{gR^2}{r}.$$

Taking this into account, we can obtain the following expression for the work needed to escape from the Earth's gravitational field, more precisely, to move a body of mass m from the surface of the Earth to an infinite distance from the center of the Earth. It is natural to take that quantity to be the limit $\lim_{r\to+\infty} A_{Rr}$.

Thus the *escape work* is

$$A = A_{R\infty} = \lim_{r\to+\infty} A_{Rr} = \lim_{r\to+\infty} m\left(\frac{gR^2}{R} - \frac{gR^2}{r}\right) = mgR.$$

6.4.6 Problems and Exercises

1. Figure 6.5 shows the graph of the dependence $F = F(x)$ of a force acting along the x-axis on a test particle located at the point x on the axis.

a) Sketch the potential for this force in the same coordinates.

b) Describe the potential of the force $-F(x)$.

c) Investigate to determine which of these two cases is such that the position x_0 of the particle is a stable equilibrium position and what property of the potential is involved in stability.

2. Based on the result of Example 10, compute the velocity a body must have in order to escape from the gravitational field of the Earth (the escape velocity for the Earth).

Fig. 6.6

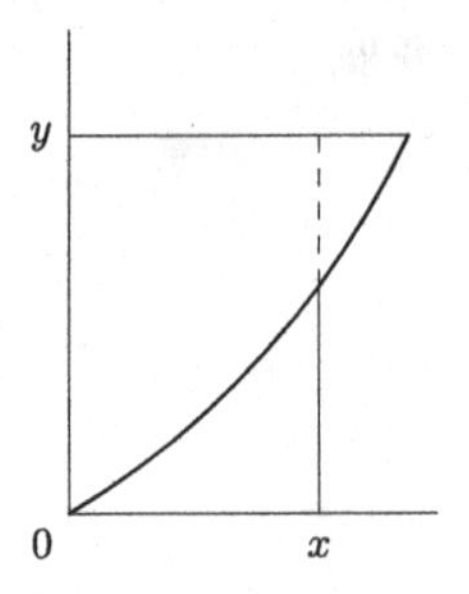

3. On the basis of Example 9

a) derive the equation $R\ddot{\varphi} = -g \sin \varphi$ for the oscillations of a mathematical pendulum;

b) assuming the oscillations are small, obtain an approximate solution of this equation;

c) from the approximate solution, determine the period of oscillation of the pendulum and compare the result with formula (6.68).

4. A wheel of radius r rolls without slipping over a horizontal plane at a uniform velocity v. Suppose at time $t = 0$ the uppermost point A of the wheel has coordinates $(0, 2r)$ in a Cartesian coordinate system whose x-axis lies in the plane and is directed along the velocity vector.

a) Write the law of motion $t \mapsto (x(t), y(t))$ of the point A.

b) Find the velocity of A as a function of time.

c) Describe graphically the trajectory of A. (This curve is called a *cycloid*.)

d) Find the length of one arch of the cycloid (the length of one period of this periodic curve).

e) The cycloid has a number of interesting properties, one of which, discovered by Huygens[10] is that the period of oscillation of a cycloidal pendulum (a ball rolling in a cycloidal well) is independent of the height to which it rises above the lowest point of the well. Try to prove this, using Example 9. (See also Problem 6 of the next section, which is devoted to improper integrals.)

5. a) Starting from Fig. 6.6, explain why, if $y = f(x)$ and $x = g(y)$ are a pair of mutually inverse continuous nonnegative functions equal to 0 at $x = 0$ and $y = 0$ respectively, then the inequality

$$xy \leq \int_0^x f(t)\, dt + \int_0^y g(t)\, dt$$

must hold.

---

[10]Ch. Huygens (1629–1695) – Dutch engineer, physicist, mathematician, and astronomer.

b) Obtain Young's inequality

$$xy \le \frac{1}{p}x^p + \frac{1}{q}y^q$$

from a) for $x, y \ge 0$, $p, q > 0$, $\frac{1}{p} + \frac{1}{q} = 1$.

c) What geometric meaning does equality have in the inequalities of a) and b)?

6. *The Buffon needle problem.*[11] The number π can be computed in the following rather surprising way.

We take a large sheet of paper, ruled into parallel lines a distance h apart and we toss a needle of length $l < h$ at random onto it. Suppose we have thrown the needle N times, and on n of those times the needle landed across one of the lines. If N is sufficiently large, then $\pi \approx \frac{2l}{ph}$, where $p = \frac{n}{N}$ is the approximate probability that the needle will land across a line.

Starting from geometric considerations connected with the computation of area, try to give a satisfactory explanation of this method of computing π.

6.5 Improper Integrals

In the preceding section we encountered the need for a somewhat broader concept of the Riemann integral. There, in studying a particular problem, we formed an idea of the direction in which this should be done. The present section is devoted to carrying out those ideas.

6.5.1 Definition, Examples, and Basic Properties of Improper Integrals

Definition 1 Suppose the function $x \mapsto f(x)$ is defined on the interval $[a, +\infty[$ and integrable on every closed interval $[a, b]$ contained in that interval.

The quantity

$$\int_a^{+\infty} f(x)\,\mathrm{d}x := \lim_{b \to +\infty} \int_a^b f(x)\,\mathrm{d}x,$$

if this limit exists, is called the *improper Riemann integral* or the *improper integral of the function f over the interval* $[a, +\infty[$.

The expression $\int_a^{+\infty} f(x)\,\mathrm{d}x$ itself is also called an improper integral, and in that case we say that the integral *converges* if the limit exists and *diverges* otherwise. Thus the question of the convergence of an improper integral is equivalent to the question whether the improper integral is defined or not.

[11] J.L.L. Buffon (1707–1788) – French experimental scientist.

Example 1 Let us investigate the values of the parameter α for which the improper integral

$$\int_1^{+\infty} \frac{\mathrm{d}x}{x^\alpha} \tag{6.69}$$

converges, or what is the same, is defined.

Since

$$\int_1^b \frac{\mathrm{d}x}{x^\alpha} = \begin{cases} \frac{1}{1-\alpha} x^{1-\alpha}\big|_1^b & \text{for } \alpha \neq 1, \\ \ln x\big|_1^b & \text{for } \alpha = 1, \end{cases}$$

the limit

$$\lim_{b\to+\infty} \int_1^b \frac{\mathrm{d}x}{x^\alpha} = \frac{1}{\alpha - 1}$$

exists only for $\alpha > 1$.

Thus,

$$\int_1^{\infty} \frac{\mathrm{d}x}{x^\alpha} = \frac{1}{\alpha - 1}, \quad \text{if } \alpha > 1,$$

and for other values of the parameter α the integral (6.69) diverges, that is, is not defined.

Definition 2 Suppose the function $x \mapsto f(x)$ is defined on the interval $[a, B[$ and integrable on any closed interval $[a, b] \subset [a, B[$. The quantity

$$\int_a^B f(x)\,\mathrm{d}x := \lim_{b\to B-0} \int_a^b f(x)\,\mathrm{d}x,$$

if this limit exists, is called the *improper integral of f over the interval* $[a, B[$.

The essence of this definition is that in any neighborhood of B the function f may happen to be unbounded.

Similarly, if a function $x \mapsto f(x)$ is defined on the interval $]A, b]$ and integrable on every closed interval $[a, b] \subset]A, b]$, then by definition we set

$$\int_A^b f(x)\,\mathrm{d}x := \lim_{a\to A+0} \int_a^b f(x)\,\mathrm{d}x$$

and also by definition we set

$$\int_{-\infty}^b f(x)\,\mathrm{d}x := \lim_{a\to-\infty} \int_a^b f(x)\,\mathrm{d}x.$$

Example 2 Let us investigate the values of the parameter α for which the integral

$$\int_0^1 \frac{dx}{x^\alpha} \tag{6.70}$$

converges.

Since for $a \in]0, 1]$

$$\int_a^1 \frac{dx}{x^\alpha} = \begin{cases} \frac{1}{1-\alpha} x^{1-\alpha}\big|_a^1, & \text{if } \alpha \neq 1, \\ \ln x\big|_a^1, & \text{if } \alpha = 1, \end{cases}$$

it follows that the limit

$$\lim_{a\to+0} \int_a^1 \frac{dx}{x^\alpha} = \frac{1}{1-\alpha}$$

exists only for $\alpha < 1$.

Thus the integral (6.70) is defined only for $\alpha < 1$.

Example 3

$$\int_{-\infty}^0 e^x\,dx = \lim_{a\to-\infty} \int_a^0 e^x\,dx = \lim_{a\to-\infty} \left(e^x\big|_a^0\right) = \lim_{a\to-\infty} \left(1 - e^a\right) = 1.$$

Since the question of the convergence of an improper integral is answered in the same way for both integrals over an infinite interval and functions unbounded near one of the endpoints of a finite interval of integration, we shall study both of these cases together from now on, introducing the following basic definition.

Definition 3 Let $[a, \omega[$ be a finite or infinite interval and $x \mapsto f(x)$ a function defined on that interval and integrable over every closed interval $[a, b] \subset [a, \omega[$. Then by definition

$$\int_a^\omega f(x)\,dx := \lim_{b\to\omega} \int_a^b f(x)\,dx, \tag{6.71}$$

if this limit exists as $b \to \omega$, $b \in [a, \omega[$.

From now on, unless otherwise stated, when studying the improper integral (6.71) we shall assume that the integrand satisfies the hypotheses of Definition 3.

Moreover, for the sake of definiteness we shall assume that the singularity ("impropriety") of the integral arises from only the upper limit of integration. The study of the case when it arises from the lower limit is carried out word for word in exactly the same way.

From Definition 3, properties of the integral, and properties of the limit, one can draw the following conclusions about properties of an improper integral.

Proposition 1 *Suppose $x \mapsto f(x)$ and $x \to g(x)$ are functions defined on an interval $[a, \omega[$ and integrable on every closed interval $[a, b] \subset [a, \omega[$. Suppose the improper integrals*

$$\int_a^\omega f(x)\,\mathrm{d}x, \tag{6.72}$$

$$\int_a^\omega g(x)\,\mathrm{d}x \tag{6.73}$$

are defined.

Then a) *if $\omega \in \mathbb{R}$ and $f \in \mathcal{R}[a, \omega]$, the values of the integral* (6.72) *are the same, whether it is interpreted as a proper or an improper integral;*

b) *for any $\lambda_1, \lambda_2 \in \mathbb{R}$ the function $(\lambda_1 f + \lambda_2 g)(x)$ is integrable in the improper sense on $[a, \omega[$ and the following equality holds:*

$$\int_a^\omega (\lambda_1 f + \lambda_2 g)(x)\,\mathrm{d}x = \lambda_1 \int_a^\omega f(x)\,\mathrm{d}x + \lambda_2 \int_a^\omega g(x)\,\mathrm{d}x;$$

c) *if $c \in [a, \omega[$, then*

$$\int_a^\omega f(x)\,\mathrm{d}x = \int_a^c f(x)\,\mathrm{d}x + \int_c^\omega f(x)\,\mathrm{d}x;$$

d) *if $\varphi : [\alpha, \gamma[\to [a, \omega[$ is a smooth strictly monotonic mapping with $\varphi(\alpha) = a$ and $\varphi(\beta) \to \omega$ as $\beta \to \gamma$, $\beta \in [\alpha, \gamma[$, then the improper integral of the function $t \mapsto (f \circ \varphi)(t)\varphi'(t)$ over $[\alpha, \gamma[$ exists and the following equality holds:*

$$\int_a^\omega f(x)\,\mathrm{d}x = \int_\alpha^\gamma (f \circ \varphi)(t)\varphi'(t)\,\mathrm{d}t.$$

Proof Part a) follows from the continuity of the function

$$\mathcal{F}(b) = \int_a^b f(x)\,\mathrm{d}x$$

on the closed interval $[a, \omega]$ on which $f \in \mathcal{R}[a, \omega]$.

Part b) follows from the fact that for $b \in [a, \omega[$

$$\int_a^b (\lambda_1 f + \lambda_2 g)(x)\,\mathrm{d}x = \lambda_1 \int_a^b f(x)\,\mathrm{d}x + \lambda_2 \int_a^b g(x)\,\mathrm{d}x.$$

Part c) follows from the equality

$$\int_a^b f(x)\,\mathrm{d}x = \int_a^c f(x)\,\mathrm{d}x + \int_c^b f(x)\,\mathrm{d}x,$$

which holds for all $b, c \in [a, \omega[$.

Part d) follows from the formula for change of variable in the definite integral:

$$\int_{a=\varphi(\alpha)}^{b=\varphi(\beta)} f(x)\,\mathrm{d}x = \int_{\alpha}^{\beta} (f\circ\varphi)(t)\varphi'(t)\,\mathrm{d}t. \qquad \square$$

Remark 1 To the properties of the improper integral expressed in Proposition 1 we should add the very useful rule for integration by parts in an improper integral, which we give in the following formulation:

If $f, g \in C^{(1)}[a, \omega[$ and the limit $\lim\limits_{\substack{x\to\omega\\ x\in[a,\omega[}} (f\cdot g)(x)$ exists, then the functions $f\cdot g'$ and $f'\cdot g$ are either both integrable or both nonintegrable in the improper sense on $[a, \omega[$, and when they are integrable the following equality holds:

$$\int_a^{\omega} (f\cdot g')(x)\,\mathrm{d}x = (f\cdot g)(x)\big|_a^{\omega} - \int_a^{\omega} (f'\cdot g)(x)\,\mathrm{d}x,$$

where

$$(f\cdot g)(x)\big|_a^{\omega} = \lim_{\substack{x\to\omega\\ x\in[a,\omega[}} (f\cdot g)(x) - (f\cdot g)(a).$$

Proof This follows from the formula

$$\int_a^{b} (f\cdot g')(x)\,\mathrm{d}x = (f\cdot g)\big|_{\alpha}^{b} - \int_a^{b} (f'\cdot g)(x)\,\mathrm{d}x$$

for integration by parts in a proper integral. $\square$

Remark 2 It is clear from part c) of Proposition 1 that the improper integrals

$$\int_a^{\omega} f(x)\,\mathrm{d}x \quad \text{and} \quad \int_c^{\omega} f(x)\,\mathrm{d}x$$

either both converge or both diverge. Thus, in improper integrals, as in series, convergence is independent of any initial piece of the series or integral.

For that reason, when posing the question of convergence of an improper integral, we sometimes omit entirely the limit of integration at which the integral does not have a singularity.

With that convention the results obtained in Examples 1 and 2 can be rewritten as follows:

the integral $\int^{+\infty} \frac{\mathrm{d}x}{x^{\alpha}}$ converges only for $\alpha > 1$;
the integral $\int_{+0} \frac{\mathrm{d}x}{x^{\alpha}}$ converges only for $\alpha < 1$.

The sign $+0$ in the last integral shows that the region of integration is contained in $x > 0$.

By a change of variable in this last integral, we immediately find that the integral $\int_{x_0+0} \frac{\mathrm{d}x}{(x-x_0)^{\alpha}}$ converges only for $\alpha < 1$.

6.5.2 Convergence of an Improper Integral

a. The Cauchy Criterion

By Definition 3, the convergence of the improper integral (6.71) is equivalent to the existence of a limit for the function

$$\mathcal{F}(b) = \int_a^b f(x)\,\mathrm{d}x \tag{6.74}$$

as $b \to \omega$, $b \in [a, \omega[$.

This relation is the reason why the following proposition holds.

Proposition 2 (Cauchy criterion for convergence of an improper integral) *If the function $x \mapsto f(x)$ is defined on the interval $[a, \omega[$ and integrable on every closed interval $[a, b] \subset [a, \omega[$, then the integral $\int_a^\omega f(x)\,\mathrm{d}x$ converges if and only if for every $\varepsilon > 0$ there exists $B \in [a, \omega[$ such that the relation*

$$\left| \int_{b_1}^{b_2} f(x)\,\mathrm{d}x \right| < \varepsilon$$

holds for any $b_1, b_2 \in [a, \omega[$ satisfying $B < b_1$ and $B < b_2$.

Proof As a matter of fact, we have

$$\int_{b_1}^{b_2} f(x)\,\mathrm{d}x = \int_a^{b_2} f(x)\,\mathrm{d}x - \int_a^{b_1} f(x)\,\mathrm{d}x = \mathcal{F}(b_2) - \mathcal{F}(b_1),$$

and therefore the condition is simply the Cauchy criterion for the existence of a limit for the function $\mathcal{F}(b)$ as $b \to \omega$, $b \in [a, \omega[$. □

b. Absolute Convergence of an Improper Integral

Definition 4 The improper integral $\int_a^\omega f(x)\,\mathrm{d}x$ converges *absolutely* if the integral $\int_a^\omega |f|(x)\,\mathrm{d}x$ converges.

Because of the inequality

$$\left| \int_{b_1}^{b_2} f(x)\,\mathrm{d}x \right| \le \left| \int_{b_1}^{b_2} |f|(x)\,\mathrm{d}x \right|$$

and Proposition 2, we can conclude that if an integral converges absolutely, then it converges.

The study of absolute convergence reduces to the study of convergence of integrals of nonnegative functions. But in this case we have the following proposition.

Proposition 3 *If a function f satisfies the hypotheses of Definition 3 and $f(x) \geq 0$ on $[a, \omega[$, then the improper integral (6.71) exists if and only if the function (6.74) is bounded on $[a, \omega[$.*

Proof Indeed, if $f(x) \geq 0$ on $[a, \omega[$, then the function (6.74) is nondecreasing on $[a, \omega[$, and therefore it has a limit as $b \to \omega$, $b \in [a, \omega[$, if and only if it is bounded. □

As an example of the use of this proposition, we consider the following corollary of it.

Corollary 1 (Integral test for convergence of a series) *If the function $x \mapsto f(x)$ is defined on the interval $[1, +\infty[$, nonnegative, nonincreasing, and integrable on each closed interval $[1, b] \subset [1, +\infty[$, then the series*

$$\sum_{n=1}^{\infty} f(n) = f(1) + f(2) + \cdots$$

and the integral

$$\int_1^{+\infty} f(x)\,\mathrm{d}x$$

either both converge or both diverge.

Proof It follows from the hypotheses that the inequalities

$$f(n+1) \leq \int_n^{n+1} f(x)\,\mathrm{d}x \leq f(n)$$

hold for any $n \in \mathbb{N}$. After summing these inequalities, we obtain

$$\sum_{n=1}^{k} f(n+1) \leq \int_1^{k+1} f(x)\,\mathrm{d}x \leq \sum_{n=1}^{k} f(n)$$

or

$$s_{k+1} - f(1) \leq \mathcal{F}(k+1) \leq s_k,$$

where $s_k = \sum_{n=1}^{k} f(n)$ and $\mathcal{F}(b) = \int_1^b f(x)\,\mathrm{d}x$. Since s_k and $\mathcal{F}(b)$ are nondecreasing functions of their arguments, these inequalities prove the proposition. □

In particular, one can say that the result of Example 1 is equivalent to the assertion that the series

$$\sum_{n=1}^{\infty} \frac{1}{n^{\alpha}}$$

converges only for $\alpha > 1$.

The most frequently used corollary of Proposition 3 is the following theorem.

Theorem 1 (Comparison theorem) *Suppose the functions $x \to f(x)$ and $x \to g(x)$ are defined on the interval $[a, \omega[$ and integrable on any closed interval $[a, b] \subset [a, \omega[$.*

If

$$0 \leq f(x) \leq g(x)$$

on $[a, \omega[$, then convergence of the integral (6.73) *implies convergence of* (6.72) *and the inequality*

$$\int_a^{\omega} f(x)\,\mathrm{d}x \leq \int_a^{\omega} g(x)\,\mathrm{d}x$$

holds. Divergence of the integral (6.72) *implies divergence of* (6.73).

Proof From the hypotheses of the theorem and the inequalities for proper Riemann integrals we have

$$\mathcal{F}(b) = \int_a^b f(x)\,\mathrm{d}x \leq \int_a^b g(x)\,\mathrm{d}x = \mathcal{G}(b)$$

for any $b \in [a, \omega[$. Since both functions $\mathcal{F}$ and $\mathcal{G}$ are nondecreasing on $[a, \omega[$, the theorem follows from this inequality and Proposition 3. □

Remark 3 If instead of satisfying the inequalities $0 \leq f(x) \leq g(x)$ the functions f and g in the theorem are known to be nonnegative and of the same order as $x \to \omega$, $x \in [a, \omega[$, that is, there are positive constants c_1 and c_2 such that

$$c_1 f(x) \leq g(x) \leq c_2 f(x),$$

then by the linearity of the improper integral and theorem just proved, in this case we can conclude that the integrals (6.72) and (6.73) either both converge or both diverge.

Example 4 The integral

$$\int^{+\infty} \frac{\sqrt{x}\,\mathrm{d}x}{\sqrt{1+x^4}}$$

converges, since

$$\frac{\sqrt{x}}{\sqrt{1+x^4}} \sim \frac{1}{x^{3/2}}$$

as $x \to +\infty$.

Example 5 The integral

$$\int_1^{+\infty} \frac{\cos x}{x^2}\,\mathrm{d}x$$

converges absolutely, since

$$\left|\frac{\cos x}{x^2}\right| \le \frac{1}{x^2}$$

for $x \ge 1$. Consequently,

$$\left|\int_1^{+\infty} \frac{\cos x}{x^2}\,\mathrm{d}x\right| \le \int_1^{+\infty} \left|\frac{\cos x}{x^2}\right| \mathrm{d}x \le \int_1^{+\infty} \frac{1}{x^2}\,\mathrm{d}x = 1.$$

Example 6 The integral

$$\int_1^{+\infty} \mathrm{e}^{-x^2}\,\mathrm{d}x$$

converges, since $\mathrm{e}^{-x^2} < \mathrm{e}^{-x}$ for $x > 1$ and

$$\int_1^{+\infty} \mathrm{e}^{-x^2}\,\mathrm{d}x < \int_1^{+\infty} \mathrm{e}^{-x}\,\mathrm{d}x = \frac{1}{\mathrm{e}}.$$

Example 7 The integral

$$\int^{+\infty} \frac{\mathrm{d}x}{\ln x}$$

diverges, since

$$\frac{1}{\ln x} > \frac{1}{x}$$

for sufficiently large values of x.

Example 8 The Euler integral

$$\int_0^{\pi/2} \ln \sin x\,\mathrm{d}x$$

converges, since

$$|\ln \sin x| \sim |\ln x| < \frac{1}{\sqrt{x}}$$

as $x \to +0$.

Example 9 The elliptic integral

$$\int_0^1 \frac{\mathrm{d}x}{\sqrt{(1-x^2)(1-k^2x^2)}}$$

converges for $0 \le k^2 < 1$, since

$$\sqrt{(1-x^2)(1-k^2x^2)} \sim \sqrt{2(1-k^2)}(1-x)^{1/2}$$

as $x \to 1-0$.

Example 10 The integral

$$\int_0^{\varphi} \frac{\mathrm{d}\theta}{\sqrt{\cos\theta - \cos\varphi}}$$

converges, since

$$\sqrt{\cos\theta - \cos\varphi} = \sqrt{2\sin\frac{\varphi+\theta}{2}\sin\frac{\varphi-\theta}{2}} \sim \sqrt{\sin\varphi}(\varphi-\theta)^{1/2}$$

as $\theta \to \varphi - 0$.

Example 11 The integral

$$T = 2\sqrt{\frac{L}{g}}\int_0^{\varphi_0} \frac{\mathrm{d}\psi}{\sqrt{\sin^2\frac{\varphi_0}{2}\sin^2\frac{\varphi}{2}}} \tag{6.75}$$

converges for $0 < \varphi_0 < \pi$ since as $\psi \to \varphi_0 - 0$ we have

$$\sqrt{\sin^2\frac{\varphi_0}{2} - \sin^2\frac{\psi}{2}} \sim \sqrt{\sin\varphi_0}(\varphi_0 - \psi)^{1/2}. \tag{6.76}$$

Relation (6.75) expresses the dependence of the period of oscillations of a pendulum on its length L and its initial angle of displacement, measured from the radius it occupies at the lowest point of its trajectory. Formula (6.75) is an elementary version of formula (6.65) of the preceding section.

A pendulum can be thought of, for example, as consisting of a weightless rod, one end of which is attached by a hinge while the other end, to which a point mass is attached, is free.

In that case one can speak of arbitrary initial angles $\varphi_0 \in [0, \pi]$. For $\varphi_0 = 0$ and $\varphi_0 = \pi$, the pendulum will not oscillate at all, being in a state of stable equilibrium in the first case and unstable equilibrium in the second.

It is interesting to note that (6.75) and (6.76) easily imply that $T \to \infty$ as $\varphi_0 \to \pi - 0$, that is, the period of oscillation of a pendulum increases without bound as its initial position approaches the upper (unstable) equilibrium position.

c. Conditional Convergence of an Improper Integral

Definition 5 If an improper integral converges but not absolutely, we say that it converges *conditionally*.

Example 12 Using Remark 1, by the formula for integration by parts in an improper integral, we find that

$$\int_{\pi/2}^{+\infty} \frac{\sin x}{x}\,\mathrm{d}x = -\frac{\cos x}{x}\bigg|_{\pi/2}^{+\infty} - \int_{\pi/2}^{+\infty} \frac{\cos x}{x^2}\,\mathrm{d}x = -\int_{\pi/2}^{+\infty} \frac{\cos x}{x^2}\,\mathrm{d}x,$$

provided the last integral converges. But, as we saw in Example 5, this integral converges, and hence the integral

$$\int_{\pi/2}^{+\infty} \frac{\sin x}{x}\,\mathrm{d}x \tag{6.77}$$

also converges.

At the same time, the integral (6.77) is not absolutely convergent. Indeed, for $b \in [\pi/2, +\infty[$ we have

$$\int_{\pi/2}^{b} \left|\frac{\sin x}{x}\right| \mathrm{d}x \geq \int_{\pi/2}^{b} \frac{\sin^2 x}{x}\,\mathrm{d}x = \frac{1}{2}\int_{\pi/2}^{b} \frac{\mathrm{d}x}{x} - \frac{1}{2}\int_{\pi/2}^{b} \frac{\cos 2x}{x}\,\mathrm{d}x. \tag{6.78}$$

The integral

$$\int_{\pi/2}^{+\infty} \frac{\cos 2x}{x}\,\mathrm{d}x,$$

as can be verified through integration by parts, is convergent, so that as $b \to +\infty$, the difference on the right-hand side of relation (6.78) tends to $+\infty$. Thus, by estimate (6.78), the integral (6.77) is not absolutely convergent.

We now give a special convergence test for improper integrals based on the second mean-value theorem and hence essentially on the same formula for integration by parts.

Proposition 4 (Abel–Dirichlet test for convergence of an integral) *Let $x \mapsto f(x)$ and $x \mapsto g(x)$ be functions defined on an interval $[a, \omega[$ and integrable on every closed interval $[a, b] \subset [a, \omega[$. Suppose that g is monotonic.*

Then a sufficient condition for convergence of the improper integral

$$\int_{a}^{\omega} (f \cdot g)(x)\,\mathrm{d}x \tag{6.79}$$

is that the one of the following pairs of conditions hold:

α_1) *the integral $\int_a^\omega f(x)\,\mathrm{d}x$ converges,*
β_1) *the function g bounded on $[a, \omega[$,*

or

α_2) *the function $\mathcal{F}(b) = \int_a^b f(x)\,\mathrm{d}x$ is bounded on $[a, \omega[$,*
β_2) *the function $g(x)$ tends to zero as $x \to \omega$, $x \in [a, \omega[$.*

Proof For any b_1 and b_2 in $[a, \omega[$ we have, by the second mean-value theorem,

$$\int_{b_1}^{b_2} (f \cdot g)(x)\,\mathrm{d}x = g(b_1) \int_{b_1}^{\xi} f(x)\,\mathrm{d}x + g(b_2) \int_{\xi}^{b_2} f(x)\,\mathrm{d}x,$$

where ξ is a point lying between b_1 and b_2. Hence by the Cauchy convergence criterion (Proposition 2), we conclude that the integral (6.79) does indeed converge if either of the two pairs of conditions holds. □

6.5.3 Improper Integrals with More than One Singularity

Up to now we have spoken only of improper integrals with one singularity caused either by the unboundedness of the function at one of the endpoints of the interval of integration or by an infinite limit of integration. In this subsection we shall show in what sense other possible variants of an improper integral can be taken.

If both limits of integration are singularities of either of these two types, then by definition

$$\int_{\omega_1}^{\omega_2} f(x)\,\mathrm{d}x := \int_{\omega_1}^{c} f(x)\,\mathrm{d}x + \int_{c}^{\omega_2} f(x)\,\mathrm{d}x, \tag{6.80}$$

where c is an arbitrary point of the open interval $]\omega_1, \omega_2[$.

It is assumed here that each of the improper integrals on the right-hand side of (6.80) converges. Otherwise we say that the integral on the left-hand side of (6.80) diverges.

By Remark 2 and the additive property of the improper integral, the definition (6.80) is unambiguous in the sense that it is independent of the choice of the point $c \in]\omega_1, \omega_2[$.

Example 13

$$\int_{-1}^{1} \frac{\mathrm{d}x}{\sqrt{1-x^2}} = \int_{-1}^{0} \frac{\mathrm{d}x}{\sqrt{1-x^2}} + \int_{0}^{1} \frac{\mathrm{d}x}{\sqrt{1-x^2}} =$$
$$= \arcsin x\big|_{-1}^{0} + \arcsin x\big|_{0}^{1} = \arcsin x\big|_{-1}^{1} = \pi.$$

Example 14 The integral

$$\int_{-\infty}^{+\infty} \mathrm{e}^{-x^2}\,\mathrm{d}x$$

is called the *Euler–Poisson integral*, and sometimes the *Gaussian integral*. It obviously converges in the sense given above. It will be shown later that its value is $\sqrt{\pi}$.

Example 15 The integral

$$\int_0^{+\infty} \frac{dx}{x^\alpha}$$

diverges, since for every α at least one of the two integrals

$$\int_0^1 \frac{dx}{x^\alpha} \quad \text{and} \quad \int_1^{+\infty} \frac{dx}{x^\alpha}$$

diverges.

Example 16 The integral

$$\int_0^{+\infty} \frac{\sin x}{x^\alpha}\, dx$$

converges if each of the integrals

$$\int_0^1 \frac{\sin x}{x^\alpha}\, dx \quad \text{and} \quad \int_1^{+\infty} \frac{\sin x}{x^\alpha}\, dx$$

converges. The first of these integrals converges if $\alpha < 2$, since

$$\frac{\sin x}{x^\alpha} \sim \frac{1}{x^{\alpha-1}}$$

as $x \to +0$. The second integral converges if $\alpha > 0$, as one can verify directly through an integration by parts similar to the one shown in Example 12, or by citing the Abel–Dirichlet test. Thus the original integral has a meaning for $0 < \alpha < 2$.

In the case when the integrand is not bounded in a neighborhood of one of the interior points ω of the closed interval of integration $[a, b]$, we set

$$\int_a^b f(x)\, dx := \int_a^\omega f(x)\, dx + \int_\omega^b f(x)\, dx, \tag{6.81}$$

requiring that both of the integrals on the right-hand side exist.

Example 17 In the sense of the convention (6.81)

$$\int_{-1}^1 \frac{dx}{\sqrt{|x|}} = 4.$$

Example 18 The integral $\int_{-1}^1 \frac{dx}{x}$ is not defined.

Besides (6.81), there is a second convention about computing the integral of a function that is unbounded in a neighborhood of an interior point ω of a closed interval of integration. To be specific, we set

$$\mathrm{PV} \int_a^b f(x)\, dx := \lim_{\delta \to +0} \left(\int_a^{\omega-\delta} f(x)\, dx + \int_{\omega+\delta}^b f(x)\, dx \right), \tag{6.82}$$

if the integral on the right-hand side exists. This limit is called, following Cauchy, the *principal value* of the integral, and, to distinguish the definitions (6.81) and (6.82), we put the letters PV in front of the second to indicate that it is the principal value.

In accordance with this convention we have

Example 19

$$\mathrm{PV}\int_{-1}^{1}\frac{\mathrm{d}x}{x}=0.$$

We also adopt the following definition:

$$\mathrm{PV}\int_{-\infty}^{+\infty} f(x)\,\mathrm{d}x := \lim_{R\to+\infty}\int_{-R}^{R} f(x)\,\mathrm{d}x. \tag{6.83}$$

Example 20

$$\mathrm{PV}\int_{-\infty}^{+\infty} x\,\mathrm{d}x = 0.$$

Finally, if there are several (finitely many) singularities of one kind or another on the interval of integration, at interior points or endpoints, then the nonsingular points of the interval are divided into a finite number of such intervals, each containing only one singularity, and the integral is computed as the sum of the integrals over the closed intervals of the partition.

It can be verified that the result of such a computation is not affected by the arbitrariness in the choice of a partition.

Example 21 The precise definition of the *logarithmic integral* can now be written as

$$\mathrm{li}\,x=\begin{cases}\int_0^x \frac{\mathrm{d}t}{\ln t}, & \text{if } 0<x<1,\\ \mathrm{PV}\int_0^x \frac{\mathrm{d}t}{\ln t}, & \text{if } 1<x.\end{cases}$$

In the last case the symbol PV refers to the only interior singularity on the interval $]0, x[$, which is located at 1. We remark that in the sense of the definition in formula (6.81) this integral is not convergent.

6.5.4 Problems and Exercises

1. Show that the following functions have the stated properties.

a) $\mathrm{Si}(x)=\int_0^x \frac{\sin t}{t}\,\mathrm{d}t$ (the *sine integral*) is defined on all of $\mathbb{R}$, is an odd function, and has a limit as $x\to+\infty$.

b) $\mathrm{si}(x)=-\int_x^{\infty} \frac{\sin t}{t}\,\mathrm{d}t$ is defined on all of $\mathbb{R}$ and differs from $\mathrm{Si}\,x$ only by a constant;

c) $\operatorname{Ci} x = -\int_x^{\infty} \frac{\cos t}{t}\,\mathrm{d}t$ (the *cosine integral*) can be computed for sufficiently large values of x by the approximate formula $\operatorname{Ci} x \approx \frac{\sin x}{x}$; estimate the region of values where the absolute error of this approximation is less than 10^{-4}.

2. Show that

a) the integrals $\int_1^{+\infty} \frac{\sin x}{x^\alpha}\,\mathrm{d}x$, $\int_1^{+\infty} \frac{\cos x}{x^\alpha}\,\mathrm{d}x$ converge only for $\alpha > 0$, and absolutely only for $\alpha > 1$;

b) the *Fresnel integrals*

$$C(x) = \frac{1}{\sqrt{2}} \int_0^{\sqrt{x}} \cos t^2\,\mathrm{d}t, \qquad S(x) = \frac{1}{\sqrt{2}} \int_0^{\sqrt{x}} \sin t^2\,\mathrm{d}t$$

are infinitely differentiable functions on the interval $]0, +\infty[$, and both have a limit as $x \to +\infty$.

3. Show that

a) the elliptic integral of first kind

$$F(k, \varphi) = \int_0^{\sin \varphi} \frac{\mathrm{d}t}{\sqrt{(1-t^2)(1-k^2t^2)}}$$

is defined for $0 \le k < 1$, $0 \le \varphi \le \frac{\pi}{2}$ and can be brought into the form

$$F(k, \varphi) = \int_0^{\varphi} \frac{\mathrm{d}\psi}{\sqrt{1 - k^2 \sin^2 \psi}};$$

b) the complete elliptic integral of first kind

$$K(k) = \int_0^{\pi/2} \frac{\mathrm{d}\psi}{\sqrt{1 - k^2 \sin^2 \psi}}$$

increases without bound as $k \to 1 - 0$.

4. Show that

a) the exponential integral $\operatorname{Ei}(x) = \int_{-\infty}^{x} \frac{e^t}{t}\,\mathrm{d}t$ is defined and infinitely differentiable for $x < 0$;

b) $-\operatorname{Ei}(-x) = \frac{e^{-x}}{x}\left(1 - \frac{1}{x} + \frac{2!}{x^2} - \cdots + (-1)^n \frac{n!}{x^n} + o\left(\frac{1}{x^n}\right)\right)$ as $x \to +\infty$;

c) the series $\sum_{n=0}^{\infty} (-1)^n \frac{n!}{x^n}$ does not converge for any value of $x \in \mathbb{R}$;

d) $\operatorname{li}(x) \sim \frac{x}{\ln x}$ as $x \to +0$. (For the definition of the logarithmic integral $\operatorname{li}(x)$ see Example 21.)

5. Show that

a) the function $\Phi(x) = \frac{1}{\sqrt{\pi}} \int_{-x}^{x} e^{-t^2}\,\mathrm{d}t$, called the *error function* and often denoted $\operatorname{erf}(x)$, is defined, even, and infinitely differentiable on $\mathbb{R}$ and has a limit as $x \to +\infty$;

b) if the limit in a) is equal to 1 (and it is), then

$$\operatorname{erf}(x) = \frac{2}{\sqrt{\pi}} \int_0^x e^{-t^2}\,dt =$$

$$= 1 - \frac{2}{\sqrt{\pi}} e^{-x^2} \left(\frac{1}{2x} - \frac{1}{2^2 x^3} + \frac{1 \cdot 3}{2^3 x^5} - \frac{1 \cdot 3 \cdot 5}{2^4 x^7} + o\left(\frac{1}{x^7}\right) \right)$$

as $x \to +\infty$.

6. Prove the following statements.

a) If a heavy particle slides under gravitational attraction along a curve given in parametric form as $x = x(\theta)$, $y = y(\theta)$, and at time $t = 0$ the particle had zero velocity and was located at the point $x_0 = x(\theta_0)$, $y_0 = y(\theta_0)$, then the following relation holds between the parameter θ defining a point on the curve and the time t at which the particle passes this point (see formula (6.63) of Sect. 6.4)

$$t = \pm \int_{\theta_0}^{\theta} \sqrt{\frac{(x'(\theta))^2 + (y'(\theta))^2}{2g(y_0 - y(\theta))}}\, d\theta,$$

in which the improper integral necessarily converges if $y'(\theta_0) \neq 0$. (The ambiguous sign is chosen positive or negative according as t and θ have the same kind of monotonicity or the opposite kind; that is, if an increasing θ corresponds to an increasing t, then one must obviously choose the positive sign.)

b) The period of oscillation of a particle in a well having cross section in the shape of a cycloid

$$\begin{aligned} x &= R(\theta + \pi + \sin\theta), \\ y &= -R(1 + \cos\theta), \end{aligned} \qquad |\theta| \leq \pi,$$

is independent of the level $y_0 = -R(1 + \cos\theta_0)$ from which it begins to slide and is equal to $4\pi\sqrt{R/g}$ (see Problem 4 of Sect. 6.4).

Chapter 7
Functions of Several Variables: Their Limits and Continuity

Up to now we have considered almost exclusively numerical-valued functions $x \mapsto f(x)$ in which the number $f(x)$ was determined by giving a single number x from the domain of definition of the function.

However, many quantities of interest depend on not just one, but many factors, and if the quantity itself and each of the factors that determine it can be characterized by some number, then this dependence reduces to the fact that a value $y = f(x^1, \dots, x^n)$ of the quantity in question is made to correspond to an ordered set $(x^1, \dots, x^n)$ of numbers, each of which describes the state of the corresponding factor. The quantity assumes this value when the factors determining this quantity are in these states.

For example, the area of a rectangle is the product of the lengths of its sides. The volume of a given quantity of gas is computed by the formula

$$V = R\frac{mT}{p},$$

where R is a constant, m is the mass, T is the absolute temperature, and p is the pressure of the gas. Thus the value of V depends on a variable ordered triple of numbers (m, T, p), or, as we say, V is a function of the three variables m, T, and p.

Our goal is to learn how to study functions of several variables just as we learned how to study functions of one variable.

As in the case of functions of one variable, the study of functions of several numerical variables begins by describing their domains of definition.

7.1 The Space $\mathbb{R}^m$ and the Most Important Classes of Its Subsets

7.1.1 The Set $\mathbb{R}^m$ and the Distance in It

We make the convention that $\mathbb{R}^m$ denotes the set of ordered m-tuples $(x^1, \dots, x^m)$ of real numbers $x^i \in \mathbb{R}$ $(i = 1, \dots, m)$.

V.A. Zorich, *Mathematical Analysis I*, Universitext,
DOI 10.1007/978-3-662-48792-1_7

Each such m-tuple will be denoted by a single letter $x = (x^1, \dots, x^m)$ and, in accordance with convenient geometric terminology, will be called a *point* of $\mathbb{R}^m$. The number x^i in the set $(x^1, \dots, x^m)$ will be called the ith *coordinate* of the point $x = (x^1, \dots, x^m)$.

The geometric analogies can be extended by introducing a distance on $\mathbb{R}^m$ between the points $x_1 = (x_1^1, \dots, x_1^m)$ and $x_2 = (x_2^1, \dots, x_2^m)$ according to the formula

$$d(x_1, x_2) = \sqrt{\sum_{i=1}^{m} (x_1^i - x_2^i)^2}. \tag{7.1}$$

The function

$$d : \mathbb{R}^m \times \mathbb{R}^m \to \mathbb{R}$$

defined by the formula (7.1) obviously has the following properties:

a) $d(x_1, x_2) \geq 0$;
b) $(d(x_1, x_2) = 0) \Leftrightarrow (x_1 = x_2)$;
c) $d(x_1, x_2) = d(x_2, x_1)$;
d) $d(x_1, x_3) \leq d(x_1, x_2) + d(x_2, x_3)$.

This last inequality (called, again because of geometric analogies, the *triangle inequality*) is a special case of Minkowski's inequality (see Sect. 5.4.2).

A function defined on pairs of points (x_1, x_2) of a set X and possessing the properties a), b), c), and d) is called a *metric* or *distance* on X.

A set X together with a fixed metric on it is called a *metric space*.

Thus we have turned $\mathbb{R}^m$ into a metric space by endowing it with the metric given by relation (7.1).

The reader can get information on arbitrary metric spaces in Chap. 9 (Part 2). Here we do not wish to become distracted from the particular metric space $\mathbb{R}^m$ that we need at the moment.

Since the space $\mathbb{R}^m$ with metric (7.1) will be our only metric space in this chapter, forming our object of study, we have no need for the general definition of a metric space at the moment. It is given only to explain the term "space" used in relation to $\mathbb{R}^m$ and the term "metric" in relation to the function (7.1).

It follows from (7.1) that for $i \in \{1, \dots, m\}$

$$|x_1^i - x_2^i| \leq d(x_1, x_2) \leq \sqrt{m} \max_{1 \leq i \leq m} |x_1^i - x_2^i|, \tag{7.2}$$

that is, the distance between the points $x_1, x_2 \in \mathbb{R}^m$ is small if and only if the corresponding coordinates of these points are close together.

It is clear from (7.2) and also from (7.1) that for $m = 1$, the set $\mathbb{R}^1$ is the same as the set of real numbers, between whose points the distance is measured in the standard way by the absolute value of the difference of the numbers.

7.1.2 Open and Closed Sets in $\mathbb{R}^m$

Definition 1 For $\delta > 0$ the set

$$B(a; \delta) = \{x \in \mathbb{R}^m \mid d(a, x) < \delta\}$$

is called the *ball with center* $a \in \mathbb{R}^m$ *of radius* δ or the δ*-neighborhood* of the point $a \in \mathbb{R}^m$.

Definition 2 A set $G \subset \mathbb{R}^m$ is *open* in $\mathbb{R}^m$ if for every point $x \in G$ there is a ball $B(x; \delta)$ such that $B(x; \delta) \subset G$.

Example 1 $\mathbb{R}^m$ is an open set in $\mathbb{R}^m$.

Example 2 The empty set $\varnothing$ contains no points at all and hence may be regarded as satisfying Definition 2, that is, $\varnothing$ is an open set in $\mathbb{R}^m$.

Example 3 A ball $B(a; r)$ is an open set in $\mathbb{R}^m$. Indeed, if $x \in B(a; r)$, that is, $d(a, x) < r$, then for $0 < \delta < r - d(a, x)$, we have $B(x; \delta) \subset B(a; r)$, since

$$\big(\xi \in B(x; \delta)\big) \Rightarrow \big(d(x, \xi) < \delta\big) \Rightarrow$$
$$\Rightarrow \big(d(a, \xi) \le d(a, x) + d(x, \xi) < d(a, x) + r - d(a, x) = r\big).$$

Example 4 A set $G = \{x \in \mathbb{R}^m \mid d(a, x) > r\}$, that is, the set of points whose distance from a fixed point $a \in \mathbb{R}^m$ is larger than r, is open. This fact is easy to verify, as in Example 3, using the triangle inequality for the metric.

Definition 3 The set $F \subset \mathbb{R}^m$ is *closed* in $\mathbb{R}^m$ if its complement $G = \mathbb{R}^m \backslash F$ is open in $\mathbb{R}^m$.

Example 5 The set $\overline{B}(a; r) = \{x \in \mathbb{R}^m \mid d(a, x) \le r\}$, $r \ge 0$, that is, the set of points whose distance from a fixed point $a \in \mathbb{R}^m$ is at most r, is closed, as follows from Definition 3 and Example 4. The set $\overline{B}(a; r)$ is called the *closed ball with center* a *of radius* r.

Proposition 1 a) *The union* $\bigcup_{\alpha \in A} G_\alpha$ *of the sets of any system* $\{G_\alpha, \alpha \in A\}$ *of open sets in* $\mathbb{R}^m$ *is an open set in* $\mathbb{R}^m$.

b) *The intersection* $\bigcap_{i=1}^n G_i$ *of a finite number of open sets in* $\mathbb{R}^m$ *is an open set in* $\mathbb{R}^m$.

a′) *The intersection* $\bigcap_{\alpha \in A} F_\alpha$ *of the sets of any system* $\{F_\alpha, \alpha \in A\}$ *of closed sets* F_α *in* $\mathbb{R}^m$ *is a closed set in* $\mathbb{R}^m$.

b′) *The union* $\bigcup_{i=1}^n F_i$ *of a finite number of closed sets in* $\mathbb{R}^m$ *is a closed set in* $\mathbb{R}^m$.

Proof a) If $x \in \bigcup_{\alpha\in A} G_\alpha$, then there exists $\alpha_0 \in A$ such that $x \in G_{\alpha_0}$, and consequently there is a δ-neighborhood $B(x;\delta)$ of x such that $B(x;\delta) \subset G_{\alpha_0}$. But then $B(x;\delta) \subset \bigcup_{\alpha\in A} G_\alpha$.

b) Let $x \in \bigcap_{i=1}^n G_i$. Then $x \in G_i$ $(i = 1,\dots,n)$. Let $\delta_1,\dots,\delta_n$ be positive numbers such that $B(x;\delta_i) \subset G_i$ $(i = 1,\dots,n)$. Setting $\delta = \min\{\delta_1,\dots,\delta_n\}$, we obviously find that $\delta > 0$ and $B(x;\delta) \subset \bigcap_{i=1}^n G_i$.

a′) Let us show that the set $C(\bigcap_{\alpha\in A} F_\alpha)$ complementary to $\bigcap_{\alpha\in A} F_\alpha$ in $\mathbb{R}^m$ is an open set in $\mathbb{R}^m$.

Indeed,

$$C\Bigl(\bigcap_{\alpha\in A} F_\alpha\Bigr) = \bigcup_{\alpha\in A}(CF_\alpha) = \bigcup_{\alpha\in A} G_\alpha,$$

where the sets $G_\alpha = CF_\alpha$ are open in $\mathbb{R}^m$. Part a′) now follows from a).

b′) Similarly, from b) we obtain

$$C\Bigl(\bigcup_{i=1}^n F_i\Bigr) = \bigcap_{i=1}^n (CF_i) = \bigcap_{i=1}^n G_i.$$

□

Example 6 The set $S(a;r) = \{x \in \mathbb{R}^m \mid d(a,x) = r\}$, $r \geq 0$, is called the *sphere of radius r with center $a \in \mathbb{R}^m$*. The complement of $S(a;r)$ in $\mathbb{R}^m$, by Examples 3 and 4, is the union of open sets. Hence by the proposition just proved it is open, and the sphere $S(a;r)$ is closed in $\mathbb{R}^m$.

Definition 4 An open set in $\mathbb{R}^m$ containing a given point is called a *neighborhood* of that point in $\mathbb{R}^m$.

In particular, as follows from Example 3, the δ-neighborhood of a point is a neighborhood of it.

Definition 5 In relation to a set $E \subset \mathbb{R}^m$ a point $x \in \mathbb{R}^m$ is

an interior point if some neighborhood of it is contained in E;
an exterior point if it is an interior point of the complement of E in $\mathbb{R}^m$;
a boundary point if it is neither an interior point nor an exterior point.

It follows from this definition that the characteristic property of a boundary point of a set is that every neighborhood of it contains both points of the set and points not in the set.

Example 7 The sphere $S(a;r)$, $r > 0$ is the set of boundary points of both the open ball $B(a;r)$ and the closed ball $\overline{B}(a;r)$.

Example 8 A point $a \in \mathbb{R}^m$ is a boundary point of the set $\mathbb{R}^m \backslash a$, which has no exterior points.

Example 9 All points of the sphere $S(a;r)$ are boundary points of it; regarded as a subset of $\mathbb{R}^m$, the sphere $S(a;r)$ has no interior points.

Definition 6 A point $a \in \mathbb{R}^m$ is a *limit point* of the set $E \subset \mathbb{R}^m$ if for any neighborhood $O(a)$ of a the intersection $E \cap O(a)$ is an infinite set.

Definition 7 The union of a set E and all its limit points in $\mathbb{R}^m$ is the *closure* of E in $\mathbb{R}^m$.

The closure of the set E is usually denoted $\overline{E}$.

Example 10 The set $\overline{B}(a;r) = B(a;r) \cup S(a;r)$ is the set of limit points of the open ball $B(a;r)$; that is why $\overline{B}(a;r)$, in contrast to $B(a;r)$, is called a closed ball.

Example 11 $\overline{S}(a;r) = S(a;r)$.

Rather than proving this last equality, we shall prove the following useful proposition.

Proposition 2 (F *is closed in* $\mathbb{R}^m$) $\Leftrightarrow$ ($F = \overline{F}$ *in* $\mathbb{R}^m$).

In other words, F is closed in $\mathbb{R}^m$ if and only if it contains all its limit points.

Proof Let F be closed in $\mathbb{R}^m$, $x \in \mathbb{R}^m$, and $x \notin F$. Then the open set $G = \mathbb{R}^m \setminus F$ is a neighborhood of x that contains no points of F. Thus we have shown that if $x \notin F$, then x is not a limit point of F.

Let $F = \overline{F}$. We shall verify that the set $G = \mathbb{R}^m \setminus \overline{F}$ is open in $\mathbb{R}^m$. If $x \in G$, then $x \notin \overline{F}$, and therefore x is not a limit point of F. Hence there is a neighborhood of x containing only a finite number of points $x_1, \dots, x_n$ of F. Since $x \notin F$, one can construct, for example, balls about x, $O_1(x), \dots, O_n(x)$ such that $x_i \notin O_i(x)$. Then $O(x) = \bigcap_{i=1}^n O_i(x)$ is an open neighborhood of x containing no points of F at all, that is, $O(x) \subset \mathbb{R}^m \setminus F$ and hence the set $\mathbb{R}^m \setminus F = \mathbb{R}^m \setminus \overline{F}$ is open. Therefore F is closed in $\mathbb{R}^m$. □

7.1.3 Compact Sets in $\mathbb{R}^m$

Definition 8 A set $K \subset \mathbb{R}^m$ is *compact* if from every covering of K by sets that are open in $\mathbb{R}^m$ one can extract a finite covering.

Example 12 A closed interval $[a,b] \subset \mathbb{R}^1$ is compact by the finite covering lemma (Heine–Borel theorem).

Example 13 A generalization to $\mathbb{R}^m$ of the concept of a closed interval is the set

$$I = \{x \in \mathbb{R}^m \mid a^i \le x^i \le b^i,\ i = 1, \dots, m\},$$

which is called an *m-dimensional interval*, or an *m-dimensional block* or an *m-dimensional parallelepiped*.

We shall show that I is compact in $\mathbb{R}^m$.

Proof Assume that from some open covering of I one cannot extract a finite covering. Bisecting each of the coordinate closed intervals $I^i = \{x^i \in \mathbb{R}\colon a^i \le x^i \le b^i\}$ $(i = 1, \dots, m)$, we break the interval I into 2^m intervals, at least one of which does not admit a covering by a finite number of sets from the open system we started with. We proceed with this interval exactly as with the original interval. Continuing this division process, we obtain a sequence of nested intervals $I = I_1 \supset I_2 \supset \dots \supset I_n \supset \cdots$, none of which admits a finite covering. If $I_n = \{x \in \mathbb{R}^m \mid a_n^i \le x^i \le b_n^i,\ i, \dots, m\}$, then for each $i \in \{1, \dots, m\}$ the coordinate closed intervals $a_n^i \le x^i \le b_n^i$ $(n = 1, 2, \dots)$ form, by construction, a system of nested closed intervals whose lengths tend to zero. By finding the point $\xi^i \in [a_n^i, b_n^i]$ common to all of these intervals for each $i \in \{1, \dots, m\}$, we obtain a point $\xi = (\xi^1, \dots, \xi^m)$ belonging to all the intervals $I = I_1, I_2, \dots, I_n, \dots$. Since $\xi \in I$, there is an open set G in the system of covering sets such that $\xi \in G$. Then for some $\delta > 0$ we also have $B(\xi; \delta) \subset G$. But by construction and the relation (7.2) there exists N such that $I_n \subset B(\xi; \delta) \subset G$ for $n > N$. We have now reached a contradiction with the fact that the intervals I_n do not admit a finite covering by sets of the given system. □

Proposition 3 *If K is a compact set in $\mathbb{R}^m$, then*

a) *K is closed in $\mathbb{R}^m$;*
b) *any closed subset of $\mathbb{R}^m$ contained in K is itself compact.*

Proof a) We shall show that any point $a \in \mathbb{R}^m$ that is a limit point of K must belong to K. Suppose $a \notin K$. For each point $x \in K$ we construct a neighborhood $G(x)$ such that a has a neighborhood disjoint from $G(x)$. The set $\{G(x)\}$, $x \in K$, consisting of all such neighborhoods forms an open covering of the compact set K, from which we can select a finite covering $G(x_1), \dots, G(x_n)$. If now $O_i(a)$ is a neighborhood of a such that $G(x_i) \cap O_i(a) = \varnothing$, then the set $O(a) = \bigcap_{i=1}^n O_i(a)$ is also a neighborhood of a, and obviously $K \cap O(a) = \varnothing$. Thus a cannot be a limit point of K.

b) Suppose F is a closed subset of $\mathbb{R}^m$ and $F \subset K$. Let $\{G_\alpha\}$, $\alpha \in A$, be a covering of F by sets that are open in $\mathbb{R}^m$. Adjoining to this collection the open set $G = \mathbb{R}^m \setminus F$, we obtain an open covering of $\mathbb{R}^m$, and in particular, an open covering of K, from which we select a finite covering of K. This finite covering of K will also cover the set F. Observing that $G \cap F = \varnothing$, one can say that if G belongs to this finite covering, we will still have a finite covering of F by sets of the original system $\{G_\alpha\}$, $\alpha \in A$, if we remove G. □

Definition 9 The *diameter* of a set $E \subset \mathbb{R}^m$ is the quantity

$$d(E) := \sup_{x_1, x_2 \in E} d(x_1, x_2).$$

$T\mathbb{R}^m_y \to T\mathbb{R}^m_x$, then the mapping $f^{-1}: V(y) \to U(x)$ is differentiable at the point $y = f(x)$, and the following equality holds:

$$(f^{-1})'(y) = \left[f'(x)\right]^{-1}.$$

Thus, mutually inverse differentiable mappings have mutually inverse tangent mappings at corresponding points.

Proof We use the following notation:

$$f(x) = y, \qquad f(x+h) = y+t, \qquad t = f(x+h) - f(x),$$

so that

$$f^{-1}(y) = x, \qquad f^{-1}(y+t) = x+h, \qquad h = f^{-1}(y+t) - f^{-1}(y).$$

We shall assume that h is so small that $x + h \in U(x)$, and hence $y + t \in V(y)$. It follows from the continuity of f at x and f^{-1} at y that

$$t = f(x+h) - f(x) \to 0 \quad \text{as } h \to 0 \tag{8.49}$$

and

$$h = f^{-1}(y+t) - f^{-1}(y) \to 0 \quad \text{as } t \to 0. \tag{8.50}$$

It follows from the differentiability of f at x that

$$t = f'(x)h + o(h) \quad \text{as } h \to 0, \tag{8.51}$$

that is, we can even assert that $t = O(h)$ as $h \to 0$ (see relations (8.17) and (8.18) of Sect. 8.1).

We shall show that if $f'(x)$ is an invertible linear mapping, then we also have $h = O(t)$ as $t \to 0$.

Indeed, we find successively by (8.51) that

$$\begin{aligned}
&\left[f'(x)\right]^{-1}t = h + \left[f'(x)\right]^{-1}o(h) && \text{as } h \to 0,\\
&\left[f'(x)\right]^{-1}t = h + o(h) && \text{as } h \to 0,\\
&\left\|\left[f'(x)\right]^{-1}t\right\| \geq \|h\| - \left\|o(h)\right\| && \text{as } h \to 0,\\
&\left\|\left[f'(x)\right]^{-1}t\right\| \geq \tfrac{1}{2}\|h\| && \text{for } \|h\| < \delta,
\end{aligned} \tag{8.52}$$

where the number $\delta > 0$ is chosen so that $\|o(h)\| < \frac{1}{2}\|h\|$ when $\|h\| < \delta$. Then, taking account of (8.50), that is, the relation $h \to 0$ as $t \to 0$, we find

$$\|h\| \leq 2\left\|\left[f'(x)\right]^{-1}t\right\| = O\left(\|t\|\right) \quad \text{as } t \to 0,$$

which is equivalent to

$$h = O(t) \quad \text{as } t \to 0.$$

From this it follows in particular that

$$o(h) = o(t) \quad \text{as } t \to 0.$$

Taking this relation into account, we find by (8.50) and (8.52) that

$$h = \big[f'(x)\big]^{-1} t + o(t) \quad \text{as } t \to 0$$

or

$$f^{-1}(y+t) - f^{-1}(y) = \big[f'(x)\big]^{-1} t + o(t) \quad \text{as } t \to 0. \qquad \square$$

It is known from algebra that if the matrix A corresponds to the linear transformation $L : \mathbb{R}^m \to \mathbb{R}^m$, then the matrix A^{-1} inverse to A corresponds to the linear transformation $L^{-1} : \mathbb{R}^m \to \mathbb{R}^m$ inverse to L. The construction of the elements of the inverse matrix is also known from algebra. Consequently, the theorem just proved provides a direct recipe for constructing the mapping $(f^{-1})'(y)$.

We remark that when $m = 1$, that is, when $\mathbb{R}^m = \mathbb{R}$, the Jacobian of the mapping $f : U(x) \to V(y)$ at the point x reduces to the single number $f'(x)$ – the derivative of the function f at x – and the linear transformation $f'(x) : T\mathbb{R}_x \to T\mathbb{R}_y$ reduces to multiplication by that number: $h \mapsto f'(x)h$. This linear transformation is invertible if and only if $f'(x) \neq 0$, and the matrix of the inverse mapping $[f'(x)]^{-1} : T\mathbb{R}_y \to T\mathbb{R}_x$ also consists of a single number, equal to $[f'(x)]^{-1}$, that is, the reciprocal of $f'(x)$. Hence Theorem 4 also subsumes the rule for finding the derivative of an inverse function proved earlier.

8.3.4 Problems and Exercises

1. a) We shall regard two paths $t \mapsto x_1(t)$ and $t \mapsto x_2(t)$ as equivalent at the point $x_0 \in \mathbb{R}^m$ if $x_1(0) = x_2(0) = x_0$ and $d(x_1(t), x_2(t)) = o(t)$ as $t \to 0$.

Verify that this relation is an equivalence relation, that is, it is reflexive, symmetric, and transitive.

b) Verify that there is a one-to-one correspondence between vectors $\mathbf{v} \in T\mathbb{R}^m_{x_0}$ and equivalence classes of smooth paths at the point x_0.

c) By identifying the tangent space $T\mathbb{R}^m_{x_0}$ with the set of equivalence classes of smooth paths at the point $x_0 \in \mathbb{R}^m$, introduce the operations of addition and multiplication by a scalar for equivalence classes of paths.

d) Determine whether the operations you have introduced depend on the coordinate system used in $\mathbb{R}^m$.

Definition 10 A set $E \subset \mathbb{R}^m$ is *bounded* if its diameter is finite.

Proposition 4 *If K is a compact set in $\mathbb{R}^m$, then K is a bounded subset of $\mathbb{R}^m$.*

Proof Take an arbitrary point $a \in \mathbb{R}^m$ and consider the sequence of open balls $\{B(a; n)\}$ $(n = 1, 2, \dots)$. They form an open covering of $\mathbb{R}^m$ and consequently also of K. If K were not bounded, it would be impossible to select a finite covering of K from this sequence. □

Proposition 5 *The set $K \subset \mathbb{R}^m$ is compact if and only if K is closed and bounded in $\mathbb{R}^m$.*

Proof The necessity of these conditions was proved in Propositions 3 and 4.

Let us verify that the conditions are sufficient. Since K is a bounded set, there exists an m-dimensional interval I containing K. As was shown in Example 13, I is compact in $\mathbb{R}^m$. But if K is a closed set contained in the compact set I, then by Proposition 3b) it is itself compact. □

7.1.4 Problems and Exercises

1. The *distance $d(E_1, E_2)$ between the sets $E_1, E_2 \subset \mathbb{R}^m$* is the quantity

$$d(E_1, E_2) := \inf_{x_1 \in E_1, x_2 \in E_2} d(x_1, x_2).$$

Give an example of closed sets E_1 and E_2 in $\mathbb{R}^m$ having no points in common for which $d(E_1, E_2) = 0$.

2. Show that

a) the closure $\overline{E}$ in $\mathbb{R}^m$ of any set $E \subset \mathbb{R}^m$ is a closed set in $\mathbb{R}^m$;
b) the set ∂E of boundary points of any set $E \subset \mathbb{R}^m$ is a closed set;
c) if G is an open set in $\mathbb{R}^m$ and F is closed in $\mathbb{R}^m$, then $G \setminus F$ is open in $\mathbb{R}^m$.

3. Show that if $K_1 \supset K_2 \supset \cdots \supset K_n \supset \cdots$ is a sequence of nested nonempty compact sets, then $\bigcap_{i=1}^{\infty} K_i \neq \varnothing$.

4. a) In the space $\mathbb{R}^k$ a two-dimensional sphere S^2 and a circle S^1 are situated so that the distance from any point of the sphere to any point of the circle is the same. Is this possible?

b) Consider problem a) for spheres S^m, S^n of arbitrary dimension in $\mathbb{R}^k$. Under what relation on m, n, and k is this situation possible?

7.2 Limits and Continuity of Functions of Several Variables

7.2.1 The Limit of a Function

In Chap. 3 we studied in detail the operation of passing to the limit for a real-valued function $f : X \to \mathbb{R}$ defined on a set in which a base $\mathcal{B}$ was fixed.

In the next few sections we shall be considering functions $f : X \to \mathbb{R}^n$ defined on subsets of $\mathbb{R}^m$ with values in $\mathbb{R} = \mathbb{R}^1$ or more generally in $\mathbb{R}^n$, $n \in \mathbb{N}$. We shall now make a number of additions to the theory of limits connected with the specifics of this class of functions.

However, we begin with the basic general definition.

Definition 1 A point $A \in \mathbb{R}^n$ is the *limit of the mapping* $f : X \to \mathbb{R}^n$ over a base $\mathcal{B}$ in X if for every neighborhood $V(A)$ of the point there exists an element $B \in \mathcal{B}$ of the base whose image $f(B)$ is contained in $V(A)$.

In brief,

$$\Bigl(\lim_{\mathcal{B}} f(x) = A\Bigr) := \bigl(\forall V(A)\ \exists B \in \mathcal{B}\ \bigl(f(B) \subset V(A)\bigr)\bigr).$$

We see that the definition of the limit of a function $f : X \to \mathbb{R}^n$ is exactly the same as the definition of the limit of a function $f : X \to \mathbb{R}$ if we keep in mind what a neighborhood $V(A)$ of a point $A \in \mathbb{R}^n$ is for every $n \in \mathbb{N}$.

Definition 2 A mapping $f : X \to \mathbb{R}^n$ is *bounded* if the set $f(X) \subset \mathbb{R}^n$ is bounded in $\mathbb{R}^n$.

Definition 3 Let $\mathcal{B}$ be a base in X. A mapping $f : X \to \mathbb{R}^n$ is *ultimately bounded* over the base $\mathcal{B}$ if there exists an element B of $\mathcal{B}$ on which f is bounded.

Taking these definitions into account and using the same reasoning that we gave in Chap. 3, one can verify without difficulty that

a function $f : X \to \mathbb{R}^n$ can have at most one limit over a given base $\mathcal{B}$ in X;

a function $f : X \to \mathbb{R}^n$ having a limit over a base $\mathcal{B}$ is ultimately bounded over that base.

Definition 1 can be rewritten in another form making explicit use of the metric in $\mathbb{R}^n$, namely

Definition 1′

$$\Bigl(\lim_{\mathcal{B}} f(x) = A \in \mathbb{R}^n\Bigr) := \bigl(\forall \varepsilon > 0\ \exists B \in \mathcal{B}\ \forall x \in B\ \bigl(d\bigl(f(x), A\bigr) < \varepsilon\bigr)\bigr)$$

or

Definition 1″

$$\Big(\lim_{\mathcal{B}} f(x) = A \in \mathbb{R}^n\Big) := \Big(\lim_{\mathcal{B}}\big(d\big(f(x), A\big)\big) = 0\Big).$$

The specific property of a mapping $f : X \to \mathbb{R}^n$ is that, since a point $y \in \mathbb{R}^n$ is an ordered n-tuple $(y^1, \dots, y^n)$ of real numbers, defining a function $f : X \to \mathbb{R}^n$ is equivalent to defining n real-valued functions $f^i : X \to \mathbb{R}$ $(i = 1, \dots, n)$, where $f^i(x) = y^i$ $(i = 1, \dots, n)$.

If $A = (A^1, \dots, A^n)$ and $y = (y^1, \dots, y^n)$, we have the inequalities

$$\big|y^i - A^i\big| \le d(y, A) \le \sqrt{n} \max_{1 \le i \le n} \big|y^i - A^i\big|, \tag{7.3}$$

from which one can see that

$$\lim_{\mathcal{B}} f(x) = A \Leftrightarrow \lim_{\mathcal{B}} f^i(x) = A^i \quad (i = 1, \dots, n), \tag{7.4}$$

that is, convergence in $\mathbb{R}^n$ is coordinatewise.

Now let $X = \mathbb{N}$ be the set of natural numbers and $\mathcal{B}$ the base $k \to \infty$, $k \in \mathbb{N}$, in X. A function $f : \mathbb{N} \to \mathbb{R}^n$ in this case is a sequence $\{y_k\}$, $k \in \mathbb{N}$, of points of $\mathbb{R}^n$.

Definition 4 A sequence $\{y_k\}$, $k \in \mathbb{N}$, of points $y_k \in \mathbb{R}^n$ is *fundamental* (a *Cauchy sequence*) if for every $\varepsilon > 0$ there exists a number $N \in \mathbb{N}$ such that $d(y_{k_1}, y_{k_2}) < \varepsilon$ for all $k_1, k_2 > N$.

One can conclude from the inequalities (7.3) that a sequence of points $y_k = (y_k^1, \dots, y_k^n) \in \mathbb{R}^n$ is a Cauchy sequence if and only if each sequence of coordinates having the same labels $\{y_k^i\}$, $k \in \mathbb{N}$, $i = 1, \dots, n$, is a Cauchy sequence.

Taking account of relation (7.4) and the Cauchy criterion for numerical sequences, one can now assert that a sequence of points $\mathbb{R}^n$ converges if and only if it is a Cauchy sequence.

In other words, the Cauchy criterion is also valid in $\mathbb{R}^n$.

Later on we shall call metric spaces in which every Cauchy sequence has a limit *complete* metric spaces. Thus we have now established that $\mathbb{R}^n$ is a complete metric space for every $n \in \mathbb{N}$.

Definition 5 The *oscillation* of a function $f : X \to \mathbb{R}^n$ on a set $E \subset X$ is the quantity

$$\omega(f; E) := d\big(f(E)\big),$$

where $d(f(E))$ is the diameter of $f(E)$.

As one can see, this is a direct generalization of the definition of the oscillation of a real-valued function, which Definition 5 becomes when $n = 1$.

The validity of the following Cauchy criterion for the existence of a limit for functions $f : X \to \mathbb{R}^n$ with values in $\mathbb{R}^n$ results from the completeness of $\mathbb{R}^n$.

Theorem 1 *Let X be a set and $\mathcal{B}$ a base in X. A function $f : X \to \mathbb{R}^n$ has a limit over the base $\mathcal{B}$ if and only if for every $\varepsilon > 0$ there exists an element $B \in \mathcal{B}$ of the base on which the oscillation of the function is less than ε.*

Thus,

$$\exists \lim_{\mathcal{B}} f(x) \Leftrightarrow \forall \varepsilon > 0 \, \exists B \in \mathcal{B} \, \big(\omega(f; B) < \varepsilon\big).$$

The proof of Theorem 1 is a verbatim repetition of the proof of the Cauchy criterion for numerical functions (Theorem 4 in Sect. 3.2), except for one minor change: $|f(x_1) - f(x_2)|$ must be replaced throughout by $d(f(x_1), f(x_2))$.

One can also verify Theorem 1 another way, regarding the Cauchy criterion as known for real-valued functions and using relations (7.4) and (7.3).

The important theorem on the limit of a composite function also remains valid for functions with values in $\mathbb{R}^n$.

Theorem 2 *Let Y be a set, $\mathcal{B}_Y$ a base in Y, and $g : Y \to \mathbb{R}^n$ a mapping having a limit over the base $\mathcal{B}_Y$.*

Let X be a set, $\mathcal{B}_X$ a base in X, and $f : X \to Y$ a mapping of X into Y such that for each $B_Y \in \mathcal{B}_Y$ there exists $B_X \in \mathcal{B}_X$ such that the image $f(B_X)$ is contained in B_Y.

Under these conditions the composition $g \circ f : X \to \mathbb{R}^n$ of the mappings f and g is defined and has a limit over the base $\mathcal{B}_X$, and

$$\lim_{\mathcal{B}_X} (g \circ f)(x) = \lim_{\mathcal{B}_Y} g(y).$$

The proof of Theorem 2 can be carried out either by repeating the proof of Theorem 5 of Sect. 3.2, replacing $\mathbb{R}$ by $\mathbb{R}^n$, or by invoking that theorem and using relation (7.4).

Up to now we have considered functions $f : X \to \mathbb{R}^n$ with values in $\mathbb{R}^n$, without specifying their domains of definition X in any way. From now on we shall primarily be interested in the case when X is a subset of $\mathbb{R}^m$.

As before, we make the following conventions.

$U(a)$ is a neighborhood of the point $a \in \mathbb{R}^m$;
$\overset{\circ}{U}(a)$ is a deleted neighborhood of $a \in \mathbb{R}^m$, that is, $\overset{\circ}{U}(a) := U(a) \backslash a$;
$U_E(a)$ is a neighborhood of a in the set $E \subset \mathbb{R}^m$, that is, $U_E(a) := E \cap U(a)$;
$\overset{\circ}{U}_E(a)$ is a deleted neighborhood of a in E, that is, $\overset{\circ}{U}_E(a) := E \cap \overset{\circ}{U}(a)$;
$x \to a$ is the base of deleted neighborhoods of a in $\mathbb{R}^m$;
$x \to \infty$ is the base of neighborhoods of infinity, that is, the base consisting of the sets $\mathbb{R}^m \backslash B(a; r)$;
$x \to a, x \in E$ or $(E \ni x \to a)$ is the base of deleted neighborhoods of a in E if a is a limit point of E;
$x \to \infty, x \in E$ or $(E \ni x \to \infty)$ is the base of neighborhoods of infinity in E consisting of the sets $E \backslash B(a; r)$, if E is an unbounded set.

In accordance with these definitions, one can, for example, give the following specific form of Definition 1 for the limit of a function when speaking of a function $f : E \to \mathbb{R}^n$ mapping a set $E \subset \mathbb{R}^m$ into $\mathbb{R}^n$:

$$\Bigl(\lim_{E\ni x\to a} f(x) = A\Bigr) := \bigl(\forall \varepsilon > 0\ \exists \mathring{U}_E(a)\ \forall x \in \mathring{U}_E(a)\ \bigl(d\bigl(f(x), A\bigr) < \varepsilon\bigr)\bigr).$$

The same thing can be written another way:

$$\Bigl(\lim_{E\ni x\to a} f(x) = A\Bigr) :=$$
$$= \bigl(\forall \varepsilon > 0\ \exists \delta > 0\ \forall x \in E\ \bigl(0 < d(x, a) < \delta \Rightarrow d\bigl(f(x), A\bigr) < \varepsilon\bigr)\bigr).$$

Here it is understood that the distances $d(x, a)$ and $d(f(x), A)$ are measured in the spaces ($\mathbb{R}^m$ and $\mathbb{R}^n$) in which these points lie.

Finally,

$$\Bigl(\lim_{x\to\infty} f(x) = A\Bigr) := \bigl(\forall \varepsilon > 0\ \exists B(a; r)\ \forall x \in \mathbb{R}^m \backslash B(a; r)\ \bigl(d\bigl(f(x), A\bigr) < \varepsilon\bigr)\bigr).$$

Let us also agree to say that, in the case of a mapping $f : X \to \mathbb{R}^n$, the phrase "$f(x) \to \infty$ in the base $\mathcal{B}$" means that for any ball $B(A; r) \subset \mathbb{R}^n$ there exists $B \in \mathcal{B}$ of the base $\mathcal{B}$ such that $f(B) \subset \mathbb{R}^n \backslash B(A; r)$.

Example 1 Let $x \mapsto \pi^i(x)$ be the mapping $\pi^i : \mathbb{R}^m \to \mathbb{R}$ assigning to each $x = (x^1, \dots, x^m)$ in $\mathbb{R}^m$ its ith coordinate x^i. Thus

$$\pi^i(x) = x^i.$$

If $a = (a^1, \dots, a^m)$, then obviously

$$\pi^i(x) \to a^i \quad \text{as } x \to a.$$

The function $x \mapsto \pi^i(x)$ does not tend to any finite value nor to infinity as $x \to \infty$ if $m > 1$.

On the other hand,

$$f(x) = \sum_{i=1}^{m} \bigl(\pi^i(x)\bigr)^2 \to \infty \quad \text{as } x \to \infty.$$

One should not think that the limit of a function of several variables can be found by computing successively the limits with respect to each of its coordinates. The following examples show why this is not the case.

Example 2 Let the function $f : \mathbb{R}^2 \to \mathbb{R}$ be defined at the point $(x, y) \in \mathbb{R}^2$ as follows:

$$f(x, y) = \begin{cases} \frac{xy}{x^2+y^2}, & \text{if } x^2 + y^2 \neq 0, \\ 0, & \text{if } x^2 + y^2 = 0. \end{cases}$$

Then $f(0, y) = f(x, 0) = 0$, while $f(x, x) = \frac{1}{2}$ for $x \neq 0$.

Hence this function has no limit as $(x, y) \to (0, 0)$.

On the other hand,

$$\lim_{y \to 0} \left(\lim_{x \to 0} f(x, y) \right) = \lim_{y \to 0} (0) = 0,$$

$$\lim_{x \to 0} \left(\lim_{y \to 0} f(x, y) \right) = \lim_{x \to 0} (0) = 0.$$

Example 3 For the function

$$f(x, y) = \begin{cases} \frac{x^2 - y^2}{x^2 + y^2}, & \text{if } x^2 + y^2 \neq 0, \\ 0, & \text{if } x^2 + y^2 = 0, \end{cases}$$

we have

$$\lim_{x \to 0} \left(\lim_{y \to 0} f(x, y) \right) = \lim_{x \to 0} \left(\frac{x^2}{x^2} \right) = 1,$$

$$\lim_{y \to 0} \left(\lim_{x \to 0} f(x, y) \right) = \lim_{y \to 0} \left(-\frac{y^2}{y^2} \right) = -1.$$

Example 4 For the function

$$f(x, y) = \begin{cases} x + y \sin \frac{1}{x}, & \text{if } x \neq 0, \\ 0, & \text{if } x = 0, \end{cases}$$

we have

$$\lim_{(x,y) \to (0,0)} f(x, y) = 0,$$

$$\lim_{x \to 0} \left(\lim_{y \to 0} f(x, y) \right) = 0,$$

yet at the same time the iterated limit

$$\lim_{y \to 0} \left(\lim_{x \to 0} f(x, y) \right)$$

does not exist at all.

Example 5 The function

$$f(x, y) = \begin{cases} \frac{x^2 y}{x^4 + y^2}, & \text{if } x^2 + y^2 \neq 0, \\ 0, & \text{if } x^2 + y^2 = 0, \end{cases}$$

has a limit of zero upon approach to the origin along any ray $x = \alpha t$, $y = \beta t$.

At the same time, the function equals $\frac{1}{2}$ at any point of the form (a, a^2), where $a \neq 0$, and so the function has no limit as $(x, y) \to (0, 0)$.

7.2.2 Continuity of a Function of Several Variables and Properties of Continuous Functions

Let E be a subset of $\mathbb{R}^m$ and $f : E \to \mathbb{R}^n$ a function defined on E with values in $\mathbb{R}^n$.

Definition 6 The function $f : E \to \mathbb{R}^n$ is *continuous* at $a \in E$ if for every neighborhood $V(f(a))$ of the value $f(a)$ that the function assumes at a, there exists a neighborhood $U_E(a)$ of a in E whose image $f(U_E(a))$ is contained in $V(f(a))$.

Thus

$$\begin{aligned}&\big(f : E \to \mathbb{R}^n \text{ is continuous at } a \in E\big) := \\ &= \big(\forall V\big(f(a)\big)\ \exists U_E(a)\ \big(f\big(U_E(a)\big) \subset V\big(f(a)\big)\big)\big).\end{aligned}$$

We see that Definition 6 has the same form as Definition 1 for continuity of a real-valued function, which we are familiar with from Sect. 4.1. As was the case there, we can give the following alternate expression for this definition:

$$\begin{aligned}&\big(f : E \to \mathbb{R}^n \text{ is continuous at } a \in E\big) := \\ &= \big(\forall \varepsilon > 0\ \exists \delta > 0\ \forall x \in E\ \big(d(x,a) < \delta \Rightarrow d\big(f(x), f(a)\big) < \varepsilon\big)\big),\end{aligned}$$

or, if a is a limit point of E,

$$\big(f : E \to \mathbb{R}^n \text{ is continuous at } a \in E\big) := \Big(\lim_{E \ni x \to a} f(x) = f(a)\Big).$$

As noted in Chap. 4, the concept of continuity is of interest precisely in connection with a point $a \in E$ that is a limit point of the set E on which the function f is defined.

It follows from Definition 6 and relation (7.4) that the mapping $f : E \to \mathbb{R}^n$ defined by the relation

$$\begin{aligned}\big(x^1, \dots, x^m\big) = x \stackrel{f}{\longmapsto} y = \big(y^1, \dots, y^n\big) = \\ = \big(f^1\big(x^1, \dots, x^m\big), \dots, f^n\big(x^1, \dots, x^m\big)\big),\end{aligned}$$

is continuous at a point if and only if each of the functions $y^i = f^i(x^1, \dots, x^m)$ is continuous at that point.

In particular, we recall that we defined a path in $\mathbb{R}^n$ to be a mapping $f : I \to \mathbb{R}^n$ of an interval $I \subset \mathbb{R}$ defined by continuous functions $f^1(x), \dots, f^n(x)$ in the form

$$x \mapsto y = \big(y^1, \dots, y^n\big) = \big(f^1(x), \dots, f^n(x)\big).$$

Thus we can now say that a *path in* $\mathbb{R}^n$ *is a continuous mapping* of an interval $I \subset \mathbb{R}$ of the real line into $\mathbb{R}^n$.

By analogy with the definition of oscillation of a real-valued function at a point, we introduce the concept of oscillation at a point for a function with values in $\mathbb{R}^n$.

Let E be a subset of $\mathbb{R}^m$, $a \in E$, and $B_E(a;r) = E \cap B(a;r)$.

Definition 7 The *oscillation of the function* $f : E \to \mathbb{R}^r$ at the point $a \in E$ is the quantity

$$\omega(f;a) := \lim_{r \to +0} \omega\bigl(f; B_E(a;r)\bigr).$$

From Definition 6 of continuity of a function, taking account of the properties of a limit and the Cauchy criterion, we obtain a set of frequently used local properties of continuous functions. We now list them.

Local Properties of Continuous Functions

a) *A mapping $f : E \to \mathbb{R}^n$ of a set $E \subset \mathbb{R}^m$ is continuous at a point $a \in E$ if and only if $\omega(f;a) = 0$.*

b) *A mapping $f : E \to \mathbb{R}^n$ that is continuous at $a \in E$ is bounded in some neighborhood $U_E(a)$ of that point.*

c) *If the mapping $g : Y \to \mathbb{R}^k$ of the set $Y \subset \mathbb{R}^n$ is continuous at a point $y_0 \in Y$ and the mapping $f : X \to Y$ of the set $X \subset \mathbb{R}^m$ is continuous at a point $x_0 \in X$ and $f(x_0) = y_0$, then the mapping $g \circ f : X \to \mathbb{R}^k$ is defined, and it is continuous at $x_0 \in X$.*

Real-valued functions possess, in addition, the following properties.

d) *If the function $f : E \to \mathbb{R}$ is continuous at the point $a \in E$ and $f(a) > 0$ (or $f(a) < 0$), there exists a neighborhood $U_E(a)$ of a in E such that $f(x) > 0$ (resp. $f(x) < 0$) for all $x \in U_E(a)$.*

e) *If the functions $f : E \to \mathbb{R}$ and $g : E \to \mathbb{R}$ are continuous at $a \in E$, then any linear combination of them $(\alpha f + \beta g) : E \to \mathbb{R}$, where $\alpha, \beta \in \mathbb{R}$, their product $(f \cdot g) : E \to \mathbb{R}$, and, if $g(x) \neq 0$ on E, their quotient $(\frac{f}{g}) : E \to \mathbb{R}$ are defined on E and continuous at a.*

Let us agree to say that the function $f : E \to \mathbb{R}^n$ is *continuous on the set E* if it is continuous at each point of the set.

The set of functions $f : E \to \mathbb{R}^n$ that are continuous on E will be denoted $C(E, \mathbb{R}^n)$ or simply $C(E)$, if the range of values of the functions is unambiguously determined from the context. As a rule, this abbreviation will be used when $\mathbb{R}^n = \mathbb{R}$.

Example 6 The functions $(x^1, \dots, x^m) \overset{\pi^i}{\longmapsto} x^i \, (i = 1, \dots, m)$, mapping $\mathbb{R}^m$ onto $\mathbb{R}$ (projections) are obviously continuous at each point $a = (a^1, \dots, a^m) \in \mathbb{R}^m$, since $\lim_{x \to a} \pi^i(x) = a^i = \pi^i(a)$.

Example 7 Any function $x \mapsto f(x)$ defined on $\mathbb{R}$, for example $x \mapsto \sin x$, can also be regarded as a function $(x, y) \overset{F}{\longmapsto} f(x)$ defined, say, on $\mathbb{R}^2$. In that case, if f was continuous as a function on $\mathbb{R}$, then the new function $(x, y) \overset{F}{\longmapsto} f(x)$ will be continuous as a function on $\mathbb{R}^2$. This can be verified either directly from the definition of continuity or by remarking that the function F is the composition $(f \circ \pi^1)(x, y)$ of continuous functions.

In particular, it follows from this, when we take account of c) and e), that the functions

$$f(x, y) = \sin x + e^{xy}, \qquad f(x, y) = \arctan\bigl(\ln\bigl(|x| + |y| + 1\bigr)\bigr),$$

for example, are continuous on $\mathbb{R}^2$.

We remark that the reasoning just used is essentially local, and the fact that the functions f and F studied in Example 7 were defined on the entire real line $\mathbb{R}$ or the plane $\mathbb{R}^2$ respectively was purely an accidental circumstance.

Example 8 The function $f(x, y)$ of Example 2 is continuous at any point of the space $\mathbb{R}^2$ except $(0, 0)$. We remark that, despite the discontinuity of $f(x, y)$ at this point, the function is continuous in either of its two variables for each fixed value of the other variable.

Example 9 If a function $f : E \to \mathbb{R}^n$ is continuous on the set E and $\tilde{E}$ is a subset of E, then the restriction $f|_{\tilde{E}}$ of f to this subset is continuous on $\tilde{E}$, as follows immediately from the definition of continuity of a function at a point.

We now turn to the global properties of continuous functions. To state them for functions $f : E \to \mathbb{R}^n$, we first give two definitions.

Definition 8 A mapping $f : E \to \mathbb{R}^n$ of a set $E \subset \mathbb{R}^m$ into $\mathbb{R}^n$ is *uniformly continuous* on E if for every $\varepsilon > 0$ there is a number $\delta > 0$ such that $d(f(x_1), f(x_2)) < \varepsilon$ for any points $x_1, x_2 \in E$ such that $d(x_1, x_2) < \delta$.

As before, the distances $d(x_1, x_2)$ and $d(f(x_1), f(x_2))$ are assumed to be measured in $\mathbb{R}^m$ and $\mathbb{R}^n$ respectively.

When $m = n = 1$, this definition is the definition of uniform continuity of numerical-valued functions that we already know.

Definition 9 A set $E \subset \mathbb{R}^m$ is *pathwise connected* if for any pair of its points x_0, x_1, there exists a path $\Gamma : I \to E$ with support in E and endpoints at these points.

In other words, it is possible to go from any point $x_0 \in E$ to any other point $x_1 \in E$ without leaving E.

Since we shall not be considering any other concept of connectedness for a set except pathwise connectedness for the time being, for the sake of brevity we shall temporarily call pathwise connected sets simply *connected*.

Definition 10 A *domain* in $\mathbb{R}^m$ is an open connected set.

Example 10 An open ball $B(a;r)$, $r>0$, in $\mathbb{R}^m$ is a domain. We already know that $B(a;r)$ is open in $\mathbb{R}^m$. Let us verify that the ball is connected. Let $x_0=(x_0^1\dots,x_0^m)$ and $x_1=(x_1^1,\dots,x_1^m)$ be two points of the ball. The path defined by the functions $x^i(t)=tx_1^i+(1-t)x_0^i$ $(i=1,\dots,m)$, defined on the closed interval $0\le t\le 1$, has x_0 and x_1 as its endpoints. In addition, its support lies in the ball $B(a;r)$, since, by Minkowski's inequality, for any $t\in[0,1]$,

$$d\big(x(t),a\big)=\sqrt{\sum_{i=1}^m\big(x^i(t)-a^i\big)^2}=\sqrt{\sum_{i=1}^m\big(t\big(x_1^i-a^i\big)+(1-t)\big(x_0^i-a^i\big)\big)^2}\le$$

$$\le\sqrt{\sum_{i=1}^m\big(t\big(x_1^i-a^i\big)\big)^2}+\sqrt{\sum_{i=1}^m\big((1-t)\big(x_0^i-a^i\big)\big)^2}=$$

$$=t\cdot\sqrt{\sum_{i=1}^m\big(x_1^i-a^i\big)^2}+(1-t)\cdot\sqrt{\sum_{i=1}^m\big(x_0^i-a^i\big)^2}<tr+(1-t)r=r.$$

Example 11 The circle (one-dimensional sphere) of radius $r>0$ is the subset of $\mathbb{R}^2$ given by the equation $(x^1)^2+(x^2)^2=r^2$. Setting $x^1=r\cos t$, $x^2=r\sin t$, we see that any two points of the circle can be joined by a path that goes along the circle. Hence a circle is a connected set. However, this set is not a domain in $\mathbb{R}^2$, since it is not open in $\mathbb{R}^2$.

We now state the basic facts about continuous functions in the large.

Global Properties of Continuous Functions

a) *If a mapping $f:K\to\mathbb{R}^n$ is continuous on a compact set $K\subset\mathbb{R}^m$, then it is uniformly continuous on K.*

b) *If a mapping $f:K\to\mathbb{R}^n$ is continuous on a compact set $K\subset\mathbb{R}^m$, then it is bounded on K.*

c) *If a function $f:K\to\mathbb{R}$ is continuous on a compact set $K\subset\mathbb{R}^m$, then it assumes its maximal and minimal values at some points of K.*

d) *If a function $f:E\to\mathbb{R}$ is continuous on a connected set E and assumes the values $f(a)=A$ and $f(b)=B$ at points $a,b\in E$, then for any C between A and B, there is a point $c\in E$ at which $f(c)=C$.*

Earlier (Sect. 4.2), when we were studying the local and global properties of functions of one variable, we gave proofs of these properties that extend to the more general case considered here. The only change that must be made in the earlier proofs is that expressions of the type $|x_1-x_2|$ or $|f(x_1)-f(x_2)|$ must be replaced by $d(x_1,x_2)$ and $d(f(x_1),f(x_2))$, where d is the metric in the space where the points in question are located. This remark applies fully to everything except the last statement d), whose proof we now give.

Proof d) Let $\Gamma : I \to E$ be a path that is a continuous mapping of an interval $[\alpha, \beta] = I \subset R$ such that $\Gamma(\alpha) = a$, $\Gamma(\beta) = b$. By the connectedness of E there exists such a path. The function $f \circ \Gamma : I \to \mathbb{R}$, being the composition of continuous functions, is continuous; therefore there is a point $\gamma \in [\alpha, \beta]$ on the closed interval $[\alpha, \beta]$ at which $f \circ \Gamma(\gamma) = C$. Set $c = \Gamma(\gamma)$. Then $c \in E$ and $f(c) = C$. □

Example 12 The sphere $S(0; r)$ defined in $\mathbb{R}^m$ by the equation

$$(x^1)^2 + \cdots + (x^m)^2 = r^2,$$

is a compact set.

Indeed, it follows from the continuity of the function

$$(x^1, \dots, x^m) \mapsto (x^1)^2 + \cdots + (x^m)^2$$

that the sphere is closed, and from the fact that $|x^i| \le r$ $(i = 1, \dots, m)$ on the sphere that it is bounded.

The function

$$(x^1, \dots, x^m) \mapsto (x^1)^2 + \cdots + (x^k)^2 - (x^{k+1})^2 - \cdots - (x^m)^2$$

is continuous on all of $\mathbb{R}^m$, so that its restriction to the sphere is also continuous, and by the global property c) of continuous functions assumes its minimal and maximal values on the sphere. At the points $(1, 0, \dots, 0)$ and $(0, \dots, 0, 1)$ this function assumes the values 1 and -1 respectively. By the connectedness of the sphere (see Problem 3 at the end of this section), global property d) of continuous functions now enables us to assert that there is a point on the sphere where this function assumes the value 0.

Example 13 The open set $\mathbb{R}^m \setminus S(0; r)$ for $r > 0$ is not a domain, since it is not connected.

Indeed, if $\Gamma : I \to \mathbb{R}^m$ is a path one end of which is at the point $x_0 = (0, \dots, 0)$ and the other at some point $x_1 = (x_1^1, \dots, x_1^m)$ such that $(x_1^1)^2 + \cdots + (x_1^m)^2 > r^2$, then the composition of the continuous functions $\Gamma : I \to \mathbb{R}^m$ and $f : \mathbb{R}^m \to \mathbb{R}$, where

$$(x^1, \dots, x^m) \overset{f}{\longmapsto} (x^1)^2 + \cdots + (x^m)^2,$$

is a continuous function on I assuming values less than r^2 at one endpoint and greater than r^2 at the other. Hence there is a point γ on I at which $(f \circ \Gamma)(\gamma) = r^2$. Then the point $x_\gamma = \Gamma(\gamma)$ in the support of the path turns out to lie on the sphere $S(0; r)$. We have thus shown that it is impossible to get out of the ball $B(0; r) \subset \mathbb{R}^m$ without intersecting its boundary sphere $S(0; r)$.

7.2.3 Problems and Exercises

1. Let $f \in C(\mathbb{R}^m; \mathbb{R})$. Show that

a) the set $E_1 = \{x \in \mathbb{R}^m \mid f(x) < c\}$ is open in $\mathbb{R}^m$;
b) the set $E_2 = \{x \in \mathbb{R}^m \mid f(x) \le c\}$ is closed in $\mathbb{R}^m$;
c) the set $E_3 = \{x \in \mathbb{R}^m \mid f(x) = c\}$ is closed in $\mathbb{R}^m$;
d) if $f(x) \to +\infty$ as $x \to \infty$, then E_2 and E_3 are compact in $\mathbb{R}^m$;
e) for any $f : \mathbb{R}^m \to \mathbb{R}$ the set $E_4 = \{x \in \mathbb{R}^m \mid \omega(f; x) \ge \varepsilon\}$ is closed in $\mathbb{R}^m$.

2. Show that the mapping $f : \mathbb{R}^m \to \mathbb{R}^n$ is continuous if and only if the preimage of every open set in $\mathbb{R}^n$ is an open set in $\mathbb{R}^m$.

3. Show that

a) the image $f(E)$ of a connected set $E \subset \mathbb{R}^m$ under a continuous mapping $f : E \to \mathbb{R}^n$ is a connected set;

b) the union of connected sets having a point in common is a connected set;

c) the hemisphere $(x^1)^2 + \cdots + (x^m)^2 = 1$, $x^m \ge 0$, is a connected set;

d) the sphere $(x^1)^2 + \cdots + (x^m)^2 = 1$ is a connected set;

e) if $E \subset \mathbb{R}$ and E is connected, then E is an interval in $\mathbb{R}$ (that is, a closed interval, a half-open interval, an open interval, or the entire real line);

f) if x_0 is an interior point and x_1 an exterior point in relation to the set $M \subset \mathbb{R}^m$, then the support of any path with endpoints x_0, x_1 intersects the boundary of the set M.

Chapter 8
The Differential Calculus of Functions of Several Variables

8.1 The Linear Structure on $\mathbb{R}^m$

8.1.1 $\mathbb{R}^m$ *as a Vector Space*

The concept of a vector space is already familiar to you from your study of algebra.

If we introduce the operation of addition of elements $x_1 = (x_1^1, \dots, x_1^m)$ and $x_2 = (x_2^1, \dots, x_2^m)$ in $\mathbb{R}^m$ by the formula

$$x_1 + x_2 = \left(x_1^1 + x_2^1, \dots, x_1^m + x_2^m\right), \tag{8.1}$$

and multiplication of an element $x = (x^1, \dots, x^m)$ by a number $\lambda \in \mathbb{R}$ via the relation

$$\lambda x = \left(\lambda x^1, \dots, \lambda x^m\right), \tag{8.2}$$

then $\mathbb{R}^m$ becomes a vector space over the field of real numbers. Its points can now be called vectors.

The vectors

$$e_i = (0, \dots, 0, 1, 0, \dots, 0) \quad (i = 1, \dots, m) \tag{8.3}$$

(where the 1 stands only in the ith place) form a maximal linearly independent set of vectors in this space, as a result of which it turns out to be an m-dimensional vector space.

Any vector $x \in \mathbb{R}^m$ can be expanded with respect to the basis (8.3), that is, represented in the form

$$x = x^1 e_1 + \dots + x^m e_m. \tag{8.4}$$

When vectors are indexed, we shall write the index as a subscript, while denoting its coordinates, as we have been doing, by superscripts. This is convenient for many

V.A. Zorich, *Mathematical Analysis I*, Universitext,
DOI 10.1007/978-3-662-48792-1_8

reasons, one of which is that, following Einstein,[1] we can make the convention of writing expressions like (8.4) briefly in the form

$$x = x^i e_i, \tag{8.5}$$

taking the simultaneous presence of subscript and superscript with the same letter to indicate summation with respect to that letter over its range of variation.

8.1.2 Linear Transformations $L : \mathbb{R}^m \to \mathbb{R}^n$

We recall that a mapping $L : X \to Y$ from a vector space X into a vector space Y is called *linear* if

$$L(\lambda_1 x_1 + \lambda_2 x_2) = \lambda_1 L(x_1) + \lambda_2 L(x_2)$$

for any $x_1, x_2 \in X$, and $\lambda_1, \lambda_2 \in \mathbb{R}$. We shall be interested in linear mappings $L : \mathbb{R}^m \to \mathbb{R}^n$.

If $\{e_1, \dots, e_m\}$ and $\{\tilde{e}_1, \dots, \tilde{e}_n\}$ are fixed bases of $\mathbb{R}^m$ and $\mathbb{R}^n$ respectively, then, knowing the expansion

$$L(e_i) = a_i^1 \tilde{e}_1 + \cdots + a_i^n \tilde{e}_n = a_i^j \tilde{e}_j \quad (i = 1, \dots, m) \tag{8.6}$$

of the images of the basis vectors under the linear mapping $L : \mathbb{R}^m \to \mathbb{R}^n$, we can use the linearity of L to find the expansion of the image $L(h)$ of any vector $h = h^1 e_1 + \cdots + h^m e_m = h^i e_i$ in the basis $\{\tilde{e}_1, \dots, \tilde{e}_n\}$. To be specific,

$$L(h) = L\big(h^i e_i\big) = h^i L(e_i) = h^i a_i^j \tilde{e}_j = a_i^j h^i \tilde{e}_j. \tag{8.7}$$

Hence, in coordinate notation:

$$L(h) = \big(a_i^1 h^i, \dots, a_i^n h^i\big). \tag{8.8}$$

For a fixed basis in $\mathbb{R}^n$ the mapping $L : \mathbb{R}^m \to \mathbb{R}^n$ can thus be regarded as a set

$$L = \big(L^1, \dots, L^n\big) \tag{8.9}$$

of n (coordinate) mappings $L^j : \mathbb{R}^m \to \mathbb{R}$.

Taking account of (8.8), we easily conclude that a mapping $L : \mathbb{R}^m \to \mathbb{R}^n$ is linear if and only if each mapping L^j in the set (8.9) is linear.

If we write (8.9) as a column, taking account of relation (8.8), we have

$$L(h) = \begin{pmatrix} L^1(h) \\ \vdots \\ L^n(h) \end{pmatrix} = \begin{pmatrix} a_1^1 & \cdots & a_m^1 \\ \vdots & \ddots & \vdots \\ a_1^n & \cdots & a_m^n \end{pmatrix} \begin{pmatrix} h^1 \\ \vdots \\ h^m \end{pmatrix}. \tag{8.10}$$

---

[1] A. Einstein (1879–1955) – greatest physicist of the twentieth century. His work in quantum theory and especially in the theory of relativity exerted a revolutionary influence on all of modern physics.

Thus, fixing bases in $\mathbb{R}^m$ and $\mathbb{R}^n$ enables us to establish a one-to-one correspondence between linear transformations $L : \mathbb{R}^m \to \mathbb{R}^n$ and $m \times n$-matrices (a_i^j) $(i = 1, \dots, m,\ j = 1, \dots, n)$. When this is done, the ith column of the matrix (a_i^j) corresponding to the transformation L consists of the coordinates of $L(e_i)$, the image of the vector $e_i \in \{e_1, \dots, e_m\}$. The coordinates of the image of an arbitrary vector $h = h^i e_i \in \mathbb{R}^m$ can be obtained from (8.10) by multiplying the matrix of the linear transformation by the column of coordinates of h.

Since $\mathbb{R}^n$ has the structure of a vector space, one can speak of linear combinations $\lambda_1 f_1 + \lambda_2 f_2$ of mappings $f_1 : X \to \mathbb{R}^n$ and $f_2 : X \to \mathbb{R}^n$, setting

$$(\lambda_1 f_1 + \lambda_2 f_2)(x) := \lambda_1 f_1(x) + \lambda_2 f_2(x). \tag{8.11}$$

In particular, a linear combination of linear transformations $L_1 : \mathbb{R}^m \to \mathbb{R}^n$ and $L_2 : \mathbb{R}^m \to \mathbb{R}^n$ is, according to the definition (8.11), a mapping

$$h \mapsto \lambda_1 L_1(h) + \lambda_2 L_2(h) = L(h),$$

which is obviously linear. The matrix of this transformation is the corresponding linear combination of the matrices of the transformations L_1 and L_2.

The composition $C = B \circ A$ of linear transformations $A : \mathbb{R}^m \to \mathbb{R}^n$ and $B : \mathbb{R}^n \to \mathbb{R}^k$ is obviously also a linear transformation, whose matrix, as follows from (8.10), is the product of the matrix of A and the matrix of B (which is multiplied on the left). Actually, the law of multiplication for matrices was defined in the way you are familiar with precisely so that the product of matrices would correspond to the composition of the transformations.

8.1.3 *The Norm in $\mathbb{R}^m$*

The quantity

$$\|x\| = \sqrt{(x^1)^2 + \cdots + (x_m)^2} \tag{8.12}$$

is called the *norm* of the vector $x = (x^1, \dots, x^m) \in \mathbb{R}^m$.

It follows from this definition, taking account of Minkowski's inequality, that

1^0 $\|x\| \ge 0$,
2^0 $(\|x\| = 0) \Leftrightarrow (x = 0)$,
3^0 $\|\lambda x\| = |\lambda| \cdot \|x\|$, where $\lambda \in \mathbb{R}$,
4^0 $\|x_1 + x_2\| \le \|x_1\| + \|x_2\|$.

In general, any function $\| \ \| : X \to \mathbb{R}$ on a vector space X satisfying conditions 1^0–4^0 is called a *norm* on the vector space. Sometimes, to be precise as to which norm is being discussed, the norm sign has a symbol attached to it to denote the space in which it is being considered. For example, we may write $\|x\|_{\mathbb{R}^m}$ or

$\|y\|_{\mathbb{R}^n}$. As a rule, however, we shall not do that, since it will always be clear from the context which space and which norm are meant.

We remark that by (8.12)

$$\|x_2 - x_1\| = d(x_1, x_2), \tag{8.13}$$

where $d(x_1, x_2)$ is the distance in $\mathbb{R}^m$ between the vectors x_1 and x_2, regarded as points of $\mathbb{R}^m$.

It is clear from (8.13) that the following conditions are equivalent:

$$x \to x_0, \qquad d(x, x_0) \to 0, \qquad \|x - x_0\| \to 0.$$

In view of (8.13), we have, in particular,

$$\|x\| = d(0, x).$$

Property 4^0 of a norm is called the *triangle inequality*, and it is now clear why.

The triangle inequality extends by induction to the sum of any finite number of terms. To be specific, the following inequality holds:

$$\|x_1 + \cdots + x_k\| \le \|x_1\| + \cdots + \|x_k\|.$$

The presence of the norm of a vector enables us to compare the size of values of functions $f : X \to \mathbb{R}^m$ and $g : X \to \mathbb{R}^n$.

Let us agree to write $f(x) = o(g(x))$ or $f = o(g)$ over a base $\mathcal{B}$ in X if $\|f(x)\|_{\mathbb{R}^m} = o(\|g(x)\|_{\mathbb{R}^n})$ over the base $\mathcal{B}$.

If $f(x) = (f^1(x), \ldots, f^m(x))$ is the coordinate representation of the mapping $f : X \to \mathbb{R}^m$, then in view of the inequalities

$$\left|f^i(x)\right| \le \left\|f(x)\right\| \le \sum_{i=1}^{m} \left|f^i(x)\right| \tag{8.14}$$

one can make the following observation, which will be useful below:

$$\bigl(f = o(g) \text{ over the base } \mathcal{B}\bigr) \Leftrightarrow \bigl(f^i = o(g) \text{ over the base } \mathcal{B};\ i = 1, \ldots, m\bigr). \tag{8.15}$$

We also make the convention that the statement $f = O(g)$ over the base $\mathcal{B}$ in X will mean that $\|f(x)\|_{\mathbb{R}^m} = O(\|g(x)\|_{\mathbb{R}^n})$ over the base $\mathcal{B}$.

We then obtain from (8.14)

$$\bigl(f = O(g) \text{ over the base } \mathcal{B}\bigr) \Leftrightarrow \bigl(f^i = O(g) \text{ over the base } \mathcal{B};\ i = 1, \ldots, m\bigr). \tag{8.16}$$

Example Consider a linear transformation $L : \mathbb{R}^m \to \mathbb{R}^n$. Let $h = h^1 e_1 + \cdots + h^m e_m$ be an arbitrary vector in $\mathbb{R}^m$. Let us estimate $\|L(h)\|_{\mathbb{R}^n}$:

$$\left\|L(h)\right\| = \left\|\sum_{i=1}^{m} h^i L(e_i)\right\| \le \sum_{i=1}^{m} \left\|L(e_i)\right\| \left|h^i\right| \le \left(\sum_{i=1}^{m} \left\|L(e_i)\right\|\right) \|h\|. \tag{8.17}$$

Thus one can assert that

$$L(h) = O(h) \quad \text{as } h \to 0. \tag{8.18}$$

In particular, it follows from this that $L(x - x_0) = L(x) - L(x_0) \to 0$ as $x \to x_0$, that is, a linear transformation $L : \mathbb{R}^m \to \mathbb{R}^n$ is continuous at every point $x_0 \in \mathbb{R}^m$. From estimate (8.17) it is even clear that a linear transformation is uniformly continuous.

8.1.4 The Euclidean Structure on $\mathbb{R}^m$

The concept of the *inner product* in a real vector space is known from algebra as a numerical function $\langle x, y\rangle$ defined on pairs of vectors of the space and possessing the properties

$$\begin{aligned}
&\langle x, x\rangle \ge 0,\\
&\langle x, x\rangle = 0 \Leftrightarrow x = 0,\\
&\langle x_1, x_2\rangle = \langle x_2, x_1\rangle,\\
&\langle \lambda x_1, x_2\rangle = \lambda\langle x_1, x_2\rangle, \quad \text{where } \lambda \in \mathbb{R},\\
&\langle x_1 + x_2, x_3\rangle = \langle x_1, x_3\rangle + \langle x_2, x_3\rangle.
\end{aligned}$$

It follows in particular from these properties that if a basis $\{e_1, \dots, e_m\}$ is fixed in the space, then the inner product $\langle x, y\rangle$ of two vectors x and y can be expressed in terms of their coordinates $(x^1, \dots, x^m)$ and $(y^1, \dots, y^m)$ as the bilinear form

$$\langle x, y\rangle = g_{ij} x^i y^j \tag{8.19}$$

(where summation over i and j is understood), in which $g_{ij} = \langle e_i, e_j\rangle$.

Vectors are said to be *orthogonal* if their inner product equals 0.

A basis $\{e_1, \dots, e_m\}$ is *orthonormal* if $g_{ij} = \delta_{ij}$, where

$$\delta_{ij} = \begin{cases} 0, & \text{if } i \ne j,\\ 1, & \text{if } i = j. \end{cases}$$

In an orthonormal basis the inner product (8.19) has the very simple form

$$\langle x, y\rangle = \delta_{ij} x^i y^j,$$

or

$$\langle x, y\rangle = x^1 \cdot y^1 + \dots + x^m \cdot y^m. \tag{8.20}$$

Coordinates in which the inner product has this form are called *Cartesian* coordinates.

We recall that the space $\mathbb{R}^m$ with an inner product defined in it is called *Euclidean space*.

Between the inner product (8.20) and the norm of a vector (8.12) there is an obvious connection

$$\langle x, x\rangle = \|x\|^2.$$

The following inequality is known from algebra:

$$\langle x, y\rangle^2 \le \langle x, x\rangle\langle y, y\rangle.$$

It shows in particular that for any pair of vectors there is an angle $\varphi \in [0, \pi]$ such that

$$\langle x, y\rangle = \|x\|\,\|y\|\cos\varphi.$$

This angle is called the *angle between the vectors x and y*. That is the reason we regard vectors whose inner product is zero as orthogonal.

We shall also find useful the following simple, but very important fact, known from algebra:

any linear function $L : \mathbb{R}^m \to \mathbb{R}$ in Euclidean space has the form

$$L(x) = \langle \xi, x\rangle,$$

where $\xi \in \mathbb{R}^m$ is a fixed vector determined uniquely by the function L.

8.2 The Differential of a Function of Several Variables

8.2.1 Differentiability and the Differential of a Function at a Point

Definition 1 A function $f : E \to \mathbb{R}^n$ defined on a set $E \subset \mathbb{R}^m$ is *differentiable at the point* $x \in E$, which is a limit point of E, if

$$\boxed{f(x+h) - f(x) = L(x)h + \alpha(x; h),} \tag{8.21}$$

where $L(x) : \mathbb{R}^m \to \mathbb{R}^n$ is a function[2] that is linear in h and $\alpha(x; h) = o(h)$ as $h \to 0$, $x + h \in E$.

The vectors

$$\Delta x(h) := (x + h) - x = h,$$

$$\Delta f(x; h) := f(x + h) - f(x)$$

---

[2]By analogy with the one-dimensional case, we allow ourselves to write $L(x)h$ instead of $L(x)(h)$. We note also that in the definition we are assuming that $\mathbb{R}^m$ and $\mathbb{R}^n$ are endowed with the norm of Sect. 8.1.

are called respectively the *increment of the argument* and the *increment of the function* (corresponding to this increment of the argument). These vectors are traditionally denoted by the symbols of the functions of h themselves Δx and $\Delta f(x)$. The linear function $L(x): \mathbb{R}^m \to \mathbb{R}^n$ in (8.21) is called the *differential, tangent mapping*, or *derivative mapping* of the function $f: E \to \mathbb{R}^n$ at the point $x \in E$.

The differential of the function $f: E \to \mathbb{R}^n$ at a point $x \in E$ is denoted by the symbols $\mathrm{d}f(x)$, $Df(x)$, or $f'(x)$.

In accordance with the notation just introduced, we can rewrite relation (8.21) as

$$f(x+h) - f(x) = f'(x)h + \alpha(x; h)$$

or

$$\Delta f(x; h) = \mathrm{d}f(x)h + \alpha(x; h).$$

We remark that the differential is defined on the displacements h from the point $x \in \mathbb{R}^m$.

To emphasize this, we attach a copy of the vector space $\mathbb{R}^m$ to the point $x \in \mathbb{R}^m$ and denote it $T_x\mathbb{R}^m$, $T\mathbb{R}^m(x)$, or $T\mathbb{R}^m_x$. The space $T\mathbb{R}^m_x$ can be interpreted as a set of vectors attached at the point $x \in \mathbb{R}^m$. The vector space $T\mathbb{R}^m_x$ is called the *tangent space to* $\mathbb{R}^m$ *at* $x \in \mathbb{R}^m$. The origin of this terminology will be explained below.

The value of the differential on a vector $h \in T\mathbb{R}^m_x$ is the vector $f'(x)h \in T\mathbb{R}^n_{f(x)}$ attached to the point $f(x)$ and approximating the increment $f(x+h) - f(x)$ of the function caused by the increment h of the argument x.

Thus $\mathrm{d}f(x)$ or $f'(x)$ is a linear transformation $f'(x): T\mathbb{R}^m_x \to T\mathbb{R}^n_{f(x)}$.

We see that, in complete agreement with the one-dimensional case that we studied, a vector-valued function of several variables is differentiable at a point if its increment $\Delta f(x; h)$ at that point is linear as a function of h up to the correction term $\alpha(x; h)$, which is infinitesimal as $h \to 0$ compared to the increment of the argument.

8.2.2 The Differential and Partial Derivatives of a Real-Valued Function

If the vectors $f(x+h)$, $f(x)$, $L(x)h$, $\alpha(x; h)$ in $\mathbb{R}^n$ are written in coordinates, Eq. (8.21) becomes equivalent to the n equalities

$$f^i(x+h) - f^i(x) = L^i(x)h + \alpha^i(x; h) \quad (i = 1, \dots, n) \tag{8.22}$$

between real-valued functions, in which, as follows from relations (8.9) and (8.15) of Sect. 8.1, $L^i(x): \mathbb{R}^m \to \mathbb{R}$ are linear functions and $\alpha^i(x; h) = o(h)$ as $h \to 0$, $x + h \in E$, for every $i = 1, \dots, n$.

Thus we have the following proposition.

Proposition 1 *A mapping $f : E \to \mathbb{R}^n$ of a set $E \subset \mathbb{R}^m$ is differentiable at a point $x \in E$ that is a limit point of E if and only if the functions $f^i : E \to \mathbb{R}$ $(i = 1, \dots, n)$ that define the coordinate representation of the mapping are differentiable at that point.*

Since relations (8.21) and (8.22) are equivalent, to find the differential $L(x)$ of a mapping $f : E \to \mathbb{R}^n$ it suffices to learn how to find the differentials $L^i(x)$ of its coordinate functions $f^i : E \to \mathbb{R}$.

Thus, let us consider a real-valued function $f : E \to \mathbb{R}$, defined on a set $E \subset \mathbb{R}^m$ and differentiable at an interior point $x \in E$ of that set. We remark that in the future we shall mostly be dealing with the case when E is a domain in $\mathbb{R}^m$. If x is an interior point of E, then for any sufficiently small displacement h from x the point $x + h$ will also belong to E, and consequently will also be in the domain of definition of the function $f : E \to \mathbb{R}$.

If we pass to the coordinate notation for the point $x = (x^1, \dots, x^m)$, the vector $h = (h^1, \dots, h^m)$, and the linear function $L(x)h = a_1(x)h^1 + \dots + a_m(x)h^m$, then the condition

$$f(x+h) - f(x) = L(x)h + o(h) \quad \text{as } h \to 0 \tag{8.23}$$

can be rewritten as

$$\begin{aligned} &f(x^1 + h^1, \dots, x^m + h^m) - f(x^1, \dots, x^m) = \\ &= a_1(x)h^1 + \dots + a_m(x)h^m + o(h) \quad \text{as } h \to 0, \end{aligned} \tag{8.24}$$

where $a_1(x), \dots, a_m(x)$ are real numbers connected with the point x.

We wish to find these numbers. To do this, instead of an arbitrary displacement h we consider the special displacement

$$h_i = h^i e_i = 0 \cdot e_1 + \dots + 0 \cdot e_{i-1} + h^i e_i + 0 \cdot e_{i+1} + \dots + 0 \cdot e_m$$

by a vector h_i collinear with the vector e_i of the basis $\{e_1, \dots, e_m\}$ in $\mathbb{R}^m$.

When $h = h_i$, it is obvious that $\|h\| = |h^i|$, and so by (8.24), for $h = h_i$ we obtain

$$\begin{aligned} &f(x^1, \dots, x^{i-1}, x^i + h^i, x^{i+1}, \dots, x^m) - f(x^1, \dots, x^i, \dots, x^m) = \\ &= a_i(x)h^i + o(h^i) \quad \text{as } h^i \to 0. \end{aligned} \tag{8.25}$$

This means that if we fix all the variables in the function $f(x^1, \dots, x^m)$ except the ith one, the resulting function of the ith variable alone is differentiable at the point x^i.

In that way, from (8.25) we find that

$$a_i(x) = \lim_{h^i \to 0} \frac{f(x^1, \dots, x^{i-1}, x^i + h^i, x^{i+1}, \dots, x^m) - f(x^1, \dots, x^i, \dots, x^m)}{h^i}. \tag{8.26}$$

Definition 2 The limit (8.26) is called the *partial derivative* of the function $f(x)$ at the point $x=(x^1,\dots,x^m)$ with respect to the variable x^i. We denote it by one of the following symbols:

$$\frac{\partial f}{\partial x^i}(x), \quad \partial_i f(x), \quad D_i f(x), \quad f'_{x^i}(x).$$

Example 1 If $f(u,v)=u^3+v^2\sin u$, then

$$\partial_1 f(u,v)=\frac{\partial f}{\partial u}(u,v)=3u^2+v^2\cos u,$$

$$\partial_2 f(u,v)=\frac{\partial f}{\partial v}(\mathrm{u},v)=2v\sin u.$$

Example 2 If $f(x,y,z)=\arctan(xy^2)+\mathrm{e}^z$, then

$$\partial_1 f(x,y,z)=\frac{\partial f}{\partial x}(x,y,z)=\frac{y^2}{1+x^2y^4},$$

$$\partial_2 f(x,y,z)=\frac{\partial f}{\partial y}(x,y,z)=\frac{2xy}{1+x^2y^4},$$

$$\partial_3 f(x,y,z)=\frac{\partial f}{\partial z}(x,y,z)=\mathrm{e}^z.$$

Thus we have proved the following result.

Proposition 2 *If a function $f:E\to\mathbb{R}$ defined on a set $E\subset\mathbb{R}^m$ is differentiable at an interior point $x\in E$ of that set, then the function has a partial derivative at that point with respect to each variable, and the differential of the function is uniquely determined by these partial derivatives in the form*

$$\mathrm{d}f(x)h=\frac{\partial f}{\partial x^1}(x)h^1+\dots+\frac{\partial f}{\partial x^m}(x)h^m. \tag{8.27}$$

Using the convention of summation on an index that appears as both a subscript and a superscript, we can write formula (8.27) succinctly:

$$\mathrm{d}f(x)h=\partial_i f(x)h^i. \tag{8.28}$$

Example 3 If we had known (as we soon will know) that the function $f(x,y,z)$ considered in Example 2 is differentiable at the point $(0,1,0)$, we could have written immediately

$$\mathrm{d}f(0,1,0)h=1\cdot h^1+0\cdot h^2+1\cdot h^3=h^1+h^3$$

and accordingly

$$f\left(h^1,1+h^2,h^3\right)-f(0,1,0)=\mathrm{d}f(0,1,0)h+o(h)$$

or

$$\arctan\big(h^1(1+h^2)^2\big)+e^{h^3}=1+h^1+h^3+o(h) \quad \text{as } h\to 0.$$

Example 4 For the function $x=(x^1,\dots,x^m)\overset{\pi^i}{\longmapsto} x^i$, which assigns to the point $x\in\mathbb{R}^m$ its ith coordinate, we have

$$\Delta\pi^i(x;h)=\big(x^i+h^i\big)-x^i=h^i,$$

that is, the increment of this function is itself a linear function in $h: h\overset{\pi^i}{\longmapsto} h^i$. Thus, $\Delta\pi^i(x;h)=\mathrm{d}\pi^i(x)h$, and the mapping $\mathrm{d}\pi^i(x)=\mathrm{d}\pi^i$ turns out to be actually independent of $x\in\mathbb{R}^m$ in the sense that $\mathrm{d}\pi^i(x)h=h^i$ at every point $x\in\mathbb{R}^m$. If we write $x^i(x)$ instead of $\pi^i(x)$, we find that $\mathrm{d}x^i(x)h=\mathrm{d}x^ih=h^i$.

Taking this fact and formula (8.28) into account, we can now represent the differential of any function as a linear combination of the differentials of the coordinates of its argument $x\in\mathbb{R}^m$. To be specific:

$$\mathrm{d}f(x)=\partial_i f(x)\,\mathrm{d}x^i=\frac{\partial f}{\partial x^1}\,\mathrm{d}x^1+\cdots+\frac{\partial f}{\partial x^m}\,\mathrm{d}x^m, \tag{8.29}$$

since for any vector $h\in T\mathbb{R}^m_x$ we have

$$\mathrm{d}f(x)h=\partial_i f(x)h^i=\partial_i f(x)\,\mathrm{d}x^i h.$$

8.2.3 Coordinate Representation of the Differential of a Mapping. The Jacobi Matrix

Thus we have found formula (8.27) for the differential of a real-valued function $f:E\to\mathbb{R}$. But then, by the equivalence of relations (8.21) and (8.22), for any mapping $f:E\to\mathbb{R}^n$ of a set $E\subset\mathbb{R}^m$ that is differentiable at an interior point $x\in E$, we can write the coordinate representation of the differential $\mathrm{d}f(x)$ as

$$\mathrm{d}f(x)h=\begin{pmatrix}\mathrm{d}f^1(x)h\\ \vdots\\ \mathrm{d}f^n(x)h\end{pmatrix}=\begin{pmatrix}\partial_i f^1(x)h^i\\ \vdots\\ \partial_i f^n(x)h^i\end{pmatrix}=$$

$$=\begin{pmatrix}\frac{\partial f^1}{\partial x^1}(x) & \cdots & \frac{\partial f^1}{\partial x^m}(x)\\ \vdots & \ddots & \vdots\\ \frac{\partial f^n}{\partial x^1}(x) & \cdots & \frac{\partial f^n}{\partial x^m}(x)\end{pmatrix}\begin{pmatrix}h^1\\ \vdots\\ h^m\end{pmatrix}. \tag{8.30}$$

Definition 3 The matrix $(\partial_i f^j(x))$ $(i = 1, \dots, m,\ j = 1, \dots, n)$ of partial derivatives of the coordinate functions of a given mapping at the point $x \in E$ is called the *Jacobi matrix*[3] or the *Jacobian*[4] of the mapping at the point.

In the case when $n = 1$, we are simply brought back to formula (8.27), and when $n = 1$ and $m = 1$, we arrive at the differential of a real-valued function of one real variable.

The equivalence of relations (8.21) and (8.22) and the uniqueness of the differential (8.27) of a real-valued function implies the following result.

Proposition 3 *If a mapping $f : E \to \mathbb{R}^n$ of a set $E \subset \mathbb{R}^m$ is differentiable at an interior point $x \in E$, then it has a unique differential $\mathrm{d}f(x)$ at that point, and the coordinate representation of the mapping $\mathrm{d}f(x) : T\mathbb{R}^m_x \to T\mathbb{R}^n_{f(x)}$ is given by relation* (8.30).

8.2.4 Continuity, Partial Derivatives, and Differentiability of a Function at a Point

We complete our discussion of the concept of differentiability of a function at a point by pointing out some connections among the continuity of a function at a point, the existence of partial derivatives of the function at that point, and differentiability at the point.

In Sect. 8.1 (relations (8.17) and (8.18)) we established that if $L : \mathbb{R}^m \to \mathbb{R}^n$ is a linear transformation, then $Lh \to 0$ as $h \to 0$. Therefore, one can conclude from relation (8.21) that a function that is differentiable at a point is continuous at that point, since

$$f(x+h) - f(x) = L(x)h + o(h) \quad \text{as } h \to 0, x+h \in E.$$

The converse, of course, is not true because, as we know, it fails even in the one-dimensional case.

Thus the relation between continuity and differentiability of a function at a point in the multidimensional case is the same as in the one-dimensional case.

The situation is completely different in regard to the relations between partial derivatives and the differential. In the one-dimensional case, that is, in the case of a real-valued function of one real variable, the existence of the differential and the existence of the derivative for a function at a point are equivalent conditions. For functions of several variables, we have shown (Proposition 2) that differentiability of a function at an interior point of its domain of definition guarantees the existence of a partial derivative with respect to each variable at that point. However, the converse is not true.

---

[3]C.G.J. Jacobi (1804–1851) – well-known German mathematician.

[4]The term *Jacobian* is more often applied to the determinant of this matrix (when it is square).

Example 5 The function

$$f(x^1,x^2)=\begin{cases}0, & \text{if } x^1x^2=0,\\ 1, & \text{if } x^1x^2\neq 0,\end{cases}$$

equals 0 on the coordinate axes and therefore has both partial derivatives at the point $(0,0)$:

$$\partial_1 f(0,0)=\lim_{h^1\to 0}\frac{f(h^1,0)-f(0,0)}{h^1}=\lim_{h^1\to 0}\frac{0-0}{h^1}=0,$$

$$\partial_2 f(0,0)=\lim_{h^2\to 0}\frac{f(0,h^2)-f(0,0)}{h^2}=\lim_{h^2\to 0}\frac{0-0}{h^2}=0.$$

At the same time, this function is not differentiable at $(0,0)$, since it is obviously discontinuous at that point.

The function given in Example 5 fails to have one of its partial derivatives at points of the coordinate axes different from $(0,0)$. However, the function

$$f(x,y)=\begin{cases}\frac{xy}{x^2+y^2}, & \text{if } x^2+y^2\neq 0,\\ 0, & \text{if } x^2+y^2=0\end{cases}$$

(which we encountered in Example 2 of Sect. 7.2) has partial derivatives at all points of the plane, but it also is discontinuous at the origin and hence not differentiable there.

Thus the possibility of writing the right-hand side of (8.27) and (8.28) still does not guarantee that this expression will represent the differential of the function we are considering, since the function may be nondifferentiable.

This circumstance might have been a serious hindrance to the entire differential calculus of functions of several variables, if it had not been determined (as will be proved below) that continuity of the partial derivatives at a point is a sufficient condition for differentiability of the function at that point.

8.3 The Basic Laws of Differentiation

8.3.1 Linearity of the Operation of Differentiation

Theorem 1 *If the mappings $f_1: E\to\mathbb{R}^n$ and $f_2: E\to\mathbb{R}^n$, defined on a set $E\subset\mathbb{R}^m$, are differentiable at a point $x\in E$, then a linear combination of them $(\lambda_1 f_1+\lambda_2 f_2): E\to\mathbb{R}^n$ is also differentiable at that point, and the following equality holds*:

$$(\lambda_1 f_1+\lambda_2 f_2)'(x)=\big(\lambda_1 f_1'+\lambda_2 f_2'\big)(x). \tag{8.31}$$

Equality (8.31) shows that the operation of differentiation, that is, forming the differential of a mapping at a point, is a linear transformation on the vector space of mappings $f : E \to \mathbb{R}^n$ that are differentiable at a given point of the set E. The left-hand side of (8.31) contains by definition the linear transformation $(\lambda_1 f_1 + \lambda_2 f_2)'(x)$, while the right-hand side contains the linear combination $(\lambda_1 f_1' + \lambda_2 f_2')(x)$ of linear transformations $f_1'(x) : \mathbb{R}^m \to \mathbb{R}^n$, and $f_2'(x) : \mathbb{R}^m \to \mathbb{R}^n$, which, as we know from Sect. 8.1, is also a linear transformation. Theorem 1 asserts that these mappings are the same.

Proof

$$\begin{aligned}
&(\lambda_1 f_1 + \lambda_2 f_2)(x+h) - (\lambda_1 f_2 + \lambda_2 f_2)(x) = \\
&\quad = \bigl(\lambda_1 f_1(x+h) + \lambda_2 f_2(x+h)\bigr) - \bigl(\lambda_1 f_1(x) + \lambda_2 f_2(x)\bigr) = \\
&\quad = \lambda_1\bigl(f_1(x+h) - f_1(x)\bigr) + \lambda_2\bigl(f_2(x+h) - f_2(x)\bigr) = \\
&\quad = \lambda_1\bigl(f_1'(x)h + o(h)\bigr) + \lambda_2\bigl(f_2'(x)h + o(h)\bigr) = \\
&\quad = \bigl(\lambda_1 f_1'(x) + \lambda_2 f_2'(x)\bigr)h + o(h).
\end{aligned}$$

□

If the functions in question are real-valued, the operations of multiplication and division (when the denominator is not zero) can also be performed. We have then the following theorem.

Theorem 2 *If the functions $f : E \to \mathbb{R}$ and $g : E \to \mathbb{R}$, defined on a set $E \subset \mathbb{R}^m$, are differentiable at the point $x \in E$, then*

a) *their product is differentiable at x and*

$$(f \cdot g)'(x) = g(x)f'(x) + f(x)g'(x); \tag{8.32}$$

b) *their quotient is differentiable at x if $g(x) \neq 0$, and*

$$\left(\frac{f}{g}\right)'(x) = \frac{1}{g^2(x)}\bigl(g(x)f'(x) - f(x)g'(x)\bigr). \tag{8.33}$$

The proof of this theorem is the same as the proof of the corresponding parts of Theorem 1 in Sect. 5.2, so that we shall omit the details.

Relations (8.31), (8.32), and (8.33) can be rewritten in the other notations for the differential. To be specific:

$$\begin{aligned}
\mathrm{d}\bigl(\lambda_1 f_1(x) + \lambda_2 f_2\bigr)(x) &= (\lambda_1 \,\mathrm{d} f_1 + \lambda_2 \,\mathrm{d} f_2)(x), \\
\mathrm{d}(f \cdot g)(x) &= g(x)\,\mathrm{d} f(x) + f(x)\,\mathrm{d} g(x), \\
\mathrm{d}\left(\frac{f}{g}\right)(x) &= \frac{1}{g^2(x)}\bigl(g(x)\,\mathrm{d} f(x) - f(x)\,\mathrm{d} g(x)\bigr).
\end{aligned}$$

Let us see what these equalities mean in the coordinate representation of the mappings. We know that if a mapping $\varphi : E \to \mathbb{R}^n$ that is differentiable at an interior point x of the set $E \subset \mathbb{R}^m$ is written in the coordinate form

$$\varphi(x) = \begin{pmatrix} \varphi^1(x^1, \dots, x^m) \\ \vdots \\ \varphi^n(x^1, \dots, x^m) \end{pmatrix},$$

then the Jacobi matrix

$$\varphi'(x) = \begin{pmatrix} \partial_1 \varphi^1 & \cdots & \partial_m \varphi^1 \\ \vdots & \ddots & \vdots \\ \partial_1 \varphi^n & \cdots & \partial_m \varphi^n \end{pmatrix} (x) = \big(\partial_1 \varphi^j\big)(x)$$

will correspond to its differential $d\varphi(x) : \mathbb{R}^m \to \mathbb{R}^n$ at this point.

For fixed bases in $\mathbb{R}^m$ and $\mathbb{R}^n$ the correspondence between linear transformations $L : \mathbb{R}^m \to \mathbb{R}^n$ and $m \times n$ matrices is one-to-one, and hence the linear transformation L can be identified with the matrix that defines it.

Even so, we shall as a rule use the symbol $f'(x)$ rather than $df(x)$ to denote the Jacobi matrix, since it corresponds better to the traditional distinction between the concepts of derivative and differential that holds in the one-dimensional case.

Thus, by the uniqueness of the differential, at an interior point x of E we obtain the following coordinate notation for (8.31), (8.32), and (8.33), denoting the equality of the corresponding Jacobi matrices:

$$\big(\partial_i(\lambda_1 f_1^j + \lambda_2 f_2^j)\big)(x) = \big(\lambda_1 \partial_i f_1^j + \lambda_2 \partial_i f_2^j\big)(x)$$
$$(i = 1, \dots, m,\ j = 1, \dots, n), \tag{8.31$'$}$$

$$\big(\partial_i(f \cdot g)\big)(x) = g(x)\partial_i f(x) + f(x)\partial_i g(x) \quad (i = 1, \dots, m), \tag{8.32$'$}$$

$$\left(\partial_i\left(\frac{f}{g}\right)\right)(x) = \frac{1}{g^2(x)}\big(g(x)\partial_i f(x) - f(x)\partial_i g(x)\big) \quad (i = 1, \dots, m). \tag{8.33$'$}$$

It follows from the elementwise equality of these matrices, for example, that the partial derivative with respect to the variable x^i of the product of real-valued functions $f(x^1, \dots, x^m)$ and $g(x^1, \dots, x^m)$ should be taken as follows:

$$\frac{\partial(f \cdot g)}{\partial x^i}\big(x^1, \dots, x^m\big) = g\big(x^1, \dots, x^m\big)\frac{\partial f}{\partial x^i}\big(x^1, \dots, x^m\big) +$$
$$+ f\big(x^1, \dots, x^m\big)\frac{\partial g}{\partial x^i}\big(x^1, \dots, x^m\big).$$

We note that both this equality and the matrix equalities (8.31′), (8.32′), and (8.33′) are obvious consequences of the definition of a partial derivative and the usual rules for differentiating real-valued functions of one real variable. However, we know that the existence of partial derivatives may still turn out to be insufficient for a function of several variables to be differentiable. For that reason, along with the important and completely obvious equalities (8.31′), (8.32′), and (8.33′), the assertions about the existence of a differential for the corresponding mapping in Theorems 1 and 2 acquire a particular importance.

We remark finally that by induction using (8.32) one can obtain the relation

$$\mathrm{d}(f_1,\dots,f_k)(x) = (f_2\cdots f_k)(x)\,\mathrm{d}f_1(x) + \cdots + (f_1\cdots f_{k-1})\,\mathrm{d}f_k(x)$$

for the differential of a product $(f_1\cdots f_k)$ of differentiable real-valued functions.

8.3.2 *Differentiation of a Composition of Mappings (Chain Rule)*

a. The Main Theorem

Theorem 3 *If the mapping $f: X \to Y$ of a set $X \subset \mathbb{R}^m$ into a set $Y \subset \mathbb{R}^n$ is differentiable at a point $x \in X$, and the mapping $f: Y \to \mathbb{R}^k$ is differentiable at the point $y = f(x) \in Y$, then their composition $g \circ f : X \to \mathbb{R}^k$ is differentiable at x and the differential $\mathrm{d}(g\circ f): T\mathbb{R}^m_x \to T\mathbb{R}^k_{g(f(x))}$ of the composition equals the composition $\mathrm{d}g(y)\circ \mathrm{d}f(x)$ of the differentials*

$$\mathrm{d}f(x): T\mathbb{R}^m_x \to T\mathbb{R}^n_{f(x)=y}, \qquad \mathrm{d}g(y): T\mathbb{R}^n_y \to T\mathbb{R}^k_{g(y)}.$$

The proof of this theorem repeats almost completely the proof of Theorem 2 of Sect. 5.2. In order to call attention to a new detail that arises in this case, we shall nevertheless carry out the proof again, without going into technical details that have already been discussed, however.

Proof Using the differentiability of the mappings f and g at the points x and $y = f(x)$, and also the linearity of the differential $g'(x)$, we can write

$$\begin{aligned}
&(g\circ f)(x+h) - (g\circ f)(x) = \\
&\quad = g\big(f(x+h)\big) - g\big(f(x)\big) = \\
&\quad = g'\big(f(x)\big)\big(f(x+h)-f(x)\big) + o\big(f(x+h)-f(x)\big) = \\
&\quad = g'(y)\big(f'(x)h + o(h)\big) + o\big(f(x+h)-f(x)\big) = \\
&\quad = g'(y)\big(f'(x)h\big) + g'(y)\big(o(h)\big) + o\big(f(x+h)-f(x)\big) = \\
&\quad = \big(g'(y)\circ f'(x)\big)h + \alpha(x;h),
\end{aligned}$$

where $g'(y)\circ f'(x)$ is a linear mapping (being a composition of linear mappings), and

$$\alpha(x;h) = g'(y)\big(o(h)\big) + o\big(f(x+h)-f(x)\big).$$

But, as relations (8.17) and (8.18) of Sect. 8.1 show,

$$g'(y)\big(o(h)\big) = o(h) \quad \text{as } h \to 0,$$
$$f(x+h) - f(x) = f'(x)h + o(h) = O(h) + o(h) = O(h) \quad \text{as } h \to 0,$$

and

$$o\big(f(x+h) - f(x)\big) = o\big(O(h)\big) = o(h) \quad \text{as } h \to 0.$$

Consequently,

$$\alpha(x;h) = o(h) + o(h) = o(h) \quad \text{as } h \to 0,$$

and the theorem is proved. □

When rewritten in coordinate form, Theorem 3 means that if x is an interior point of the set X and

$$f'(x) = \begin{pmatrix} \partial_1 f^1(x) & \cdots & \partial_m f^1(x) \\ \vdots & \ddots & \vdots \\ \partial_1 f^n(x) & \cdots & \partial_m f^n(x) \end{pmatrix} = \big(\partial_i f^j\big)(x),$$

and $y = f(x)$ is an interior point of the set Y and

$$g'(y) = \begin{pmatrix} \partial_1 g^1(y) & \cdots & \partial_n g^1(y) \\ \vdots & \ddots & \vdots \\ \partial_1 g^k(y) & \cdots & \partial_n g^k(y) \end{pmatrix} = \big(\partial_j g^k\big)(y),$$

then

$$(g \circ f)'(x) = \begin{pmatrix} \partial_1 (g^1 \circ f)(x) & \cdots & \partial_m (g^1 \circ f)(x) \\ \vdots & \ddots & \vdots \\ \partial_1 (g^k \circ f)(x) & \cdots & \partial_m (g^k \circ f)(x) \end{pmatrix} = \big(\partial_i (g^l \circ f)\big)(x) =$$
$$= \begin{pmatrix} \partial_1 g^1(y) & \cdots & \partial_n g^1(y) \\ \vdots & \ddots & \vdots \\ \partial_1 g^k(y) & \cdots & \partial_n g^k(y) \end{pmatrix} \begin{pmatrix} \partial_1 f^1(x) & \cdots & \partial_m f^1(x) \\ \vdots & \ddots & \vdots \\ \partial_1 f^n(x) & \cdots & \partial_m f^n(x) \end{pmatrix} =$$
$$= \big(\partial_j g^l(y) \cdot \partial_i f^j(x)\big).$$

In the equality

$$\big(\partial_i (g^l \circ f)\big)(x) = \big(\partial_j g^l\big(f(x)\big) \cdot \partial_i f^j(x)\big) \tag{8.34}$$

summation is understood on the right-hand side with respect to the index j over its interval of variation, that is, from 1 to n.

In contrast to Eqs. (8.31′), (8.32′), and (8.33′), relation (8.34) is nontrivial even in the sense of elementwise equality of the matrices occurring in it.

Let us now consider some important cases of the theorem just proved.

b. The Differential and Partial Derivatives of a Composite Real-Valued Function

Let $z = g(y^1, \dots, y^n)$ be a real-valued function of the real variables $y^1, \dots, y^n$, each of which in turn is a function $y^j = f^j(x^1, \dots, x^m)$ $(j = 1, \dots, n)$ of the variables $x^1, \dots, x^m$. Assuming that the functions g and f^j are differentiable $(j = 1, \dots, n)$, let us find the partial derivative $\frac{\partial (g \circ f)}{\partial x^i}(x)$ of the composition of the mappings $f : X \to Y$ and $g : Y \to \mathbb{R}$.

According to formula (8.34), in which $l = 1$ under the present conditions, we find

$$\partial_i (g \circ f)(x) = \partial_j g\bigl(f(x)\bigr) \cdot \partial_i f^j(x), \tag{8.35}$$

or, in notation that shows more detail,

$$\begin{aligned} \frac{\partial z}{\partial x^i}(x) = \frac{\partial (g \circ f)}{\partial x^i}\bigl(x^1, \dots, x^m\bigr) &= \frac{\partial g}{\partial y^1} \cdot \frac{\partial y^1}{\partial x^i} + \cdots + \frac{\partial g}{\partial y^n} \cdot \frac{\partial y^n}{\partial x^i} = \\ &= \partial_1 g\bigl(f(x)\bigr) \cdot \partial_i f^1(x) + \cdots + \partial_n g\bigl(f(x)\bigr) \cdot \partial_i f^n(x). \end{aligned}$$

c. The Derivative with Respect to a Vector and the Gradient of a Function at a Point

Consider the stationary flow of a liquid or gas in some domain G of $\mathbb{R}^3$. The term "stationary" means that the velocity of the flow at each point of G does not vary with time, although of course it may vary from one point of G to another. Suppose, for example, $f(x) = f(x^1, x^2, x^3)$ is the pressure in the flow at the point $x = (x^1, x^2, x^3) \in G$. If we move about in the flow according to the law $x = x(t)$, where t is time, we shall record a pressure $(f \circ x)(t) = f(x(t))$ at time t. The rate of variation of pressure over time along our trajectory is obviously the derivative $\frac{\mathrm{d}(f \circ x)}{\mathrm{d}t}(t)$ of the function $(f \circ x)(t)$ with respect to time. Let us find this derivative, assuming that $f(x^1, x^2, x^3)$ is a differentiable function in the domain G. By the rule for differentiating composite functions, we find

$$\frac{\mathrm{d}(f \circ x)}{\mathrm{d}t}(t) = \frac{\partial f}{\partial x^1}\bigl(x(t)\bigr)\dot{x}^1(t) + \frac{\partial f}{\partial x^2}\bigl(x(t)\bigr)\dot{x}^2(t) + \frac{\partial f}{\partial x^3}\bigl(x(t)\bigr)\dot{x}^3(t), \tag{8.36}$$

where $\dot{x}^i(t) = \frac{\mathrm{d}x^i}{\mathrm{d}t}(t)$ $(i = 1, 2, 3)$.

Since the vector $(\dot{x}^1, \dot{x}^2, \dot{x}^3) = v(t)$ is the velocity of our displacement at time t and $(\partial_1 f, \partial_2 f, \partial_3 f)(x)$ is the coordinate notation for the differential $\mathrm{d}f(x)$ of the function f at the point x, Eq. (8.36) can also be rewritten as

$$\frac{\mathrm{d}(f \circ x)}{\mathrm{d}t}(t) = \mathrm{d}f\big(x(t)\big)\,v(t), \tag{8.37}$$

that is, the required quantity is the value of the differential $\mathrm{d}f(x(t))$ of the function $f(x)$ at the point $x(t)$ evaluated at the velocity vector $v(t)$ of the motion.

In particular, if we were at the point $x_0 = x(0)$ at time $t = 0$, then

$$\frac{\mathrm{d}(f \circ x)}{\mathrm{d}t}(0) = \mathrm{d}f(x_0)\,v, \tag{8.38}$$

where $v = v(0)$ is the velocity vector at time $t = 0$.

The right-hand side of (8.38) depends only on the point $x_0 \in G$ and the velocity vector v that we have at that point; it is independent of the specific form of the trajectory $x = x(t)$, provided the condition $\dot{x}(0) = v$ holds. That means that the value of the left-hand side of Eq. (8.38) is the same on any trajectory of the form

$$x(t) = x_0 + vt + \alpha(t), \tag{8.39}$$

where $\alpha(t) = o(t)$ as $t \to 0$, since this value is completely determined by giving the point x_0 and the vector $v \in T\mathbb{R}^3_{x_0}$ attached at that point. In particular, if we wished to compute the value of the left-hand side of Eq. (8.38) directly (and hence also the right-hand side), we could choose the law of motion to be

$$x(t) = x_0 + vt, \tag{8.40}$$

corresponding to a uniform motion at velocity v under which we are at the point $x(0) = x_0$ at time $t = 0$.

We now give the following

Definition 1 If the function $f(x)$ is defined in a neighborhood of the point $x_0 \in \mathbb{R}^m$ and the vector $v \in T\mathbb{R}^m_{x_0}$ is attached at the point x_0, then the quantity

$$\boxed{D_v f(x_0) := \lim_{t \to 0} \frac{f(x_0 + vt) - f(x_0)}{t}} \tag{8.41}$$

(if the indicated limit exists) is called the *derivative of f at the point x_0 with respect to the vector v* or *the derivative along the vector v at the point x_0*.

It follows from these considerations that if the function f is differentiable at the point x_0, then the following equality holds for any function $x(t)$ of the form (8.39), and in particular, for any function of the form (8.40):

$$D_v f(x_0) = \frac{\mathrm{d}(f \circ x)}{\mathrm{d}t}(0) = \mathrm{d}f(x_0)\,v. \tag{8.42}$$

In coordinate notation, this equality says

$$D_v f(x_0) = \frac{\partial f}{\partial x^1}(x_0)v^1 + \cdots + \frac{\partial f}{\partial x^m}(x_0)v^m. \tag{8.43}$$

In particular, for the basis vectors $e_1 = (1, 0, \ldots, 0), \ldots, e_m = (0, \ldots, 0, 1)$ this formula implies

$$D_{e_i} f(x_0) = \frac{\partial f}{\partial x^i}(x_0) \quad (i = 1, \ldots, m).$$

By virtue of the linearity of the differential $\mathrm{d}f(x_0)$, we deduce from Eq. (8.42) that if f is differentiable at the point x_0, then for any vectors $v_1, v_2 \in T\mathbb{R}^m_{x_0}$ and any $\lambda_1, \lambda_2 \in \mathbb{R}$ the function has a derivative at the point x_0 with respect to the vector $(\lambda_1 v_1 + \lambda_2 v_2) \in T\mathbb{R}^m_{x_0}$, and that

$$D_{\lambda_1 v_1 + \lambda_2 v_2} f(x_0) = \lambda_1 D_{v_1} f(x_0) + \lambda_2 D_{v_2} f(x_0). \tag{8.44}$$

If $\mathbb{R}^m$ is regarded as a Euclidean space, that is, as a vector space with an inner product, then (see Sect. 8.1) it is possible to write any linear functional $L(v)$ as the inner product $\langle \xi, v \rangle$ of a fixed vector $\xi = \xi(L)$ and the variable vector v.

In particular, there exists a vector ξ such that

$$\mathrm{d}f(x_0)v = \langle \xi, v \rangle. \tag{8.45}$$

Definition 2 The vector $\xi \in T\mathbb{R}^m_{x_0}$ corresponding to the differential $\mathrm{d}f(x_0)$ of the function f at the point x_0 in the sense of Eq. (8.45) is called the *gradient* of the function at that point and is denoted $\operatorname{grad} f(x_0)$.

Thus, by definition

$$\boxed{\mathrm{d}f(x_0)v = \big\langle \operatorname{grad} f(x_0), v \big\rangle.} \tag{8.46}$$

If a Cartesian coordinate system has been chosen in $\mathbb{R}^m$, then, by comparing relations (8.42), (8.43), and (8.46), we conclude that the gradient has the following representation in such a coordinate system:

$$\operatorname{grad} f(x_0) = \left(\frac{\partial f}{\partial x^1}, \ldots, \frac{\partial f}{\partial x^m}\right)(x_0). \tag{8.47}$$

We shall now explain the geometric meaning of the vector $\operatorname{grad} f(x_0)$.

Let $e \in T\mathbb{R}^m_{x_0}$ be a unit vector. Then by (8.46)

$$D_e f(x_0) = \big|\operatorname{grad} f(x_0)\big| \cos\varphi, \tag{8.48}$$

where φ is the angle between the vectors e and $\operatorname{grad} f(x_0)$.

Thus if $\operatorname{grad} f(x_0) \neq 0$ and $e = \|\operatorname{grad} f(x_0)\|^{-1} \operatorname{grad} f(x_0)$, the derivative $D_e f(x_0)$ assumes a maximum value. That is, the rate of increase of the function f (expressed in the units of f relative to a unit length in $\mathbb{R}^m$) is maximal and equal to $\|\operatorname{grad} f(x_0)\|$ for motion from the point x_0 precisely when the displacement is in the direction of the vector grad $f(x_0)$. The value of the function decreases most sharply under displacement in the opposite direction, and the rate of variation of the function is zero in a direction perpendicular to the vector grad $f(x_0)$.

The derivative with respect to a unit vector in a given direction is usually called the *directional derivative in that direction.*

Since a unit vector in Euclidean space is determined by its direction cosines

$$e = (\cos\alpha_1, \dots, \cos\alpha_m),$$

where α_i is the angle between the vector e and the basis vector e_i in a Cartesian coordinate system, it follows that

$$D_e f(x_0) = \langle \operatorname{grad} f(x_0), e\rangle = \frac{\partial f}{\partial x^1}(x_0)\cos\alpha_1 + \cdots + \frac{\partial f}{\partial x^m}(x_0)\cos\alpha_m.$$

The vector grad $f(x_0)$ is encountered very frequently and has numerous applications. For example the so-called *gradient methods* for finding extrema of functions of several variables numerically (using a computer) are based on the geometric property of the gradient just noted. (In this connection, see Problem 2 at the end of this section.)

Many important vector fields, such as, for example, a Newtonian gravitational field or the electric field due to charge, are the gradients of certain scalar-valued functions, known as the potentials of the fields (see Problem 3).

Many physical laws use the vector grad f in their very statement. For example, in the mechanics of continuous media the equivalent of Newton's basic law of dynamics $ma = F$ is the relation

$$\rho\mathbf{a} = -\operatorname{grad} p,$$

which connects the acceleration $\mathbf{a} = \mathbf{a}(x,t)$ in the flow of an ideal liquid or gas free of external forces at the point x and time t with the density of the medium $\rho = \rho(x,t)$ and the gradient of the pressure $p = p(x,t)$ at the same point and time (see Problem 4).

We shall discuss the vector grad f again later, when we study vector analysis and the elements of field theory.

8.3.3 Differentiation of an Inverse Mapping

Theorem 4 *Let $f: U(x) \to V(y)$ be a mapping of a neighborhood $U(x) \subset \mathbb{R}^m$ of the point x onto a neighborhood $V(y) \subset \mathbb{R}^m$ of the point $y = f(x)$. Assume that f is continuous at the point x and has an inverse mapping $f^{-1}: V(y) \to U(x)$ that is continuous at the point y.*

Given these assumptions, if the mapping f is differentiable at x and the tangent mapping $f'(x): T\mathbb{R}^m_x \to T\mathbb{R}^m_y$ to f at the point x has an inverse $[f'(x)]^{-1}$:

2. a) Draw the graph of the function $z = x^2 + 4y^2$, where (x, y, z) are Cartesian coordinates in $\mathbb{R}^3$.

b) Let $f : G \to \mathbb{R}$ be a numerically valued function defined on a domain $G \subset \mathbb{R}^m$. A *level set* (c-level) of the function is a set $E \subset G$ on which the function assumes only one value ($f(E) = c$). More precisely, $E = f^{-1}(c)$. Draw the level sets in $\mathbb{R}^2$ for the function given in part a).

c) Find the gradient of the function $f(x, y) = x^2 + 4y^2$, and verify that at any point (x, y) the vector $\operatorname{grad} f$ is orthogonal to the level curve of the function f passing through the point.

d) Using the results of a), b), and c), lay out what appears to be the shortest path on the surface $z = x^2 + 4y^2$ descending from the point $(2, 1, 8)$ to the lowest point on the surface $(0, 0, 0)$.

e) What algorithm, suitable for implementation on a computer, would you propose for finding the minimum of the function $f(x, y) = x^2 + 4y^2$?

3. We say that a *vector field* is defined in a domain G of $\mathbb{R}^m$ if a vector $\mathbf{v}(x) \in T\mathbb{R}^m_x$ is assigned to each point $x \in G$. A vector field $\mathbf{v}(x)$ in G is called a *potential field* if there is a numerical-valued function $U : G \to \mathbb{R}$ such that $\mathbf{v}(x) = \operatorname{grad} U(x)$. The function $U(x)$ is called the *potential* of the field $\mathbf{v}(x)$. (In physics it is the function $-U(x)$ that is usually called the potential, and the function $U(x)$ is called the *force function* when a field of force is being discussed.)

a) On a plane with Cartesian coordinates (x, y) draw the field $\operatorname{grad} f(x, y)$ for each of the following functions: $f_1(x, y) = x^2 + y^2$; $f_2(x, y) = -(x^2 + y^2)$; $f_3(x, y) = \arctan(x/y)$ in the domain $y > 0$; $f_4(x, y) = xy$.

b) By Newton's law a particle of mass m at the point $0 \in \mathbb{R}^3$ attracts a particle of mass 1 at the point $x \in \mathbb{R}^3$ $(x \neq 0)$ with force $\mathbf{F} = -m|\mathbf{r}|^{-3}\mathbf{r}$, where $\mathbf{r}$ is the vector $\overrightarrow{Ox}$ (we have omitted the dimensional constant G_0). Show that the vector field $\mathbf{F}(x)$ in $\mathbb{R}^3 \backslash 0$ is a potential field.

c) Verify that masses m_i $(i = 1, \dots, n)$ located at the points (ξ_i, η_i, ζ_i) $(i = 1, \dots, n)$ respectively, create a Newtonian force field except at these points and that the potential is the function

$$U(x, y, z) = \sum_{i=1}^{n} \frac{m_i}{\sqrt{(x - \xi_i)^2 + (y - \eta_i)^2 + (z - \zeta_i)^2}}.$$

d) Find the potential of the electrostatic field created by point charges q_i $(i = 1, \dots, n)$ located at the points (ξ_i, η_i, ζ_i) $(i = 1, \dots, n)$ respectively.

4. Consider the motion of an ideal incompressible liquid in a space free of external forces (in particular, free of gravitational forces).

Let $\mathbf{v} = \mathbf{v}(x, y, z, t)$, $\mathbf{a} = \mathbf{a}(x, y, z, t)$, $\rho = \rho(x, y, z, t)$, and $p = p(x, y, z, t)$ be respectively the velocity, acceleration, density, and pressure of the fluid at the point (x, y, z) of the medium at time t.

An ideal liquid is one in which the pressure is the same in all directions at each point.

a) Distinguish a volume of the liquid in the form of a small parallelepiped, one of whose edges is parallel to the vector $\operatorname{grad} p(x, y, z, t)$ (where $\operatorname{grad} p$ is taken with respect to the spatial coordinates). Estimate the force acting on this volume due to the pressure drop, and give an approximate formula for the acceleration of the volume, assuming the fluid is incompressible.

b) Determine whether the result you obtained in a) is consistent with *Euler's equation*

$$\rho \mathbf{a} = -\operatorname{grad} p.$$

c) A curve whose tangent at each point has the direction of the velocity vector at that point is called a *streamline*. The motion is called *stationary* if the functions $\mathbf{v}$, $\mathbf{a}$, ρ, and p are independent of t. Using b), show that along a streamline in the stationary flow of an incompressible liquid the quantity $\frac{1}{2}\|\mathbf{v}\|^2 + p/\rho$ is constant (*Bernoulli's law*[5]).

d) How do the formulas in a) and b) change if the motion takes place in the gravitational field near the surface of the earth? Show that in this case

$$\rho \mathbf{a} = -\operatorname{grad}(gz + p).$$

so that now the quantity $\frac{1}{2}\|\mathbf{v}\|^2 + gz + p/\rho$ is constant along each streamline of the stationary motion of an incompressible liquid, where g is the gravitational acceleration and z is the height of the streamline measured from some zero level.

e) Explain, on the basis of the preceding results, why a load-bearing wing has a characteristic convex-upward profile.

f) An incompressible ideal liquid of density ρ was used to fill a cylindrical glass with a circular base of radius R to a depth h. The glass was then revolved about its axis with angular velocity ω. Using the incompressibility of the liquid, find the equation $z = f(x, y)$ of its surface in stationary mode (see also Problem 3 of Sect. 5.1).

g) From the equation $z = f(x, y)$ found in part f) for the surface, write a formula $p = p(x, y, z)$ for the pressure at each point (x, y, z) of the volume filled by the rotating liquid. Check to see whether the equation $\rho \mathbf{a} = -\operatorname{grad}(gz + p)$ of part d) holds for the formula that you found.

h) Can you now explain why tea leaves sink (although not very rapidly!) and why they accumulate at the center of the bottom of the cup, rather than its side, when the tea is stirred?

5. *Estimating the errors in computing the values of a function.*

a) Using the definition of a differentiable function and the approximate equality $\Delta f(x; h) \approx \mathrm{d}f(x)h$, show that the relative error $\delta = \delta(f(x); h)$ in the value of the product $f(x) = x^1 \cdots x^m$ of m nonzero factors due to errors in determining the factors themselves can be found in the form $\delta \approx \sum_{i=1}^{m} \delta_i$, where δ_i is the relative error in the determination of the ith factor.

---

[5]Daniel Bernoulli (1700–1782) – Swiss scholar, one of the outstanding physicists and mathematicians of his time.

b) Using the equality $\mathrm{d}\ln f(x) = \frac{1}{f(x)}\,\mathrm{d}f(x)$, obtain the result of part a) again and show that in general the relative error in a fraction

$$\frac{f_1 \cdots f_n}{g_1 \cdots g_k}(x_1, \dots, x_m)$$

can be found as the sum of the relative errors of the values of the functions $f_1, \dots, f_n, g_1, \dots, g_k$.

6. *Homogeneous functions and Euler's identity.* A function $f : G \to \mathbb{R}$ defined in some domain $G \subset \mathbb{R}^m$ is called *homogeneous* (resp. *positive-homogeneous*) *of degree n* if the equality

$$f(\lambda x) = \lambda^n f(x) \quad \bigl(\text{resp. } f(\lambda x) = |\lambda|^n f(x)\bigr)$$

holds for any $x \in \mathbb{R}^m$ and $\lambda \in \mathbb{R}$ such that $x \in G$ and $\lambda x \in G$.

A function is *locally homogeneous* of degree n in the domain G if it is a homogeneous function of degree n in some neighborhood of each point of G.

a) Prove that in a convex domain every locally homogeneous function is homogeneous.

b) Let G be the plane $\mathbb{R}^2$ with the ray $L = \{(x, y) \in \mathbb{R}^2 \mid x = 2 \wedge y \ge 0\}$ removed. Verify that the function

$$f(x, y) = \begin{cases} y^4/x, & \text{if } x > 2 \wedge y > 0, \\ y^3, & \text{at other points of the domain,} \end{cases}$$

is locally homogeneous in G, but is not a homogeneous function in that domain.

c) Determine the degree of homogeneity or positive homogeneity of the following functions with their natural domains of definition:

$$f_1(x^1, \dots, x^m) = x^1x^2 + x^2x^3 + \cdots + x^{m-1}x^m;$$

$$f_2(x^1, x^2, x^3, x^4) = \frac{x^1x^2 + x^3x^4}{x^1x^2x^3 + x^2x^3x^4};$$

$$f_3(x^1, \dots, x^m) = |x^1 \cdots x^m|^l.$$

d) By differentiating the equality $f(tx) = t^n f(x)$ with respect to t, show that if a differentiable function $f : G \to \mathbb{R}$ is locally homogeneous of degree n in a domain $G \subset \mathbb{R}^m$, it satisfies the following *Euler identity for homogeneous functions*:

$$x^1 \frac{\partial f}{\partial x^1}(x^1, \dots, x^m) + \cdots + x^m \frac{\partial f}{\partial x^m}(x^1, \dots, x^m) \equiv nf(x^1, \dots, x^m).$$

e) Show that if Euler's identity holds for a differentiable function $f : G \to \mathbb{R}$ in a domain G, then that function is locally homogeneous of degree n in G.

Hint: Verify that the function $\varphi(t) = t^{-n} f(tx)$ is defined for every $x \in G$ and is constant in some neighborhood of 1.

7. *Homogeneous functions and the dimension method.*

1^0. *The dimension of a physical quantity and the properties of functional relations between physical quantities.*

Physical laws establish interconnections between physical quantities, so that if certain units of measurement are adopted for some of these quantities, then the units of measurement of the quantities connected with them can be expressed in a certain way in terms of the units of measurement of the fixed quantities. That is how the basic and derived units of different systems of measurement arise.

In the International System, the basic mechanical units of measurement are taken to be the unit of length (the meter, denoted m), mass (the kilogram, denoted kg), and time (the second, denoted s).

The expression of a derived unit of measurement in terms of the basic mechanical units is called its *dimension*. This definition will be made more precise below.

The dimension of any mechanical quantity is written symbolically as a formula expressing it in terms of the symbols L, M, and T proposed by Maxwell[6] as the dimensions of the basic units mentioned above. For example, the dimensions of velocity, acceleration, and force have respectively the forms

$$[v] = LT^{-1}, \qquad [a] = LT^{-2}, \qquad [F] = MLT^{-2}.$$

If physical laws are to be independent of the choice of units of measurement, one expression of that invariance should be certain properties of the functional relation

$$x_0 = f(x_1, \dots, x_k, x_{k+1}, \dots, x_n) \tag{*}$$

between the numerical characteristics of the physical quantities.

Consider, for example, the relation $c = f(a, b) = \sqrt{a^2 + b^2}$ between the lengths of the legs and the length of the hypotenuse of a right triangle. Any change of scale should affect all the lengths equally, so that for all admissible values of a and b the relation $f(\alpha a, \alpha b) = \varphi(\alpha) f(a, b)$ should hold, and in the present case $\varphi(\alpha) = \alpha$.

A basic (and, at first sight, obvious) presupposition of dimension theory is that *a relation* (*) *claiming physical significance must be such that when the scales of the basic units of measurement are changed, the numerical values of all terms of the same type occurring in the formula must be multiplied by the same factor.*

In particular, if x_1, x_2, x_3 are basic independent physical quantities and the relation $(x_1, x_2, x_3) \mapsto f(x_1, x_2, x_3)$ expresses the way a fourth physical quantity depends on them, then, by the principle just stated, for any admissible values of x_1, x_2, x_3 the equality

$$f(\alpha_1 x_1, \alpha_2 x_2, \alpha_3 x_3) = \varphi(\alpha_1, \alpha_{2)} \alpha_3) f(x_2, x_2, x_3), \tag{**}$$

must hold with some particular function φ.

[6] J.C. Maxwell (1831–1879) – outstanding British physicist. He created the mathematical theory of the electromagnetic field, and is also famous for his research in the kinetic theory of gases, optics, and mechanics.

The function φ in (**) characterizes completely the dependence of the numerical value of the physical quantity in question on a change in the scale of the basic fixed physical quantities. Thus, this function should be regarded as the *dimension* of that physical quantity relative to the fixed basic units of measurement.

We now make the form of the dimension function more precise.

a) Let $x \mapsto f(x)$ be a function of one variable satisfying the condition $f(\alpha x) = \varphi(\alpha) f(x)$, where f and φ are differentiable functions.

Show that $\varphi(\alpha) = \alpha^d$.

b) Show that the dimension function φ in Eq. (**) always has the form $\alpha_1^{d_1} \cdot \alpha_2^{d_2} \cdot \alpha_3^{d_3}$, where the exponents d_1, d_2, d_3 are certain real numbers. Thus if, for example, the basic units of L, M, and T are fixed, then the set (d_1, d_2, d_3) of exponents expressed in the power representation $L^{d_1} M^{d_2} T^{d_3}$ can also be regarded as the *dimension* of the given physical quantity.

c) In part b) it was found that the dimension function is always a power function, that is, it is a homogeneous function of a certain degree with respect to each of the basic units of measurement. What does it mean if the degree of homogeneity of the dimension function of a certain physical quantity relative to one of the basic units of measurement is zero?

2^0 *The Π-theorem and the dimension method.*

Let $[x_i] = X_i$ $(i = 0, 1, \dots, n)$ be the dimensions of the physical quantities occurring in the law (*).

Assume that the dimensions of $x_0, x_{k+1}, \dots, x_n$ can be expressed in terms of the dimensions of $x_1, \dots, x_k$, that is,

$$[x_0] = X_0 = X_1^{p_0^1} \cdots X_k^{p_0^k},$$

$$[x_{k+i}] = X_{k+i} = X_1^{p_i^1} \cdots X_k^{p_i^k} \quad (i = 1, \dots, n-k).$$

d) Show that the following relation must then hold, along with (*):

$$\alpha_1^{p_0^1} \cdots \alpha_k^{p_0^k} x_0 = f\big(\alpha_1 x_1, \dots, \alpha_k x_k, \alpha_1^{p_1^1} \cdots \alpha_k^{p_1^k} x_{k+1}, \dots, \alpha_1^{p_{n-k}^1} \cdots \alpha_k^{p_{n-k}^k} x_n\big). \tag{***}$$

e) If $x_1, \dots, x_k$ are independent, we set $\alpha_1 = x_1^{-1}, \dots, \alpha_k = x_k^{-1}$ in (***). Verify that when this is done, (***) yields the equality

$$\frac{x_0}{x_1^{p_0^1} \cdots x_k^{p_0^k}} = f\left(1, \dots, 1, \frac{x_{k+1}}{x_1^{p_1^1} \cdots x_k^{p_1^k}}, \dots, \frac{x_n}{x_1^{p_{n-k}^1} \cdots x_k^{p_{n-k}^k}}\right),$$

which is a relation

$$\Pi = f(1, \dots, 1, \Pi_1, \dots, \Pi_{n-k}) \tag{****}$$

involving the dimensionless quantities $\Pi, \Pi_1, \dots, \Pi_{n-k}$.

Thus we obtain the following

***Π*-theorem of dimension theory** *If the quantities $x_1, \dots, x_k$ in relation (*) are independent, this relation can be reduced to the function (****) of $n-k$ dimensionless parameters.*

f) Verify that if $k=n$, the function f in relation (*) can be determined up to a numerical multiple by using the Π-theorem. Use this method to find the expression $c(\varphi_0)\sqrt{l/g}$ for the period of oscillation of a pendulum (that is, a mass m suspended by a thread of length l and oscillating near the surface of the earth, where φ_0 is the initial displacement angle).

g) Find a formula $P = c\sqrt{mr/F}$ for the period of revolution of a body of mass m held in a circular orbit by a central force of magnitude F.

h) Use Kepler's law $(P_1/P_2)^2 = (r_1/r_2)^3$, which establishes for circular orbits a connection between the ratio of the periods of revolution of planets (or satellites) and the ratio of the radii of their orbits, to find, as Newton did, the exponent α in the law of universal gravitation $F = G\frac{m_1 m_2}{r^\alpha}$.

8.4 The Basic Facts of Differential Calculus of Real-Valued Functions of Several Variables

8.4.1 The Mean-Value Theorem

Theorem 1 *Let $f : G \to \mathbb{R}$ be a real-valued function defined in a region $G \subset \mathbb{R}^m$, and let the closed line segment $[x, x+h]$ with endpoints x and $x+h$ be contained in G. If the function f is continuous at the points of the closed line segment $[x, x+h]$ and differentiable at points of the open interval $]x, x+h[$, then there exists a point $\xi \in]x, x+h[$ such that the following equality holds:*

$$\boxed{f(x+h) - f(x) = f'(\xi)h.} \tag{8.53}$$

Proof Consider the auxiliary function

$$F(t) = f(x+th)$$

defined on the closed interval $0 \le t \le 1$. This function satisfies all the hypotheses of Lagrange's theorem: it is continuous on $[0, 1]$, being the composition of continuous mappings, and differentiable on the open interval $]0, 1[$, being the composition of differentiable mappings. Consequently, there exists a point $\theta \in]0, 1[$ such that

$$F(1) - F(0) = F'(\theta) \cdot 1.$$

But $F(1) = f(x+h)$, $F(0) = f(x)$, $F'(\theta) = f'(x+\theta h)h$, and hence the equality just written is the same as the assertion of the theorem. □

We now give the coordinate form of relation (8.53).

If $x = (x^1, \dots, x^m)$, $h = (h^1, \dots, h^m)$, and $\xi = (x^1 + \theta h^1, \dots, x^m + \theta h^m)$, Eq. (8.53) means that

$$
\begin{aligned}
f(x+h) - f(x) &= f(x^1 + h^1, \dots, x^m + h^m) - f(x^1, \dots, x^m) = \\
&= f'(\xi)h = \left(\frac{\partial f}{\partial x^1}(\xi), \dots, \frac{\partial f}{\partial x^m}(\xi)\right)\begin{pmatrix} h^1 \\ \cdots \\ h^m \end{pmatrix} = \\
&= \partial_1 f(\xi)h^1 + \cdots + \partial_m f(\xi)h^m = \\
&= \sum_{i=1}^{m} \partial_i f(x^1 + \theta h^1, \dots, x^m + \theta h^m)h^i.
\end{aligned}
$$

Using the convention of summation on an index that appears as both superscript and subscript, we can finally write

$$
\begin{aligned}
f(x^1 + h^1, \dots, x^m + h^m) - f(x^1, \dots, x^m) = \\
= \partial_i f(x^1 + \theta h^1, \dots, x^m + \theta h^m)h^i, \qquad (8.54)
\end{aligned}
$$

where $0 < \theta < 1$ and θ depends on both x and h.

Remark Theorem 1 is called the mean-value theorem because there exists a certain "average" point $\xi \in]x, x+h[$ at which Eq. (8.53) holds. We have already noted in our discussion of Lagrange's theorem (Sect. 5.3.1) that the mean-value theorem is specific to real-valued functions. A general finite-increment theorem for mappings will be proved in Chap. 10 (Part 2).

The following proposition is a useful corollary of Theorem 1.

Corollary *If the function $f : G \to \mathbb{R}$ is differentiable in the domain $G \subset \mathbb{R}^m$ and its differential equals zero at every point $x \in G$, then f is constant in the domain G.*

Proof The vanishing of a linear transformation is equivalent to the vanishing of all the elements of the matrix corresponding to it. In the present case

$$
df(x)h = (\partial_1 f, \dots, \partial_m f)(x)h,
$$

and therefore $\partial_1 f(x) = \cdots = \partial_m f(x) = 0$ at every point $x \in G$.

By definition, a domain is an open connected set. We shall make use of this fact.

We first show that if $x \in G$, then the function f is constant in a ball $B(x; r) \subset G$. Indeed, if $(x+h) \in B(x; r)$, then $[x, x+h] \subset B(x; r) \subset G$. Applying relation (8.53) or (8.54), we obtain

$$
f(x+h) - f(x) = f'(\xi)h = 0 \cdot h = 0,
$$

that is, $f(x+h)=f(x)$, and the values of f in the ball $B(x;r)$ are all equal to the value at the center of the ball.

Now let $x_0, x_1 \in G$ be arbitrary points of the domain G. By the connectedness of G, there exists a path $t \mapsto x(t) \in G$ such that $x(0)=x_0$ and $x(1)=x_1$. We assume that the continuous mapping $t \mapsto x(t)$ is defined on the closed interval $0 \le t \le 1$. Let $B(x_0;r)$ be a ball with center at x_0 contained in G. Since $x(0)=x_0$ and the mapping $t \mapsto x(t)$ is continuous, there is a positive number δ such that $x(t) \in B(x_0;r) \subset G$ for $0 \le t \le \delta$. Then, by what has been proved, $(f \circ x)(t) \equiv f(x_0)$ on the interval $[0,\delta]$.

Let $l = \sup \delta$, where the upper bound is taken over all numbers $\delta \in [0,1]$ such that $(f \circ x)(t) \equiv f(x_0)$ on the interval $[0,\delta]$. By the continuity of the function $f(x(t))$ we have $f(x(l)) = f(x_0)$. But then $l=1$. Indeed, if that were not so, we could take a ball $B(x(l);r) \subset G$, in which $f(x) = f(x(l)) = f(x_0)$, and then by the continuity of the mapping $t \mapsto x(t)$ find $\Delta > 0$ such that $x(t) \in B(x(l);r)$ for $l \le t \le l+\Delta$. But then $(f \circ x)(t) = f(x(l)) = f(x_0)$ for $0 \le t \le l+\Delta$, and so $l \ne \sup \delta$.

Thus we have shown that $(f \circ x)(t) = f(x_0)$ for any $t \in [0,1]$. In particular $(f \circ x)(1) = f(x_1) = f(x_0)$, and we have verified that the values of the function $f : G \to \mathbb{R}$ are the same at any two points $x_0, x_1 \in G$. □

8.4.2 A Sufficient Condition for Differentiability of a Function of Several Variables

Theorem 2 *Let $f : U(x) \to \mathbb{R}$ be a function defined in a neighborhood $U(x) \subset \mathbb{R}^m$ of the point $x = (x^1, \dots, x^m)$.*

If the function f has all partial derivatives $\frac{\partial f}{\partial x^1}, \dots, \frac{\partial f}{\partial x^m}$ at each point of the neighborhood $U(x)$ and they are continuous at x, then f is differentiable at x.

Proof Without loss of generality we shall assume that $U(x)$ is a ball $B(x;r)$. Then, together with the points $x = (x^1, \dots, x^m)$ and $x+h = (x^1+h^1, \dots, x^m+h^m)$, the points $(x^1, x^2+h^2, \dots, x^m+h^m), \dots, (x^1, x^2, \dots, x^{m-1}, x^m+h^m)$ and the lines connecting them must also belong to the domain $U(x)$. We shall use this fact, applying the Lagrange theorem for functions of one variable in the following computation:

$$
\begin{aligned}
f(x+h) - f(x) &= f(x^1+h^1, \dots, x^m+h^m) - f(x^1, \dots, x^m) = \\
&= f(x^1+h^1, \dots, x^m+h^m) - f(x^1, x^2+h^2, \dots, x^m+h^m) + \\
&\quad + f(x^1, x^2+h^2, \dots, x^m+h^m) - \\
&\quad - f(x^1, x^2, x^3+h^3, \dots, x^m+h^m) + \dots + \\
&\quad + f(x^1, x^2, \dots, x^{m-1}, x^m+h^m) - f(x^1, \dots, x^m) =
\end{aligned}
$$

$$= \partial_1 f(x^1 + \theta^1 h^1, x^2 + h^2, \dots, x^m + h^m)h^1 +$$
$$+ \partial_2 f(x^1, x^2 + \theta^2 h^2, x^3 + h^3, \dots, x^m + h^m)h^2 + \dots +$$
$$+ \partial_m f(x^1, x^2, \dots, x^{m-1}, x^m + \theta^m h^m)h^m.$$

So far we have used only the fact that the function f has partial derivatives with respect to each of its variables in the domain $U(x)$.

We now use the fact that these partial derivatives are continuous at x. Continuing the preceding computation, we obtain

$$f(x+h) - f(x) = \partial_1 f(x^1, \dots, x^m)h^1 + \alpha^1 h^1 +$$
$$+ \partial_2 f(x^1, \dots, x^m)h^2 + \alpha^2 h^2 + \dots +$$
$$+ \partial_m f(x^1, \dots, x^m)h^m + \alpha^m h^m,$$

where the quantities $\alpha_1, \dots, \alpha_m$ tend to zero as $h \to 0$ by virtue of the continuity of the partial derivatives at the point x.

But this means that

$$f(x+h) - f(x) = L(x)h + o(h) \quad \text{as } h \to 0,$$

where $L(x)h = \partial_1 f(x^1, \dots, x^m)h^1 + \dots + \partial_m f(x^1, \dots, x^m)h^m$. □

It follows from Theorem 2 that if the partial derivatives of a function $f : G \to \mathbb{R}$ are continuous in the domain $G \subset \mathbb{R}^m$, then the function is differentiable at that point of the domain.

Let us agree from now on to use the symbol $C^{(1)}(G; \mathbb{R})$, or, more simply, $C^{(1)}(G)$ to denote the set of functions having continuous partial derivatives in the domain G.

8.4.3 Higher-Order Partial Derivatives

If a function $f : G \to \mathbb{R}$ defined in a domain $G \subset \mathbb{R}^m$ has a partial derivative $\frac{\partial f}{\partial x^i}(x)$ with respect to one of the variables $x^1, \dots, x^m$, this partial derivative is a function $\partial_i f : G \to \mathbb{R}$, which in turn may have a partial derivative $\partial_j(\partial_i f)(x)$ with respect to a variable x^j.

The function $\partial_j(\partial_i f) : G \to \mathbb{R}$ is called the *second partial derivative of f with respect to the variables x^i and x^j* and is denoted by one of the following symbols:

$$\partial_{ji} f(x), \quad \frac{\partial^2 f}{\partial x^j \partial x^i}(x).$$

The order of the indices indicates the order in which the differentiation is carried out with respect to the corresponding variables.

We have now defined partial derivatives of second order.

If a partial derivative of order k

$$\partial_{i_1\cdots i_k} f(x) = \frac{\partial^k f}{\partial x^{i_1} \cdots \partial x^{i_k}}(x)$$

has been defined, we define by induction the partial derivative of order $k+1$ by the relation

$$\partial_{i i_1\cdots i_k} f(x) := \partial_i(\partial_{i_1\cdots i_k} f)(x).$$

At this point a question arises that is specific for functions of several variables: Does the order of differentiation affect the partial derivative computed?

Theorem 3 *If the function $f : G \to \mathbb{R}$ has partial derivatives*

$$\frac{\partial^2 f}{\partial x^i \partial x^j}(x), \quad \frac{\partial^2 f}{\partial x^j \partial x^i}(x)$$

in a domain G, then at every point $x \in G$ at which both partial derivatives are continuous, their values are the same.

Proof Let $x \in G$ be a point at which both functions $\partial_{ij} f : G \to \mathbb{R}$ and $\partial_{ji} f : G \to \mathbb{R}$ are continuous. From this point on all of our arguments are carried out in the context of a ball $B(x; r) \subset G, r > 0$, which is a convex neighborhood of the point x. We wish to verify that

$$\frac{\partial^2 f}{\partial x^i \partial x^j}(x^1, \dots, x^m) = \frac{\partial^2 f}{\partial x^j \partial x^i}(x^1, \dots, x^m).$$

Since only the variables x^i and x^j will be changing in the computations to follow, we shall assume for the sake of brevity that f is a function of two variables $f(x^1, x^2)$, and we need to verify that

$$\frac{\partial^2 f}{\partial x^1 \partial x^2}(x^1, x^2) = \frac{\partial^2 f}{\partial x^2 \partial x^1}(x^1, x^2),$$

if the two functions are both continuous at the point (x^1, x^2).

Consider the auxiliary function

$$F(h^1, h^2) = f(x^1 + h^1, x^2 + h^2) - f(x^1 + h^1, x^2) - f(x^1, x^2 + h^2) + f(x^1, x^2),$$

where the displacement $h = (h^1, h^2)$ is assumed to be sufficiently small, namely so small that $x + h \in B(x; r)$.

If we regard $F(h^1, h^2)$ as the difference

$$F(h^1, h^2) = \varphi(1) - \varphi(0),$$

where $\varphi(t) = f(x^1 + th^1, x^2 + h^2) - f(x^1 + th^1, x^2)$, we find by Lagrange's theorem that

$$F(h^1, h^2) = \varphi'(\theta_1) = \left(\partial_1 f(x^1 + \theta_1 h^1, x^2 + h^2) - \partial f(x^1 + \theta_1 h^1, x^2)\right)h^1.$$

Again applying Lagrange's theorem to this last difference, we find that

$$F(h^1, h^2) = \partial_{21} f(x^1 + \theta_1 h^1, x^2 + \theta_2 h^2)h^2 h^1. \tag{8.55}$$

If we now represent $F(h^1, h^2)$ as the difference

$$F(h^1, h^2) = \tilde{\varphi}(1) - \tilde{\varphi}(0),$$

where $\tilde{\varphi}(t) = f(x^1 + h^1, x^2 + th^2) - f(x^1, x^2 + th^2)$, we find similarly that

$$F(h^1, h^2) = \partial_{12} f(x^1 + \tilde{\theta}_1 h^1, x^2 + \tilde{\theta}_2 h^2)h^1 h^2. \tag{8.56}$$

Comparing (8.55) and (8.56), we conclude that

$$\partial_{21} f(x^1 + \theta_1 h^1, x^2 + \theta_2 h^2) = \partial_{12} f(x^1 + \tilde{\theta}_1 h^1, x^2 + \tilde{\theta}_2 h^2), \tag{8.57}$$

where $\theta_1, \theta_2, \tilde{\theta}_1, \tilde{\theta}_2 \in \,]0, 1[$. Using the continuity of the partial derivatives at the point (x^1, x^2), as $h \to 0$, we get the equality we need as a consequence of (8.57).

$$\partial_{21} f(x^1, x^2) = \partial_{12} f(x^1, x^2).$$ □

We remark that without additional assumptions we cannot say in general that $\partial_{ij} f(x) = \partial_{ji} f(x)$ if both of the partial derivatives are defined at the point x (see Problem 2 at the end of this section).

Let us agree to denote the set of functions $f : G \to \mathbb{R}$ all of whose partial derivatives up to order k inclusive are defined and continuous in the domain $G \subset \mathbb{R}^m$ by the symbol $C^{(k)}(G; \mathbb{R})$ or $C^{(k)}(G)$.

As a corollary of Theorem 3, we obtain the following.

Proposition 1 *If $f \in C^{(k)}(G; \mathbb{R})$, the value $\partial_{i_1 \ldots i_k} f(x)$ of the partial derivative is independent of the order $i_1, \ldots, i_k$ of differentiation, that is, remains the same for any permutation of the indices $i_1, \ldots, i_k$.*

Proof In the case $k = 2$ this proposition is contained in Theorem 3.

Let us assume that the proposition holds up to order n inclusive. We shall show that then it also holds for order $n + 1$.

But $\partial_{i_1 i_2 \cdots i_{n+1}} f(x) = \partial_{i_1}(\partial_{i_2 \cdots i_{n+1}} f)(x)$. By the induction assumption the indices $i_2, \ldots, i_{n+1}$ can be permuted without changing the function $\partial_{i_2 \cdots i_{n+1}} f(x)$, and hence without changing $\partial_{i_1 \cdots i_{n+1}} f(x)$. For that reason it suffices to verify that one can also permute, for example, the indices i_1 and i_2 without changing the value of the derivative $\partial_{i_1 i_2 \cdots i_{n+1}} f(x)$.

Since

$$\partial_{i_1 i_2 \cdots i_{n+1}} f(x) = \partial_{i_1 i_2} (\partial_{i_3 \cdots i_{n+1}} f)(x),$$

the possibility of this permutation follows immediately from Theorem 3. By the induction principle Proposition 1 is proved. □

Example 1 Let $f(x) = f(x^1, x^2)$ be a function of class $C^{(k)}(G; \mathbb{R})$.

Let $h = (h^1, h^2)$ be such that the closed interval $[x, x + h]$ is contained in the domain G. We shall show that the function

$$\varphi(t) = f(x + th),$$

which is defined on the closed interval [0, 1], belongs to class $C^{(k)}[0, 1]$ and find its derivative of order k with respect to t.

We have

$$\varphi'(t) = \partial_1 f\big(x^1 + th^1, x^2 + th^2\big)h^1 + \partial_2 f\big(x^1 + th^1,\ x^2 + th^2\big)h^2,$$

$$\begin{aligned}\varphi''(t) &= \partial_{11} f(x + th)h^1 h^1 + \partial_{21} f(x + th)h^2 h^1 + \\ &\quad + \partial_{12} f(x + th)h^1 h^2 + \partial_{22} f(x + th)h^2 h^2 = \\ &= \partial_{11} f(x + th)\big(h^1\big)^2 + 2\partial_{12} f(x + th)h^1 h^2 + \partial_{22} f(x + th)\big(h^2\big)^2.\end{aligned}$$

These relations can be written as the action of the operator $(h^1 \partial_1 + h^2 \partial_2)$:

$$\varphi'(t) = \big(h^1 \partial_1 + h^2 \partial_2\big) f(x + th) = h^i \partial_i f(x + th),$$

$$\varphi''(t) = \big(h^1 \partial_1 + h^2 \partial_2\big)^2 f(x + th) = h^{i_1} h^{i_2} \partial_{i_1 i_2} f(x + th).$$

By induction we obtain

$$\varphi^{(k)}(t) = \big(h^1 \partial_1 + h^2 \partial_2\big)^k f(x + th) = h^{i_1} \cdots h^{i_k} \partial_{i_1 \cdots i_k} f(x + th)$$

(summation over all sets $i_1, \ldots, i_k$ of k indices, each assuming the values 1 and 2, is meant).

Example 2 If $f(x) = f(x^1, \ldots, x^m)$ and $f \in C^{(k)}(G; \mathbb{R})$, then, under the assumption that $[x, x + h] \subset G$, for the function $\varphi(t) = f(x + th)$ defined on the closed interval [0, 1] we obtain

$$\varphi^{(k)}(t) = h^{i_1} \cdots h^{i_k} \partial_{i_1 \cdots i_k} f(x + th), \tag{8.58}$$

where summation over all sets of indices $i_1, \ldots, i_k$, each assuming all values from 1 to m inclusive, is meant on the right.

We can also write formula (8.58) as

$$\varphi^{(k)}(t) = \big(h^1 \partial_1 + \cdots + h^m \partial_m\big)^k f(x + th). \tag{8.59}$$

8.4.4 Taylor's Formula

Theorem 4 *If the function $f : U(x) \to \mathbb{R}$ is defined and belongs to class $C^{(n)}(U(x); \mathbb{R})$ in a neighborhood $U(x) \subset \mathbb{R}^m$ of the point $x \in \mathbb{R}^m$, and the closed interval $[x, x+h]$ is completely contained in $U(x)$, then the following equality holds:*

$$f(x^1+h^1,\dots,x^m+h^m) - f(x^1,\dots,x^m) = \\ = \sum_{k=1}^{n-1} \frac{1}{k!}\left(h^1\partial_1 + \dots + h^m\partial_m\right)^k f(x) + r_{n-1}(x;h), \tag{8.60}$$

where

$$r_{n-1}(x;h) = \int_0^1 \frac{(1-t)^{n-1}}{(n-1)!}\left(h^1\partial_1 + \dots + h^m\partial_m\right)^n f(x+th)\,\mathrm{d}t. \tag{8.61}$$

Equality (8.60), together with (8.61), is called *Taylor's formula with integral form of the remainder.*

Proof Taylor's formula follows immediately from the corresponding Taylor formula for a function of one variable. In fact, consider the auxiliary function

$$\varphi(t) = f(x+th),$$

which, by the hypotheses of Theorem 4, is defined on the closed interval $0 \le t \le 1$ and (as we have verified above) belongs to the class $C^{(n)}[0,1]$.

Then for $\tau \in [0,1]$, by Taylor's formula for functions of one variable, we can write that

$$\varphi(\tau) = \varphi(0) + \frac{1}{1!}\varphi'(0)\tau + \dots + \frac{1}{(n-1)!}\varphi^{(n-1)}(0)\tau^{n-1} + \\ + \int_0^1 \frac{(1-t)^{n-1}}{(n-1)!}\varphi^{(n)}(t\tau)\tau^n\,\mathrm{d}t.$$

Setting $\tau = 1$ here, we obtain

$$\varphi(1) = \varphi(0) + \frac{1}{1!}\varphi'(0) + \dots + \frac{1}{(n-1)!}\varphi^{(n-1)}(0) + \\ + \int_0^1 \frac{(1-t)^{n-1}}{(n-1)!}\varphi^{(n)}(t)\,\mathrm{d}t. \tag{8.62}$$

Substituting the values

$$\varphi^{(k)}(0) = \left(h^1\partial_1 + \dots + h^m\partial_m\right)^k f(x) \quad (k = 0,\dots,n-1),$$

$$\varphi^{(n)}(t) = \left(h^1\partial_1 + \cdots + h^m\partial_m\right)^n f(x+th),$$

into this equality in accordance with formula (8.59), we find what Theorem 4 asserts. □

Remark If we write the remainder term in relation (8.62) in the Lagrange form rather than the integral form, then the equality

$$\varphi(1) = \varphi(0) + \frac{1}{1!}\varphi'(0) + \cdots + \frac{1}{(n-1)!}\varphi^{(n-1)}(0) + \frac{1}{n!}\varphi^{(n)}(\theta),$$

where $0 < \theta < 1$, implies Taylor's formula (8.60) with remainder term

$$r_{n-1}(x;h) = \frac{1}{n!}\left(h^1\partial_1 + \cdots + h^m\partial_m\right)^n f(x+\theta h). \tag{8.63}$$

This form of the remainder term, as in the case of functions of one variable, is called the *Lagrange form of the remainder term in Taylor's formula*.

Since $f \in C^{(n)}(U(x);\mathbb{R})$, it follows from (8.63) that

$$r_{n-1}(x;h) = \frac{1}{n!}\left(h^1\partial_1 + \cdots + h^m\partial_m\right)^n f(x) + o\left(\|h\|^n\right) \quad \text{as } h \to 0,$$

and so we have the equality

$$f\left(x^1+h^1,\dots,x^m+h^m\right) - f\left(x^1,\dots,x^m\right) =$$

$$= \sum_{k=1}^{n} \frac{1}{k!}\left(h^1\partial_1 + \cdots + h^m\partial_m\right)^k f(x) + o\left(\|h\|^n\right) \quad \text{as } h \to 0, \tag{8.64}$$

called *Taylor's formula with the remainder term in Peano form*.

8.4.5 Extrema of Functions of Several Variables

One of the most important applications of differential calculus is its use in finding extrema of functions.

Definition 1 A function $f: E \to \mathbb{R}$ defined on a set $E \subset \mathbb{R}^m$ has a *local maximum* (resp. *local minimum*) at an interior point x_0 of E if there exists a neighborhood $U(x_0) \subset E$ of the point x_0 such that $f(x) \le f(x_0)$ (resp. $f(x) \ge f(x_0)$) for all $x \in U(x_0)$.

If the strict inequality $f(x) < f(x_0)$ holds for $x \in U(x_0) \setminus x_0$ (or, respectively, $f(x) > f(x_0)$), the function has a *strict local maximum* (resp. *strict local minimum*) at x_0.

Definition 2 The local minima and maxima of a function are called its *local extrema.*

Theorem 5 *Suppose a function $f : U(x_0) \to \mathbb{R}$ defined in a neighborhood $U(x_0) \subset \mathbb{R}^m$ of the point $x_0 = (x_0^1, \dots, x_0^m)$ has partial derivatives with respect to each of the variables $x^1, \dots, x^m$ at the point x_0.*

Then a necessary condition for the function to have a local extremum at x_0 is that the following equalities hold at that point:

$$\frac{\partial f}{\partial x^1}(x_0) = 0, \quad \dots, \quad \frac{\partial f}{\partial x^m}(x_0) = 0. \tag{8.65}$$

Proof Consider the function $\varphi(x^1) = f(x^1, x_0^2, \dots, x_0^m)$ of one variable defined, according to the hypotheses of the theorem, in some neighborhood of the point x_0^1 on the real line. At x_0^1 the function $\varphi(x^1)$ has a local extremum, and since

$$\varphi'(x_0^1) = \frac{\partial f}{\partial x^1}(x_0^1, x_0^2, \dots, x_0^m),$$

it follows that $\frac{\partial f}{\partial x^1}(x_0) = 0$.

The other equalities in (8.65) are proved similarly. □

We call attention to the fact that relations (8.65) give only necessary but not sufficient conditions for an extremum of a function of several variables. An example that confirms this is any example constructed for this purpose for functions of one variable. Thus, where previously we spoke of the function $x \mapsto x^3$, whose derivative is zero at zero, but has no extremum there, we can now consider the function

$$f(x^1, \dots, x^m) = (x^1)^3,$$

all of whose partial derivatives are zero at $x_0 = (0, \dots, 0)$, while the function obviously has no extremum at that point.

Theorem 5 shows that if the function $f : G \to \mathbb{R}$ is defined on an open set $G \subset \mathbb{R}^m$, its local extrema are found either among the points at which f is not differentiable or at the points where the differential $df(x_0)$ or, what is the same, the tangent mapping $f'(x_0)$, vanishes.

We know that if a mapping $f : U(x_0) \to \mathbb{R}^n$ defined in a neighborhood $U(x_0) \subset \mathbb{R}^m$ of the point $x_0 \in \mathbb{R}^m$ is differentiable at x_0, then the matrix of the tangent mapping $f'(x_0) : \mathbb{R}^m \to \mathbb{R}^n$ has the form

$$\begin{pmatrix} \partial^1 f^1(x_0) & \cdots & \partial_m f^1(x_0) \\ \vdots & \ddots & \vdots \\ \partial_1 f^n(x_0) & \cdots & \partial_m f^n(x_0) \end{pmatrix}. \tag{8.66}$$

Definition 3 The point x_0 is a *critical point of the mapping* $f : U(x_0) \to \mathbb{R}^n$ if the rank of the Jacobi matrix (8.66) of the mapping at that point is less than $\min\{m, n\}$, that is, smaller than the maximum possible value it can have.

In particular, if $n = 1$, the point x_0 is critical if condition (8.65) holds, that is, all the partial derivatives of the function $f : U(x_0) \to \mathbb{R}$ vanish.

The critical points of real-valued functions are also called the *stationary* points of these functions.

After the critical points of a function have been found by solving the system (8.65), the subsequent analysis to determine whether they are extrema or not can often be carried out using Taylor's formula and the following sufficient conditions for the presence or absence of an extremum provided by that formula.

Theorem 6 *Let $f : U(x_0) \to \mathbb{R}$ be a function of class $C^{(2)}(U(x_0); \mathbb{R})$ defined in a neighborhood $U(x_0) \subset \mathbb{R}^m$ of the point $x_0 = (x_0^1, \dots, x_0^m) \in \mathbb{R}^m$, and let x_0 be a critical point of the function f.*

If, in the Taylor expansion of the function at the point x_0

$$f(x_0^1 + h^1, \dots, x_0^m + h^m) =$$
$$= f(x_0^1, \dots, x_0^m) + \frac{1}{2!} \sum_{i,j=1}^{m} \frac{\partial^2 f}{\partial x^i \partial x^j}(x_0) h^i h^j + o(\|h\|^2) \tag{8.67}$$

the quadratic form

$$\sum_{i,j=1}^{m} \frac{\partial^2 f}{\partial x^i \partial x^j}(x_0) h^i h^j \equiv \partial_{ij} f(x_0) h^i h^j \tag{8.68}$$

a) *is positive-definite or negative-definite, then the point x_0 has a local extremum at x_0, which is a strict local minimum if the quadratic form* (8.68) *is positive-definite and a strict local maximum if it is negative-definite;*

b) *assumes both positive and negative values, then the function does not have an extremum at x_0.*

Proof Let $h \neq 0$ and $x_0 + h \in U(x_0)$. Let us represent (8.67) in the form

$$f(x_0 + h) - f(x_0) = \frac{1}{2!} \|h\|^2 \left[\sum_{i,j=1}^{m} \frac{\partial^2 f}{\partial x^i \partial x^j}(x_0) \frac{h^i}{\|h\|} \frac{h^j}{\|h\|} + o(1) \right], \tag{8.69}$$

where $o(1)$ is infinitesimal as $h \to 0$.

It is clear from (8.69) that the sign of the difference $f(x_0 + h) - f(x_0)$ is completely determined by the sign of the quantity in brackets. We now undertake to study this quantity.

The vector $e = (h^1/\|h\|, \dots, h^m/\|h\|)$ obviously has norm 1. The quadratic form (8.68) is continuous as a function $h \in \mathbb{R}^m$, and therefore its restriction to the unit sphere $S(0; 1) = \{x \in \mathbb{R}^m \mid \|x\| = 1\}$ is also continuous on $S(0; 1)$. But the sphere S is a closed bounded subset in $\mathbb{R}^m$, that is, it is compact. Consequently, the form (8.68) has both a minimum point and a maximum point on S, at which it assumes respectively the values m and M.

If the form (8.68) is positive-definite, then $0 < m \le M$, and there is a number $\delta > 0$ such that $|o(1)| < m$ for $\|h\| < \delta$. Then for $\|h\| < \delta$ the bracket on the right-hand side of (8.69) is positive, and consequently $f(x_0 + h) - f(x_0) > 0$ for $0 < \|h\| < \delta$. Thus, in this case the point x_0 is a strict local minimum of the function.

One can verify similarly that when the form (8.68) is negative-definite, the function has a strict local maximum at the point x_0.

Thus a) is now proved.

We now prove b).

Let e_m and e_M be points of the unit sphere at which the form (8.68) assumes the values m and M respectively, and let $m < 0 < M$.

Setting $h = te_m$, where t is a sufficiently small positive number (so small that $x_0 + te_m \in U(x_0)$), we find by (8.69) that

$$f(x_0 + te_m) - f(x_0) = \frac{1}{2!}t^2\big(m + o(1)\big),$$

where $o(1) \to 0$ as $t \to 0$. Starting at some time (that is, for all sufficiently small values of t), the quantity $m + o(1)$ on the right-hand side of this equality will have the sign of m, that is, it will be negative. Consequently, the left-hand side will also be negative.

Similarly, setting $h = te_M$, we obtain

$$f(x_0 + te_M) - f(x_0) = \frac{1}{2!}t^2\big(M + o(1)\big),$$

and consequently for all sufficiently small t the difference $f(x_0 + te_M) - f(x_0)$ is positive.

Thus, if the quadratic form (8.68) assumes both positive and negative values on the unit sphere, or, what is obviously equivalent, in $\mathbb{R}^m$, then in any neighborhood of the point x_0 there are both points where the value of the function is larger than $f(x_0)$ and points where the value is smaller than $f(x_0)$. Hence, in that case x_0 is not a local extremum of the function. □

We now make a number of remarks in connection with this theorem.

Remark 1 Theorem 6 says nothing about the case when the form (8.68) is semi-definite, that is, nonpositive or nonnegative. It turns out that in this case the point may be an extremum, or it may not. This can be seen, in particular from the following example.

Example 3 Let us find the extrema of the function $f(x, y) = x^4 + y^4 - 2x^2$, which is defined in $\mathbb{R}^2$.

In accordance with the necessary conditions (8.65) we write the system of equations

$$\begin{cases} \dfrac{\partial f}{\partial x}(x, y) = 4x^3 - 4x = 0, \\ \dfrac{\partial f}{\partial y}(x, y) = 4y^3 = 0, \end{cases}$$

from which we find three critical points: $(-1, 0)$, $(0, 0)$, $(1, 0)$.

Since

$$\frac{\partial^2 f}{\partial x^2}(x, y) = 12x^2 - 4, \qquad \frac{\partial^2 f}{\partial x \partial y}(x, y) \equiv 0, \qquad \frac{\partial^2 f}{\partial y^2}(x, y) = 12y^2,$$

at the three critical points the quadratic form (8.68) has respectively the form

$$8(h^1)^2, \quad -4(h^1)^2, \quad 8(h^1)^2.$$

That is, in all cases it is positive semi-definite or negative semi-definite. Theorem 6 is not applicable, but since $f(x, y) = (x^2 - 1)^2 + y^4 - 1$, it is obvious that the function $f(x, y)$ has a strict minimum -1 (even a global minimum) at the points $(-1, 0)$, and $(1, 0)$, while there is no extremum at $(0, 0)$, since for $x = 0$ and $y \neq 0$, we have $f(0, y) = y^4 > 0$, and for $y = 0$ and sufficiently small $x \neq 0$ we have $f(x, 0) = x^4 - 2x^2 < 0$.

Remark 2 After the quadratic form (8.68) has been obtained, the study of its definiteness can be carried out using the Sylvester[7] criterion. We recall that by the Sylvester criterion, a quadratic form $\sum_{ij=1}^{m} a_{ij} x^i x^j$ with symmetric matrix

$$\begin{pmatrix} a_{11} & \cdots & a_{1m} \\ \vdots & \ddots & \vdots \\ a_{m1} & \cdots & a_{mm} \end{pmatrix}$$

is positive-definite if and only if all its principal minors are positive; the form is negative-definite if and only if $a_{11} < 0$ and the sign of the principal minor reverses each time its order increases by one.

Example 4 Let us find the extrema of the function

$$f(x, y) = xy \ln(x^2 + y^2),$$

which is defined everywhere in the plane $\mathbb{R}^2$ except at the origin.

[7] J.J. Sylvester (1814–1897) – British mathematician. His best-known works were on algebra.

Solving the system of equations

$$\begin{cases} \dfrac{\partial f}{\partial x}(x,y) = y\ln(x^2+y^2) + \dfrac{2x^2y}{x^2+y^2} = 0, \\ \dfrac{\partial f}{\partial y}(x,y) = x\ln(x^2+y^2) + \dfrac{2xy^2}{x^2+y^2} = 0, \end{cases}$$

we find all the critical points of the function

$$(0,\pm 1); \quad (\pm 1, 0); \quad \left(\pm\frac{1}{\sqrt{2e}}, \pm\frac{1}{\sqrt{2e}}\right); \quad \left(\pm\frac{1}{\sqrt{2e}}, \mp\frac{1}{\sqrt{2e}}\right).$$

Since the function is odd with respect to each of its arguments individually, the points $(0,\pm 1)$ and $(\pm 1, 0)$ are obviously not extrema of the function.

It is also clear that this function does not change its value when the signs of both variables x and y are changed. Thus by studying only one of the remaining critical points, for example, $(\frac{1}{\sqrt{2e}}, \frac{1}{\sqrt{2e}})$ we will be able to draw conclusions on the nature of the others.

Since

$$\frac{\partial^2 f}{\partial x^2}(x,y) = \frac{6xy}{x^2+y^2} - \frac{4x^3y}{(x^2+y^2)^2},$$

$$\frac{\partial^2 f}{\partial x \partial y}(x,y) = \ln(x^2+h^2) + 2 - \frac{4x^2y^2}{(x^2+y^2)^2},$$

$$\frac{\partial^2 f}{\partial y^2}(x,y) = \frac{6xy}{x^2+y^2} - \frac{4xy^3}{(x^2+y^2)^2},$$

at the point $(\frac{1}{\sqrt{2e}}, \frac{1}{\sqrt{2e}})$ the quadratic form $\partial_{ij} f(x_0)h^i h^j$ has the matrix

$$\begin{pmatrix} 2 & 0 \\ 0 & 2 \end{pmatrix},$$

that is, it is positive-definite, and consequently at that point the function has a local minimum

$$f\left(\frac{1}{\sqrt{2e}}, \frac{1}{\sqrt{2e}}\right) = -\frac{1}{2e}.$$

By the observations made above on the properties of this function, one can conclude immediately that

$$f\left(-\frac{1}{\sqrt{2e}}, -\frac{1}{\sqrt{2e}}\right) = -\frac{1}{2e}$$

is also a local minimum and

$$f\left(\frac{1}{\sqrt{2e}}, -\frac{1}{\sqrt{2e}}\right) = f\left(-\frac{1}{\sqrt{2e}}, \frac{1}{\sqrt{2e}}\right) = \frac{1}{2e}$$

are local maxima of the function. This, however, could have been verified directly, by checking the definiteness of the corresponding quadratic form. For example, at the point $(-\frac{1}{\sqrt{2e}}, \frac{1}{\sqrt{2e}})$ the matrix of the quadratic form (8.68) has the form

$$\begin{pmatrix} -2 & 0 \\ 0 & -2 \end{pmatrix},$$

from which it is clear that it is negative-definite.

Remark 3 It should be kept in mind that we have given necessary conditions (Theorem 5) and sufficient conditions (Theorem 6) for an extremum of a function only at an interior point of its domain of definition. Thus in seeking the absolute maximum or minimum of a function, it is necessary to examine the boundary points of the domain of definition along with the critical interior points, since the function may assume its maximal or minimal value at one of these boundary points.

The general principles of studying noninterior extrema will be considered in more detail later (see the section devoted to extrema with constraint). It is useful to keep in mind that in searching for minima and maxima one may use certain simple considerations connected with the nature of the problem along with the formal techniques, and sometimes even instead of them. For example, if a differentiable function being studied in $\mathbb{R}^m$ must have a minimum because of the nature of the problem and turns out to be unbounded above, then if the function has only one critical point, one can assert without further investigation that that point is the minimum.

Example 5 (Huygens' problem) On the basis of the laws of conservation of energy and momentum of a closed mechanical system one can show by a simple computation that when two perfectly elastic balls having mass m_1 and m_2 and initial velocities v_1 and v_2 collide, their velocities after a central collision (when the velocities are directed along the line joining the centers) are determined by the relations

$$\tilde{v}_1 = \frac{(m_1 - m_2)v_1 + 2m_2v_2}{m_1 + m_2},$$

$$\tilde{v}_2 = \frac{(m_2 - m_1)v_2 + 2m_1v_1}{m_1 + m_2}.$$

In particular, if a ball of mass M moving with velocity V strikes a motionless ball of mass m, then the velocity v acquired by the latter can be found from the formula

$$v = \frac{2M}{m + M}V, \tag{8.70}$$

from which one can see that if $0 \le m \le M$, then $V \le v \le 2V$.

How can a significant part of the kinetic energy of a larger mass be communicated to a body of small mass? To do this, for example, one can insert balls with intermediate masses between the balls of small and large mass: $m < m_1 < m_2 < \cdots < m_n < M$. Let us compute (after Huygens) how the masses $m_1, m_2, \dots, m_n$

should be chosen to that the body m will acquire maximum velocity after successive central collisions.

In accordance with formula (8.70) we obtain the following expression for the required velocity as a function of the variables $m_1, m_2, \ldots, m_n$:

$$v = \frac{m_1}{m + m_1} \cdot \frac{m_2}{m_1 + m_2} \cdot \cdots \cdot \frac{m_n}{m_{n-1} + m_n} \cdot \frac{M}{m_n + M} \cdot 2^{n+1} V. \tag{8.71}$$

Thus Huygens' problem reduces to finding the maximum of the function

$$f(m_1, \ldots, m_n) = \frac{m_1}{m + m_1} \cdot \cdots \cdot \frac{m_n}{m_{n-1} + m_n} \cdot \frac{M}{m_n + M}.$$

The system of equations (8.65), which gives the necessary conditions for an interior extremum, reduces to the following system in the present case:

$$\begin{cases} m \cdot m_2 - m_1^2 = 0, \\ m_1 \cdot m_3 - m_2^2 = 0, \\ \qquad \vdots \\ m_{n-1} \cdot M - m_n^2 = 0, \end{cases}$$

from which it follows that the numbers $m, m_1, \ldots, m_n, M$ form a geometric progression with ratio q equal to $\sqrt[n+1]{M/m}$.

The value of the velocity (8.71) that results from this choice of masses is given by

$$v = \left(\frac{2q}{1+q}\right)^{n+1} V, \tag{8.72}$$

which agrees with (8.70) if $n = 0$.

It is clear from physical considerations that formula (8.72) gives the maximal value of the function (8.71). However, this can also be verified formally (without invoking the cumbersome second derivatives. See Problem 9 at the end of this section).

We remark that it is clear from (8.72) that if $m \to 0$, then $v \to 2^{n+1} V$. Thus the intermediate masses do indeed significantly increase the portion of the kinetic energy of the mass M that is transmitted to the small mass m.

8.4.6 Some Geometric Images Connected with Functions of Several Variables

a. The Graph of a Function and Curvilinear Coordinates

Let x, y, and z be Cartesian coordinates of a point in $\mathbb{R}^3$ and let $z = f(x, y)$ be a continuous function defined in some domain G of the plane $\mathbb{R}^2$ of the variables x and y.

By the general definition of the graph of a function, the graph of the function $f: G \to \mathbb{R}$ in our case is the set $S = \{(x, y, z) \in \mathbb{R}^3 \mid (x, y) \in G, z = f(x, y)\}$ in the space $\mathbb{R}^3$.

It is obvious that the mapping $G \xrightarrow{F} S$ defined by the relation $(x, y) \mapsto (x, y, f(x, y))$ is a continuous one-to-one mapping of G onto S, by which one can determine every point of S by exhibiting the point of G corresponding to it, or, what is the same, giving the coordinates (x, y) of this point of G.

Thus the pairs of numbers $(x, y) \in G$ can be regarded as certain coordinates of the points of a set S – the graph of the function $z = f(x, y)$. Since the points of S are given by pairs of numbers, we shall conditionally agree to call S a *two-dimensional surface in* $\mathbb{R}^3$. (The general definition of a surface will be given later.)

If we define a path $\Gamma : I \to G$ in G, then a path $F \circ \Gamma : I \to S$ automatically appears on the surface S. If $x = x(t)$ and $y = y(t)$ is a parametric definition of the path Γ, then the path $F \circ \Gamma$ on S is given by the three functions $x = x(t)$, $y = y(t)$, $z = z(t) = f(x(t), y(t))$. In particular, if we set $x = x_0 + t$, $y = y_0$, we obtain a curve $x = x_0 + t$, $y = y_0$, $z = f(x_0 + t, y_0)$ on the surface S along which the coordinate $y = y_0$ of the points of S does not change. Similarly one can exhibit a curve $x = x_0$, $y = y_0 + t$, $z = f(x_0, y_0 + t)$ on S along which the first coordinate x_0 of the points of S does not change. By analogy with the planar case these curves on S are naturally called *coordinate lines* on the surface S. However, in contrast to the coordinate lines in $G \subset \mathbb{R}^2$, which are pieces of straight lines, the coordinate lines on S are in general curves in $\mathbb{R}^3$. For that reason, the coordinates (x, y) of points of the surface S are often called *curvilinear coordinates* on S.

Thus the graph of a continuous function $z = f(x, y)$, defined in a domain $G \subset R^2$ is a two-dimensional surface S in $\mathbb{R}^3$ whose points can be defined by curvilinear coordinates $(x, y) \in G$.

At this point we shall not go into detail on the general definition of a surface, since we are interested only in a special case of a surface – the graph of a function. However, we assume that from the course in analytic geometry the reader is well acquainted with some important particular surfaces in $\mathbb{R}^3$ (such as a plane, an ellipsoid, paraboloids, and hyperboloids).

b. The Tangent Plane to the Graph of a Function

Differentiability of a function $z = f(x, y)$ at the point $(x_0, y_0) \in G$ means that

$$f(x, y) = f(x_0, y_0) + A(x - x_0) + B(y - y_0) + \\ + o\left(\sqrt{(x - x_0)^2 + (y - y_0)^2}\right) \quad \text{as } (x, y) \to (x_0, y_0), \tag{8.73}$$

where A and B are certain constants.

In $\mathbb{R}^3$ let us consider the plane

$$z = z_0 + A(x - x_0) + B(y - y_0), \tag{8.74}$$

where $z_0 = f(x_0, y_0)$. Comparing equalities (8.73) and (8.74), we see that the graph of the function is well approximated by the plane (8.74) in a neighborhood of the point (x_0, y_0, z_0). More precisely, the point $(x, y, f(x, y))$ of the graph of the function differs from the point $(x, y, z(x, y))$ of the plane (8.74) by an amount that is infinitesimal in comparison with the magnitude $\sqrt{(x - x_0)^2 + (y - y_0)^2}$ of the displacement of its curvilinear coordinates (x, y) from the coordinates (x_0, y_0) of the point (x_0, y_0, z_0).

By the uniqueness of the differential of a function, the plane (8.74) possessing this property is unique and has the form

$$z = f(x_0, y_0) + \frac{\partial f}{\partial x}(x_0, y_0)(x - x_0) + \frac{\partial f}{\partial y}(x_0, y_0)(y - y_0). \tag{8.75}$$

This plane is called the *tangent plane to the graph of the function* $z = f(x, y)$ *at the point* $(x_0, y_0, f(x_0, y_0))$.

Thus, the differentiability of a function $z = f(x, y)$ at the point (x_0, y_0) and the existence of a tangent plane to the graph of this function at the point $(x_0, y_0, f(x_0, y_0))$ are equivalent conditions.

c. The Normal Vector

Writing Eq. (8.75) for the tangent plane in the canonical form

$$\frac{\partial f}{\partial x}(x_0, y_0)(x - x_0) + \frac{\partial f}{\partial y}(x_0, y_0)(y - y_0) - \big(z - f(x_0, y_0)\big) = 0,$$

we conclude that the vector

$$\left(\frac{\partial f}{\partial x}(x_0, y_0), \frac{\partial f}{\partial y}(x_0, y_0), -1\right) \tag{8.76}$$

is the *normal vector to the tangent plane*. Its direction is considered to be the direction normal or orthogonal to the surface S (the graph of the function) at the point $(x_0, y_0, f(x_0, y_0))$.

In particular, if (x_0, y_0) is a critical point of the function $f(x, y)$, then the normal vector to the graph at the point $(x_0, y_0, f(x_0, y_0))$ has the form $(0, 0, -1)$ and consequently, the tangent plane to the graph of the function at such a point is horizontal (parallel to the xy-plane).

The three graphs in Fig. 8.1 illustrate what has just been said.

Figures 8.1a and c depict the location of the graph of a function with respect to the tangent plane in a neighborhood of a local extremum (minimum and maximum respectively), while Fig. 8.1b shows the graph in the neighborhood of a so-called *saddle point*.

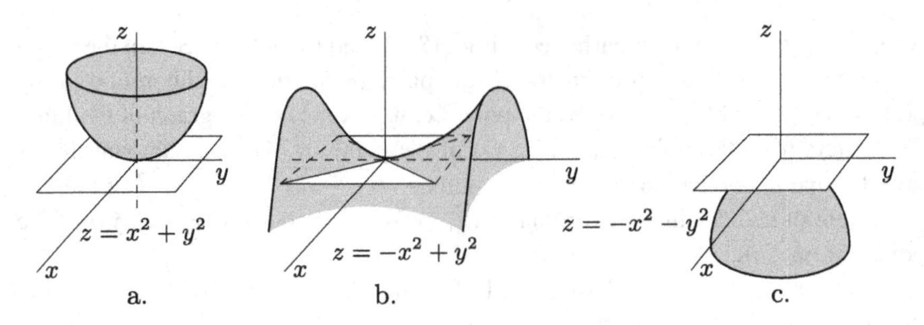

Fig. 8.1

d. Tangent Planes and Tangent Vectors

We know that if a path $\Gamma : I \to \mathbb{R}^3$ in $\mathbb{R}^3$ is given by differentiable functions $x = x(t)$, $y = y(t)$, $z = z(t)$, then the vector $(\dot{x}(0), \dot{y}(0), \dot{z}(0))$ is the velocity vector at time $t = 0$. It is a direction vector of the tangent at the point $x_0 = x(0)$, $y_0 = y(0)$, $z_0 = z(0)$ to the curve in $\mathbb{R}^3$ that is the support of the path Γ.

Now let us consider a path $\Gamma : I \to S$ on the graph of a function $z = f(x, y)$ given in the form $x = x(t)$, $y = y(t)$, $z = f(x(t), y(t))$. In this particular case we find that

$$\big(\dot{x}(0), \dot{y}(0), \dot{z}(0)\big) = \left(\dot{x}(0), \dot{y}(0), \frac{\partial f}{\partial x}(x_0, y_0)\dot{x}(0) + \frac{\partial f}{\partial y}(x_0, y_0)\dot{y}(0)\right),$$

from which it can be seen that this vector is orthogonal to the vector (8.76) normal to the graph S of the function at the point $(x_0, y_0, f(x_0, y_0))$. Thus we have shown that if a vector (ξ, η, ζ) is tangent to a curve on the surface S at the point $(x_0, y_0, f(x_0, y_0))$, then it is orthogonal to the vector (8.76) and (in this sense) lies in the plane (8.75) tangent to the surface S at the point in question. More precisely we could say that the whole line $x = x_0 + \xi t$, $y = y_0 + \eta t$, $z = f(x_0, y_0) + \zeta t$ lies in the tangent plane (8.75).

Let us now show that the converse is also true, that is, if a line $x = x_0 + \xi t$, $y = y_0 + \eta t$, $z = f(x_0, y_0) + \zeta t$, or what is the same, the vector (ξ, η, ζ), lies in the plane (8.75), then there is a path on S for which the vector (ξ, η, ζ) is the velocity vector at the point $(x_0, y_0, f(x_0, y_0))$.

The path can be taken, for example, to be

$$x = x_0 + \xi t, \qquad y = y_0 + \eta t, \qquad z = f(x_0 + \xi t, y_0 + \eta t).$$

In fact, for this path,

$$\dot{x}(0) = \xi, \qquad \dot{y}(0) = \eta, \qquad \dot{z}(0) = \frac{\partial f}{\partial x}(x_0, y_0)\xi + \frac{\partial f}{\partial y}(x_0, y_0)\eta.$$

In view of the equality

$$\frac{\partial f}{\partial x}(x_0, y_0)\dot{x}(0) + \frac{\partial f}{\partial y}(x_0, y_0)\dot{y}(0) - \dot{z}(0) = 0$$

Fig. 8.2

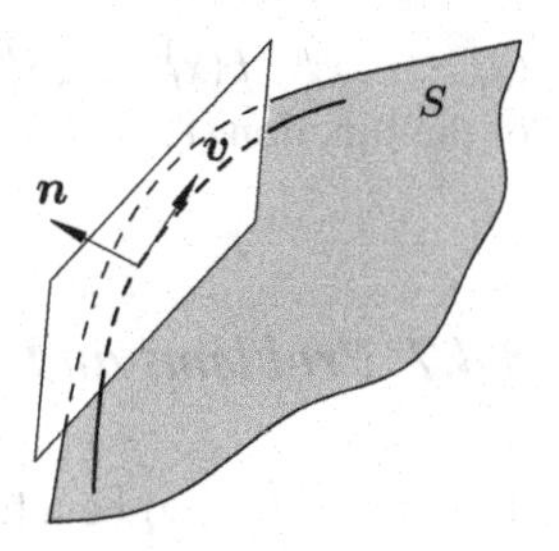

and the hypothesis that

$$\frac{\partial f}{\partial x}(x_0, y_0)\xi + \frac{\partial f}{\partial y}(x_0, y_0)\eta - \zeta = 0.$$

We conclude that

$$\big(\dot{x}(0), \dot{y}(0), \dot{z}(0)\big) = (\xi, \eta, \zeta).$$

Hence the tangent plane to the surface S at the point (x_0, y_0, z_0) is formed by the vectors that are tangents at the point (x_0, y_0, z_0) to curves on the surface S passing through the point (see Fig. 8.2).

This is a more geometric description of the tangent plane. In any case, one can see from it that if the tangent to a curve is invariantly defined (with respect to the choice of coordinates), then the tangent plane is also invariantly defined.

We have been considering functions of two variables for the sake of visualizability, but everything that was said obviously carries over to the general case of a function

$$y = f(x^1, \dots, x^m) \tag{8.77}$$

of m variables, where $m \in \mathbb{N}$.

At the point $(x_0^1, \dots, x_0^m, f(x_0^1, \dots, x_0^m))$ the plane tangent to the graph of such a function can be written in the form

$$y = f(x_0^1, \dots, x_0^m) + \sum_{i=1}^{m} \frac{\partial f}{\partial x^i}(x_0^1, \dots, x_0^m)(x^i - x_0^i); \tag{8.78}$$

the vector

$$\left(\frac{\partial f}{\partial x^1}(x_0), \dots, \frac{\partial f}{\partial x^m}(x_0), -1\right)$$

is the normal vector to the plane (8.78). This plane itself, like the graph of the function (8.77), has dimension m, that is, any point is now given by a set $(x^1, \dots, x^m)$ of m coordinates.

Thus, Eq. (8.78) defines a hyperplane in $\mathbb{R}^{m+1}$.

Repeating verbatim the reasoning above, one can verify that the tangent plane (8.78) consists of vectors that are tangent to curves passing through the point

$(x_0^1, \dots, x_0^m, f(x_0^1, \dots, x_0^m))$ and lying on the m-dimensional surface S – the graph of the function (8.77).

8.4.7 Problems and Exercises

1. Let $z = f(x, y)$ be a function of class $C^{(1)}(G; \mathbb{R})$.

a) If $\frac{\partial f}{\partial y}(x, y) \equiv 0$ in G, can one assert that f is independent of y in G?

b) Under what condition on the domain G does the preceding question have an affirmative answer?

2. a) Verify that for the function

$$f(x, y) = \begin{cases} xy\frac{x^2-y^2}{x^2+y^2}, & \text{if } x^2 + y^2 \neq 0, \\ 0, & \text{if } x^2 + y^2 = 0, \end{cases}$$

the following relations hold:

$$\frac{\partial f}{\partial x \partial y}(0, 0) = 1 \neq -1 = \frac{\partial^2 f}{\partial y \partial x}(0, 0).$$

b) Prove that if the function $f(x, y)$ has partial derivatives $\frac{\partial f}{\partial x}$ and $\frac{\partial f}{\partial y}$ in some neighborhood U of the point (x_0, y_0), and if the mixed derivative $\frac{\partial^2 f}{\partial x \partial y}$ (or $\frac{\partial^2 f}{\partial y \partial x}$) exists in U and is continuous at (x_0, y_0), then the mixed derivative $\frac{\partial^2 f}{\partial y \partial x}$ (resp. $\frac{\partial^2 f}{\partial x \theta y}$) also exists at that point and the following equality holds:

$$\frac{\partial^2 f}{\partial x \partial y}(x_0, y_0) = \frac{\partial^2 f}{\partial y \partial x}(x_0, y_0).$$

3. Let $x^1, \dots, x^m$ be Cartesian coordinates in $\mathbb{R}^m$. The differential operator

$$\Delta = \sum_{i=1}^{m} \frac{\partial^2}{\partial x^{i2}},$$

acting on functions $f \in C^{(2)}(G; \mathbb{R})$ according to the rule

$$\Delta f = \sum_{i=1}^{m} \frac{\partial^2 f}{\partial x^{i2}}(x^1, \dots, x^m),$$

is called the *Laplacian*.

The equation $\Delta f = 0$ for the function f in the domain $G \subset \mathbb{R}^m$ is called *Laplace's equation*, and its solutions are called *harmonic functions in the domain G*.

a) Show that if $x = (x^1, \dots, x^m)$ and

$$\|x\| = \sqrt{\sum_{i=1}^{m} (x^i)^2},$$

then for $m > 2$ the function

$$f(x) = \|x\|^{-\frac{2-m}{2}}$$

is harmonic in the domain $\mathbb{R}^m \setminus 0$, where $0 = (0, \dots, 0)$.

b) Verify that the function

$$f(x^1, \dots, x^m, t) = \frac{1}{(2a\sqrt{\pi t})^m} \cdot \exp\left(-\frac{\|x\|^2}{4a^2 t}\right),$$

which is defined for $t > 0$ and $x = (x^1, \dots, x^m) \in \mathbb{R}^m$, satisfies the *heat equation*

$$\frac{\partial f}{\partial t} = a^2 \Delta f,$$

that is, verify that $\frac{\partial f}{\partial t} = a^2 \sum_{i=1}^{m} \frac{\partial^2 f}{\partial x^{i2}}$ at each point of the domain of definition of the function.

4. *Taylor's formula in multi-index notation*. The symbol $\alpha := (\alpha_1, \dots, \alpha_m)$ consisting of nonnegative integers α_i, $i = 1, \dots, m$, is called the *multi-index* α.

The following notation is conventional:

$$|\alpha| := \alpha_1 + \dots + \alpha_m,$$

$$\alpha! := \alpha_1! \cdots \alpha_m!;$$

finally, if $a = (a_1, \dots, a_m)$, then

$$a^\alpha := a_1^{\alpha_1} \cdots a_m^{\alpha_m}.$$

a) Verify that if $k \in \mathbb{N}$, then

$$(a_1 + \dots + a_m)^k = \sum_{|\alpha|=k} \frac{k!}{\alpha_1! \cdots \alpha_m!} a_1^{\alpha_1} \cdots a_m^{\alpha_m},$$

or

$$(a_1 + \dots + a_m)^k = \sum_{|\alpha|=k} \frac{k!}{\alpha!} a^\alpha,$$

where the summation extends over all sets $\alpha = (\alpha_1, \dots, \alpha_m)$ of nonnegative integers such that $\sum_{i=1}^{m} \alpha_i = k$.

b) Let

$$D^\alpha f(x) := \frac{\partial^{|\alpha|} f}{(\partial x^1)^{\alpha_1} \cdots (\partial x^m)^{\alpha_m}}(x).$$

Show that if $f \in C^{(k)}(G; \mathbb{R})$, then the equality

$$\sum_{i_1+\cdots+i_m=k} \partial_{i_1 \cdots i_k} f(x) h^{i_1} \cdots h^{i_k} = \sum_{|\alpha|=k} \frac{k!}{\alpha!} D^\alpha f(x) h^\alpha,$$

where $h = (h^1, \ldots, h^m)$, holds at any point $x \in G$.

c) Verify that in multi-index notation Taylor's theorem with the Lagrange form of the remainder, for example, can be written as

$$f(x+h) = \sum_{|\alpha|=0}^{n-1} \frac{1}{\alpha!} D^\alpha f(x) h^\alpha + \sum_{|\alpha|=n} \frac{1}{\alpha!} D^\alpha f(x + \theta h) h^\alpha.$$

d) Write Taylor's formula in multi-index notation with the integral form of the remainder (Theorem 4).

5. a) Let $I^m = \{x = (x^1, \ldots, x^m) \in \mathbb{R}^m \mid |x^i| \le c^i, i = 1, \ldots, m\}$ be an m-dimensional closed interval and I a closed interval $[a, b] \subset \mathbb{R}$. Show that if the function $f(x, y) = f(x^1, \ldots, x^m, y)$ is defined and continuous on the set $I^m \times I$, then for any positive number $\varepsilon > 0$ there exists a number $\delta > 0$ such that $|f(x, y_1) - f(x, y_2)| < \varepsilon$ if $x \in I^m$, $y_1, y_2 \in I$, and $|y_1 - y_2| < \delta$.

b) Show that the function

$$F(x) = \int_a^b f(x, y)\, dy$$

is defined and continuous on the closed interval I^m.

c) Show that if $f \in C(I^m; \mathbb{R})$, then the function

$$\mathcal{F}(x, t) = f(tx)$$

is defined and continuous on $I^m \times I^1$, where $I^1 = \{t \in \mathbb{R} \mid |t| \le 1\}$.

d) Prove *Hadamard's lemma*:

If $f \in C^{(1)}(I^m; \mathbb{R})$ and $f(0) = 0$, there exist functions $g_1, \ldots, g_m \in C(I^m; \mathbb{R})$ such that

$$f(x^1, \ldots, x^m) = \sum_{i=1}^m x^i g_i(x^1, \ldots, x^m)$$

in I^m, and in addition

$$g_i(0) = \frac{\partial f}{\partial x^i}(0), \quad i = 1, \ldots, m.$$

6. Prove the following *generalization of Rolle's theorem for functions of several variables.*

If the function f is continuous in a closed ball $\overline{B}(0; r)$, equal to zero on the boundary of the ball, and differentiable in the open ball $B(0; r)$, then at least one of the points of the open ball is a critical point of the function.

7. Verify that the function

$$f(x, y) = \bigl(y - x^2\bigr)\bigl(y - 3x^2\bigr)$$

does not have an extremum at the origin, even though its restriction to each line passing through the origin has a strict local minimum at that point.

8. *The method of least squares.* This is one of the commonest methods of processing the results of observations. It consists of the following. Suppose it is known that the physical quantities x and y are linearly related:

$$y = ax + b \tag{8.79}$$

or suppose an empirical formula of this type has been constructed on the basis of experimental data.

Let us assume that n observations have been made, in each of which both x and y were measured, resulting in n pairs of values $x_1, y_1; \dots; x_n, y_n$. Since the measurements have errors, even if the relation (8.79) is exact, the equalities

$$y_k = ax_k + b$$

may fail to hold for some of the values of $k \in \{1, \dots, n\}$, no matter what the coefficients a and b are.

The problem is to determine the unknown coefficients a and b in a reasonable way from these observational results.

Basing his argument on analysis of the probability distribution of the magnitude of observational errors, Gauss established that the most probable values for the coefficients a and b with a given set of observational results should be sought by use of the following *least-squares principle*:

If $\delta_k = (ax_k + b) - y_k$ is the discrepancy in the kth observation, then a and b should be chosen so that the quantity

$$\Delta = \sum_{k=1}^{n} \delta_k^2,$$

that is, the sum of the squares of the discrepancies, has a minimum.

a) Show that the least-squares principle for relation (8.79) leads to the following system of linear equations

$$\begin{cases} [x_k, x_k]a + [x_k, 1]b = [x_k, y_k], \\ [1, x_k]a + [1, 1]b = [1, y_k], \end{cases}$$

Table 8.1

Temperature, °C	Frequency, %	Temperature, °C	Frequency, %
0	39	20	136
5	54	25	182
10	74	30	254
15	100		

for determining the coefficients a and b. Here, following Gauss, we write $[x_k, x_k] := x_1x_1 + \cdots + x_nx_n$, $[x_k, 1] := x_1 \cdot 1 + \cdots + x_n \cdot 1$, $[x_k, y_k] := x_1y_1 + \cdots + x_ny_n$, and so forth.

b) Write the system of equations for the numbers $a_1, \dots, a_m, b$ to which the least-squares principle leads when Eq. (8.79) is replaced by the relation

$$y = \sum_{i=1}^{m} a_i x^i + b,$$

(or, more briefly, $y = a_i x^i + b$) between the quantities $x^1, \dots, x^m$ and y.

c) How can the method of least squares be used to find empirical formulas of the form

$$y = c x_1^{\alpha_1} \cdots x_n^{\alpha_n}$$

connecting physical quantities $x_1, \dots, x_m$ with the quantity y?

d) (M. Germain) The frequency R of heart contractions was measured at different temperatures T in several dozen specimens of *Nereis diversicolor*. The frequencies were expressed in percents relative to the contraction frequency at 15 °C. The results are given in Table 8.1.

The dependence of R on T appears to be exponential. Assuming $R = Ae^{bT}$, find the values of the constants A and b that best fit the experimental results.

9. a) Show that in Huygens' problem, studied in Example 5, the function (8.71) tends to zero if at least one of the variables $m_1, \dots, m_n$ tends to infinity.

b) Show that the function (8.71) has a maximum point in $\mathbb{R}^n$ and hence the unique critical point of that function in $\mathbb{R}^n$ must be its maximum.

c) Show that the quantity v defined by formula (8.72) is monotonically increasing as n increases and find its limit as $n \to \infty$.

10. a) During so-called exterior disk grinding the grinding tool – a rapidly rotating grinding disk (with an abrasive rim) that acts as a file – is brought into contact with the surface of a circular machine part that is rotating slowly compared with the disk (see Fig. 8.3).

The disk K is gradually pressed against the machine part D, causing a layer H of metal to be removed, reducing the part to the required size and producing a smooth working surface for the device. In the machine where it will be placed this surface will usually be a working surface. In order to extend its working life, the metal of the machine part is subjected to a preliminary annealing to harden the steel. However,

Fig. 8.3

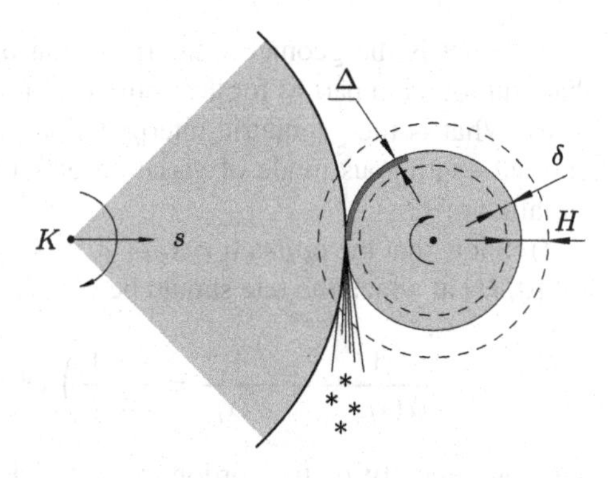

because of the high temperature in the contact zone between the machine part and the grinding disk, structural changes can (and frequently do) occur in a certain layer Δ of metal in the machine part, resulting in decreased hardness of the steel in that layer. The quantity Δ is a monotonic function of the rate s at which the disk is applied to the machine part, that is, $\Delta = \varphi(s)$. It is known that there is a certain critical rate $s_0 > 0$ at which the relation $\Delta = 0$ still holds, while $\Delta > 0$ whenever $s > s_0$. For the following discussion it is convenient to introduce the relation

$$s = \psi(\Delta)$$

inverse to the one just given. This new relation is defined for $\Delta > 0$.

Here ψ is a monotonically increasing function known experimentally, defined for $\Delta \geq 0$, and $\psi(0) = s_0 > 0$.

The grinding process must be carried out in such a way that there are no structural changes in the metal on the surface eventually produced.

In terms of rapidity, the optimal grinding mode under these conditions would obviously be a set of variations in the rate s of application of the grinding disk for which

$$s = \psi(\delta),$$

where $\delta = \delta(t)$ is the thickness of the layer of metal not yet removed up to time t, or, what is the same, the distance from the rim of the disk at time t to the final surface of the device being produced. Explain this.

b) Find the time needed to remove a layer of thickness H when the rate of application of the disk is optimally adjusted.

c) Find the dependence $s = s(t)$ of the rate of application of the disk on time in the optimal mode under the condition that the function $\Delta \overset{\psi}{\longmapsto} s$ is linear: $s = s_0 + \lambda\Delta$.

Due to the structural properties of certain kinds of grinding lathes, the rate s can undergo only discrete changes. This poses the problem of optimizing the productivity of the process under the additional condition that only a fixed number n of switches in the rate s are allowed. The answers to the following questions give a picture of the optimal mode.

d) What is the geometric interpretation of the grinding time $t(H) = \int_0^H \frac{d\delta}{\psi(\delta)}$ that you found in part b) for the optimal continuous variation of the rate s?

e) What is the geometric interpretation of the time lost in switching from the optimal continuous mode of variation of s to the time-optimal stepwise mode of variation of s?

f) Show that the points $0 = x_{n+1} < x_n < \cdots < x_1 < x_0 = H$ of the closed interval $[0, H]$ at which the rate should be switched must satisfy the conditions

$$\frac{1}{\psi(x_{i+1})} - \frac{1}{\psi(x_i)} = -\left(\frac{1}{\psi}\right)'(x_i)(x_i - x_{i-1}) \quad (i = 1, \dots, n)$$

and consequently, on the portion from x_i to x_{i+1}, the rate of application of the disk has the form $s = \psi(x_{i+1})$ $(i = 0, \dots, n)$.

g) Show that in the linear case, when $\psi(\Delta) = s_0 + \lambda\Delta$, the points x_i (in part f)) on the closed interval $[0, H]$ are distributed so that the numbers

$$\frac{s_0}{\lambda} < \frac{s_0}{\lambda} + x_n < \cdots < \frac{s_0}{\lambda} + x_1 < \frac{s_0}{\lambda} + H$$

form a geometric progression.

11. a) Verify that the tangent to a curve $\Gamma : I \to \mathbb{R}^m$ is defined invariantly relative to the choice of coordinate system in $\mathbb{R}^m$.

b) Verify that the tangent plane to the graph S of a function $y = f(x^1, \dots, x^m)$ is defined invariantly relative to the choice of coordinate system in $\mathbb{R}^m$.

c) Suppose the set $S \subset \mathbb{R}^m \times \mathbb{R}^1$ is the graph of a function $y = f(x^1, \dots, x^m)$ in coordinates $(x^1, \dots, x^m, y)$ in $\mathbb{R}^m \times \mathbb{R}^1$ and the graph of a function $\tilde{y} = \tilde{f}(\tilde{x}^1, \dots, \tilde{x}^m)$ in coordinates $(\tilde{x}^1, \dots, \tilde{x}^m, \tilde{y})$ in $\mathbb{R}^m \times \mathbb{R}^1$. Verify that the tangent plane to S is invariant relative to a linear change of coordinates in $\mathbb{R}^m \times \mathbb{R}^1$.

d) Verify that the Laplacian $\Delta f = \sum_{i=1}^m \frac{\partial^2 f}{\partial x^{i2}}(x)$ is defined invariantly relative to orthogonal coordinate transformations in $\mathbb{R}^m$.

8.5 The Implicit Function Theorem

8.5.1 Statement of the Problem and Preliminary Considerations

In this section we shall prove the implicit function theorem, which is important both intrinsically and because of its numerous applications.

Let us begin by explaining the problem. Suppose, for example, we have the relation

$$x^2 + y^2 - 1 = 0 \tag{8.80}$$

between the coordinates x, y of points in the plane $\mathbb{R}^2$. The set of all points of $\mathbb{R}^2$ satisfying this condition is the unit circle (Fig. 8.4).

Fig. 8.4

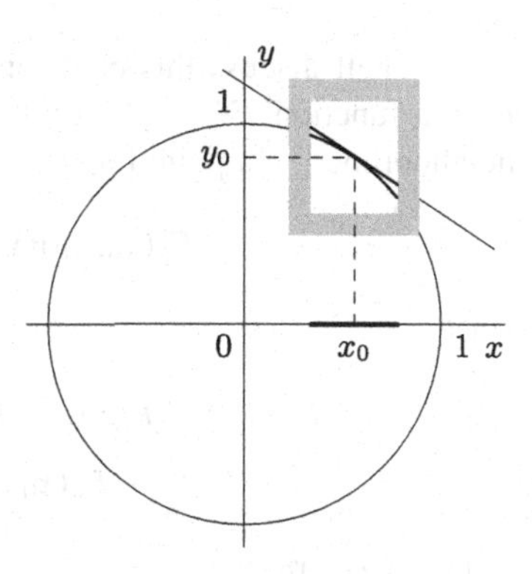

The presence of the relation (8.80) shows that after fixing one of the coordinates, for example, x, we can no longer choose the second coordinate arbitrarily. Thus relation (8.80) determines the dependence of y on x. We are interested in the question of the conditions under which the implicit relation (8.80) can be solved as an explicit functional dependence $y = y(x)$.

Solving Eq. (8.80) with respect to y, we find that

$$y = \pm\sqrt{1 - x^2}, \tag{8.81}$$

that is, to each value of x such that $|x| < 1$, there are actually two admissible values of y. In forming a functional relation $y = y(x)$ satisfying relation (8.80) one cannot give preference to either of the values (8.81) without invoking additional requirements. For example, the function $y(x)$ that assumes the value $+\sqrt{1 - x^2}$ at rational points of the closed interval $[-1, 1]$ and the value $-\sqrt{1 - x^2}$ at irrational points obviously satisfies (8.80).

It is clear that one can create infinitely many functional relations satisfying (8.80) by varying this example.

The question whether the set defined in $\mathbb{R}^2$ by (8.80) is the graph of a function $y = y(x)$ obviously has a negative answer, since from the geometric point of view it is equivalent to the question whether it is possible to establish a one-to-one direct projection of a circle into a line.

But observation (see Fig. 8.4) suggests that nevertheless, in a neighborhood of a particular point (x_0, y_0) the arc projects in a one-to-one manner into the x-axis, and that it can be represented uniquely as $y = y(x)$, where $x \mapsto y(x)$ is a continuous function defined in a neighborhood of the point x_0 and assuming the value y_0 at x_0. In this aspect, the only bad points are $(-1, 0)$ and $(1, 0)$, since no arc of the circle having them as interior points projects in a one-to-one manner into the x-axis. Even so, neighborhoods of these points on the circle are well situated relative to the y-axis, and can be represented as the graph of a function $x = x(y)$ that is continuous in a neighborhood of the point 0 and assumes the value -1 or 1 according as the arc in question contains the point $(-1, 0)$ or $(1, 0)$.

How is it possible to find out analytically when a geometric locus of points defined by a relation of the type (8.80) can be represented in the form of an explicit function $y = y(x)$ or $x = x(y)$ in a neighborhood of a point (x_0, y_0) on the locus?

We shall discuss this question using the following, now familiar, method. We have a function $F(x, y) = x^2 + y^2 - 1$. The local behavior of this function in a neighborhood of a point (x_0, y_0) is well described by its differential

$$F'_x(x_0, y_0)(x - x_0) + F'_y(x_0, y_0)(y - y_0),$$

since

$$F(x, y) = F(x_0, y_0) + F'_x(x_0, y_0)(x - x_0) + \\ + F'_y(x_0, y_0)(y - y_0) + o\big(|x - x_0| + |y - y_0|\big)$$

as $(x, y) \to (x_0, y_0)$.

If $F(x_0, y_0) = 0$ and we are interested in the behavior of the level curve

$$F(x, y) = 0$$

of the function in a neighborhood of the point (x_0, y_0), we can judge that behavior from the position of the (tangent) line

$$F'_x(x_0, y_0)(x - x_0) + F'_y(x_0, y_0)(y - y_0) = 0. \tag{8.82}$$

If this line is situated so that its equation can be solved with respect to y, then, since the curve $F(x, y) = 0$ differs very little from this line in a neighborhood of the point (x_0, y_0), we may hope that it also can be written in the form $y = y(x)$ in some neighborhood of the point (x_0, y_0).

The same can be said about local solvability of $F(x, y) = 0$ with respect to x.

Writing Eq. (8.82) for the specific relation (8.80), we obtain the following equation for the tangent line:

$$x_0(x - x_0) + y_0(y - y_0) = 0.$$

This equation can always be solved for y when $y_0 \neq 0$, that is, at all points of the circle (8.80) except $(-1, 0)$ and $(1, 0)$. It is solvable with respect to x at all points of the circle except $(0, -1)$ and $(0, 1)$.

8.5.2 An Elementary Version of the Implicit Function Theorem

In this section we shall obtain the implicit function theorem by a very intuitive, but not very constructive method, one that is adapted only to the case of real-valued functions of real variables. The reader can become familiar with another method of obtaining this theorem, one that is in many ways preferable, and with a more detailed analysis of its structure in Chap. 10 (Part 2), and also in Problem 4 at the end of the section.

The following proposition is an elementary version of the implicit function theorem.

Proposition 1 *If the function* $F : U(x_0, y_0) \to \mathbb{R}$ *defined in a neighborhood* $U(x_0, y_0)$ *of the point* $(x_0, y_0) \in \mathbb{R}^2$ *is such that*

1^0 $F \in C^{(p)}(U; \mathbb{R})$, *where* $p \geq 1$,
2^0 $F(x_0, y_0) = 0$,
3^0 $F'_y(x_0, y_0) \neq 0$,

then there exist a two-dimensional interval $I = I_x \times I_y$ *where*

$$I_x = \{x \in \mathbb{R} \mid |x - x_0| < \alpha\}, \qquad I_y = \{y \in \mathbb{R} \mid |y - y_0| < \beta\},$$

that is a neighborhood of the point (x_0, y_0) *contained in* $U(x_0, y_0)$, *and a function* $f \in C^{(p)}(I_x; I_y)$ *such that*

$$F(x, y) = 0 \Leftrightarrow y = f(x), \tag{8.83}$$

for any point $(x, y) \in I_x \times I_y$ *and the derivative of the function* $y = f(x)$ *at the points* $x \in I_x$ *can be computed from the formula*

$$f'(x) = -\big[F'_y\big(x, f(x)\big)\big]^{-1}\big[F'_x\big(x, f(x)\big)\big]. \tag{8.84}$$

Before taking up the proof, we shall give several possible reformulations of the conclusion (8.83), which should bring out the meaning of the relation itself.

Proposition 1 says that under hypotheses 1^0, 2^0, and 3^0 the portion of the set defined by the relation $F(x, y) = 0$ that belongs to the neighborhood $I_x \times I_y$ of the point (x_0, y_0) is the graph of a function $f : I_x \to I_y$ of class $C^{(p)}(I_x; I_y)$.

In other words, one can say that inside the neighborhood I of the point (x_0, y_0) the equation $F(x, y) = 0$ has a unique solution for y, and the function $y = f(x)$ is that solution, that is, $F(x, f(x)) \equiv 0$ on I_x.

It follows in turn from this that if $y = \tilde{f}(x)$ is a function defined on I_x that is known to satisfy the relation $F(x, \tilde{f}(x)) \equiv 0$ on I_x, $\tilde{f}(x_0) = y_0$, and this function is continuous at the point $x_0 \in I_x$, then there exists a neighborhood $\Delta \subset I_x$ of x_0 such that $\tilde{f}(\Delta) \subset I_y$, and then $\tilde{f}(x) \equiv f(x)$ for $x \in \Delta$.

Without the assumption that the function $\tilde{f}$ is continuous at the point x_0 and the condition $\tilde{f}(x_0) = y_0$, this last conclusion could turn out to be incorrect, as can be seen from the example of the circle already studied.

Let us now prove Proposition 1.

Proof Suppose for definiteness that $F'_y(x_0, y_0) > 0$. Since $F \in C^{(1)}(U; \mathbb{R})$, it follows that $F'_y(x, y) > 0$ also in some neighborhood of (x_0, y_0). In order to avoid introducing new notation, we can assume without loss of generality that $F'_y(x, y) > 0$ at every point of the original neighborhood $U(x_0, y_0)$.

Moreover, shrinking the neighborhood $U(x_0, y_0)$ if necessary, we can assume that it is a disk of radius $r = 2\beta > 0$ with center at (x_0, y_0).

Since $F'_y(x, y) > 0$ in U, the function $F(x_0, y)$ is defined and monotonically increasing as a function of y on the closed interval $y_0 - \beta \leq y \leq y_0 + \beta$. Consequently,

$$F(x_0, y_0 - \beta) < F(x_0, y_0) = 0 < F(x_0, y_0 + \beta).$$

By the continuity of the function F in U, there exists a positive number $\alpha < \beta$ such that the relations

$$F(x, y_0 - \beta) < 0 < F(x, y_0 + \beta)$$

hold for $|x - x_0| \le \alpha$.

We shall now show that the rectangle $I = I_x \times I_y$, where

$$I_x = \{x \in \mathbb{R} \mid |x - x_0| < \alpha\}, \qquad I_y = \{y \in \mathbb{R} \mid |y - y_0| < \beta\},$$

is the required two-dimensional interval in which relation (8.83) holds.

For each $x \in I_x$ we fix the vertical closed interval with endpoints $(x, y_0 - \beta)$, $(x, y_0 + \beta)$. Regarding $F(x, y)$ as a function of y on that closed interval, we obtain a strictly increasing continuous function that assumes values of opposite sign at the endpoints of the interval. Consequently, for each $x \in I_x$, there is a unique point $y(x) \in I_y$ such that $F(x, y(x)) = 0$. Setting $y(x) = f(x)$, we arrive at relation (8.83).

We now establish that $f \in C^{(p)}(I_x; I_y)$.

We begin by showing that the function f is continuous at x_0 and that $f(x_0) = y_0$. This last equality obviously follows from the fact that for $x = x_0$ there is a unique point $y(x_0) \in I_y$ such that $F(x_0, y(x_0)) = 0$. At the same time, $F(x_0, y_0) = 0$, and so $f(x_0) = y_0$.

Given a number ε, $0 < \varepsilon < \beta$, we can repeat the proof of the existence of the function $f(x)$ and find a number δ, $0 < \delta < \alpha$ such that in the two-dimensional interval $\tilde{I} = \tilde{I}_x \times \tilde{I}_y$, where

$$\tilde{I}_x = \{x \in \mathbb{R} \mid |x - x_0| < \delta\}, \qquad \tilde{I}_y = \{y \in \mathbb{R} \mid |y - y_0| < \varepsilon\},$$

the relation

$$\bigl(F(x, y) = 0 \text{ in } \tilde{I}\bigr) \Leftrightarrow \bigl(y = \tilde{f}(x), x \in \tilde{I}_x\bigr) \tag{8.85}$$

holds with a new function $\tilde{f} : \tilde{I}_x \to \tilde{I}_y$.

But $\tilde{I}_x \subset I_x$, $\tilde{I}_y - \subset I_y$, and $\tilde{I} \subset I$, and therefore it follows from (8.83) and (8.85) that $\tilde{f}(x) \equiv f(x)$ for $x \in \tilde{I}_x \subset I_x$. We have thus verified that $|f(x) - f(x_0)| = |f(x) - y_0| < \varepsilon$ for $|x - x_0| < \delta$.

We have now established that the function f is continuous at the point x_0. But any point $(x, y) \in I$ at which $F(x, y) = 0$ can also be taken as the initial point of the construction, since conditions 2^0 and 3^0 hold at that point. Carrying out that construction inside the interval I, we would once again arrive via (8.83) at the corresponding part of the function f considered in a neighborhood of x. Hence the function f is continuous at x. Thus we have established that $f \in C(I_x; I_y)$.

We shall now show that $f \in C^{(1)}(I_x; I_y)$ and establish formula (8.84).

Let the number Δx be such that $x + \Delta x \in I_x$. Let $y = f(x)$ and $y + \Delta y = f(x + \Delta x)$. Applying the mean-value theorem to the function $F(x, y)$ inside the interval I, we find that

$$
\begin{aligned}
0 &= F\big(x+\Delta x, f(x+\Delta x)\big) - F\big(x, f(x)\big) = \\
&= F(x+\Delta x, y+\Delta y) - F(x,y) = \\
&= F'_x(x+\theta\Delta x, y+\theta\Delta y)\Delta x + F'_y(x+\theta\Delta x, y+\theta\Delta y)\Delta y \quad (0<\theta<1),
\end{aligned}
$$

from which, taking account of the relation $F'_y(x,y) \neq 0$ in I, we obtain

$$
\frac{\Delta y}{\Delta x} = -\frac{F'_x(x+\theta\Delta x, y+\theta\Delta y)}{F'_y(x+\theta\Delta x, y+\theta\Delta y)}. \tag{8.86}
$$

Since $f \in C(I_x; I_y)$, it follows that $\Delta y \to 0$ as $\Delta x \to 0$, and, taking account of the relation $F \in C^{(1)}(U; \mathbb{R})$, as $\Delta x \to 0$ in (8.86), we obtain

$$
f'(x) = -\frac{F'_x(x,y)}{F'_y(x,y)},
$$

where $y = f(x)$. Thus formula (8.84) is now established.

By the theorem on continuity of composite functions, it follows from formula (8.84) that $f \in C^{(1)}(I_x; I_y)$.

If $F \in C^{(2)}(U; \mathbb{R})$, the right-hand side of formula (8.84) can be differentiated with respect to x, and we find that

$$
f''(x) = -\frac{[F''_{xx} + F''_{xy} \cdot f'(x)]F'_y - F'_x[F''_{xy} + F''_{yy} \cdot f'(x)]}{(F'_y)^2}, \tag{8.84'}
$$

where F'_x, F'_y, F''_{xx}, F''_{xy}, and F''_{yy} are all computed at the point $(x, f(x))$.

Thus $f \in C^{(2)}(I_x; I_y)$ if $F \in C^{(2)}(U; \mathbb{R})$. Since the order of the derivatives of f on the right-hand side of (8.84), (8.84′), and so forth, is one less than the order on the left-hand side of the equality, we find by induction that $f \in C^{(p)}(I_x; I_y)$ if $F \in C^{(p)}(U, \mathbb{R})$. □

Example 1 Let us return to relation (8.80) studied above, which defines a circle in $\mathbb{R}^2$, and verify Proposition 1 on this example.

In this case

$$
F(x,y) = x^2 + y^2 - 1,
$$

and it is obvious that $F \in C^{(\infty)}(\mathbb{R}^2; \mathbb{R})$. Next,

$$
F'_x(x,y) = 2x, \qquad F'_y(x,y) = 2y,
$$

so that $F'_y(x,y) \neq 0$ if $y \neq 0$. Thus, by Proposition 1, for any point (x_0, y_0) of this circle different from the points $(-1,0)$ and $(1,0)$ there is a neighborhood such that the arc of the circle contained in that neighborhood can be written in the form $y = f(x)$. Direct computation confirms this, and $f(x) = \sqrt{1-x^2}$ or $f(x) = -\sqrt{1-x^2}$.

Next, by Proposition 1,

$$f'(x_0) = -\frac{F'_x(x_0, y_0)}{F'_y(x_0, y_0)} = -\frac{x_0}{y_0}. \tag{8.87}$$

Direct computation yields

$$f'(x) = \begin{cases} -\frac{x}{\sqrt{1-x^2}}, & \text{if } f(x) = \sqrt{1-x^2}, \\ \frac{x}{\sqrt{1-x^2}}, & \text{if } f(x) = -\sqrt{1-x^2}, \end{cases}$$

which can be written as the single expression

$$f'(x) = -\frac{x}{f(x)} = -\frac{x}{y},$$

and computation with it leads to the same result,

$$f'(x_0) = -\frac{x_0}{y_0},$$

as computation from formula (8.87) obtained from Proposition 1.

It is important to note that formula (8.84) or (8.87) makes it possible to compute $f'(x)$ without even having an explicit expression for the relation $y = f(x)$, if only we know that $f(x_0) = y_0$. The condition $y_0 = f(x_0)$ must be prescribed, however, in order to distinguish the portion of the level curve $F(x, y) = 0$ that we intend to describe in the form $y = f(x)$.

It is clear from the example of the circle that giving only the coordinate x_0 does not determine an arc of the circle, and only after fixing y_0 have we distinguished one of the two possible arcs in this case.

8.5.3 Transition to the Case of a Relation $F(x^1, \dots, x^m, y) = 0$

The following proposition is a simple generalization of Proposition 1 to the case of a relation $F(x^1, \dots, x^m, y) = 0$.

Proposition 2 *If a function $F : U \to \mathbb{R}$ defined in a neighborhood $U \subset \mathbb{R}^{m+1}$ of the point $(x_0, y_0) = (x_0^1, \dots, x_0^m, y_0) \in \mathbb{R}^{m+1}$ is such that*

1^0 $F \in C^{(p)}(U; \mathbb{R})$, $p \geq 1$,
2^0 $F(x_0, y_0) = F(x_0^1, \dots, x_0^m, y_0) = 0$,
3^0 $F'_y(x_0, y_0) = F'_y(x_0^1, \dots, x_0^m, y_0) \neq 0$,

then there exists an $(m+1)$-dimensional interval $I = I_x^m \times I_y^1$, where

$$I_x^m = \{x = (x^1, \dots, x^m) \in \mathbb{R}^m \mid |x^i - x_0^i| < \alpha^i,\ i = 1, \dots, m\},$$

$$I_y^1 = \{y \in \mathbb{R} \mid |y - y_0| < \beta\},$$

which is a neighborhood of the point (x_0, y_0) *contained in* U, *and a function* $f \in C^{(p)}(I_x^m; I_y^1)$ *such that for any point* $(x, y) \in I_x^m \times I_y^1$

$$F(x^1, \ldots, x^m, y) = 0 \Leftrightarrow y = f(x^1, \ldots, x^m), \tag{8.88}$$

and the partial derivatives of the function $y \in f(x^1, \ldots, x^m)$ *at the points of* I_x *can be computed from the formula*

$$\frac{\partial f}{\partial x^i}(x) = -\left[F'_y(x, f(x))\right]^{-1}\left[F'_x(x, f(x))\right]. \tag{8.89}$$

Proof The proof of the existence of the interval $I^{m+1} = I_x^m \times I_y^1$ and the existence of the function $y = f(x) = f(x^1, \ldots, x^m)$ and its continuity in I_x^m is a verbatim repetition of the corresponding part of the proof of Proposition 1, with only a single change, which reduces to the fact that the symbol x must now be interpreted as $(x^1, \ldots, x^m)$ and α as $(\alpha^1, \ldots, \alpha^m)$.

If we now fix all the variables in the functions $F(x^1, \ldots, x^m, y)$ and $f(x^1, \ldots, x^m)$ except x^i and y, we have the hypotheses of Proposition 1, where now the role of x is played by the variable x^i. Formula (8.89) follows from this. It is clear from this formula that $\frac{\partial f}{\partial x^i} \in C(I_x^m; I_y^1)$ $(i = 1, \ldots, m)$, that is, $f \in C^{(1)}(I_x^m; I_y^1)$. Reasoning as in the proof of Proposition 1, we establish by induction that $f \in C^{(p)}(I_x^m; I_y^1)$ when $F \in C^{(p)}(U; \mathbb{R})$. □

Example 2 Assume that the function $F : G \to \mathbb{R}$ is defined in a domain $G \subset \mathbb{R}^m$ and belongs to the class $C^{(1)}(G; \mathbb{R})$; $x_0 = (x_0^1, \ldots, x_0^m) \in G$ and $F(x_0) = F(x_0^1, \ldots, x_0^m) = 0$. If x_0 is not a critical point of F, then at least one of the partial derivatives of F at x_0 is nonzero. Suppose, for example, that $\frac{\partial F}{\partial x^m}(x_0) \neq 0$.

Then, by Proposition 2, in some neighborhood of x_0 the subset of $\mathbb{R}^m$ defined by the equation $F(x^1, \ldots, x^m) = 0$ can be defined as the graph of a function $x^m = f(x^1, \ldots, x^{m-1})$, defined in a neighborhood of the point $(x_0^1, \ldots, x_0^{m-1}) \in \mathbb{R}^{m-1}$ that is continuously differentiable in this neighborhood and such that $f(x_0^1, \ldots, x_0^{m-1}) = x_0^m$.

Thus, in a neighborhood of a noncritical point x_0 of F the equation

$$F(x^1, \ldots, x^m) = 0$$

defines an $(m - 1)$-dimensional surface.

In particular, in the case of $\mathbb{R}^3$ the equation

$$F(x, y, z) = 0$$

defines a two-dimensional surface in a neighborhood of a noncritical point (x_0, y_0, z_0) satisfying the equation, which, when the condition $\frac{\partial F}{\partial z}(x_0, y_0, z_0) \neq 0$

holds, can be locally written in the form

$$z = f(x, y).$$

As we know, the equation of the plane tangent to the graph of this function at the point (x_0, y_0, z_0) has the form

$$z - z_0 = \frac{\partial f}{\partial x}(x_0, y_0)(x - x_0) + \frac{\partial f}{\partial y}(x_0, y_0)(y - y_0).$$

But by formula (8.89)

$$\frac{\partial f}{\partial x}(x_0, y_0) = -\frac{F'_x(x_0, y_0, z_0)}{F'_z(x_0, y_0, z_0)}, \qquad \frac{\partial f}{\partial y}(x_0, y_0) = -\frac{F'_y(x_0, y_0, z_0)}{F'_z(x_0, y_0, z_0)},$$

and therefore the equation of the tangent plane can be rewritten as

$$F'_x(x_0, y_0, z_0)(x - x_0) + F'_y(x_0, y_0, z_0)(y - y_0) + F'_z(x_0, y_0, z_0)(z - z_0) = 0,$$

which is symmetric in the variables x, y, z.

Similarly, in the general case we obtain the equation

$$\sum_{i=1}^{m} F'_{x^i}(x_0)\left(x^i - x_0^i\right) = 0$$

of the hyperplane in $\mathbb{R}^m$ tangent at the point $x_0 = (x_0^1, \dots, x_0^m)$ to the surface given by the equation $F(x^1, \dots, x^m) = 0$ (naturally, under the assumptions that $F(x_0) = 0$ and that x_0 is a noncritical point of F).

It can be seen from these equations that, given the Euclidean structure on $\mathbb{R}^m$, one can assert that the vector

$$\operatorname{grad} F(x_0) = \left(\frac{\partial F}{\partial x^1}, \dots, \frac{\partial F}{\partial x^m}\right)(x_0)$$

is orthogonal to the r-level surface $F(x) = r$ of the function F at a corresponding point $x_0 \in \mathbb{R}^m$.

For example, for the function

$$F(x, y, z) = \frac{x^2}{a^2} + \frac{y^2}{b^2} + \frac{z^2}{c^2},$$

defined in $\mathbb{R}^3$, the r-level is the empty set if $r < 0$, a single point if $r = 0$, and the ellipsoid

$$\frac{x^2}{a^2} + \frac{y^2}{b^2} + \frac{z^2}{c^2} = r$$

if $r > 0$. If (x_0, y_0, z_0) is a point on this ellipsoid, then by what has been proved, the vector

$$\operatorname{grad} F(x_0, y_0, z_0) = \left(\frac{2x_0}{a^2}, \frac{2y_0}{b^2}, \frac{2z_0}{c^2}\right)$$

is orthogonal to this ellipsoid at the point (x_0, y_0, z_0), and the tangent plane to it at this point has the equation

$$\frac{x_0(x - x_0)}{a^2} + \frac{y_0(y - y_0)}{b^2} + \frac{z_0(z - z_0)}{c^2} = 0,$$

which, when we take account of the fact that the point (x_0, y_0, z_0) lies on the ellipsoid, can be rewritten as

$$\frac{x_0 x}{a^2} + \frac{y_0 y}{b^2} + \frac{z_0 z}{c^2} = r.$$

8.5.4 The Implicit Function Theorem

We now turn to the general case of a system of equations

$$\begin{cases} F^1(x^1, \dots, x^m, y^1, \dots y^n) = 0, \\ \vdots \\ F^n(x^1, \dots, x^m, y^1, \dots, y^n) = 0, \end{cases} \tag{8.90}$$

which we shall solve with respect to $y^1, \dots, y^n$, that is, find a system of functional relations

$$\begin{cases} y^1 = f^1(x^1, \dots, x^m), \\ \vdots \\ y^n = f^n(x^1, \dots, x^m), \end{cases} \tag{8.91}$$

locally equivalent to the system (8.90).

For the sake of brevity, convenience in writing, and clarity of statement, let us agree that $x = (x^1, \dots, x^m)$, $y = (y^1, \dots, y^n)$. We shall write the left-hand side of the system (8.90) as $F(x, y)$, the system of equations (8.90) as $F(x, y) = 0$, and the mapping (8.91) as $y = f(x)$.

If

$$x_0 = (x_0^1, \dots, x_0^m), \qquad y_0 = (y_0^1, \dots, y_0^n),$$
$$\alpha = (\alpha^1, \dots, \alpha^m), \qquad \beta = (\beta^1, \dots, \beta^n),$$

the notation $|x - x_0| < \alpha$ or $|y - y_0| < \beta$ will mean that $|x^i - x_0^i| < \alpha^i$ $(i = 1, \dots, m)$ or $|y^j - y_0^j| < \beta^j$ $(j = 1, \dots, n)$ respectively.

We next set

$$f'(x) = \begin{pmatrix} \frac{\partial f^1}{\partial x^1} & \cdots & \frac{\partial f^1}{\partial x^m} \\ \vdots & \ddots & \vdots \\ \frac{\partial f^n}{\partial x^1} & \cdots & \frac{\partial f^n}{\partial x^m} \end{pmatrix}(x), \tag{8.92}$$

$$F'_x(x, y) = \begin{pmatrix} \frac{\partial F^1}{\partial x^1} & \cdots & \frac{\partial F^1}{\partial x^m} \\ \vdots & \ddots & \vdots \\ \frac{\partial F^n}{\partial x^1} & \cdots & \frac{\partial F^n}{\partial x^m} \end{pmatrix}(x, y), \tag{8.93}$$

$$F'_y(x, y) = \begin{pmatrix} \frac{\partial F^1}{\partial y^1} & \cdots & \frac{\partial F^1}{\partial y^n} \\ \vdots & \ddots & \vdots \\ \frac{\partial F^n}{\partial y^1} & \cdots & \frac{\partial F^n}{\partial y^n} \end{pmatrix}(x, y). \tag{8.94}$$

We remark that the matrix $F'_y(x, y)$ is square and hence invertible if and only if its determinant is nonzero. In the case $n = 1$, it reduces to a single element, and in that case the invertibility of $F'_y(x, y)$ is equivalent to the condition that that single element is nonzero. As usual, we shall denote the matrix inverse to $F'_y(x, y)$ by $[F'_y(x, y)]^{-1}$.

We now state the main result of the present section.

Theorem 1 (Implicit function theorem) *If the mapping $F : U \to \mathbb{R}^n$ defined in a neighborhood U of the point $(x_0, y_0) \in \mathbb{R}^{m+n}$ is such that*

1^0 $F \in C^{(p)}(U; \mathbb{R}^n)$, $p \ge 1$,
2^0 $F(x_0, y_0) = 0$,
3^0 $F'_y(x_0, y_0)$ *is an invertible matrix,*

then there exists an $(m + n)$-dimensional interval $I = I^m_x \times I^n_y \subset U$, where

$$I^m_x = \{x \in \mathbb{R}^m \mid |x - x_0| < \alpha\}, \qquad I^n_y = \{y \in \mathbb{R}^n \mid |y - y_0| < \beta\},$$

and a mapping $f \in C^{(p)}(I^m_x; I^n_y)$ such that

$$F(x, y) = 0 \Leftrightarrow y = f(x), \tag{8.95}$$

for any point $(x, y) \in I^m_x \times I^n_y$ and

$$f'(x) = -\big[F'_y\big(x, f(x)\big)\big]^{-1}\big[F'_x\big(x, f(x)\big)\big]. \tag{8.96}$$

Proof The proof of the theorem will rely on Proposition 2 and the elementary properties of determinants. We shall break it into stages, reasoning by induction.

For $n = 1$, the theorem is the same as Proposition 2 and is therefore true.

Suppose the theorem is true for dimension $n - 1$. We shall show that it is then valid for dimension n.

a) By hypothesis 3^0, the determinant of the matrix (8.94) is nonzero at the point $(x_0, y_0) \in \mathbb{R}^{m+n}$ and hence in some neighborhood of the point (x_0, y_0). Consequently at least one element of the last row of this matrix is nonzero. Up to a change in the notation, we may assume that the element $\frac{\partial F^n}{\partial y^n}$ is nonzero.

b) Then applying Proposition 2 to the relation

$$F^n(x^1, \dots, x^m, y^1, \dots, y^n) = 0,$$

we find an interval $\tilde{I}^{m+n} = (\tilde{I}_x^m \times \tilde{I}_y^{n-1}) \times I_y^1 \subset U$ and a function $\tilde{f} \in C^{(p)}(\tilde{I}_x^m \times \tilde{I}_y^{n-1}; I_y^1)$ such that

$$\begin{aligned}
&\left(F^n(x^1, \dots, x^m, y^1, \dots, y^n) = 0 \text{ in } \tilde{I}^{m+n}\right) \Leftrightarrow \\
&\quad \Leftrightarrow \left(y^n = \tilde{f}(x^1, \dots, x^m, y^1, \dots, y^{n-1}),\right. \\
&\quad \left.(x^1, \dots, x^m) \in \tilde{I}_x^m, (y^1, \dots, y^{n-1}) \in \tilde{I}_y^{n-1}\right).
\end{aligned} \tag{8.97}$$

c) Substituting the resulting expression $y^n = \tilde{f}(x, y^1, \dots, y^{n-1})$ for the variable y^n in the first $n-1$ equations of (8.90), we obtain $n-1$ relations

$$\begin{cases}
\Phi^1(x^1, \dots, x^m, y^1, \dots, y^{n-1}) := \\
\quad = F^1(x^1, \dots, x^m, y^1, \dots, y^{n-1}, \tilde{f}(x^1, \dots, x^m, y^1, \dots, y^{n-1})) = 0, \\
\vdots \\
\Phi^{n-1}(x^1, \dots, x^m, y^1, \dots, y^{n-1}) := \\
\quad = F^{n-1}(x^1, \dots, x^m, y^1, \dots, y^{n-1}, \tilde{f}(x^1, \dots, x^m, y^1, \dots, y^{n-1})) = 0.
\end{cases} \tag{8.98}$$

It is clear that $\Phi^i \in C^{(p)}(\tilde{I}_x^m \times \tilde{I}_y^{n-1}; \mathbb{R})$ $(i = 1, \dots, n-1)$, and

$$\Phi^i(x_0^1, \dots, x_0^m; y_0^1, \dots, y_0^{n-1}) = 0 \quad (i, \dots, n-1),$$

since $\tilde{f}(x_0^1, \dots, x_0^m, y_0^1, \dots, y_0^{n-1}) = y_0^n$ and $F^i(x_0, y_0) = 0$ $(i = 1, \dots, n)$.

By definition of the functions Φ^k $(k = 1, \dots, n-1)$,

$$\frac{\partial \Phi^k}{\partial y^i} = \frac{\partial F^k}{\partial y^i} + \frac{\partial F^k}{\partial y^n} \cdot \frac{\partial \tilde{f}}{\partial y^i} \quad (i, k = 1, \dots, n-1). \tag{8.99}$$

Further setting

$$\begin{aligned}
&\Phi^n(x^1, \dots, x^m, y_1, \dots, y^{n-1}) := \\
&\quad = F^n(x^1, \dots, x^m, y^1, \dots, y^{n-1}, \tilde{f}(x^1, \dots, x^m, y^1, \dots, y^{n-1})),
\end{aligned}$$

we find by (8.97) that $\Phi^n \equiv 0$ in its domain of definition, and therefore

$$\frac{\partial \Phi^n}{\partial y^i} = \frac{\partial F^n}{\partial y^i} + \frac{\partial F^n}{\partial y^n} \cdot \frac{\partial \tilde{f}}{\partial y^i} \equiv 0 \quad (i = 1, \dots, n-1). \tag{8.100}$$

Taking account of relations (8.99) and (8.100) and the properties of determinants, we can now observe that the determinant of the matrix (8.94) equals the determinant of the matrix

$$\begin{pmatrix} \frac{\partial F^1}{\partial y^1} + \frac{\partial F^1}{\partial y^n} \cdot \frac{\partial \tilde{f}}{\partial y^1} & \cdots & \frac{\partial F^1}{\partial y^{n-1}} + \frac{\partial F^1}{\partial y^n} \cdot \frac{\partial \tilde{f}}{\partial y^{n-1}} & \frac{\partial F^1}{\partial y^n} \\ \vdots & \ddots & \vdots & \vdots \\ \frac{\partial F^n}{\partial y^1} + \frac{\partial F^n}{\partial y^n} \cdot \frac{\partial \tilde{f}}{\partial y^1} & \cdots & \frac{\partial F^n}{\partial y^{n-1}} + \frac{\partial F^n}{\partial y^n} \cdot \frac{\partial \tilde{f}}{\partial y^{n-1}} & \frac{\partial F^n}{\partial y^n} \end{pmatrix} =$$

$$= \begin{pmatrix} \frac{\partial \Phi^1}{\partial y^1} & \cdots & \frac{\partial \Phi^1}{\partial y^{n-1}} & \frac{\partial F^1}{\partial y^n} \\ \vdots & \ddots & \vdots & \vdots \\ \frac{\partial \Phi^{n-1}}{\partial y^1} & \cdots & \frac{\partial \Phi^{n-1}}{\partial y^{n-1}} & \frac{\partial F^{n-1}}{\partial y^n} \\ 0 & \cdots & 0 & \frac{\partial F^n}{\partial y^n} \end{pmatrix}.$$

By assumption, $\frac{\partial F^n}{\partial y^n} \neq 0$, and the determinant of the matrix (8.94) is nonzero. Consequently, in some neighborhood of $(x_0^1, \dots, x_0^m, y_0^1, \dots y_0^{n-1})$ the determinant of the matrix

$$\begin{pmatrix} \frac{\partial \Phi^1}{\partial y^1} & \cdots & \frac{\partial \Phi^1}{\partial y^{n-1}} \\ \vdots & \ddots & \vdots \\ \frac{\partial \Phi^{n-1}}{\partial y^1} & \cdots & \frac{\partial \Phi^{n-1}}{\partial y^{n-1}} \end{pmatrix} (x^1, \dots, x^m, y^1, \dots, y^{n-1})$$

is nonzero.

Then by the induction hypothesis there exist an interval $I^{m+n-1} = I_x^m \times I_y^{n-1} \subset \tilde{I}_x^m \times \tilde{I}_y^{n-1}$, which is a neighborhood of $(x_0^1, \dots, x_0^m, y_0^1, \dots, y_0^{n-1})$ in $\mathbb{R}^{m-1}$, and a mapping $f \in C^{(p)}(I_x^m; I_y^{n-1})$ such that the system (8.98) is equivalent on the interval $I^{m+n-1} = I_x^m \times I_y^{n-1}$ to the relations

$$\begin{cases} y^1 = f^1(x^1, \dots, x^m), \\ \vdots \\ y^{n-1} = f^{n-1}(x^1, \dots, x^m). \end{cases} \quad x \in I_x^m. \tag{8.101}$$

d) Since $I_y^{n-1} \subset \tilde{I}_y^{n-1}$, and $I_x^m \subset \tilde{I}_x^m$, substituting $f^1, \dots, f^{n-1}$ from (8.101) in place of the corresponding variables in the function

$$y^n = \tilde{f}(x^1, \dots, x^m, y^1, \dots, y^{n-1})$$

from (8.97) we obtain a relation

$$y^n = f^n(x^1, \dots, x^m) \tag{8.102}$$

between y^n and $(x^1, \dots, x^m)$.

e) We now show that the system

$$\begin{cases} y^1 = f^1(x^1, \dots, x^m), \\ \vdots \\ y^n = f^n(x^1, \dots, x^m), \end{cases} \qquad x \in I_x^m, \tag{8.103}$$

which defines a mapping $f \in C^{(p)}(I_x^m; I_y^n)$, where $I_y^n = I_y^{n-1} \times I_y^1$, is equivalent to the system of equations (8.90) in the neighborhood $I^{m+n} = I_x^m \times I_y^n$.

In fact, inside $\tilde{I}^{m+n} = (\tilde{I}_x^m \times \tilde{I}_y^{n-1}) \times I_y^1$ we began by replacing the last equation of the original system (8.90) with the equality $y^n = \tilde{f}(x, y^1, \dots, y^{n-1})$, which is equivalent to it by virtue of (8.97). From the second system so obtained, we passed to a third system equivalent to it by replacing the variable y^n in the first $n-1$ equations with $\tilde{f}(x, y^1, \dots, y^{n-1})$. We then replaced the first $n-1$ equations (8.98) of the third system inside $I_x^m \times I_y^{n-1} \subset \tilde{I}_x^m \times \tilde{I}_y^{n-1}$ with relations (8.101), which are equivalent to them. In that way, we obtained a fourth system, after which we passed to the final system (8.103), which is equivalent to it inside $I_x^m \times I_y^{n-1} \times I_y^1 = I^{m+n}$, by replacing the variables $y^1, \dots, y^{n-1}$ with their expressions (8.101) in the last equation $y^n = \tilde{f}(x^1, \dots, x^m, y^1, \dots, y^{n-1})$ of the fourth system, obtaining (8.102) as the last equation.

f) To complete the proof of the theorem it remains only to verify formula (8.96).

Since the systems (8.90) and (8.91) are equivalent in the neighborhood $I_x^m \times I_y^n$ of the point (x_0, y_0), it follows that

$$F\big(x, f(x)\big) \equiv 0, \quad \text{if } x \in I_x^m.$$

In coordinates this means that in the domain I_x^m

$$F^k\big(x^1, \dots, x^m, f^1(x^1, \dots, x^m), \dots, f^n(x^1, \dots, x^m)\big) \equiv 0 \quad (k = 1, \dots, n). \tag{8.104}$$

Since $f \in C^{(p)}(I_x^m; I_y^n)$ and $F \in C^{(p)}(U; \mathbb{R}^n)$, where $p \geq 1$, it follows that $F(\cdot, f(\cdot)) \in C^{(p)}(I_x^m; \mathbb{R}^n)$ and, differentiating the identity (8.104), we obtain

$$\frac{\partial F^k}{\partial x^i} + \sum_{j=1}^{n} \frac{\partial F^k}{\partial y^j} \cdot \frac{\partial f^j}{\partial x^i} = 0 \quad (k = 1, \dots, n; i = 1, \dots, m). \tag{8.105}$$

Relations (8.105) are obviously equivalent to the single matrix equality

$$F_x'(x, y) + F_y'(x, y) \cdot f'(x) = 0,$$

in which $y = f(x)$.

Taking account of the invertibility of the matrix $F_y'(x, y)$ in a neighborhood of the point (x_0, y_0), we find by this equality that

$$f'(x) = -\big[F_y'\big(x, f(x)\big)\big]^{-1}\big[F_x'\big(x, f(x)\big)\big],$$

and the theorem is completely proved. □

8.5.5 Problems and Exercises

1. On the plane $\mathbb{R}^2$ with coordinates x and y a curve is defined by the relation $F(x, y) = 0$, where $F \in C^{(2)}(\mathbb{R}^2, \mathbb{R})$. Let (x_0, y_0) be a noncritical point of the function $F(x, y)$ lying on the curve.

a) Write the equation of the tangent to this curve at this point (x_0, y_0).

b) Show that if (x_0, y_0) is a point of inflection of the curve, then the following equality holds:

$$\left(F''_{xx} F'^2_y - 2F''_{xy} F'_x F'_y + F''_{yy} F'^2_x\right)(x_0, y_0) = 0.$$

c) Find a formula for the curvature of the curve at the point (x_0, y_0).

2. *The Legendre transform in m variables.* The *Legendre transform* of $x^1, \dots, x^m$ and the function $f(x^1, \dots, x^m)$ is the transformation to the new variables $\xi_1, \dots, \xi_m$ and function $f^*(\xi_1, \dots, \xi_m)$ defined by the relations

$$\begin{cases} \xi_i = \dfrac{\partial f}{\partial x^i}(x^1, \dots, x^m) \quad (i = 1, \dots, m), \\ f^*(\xi_1, \dots, \xi_m) = \displaystyle\sum_{i=1}^{m} \xi_i x^i - f(x^1, \dots, x^m). \end{cases} \tag{8.106}$$

a) Give a geometric interpretation of the Legendre transform (8.106) as the transition from the coordinates $(x^1, \dots, x^m, f(x^1, \dots, x^m))$ of a point on the graph of the function $f(x)$ to the parameters $(\xi_1, \dots, \xi_m, f^*(\xi_1, \dots, \xi_m))$ defining the equation of the plane tangent to the graph at that point.

b) Show that the Legendre transform is guaranteed to be possible locally if $f \in C^{(2)}$ and $\det(\frac{\partial^2 f}{\partial x^i \partial x^j}) \neq 0$.

c) Using the same definition of convexity for a function $f(x) = f(x^1, \dots, x^m)$ as in the one-dimensional case (taking x to be the vector $(x^1, \dots, x^m) \in \mathbb{R}^m$), show that the Legendre transform of a convex function is a convex function.

d) Show that

$$df^* = \sum_{i=1}^{m} x^i \, d\xi_i + \sum_{i=1}^{m} \xi_i \, dx^i - df = \sum_{i=1}^{m} x^i \, d\xi_i,$$

and deduce from this relation that the Legendre transform is involutive, that is, verify the equality

$$(f^*)^*(x) = f(x).$$

e) Taking account of d), write the transform (8.106) in the following form, which is symmetric in the variables:

$$\begin{cases} f^*(\xi_1, \dots, \xi_m) + f(x^1, \dots, x^m) = \sum_{i=1}^{m} \xi_i x^i, \\ \xi_i = \frac{\partial f}{\partial x^i}(x^1, \dots, x^m), \quad x^i = \frac{\partial f^*}{\partial \xi_i}(\xi_1, \dots, \xi_m) \end{cases} \tag{8.107}$$

or, more briefly, in the form

$$f^*(\xi) + f(x) = \xi x, \qquad \xi = \nabla f(x), \qquad x = \nabla f^*(\xi),$$

where

$$\nabla f(x) = \left(\frac{\partial f}{\partial x^1}, \dots, \frac{\partial f}{\partial x^m}\right)(x), \qquad \nabla f^*(\xi) = \left(\frac{\partial r^*}{\partial \xi_1}, \dots, \frac{\partial f^*}{\partial \xi_m}\right)(\xi),$$

$$\xi x = \xi_i x^i = \sum_{i=1}^{m} \xi_i x^i.$$

f) The matrix formed from the second-order partial derivatives of a function (and sometimes the determinant of this matrix) is called the *Hessian* of the function at a given point.

Let d_{ij} and d^*_{ij} be the co-factors of the elements $\frac{\partial^2 f}{\partial x^i \partial x^j}$ and $\frac{\partial^2 f^*}{\partial \xi_i \partial \xi_j}$ of the Hessians

$$\begin{pmatrix} \frac{\partial^2 f}{\partial x^1 \partial x^1} & \cdots & \frac{\partial^2 f}{\partial x^1 \partial x^m} \\ \vdots & \ddots & \vdots \\ \frac{\partial^2 f}{\partial x^m \partial x^1} & \cdots & \frac{\partial^2 f}{\partial x^m \partial x^m} \end{pmatrix}(x), \qquad \begin{pmatrix} \frac{\partial^2 f^*}{\partial \xi_1 \partial \xi_1} & \cdots & \frac{\partial^2 f^*}{\partial \xi_1 \partial \xi_m} \\ \vdots & \ddots & \vdots \\ \frac{\partial^2 f^*}{\partial \xi_m \partial \xi_1} & \cdots & \frac{\partial^2 f^*}{\partial \xi_m \partial \xi_m} \end{pmatrix}(\xi)$$

of the functions $f(x)$ and $f^*(\xi)$, and let d and d^* be the determinants of these matrices. Assuming that $d \neq 0$, show that $d \cdot d^* = 1$ and that

$$\frac{\partial^2 f}{\partial x^i \partial x^j}(x) = \frac{d^*_{ij}}{d^*}(\xi), \qquad \frac{\partial^2 f^*}{\partial \xi_i \partial \xi_j}(\xi) = \frac{d_{ij}}{d}(x).$$

g) A soap film spanning a wire frame forms a so-called *minimal surface*, having minimal area among all the surfaces spanning the contour.

If that surface is locally defined as the graph of a function $z = f(x, y)$, it turns out that the function f must satisfy the following equation for minimal surfaces:

$$\left(1 + f_y'^2\right) f_{xx}'' - 2 f_x' f_y' f_{xy}'' + \left(1 + f_x'^2\right) f_{yy}'' = 0.$$

Show that after a Legendre transform is performed this equation is brought into the form

$$\left(1 + \eta^2\right) f^{*\,\prime\prime}_{\eta\eta} + 2\xi\eta f^{*\,\prime\prime}_{\xi\eta} + \left(1 + \xi^2\right) f^{*\,\prime\prime}_{\xi\xi} = 0.$$

3. *Canonical variables and the Hamilton equations.*[8]

a) In the calculus of variations and the fundamental principles of classical mechanics the following system of equations, due to Euler and Lagrange, plays an important role:

$$\begin{cases} \left(\dfrac{\partial L}{\partial x} - \dfrac{\mathrm{d}}{\mathrm{d}t}\dfrac{\partial L}{\partial v}\right)(t, x, v) = 0, \\ v = \dot{x}(t), \end{cases} \tag{8.108}$$

where $L(t, x, v)$ is a given function of the variables t, x, v, of which t is usually time, x the coordinate, and v the velocity.

The system (8.108) consists of two relations in three variables. Usually we wish to determine $x = x(t)$ and $v = v(t)$ from (8.108), which essentially reduces to determining the relation $x = x(t)$, since $v = \frac{\mathrm{d}x}{\mathrm{d}t}$.

Write the first equation of (8.108) in more detail, expanding the derivative $\frac{\mathrm{d}}{\mathrm{d}t}$ taking account of the equalities $x = x(t)$ and $v = v(t)$.

b) Show that if we change from the coordinates t, x, v, L to the so-called *canonical coordinates* t, x, p, H by performing the Legendre transform (see Problem 2)

$$\begin{cases} p = \dfrac{\partial L}{\partial v}, \\ H = pv - L \end{cases}$$

with respect to the variables v and L to replace them with p and H, then the Euler–Lagrange system (8.108) assumes the symmetric form

$$\dot{p} = -\frac{\partial H}{\partial x}, \qquad \dot{x} = \frac{\partial H}{\partial p}, \tag{8.109}$$

in which it is called *system of Hamilton equations*.

c) In the multidimensional case, when $L = L(t, x^1, \dots, x^m, v^1, \dots, v^m)$ the Euler–Lagrange system has the form

$$\begin{cases} \left(\dfrac{\partial L}{\partial x^i} - \dfrac{\mathrm{d}}{\mathrm{d}t}\dfrac{\partial L}{\partial v^i}\right)(t, x, v) = 0, \\ v^i = \dot{x}^i(t) \quad (i = 1, \dots, m), \end{cases} \tag{8.110}$$

where for brevity we have set $x = (x^1, \dots, x^m)$, $v = (v^1, \dots, v^m)$.

By performing a Legendre transform with respect to the variables $v^1, \dots, v^m, L$, change from the variables $t, x^1, \dots, x^m, v^1, \dots, v^m, L$ to the canonical variables

[8] W.R. Hamilton (1805–1865) – famous Irish mathematician and specialist in mechanics. He stated a variational principle (Hamilton's principle), constructed a phenomenological theory of optic phenomena, and was the creator of quaternions and the founder of vector analysis (in fact, the term "vector" is due to him).

$t, x^1, \dots, x^m, p_1, \dots, p_m, H$ and show that in these variables the system (8.110) becomes the following system of Hamilton equations:

$$\dot{p}_i = -\frac{\partial H}{\partial x^i}, \quad \dot{x}^i = \frac{\partial H}{\partial p_i} \quad (i = 1, \dots, m). \tag{8.111}$$

4. *The implicit function theorem.*
The solution of this problem gives another proof of the fundamental theorem of this section, perhaps less intuitive and constructive than the one given above, but shorter.

a) Suppose the hypotheses of the implicit function theorem are satisfied, and let

$$F_y^i(x, y) = \left(\frac{\partial F^i}{\partial y^1}, \dots, \frac{\partial F^i}{\partial y^n}\right)(x, y)$$

be the ith row of the matrix $F_y'(x, y)$.

Show that the determinant of the matrix formed from the vectors $F_y^i(x_i, y_i)$ is nonzero if all the points (x_i, y_i) $(i = 1, \dots, n)$ lie in some sufficiently small neighborhood $U = I_x^m \times I_y^n$ of (x_0, y_0).

b) Show that, if for $x \in I_x^m$ there are points $y_1, y_2 \in I_y^n$ such that $F(x, y_1) = 0$ and $F(x, y_2) = 0$, then for each $i \in \{1, \dots, n\}$ there is a point (x, y_i) lying on the closed interval with endpoints (x, y_1) and (x, y_2) such that

$$F_y^i(x, y_i)(y_2 - y_1) = 0 \quad (i = 1, \dots, n).$$

Show that this implies that $y_1 = y_2$, that is, if the implicit function $f : I_x^m \to I_y^n$ exists, it is unique.

c) Show that if the open ball $B(y_0; r)$ is contained in I_y^n, then $F(x_0, y) \neq 0$ for $\|y - y_0\|_{\mathbb{R}^n} = r > 0$.

d) The function $\|F(x_0, y)\|_{\mathbb{R}^n}^2$ is continuous and has a positive minimum value μ on the sphere $\|y - y_0\|_{\mathbb{R}^n} = r$.

e) There exists $\delta > 0$ such that for $\|x - x_0\|_{\mathbb{R}^m} < \delta$ we have

$$\|F(x, y)\|_{\mathbb{R}^n}^2 \geq \frac{1}{2}\mu, \quad \text{if } \|y - y_0\|_{\mathbb{R}^n} = r,$$

$$\|F(x, y)\|_{\mathbb{R}^n}^2 < \frac{1}{2}\mu, \quad \text{if } y = y_0.$$

f) For any fixed x such that $\|x - x_0\| < \delta$ the function $\|F(x, y)\|_{\mathbb{R}^n}^2$ attains a minimum at some interior point $y = f(x)$ of the open ball $\|y - y_0\|_{\mathbb{R}^n} \leq r$, and since the matrix $F_y'(x, f(x))$ is invertible, it follows that $F(x, f(x)) = 0$. This establishes the existence of the implicit function $f : B(x_0; \delta) \to B(y_0; r)$.

g) If $\Delta y = f(x + \Delta x) - f(x)$, then

$$\Delta y = -\left[\tilde{F}_y'\right]^{-1} \cdot \left[\tilde{F}_x'\right]\Delta x,$$

where $\tilde{F}'_y$ is the matrix whose rows are the vectors $F^i_y(x_i, y_i)$ $(i = 1, \dots, n)$, (x_i, y_i) being a point on the closed interval with endpoints (x, y) and $(x + \Delta x, y + \Delta y)$. The symbol $\tilde{F}'_x$ has a similar meaning.

Show that this relation implies that the function $y = f(x)$ is continuous.

h) Show that

$$f'(x) = -\left[\tilde{F}'_y(x, f(x))\right]^{-1} \cdot \left[\tilde{F}'_x(x, f(x))\right].$$

5. "If $f(x, y, z) = 0$, then $\frac{\partial z}{\partial y} \cdot \frac{\partial y}{\partial x} \cdot \frac{\partial x}{\partial z} = -1$."

a) Give a precise meaning to this statement.

b) Verify that it holds in the example of Clapeyron's ideal gas equation

$$\frac{P \cdot V}{T} = \text{const}$$

and in the general case of a function of three variables.

c) Write the analogous statement for the relation $f(x^1, \dots, x^m) = 0$ among m variables. Verify that it is correct.

6. Show that the roots of the equation

$$z^n + c_1 z^{n-1} + \dots + c_n = 0$$

are smooth functions of the coefficients, at least when they are all distinct.

8.6 Some Corollaries of the Implicit Function Theorem

8.6.1 The Inverse Function Theorem

Definition 1 A mapping $f : U \to V$, where U and V are open subsets of $\mathbb{R}^m$, is a $C^{(p)}$*-diffeomorphism* or a *diffeomorphism of smoothness* p $(p = 0, 1, \dots)$, if

1) $f \in C^{(p)}(U; V)$;
2) f is a bijection;
3) $f^{-1} \in C^{(p)}(V; U)$.

A $C^{(0)}$-diffeomorphism is called a *homeomorphism*.

As a rule, in this book we shall consider only the smooth case, that is, the case $p \in \mathbb{N}$ or $p = \infty$.

The basic idea of the following frequently used theorem is that if the differential of a mapping is invertible at a point, then the mapping itself is invertible in some neighborhood of the point.

Theorem 1 (Inverse function theorem) *If a mapping $f : G \to \mathbb{R}^m$ of a domain $G \subset \mathbb{R}^m$ is such that*

1^0 $f \in C^{(p)}(G; \mathbb{R}^m)$, $p \geq 1$,
2^0 $y_0 = f(x_0)$ *at* $x_0 \in \mathbb{G}$,
3^0 $f'(x_0)$ *is invertible,*

then there exists a neighborhood $U(x_0) \subset G$ of x_0 and a neighborhood $V(y_0)$ of y_0 such that $f : U(x_0) \to V(y_0)$ is a $C^{(p)}$-diffeomorphism. Moreover, if $x \in U(x_0)$ and $y = f(x) \in V(y_0)$, then

$$(f^{-1})'(y) = (f'(x))^{-1}.$$

Proof We rewrite the relation $y = f(x)$ in the form

$$F(x, y) = f(x) - y = 0. \tag{8.112}$$

The function $F(x, y) = f(x) - y$ is defined for $x \in G$ and $y \in \mathbb{R}^m$, that is it is defined in the neighborhood $G \times \mathbb{R}^m$ of the point $(x_0, y_0) \in \mathbb{R}^m \times \mathbb{R}^m$.

We wish to solve Eq. (8.112) with respect to x in some neighborhood of (x_0, y_0). By hypotheses 1^0, 2^0, 3^0 of the theorem the mapping $F(x, y)$ has the property that

$$\begin{aligned}
&F \in C^{(p)}(G \times \mathbb{R}^m; \mathbb{R}^m), \quad p \geq 1,\\
&F(x_0, y_0) = 0,\\
&F'_x(x_0, y_0) = f'(x_0) \quad \text{is invertible.}
\end{aligned}$$

By the implicit function theorem there exist a neighborhood $I_x \times I_y$ of (x_0, y_0) and a mapping $g \in C^{(p)}(I_y; I_x)$ such that

$$f(x) - y = 0 \Leftrightarrow x = g(y) \tag{8.113}$$

for any point $(x, y) \in I_x \times I_y$ and

$$g'(y) = -[F'_x(x, y)]^{-1}[F'_y(x, y)].$$

In the present case

$$F'_x(x, y) = f'(x), \qquad F'_y(x, y) = -E,$$

where E is the identity matrix; therefore

$$g'(y) = (f'(x))^{-1}. \tag{8.114}$$

If we set $V = I_y$ and $U = g(V)$, relation (8.113) shows that the mappings $f : U \to V$ and $g : V \to U$ are mutually inverse, that is, $g = f^{-1}$ on V.

Fig. 8.5

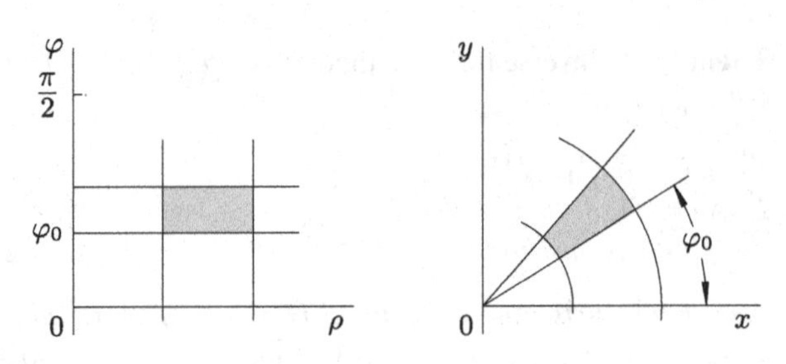

Since $V = I_y$, it follows that V is a neighborhood of y_0. This means that under hypotheses 1^0, 2^0, and 3^0 the image $y_0 = f(x_0)$ of $x_0 \in G$, which is an interior point of G, is an interior point of the image $f(G)$ of G. By formula (8.114) the matrix $g'(y_0)$ is invertible. Therefore the mapping $g : V \to U$ has properties 1^0, 2^0, and 3^0 relative to the domain V and the point $y_0 \in V$. Hence by what has already been proved $x_0 = g(y_0)$ is an interior point of $U = g(V)$.

Since by (8.114) hypotheses 1^0, 2^0, and 3^0 obviously hold at any point $y \in V$, any point $x = g(y)$ is an interior point of U. Thus U is an open (and obviously even connected) neighborhood of $x_0 \in \mathbb{R}^m$.

We have now verified that the mapping $f : U \to V$ satisfies all the conditions of Definition 1 and the assertion of Theorem 1. $\square$

We shall now give several examples that illustrate Theorem 1.

The inverse function theorem is very often used in converting from one coordinate system to another. The simplest version of such a change of coordinates was studied in analytic geometry and linear algebra and has the form

$$\begin{pmatrix} y^1 \\ \vdots \\ y^m \end{pmatrix} = \begin{pmatrix} a_1^1 & \cdots & a_m^1 \\ \vdots & \ddots & \vdots \\ a_1^m & \cdots & a_m^m \end{pmatrix} \begin{pmatrix} x^1 \\ \vdots \\ x^m \end{pmatrix}$$

or, in compact notation, $y^j = a_i^j x^i$. This linear transformation $A : \mathbb{R}_x^m \to \mathbb{R}_y^m$ has an inverse $A^{-1} : \mathbb{R}_y^m \to \mathbb{R}_x^m$ defined on the entire space $\mathbb{R}_y^m$ if and only if the matrix (a_i^j) is invertible, that is, $\det(a_i^j) \neq 0$.

The inverse function theorem is a local version of this proposition, based on the fact that in a neighborhood of a point a smooth mapping behaves approximately like its differential at the point.

Example 1 (Polar coordinates) The mapping $f : \mathbb{R}_+^2 \to \mathbb{R}^2$ of the half-plane $\mathbb{R}_+^2 = \{(\rho, \varphi) \in \mathbb{R}^2 \mid \rho \geq 0\}$ onto the plane $\mathbb{R}^2$ defined by the formula

$$\begin{aligned} x &= \rho \cos \varphi, \\ y &= \rho \sin \varphi, \end{aligned} \qquad (8.115)$$

is illustrated in Fig. 8.5.

Fig. 8.6

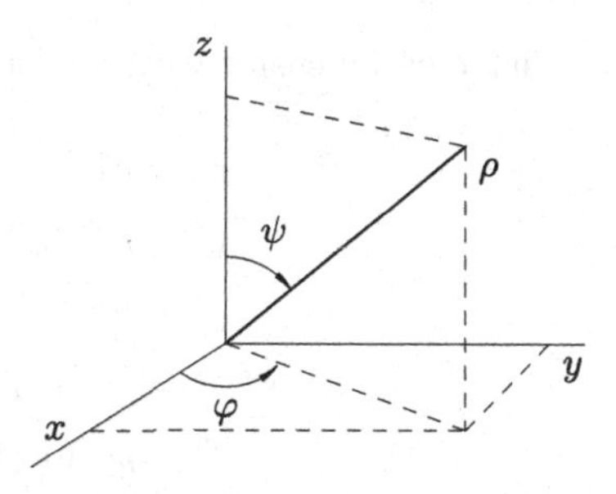

The Jacobian of this mapping, as can be easily computed, is ρ, that is, it is nonzero in a neighborhood of any point (ρ, φ), where $\rho > 0$. Therefore formulas (8.115) are locally invertible and hence locally the numbers ρ and φ can be taken as new coordinates of the point previously determined by the Cartesian coordinates x and y.

The coordinates (ρ, φ) are a well known system of curvilinear coordinates on the plane – polar coordinates. Their geometric interpretation is shown in Fig. 8.5. We note that by the periodicity of the functions $\cos\varphi$ and $\sin\varphi$ the mapping (8.115) is only locally a diffeomorphism when $\rho > 0$; it is not bijective on the entire plane. That is the reason that the change from Cartesian to polar coordinates always involves a choice of a branch of the argument φ (that is, an indication of its range of variation).

Polar coordinates (ρ, ψ, φ) in three-dimensional space $\mathbb{R}^3$ are called *spherical coordinates*. They are connected with Cartesian coordinates by the formulas

$$\begin{aligned} z &= \rho\cos\psi, \\ y &= \rho\sin\psi\sin\varphi, \\ x &= \rho\sin\psi\cos\varphi. \end{aligned} \tag{8.116}$$

The geometric meaning of the parameters ρ, ψ, and φ is shown in Fig. 8.6.

The Jacobian of the mapping (8.116) is $\rho^2\sin\psi$, and so by Theorem 1 the mapping is invertible in a neighborhood of each point (ρ, ψ, φ) at which $\rho > 0$ and $\sin\psi \neq 0$.

The sets where $\rho = \text{const}$, $\varphi = \text{const}$, or $\psi = \text{const}$ in (x, y, z)-space obviously correspond to a spherical surface (a sphere of radius ρ), a half-plane passing through the z-axis, and the surface of a cone whose axis is the z-axis respectively.

Thus in passing from coordinates (x, y, z) to coordinates (ρ, ψ, φ), for example, the spherical surface and the conical surface are flattened; they correspond to pieces of the planes $\rho = \text{const}$ and $\psi = \text{const}$ respectively. We observed a similar phenomenon in the two-dimensional case, where an arc of a circle in the (x, y)-plane corresponded to a closed interval on the line in the plane with coordinates (ρ, φ) (see Fig. 8.5). Please note that this is a local straightening.

In the m-dimensional case polar coordinates are introduced by the relations

$$\begin{aligned} x^1 &= \rho \cos \varphi_1, \\ x^2 &= \rho \sin \varphi_1 \cos \varphi_2, \\ &\vdots \\ x^{m-1} &= \rho \sin \varphi_1 \sin \varphi_2 \cdots \sin \varphi_{m-2} \cos \varphi_{m-1}, \\ x^m &= \rho \sin \varphi_1 \sin \varphi_2 \cdots \sin \varphi_{m-2} \sin \varphi_{m-1}. \end{aligned} \tag{8.117}$$

The Jacobian of this transformation is

$$\rho^{m-1} \sin^{m-2} \varphi_1 \sin^{m-3} \varphi_2 \cdots \sin \varphi_{m-2}, \tag{8.118}$$

and by Theorem 1 it is also locally invertible everywhere where this Jacobian is nonzero.

Example 2 (The general idea of local rectification of curves) New coordinates are usually introduced for the purpose of simplifying the analytic expression for the objects that occur in a problem and making them easier to visualize in the new notation.

Suppose for example, a curve in the plane $\mathbb{R}^2$ is defined by the equation

$$F(x, y) = 0.$$

Assume that F is a smooth function, that the point (x_0, y_0) lies on the curve, that is, $F(x_0, y_0) = 0$, and that this point is not a critical point of F. For example, suppose $F'_y(x, y) \neq 0$.

Let us try to choose coordinates ξ, η so that in these coordinates a closed interval of a coordinate line, for example, the line $\eta = 0$, corresponds to an arc of this curve.

We set

$$\xi = x - x_0, \qquad \eta = F(x, y).$$

The Jacobi matrix

$$\begin{pmatrix} 1 & 0 \\ F'_x & F'_y \end{pmatrix} (x, y)$$

of this transformation has as its determinant the number $F'_y(x, y)$, which by assumption is nonzero at (x_0, y_0). Then by Theorem 1, this mapping is a diffeomorphism of a neighborhood of (x_0, y_0) onto a neighborhood of the point $(\xi, \eta) = (0, 0)$. Hence, inside this neighborhood, the numbers ξ and η can be taken as new coordinates of points lying in a neighborhood of (x_0, y_0). In the new coordinates, the curve obviously has the equation $\eta = 0$, and in this sense we have indeed achieved a local rectification of it (see Fig. 8.7).

Fig. 8.7

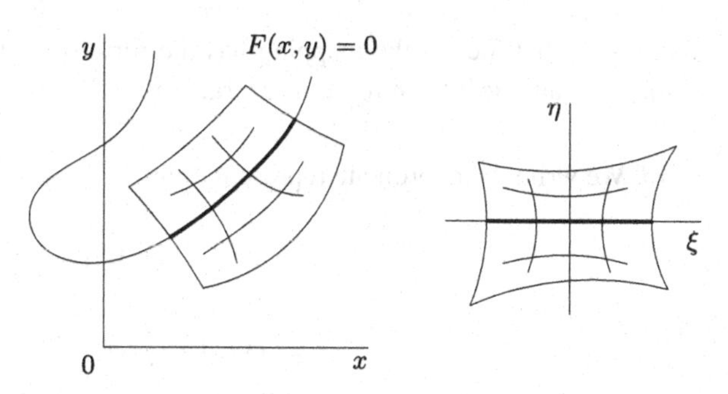

Fig. 8.8

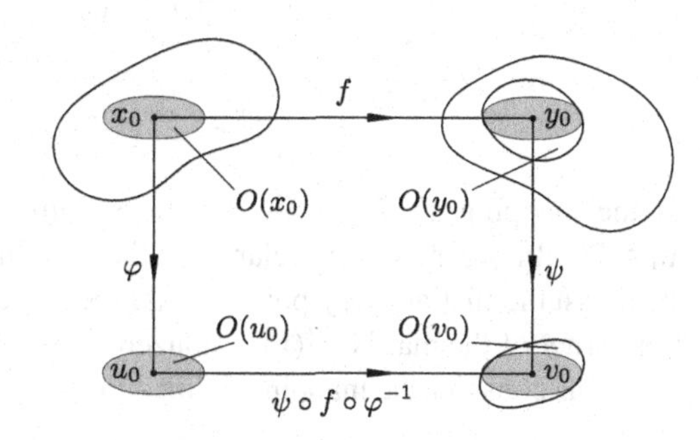

8.6.2 Local Reduction of a Smooth Mapping to Canonical Form

In this subsection we shall consider only one question of this type. To be specific, we shall exhibit a canonical form to which one can locally reduce any smooth mapping of constant rank by means of a suitable choice of coordinates.

We recall that the *rank* of a mapping $f : U \to \mathbb{R}^n$ of a domain $U \subset \mathbb{R}^m$ at a point $x \in U$ is the rank of the linear transformation tangent to it at the point, that is, the rank of the matrix $f'(x)$. The rank of a mapping at a point is usually denoted rank $f(x)$.

Theorem 2 (The rank theorem) *Let $f : U \to \mathbb{R}^n$ be a mapping defined in a neighborhood $U \subset \mathbb{R}^m$ of a point $x_0 \in \mathbb{R}^m$. If $f \in C^{(p)}(U; \mathbb{R}^n)$, $p \geq 1$, and the mapping f has the same rank k at every point $x \in U$, then there exist neighborhoods $O(x_0)$ of x_0 and $O(y_0)$ of $y_0 = f(x_0)$ and diffeomorphisms $u = \varphi(x)$, $v = \psi(y)$ of those neighborhoods, of class $C^{(p)}$, such that the mapping $v = \psi \circ f \circ \varphi^{-1}(u)$ has the coordinate representation*

$$\left(u^1, \dots, u^k, \dots, u^m\right) = u \mapsto v = \left(v^1, \dots, v^n\right) = \left(u^1, \dots, u^k, 0, \dots, 0\right) \quad (8.119)$$

in the neighborhood $O(u_0) = \varphi(O(x_0))$ of $u_0 = \varphi(x_0)$.

In other words, the theorem asserts (see Fig. 8.8) that one can choose coordinates $(u^1, \dots, u^m)$ in place of $(x^1, \dots, x^m)$ and $(v^1, \dots, v^n)$ in place of $(y^1, \dots, y^n)$ in such a way that locally the mapping has the form (8.119) in the new coordinates, that is, the canonical form for a linear transformation of rank k.

Proof We write the coordinate representation

$$\begin{aligned} y^1 &= f^1(x^1,\dots,x^m),\\ &\vdots\\ y^k &= f^k(x^1,\dots,x^m),\\ y^{k+1} &= f^{k+1}(x^1,\dots,x^m),\\ &\vdots\\ y^n &= f^n(x^1,\dots,x^m) \end{aligned} \tag{8.120}$$

of the mapping $f : U \to \mathbb{R}^n_y$, which is defined in a neighborhood of the point $x_0 \in \mathbb{R}^m_x$. In order to avoid relabeling the coordinates and the neighborhood U, we shall assume that at every point $x \in U$, the principal minor of order k in the upper left corner of the matrix $f'(x)$ is nonzero.

Let us consider the mapping defined in a neighborhood U of x_0 by the equalities

$$\begin{aligned} u^1 &= \varphi^1(x^1,\dots,x^m) = f^1(x^1,\dots,x^m),\\ &\vdots\\ u^k &= \varphi^k(x^1,\dots,x^m) = f^k(x^1,\dots,x^m),\\ u^{k+1} &= \varphi^{k+1}(x^1,\dots,x^m) = x^{k+1},\\ &\vdots\\ u^m &= \varphi^m(x^1,\dots,x^m) = x^m. \end{aligned} \tag{8.121}$$

The Jacobi matrix of this mapping has the form

$$\begin{pmatrix} \frac{\partial f^1}{\partial x^1} & \cdots & \frac{\partial f^1}{\partial x^k} & \vdots & \frac{\partial f^1}{\partial x^{k+1}} & \cdots & \frac{\partial f^1}{\partial x^m}\\ \vdots & \ddots & \vdots & \vdots & \vdots & \ddots & \vdots\\ \frac{\partial f^k}{\partial x^1} & \cdots & \frac{\partial f^k}{\partial x^k} & \vdots & \frac{\partial f^k}{\partial x^{k+1}} & \cdots & \frac{\partial f^k}{\partial x^m}\\ \vdots & \ddots & \vdots & \vdots & \vdots & \ddots & \vdots\\ & & & \vdots & 1 & & 0\\ & 0 & & \vdots & & \ddots & \\ & & & \vdots & 0 & & 1 \end{pmatrix},$$

and by assumption its determinant is nonzero in U.

By the inverse function theorem, the mapping $u=\varphi(x)$ is a diffeomorphism of smoothness p of some neighborhood $\tilde{O}(x_0)\subset U$ of x_0 onto a neighborhood $\tilde{O}(u_0)=\varphi(\tilde{O}(x_0))$ of $u_0=\varphi(x_0)$.

Comparing relations (8.120) and (8.121), we see that the composite function $g=f\circ\varphi^{-1}:\tilde{O}(u_0)\to\mathbb{R}^n_y$ has the coordinate representation

$$\begin{aligned}
y^1 &= f^1\circ\varphi^{-1}\big(u^1,\dots,u^m\big) = u^1,\\
&\ \ \vdots\\
y^k &= f^k\circ\varphi^{-1}\big(u^1,\dots,u^m\big) = u^k,\\
y^{k+1} &= f^{k+1}\circ\varphi^{-1}\big(u^1,\dots,u^m\big) = g^{k+1}\big(u^1,\dots,u^m\big),\\
&\ \ \vdots\\
y^n &= f^n\circ\varphi^{-1}\big(u^1,\dots,u^m\big) = g^n\big(u^1,\dots,u^m\big).
\end{aligned}\tag{8.122}$$

Since the mapping $\varphi^{-1}:\tilde{O}(u_0)\to\tilde{O}(x_0)$ has maximal rank m at each point $u\in\tilde{O}(u_0)$, and the mapping $f:\tilde{O}(x_0)\to\mathbb{R}^n_y$ has rank k at every point $x\in\tilde{O}(x_0)$, it follows, as is known from linear algebra, that the matrix $g'(u)=f'(\varphi^{-1}(u))(\varphi^{-1})'(u)$ has rank k at every point $u\in\tilde{O}(u_0)$.

Direct computation of the Jacobi matrix of the mapping (8.122) yields

$$\left(\begin{array}{ccc|ccc}
1 & & 0 & & & \\
 & \ddots & & & 0 & \\
0 & & 1 & & & \\
\vdots & \ddots & \vdots & \vdots & \ddots & \vdots \\
\frac{\partial g^{k+1}}{\partial u^1} & \cdots & \frac{\partial g^{k+1}}{\partial u^k} & \frac{\partial g^{k+1}}{\partial u^{k+1}} & \cdots & \frac{\partial g^{k+1}}{\partial u^m}\\
\vdots & \ddots & \vdots & \vdots & \ddots & \vdots \\
\frac{\partial g^n}{\partial u^1} & \cdots & \frac{\partial g^n}{\partial u^k} & \frac{\partial g^n}{\partial u^{k+1}} & \cdots & \frac{\partial g^n}{\partial u^m}
\end{array}\right).$$

Hence at each point $u\in\tilde{O}(u_0)$ we obtain $\frac{\partial g^j}{\partial u^i}(\mathrm{u})=0$ for $i=k+1,\dots,m$; $j=k+1,\dots,n$. Assuming that the neighborhood $\tilde{O}(u_0)$ is convex (which can be achieved by shrinking $\tilde{O}(u_0)$ to a ball with center at u_0, for example), we can conclude from this that the functions g^j, $j=k+1,\dots,n$, really are independent of the variables $u^{k+1},\dots,u^m$.

After this decisive observation, we can rewrite the mapping (8.122) as

$$\begin{aligned}
y^1 &= u^1,\\
&\vdots\\
y^k &= u^k,\\
y^{k+1} &= g^{k+1}\left(u^1,\dots,u^k\right),\\
&\vdots\\
y^n &= g^n\left(u^1,\dots,u^k\right).
\end{aligned}\tag{8.123}$$

At this point we can exhibit the mapping ψ. We set

$$\begin{aligned}
v^1 = y^1 &=: \psi^1(y),\\
&\vdots\\
v^k = y^k &=: \psi^k(y),\\
v^{k+1} = y^{k+1} - g^{k+1}\left(y^1,\dots,y^k\right) &=: \psi^{k+1}(y),\\
&\vdots\\
v^n = y^n - g^n\left(y^1,\dots,y^k\right) &=: \psi^n(y).
\end{aligned}\tag{8.124}$$

It is clear from the construction of the functions g^j $(j = k+1,\dots,n)$ that the mapping ψ is defined in a neighborhood of y_0 and belongs to class $C^{(p)}$ in that neighborhood.

The Jacobi matrix of the mapping (8.124) has the form

$$\left(\begin{array}{ccc|ccc}
1 & & 0 & & & \\
 & \ddots & & & 0 & \\
0 & & 1 & & & \\
\vdots & \ddots & \vdots & & \ddots & \\
-\frac{\partial g^{k+1}}{\partial y^1} & \cdots & -\frac{\partial g^{k+1}}{\partial y^k} & 1 & & 0\\
\vdots & \ddots & \vdots & & \ddots & \\
-\frac{\partial g^n}{\partial y^1} & \cdots & -\frac{\partial g^n}{\partial y^k} & 0 & & 1
\end{array}\right).$$

Its determinant equals 1, and so by Theorem 1 the mapping ψ is a diffeomorphism of smoothness p of some neighborhood $\tilde{O}(y_0)$ of $y_0 \in \mathbb{R}^n_y$ onto a neighborhood $\tilde{O}(v_0) = \psi(\tilde{O}(y_0))$ of $v_0 \in \mathbb{R}^n_v$.

Comparing relations (8.123) and (8.124), we see that in a neighborhood $O(u_0) \subset \tilde{O}(u_0)$ of u_0 so small that $g(O(u_0)) \subset \tilde{O}(y_0)$, the mapping $\psi \circ f \circ \varphi^{-1} : O(u_0) \to$

$\mathbb{R}^n_y$ is a mapping of smoothness p from this neighborhood onto some neighborhood $O(v_0) \subset \tilde{O}(v_0)$ of $v_0 \in \mathbb{R}^n_v$ and that it has the canonical form

$$\begin{aligned} v^1 &= u^1, \\ &\vdots \\ v^k &= u^k, \\ v^{k+1} &= 0, \\ &\vdots \\ v^n &= 0. \end{aligned} \tag{8.125}$$

Setting $\varphi^{-1}(O(u_0)) = O(x_0)$ and $\psi^{-1}(O(v_0)) = O(y_0)$, we obtain the neighborhoods of x_0 and y_0 whose existence is asserted in the theorem. The proof is now complete. □

Theorem 2, like Theorem 1, is obviously a local version of the corresponding theorem from linear algebra.

In connection with the proof just given of Theorem 2, we make the following remarks, which will be useful in what follows.

Remark 1 If the rank of the mapping $f : U \to \mathbb{R}^n$ is n at every point of the original neighborhood $U \subset \mathbb{R}^m$, then the point $y_0 = f(x_0)$, where $x_0 \in U$, is an interior point of $f(U)$, that is, $f(U)$ contains a neighborhood of this point.

Proof Indeed, from what was just proved, the mapping $\psi \circ f \circ \varphi^{-1} : O(u_0) \to O(v_0)$ has the form

$$\left(u^1, \dots, u^n, \dots, u^m\right) = u \mapsto v = \left(v^1, \dots, v^n\right) = \left(u^1, \dots, u^n\right),$$

in this case, and so the image of a neighborhood of $u_0 = \varphi(x_0)$ contains some neighborhood of $v_0 = \psi \circ f \circ \varphi^{-1}(u_0)$.

But the mappings $\varphi : O(x_0) \to O(u_0)$ and $\psi : O(y_0) \to O(v_0)$ are diffeomorphisms, and therefore they map interior points to interior points. Writing the original mapping f as $f = \psi^{-1} \circ (\psi \circ f \circ \varphi^{-1}) \circ \varphi$, we conclude that $y_0 = f(x_0)$ is an interior point of the image of a neighborhood of x_0. □

Remark 2 If the rank of the mapping $f : U \to \mathbb{R}^n$ is k at every point of a neighborhood U and $k < n$, then, by Eqs. (8.120), (8.124), and (8.125), in some neighborhood of $x_0 \in U \subset \mathbb{R}^m$ the following $n - k$ relations hold;

$$f^i\left(x^1, \dots, x^m\right) = g^i\left(f^1\left(x^1, \dots, x^m\right), \dots, f^k\left(x^1, \dots, x^m\right)\right) \quad (i = k+1, \dots, n). \tag{8.126}$$

These relations are written under the assumption we have made that the principal minor of order k of the matrix $f'(x_0)$ is nonzero, that is, the rank k is realized on the set of functions $f^1, \dots, f^k$. Otherwise one may relabel the functions $f^1, \dots, f^n$ and again have this situation.

8.6.3 Functional Dependence

Definition 2 A system of continuous functions $f^i(x) = f^i(x^1, \dots, x^m)$ $(i = 1, \dots, n)$ is *functionally independent* in a neighborhood of a point $x_0 = (x_0^1, \dots, x_0^m)$ if for any continuous function $F(y) = F(y^1, \dots, y^n)$ defined in a neighborhood of $y_0 = (y_0^1, \dots, y_0^n) = (f^1(x_0), \dots, f^n(x_0)) = f(x_0)$, the relation

$$F\left(f^1\left(x^1, \dots, x^m\right), \dots, f^n\left(x^1, \dots, x^m\right)\right) \equiv 0$$

is possible at all points of a neighborhood of x_0 only when $F(y^1, \dots, y^n) \equiv 0$ in a neighborhood of y_0.

The linear independence studied in algebra is independence with respect to linear relations

$$F\left(y^1, \dots, y^n\right) = \lambda_1 y^1 + \dots + \lambda_n y^n.$$

If a system is not functionally independent, it is said to be *functionally dependent*.

When vectors are linearly dependent, one of them obviously is a linear combination of the others. A similar situation holds in the relation of functional dependence of a system of smooth functions.

Proposition 1 *If a system $f^i(x^1, \dots, x^m)$ $(i = 1, \dots, n)$ of smooth functions defined on a neighborhood $U(x_0)$ of the point $x_0 \in \mathbb{R}^m$ is such that the rank of the matrix*

$$\begin{pmatrix} \frac{\partial f^1}{\partial x^1} & \cdots & \frac{\partial f^1}{\partial x^m} \\ \vdots & \ddots & \vdots \\ \frac{\partial f^n}{\partial x^1} & \cdots & \frac{\partial f^n}{\partial x^m} \end{pmatrix}(x)$$

is equal to the same number k at every point $x \in U$, then

a) *when $k = n$, the system is functionally independent in a neighborhood of x_0;*

b) *when $k < n$, there exist a neighborhood of x_0 and k functions of the system, say $f^1, \dots, f^k$ such that the other $n - k$ functions can be represented as*

$$f^i\left(x^1, \dots, x^m\right) = g^i\left(f^1\left(x^1, \dots, x^m\right), \dots, f^k\left(x^1, \dots, x^m\right)\right)$$

in this neighborhood, where $g^i(y^1, \dots, y^k)$, $(i = k+1, \dots, n)$ are smooth functions defined in a neighborhood of $y_0 = (f^1(x_0), \dots, f^n(x_0))$ and depending only on k coordinates of the variable point $y = (y^1, \dots, y^n)$.

Proof In fact, if $k = n$, then by Remark 1 after the rank theorem, the image of a neighborhood of the point x_0 under the mapping

$$\begin{aligned} y^1 &= f^1(x^1, \dots, x^m), \\ &\vdots \\ y^n &= f^n(x^1, \dots, x^m) \end{aligned} \tag{8.127}$$

contains a neighborhood of $y_0 = f(x_0)$. But then the relation

$$F\big(f^1(x^1, \dots, x^m), \dots, f^n(x^1, \dots, x^m)\big) \equiv 0$$

can hold in a neighborhood of x_0 only if

$$F\big(y^1, \dots, y^n\big) \equiv 0$$

in a neighborhood of y_0. This proves assertion a).

If $k < n$ and the rank k of the mapping (8.127) is realized on the functions $f^1, \dots, f^k$, then by Remark 2 after the rank theorem, there exists a neighborhood of $y_0 = f(x_0)$ and $n - k$ functions $g^i(y) = g^i(y^1, \dots, y^k)$ $(i = k+1, \dots, n)$, defined on that neighborhood, having the same order of smoothness as the functions of the original system, and such that relations (8.126) hold in some neighborhood of x_0. This proves b). □

We have now shown that if $k < n$ there exist $n - k$ special functions $F^i(y) = y^i - g^i(y^1, \dots, y^k)$ $(i = k+1, \dots, n)$ that establish the relations

$$F^i\big(f^1(x), \dots, f^k(x), f^i(x)\big) \equiv 0 \quad (i = k+1, \dots, n)$$

between the functions of the system $f^1, \dots, f^k, \dots, f^n$ in a neighborhood of the point x_0.

8.6.4 Local Resolution of a Diffeomorphism into a Composition of Elementary Ones

In this subsection we shall show how, using the inverse function theorem, one can represent a diffeomorphic mapping locally as a composition of diffeomorphisms, each of which changes only one coordinate.

Definition 3 A diffeomorphism $g : U \to \mathbb{R}^m$ of an open set $U \subset \mathbb{R}^m$ will be called *elementary* if its coordinate representation is

$$\begin{cases} y^i = x^i, & i \in \{1, \dots, m\},\ i \neq j, \\ y^j = g^j(x^1, \dots, x^m), \end{cases}$$

that is, under the diffeomorphism $g : U \to \mathbb{R}^m$ only one coordinate of the point being mapped is changed.

Proposition 2 *If $f : G \to \mathbb{R}^m$ is a diffeomorphism of an open set $G \subset \mathbb{R}^m$, then for any point $x_0 \in G$ there is a neighborhood of the point in which the representation $f = g_1 \circ \cdots \circ g_n$ holds, where $g_1, \ldots, g_n$ are elementary diffeomorphisms.*

Proof We shall verify this by induction.

If the original mapping f is itself elementary, the proposition holds trivially for it.

Assume that the proposition holds for diffeomorphisms that alter at most $(k - 1)$ coordinates, where $k - 1 < n$. Now consider a diffeomorphism $f : G \to \mathbb{R}^m$ that alters k coordinates:

$$\begin{aligned} y^1 &= f^1(x^1, \ldots, x^m), \\ &\vdots \\ y^k &= f^k(x^1, \ldots, x^m), \\ y^{k+1} &= x^{k+1}, \\ &\vdots \\ y^m &= x^m. \end{aligned} \tag{8.128}$$

We have assumed that it is the first k coordinates that are changed, which can be achieved by linear changes of variable. Hence this assumption causes no loss in generality.

Since f is a diffeomorphism, its Jacobi matrix $f'(x)$ is nondegenerate at each point, for

$$(f^{-1})'(f(x)) = [f'(x)]^{-1}.$$

Let us fix $x_0 \in G$ and compute the determinant of $f'(x_0)$:

$$\left|\begin{array}{ccc:ccc} \frac{\partial f^1}{\partial x^1} & \cdots & \frac{\partial f^1}{\partial x^k} & \frac{\partial f^1}{\partial x^{k+1}} & \cdots & \frac{\partial f^1}{\partial x^m} \\ \vdots & \ddots & \vdots & \vdots & \ddots & \vdots \\ \frac{\partial f^k}{\partial x^1} & \cdots & \frac{\partial f^k}{\partial x^k} & \frac{\partial f^k}{\partial x^{k+1}} & \cdots & \frac{\partial f^k}{\partial x^m} \\ \vdots & \ddots & \vdots & \vdots & \ddots & \vdots \\ & & & 1 & & 0 \\ & 0 & & & \ddots & \\ & & & 0 & & 1 \end{array}\right|(x_0) = \begin{vmatrix} \frac{\partial f^1}{\partial x^1} & \cdots & \frac{\partial f^1}{\partial x^k} \\ \vdots & \ddots & \vdots \\ \frac{\partial f^k}{\partial x^1} & \cdots & \frac{\partial f^k}{\partial x^k} \end{vmatrix}(x_0) \neq 0.$$

Thus one of the minors of order $k - 1$ of this last determinant must be nonzero. Again, for simplicity of notation, we shall assume that the principal minor of order $k - 1$ is nonzero. Now consider the auxiliary mapping $g : G \to \mathbb{R}^m$ defined by the equalities

$$\begin{aligned} u^1 &= f^1(x^1, \dots, x^m), \\ &\vdots \\ u^{k-1} &= f^{k-1}(x^1, \dots, x^m), \\ u^k &= x^k, \\ &\vdots \\ u^m &= x^m. \end{aligned} \tag{8.129}$$

Since the Jacobian

$$\begin{vmatrix} \frac{\partial f^1}{\partial x^1} & \cdots & \frac{\partial f^1}{\partial x^{k-1}} & \vdots & \frac{\partial f^1}{\partial x^k} & \cdots & \frac{\partial f^1}{\partial x^m} \\ \vdots & \ddots & \vdots & \vdots & \vdots & \ddots & \vdots \\ \frac{\partial f^{k-1}}{\partial x^1} & \cdots & \frac{\partial f^{k-1}}{\partial x^{k-1}} & \vdots & \frac{\partial f^{k-1}}{\partial x^k} & \cdots & \frac{\partial f^{k-1}}{\partial x^m} \\ \vdots & \ddots & \vdots & \vdots & \vdots & \ddots & \vdots \\ & & & \vdots & 1 & & 0 \\ & 0 & & \vdots & & \ddots & \\ & & & \vdots & 0 & & 1 \end{vmatrix} (x_0) = \begin{vmatrix} \frac{\partial f^1}{\partial x^1} & \cdots & \frac{\partial f^1}{\partial x^{k-1}} \\ \vdots & \ddots & \vdots \\ \frac{\partial f^{k-1}}{\partial x^1} & \cdots & \frac{\partial f^{k-1}}{\partial x^{k-1}} \end{vmatrix} (x_0) \neq 0$$

of the mapping $g : G \to \mathbb{R}^m$ is nonzero at $x_0 \in G$, the mapping g is a diffeomorphism in some neighborhood of x_0.

Then, in some neighborhood of $u_0 = g(x_0)$ the mapping inverse to g, $x = g^{-1}(u)$, is defined, making it possible to introduce new coordinates $(u^1, \dots, u^m)$ in a neighborhood of x_0.

Let $h = f \circ g^{-1}$. In other words, the mapping $y = h(u)$ is the mapping (8.128) $y = f(x)$ written in u-coordinates. The mapping h, being the composition of diffeomorphisms, is a diffeomorphism of some neighborhood of u_0. Its coordinate expression obviously has the form

$$\begin{aligned} y^1 &= f^1 \circ g^{-1}(u) = u^1, \\ &\vdots \\ y^{k-1} &= f^{k-1} \circ g^{-1}(u) = u^{k-1}, \\ y^k &= f^k \circ g^{-1}(u), \\ y^{k+1} &= u^{k+1}, \\ &\vdots \\ y^m &= u^m, \end{aligned}$$

that is, h is an elementary diffeomorphism.

But $f = h \circ g$, and by the induction hypothesis the mapping g defined by (8.129) can be resolved into a composition of elementary diffeomorphisms. Thus, the diffeomorphism f, which alters k coordinates, can also be resolved into a composition of elementary diffeomorphisms in a neighborhood of x_0, which completes the induction. □

8.6.5 Morse's Lemma

This same circle of ideas contains an intrinsically beautiful lemma of Morse[9] on the local reduction of smooth real-valued functions to canonical form in a neighborhood of a nondegenerate critical point. This lemma is also important in applications.

Definition 4 Let x_0 be a critical point of the function $f \in C^{(2)}(U;\mathbb{R})$ defined in a neighborhood U of this point.

The critical point x_0 is a *nondegenerate critical point of* f if the Hessian of the function at that point (that is, the matrix $\frac{\partial^2 f}{\partial x^i \partial x^j}(x_0)$ formed from the second-order partial derivatives) has a nonzero determinant.

If x_0 is a critical point of the function, that is, $f'(x_0) = 0$, then by Taylor's formula

$$f(x) - f(x_0) = \frac{1}{2!}\sum_{i,j} \frac{\partial^2 f}{\partial x^i \partial x^j}(x_0)\left(x^i - x_0^i\right)\left(x^j - x_0^j\right) + o\left(\|x - x_0\|^2\right). \quad (8.130)$$

Morse's lemma asserts that one can make a local change of coordinates $x = g(y)$ such that the function will have the form

$$(f \circ g)(y) - f(x_0) = -\left(y^1\right)^2 - \cdots - \left(y^k\right)^2 + \left(y^{k+1}\right)^2 + \cdots + \left(y^m\right)^2$$

when expressed in y-coordinates.

If the remainder term $o(\|x - x_0\|^2)$ were not present on the right-hand side of Eq. (8.130), that is, the difference $f(x) - f(x_0)$ were a simple quadratic form, then, a is known from algebra, it could be brought into the indicated canonical form by a linear transformation. Thus the assertion we are about to prove is a local version of the theorem on reduction of a quadratic form to canonical form. The proof will use the idea of the proof of this algebraic theorem. We shall also rely on the inverse function theorem and the following proposition.

Hadamard's lemma *Let $f : U \to \mathbb{R}$ be a function of class $C^{(p)}(U;\mathbb{R})$, $p \geq 1$, defined in a convex neighborhood U of the point $0 = (0, \ldots, 0) \in \mathbb{R}^m$ and such that*

---

[9] H.C.M. Morse (1892–1977) – American mathematician; his main work was devoted to the application of topological methods in various areas of analysis.

$f(0)=0$. Then there exist functions $g_i \in C^{(p-1)}(U;\mathbb{R})$ $(i=1,\dots,m)$ such that the equality

$$f\left(x^1,\dots,x^m\right)=\sum_{i=1}^m x^i g_i\left(x^1,\dots,x^m\right) \tag{8.131}$$

holds in U, and $g_i(0)=\frac{\partial f}{\partial x^i}(0)$.

Proof Equality (8.131) is essentially another useful expression for Taylor's formula with the integral form of the remainder term. It follows from the equalities

$$f\left(x^1,\dots,x^m\right)=\int_0^1 \frac{\mathrm{d}f(tx^1,\dots,tx^m)}{\mathrm{d}t}\,\mathrm{d}t=\sum_{i=1}^m x^i\int_0^1 \frac{\partial f}{\partial x^i}\left(tx^1,\dots,tx^m\right)\mathrm{d}t,$$

if we set

$$g_i\left(x^1,\dots,x^m\right)=\int_0^1 \frac{\partial f}{\partial x^i}\left(tx^1,\dots,tx^m\right)\mathrm{d}t \quad (i=1,\dots,m).$$

The fact that $g_i(0)=\frac{\partial f}{\partial x^i}(0)$ $(i=1,\dots,m)$ is obvious, and it is also not difficult to verify that $g_i\in C^{(p-1)}(U;\mathbb{R})$. However, we shall not undertake the verification just now, since we shall later give a general rule for differentiating an integral depending on a parameter, from which the property we need for the functions g_i will follow immediately.

Thus, up to this verification, Hadamard's formula (8.131) is proved. □

Morse's lemma *If $f: G\to\mathbb{R}$ is a function of class $C^{(3)}(G;\mathbb{R})$ defined on an open set $G\subset\mathbb{R}^m$ and $x_0\in G$ is a nondegenerate critical point of that function, then there exists a diffeomorphism $g: V\to U$ of some neighborhood of the origin 0 in $\mathbb{R}^m$ onto a neighborhood U of x_0 such that*

$$(f\circ g)(y)=f(x_0)-\left[\left(y^1\right)^2+\dots+\left(y^k\right)^2\right]+\left[\left(y^{k+1}\right)^2+\dots+\left(y^m\right)^2\right]$$

for all $y\in V$.

Proof By linear changes of variable we can reduce the problem to the case when $x_0=0$ and $f(x_0)=0$, and from now on we shall assume that these conditions hold.

Since $x_0=0$ is a critical point of f, we have $g_i(0)=0$ in formula (8.131) $(i=1,\dots,m)$. Then, also by Hadamard's lemma,

$$g_i\left(x^1,\dots,x^m\right)=\sum_{j=1}^m x^j h_{ij}\left(x^1,\dots,x^m\right),$$

where h_{ij} are smooth functions in a neighborhood of 0 and consequently

$$f(x^1, \dots, x^m) = \sum_{i,j=1}^{m} x^i x^j h_{ij}(x^1, \dots, x^m). \tag{8.132}$$

By making the substitution $\tilde{h}_{ij} = \frac{1}{2}(h_{ij} + h_{ji})$ if necessary, we can assume that $h_{ij} = h_{ji}$. We remark also that, by the uniqueness of the Taylor expansion, the continuity of the functions h_{ij} implies that $h_{ij}(0) = \frac{\partial^2 f}{\partial x^i \partial x^j}(0)$ and hence the matrix $(h_{ij}(0))$ is nondegenerate.

The function f has now been written in a manner that resembles a quadratic form, and we wish, so to speak, to reduce it to diagonal form.

As in the classical case, we proceed by induction.

Assume that there exist coordinates $u^1, \dots, u^m$ in a neighborhood U_1 of $0 \in \mathbb{R}^m$, that is, a diffeomorphism $x = \varphi(u)$, such that

$$(f \circ \varphi)(u) = \pm(u^1)^2 \pm \cdots \pm (u^{r-1})^2 + \sum_{i,j=r}^{m} u^i u^j H_{ij}(u^1, \dots, u^m) \tag{8.133}$$

in the coordinates $u^1, \dots, u^m$, where $r \ge 1$ and $H_{ij} = H_{ji}$.

We observe that relation (8.133) holds for $r = 1$, as one can see from (8.132), where $H_{ij} = h_{ij}$.

By the hypothesis of the lemma the quadratic form $\sum_{i,j=1}^m x^i x^j h_{ij}(0)$ is nondegenerate, that is, $\det(h_{ij}(0)) \ne 0$. The change of variable $x = \varphi(u)$ is carried out by a diffeomorphism, so that $\det \varphi'(0) \ne 0$. But then the matrix of the quadratic form $\pm(u^1)^2 \pm \cdots \pm (u^{r-1})^2 + \sum_{i,j=r}^m u^i u^j H_{ij}(0)$ obtained from the matrix $(h_{ij}(0))$ through right-multiplication by the matrix $\varphi'(0)$ and left-multiplication by the transpose of $\varphi'(0)$ is also nondegenerate. Consequently, at least one of the numbers $H_{ij}(0)$ $(i, j = r, \dots, m)$ is nonzero. By a linear change of variable we can bring the form $\sum_{i,j=r}^m u^i u^j H_{ij}(0)$ to diagonal form, and so we may assume that $H_{rr}(0) \ne 0$ in Eq. (8.133). By the continuity of the functions $H_{ij}(u)$ the inequality $H_{rr}(u) \ne 0$ will also hold in some neighborhood of $u = 0$.

Let us set $\psi(u^1, \dots, u^m) = \sqrt{|H_{rr}(u)|}$. Then the function ψ belongs to the class $C^{(1)}(U_2; \mathbb{R})$ in some neighborhood $U_2 \subset U_1$ of $u = 0$. We now change to coordinates $(v^1, \dots, v^m)$ by the formulas

$$\begin{aligned} v^i &= u^i, \quad i \ne r, \\ v^r &= \psi(u^1, \dots, u^m)\left(u^r + \sum_{i>r} \frac{u^i H_{ir}(u^1, \dots, u^m)}{H_{rr}(u^1, \dots, u^m)}\right). \end{aligned} \tag{8.134}$$

The Jacobian of the transformation (8.134) at $u = 0$ is obviously equal to $\psi(0)$, that is, it is nonzero. Then by the inverse function theorem we can assert that in some neighborhood $U_3 \subset U_2$ of $u = 0$ the mapping $v = \psi(u)$ defined by (8.134) is a diffeomorphism of class $C^{(1)}(U_3; \mathbb{R}^m)$ and therefore the variables $(v^1, \dots, v^m)$ can indeed serve as coordinates of points in U_3.

We now separate off in Eq. (8.133) all terms

$$u^r u^r H_{rr}(u^1, \dots, u^m) + 2 \sum_{j=r+1}^{m} u^r u^j H_{rj}(u^1, \dots, u^m), \tag{8.135}$$

containing u^r. In the expression (8.135) for the sum of these terms we have used the fact that $H_{ij} = H_{ji}$.

Comparing (8.134) and (8.135), we see that we can rewrite (8.135) in the form

$$\pm v^r v^r - \frac{1}{H_{rr}} \Big(\sum_{i>r} u^i H_{ir}(u^1, \dots, u^m) \Big)^2 .$$

The ambiguous sign $\pm$ appears in front of $v^r v^r$ because $H_{rr} = \pm(\psi)^2$, the positive sign being taken if $H_{rr} > 0$ and the negative sign if $H_{rr} < 0$.

Thus, after the substitution $v = \psi(u)$, the expression (8.133) becomes the equality

$$(f \circ \varphi \circ \psi^{-1})(v) = \sum_{i=1}^{r} [\pm (v^i)^2] + \sum_{i,j>r} v^i v^j \tilde{H}_{ij}(v^1, \dots, v^m),$$

where $\tilde{H}_{ij}$ are new smooth functions that are symmetric with respect to the indices i and j. The mapping $\varphi \circ \psi^{-1}$ is a diffeomorphism. Thus the induction from $r-1$ to r is now complete, and Morse's lemma is proved. □

8.6.6 Problems and Exercises

1. Compute the Jacobian of the change of variable (8.118) from polar coordinates to Cartesian coordinates in $\mathbb{R}^m$.

2. a) Let x_0 be a noncritical point of a smooth function $F : U \to \mathbb{R}$ defined in a neighborhood U of $x_0 = (x_0^1, \dots, x_0^m) \in \mathbb{R}^m$. Show that in some neighborhood $\tilde{U} \subset U$ of x_0 one can introduce curvilinear coordinates $(\xi^1, \dots, \xi^m)$ such that the set of points defined by the condition $F(x) = F(x_0)$ will be given by the equation $\xi^m = 0$ in these new coordinates.

b) Let $\varphi, \psi \in C^{(k)}(D; \mathbb{R})$, and suppose that $(\varphi(x) = 0) \Rightarrow (\psi(x) = 0)$ in the domain D. Show that if $\operatorname{grad} \varphi \neq 0$, then there is a decomposition $\psi = \theta \cdot \varphi$ in D, where $\theta \in C^{(k-1)}(D; \mathbb{R})$.

3. Let $f : \mathbb{R}^2 \to \mathbb{R}^2$ be a smooth mapping satisfying the Cauchy–Riemann equations

$$\frac{\partial f^1}{\partial x^1} = \frac{\partial f^2}{\partial x^2}, \qquad \frac{\partial f^1}{\partial x^2} = -\frac{\partial f^2}{\partial x^1}.$$

a) Show that the Jacobian of such a mapping is zero at a point if and only if $f'(x)$ is the zero matrix at that point.

b) Show that if $f'(x) \neq 0$, then the inverse f^{-1} to the mapping f is defined in a neighborhood of f and also satisfies the Cauchy–Riemann equations.

4. *Functional dependence* (direct proof).

a) Show that the functions $\pi^i(x) = x^i$ $(i = 1, \dots, m)$, regarded as functions of the point $x = (x^1, \dots, x^m) \in \mathbb{R}^m$, form an independent system of functions in a neighborhood of any point of $\mathbb{R}^m$.

b) Show that, for any function $f \in C(\mathbb{R}^m; \mathbb{R})$ the system $\pi^1, \dots, \pi^m, f$ is functionally dependent.

c) If the system of smooth functions $f^1, \dots, f^k$, $k < m$, is such that the rank of the mapping $f = (f^1, \dots, f^k)$ equals k at a point $x_0 = (x_0^1, \dots, x_0^m) \in \mathbb{R}^m$, then in some neighborhood of this point one can complete it to an independent system $f^1, \dots, f^m$ consisting of m smooth functions.

d) If the system

$$\xi^i = f^i(x^1, \dots, x^m) \quad (i = 1, \dots, m)$$

of smooth functions is such that the mapping $f = (f^1, \dots, f^m)$ has rank m at the point $x_0 = (x_0^1, \dots, x_0^m)$, then the variables $(\xi^1, \dots, \xi^m)$ can be used as curvilinear coordinates in some neighborhood $U(x_0)$ of x_0, and any function $\varphi : U(x_0) \to \mathbb{R}$ can be written as $\varphi(x) = F(f^1(x), \dots, f^m(x))$, where $F = \varphi \circ f^{-1}$.

e) The rank of the mapping provided by a system of smooth functions is also called the *rank* of the system. Show that if the rank of a system of smooth functions $f^i(x^1, \dots, x^m)$ $(i = 1, \dots, k)$ is k and the rank of the system $f^1, \dots, f^m, \varphi$ is also k at some point $x_0 \in \mathbb{R}^m$, then $\varphi(x) = F(f^1(x), \dots, f^k(x))$ in a neighborhood of the point.

Hint: Use c) and d) and show that

$$F(f^1, \dots, f^m) = F(f^1, \dots, f^k).$$

5. Show that the rank of a smooth mapping $f : \mathbb{R}^m \to \mathbb{R}^n$ is a lower semicontinuous function, that is rank $f(x) \geq$ rank $f(x_0)$ in a neighborhood of a point $x_0 \in \mathbb{R}^m$.

6. a) Give a direct proof of Morse's lemma for functions $f : \mathbb{R} \to \mathbb{R}$.

b) Determine whether Morse's lemma is applicable at the origin to the following functions:

$$f(x) = x^3; \qquad f(x) = x \sin \frac{1}{x}; \qquad f(x) = \mathrm{e}^{-1/x^2} \sin^2 \frac{1}{x};$$

$$f(x, y) = x^3 - 3xy^2; \qquad f(x, y) = x^2.$$

c) Show that nondegenerate critical points of a function $f \in C^{(3)}(\mathbb{R}^m; \mathbb{R})$ are isolated: each of them has a neighborhood in which it is the only critical point of f.

d) Show that the number k of negative squares in the canonical representation of a function in the neighborhood of a nondegenerate critical point is independent of the reduction method, that is, independent of the coordinate system in which the function has canonical form. This number is called the *index of the critical point.*

8.7 Surfaces in $\mathbb{R}^n$ and the Theory of Extrema with Constraint

To acquire an informal understanding of the theory of extrema with constraint, which is important in applications, it is useful to have some elementary information on surfaces (manifolds) in $\mathbb{R}^n$.

8.7.1 k-Dimensional Surfaces in $\mathbb{R}^n$

Generalizing the concept of a law of motion of a point mass $x = x(t)$, we have previously introduced the concept of a path in $\mathbb{R}^n$ as a continuous mapping $\Gamma : I \to \mathbb{R}^n$ of an interval $I \subset \mathbb{R}$. The degree of smoothness of the path was defined a the degree of smoothness of this mapping. The support $\Gamma(I) \subset \mathbb{R}^n$ of a path can be a rather peculiar set in $\mathbb{R}^n$, which it would be a great stretch to call a curve in some instances. For example, the support of a path might be a single point.

Similarly, a continuous or smooth mapping $f : I^k \to \mathbb{R}^n$ of a k-dimensional interval $I^k \subset \mathbb{R}^k$, called a *singular k-cell* in $\mathbb{R}^n$, may have as its image $f(I^k)$ not at all what one would like to call a k-dimensional surface in $\mathbb{R}^n$. For example, it might again be simply a point.

In order for a smooth mapping $f : G \to \mathbb{R}^n$ of a domain $G \subset \mathbb{R}^k$ to define a k-dimensional geometric figure in $\mathbb{R}^n$ whose points are described by k independent parameters $(t^1, \dots, t^k) \in G$, it suffices, as we know from the preceding section, to require that the rank of the mapping $f : G \to \mathbb{R}^n$ be k at each point $t \in G$ (naturally, $k \le n$). In that case the mapping $f : G \to f(G)$ is locally one-to-one (that is, in a neighborhood of each point $t \in G$).

Indeed, suppose rank $f(t_0) = k$ and this rank is realized, for example, on the first k of the n functions

$$\begin{cases} x^1 = f^1(t^1, \dots, t^k), \\ \vdots \\ x^n = f^n(t^1, \dots, t^k) \end{cases} \tag{8.136}$$

that define the coordinate expressions for the mapping $f : G \to \mathbb{R}^n$.

Then, by the inverse function theorem the variables $t^1, \dots, t^k$ can be expressed in terms of $x^1, \dots, x^k$ in some neighborhood $U(t_0)$ of t_0. It follows that the set $f(U(t_0))$ can be written as

$$x^{k+1} = \varphi^{k+1}(x^1, \dots, x^k), \dots, x^n = \varphi^n(x^1, \dots, x^k)$$

(that is, it projects in a one-to-one manner onto the coordinate plane of $x^1, \dots, x^k$), and therefore the mapping $f : U(t_0) \to f(U(t_0))$ is indeed one-to-one.

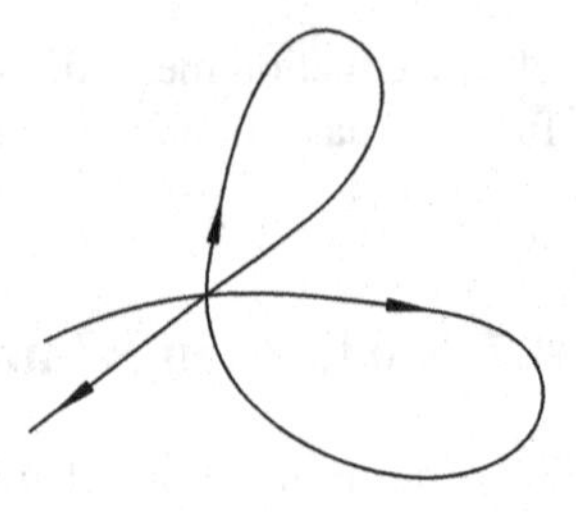

Fig. 8.9

However, even the simple example of a smooth one-dimensional path (Fig. 8.9) makes it clear that the local injectivity of the mapping $f : G \to \mathbb{R}^n$ from the parameter domain G into $\mathbb{R}^n$ is by no means necessarily a global injectivity. The trajectory may have multiple self-intersections, so that if we wish to define a smooth k-dimensional surface in $\mathbb{R}^n$ and picture it as a set that has the structure of a slightly deformed piece of a k-dimensional plane (a k-dimensional subspace of $\mathbb{R}^n$) near each of its points, it is not enough merely to map a canonical piece $G \subset \mathbb{R}^k$ of a k-dimensional plane in a regular manner into $\mathbb{R}^n$. It is also necessary to be sure that it happens to be globally imbedded in this space.

Definition 1 We shall call a set $S \subset \mathbb{R}^n$ a *k-dimensional smooth surface* in $\mathbb{R}^n$ (or a *k-dimensional submanifold of* $\mathbb{R}^n$) if for every point $x_0 \in S$ there exist a neighborhood $U(x_0)$ in $\mathbb{R}^n$ and a diffeomorphism $\varphi : U(x_0) \to I^n$ of this neighborhood onto the standard n-dimensional cube $I^n = \{t \in \mathbb{R}^n \mid |t^i| < 1, i = 1, \dots, n\}$ of the space $\mathbb{R}^n$ under which the image of the set $S \cap U(x_0)$ is the portion of the k-dimensional plane in $\mathbb{R}^n$ defined by the relations $t^{k+1} = 0, \dots, t^n = 0$ lying inside I^n (Fig. 8.10).

We shall measure the degree of smoothness of the surface S by the degree of smoothness of the diffeomorphism φ.

If we regard the variables $t^1, \dots, t^n$ as new coordinates in a neighborhood of $U(x_0)$, Definition 1 can be rewritten briefly as follows: the set $S \subset \mathbb{R}^n$ is a k-

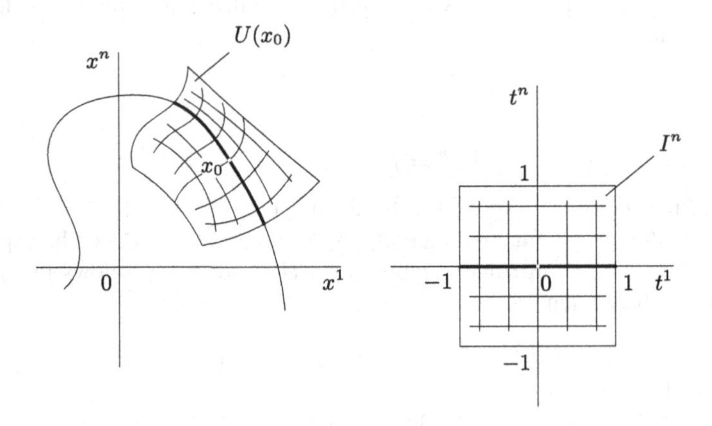

Fig. 8.10

dimensional surface (k-dimensional submanifold) in $\mathbb{R}^n$ if for every point $x_0 \in S$ there is a neighborhood $U(x_0)$ and coordinates $t^1, \dots, t^n$ in $U(x_0)$ such that in these coordinates the set $S \cap U(x_0)$ is defined by the relations

$$t^{k+1} = \cdots = t^n = 0.$$

The role of the standard n-dimensional cube in Definition 1 is rather artificial and approximately the same as the role of the standard size and shape of a page in a geographical atlas. The canonical location of the interval in the coordinate system $t^1, \dots, t^n$ is also a matter of convention and nothing more, since any cube in $\mathbb{R}^n$ can always be transformed into the standard n-dimensional cube by an additional linear diffeomorphism.

We shall often use this remark when abbreviating the verification that a set $S \subset \mathbb{R}^n$ is a surface in $\mathbb{R}^n$.

Let us consider some examples.

Example 1 The space $\mathbb{R}^n$ itself is an n-dimensional surface of class $C^{(\infty)}$. As the mapping $\varphi : \mathbb{R}^n \to I^n$ here, one can take, for example, the mapping

$$\xi^i = \frac{2}{\pi} \arctan x^i \quad (i = 1, \dots, n).$$

Example 2 The mapping constructed in Example 1 also establishes that the subspace of the vector space $\mathbb{R}^n$ defined by the conditions $x^{k+1} = \cdots = x^n = 0$ is a k-dimensional surface in $\mathbb{R}^n$ (or a k-dimensional submanifold of $\mathbb{R}^n$).

Example 3 The set in $\mathbb{R}^n$ defined by the system of relations

$$\begin{cases} a_1^1 x^1 + \cdots + a_k^1 x^k + a_{k+1}^1 x^{k+1} + \cdots + a_n^1 x^n = 0, \\ \vdots \\ a_1^{n-k} x^1 + \cdots + a_k^{n-k} x^k + a_{k+1}^{n-k} x^{k+1} + \cdots + a_n^{n-k} x^n = 0, \end{cases}$$

provided this system has rank $n - k$, is a k-dimensional submanifold of $\mathbb{R}^n$.

Indeed, suppose for example that the determinant

$$\begin{vmatrix} a_{k+1}^1 & \cdots & a_n^1 \\ \vdots & \ddots & \vdots \\ a_{k+1}^{n-k} & \cdots & a_n^{n-k} \end{vmatrix}$$

is nonzero. Then the linear transformation

$$\begin{aligned} t^1 &= x^1, \\ &\vdots \\ t^k &= x^k, \end{aligned}$$

$$t^{k+1} = a_1^1 x^1 + \cdots + a_n^1 x^n,$$

$$\vdots$$

$$t^n = a_1^{n-k} x^1 + \cdots + a_n^{n-k} x^n,$$

is obviously nondegenerate. In the coordinates $t^1, \dots, t^n$ the set is defined by the conditions $t^{k+1} = \cdots = t^n = 0$, already considered in Example 2.

Example 4 The graph of a smooth function $x^n = f(x^1, \dots, x^{n-1})$ defined in a domain $G \subset \mathbb{R}^{n-1}$ is a smooth $(n-1)$-dimensional surface in $\mathbb{R}^n$.

Indeed, setting

$$\begin{cases} t^i = x^i \quad (i = 1, \dots, n-1), \\ t^n = x^n - f\left(x^1, \dots, x^{n-1}\right), \end{cases}$$

we obtain a coordinate system in which the graph of the function has the equation $t^n = 0$.

Example 5 The circle $x^2 + y^2 = 1$ in $\mathbb{R}^2$ is a one-dimensional submanifold of $\mathbb{R}^2$, as is established by the locally invertible conversion to polar coordinates (ρ, φ) studied in the preceding section. In these coordinates the circle has equation $\rho = 1$.

Example 6 This example is a generalization of Example 3 and at the same time, as can be seen from Definition 1, gives a general form for the coordinate expression of submanifolds of $\mathbb{R}^n$.

Let $F^i(x^1, \dots, x^n)$ $(i = 1, \dots, n-k)$ be a system of smooth functions of rank $n-k$. We shall show that the relations

$$\begin{cases} F^1\left(x^1, \dots, x^k, x^{k+1}, \dots, x^n\right) = 0, \\ \vdots \\ F^{n-k}\left(x^1, \dots, x^k, x^{k+1}, \dots, x^n\right) = 0 \end{cases} \tag{8.137}$$

define a k-dimensional submanifold S in $\mathbb{R}^n$.

Suppose the condition

$$\begin{vmatrix} \frac{\partial F^1}{\partial x^{k+1}} & \cdots & \frac{\partial F^1}{\partial x^n} \\ \vdots & \ddots & \vdots \\ \frac{\partial F^{n-k}}{\partial x^{k+1}} & \cdots & \frac{\partial F^{n-k}}{\partial x^n} \end{vmatrix} (x_0) \neq 0 \tag{8.138}$$

holds at a point $x_0 \in S$. Then by the inverse function theorem the transformation

$$\begin{cases} t^i = x^i \quad (i = 1, \dots, k), \\ t^i = F^{i-k}\left(x^1, \dots, x^n\right) \quad (i = k+1, \dots, n) \end{cases}$$

is a diffeomorphism of a neighborhood of this point.

In the new coordinates $t^1, \dots, t^n$ the original system will have the form $t^{k+1} = \dots = t^n = 0$; thus, S is a k-dimensional smooth surface in $\mathbb{R}^n$.

Example 7 The set E of points of the plane $\mathbb{R}^2$ satisfying the equation $x^2 - y^2 = 0$ consists of two lines that intersect at the origin. This set is not a one-dimensional submanifold of $\mathbb{R}^2$ (verify this!) precisely because of this point of intersection.

If the origin $0 \in \mathbb{R}^2$ is removed from E, then the set $E \backslash 0$ will now obviously satisfy Definition 1. We remark that the set $E \backslash 0$ is not connected. It consists of four pairwise disjoint rays.

Thus a k-dimensional surface in $\mathbb{R}^n$ satisfying Definition 1 may happen to be a disconnected subset consisting of several connected components (and these components are connected k-dimensional surfaces). A surface in $\mathbb{R}^n$ is often taken to mean a connected k-dimensional surface. Just now we shall be interested in the problem of finding extrema of functions defined on surfaces. These are local problems, and therefore connectivity will not manifest itself in them.

Example 8 If a smooth mapping $f : G \to \mathbb{R}^n$ of the domain $G \subset \mathbb{R}^n$ defined in coordinate form by (8.136) has rank k at the point $t_0 \in G$, then there exists a neighborhood $U(t_0) \subset G$ of this point whose image $f(U(t_0)) \subset \mathbb{R}^n$ is a smooth surface in $\mathbb{R}^n$.

Indeed, as already noted above, in this case relations (8.136) can be replaced by the equivalent system

$$\begin{cases} x^{k+1} = \varphi^{k+1}(x^1, \dots, x^k), \\ \vdots \\ x^n = \varphi^n(x^1, \dots, x^k) \end{cases} \tag{8.139}$$

in some neighborhood $U(t_0)$ of $t_0 \in G$. (For simplicity of notation, we assume that the system $f^1, \dots, f^k$ has rank k.) Setting

$$F^i(x^1, \dots, x^n) = x^{k+i} - \varphi^{k+1}(x^1, \dots, x^k) \quad (i = 1, \dots, n-k),$$

we write the system (8.139) in the form (8.137). Since relations (8.138) are satisfied, Example 6 guarantees that the set $f(U(t_0))$ is indeed a k-dimensional smooth surface in $\mathbb{R}^n$.

8.7.2 The Tangent Space

In studying the law of motion $x = x(t)$ of a point mass in $\mathbb{R}^3$, starting from the relation

$$x(t) = x(0) + x'(0)t + o(t) \quad \text{as } t \to 0 \tag{8.140}$$

and assuming that the point $t = 0$ is not a critical point of the mapping $\mathbb{R} \ni t \mapsto x(t) \in \mathbb{R}^3$, that is, $x'(0) \neq 0$, we defined the line tangent to the trajectory at the point

$x(0)$ as the linear subset of $\mathbb{R}^3$ given in parametric form by the equation

$$x - x_0 = x'(0)t \tag{8.141}$$

or the equation

$$x - x_0 = \xi \cdot t, \tag{8.142}$$

where $x_0 = x(0)$ and $\xi = x'(0)$ is a direction vector of the line.

In essence, we did a similar thing in defining the tangent plane to the graph of a function $z = f(x, y)$ in $\mathbb{R}^3$. Indeed, supplementing the relation $z = f(x, y)$ with the trivial equalities $x = x$ and $y = y$, we obtain a mapping $\mathbb{R}^2 \ni (x, y) \mapsto (x, y, f(x, y)) \in \mathbb{R}^3$ to which the tangent at the point (x_0, y_0) is the linear mapping

$$\begin{pmatrix} x - x_0 \\ y - y_0 \\ z - z_0 \end{pmatrix} = \begin{pmatrix} 1 & 0 \\ 0 & 1 \\ f'_x(x_0, y_0) & f'_y(x_0, y_0) \end{pmatrix} \begin{pmatrix} x - x_0 \\ y - y_0 \end{pmatrix}, \tag{8.143}$$

where $z_0 = f(x_0, y_0)$.

Setting $t = (x - x_0, y - y_0)$ and $x = (x - x_0, y - y_0, z - z_0)$ here, and denoting the Jacobi matrix in (8.143) for this transformation by $x'(0)$, we remark that its rank is two and that in this notation relation (8.143) has the form (8.141).

The peculiarity of relation (8.143) is that only the last equality in the set of three equalities

$$\begin{cases} x - x_0 = x - x_0, \\ y - y_0 = y - y_0, \\ z - z_0 = f'_x(x_0, y_0)(x - x_0) + f'_y(x_0, y - 0)(y - y_0), \end{cases} \tag{8.144}$$

to which it is equivalent is a nontrivial relation. That is precisely the reason it is retained as the equation defining the plane tangent to the graph of $z = f(x, y)$ at (x_0, y_0, z_0).

This observation can now be used to give the definition of the k-dimensional plane tangent to a k-dimensional smooth surface $S \subset \mathbb{R}^n$.

It can be seen from Definition 1 of a surface that in a neighborhood of each of its points $x_0 \in S$ a k-dimensional surface S can be defined parametrically, that is, using mappings $I^k \ni (t^1, \dots, t^k) \mapsto (x^1, \dots, x^n) \in S$. Such a parametrization can be taken to be the restriction of the mapping $\varphi^{-1} : I^n \to U(x_0)$ to the k-dimensional plane $t^{k+1} = \dots = t^n = 0$ (see Fig. 8.10).

Since φ^{-1} is a diffeomorphism, the Jacobian of the mapping $\varphi^{-1} : I^n \to U(x_0)$ is nonzero at each point of the cube I^n. But then the mapping $I^k \ni (t^1, \dots, t^k) \mapsto (x^1, \dots, x^n) \in S$ obtained by restricting φ^{-1} to this plane must also have rank k at each point of I^k.

Now setting $(t^1, \dots, t^k) = t \in I^k$ and denoting the mapping $I^k \ni t \mapsto x \in S$ by $x = x(t)$, we obtain a local parametric representation of the surface S possessing the property expressed by (8.140), on the basis of which we take Eq. (8.141) as the equation of the tangent space or tangent plane to the surface $S \subset \mathbb{R}^n$ at $x_0 \in S$.

Thus we adopt the following definition.

Definition 2 If a k-dimensional surface $S \subset \mathbb{R}^n$, $1 \le k \le n$, is defined parametrically in a neighborhood of $x_0 \in S$ by means of a smooth mapping $(t^1, \dots, t^k) = t \mapsto x = (x^1, \dots, x^n)$ such that $x_0 = x(0)$ and the matrix $x'(0)$ has rank k, then the k-dimensional surface in $\mathbb{R}^n$ defined parametrically by the matrix equality (8.141) is called the *tangent plane* or *tangent space to the surface S at $x_0 \in S$*.

In coordinate form the following system of equations corresponds to Eq. (8.141):

$$\begin{cases} x^1 - x_0^1 = \dfrac{\partial x^1}{\partial t^1}(0)t^1 + \cdots + \dfrac{\partial x^1}{\partial t^k}(0)t^k, \\ \vdots \\ x^n - x_0^n = \dfrac{\partial x^n}{\partial t^1}(0)t^1 + \cdots + \dfrac{\partial x^n}{\partial t^k}(0)t^k. \end{cases} \tag{8.145}$$

We shall denote[10] the tangent space to the surface S at $x \in S$, as before, by TS_x.

An important and useful exercise, which the reader can do independently, is to prove the invariance of the definition of the tangent space and the verification that the linear mapping $t \mapsto x'(0)t$ tangent to the mapping $t \mapsto x(t)$, which defines the surface S locally, provides a mapping of the space $\mathbb{R}^k = T\mathbb{R}^k_0$ onto the plane $TS_{x(0)}$ (see Problem 3 at the end of this section).

Let us now determine the form of the equation of the tangent plane to the k-dimensional surface S defined in $\mathbb{R}^n$ by the system (8.137). For definiteness we shall assume that condition (8.138) holds in a neighborhood of the point $x_0 \in S$.

Setting $(x^1, \dots, x^k) = u$, $(x^{k+1}, \dots, x^n) = v$, $(F^1, \dots, F^{n-k}) = F$, we write the system (8.137) in the form

$$F(u, v) = 0, \tag{8.146}$$

and (8.138) as

$$\det F'_v(u, v) \ne 0. \tag{8.147}$$

Using the implicit function theorem, in a neighborhood of the point $(u_0, v_0) = (x_0^1, \dots, x_0^k, x_0^{k+1}, \dots, x_0^n)$ we pass from relation (8.146) to the equivalent relation

$$v = f(u), \tag{8.148}$$

which, when we supplement it with the identity $u = u$, yields the parametric representation of the surface S in a neighborhood of $x_0 \in S$:

$$\begin{cases} u = u, \\ v = f(u). \end{cases} \tag{8.149}$$

[10]This is a slight departure from the usual notation $T_x S$ or $T_x(S)$.

On the basis of Definition 2 we obtain from (8.149) the parametric equation

$$\begin{cases} u - u_0 = E \cdot t, \\ v - v_0 = f'(u_0) \cdot t \end{cases} \tag{8.150}$$

of the tangent plane; here E is the identity matrix and $t = u - u_0$.

Just as was done in the case of the system (8.144), we retain in the system (8.150) only the nontrivial equation

$$v - v_0 = f'(u_0)(u - u_0), \tag{8.151}$$

which contains the connection of the variables $x^1, \dots, x^k$ with the variables $x^{k+1}, \dots, x^n$ that determine the tangent space.

Using the relation

$$f'(u_0) = -\big[F'_v(u_0, v_0)\big]^{-1}\big[F'_u(u_0, v_0)\big],$$

which follows from the implicit function theorem, we rewrite (8.151) as

$$F'_u(u_0, v_0)(u - u_0) + F'_v(u_0, v_0)(v - v_0) = 0,$$

from which, after returning to the variables $(x^1, \dots, x^n) = x$, we obtain the equation we are seeking for the tangent space $TS_{x_0} \subset \mathbb{R}^n$, namely

$$F'_x(x_0)(x - x_0) = 0. \tag{8.152}$$

In coordinate representation Eq. (8.152) is equivalent to the system of equations

$$\begin{cases} \dfrac{\partial F^1}{\partial x^1}(x_0)\big(x^1 - x_0^1\big) + \dots + \dfrac{\partial F^1}{\partial x^n}(x_0)\big(x^n - x_0^n\big) = 0, \\ \vdots \\ \dfrac{\partial F^{n-k}}{\partial x^1}(x_0)\big(x^1 - x_0^1\big) + \dots + \dfrac{\partial F^{n-k}}{\partial x^n}(x_0)\big(x^n - x_0^n\big) = 0. \end{cases} \tag{8.153}$$

By hypothesis the rank of this system is $n - k$, and hence it defines a k-dimensional plane in $\mathbb{R}^n$.

The affine equation (8.152) is equivalent (given the point x_0) to the vector equation

$$F'_x(x_0) \cdot \xi = 0, \tag{8.154}$$

in which $\xi = x - x_0$.

Hence the vector ξ lies in the plane TS_{x_0} tangent at $x_0 \in S$ to the surface $S \subset \mathbb{R}^n$ defined by the equation $F(x) = 0$ if and only if it satisfies condition (8.154). Thus TS_{x_0} can be regarded as the vector space consisting of the vectors ξ that satisfy (8.154).

It is this fact that motivates the use of the term *tangent space*.

Let us now prove the following proposition, which we have already encountered in a special case (see Sect. 6.4).

Proposition *The space TS_{x_0} tangent to a smooth surface $S \subset \mathbb{R}^n$ at a point $x_0 \in S$ consists of the vectors tangent to smooth curves lying on the surface S and passing through the point x_0.*

Proof Let the surface S be defined in a neighborhood of the point $x_0 \in S$ by a system of equations (8.137), which we write briefly as

$$F(x) = 0, \tag{8.155}$$

where $F = (F^1, \dots, F^{n-k})$, $x = (x^1, \dots, x^n)$. Let $\Gamma : I \to S$ be an arbitrary smooth path with support on S. Taking $I = \{t \in \mathbb{R} \mid |t| < 1\}$, we shall assume that $x(0) = x_0$. Since $x(t) \in S$ for $t \in I$, after substituting $x(t)$ into Eq. (8.155), we obtain

$$F\big(x(t)\big) \equiv 0 \tag{8.156}$$

for $t \in I$. Differentiating this identity with respect to t, we find that

$$F'_x\big(x(t)\big) \cdot x'(t) \equiv 0.$$

In particular, when $t = 0$, setting $\xi = x'(0)$, we obtain

$$F'_x(x_0)\xi = 0,$$

that is, the vector ξ tangent to the trajectory at x_0 (at time $t = 0$) satisfies Eq. (8.154) of the tangent space TS_{x_0}.

We now show that for every vector ξ satisfying Eq. (8.154) there exists a smooth path $\Gamma : I \to S$ that defines a curve on S passing through x_0 at $t = 0$ and having the velocity vector ξ at time $t = 0$.

This will simultaneously establish the existence of smooth curves on S passing through x_0, which we assumed implicitly in the proof of the first part of the proposition.

Suppose for definiteness that condition (8.138) holds. Then, knowing the first k coordinates $\xi^1, \dots, \xi^k$ of the vector $\xi = (\xi^1, \dots, \xi^k, \xi^{k+1}, \dots, \xi^n)$, we determine the other coordinates $\xi^{k+1}, \dots, \xi^n$ uniquely from Eq. (8.154) (which is equivalent to the system (8.153)). Thus, if we establish that a vector $\tilde{\xi} = (\xi^1, \dots, \xi^k, \tilde{\xi}^{k+1}, \dots, \tilde{\xi}^n)$ satisfies Eq. (8.154), we can conclude that $\tilde{\xi} = \xi$. We shall make use of this fact.

Again, as was done above, we introduce for convenience the notation $u = (x^1, \dots, x^k)$, $v = (x^{k+1}, \dots, x^n)$, $x = (x^1, \dots, x^n) = (u, v)$, and $F(x) = F(u, v)$. Then Eq. (8.155) will have the form (8.146) and condition (8.138) will have the form (8.147). In the subspace $\mathbb{R}^k \subset \mathbb{R}^n$ of the variables $x^1, \dots, x^k$ we choose a parametrically defined line

$$\begin{cases} x^1 - x_0^1 = \xi^1 t, \\ \vdots \\ x^k - x_0^k = \xi^k t, \end{cases} \qquad t \in \mathbb{R},$$

having direction vector $(\xi^1, \dots, \xi^k)$, which we denote ξ_u. In more abbreviated notation this line can be written as

$$u = u_0 + \xi_u t. \tag{8.157}$$

Solving Eq. (8.146) for v, by the implicit function theorem we obtain a smooth function (8.148), which, when the right-hand side of Eq. (8.157) is substituted as its argument and (8.157) is taken account of, yields a smooth curve in $\mathbb{R}^n$ defined as follows:

$$\begin{cases} u = u_0 + \xi_u t, \\ v = f(u_0 + \xi_u t), \end{cases} \quad t \in U(0) \in \mathbb{R}. \tag{8.158}$$

Since $F(u, f(u)) \equiv 0$, this curve obviously lies on the surface S. Moreover, it is clear from Eqs. (8.158) that at $t = 0$ the curve passes through the point $(u_0, v_0) = (x_0^1, \dots, x_0^k, x_0^{k+1}, \dots, x_0^n) = x_0 \in S$.

Differentiating the identity

$$F\bigl(u(t), v(t)\bigr) = F\bigl(u_0 + \xi_u t, f(u_0 + \xi_u t)\bigr) \equiv 0$$

with respect to t, we obtain for $t = 0$

$$F'_u(u_0, v_0)\xi_u + F'_v(u_0, v_0)\tilde{\xi}_v = 0,$$

where $\tilde{\xi}_u = v'(0) = (\tilde{\xi}^{k+1}, \dots, \tilde{\xi}^n)$. This equality shows that the vector $\tilde{\xi} = (\xi_u, \tilde{\xi}_v) = (\xi^1, \dots, \xi^k, \tilde{\xi}^{k+1}, \dots, \tilde{\xi}^n)$ satisfies Eq. (8.154). Thus by the remark made above, we conclude that $\xi = \tilde{\xi}$. But the vector $\tilde{\xi}$ is the velocity vector at $t = 0$ for the trajectory (8.158). The proposition is now proved. □

8.7.3 *Extrema with Constraint*

a. Statement of the Problem

One of the most brilliant and well-known achievements of differential calculus is the collection of recipes it provides for finding the extrema of functions. The necessary conditions and sufficient differential tests for an extremum that we obtained from Taylor's theorem apply, as we have noted, to interior extrema.

In other words, these results are applicable only to the study of the behavior of functions $\mathbb{R}^n \ni x \mapsto f(x) \in \mathbb{R}$ in a neighborhood of a point $x_0 \in \mathbb{R}^n$, when the argument x can assume any value in some neighborhood of x_0 in $\mathbb{R}^n$.

Frequently a situation that is more complicated and from the practical point of view even more interesting arises, in which one seeks an extremum of a function under certain constraints that limit the domain of variation of the argument. A typical example is the isoperimetric problem, in which we seek a body of maximal volume subject to the condition that its boundary surface has a fixed area. To obtain a mathematical expression for such a problem that will be accessible to us, we shall simplify the statement and assume that the problem is to choose from the set of rectangles having a fixed perimeter $2p$ the one having the largest area σ. Denoting the lengths of the sides of the rectangle by x and y, we write

$$\sigma(x,h) = x \cdot y,$$
$$x + y = p.$$

Thus we need to find an extremum of the function $\sigma(x, y)$ under the condition that the variables x and y are connected by the equation $x + y = p$. Therefore, the extremum is being sought only on the set of points of $\mathbb{R}^2$ satisfying this relation. This particular problem, of course, can be solved without difficulty: it suffices to write $y = p - x$ and substitute this expression into the formula for $\sigma(x, y)$, then find the maximum of the function $x(p - x)$ by the usual methods. We needed this example only to explain the statement of the problem itself.

In general the problem of an extremum with constraint usually amounts to finding an extremum for a real-valued function

$$y = f\left(x^1, \dots, x^n\right) \tag{8.159}$$

of n variables under the condition that these variables must satisfy a system of equations

$$\begin{cases} F^1\left(x^1, \dots, x^n\right) = 0, \\ \vdots \\ F^m\left(x^1, \dots, x^n\right) = 0. \end{cases} \tag{8.160}$$

Since we are planning to obtain differential conditions for an extremum, we shall assume that all these functions are differentiable and even continuously differentiable. If the rank of the system of functions $F^1, \dots, F^m$ is $n - k$, conditions (8.160) define a k-dimensional smooth surface S in $\mathbb{R}^n$, and from the geometric point of view the problem of extremum with constraint amounts to finding an extremum of the function f on the surface S. More precisely, we are considering the restriction $f|_S$ of the function f to the surface S and seeking an extremum of that function.

The meaning of the concept of a local extremum itself here, of course, remains the same as before, that is, a point $x_0 \in S$ is a local extremum of f on S, or, more briefly $f|_S$, if there is a neighborhood[11] $U_S(x_0)$ of x_0 in $S \subset \mathbb{R}^n$ such that $f(x) \geq f(x_0)$ for any point $x \in U_S(x_0)$ (in which case x_0 is a local minimum) or $f(x) \leq f(x_0)$ (and then x_0 is a local maximum). If these inequalities are strict for $x \in U_S(x_0) \backslash x_0$, then the extremum, as before, will be called strict.

[11] We recall that $U_S(x_0) = S \cap U(x_0)$, where $U(x_0)$ is a neighborhood of x_0 in $\mathbb{R}^n$.

b. A Necessary Condition for an Extremum with Constraint

Theorem 1 *Let $f : D \to \mathbb{R}$ be a function defined on an open set $D \subset \mathbb{R}^n$ and belonging to $C^{(1)}(D; \mathbb{R})$. Let S be a smooth surface in D.*

A necessary condition for a point $x_0 \in S$ that is noncritical for f to be a local extremum of $f|_S$ is that

$$\boxed{TS_{x_0} \subset TN_{x_0},} \tag{8.161}$$

where TS_{x_0} is the tangent space to the surface S at x_0 and TN_{x_0} is the tangent space to the level surface $N = \{x \in D \mid f(x) = f(x_0)\}$ of f to which x_0 belongs.

We begin by remarking that the requirement that the point x_0 be non-critical for f is not an essential restriction in the context of the problem of finding an extremum with constraint, which we are discussing. Indeed, even if the point $x_0 \in D$ were a critical point of the function $f : D \to \mathbb{R}$ or an extremum of the function, it is clear that it would still be a possible or actual extremum respectively for the function $f|_S$. Thus, the new element in this problem is precisely that the function $f|_S$ may have critical points and extrema that are different from those of f.

Proof We choose an arbitrary vector $\xi \in TS_{x_0}$ and a smooth path $x = x(t)$ on S that passes through this point at $t = 0$ and for which the vector ξ is the velocity at $t = 0$, that is,

$$\frac{\mathrm{d}x}{\mathrm{d}t}(0) = \xi. \tag{8.162}$$

If x_0 is an extremum of the function $f|_S$, the smooth function $f(x(t))$ must have an extremum at $t = 0$. By the necessary condition for an extremum, its derivative must vanish at $t = 0$, that is, we must have

$$f'(x_0) \cdot \xi = 0, \tag{8.163}$$

where

$$f'(x_0) = \left(\frac{\partial f}{\partial x^1}, \dots, \frac{\partial f}{\partial x^n}\right), \qquad \xi = (\xi^1, \dots, \xi^n).$$

Since x_0 is a noncritical point of f, condition (8.163) is equivalent to the condition that $\xi \in TN_{x_0}$, for relation (8.163) is precisely the equation of the tangent space TN_{x_0}.

Thus we have proved that $TS_{x_0} \subset TN_{x_0}$. □

If the surface S is defined by the system of equations (8.160) in a neighborhood of x_0, then the space TS_{x_0}, as we know, is defined by the system of linear equations

$$\begin{cases} \frac{\partial F^1}{\partial x^1}(x_0)\xi^1 + \cdots + \frac{\partial F^1}{\partial x^n}(x_0)\xi^n = 0, \\ \vdots \\ \frac{\partial F^m}{\partial x^1}(x_0)\xi^1 + \cdots + \frac{\partial F^m}{\partial x^n}(x_0)\xi^n = 0. \end{cases} \tag{8.164}$$

The space TN_{x_0} is defined by the equation

$$\frac{\partial f}{\partial x^1}(x_0)\xi^1 + \cdots + \frac{\partial f}{\partial x^n}(x_0)\xi^n = 0, \tag{8.165}$$

and, since every solution of (8.164) is a solution of (8.165), the latter equation is a consequence of (8.163).

It follows from these considerations that the relation $TS_{x_0} \subset TN_{x_0}$ is equivalent to the analytic statement that the vector $\operatorname{grad} f(x_0)$ is a linear combination of the vectors $\operatorname{grad} F^i(x_0)$ $(i = 1, \dots, m)$, that is,

$$\boxed{\operatorname{grad} f(x_0) = \sum_{i=1}^{m} \lambda_i \operatorname{grad} F^i(x_0).} \tag{8.166}$$

Taking account of this way of writing the necessary condition for an extremum of a function (8.159) whose variables are connected by (8.160), Lagrange proposed using the following auxiliary function when seeking a constrained extremum:

$$L(x, \lambda) = f(x) - \sum_{i=1}^{m} \lambda_i F^i(x) \tag{8.167}$$

in $n + m$ variables $(x, \lambda) = (x^1, \dots, x^m, \lambda_1, \dots, \lambda_n)$.

This function is called the *Lagrange function* and the method of using it is the *method of Lagrange multipliers*.

The function (8.167) is convenient because the necessary conditions for an extremum of it, regarded as a function of $(x, \lambda) = (x^1, \dots, x^m, \lambda_1, \dots, \lambda_n)$, are precisely (8.166) and (8.160).

Indeed,

$$\begin{cases} \frac{\partial L}{\partial x^j}(x, \lambda) = \frac{\partial f}{\partial x^j}(x) - \sum_{i=1}^{m} \lambda_i \frac{\partial F^i}{\partial x^j}(x) = 0 \quad (j = 1, \dots, n), \\ \frac{\partial L}{\partial \lambda_i}(x, \lambda) = -F^i(x) = 0 \quad (i = 1, \dots, m). \end{cases} \tag{8.168}$$

Thus, in seeking an extremum of a function (8.159) whose variables are subject to the constraints (8.160), one can write the Lagrange function (8.167) with undetermined multipliers and look for its critical points. If it is possible to find $x_0 = (x_0^1, \dots, x_0^n)$ from the system (8.168) without finding $\lambda = (\lambda_1, \dots, \lambda_m)$, then, as far as the original problem is concerned, that is what should be done.

As can be seen from (8.166), the multipliers λ_i $(i = 1, \dots, m)$ are uniquely determined if the vectors $\operatorname{grad} F^i(x_0)$ $(i = 1, \dots, m)$ are linearly independent. The independence of these vectors is equivalent to the statement that the rank of the system (8.164) is m, that is, that all the equations in this system are essential (none of them is a consequence of the others).

This is usually the case, since it is assumed that all the relations (8.160) are independent, and the rank of the system of functions $F^1, \dots, F^m$ is m at every point $x \in X$.

The Lagrange function is often written as

$$L(x, \lambda) = f(x) + \sum_{i=1}^{m} \lambda_i F^i(x),$$

which differs from the preceding expression only in the inessential replacement of λ_i by $-\lambda_i$.[12]

Example 9 Let us find the extrema of a symmetric quadratic form

$$f(x) = \sum_{i,j=1}^{n} a_{ij} x^i x^j \quad (a_{ij} = a_{ji}) \tag{8.169}$$

on the sphere

$$F(x) = \sum_{i=1}^{n} (x^i)^2 - 1 = 0. \tag{8.170}$$

Let us write the Lagrange function for this problem

$$L(x, \lambda) = \sum_{i,j=1}^{n} a_{ij} x^i x^j - \lambda \left(\sum_{i=1}^{n} (x^i)^2 - 1 \right),$$

and the necessary conditions for an extremum of $L(x, \lambda)$, taking account of the relation $a_{ij} = a_{ji}$:

$$\begin{cases} \dfrac{\partial L}{\partial x^i}(x, \lambda) = 2 \left(\displaystyle\sum_{j=1}^{n} a_{ij} x^j - \lambda x^i \right) = 0 \quad (i = 1, \dots, n), \\ \dfrac{\partial L}{\partial \lambda}(x, \lambda) = -\left(\displaystyle\sum_{i=1}^{n} (x^i)^2 - 1 \right) = 0. \end{cases} \tag{8.171}$$

---

[12] In regard to the necessary criterion for an extremum with constraint, see also Problem 6 in Sect. 10.7 (Part 2).

Multiplying the first equation by x^i and summing the first relation over i, we find, taking account of the second relation, that the equality

$$\sum_{i,j=1}^{n} a_{ij}x^i x^j - \lambda = 0 \tag{8.172}$$

must hold at an extremum.

The system (8.171) minus the last equation can be rewritten as

$$\sum_{i=1}^{n} a_{ij}x^j = \lambda x^i \quad (i = 1, \dots, n), \tag{8.173}$$

from which it follows that λ is an eigenvalue of the linear operator A defined by the matrix (a_{ij}), and $x = (x^1, \dots, x^n)$ is an eigenvector of this operator corresponding to this eigenvalue.

Since the function (8.169), which is continuous on the compact set $S = \{x \in \mathbb{R}^n \mid \sum_{i=1}^{n}(x^i)^2 = 1\}$, must assume its maximal value at some point, the system (8.171), and hence also (8.173), must have a solution. Thus we have established along the way that every real symmetric matrix (a_{ij}) has at least one real eigenvalue. This is a result well-known from linear algebra, and is fundamental in the proof of the existence of a basis of eigenvectors for a symmetric operator.

To show the geometric meaning of the eigenvalue λ, we remark that if $\lambda > 0$, then, passing to the coordinates $t^i = x^i/\sqrt{\lambda}$ we find, instead of (8.172),

$$\sum_{i,j=1}^{n} a_{ij}t^i t^j = 1, \tag{8.174}$$

and, instead of (8.170),

$$\sum_{i=1}^{n} (t^i)^2 = \frac{1}{\lambda}. \tag{8.175}$$

But $\sum_{i=1}^{n}(t^i)^2$ is the square of the distance from origin to the point $t = (t^1, \dots, t^n)$, which lies on the quadric surface (8.174). Thus if (8.174) represents, for example, an ellipsoid, then the reciprocal $1/\lambda$ of the eigenvalue λ is the square of the length of one of its semi-axes.

This is a useful observation. It shows in particular that relations (8.171), which are necessary conditions for an extremum with constraint, are still not sufficient. After all, an ellipsoid in $\mathbb{R}^3$ has, besides its largest and smallest semi-axes, a third semi-axis whose length is intermediate between the two, in any neighborhood of whose endpoint there are both points nearer to the origin and points farther away from the origin than the endpoint. This last becomes completely obvious if we consider the ellipses obtained by taking a section of the original ellipsoid by two planes passing through the intermediate-length semi-axis, one passing through the smallest semi-axis and the other through the largest. In one of these cases the intermediate

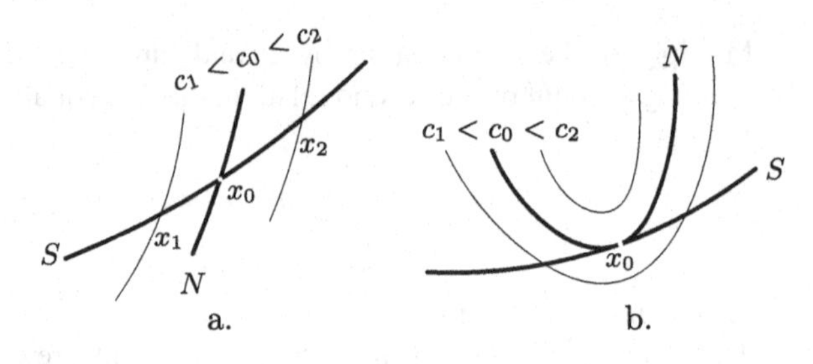

Fig. 8.11

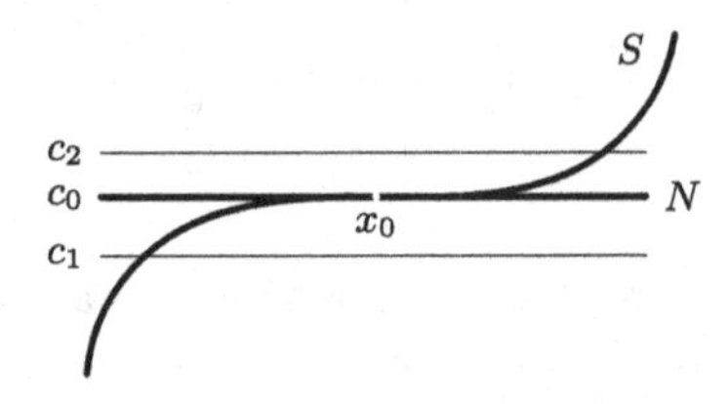

Fig. 8.12

axis will be the major semi-axis of the ellipse of intersection. In the other it will be the minor semi-axis.

To what has just been said we should add that if $1/\sqrt{\lambda}$ is the length of this intermediate semi-axis, then, as can be seen from the canonical equation of an ellipsoid, λ will be an eigenvalue of the operator A. Therefore the system (8.171), which expresses the necessary conditions for an extremum of the function $f|_S$, will indeed have a solution that does not give an extremum of the function.

The result obtained in Theorem 1 (the necessary condition for an extremum with constraint) is illustrated in Fig. 8.11a and 8.11b.

Figure 8.11a explains why the point x_0 of the surface S cannot be an extremum of $f|_S$ if S is not tangent to the surface $N = \{x \in \mathbb{R}^n \mid f(x) = f(x_0) = c_0\}$ at x_0. It is assumed here that $\operatorname{grad} f(x_0) \neq 0$. This last condition guarantees that in a neighborhood of x_0 there are points of a higher, c_2-level surface of the function f and also points of a lower, c_1-level surface of the function.

Since the smooth surface S intersects the surface N, that is, the c_0-level surface of the smooth function f, it follows that S will intersect both higher and lower level surfaces of f in a neighborhood of x_0. But this means that x_0 cannot be an extremum of $f|_S$.

Figure 8.11b shows why, when N is tangent to S at x_0, this point may turn out to be an extremum. In the figure x_0 is a local maximum of $f|_S$.

These same considerations make it possible to sketch a picture whose analytic expression can show that the necessary criterion for an extremum is not sufficient.

Indeed, in accordance with Fig. 8.12, we set, for example,

$$f(x, y) = y, \qquad F(x, y) = x^3 - y = 0.$$

It is then obvious that y has no extremum at the point $(0, 0)$ on the curve $S \subset \mathbb{R}^2$ defined by the equation $y = x^3$, even though this curve is tangent to the level line $f(x, y) = 0$ of the function f at that point. We remark that $\operatorname{grad} f(0, 0) = (0, 1) \neq 0$.

It is obvious that this is essentially the same example that served earlier to illustrate the difference between the necessary and sufficient conditions for a classical interior extremum of a function.

c. A Sufficient Condition for a Constrained Extremum

We now prove the following sufficient condition for the presence or absence of a constrained extremum.

Theorem 2 *Let $f : D \to \mathbb{R}$ be a function defined on an open set $D \subset \mathbb{R}^n$ and belonging to the class $C^{(2)}(D; \mathbb{R})$; let S be the surface in D defined by Eqs.* (8.160), *where $F^i \in C^{(2)}(D; \mathbb{R})$ $(i = 1, \dots, m)$ and the rank of the system of functions $\{F^1, \dots, F^m\}$ at each point of D is m.*

Suppose that the parameters $\lambda_1, \dots, \lambda_m$ in the Lagrange function

$$L(x) = L(x; \lambda) = f(x^1, \dots, x^n) - \sum_{i=1}^{m} \lambda_i F^i(x^1, \dots, x^n)$$

have been chosen in accordance with the necessary criterion (8.166) *for a constrained extremum of the function $f|_S$ at $x_0 \in S$.*[13]

A sufficient condition for the point x_0 to be an extremum of the function $f|_S$ is that the quadratic form

$$\frac{\partial^2 L}{\partial x^i \partial x^j}(x_0)\xi^i \xi^j \tag{8.176}$$

be either positive-definite or negative-definite for vectors $\xi \in TS_{x_0}$.

If the quadratic form (8.176) *is positive-definite on TS_{x_0}, then x_0 is a strict local minimum of $f|_S$; if it is negative-definite, then x_0 is a strict local maximum.*

A sufficient condition for the point x_0 not to be an extremum of $f|_S$ is that the form (8.176) *assume both positive and negative values on TS_{x_0}.*

Proof We first note that $L(x) \equiv f(x)$ for $x \in S$, so that if we show that $x_0 \in S$ is an extremum of the function $L|_S$, we shall have shown simultaneously that it is an extremum of $f|_S$.

By hypothesis, the necessary criterion (8.166) for an extremum of $f|_S$ at x_0 is fulfilled, so that $\operatorname{grad} L(x_0) = 0$ at this point. Hence the Taylor expansion of $L(x)$ in a neighborhood of $x_0 = (x_0^1, \dots, x_0^n)$ has the form

$$L(x) - L(x_0) = \frac{1}{2!}\frac{\partial^2 L}{\partial x^i \partial x^j}(x_0)\left(x^i - x_0^i\right)\left(x^j - x_0^j\right) + o\left(\|x - x_0\|^2\right) \tag{8.177}$$

as $x \to x_0$.

---

[13]When we keep λ fixed, $L(x; \lambda)$ becomes a function depending only on x; we allow ourselves to denote this function $L(x)$.

We now recall that, in motivating Definition 2, we noted the possibility of a local (for example, in a neighborhood of $x_0 \in S$) parametric definition of a smooth k-dimensional surface S (in the present case, $k = n - m$).

In other words, there exists a smooth mapping

$$\mathbb{R} \ni (t^1, \dots, t^k) = t \mapsto x = (x^1, \dots, x^n) \in \mathbb{R}^n$$

(as before, we shall write it in the form $x = x(t)$), under which a neighborhood of the point $0 = (0, \dots, 0) \in \mathbb{R}^k$ maps bijectively to some neighborhood of x_0 on S, and $x_0 = x(0)$.

We remark that the relation

$$x(t) - x(0) = x'(0)t + o(\|t\|) \quad \text{as } t \to 0,$$

which expresses the differentiability of the mapping $t \mapsto x(t)$ at $t = 0$, is equivalent to the n coordinate equalities

$$x^i(t) - x^i(0) = \frac{\partial x^i}{\partial t^\alpha}(0)t^\alpha + o(\|t\|) \quad (i = 1, \dots, n), \tag{8.178}$$

in which the index α ranges over the integers from 1 to k and the summation is over this index.

It follows from these numerical equalities that

$$|x^i(t) - x^i(0)| = O(\|t\|) \quad \text{as } t \to 0$$

and hence

$$\|x(t) - x(0)\|_{\mathbb{R}^n} = O(\|t\|_{\mathbb{R}^k}) \quad \text{as } t \to 0. \tag{8.179}$$

Using relations (8.178), (8.179), and (8.177), we find that as $t \to 0$

$$L(x(t)) - L(x(0)) = \frac{1}{2!}\partial_{ij}L(x_0)\partial_\alpha x^i(0)\partial_\beta x(0)t^\alpha t^\beta + o(\|t\|^2). \tag{8.177'}$$

Hence under the assumption of positive- or negative-definiteness of the form

$$\partial_{ij}L(x_0)\partial_\alpha x^i(0)\partial_\beta x^j(0)t^\alpha t^\beta \tag{8.180}$$

it follows that the function $L(x(t))$ has an extremum at $t = 0$. If the form (8.180) assumes both positive and negative values, then $L(x(t))$ has no extremum at $t = 0$. But, since some neighborhood of the point $0 \in \mathbb{R}^k$ maps to a neighborhood of $x(0) = x_0 \in S$ on the surface S under the mapping $t \mapsto x(t)$, we can conclude that the function $L|_S$ also will either have an extremum at x_0 of the same nature as the function $L(x(t))$ or, like $L(x(t))$, will not have an extremum.

Thus, it remains to verify that for vectors $\xi \in TS_{x_0}$ the expressions (8.176) and (8.180) are merely different notations for the same object.

Indeed, setting

$$\xi = x'(0)t,$$

we obtain a vector ξ tangent to S at x_0, and if $\xi = (\xi^1, \dots, \xi^n)$, $x(t) = (x^1, \dots, x^n)(t)$, and $t = (t^1, \dots, t^k)$, then

$$\xi = \partial_\beta x^j(0) t^\beta \quad (j = 1, \dots, n),$$

from which it follows that the quantities (8.176) and (8.180) are the same. □

We note that the practical use of Theorem 2 is hindered by the fact that only $k = n - m$ of the coordinates of the vector $\xi = (\xi^1, \dots, \xi^n) \in TS_{x_0}$ are independent, since the coordinates of ξ must satisfy the system (8.164) defining the space TS_{x_0}. Thus a direct application of the Sylvester criterion to the quadratic form (8.176) generally yields nothing in the present case: the form (8.176) may not be positive- or negative-definite on $T\mathbb{R}^n_{x_0}$ and yet be definite on TS_{x_0}. But if we express m coordinates of the vector ξ in terms of the other k coordinates by relations (8.164) and then substitute the resulting linear forms into (8.176), we arrive at a quadratic form in k variables whose positive- or negative-definiteness can be investigated using the Sylvester criterion.

Let us clarify what has just been said by some elementary examples.

Example 10 Suppose we are given the function

$$f(x, y, z) = x^2 - y^2 + z^2$$

in the space $\mathbb{R}^3$ with coordinates x, y, z. We seek an extremum of this function on the plane S defined by the equation

$$F(x, y, z) = 2x - y - 3 = 0.$$

Writing the Lagrange function

$$L(x, y, z) = \left(x^2 - y^2 + z^2\right) - \lambda(2x - y - 3)$$

and the necessary conditions for an extremum

$$\begin{cases} \dfrac{\partial L}{\partial x} = 2x - 2\lambda = 0, \\ \dfrac{\partial L}{\partial y} = -2y + \lambda = 0, \\ \dfrac{\partial L}{\partial z} = 2z = 0, \\ \dfrac{\partial L}{\partial \lambda} = -(2x - y - 3) = 0, \end{cases}$$

we find the possible extremum $p = (2, 1, 0)$.

Next we find the form (8.176):

$$\frac{1}{2}\partial_{ij} L \xi^i \xi^j = \left(\xi^1\right)^2 - \left(\xi^2\right)^2 + \left(\xi^3\right)^2. \tag{8.181}$$

We note that in this case the parameter λ did not occur in this quadratic form, and so we did not compute it.

We now write the condition $\xi \in TS_p$:

$$2\xi^1 - \xi^2 = 0. \tag{8.182}$$

From this equality we find $\xi^2 = 2\xi^1$ and substitute it into the form (8.181), after which it assumes the form

$$-3(\xi^1)^2 + (\xi^3)^2,$$

where this time ξ^1 and ξ^3 are independent variables.

This last form may obviously assume both positive and negative values, and therefore the function $f|_S$ has no extremum at $p \in S$.

Example 11 Under the hypotheses of Example 10 we replace $\mathbb{R}^3$ by $\mathbb{R}^2$ and the function f by

$$f(x, y) = x^2 - y^2,$$

retaining the condition

$$2x - y - 3 = 0,$$

which now defines a line S in the plane $\mathbb{R}^2$.

We find $p = (2, 1)$ as a possible extremum.

Instead of the form (8.181) we obtain the form

$$(\xi^1)^2 - (\xi^2)^2 \tag{8.183}$$

with the previous relation (8.182) between ξ^1 and ξ^2.

Thus the form (8.183) now has the form

$$-3(\xi^1)^2$$

on TS_p, that is, it is negative-definite. We conclude from this that the point $p = (2, 1)$ is a local maximum of $f|_S$.

The following simple examples are instructive in many respects. On them one can distinctly trace the working of both the necessary and the sufficient conditions for constrained extrema, including the role of the parameter and the informal role of the Lagrange function itself.

Example 12 On the plane $\mathbb{R}^2$ with Cartesian coordinates (x, y) we are given the function

$$f(x, y) = x^2 + y^2.$$

Let us find the extremum of this function on the ellipse given by the canonical relation

$$F(x, y) = \frac{x^2}{a^2} + \frac{y^2}{b^2} - 1 = 0,$$

where $0 < a < b$.

It is obvious from geometric considerations that $\min f|_S = a^2$ and $\max f|_S = b^2$. Let us obtain this result on the basis of the procedures recommended by Theorems 1 and 2.

By writing the Lagrange function

$$L(x, y, \lambda) = (x^2 + y^2) - \lambda\left(\frac{x^2}{a^2} + \frac{y^2}{b^2} - 1\right)$$

and solving the equation $\mathrm{d}L = 0$, that is, the system $\frac{\partial L}{\partial x} = \frac{\partial L}{\partial y} = \frac{\partial L}{\partial \lambda} = 0$, we find the solutions

$$(x, y, \lambda) = (\pm a, 0, a^2), \quad (0, \pm b, b^2).$$

Now in accordance with Theorem 2 we write and study the quadratic form $\frac{1}{2}\,\mathrm{d}^2 L\boldsymbol{\xi}^2$, the second term of the Taylor expansion of the Lagrange function in a neighborhood of the corresponding points:

$$\frac{1}{2}\,\mathrm{d}^2 L\boldsymbol{\xi}^2 = \left(1 - \frac{\lambda}{a^2}\right)(\xi^1)^2 + \left(1 - \frac{\lambda}{b^2}\right)(\xi^2)^2.$$

At the points $(\pm a, 0)$ of the ellipse S the tangent vector $\boldsymbol{\xi} = (\xi^1, \xi^2)$ has the form $(0, \xi^2)$, and for $\lambda = a^2$ the quadratic form assumes the form

$$\left(1 - \frac{a^2}{b^2}\right)(\xi^2)^2.$$

Taking account of the condition $0 < a < b$, we conclude that this form is positive-definite and hence at the points $(\pm a, 0) \in S$ there is a strict local (and in this case obviously also global) minimum of the function $f|_S$, that is, $\min f|_S = a^2$.

Similarly we find the form

$$\left(1 - \frac{b^2}{a^2}\right)(\xi^1)^2,$$

which corresponds to the points $(0, \pm b) \in S$, and we find $\max f|_S = b^2$.

Remark Note the role of the Lagrange function here compared with the role of the function f. At the corresponding points on these tangent vectors the differential of f (like the differential of L) vanishes, and the quadratic form $\frac{1}{2}\,\mathrm{d}^2 f\boldsymbol{\xi}^2 = (\xi^1)^2 + (\xi^2)^2$ is positive definite at whichever of these points it is computed. Nevertheless, the function $f|_S$ has a strict minimum at the points $(\pm a, 0)$ and a strict maximum at the points $(0, \pm b)$.

To understand what *is* going on here, look again at the proof of Theorem 2 and try to obtain relation (8.176) by substituting f for L in (8.177). Note that an additional term containing $x''(0)$ arises here. The reason it does not vanish is that, in contrast to $\mathrm{d}L$ the differential $\mathrm{d}f$ of f is not identically zero at the corresponding points, even though its values are indeed zero on the tangent vectors (of the form $x'(0)$).

Example 13 Let us find the extrema of the function

$$f(x,y,z)=x^2+y^2+z^2$$

on the ellipsoid S defined by the relation

$$F(x,y,z)=\frac{x^2}{a^2}+\frac{y^2}{b^2}+\frac{z^2}{c^2}-1=0,$$

where $0<a<b<c$.

By writing the Lagrange function

$$L(x,y,z,\lambda)=\left(x^2+y^2+z^2\right)-\lambda\left(\frac{x^2}{a^2}+\frac{y^2}{b^2}+\frac{z^2}{c^2}-1\right),$$

in accordance with the necessary criterion for an extremum, we find the solutions of the equation $\mathrm{d}L=0$, that is, the system $\frac{\partial L}{\partial x}=\frac{\partial L}{\partial y}=\frac{\partial L}{\partial z}=\frac{\partial L}{\partial \lambda}=0$:

$$(x,y,z,\lambda)=\left(\pm a,0,0,a^2\right),\quad \left(0,\pm b,0,b^2\right),\quad \left(0,0,\pm c,c^2\right).$$

On each respective tangent plane the quadratic form

$$\frac{1}{2}\mathrm{d}^2L\xi^2=\left(1-\frac{\lambda}{a^2}\right)\left(\xi^1\right)^2+\left(1-\frac{\lambda}{b^2}\right)\left(\xi^2\right)^2+\left(1-\frac{\lambda}{c^2}\right)\left(\xi^3\right)^2$$

in each of these cases has the form

$$\left(1-\frac{a^2}{b^2}\right)\left(\xi^2\right)^2+\left(1-\frac{a^2}{c^2}\right)\left(\xi^3\right)^2, \qquad (a)$$

$$\left(1-\frac{b^2}{a^2}\right)\left(\xi^1\right)^2+\left(1-\frac{b^2}{c^2}\right)\left(\xi^3\right)^2, \qquad (b)$$

$$\left(1-\frac{c^2}{a^2}\right)\left(\xi^1\right)^2+\left(1-\frac{c^2}{b^2}\right)\left(\xi^2\right)^2. \qquad (c)$$

Since $0<a<b<c$, it follows from Theorem 2, which gives a sufficient criterion for the presence or absence of a constrained extremum, that one can conclude that in cases (*a*) and (*c*), we have found respectively $\min f|_S=a^2$ and $\max f|_S=c^2$, while at the points $(0,\pm b,0)\in S$ corresponding to case (*b*) the function $f|_S$ has no extremum. This is in complete agreement with the obvious geometric considerations stated in the discussion of the necessary criterion for a constrained extremum.

Certain other aspects of the concepts of analysis and geometry encountered in this section, which are sometimes quite useful, including the physical interpretation of the problem of a constrained extremum itself, as well as the necessary criterion (8.166) for it as the resolution of forces at an equilibrium point and the interpretation of the Lagrange multipliers as the magnitude of the reaction of ideal constraints, are presented in the problems and exercises that follow.

8.7.4 Problems and Exercises

1. *Paths and surfaces.*

a) Let $f : I \to \mathbb{R}^2$ be a mapping of class $C^{(1)}(I; \mathbb{R}^2)$ of the interval $I \subset \mathbb{R}$. Regarding this mapping as a path in $\mathbb{R}^2$, show by example that its support $f(I)$ may fail to be a submanifold of $\mathbb{R}^2$, while the graph of this mapping in $\mathbb{R}^3 = \mathbb{R}^1 \times \mathbb{R}^2$ is always a one-dimensional submanifold of $\mathbb{R}^3$ whose projection into $\mathbb{R}^2$ is the support $f(I)$ of the path.

b) Solve problem a) in the case when I is an interval in $\mathbb{R}^k$ and $f \in C^{(1)}(I; \mathbb{R}^n)$. Show that in this case the graph of the mapping $f : I \to \mathbb{R}^n$ is a smooth k-dimensional surface in $\mathbb{R}^k \times \mathbb{R}^n$ whose projection on the subspace $\mathbb{R}^n$ equals $f(I)$.

c) Verify that if $f_1 : I_1 \to S$ and $f_2 : I_2 \to S$ are two smooth parametrizations of the same k-dimensional surface $S \subset \mathbb{R}^n$, f_1 having no critical points in I_1 and f_2 having no critical points in I_2, then the mappings

$$f_1^{-1} \circ f_2 : I_2 \to I_1, \qquad f_2^{-1} \circ f_1 : I_1 \to I_2$$

are smooth.

2. *The sphere in $\mathbb{R}^n$.*

a) On the sphere $S^2 = \{x \in \mathbb{R}^3 \mid \|x\| = 1\}$ exhibit a maximal domain of validity for the curvilinear coordinates (φ, ψ) obtained from polar coordinates in $\mathbb{R}^3$ (see formula (8.116) of the preceding section) when $\rho = 1$.

b) Answer question a) in the case of the $(m-1)$-dimensional sphere

$$S^{m-1} = \{x \in \mathbb{R}^m \mid \|x\| = 1\}$$

in $\mathbb{R}^m$ and the coordinates $(\varphi_1, \dots, \varphi_{m-1})$ on it obtained from polar coordinates in $\mathbb{R}^m$ (see formula (8.118) of the preceding section) at $\rho = 1$.

c) Can the sphere $S^k \subset \mathbb{R}^{k+1}$ be defined by a single coordinate system $(t^1, \dots, t^k)$, that is, a single diffeomorphism $f : G \to \mathbb{R}^{k+1}$ of a domain $G \subset \mathbb{R}^k$?

d) What is the smallest number of maps needed in an atlas of the Earth's surface?

e) Let us measure the distance between points of the sphere $S^2 \subset \mathbb{R}^3$ by the length of the shortest curve lying on the sphere S^2 and joining these points. Such a curve is the arc of a suitable great circle. Can there exist a local flat map of the sphere such that all the distances between points of the sphere are proportional (with the same coefficient of proportionality) to the distances between their images on the map?

Fig. 8.13

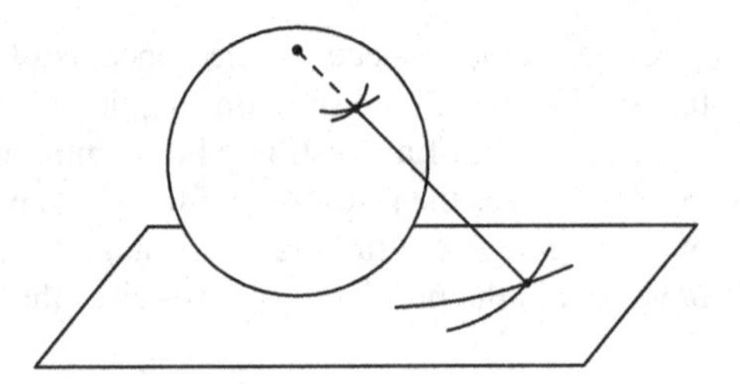

f) The angle between curves (whether lying on the sphere or not) at their point of intersection is defined as the angle between the tangents to these curves at this point.

Show that there exist local flat maps of a sphere at which the angles between the curves on the sphere and the corresponding curves on the map are the same (see Fig. 8.13, which depicts the so-called *stereographic projection*).

3. *The tangent space.*

a) Verify by direct computation that the tangent manifold TS_{x_0} to a smooth k-dimensional surface $S \subset \mathbb{R}^n$ at a point $x_0 \in S$ is independent of the choice of the coordinate system in $\mathbb{R}^n$.

b) Show that if a smooth surface $S \subset D$ maps to a smooth surface $S' \subset D'$ under a diffeomorphism $f : D \to D'$ of the domain $D \subset \mathbb{R}^n$ onto the domain $D' \subset \mathbb{R}^n$ and the point $x_0 \in S$ maps to $x'_0 \in S'$, then under the linear mapping $f'(x_0) : \mathbb{R}^n \to \mathbb{R}^n$ tangent to f at $x_0 \in D$ the vector space TS_{x_0} maps isomorphically to the vector space $TS'_{x'_0}$.

c) If under the conditions of the preceding problem the mapping $f : D \to D'$ is any mapping of class $C^{(1)}(D; D')$ under which $f(S) \subset S'$, then $f'(TS_{x_0}) \subset TS'_{x'_0}$.

d) Show that the orthogonal projection of a smooth k-dimensional surface $S \subset \mathbb{R}^n$ on to the k-dimensional tangent plane TS_{x_0} to it at $x_0 \in S$ is one-to-one in some neighborhood of the point of tangency x_0.

e) Suppose, under the conditions of the preceding problem, that $\xi \in TS_{x_0}$ and $\|\xi\| = 1$.

The equation $x - x_0 = \xi t$ of a line in $\mathbb{R}^n$ lying in TS_{x_0} can be used to characterize each point $x \in TS_{x_0} \backslash x_0$ by the pair (t, ξ). These are essentially polar coordinates in TS_{x_0}.

Show that smooth curves on the surface S intersecting only at the point x_0 correspond to the lines $x - x_0 = \xi t$ in a neighborhood of x_0. Verify that, retaining t as the parameter on these curves, we obtain paths along which the velocity at $t = 0$ is the vector $\xi \in TS_{x_0}$ that determines the line $x - x_0 = \xi t$, from which the given curve on S is obtained.

Thus the pairs (t, ξ) where $\xi \in TS_{x_0}$, $\|\xi\| = 1$, and t are real numbers from some neighborhood $U(0)$ of zero in $\mathbb{R}$, can serve as the analogue of polar coordinates in a neighborhood of $x_0 \in S$.

4. Let the function $F \in C^{(1)}(\mathbb{R}^n; \mathbb{R})$ having no critical points be such that the equation $F(x^1, \dots, x^n) = 0$ defines a compact surface S in $\mathbb{R}^n$ (that is, S is compact as a subset of $\mathbb{R}^n$). For any point $x \in S$ we find a vector $\eta(x) = \operatorname{grad} F(x)$ normal to S at x. If we force each point $x \in S$ to move uniformly with velocity $\eta(x)$, a mapping $S \ni x \mapsto x + \eta(x)t \in \mathbb{R}^n$ arises.

a) Show that for values of t sufficiently close to zero, this mapping is bijective and for each such value of t a smooth surface $\tilde{S}_t$ results from S.

b) Let E be a set in $\mathbb{R}^n$; we define the δ-neighborhood of the set E to be the set of points in $\mathbb{R}^n$ whose distance from E is less than δ.

Show that for values of t close to zero, the equation

$$F(x^1, \dots, x^n) = t$$

defines a compact surface $S_t \subset \mathbb{R}^n$, and show that the surface $\tilde{S}_t$ lies in the $\delta(t)$-neighborhood of the surface S_t, where $\delta(t) = O(t)$ as $t \to 0$.

c) With each point x of the surface $S = S_0$ we associate a unit normal vector

$$\mathbf{n}(x) = \frac{\eta(x)}{\|\eta(x)\|}$$

and consider the new mapping $S \ni x \mapsto x + \mathbf{n}(x)t \in \mathbb{R}^n$.

Show that for all values of t sufficiently close to zero this mapping is bijective, that the surface $\approx{S}_t$ obtained from S at the specific value of t is smooth, and if $t_1 \neq t_2$, then $\approx{S}_{t_1} = \cap \approx{S}_{t_2} = \varnothing$.

d) Relying on the result of the preceding exercise, show that there exists $\delta > 0$ such that there is a one-to-one correspondence between the points of the δ-neighborhood of the surface S and the pairs (t, x), where $t \in]-\delta, \delta[\subset \mathbb{R}$, $x \in S$. If $(t^1, \dots, t^k)$ are local coordinates on the surface S in the neighborhood $U_S(x_0)$ of x_0, then the quantities $(t, t^1, \dots, t^k)$ can serve as local coordinates in a neighborhood $U(x_0)$ of $x_0 \in \mathbb{R}^n$.

e) Show that for $|t| < \delta$ the point $x \in S$ is the point of the surface S closest to $(x + \mathbf{n}(x)t) \in \mathbb{R}^n$. Thus for $|t| < \delta$ the surface $\approx{S}_t$ is the geometric locus of the points of $\mathbb{R}^n$ at distance $|t|$ from S.

5. a) Let $d_p : S \to \mathbb{R}$ be the function on the smooth k-dimensional surface $S \subset \mathbb{R}^n$ defined by the equality $d_p(x) = \|p - x\|$, where p is a fixed point of $\mathbb{R}^n$, x is a point of S, and $\|p - x\|$ is the distance between these points in $\mathbb{R}^n$.

Show that at the extrema of the function $d_p(x)$ the vector $p - x$ is orthogonal to the surface S.

b) Show that on any line that intersects the surface S orthogonally at the point $q \in S$, there are at most k points p such that the function $d_p(x)$ has q as a degenerate critical point (that is, a point at which the Hessian of the function vanishes).

c) Show that in the case of a curve S ($k = 1$) in the plane $\mathbb{R}^2$ ($n = 2$) the point p for which the point $q \in S$ is a degenerate critical point of $d_p(x)$ is the center of curvature of the curve S at the point $q \in S$.

6. In the plane $\mathbb{R}^2$ with Cartesian coordinates x, y construct the level curves of the function $f(x, y) = xy$ and the curve

$$S = \{(x, y) \in \mathbb{R}^2 \mid x^2 + y^2 = 1\}.$$

Using the resulting picture, carry out a complete investigation of the problem of extrema of the function $f|_S$.

7. The following functions of class $C^{(\infty)}(\mathbb{R}^2; \mathbb{R})$ are defined on the plane $\mathbb{R}^2$ with Cartesian coordinates x, y:

$$f(x, y) = x^2 - y; \qquad F(x, y) = \begin{cases} x^2 - y + e^{-1/x^2} \sin \frac{1}{x}, & \text{if } x \neq 0, \\ x^2 - y, & \text{if } x = 0. \end{cases}$$

a) Draw the level curves of the function $f(x, y)$ and the curve S defined by the relation $F(x, y) = 0$.

b) Investigate the function $f|_S$ for extrema.

c) Show that the condition that the form $\partial_{ij} f(x_0)\xi^i\xi^j$ be positive-definite or negative-definite on TS_{x_0}, in contrast to the condition for the form $\partial_{ij} L(x_0)\xi^i\xi^j$ on TS_{x_0} given in Theorem 2, is still not sufficient for the possible extremum $x_0 \in S$ to be an actual extremum of the function $f|_S$.

d) Check to see whether the point $x_0 = (0, 0)$ is critical for the function f and whether one can study the behavior of f in a neighborhood of this point using only the second (quadratic) term of Taylor's formula, as was assumed in c).

e) Consider the pair of functions $F(x, y) = 2x^2 + y$, $f(x, y) = x^2 + y$ and show that f may have a strict maximum at a point of the curve $F(x, y) = 0$, although on the tangent to the curve at this point the function f has a strict minimum. This again emphasizes the role of the function of Lagrange in the statement proved above for the sufficient condition for a relative extremum.

f) Consider the pair of functions $F(x, y) = x^2 - y^3$, $f(x, y) = y$ and show that for $L(x, y) = f(x, y) + \lambda F$ the equation $\mathrm{d}L = 0$ may not have solution, when the extremum of f is attained at a singular point of the curve $F(x, y) = 0$. This can be taken into consideration studying instead of L, the function $\tilde{L}(x, y) = \lambda_0 f + \lambda F$, allowing the possibility of $\lambda_0 = 0$.

8. In determining principal curvatures and principal directions in differential geometry it is useful to know how to find an extremum of a quadratic form $h_{ij}u^i u^j$ under the hypothesis that another (positive-definite) form $g_{ij}u^i u^j$ is constant. Solve this problem by analogy with Example 9 which was discussed above.

9. Let $A = [a^i_j]$ be a square matrix of order n such that

$$\sum_{i=1}^{n} (a^i_j)^2 = H_j \quad (j = 1, \dots, n),$$

where $H_1, \dots, H_n$ is a fixed set of n nonnegative real numbers.

a) Show that $\det^2 A$ can have an extremum under these conditions only if the rows of the matrix A are pairwise orthogonal vectors in $\mathbb{R}^n$.

b) Starting from the equality

$$\det^2 A = \det A \cdot \det A^*,$$

where A^* is the transpose of A, show that under the conditions above

$$\max_A \det^2 A = H_1 \cdots H_n.$$

c) Prove *Hadamard's inequality* for any matrix $[a^i_j]$:

$$\det^2(a^i_j) \le \prod_{j=1}^n \left(\sum_{i=1}^n (a^i_j)^2 \right).$$

d) Give an intuitive-geometric interpretation of Hadamard's inequality.

10. a) Draw the level surfaces of the function f and the plane S in Example 10. Explain the result obtained in this example on the figure.

b) Draw the level curves of the function f and the line S in Example 11. Explain the result obtained in this example on the figure.

11. In Example 6 of Sect. 5.4, starting from Fermat's principle we obtained Snell's law for refraction of light at the interface of two media when the interface is a plane. Does this law remain valid for an arbitrary smooth interface?

12. a) A point mass in a potential force field can be in an equilibrium position (also called a state of rest or a stationary state) only at critical (stationary) points of the potential. In this situation a position of stable equilibrium corresponds to a strict local minimum of the potential and an unstable equilibrium point to a local maximum. Verify this.

b) To which constrained extremal problem (solved by Lagrange) does the problem of the equilibrium position reduce for a point mass in a potential force field (for example, gravitational) with ideal constraints (for example, a point may be confined to a smooth surface or a bead may be confined to a smooth thread or a ball to a track)? The constraint is ideal (there is no friction); this means that its effect on the point (the *reactive force of the constraint*) is always normal to the constraint.

c) What physical (mechanical) meaning do the expansion (8.166), the necessary criterion for a constrained extremum, and the Lagrange multiplier have in this case.

Note that each of the functions of the system (8.160) can be divided by the absolute value of its gradient, which obviously leads to an equivalent system (if its rank is equal to m everywhere). Hence all the vectors $\operatorname{grad} F^i(x_0)$ on the right-hand side of (8.166) can be regarded as unit normal vectors to the corresponding surface.

d) Do you agree that the Lagrange method of finding a constrained extremum becomes obvious and natural after the physical interpretation just given?

Some Problems from the Midterm Examinations

1 Introduction to Analysis (Numbers, Functions, Limits)

Problem 1 The length of a hoop girdling the Earth at the equator is increased by 1 meter, leaving a gap between the Earth and the hoop. Could an ant crawl through this gap? How big would the absolute and relative increases in the radius of the Earth be if the equator were lengthened by this amount? (The radius of the Earth is approximately 6400 km.)

Problem 2 How are the completeness (continuity) of the real numbers, the unboundedness of the series of natural numbers, and Archimedes' principle related? Why is it possible to approximate every real number arbitrarily closely by rational numbers? Explain using the model of rational fractions (rational functions) that Archimedes' principle may fail, and that in such number systems the sequence of natural numbers is bounded and there exist infinitely small numbers.

Problem 3 Four bugs sitting at the corners of the unit square begin to chase one another with unit speed, each maintaining a course in the direction of the one pursued. Describe the trajectories of their motions. What is the length of each trajectory? What is the law of motion (in Cartesian or polar coordinates)?

Problem 4 Draw a flow chart for computing $\sqrt{a}$ $(a > 0)$ by the recursive procedure

$$x_{n+1} = \frac{1}{2}\left(x_n + \frac{a}{x_n}\right).$$

How is equation solving related to finding fixed points? How do you find $\sqrt[n]{a}$?

Problem 5 Let $g(x) = f(x) + o(f(x))$ as $x \to \infty$. Is it also true that $f(x) = g(x) + o(g(x))$ as $x \to \infty$?

V.A. Zorich, *Mathematical Analysis I*, Universitext,
DOI 10.1007/978-3-662-48792-1

Problem 6 By the method of undetermined coefficients (or otherwise) find the first few (or all) coefficients of the power series for $(1+x)^\alpha$ with $\alpha = -1, -\frac{1}{2}, 0, \frac{1}{2}, 1, \frac{3}{2}$. (By interpolating the coefficients of like powers of x in such expansions, Newton wrote out the law for forming the coefficients with any $\alpha \in \mathbb{R}$. This result is known as Newton's binomial theorem.)

Problem 7 Knowing the power-series expansion of the function e^x, find by the method of undetermined coefficients (or otherwise) the first few (or all) terms of the power-series expansion of the function $\ln(1+x)$.

Problem 8 Compute $\exp A$ when A is one of the matrices

$$\begin{pmatrix} 0 & 0 \\ 0 & 0 \end{pmatrix}, \quad \begin{pmatrix} 0 & 1 \\ 0 & 0 \end{pmatrix}, \quad \begin{pmatrix} 0 & 1 & 0 \\ 0 & 0 & 1 \\ 0 & 0 & 0 \end{pmatrix}, \quad \begin{pmatrix} 1 & 0 & 0 \\ 0 & 2 & 0 \\ 0 & 0 & 3 \end{pmatrix}.$$

Problem 9 How many terms of the series for e^x must one take in order to obtain a polynomial that makes it possible to compute e^x on the interval $[-3, 5]$ within 10^{-2}?

Problem 10 If we know the power expansion of the functions $\sin x$ and $\cos x$, find with the method of the undetermined coefficients (or another one) some few first terms (or all) the power expansion of the function $\tan x$ in a neighborhood of the point $x = 0$.

Problem 11 The length of a band tightening the Earth at the Equator have increased by 1 meter, and after that the band was pulled propping up a vertical column. What is, roughly speaking, the height of the column if the radius of the Earth $\approx$6400 km?

Problem 12 Calculate

$$\lim_{x\to\infty} \left(\mathrm{e}\left(1 + \frac{1}{x}\right)^{-x}\right)^x.$$

Problem 13 Sketch the graphs of the following functions:

$$\text{a) } \log_{\cos x} \sin x; \qquad \text{b) } \arctan \frac{x^3}{(1-x)(1+x)^2}.$$

2 One-Variable Differential Calculus

Problem 1 Show that if the acceleration vector $\mathbf{a}(t)$ is orthogonal to the vector $\mathbf{v}(t)$ at each instant of time t, the magnitude $|\mathbf{v}(t)|$ remains constant.

Problem 2 Let (x, t) and $(\tilde{x}, \tilde{t})$ be respectively the coordinate of a moving point and the time in two systems of measurement. Assuming the formulas $\tilde{x} = \alpha x + \beta t$

and $\tilde{t} = \gamma x + \delta t$ for transition from one system to the other are known, find the formula for the transformation of velocities, that is, the connection between $v = \frac{dx}{dt}$ and $\tilde{v} = \frac{d\tilde{x}}{d\tilde{t}}$.

Problem 3 The function $f(x) = x^2 \sin \frac{1}{x}$ for $x \neq 0$, $f(0) = 0$ is differentiable on $\mathbb{R}$, but f' is discontinuous at $x = 0$ (verify this). We shall "prove", however, that if $f : \mathbb{R} \to \mathbb{R}$ is differentiable on $\mathbb{R}$, then f' is continuous at every point $a \in \mathbb{R}$. By Lagrange's theorem

$$\frac{f(x) - f(a)}{x - a} = f'(\xi),$$

where ξ is a point between a and x. Then if $x \to a$, it follows that $\xi \to a$. By definition,

$$\lim_{x \to a} \frac{f(x) - f(a)}{x - a} = f'(a),$$

and since this limit exists, the right-hand side of Lagrange's formula has a limit equal to it. That is, $f'(\xi) \to f'(a)$ as $\xi \to a$. The continuity of f' at a is now "proved". Where is the error?

Problem 4 Suppose the function f has $n+1$ derivatives at the point x_0, and let $\xi = x_0 + \theta_x(x - x_0)$ be the intermediate point in Lagrange's formula for the remainder term $\frac{1}{n!} f^{(n)}(\xi)(x - x_0)^n$, so that $0 < \theta_x < 1$. Show that $\theta_x \to \frac{1}{n+1}$ as $x \to x_0$ if $f^{(n+1)}(x_0) \neq 0$.

Problem 5

a) If the function $f \in C^{(n)}([a,b], \mathbb{R})$ vanishes at $n+1$ points of the interval $[a,b]$, then there exists in this interval at least one zero of the function $f^{(n)}$, the derivative of f of order n.

b) Show that the polynomial $P_n(x) = \frac{d^n (x^2-1)^n}{dx^n}$ has n roots on the interval $[-1, 1]$. (Hint: $x^2 - 1 = (x-1)(x+1)$ and $P_n^{(k)}(-1) = P_n^{(k)}(1) = 0$, for $k = 0, \dots, n-1$.)

Problem 6 Recall the geometric meaning of the derivative and show that if the function f is defined and differentiable on an interval I and $[a,b] \subset I$, then the function f' (not even necessarily continuous!) takes all the values between $f'(a)$ and $f'(b)$ on the interval $[a,b]$.

Problem 7 Prove the inequality

$$a_1^{\alpha_1} \cdots a_n^{\alpha_n} \leq \alpha_1 a_1 + \cdots + \alpha_n a_n,$$

where $a_1, \dots, a_n, \alpha_1, \dots, \alpha_n$ are nonnegative and $\alpha_1 + \cdots + \alpha_n = 1$.

Problem 8 Show that

$$\lim_{n\to\infty}\left(1+\frac{z}{n}\right)^n = \mathrm{e}^x(\cos y + i\sin y) \quad (z = x+iy),$$

so that it is natural to suppose that $\mathrm{e}^{iy} = \cos y + i \sin y$ (Euler's formula) and

$$\mathrm{e}^z = \mathrm{e}^x \mathrm{e}^{iy} = \mathrm{e}^x(\cos y + i \sin y).$$

Problem 9 Find the shape of the surface of a liquid rotating at uniform angular velocity in a glass.

Problem 10 Show that the tangent to the ellipse $\frac{x^2}{a^2}+\frac{y^2}{b^2}=1$ at the point (x_0, y_0) has the equation $\frac{xx_0}{a^2}+\frac{yy_0}{b^2}=1$, and that light rays from a source situated at one of the foci $F_1 = (-\sqrt{a^2-b^2}, 0)$, $F_2 = (\sqrt{a^2-b^2}, 0)$ of an ellipse with semiaxes $a > b > 0$ are reflected by an elliptical mirror to the other focus.

Problem 11 A particle subject to gravity, without any initial boost, begins to slide from the top of an iceberg of elliptic cross-section. The equation of the cross section is $x^2 + 5y^2 = 1$, $y \geq 0$. Compute the trajectory of the motion of the particle until it reaches the ground.

Problem 12 The value

$$s_\alpha(x_1, x_2, \ldots, x_n) = \left(\frac{x_1^\alpha + x_2^\alpha + \cdots + x_n^\alpha}{n}\right)^{1/\alpha}$$

is called the mean of order α of the numbers $x_1, x_2, \ldots, x_n$. In particular, for $\alpha = 1, 2, -1$, we obtain the *arithmetic mean, the mean square, and the harmonic mean*, respectively, of these numbers. We will assume that the numbers $x_1, x_2, \ldots, x_n$ are nonnegative and if the exponent α is less than 0, then we will suppose even that they are positive.

a) Show, using Hölder's inequality, that if $\alpha < \beta$, then

$$s_\alpha(x_1, x_2, \ldots, x_n) \leq s_\beta(x_1, x_2, \ldots, x_n),$$

and equality holds only when $x_1 = x_2 = \cdots = x_n$.

b) Show that if α tends to zero, then the value $s_\alpha(x_1, x_2, \ldots, x_n)$ tends to $\sqrt[n]{x_1 x_2 \cdots x_n}$, i.e., to the *harmonic mean* of these numbers. In view of the result of problem a), from here, for example, follows the classical inequality between the geometric and the arithmetic means of nonnegative numbers (write it down).

c) If $\alpha \to \infty$, then $s_\alpha(x_1, x_2, \ldots, x_n) \to \max\{x_1, x_2, \ldots, x_n\}$, and for $\alpha \to -\infty$, the value $s_\alpha(x_1, x_2, \ldots, x_n)$ tends to the lowest of the considered numbers, i.e., to $\min\{x_1, x_2, \ldots, x_n\}$. Prove this.

Problem 13 Let $\mathbf{r} = \mathbf{r}(t)$ denote the law of motion of a point (i.e., its radius vectors as a function of time). We suppose that it is a continuously differentiable function on the interval $a \leq t \leq b$.

a) Is it possible, according to Lagrange's mean value theorem, to claim that there exists a moment ξ on the interval $[a, b]$ such that $\mathbf{r}(b) - \mathbf{r}(a) = \mathbf{r}'(\xi) \cdot (b - a)$? Explain your answer with examples.

b) Let Convex$\{\mathbf{r}'\}$ be the convex hull (of all ends) of the vectors $\mathbf{r}'(t)$, $t \in [a, b]$. Show that there exists a vector $\mathbf{v} \in$ Convex$\{\mathbf{r}'\}$ such that $\mathbf{r}(b) - \mathbf{r}(a) = \mathbf{v} \cdot (b - a)$.

c) The relation $|\mathbf{r}(b) - \mathbf{r}(a)| \leq \sup |\mathbf{r}'(t)| \cdot |b - a|$, where the upper bound is taken over $t \in [a, b]$ has an obvious physical sense. What is it? Prove this inequality as a general mathematical fact, developing the classical Lagrange theorem on finite increments.

3 Integration and Introduction to Several Variables

Problem 1 Knowing the inequalities of Hölder, Minkowski, and Jensen for sums, obtaining the corresponding inequalities for integrals.

Problem 2 Compute the integral $\int_0^1 \mathrm{e}^{-x^2}\,\mathrm{d}x$ with a relative error of less than 10 %.

Problem 3 The function $\operatorname{erf}(x) = \frac{1}{\sqrt{\pi}} \int_{-x}^{x} \mathrm{e}^{-t^2}\,\mathrm{d}t$, called the *probability error integral*, has limit 1 as $x \to +\infty$. Draw the graph of this function and find its derivative. Show that as $x \to +\infty$

$$\operatorname{erf}(x) = 1 - \frac{2}{\sqrt{\pi}} \mathrm{e}^{-x^2} \left(\frac{1}{2x} - \frac{1}{2^2 x^3} + \frac{1 \cdot 3}{2^3 x^5} - \frac{1 \cdot 3 \cdot 5}{2^4 x^7} + o\left(\frac{1}{x^7}\right) \right).$$

How can this asymptotic formula be extended to a series? Are there any values of $x \in \mathbb{R}$ for which this series converges?

Problem 4 Does the length of a path depend on the law of motion (the parametrization)?

Problem 5 You are holding one end of a rubber band of length 1 km. A beetle is crawling toward you from the other end, which is clamped, at a rate of 1 cm/s. Each time it crawls 1 cm you lengthen the band by 1 km. Will the beetle ever reach your hand? If so, approximately how much time will it require? (A problem of L.B. Okun', proposed to A.D. Sakharov.)

Problem 6 Calculate the work done in moving a mass in the gravitational field of the Earth and show that this work depends only on the elevation of the initial and terminal positions. Find the work done in escaping from the Earth's gravitational field and the corresponding escape velocity.

Problem 7 Using the example of a pendulum and a double pendulum explain how it is possible to introduce local coordinate systems and neighborhoods into the set of corresponding configurations and how a natural topology thereby arises making it into the configuration space of a mechanical system. Is this space metrizable under these conditions?

Problem 8 Is the unit sphere in $\mathbb{R}^n$, $\mathbb{R}_0^\infty$ or $C[a,b]$ compact?

Problem 9 A subset of a given set is called an ε-*grid* if any point of the set lies at a distance less than ε from some point of the set. Denote by $N(\varepsilon)$ the smallest possible number of points in an ε-grid for a given set. Estimate the ε-*entropy* $\log_2 N(\varepsilon)$ of a closed line segment, a square, a cube, and a bounded region in $\mathbb{R}^n$. Does the quantity $\frac{\log_2 N(\varepsilon)}{\log_2(1/\varepsilon)}$ as $\varepsilon \to 0$ give a picture of the dimension of the space under consideration? Can such a dimension be equal, for example, to 0.5?

Problem 10 On the surface of the unit sphere S in $\mathbb{R}^3$ the temperature T varies continuously as a function of a point. Must there be points on the sphere where the temperature reaches a minimum or a maximum? If there are points where the temperature assumes two given values, must there be points where it assumes intermediate values? How much of this is valid when the unit sphere is taken in the space $C[a,b]$ and the temperature at the point $f \in S$ is given as

$$T(f) = \left(\int_a^b |f|(x)\,dx\right)^{-1}?$$

Problem 11

a) Taking 1.5 as an initial approximation to $\sqrt{2}$, carry out two iterations using Newton's method and observe how many decimal places of accuracy you obtain at each step.

b) By a recursive procedure find a function f satisfying the equation

$$f(x) = x + \int_0^x f(t)\,dt.$$

4 Differential Calculus of Several Variables

Problem 1 Local linearization. Consider and prove that local linearization is applicable to the following examples: instantaneous velocity and displacement; simplification of the equation of motion when the oscillations of a pendulum are small; computation of linear corrections to the values of $\exp(A)$, A^{-1}, $\det(E)$, $\langle a,b\rangle$ under small changes in the arguments (here A is an invertible matrix, E is the identity matrix, a and b are vectors, and $\langle\cdot,\cdot\rangle$ is the inner product).

Problem 2

a) What is the relative error $\delta = \frac{|\Delta f|}{|f|}$ in computing the value of a function $f(x, y, z)$ at a point (x, y, z) whose coordinates have absolute errors Δx, Δy, and Δz respectively?

b) What is the relative error in computing the volume of a room whose dimensions are as follows: length $x = 5 \pm 0.05$ m, width $y = 4 \pm 0.04$ m, height $z = 3 \pm 0.03$ m?

c) Is it true that the relative error of the value of a linear function coincides with the relative error of the value of its argument?

d) Is it true that the differential of a linear function coincides with the function itself?

e) Is it true that the relation $f' = f$ holds for a linear function f?

Problem 3

a) One of the partial derivatives of a function of two variables defined in a disk equals zero at every point. Does that mean that the function is independent of the corresponding variable in that disk?

b) Does the answer change if the disk is replaced by an arbitrary convex region?

c) Does the answer change if the disk is replaced by an arbitrary region?

d) Let $\mathbf{x} = \mathbf{x}(t)$ be the law of motion of a point in the plane (or in $\mathbb{R}^n$) in the time interval $t \in [a, b]$. Let $\mathbf{v}(t)$ be its velocity as a function of time and $C = \operatorname{conv}\{\mathbf{v}(t) \mid t \in [a, b]\}$ the smallest convex set containing all the vectors $\mathbf{v}(t)$ (usually called the *convex hull* of a set that spans it). Show that there is a vector $\mathbf{v}$ in C such that $\mathbf{x}(b) - \mathbf{x}(a) = \mathbf{v} \cdot (b - a)$.

Problem 4

a) Let $F(x, y, z) = 0$. Is it true that $\frac{\partial z}{\partial y} \cdot \frac{\partial y}{\partial x} \cdot \frac{\partial x}{\partial z} = -1$? Verify this for the relation $\frac{xy}{z} - 1 = 0$ (corresponding to the Clapeyron equation of state of an ideal gas: $\frac{PV}{T} = R$).

b) Now let $F(x, y) = 0$. Is it true that $\frac{\partial y}{\partial x} \frac{\partial x}{\partial y} = 1$?

c) What can you say in general about the relation $F(x_1, \dots, x_n) = 0$?

d) How can you find the first few terms of the Taylor expansion of the implicit function $y = f(x)$ defined by an equation $F(x, y) = 0$ in a neighborhood of a point (x_0, y_0), knowing the first few terms of the Taylor expansion of the function $F(x, y)$ in a neighborhood of (x_0, y_0), where $F(x_0, y_0) = 0$ and $F'_y(x_0, y_0)$ is invertible?

Problem 5

a) Verify that the plane tangent to the ellipsoid $\frac{x^2}{a^2} + \frac{y^2}{b^2} + \frac{z^2}{c^2} = 1$ at the point (x_0, y_0, z_0) can be defined by the equation $\frac{xx_0}{a^2} + \frac{yy_0}{b^2} + \frac{zz_0}{c^2} = 1$.

b) The point $P(t) = (\frac{a}{\sqrt{3}}, \frac{b}{\sqrt{3}}, \frac{c}{\sqrt{3}}) \cdot t$ emerged from the ellipsoid $\frac{x^2}{a^2} + \frac{y^2}{b^2} + \frac{z^2}{c^2} = 1$ at time $t = 1$. Let $p(t)$ be the point of the same ellipsoid closest to $P(t)$ at time t. Find the limiting position of $p(t)$ as $t \to +\infty$.

Problem 6

a) In the plane $\mathbb{R}^2$ with Cartesian coordinates (x, y) construct the level curves of the function $f(x, y) = xy$ and the curve $S = \{(x, y) \in \mathbb{R}^2 \mid x^2 + y^2 = 1\}$. Using the resulting picture, carry out a complete study of the extremal problem for $f|_S$, the restriction of f to the circle S.

b) What is the physical meaning of the Lagrange multiplier in Lagrange's method of finding extrema with constraints when an equilibrium position is sought for a point mass in a gravitational field if the motion of the point is constrained by ideal relations (for example, relations of the form $F_1(x, y, z) = 0$, $F_2(x, y, z) = 0$)?

Problem 7 If in a vector space V one has a nondegenerate bilinear form $B(x, y)$, then to every linear function $g^* \in V^*$ in this space there corresponds a unique vector g such that $g^*(v) = B(g, v)$, for every vector $v \in V$.

a) Show that if $V = \mathbb{R}^n$, $B(x, y) = b_{ij}x^i x^j$, and $g^* v = g_i v^i$, then the vector g has coordinates $g^j = b^{ij} g_i$, where b^{ij} is the inverse of the matrix (b_{ij}).

A symmetric scalar product $\langle \cdot, \cdot \rangle$ in the Euclidean geometry or a skew-scalar product $\omega(\cdot, \cdot)$ (when the form B is skew-symmetric) in the symplectic geometry appears most often as a bilinear form $B(\cdot, \cdot)$.

b) Let $B(v_1, v_2) = \begin{vmatrix} v_1^1 & v_1^2 \\ v_2^1 & v_2^2 \end{vmatrix}$ be the oriented area of the parallelogram spanned by the vectors $v_1, v_2 \in \mathbb{R}^2$. Find the vector $g = (g^1, g^1)$ corresponding to the linear function $g^* = (g_1, g_2)$ and with respect to the form B if we know the coefficients of g^*.

c) The vector corresponding to the differential of a function $f : \mathbb{R}^n \to \mathbb{R}$ at the point x relative to the scalar product $\langle \cdot, \cdot \rangle$ of the Euclidean space $\mathbb{R}^n$ is called as usual the gradient of the function f at this point and denoted by $\operatorname{grad} f$. Thus, $df(x)v := \langle \operatorname{grad} f, v \rangle$ for every vector $v \in T_x\mathbb{R}^n \sim \mathbb{R}^n$, applied at x.

Therefore,

$$f'(x)v = \frac{\partial f}{\partial x^1}(x)v^1 + \cdots + \frac{\partial f}{\partial x^n}(x)v^n = \big\langle \operatorname{grad} f(x), v \big\rangle = \big|\operatorname{grad} f(x)\big| \cdot |v| \cos \varphi.$$

c_1) Show that in the standard orthonormal basis, i.e., in Cartesian coordinates, $\operatorname{grad} f(x) = (\frac{\partial f}{\partial x^1}, \dots, \frac{\partial f}{\partial x^n})(x)$.

c_2) Show that the rate of growth of the function f under a motion from the point x with unit velocity is maximal when the direction of the movement coincides with the direction of the gradient of the function f at this point and is equal to $|\operatorname{grad} f(x)|$. When the movement has a perpendicular direction to the vector $\operatorname{grad} f(x)$, the function does not change.

c_3) How do the coordinates of the vector $\operatorname{grad} f(x)$ in $\mathbb{R}^2$ change if instead of considering the canonical basis (e_1, e_2), we take an orthogonal basis $(\tilde{e}_1, \tilde{e}_2) = (\lambda_1 e_1, \lambda e_2)$?

c_4) How do we calculate $\operatorname{grad} f$ in polar coordinates? Answer: $(\frac{\partial f}{\partial r}, \frac{1}{r}\frac{\partial f}{\partial \varphi})$.

d) In exercise b) above, we considered a skew-symmetric form $B(v_1, v_2)$ of the oriented area of the parallelogram in $\mathbb{R}^2$.

Since the vector corresponding to $\mathrm{d}f(x)$ relative to the symmetric form $\langle\cdot,\cdot\rangle$ is called the gradient $\operatorname{grad} f(x)$, the vector corresponding to $\mathrm{d}f(x)$ relative to the skew-symmetric form B is called *skew-gradient* and denoted by $\operatorname{sgrad} f(x)$. Write down $\operatorname{grad} f(x)$ and $\operatorname{sgrad} f(x)$ in Cartesian coordinates.

Problem 8

a) Show that in $\mathbb{R}^3$ (and in general in $\mathbb{R}^{2n+1}$), there are no nondegenerate skew-symmetric bilinear forms.

b) In the oriented space $\mathbb{R}^2$ there is a nondegenerate skew-symmetric bilinear form (the oriented area of the parallelogram), as we have seen. In $\mathbb{R}^{2n}$ with coordinates $(x^1, \dots, x^n, \dots, x^{2n}) = (p^1, \dots, p^n, q^1, \dots, q^n)$, such a form ω also exists: if $v_i = (p_i^1, \dots, p_i^n, q_i^1, \dots, q_i^n)$ $(i = 1, 2)$, then

$$\omega(v_1, v_2) = \begin{vmatrix} p_1^1 & q_1^1 \\ p_2^1 & q_2^1 \end{vmatrix} + \dots + \begin{vmatrix} p_1^n & q_1^n \\ p_2^n & q_2^n \end{vmatrix}.$$

That means that $\omega(v_1, v_2)$ is the sum of the oriented areas of parallelograms spanned by the projections of the vectors v_1, v_2 in the coordinate plane (p^j, q^j) $(j = 1, \dots, n)$.

b$_1$) Let g^* be a linear function in $\mathbb{R}^{2n}$, given with its coefficients $g^* = (p_1, \dots, p_n, q_1, \dots, q_n)$. Find the coordinates for the vector g mapped by the function g^* through the form ω.

b$_2$) The differential of the function $f : \mathbb{R}^{2n} \to \mathbb{R}$ at the point $x \in \mathbb{R}^{2n}$ through the skew-symmetric form ω is associated with a vector called the skew-gradient of the function f, as we already said, at this point and denoted by $\operatorname{sgrad} f(x)$. Find the expression for $\operatorname{sgrad} f(x)$ in the canonical Cartesian coordinates in the space $\mathbb{R}^{2n}$.

b$_3$) Find the scalar product $\langle \operatorname{grad} f(x), \operatorname{sgrad} f(x) \rangle$.

b$_4$) Show that the vector $\operatorname{sgrad} f(x)$ is directed along the level surface of the function f.

b$_5$) The law of motion $x = x(t)$ of a point x in the space $\mathbb{R}^{2n}$ is such that $\dot{x}(t) = \operatorname{sgrad} f(x(t))$. Show that $f(x(t)) = \text{const}$.

b$_6$) Write down the equation $\dot{x} = \operatorname{sgrad} f(x)$ in the canonical Cartesian notation $(p^1, \dots, p^n, q^1, \dots, q^n)$ for the coordinates and $H = H(p, q)$ for the function f. The resulting system, called a system of Hamilton's equations, is one of the central objects of mechanics.

Problem 9 Canonical variables and Hamilton's system of equations.

a) In the calculus of variations and in the fundamental variational principles of classical mechanics, the following *system of Euler–Lagrange equations* plays an important role:

$$\begin{cases} \left(\dfrac{\partial L}{\partial x} - \dfrac{\mathrm{d}}{\mathrm{d}t}\dfrac{\partial L}{\partial v}\right)(t,x,v) = 0, \\ v = \dot{x}(t), \end{cases}$$

where $L(t,x,v)$ is a given function in the variables t, x, v, where t usually denotes time, x the coordinate, and v the velocity. This is a system of two equations in three variables. From it, one usually wants to find the dependence relations $x = x(t)$ and $v = v(t)$, which essentially boils down to finding the law of motion $x = x(t)$, since $v = \dot{x}(t)$. Write down the first equation of the system in detail, revealing the derivative $\frac{\mathrm{d}}{\mathrm{d}t}$ given that $x = x(t)$ and $v = v(t)$.

b) Show that in transition from the variables t, x, v, L to the *canonical* variables t, x, p, H by taking the Legendre transform of

$$\begin{cases} p = \dfrac{\partial L}{\partial v}, \\ H = pv - L, \end{cases}$$

with respect to the variables v, L, interchanging them with the variables p, H, the Euler–Lagrange system acquires the symmetric form:

$$\dot{p} = -\frac{\partial H}{\partial x}, \qquad \dot{x} = \frac{\partial H}{\partial p}.$$

c) In mechanics, we often use the notation q and $\dot{q}$, instead of x and v. In many cases when $L(t,q,\dot{q}) = L(t,q^1,\dots,q^m,\dot{q}^1,\dots,\dot{q}^m)$, the Euler–Lagrange system of equations has the form

$$\left(\frac{\partial L}{\partial q^i} - \frac{\mathrm{d}}{\mathrm{d}t}\frac{\partial L}{\partial \dot{q}^i}\right)(t,q,\dot{q}) = 0 \quad (i = 1,\dots,m).$$

Take the Legendre transform with respect to the variables $\dot{q}$ and L, and go from the variables t, q, $\dot{q}$, L to the canonical variables t, q, p, H and show that this Euler–Lagrange system transforms into the following Hamilton equations:

$$\dot{p}_i = -\frac{\partial H}{\partial q^i}, \quad \dot{q}_i = \frac{\partial H}{\partial p^i} \quad (i = 1,\dots,m).$$

Examination Topics

1 First Semester

1.1 Introduction to Analysis and One-Variable Differential Calculus

1. Real numbers. Bounded (from above or below) numerical sets. The axiom of completeness and the existence of a least upper (greatest lower) bound of a set. Unboundedness of the set of natural numbers.
2. Fundamental lemmas connected with the completeness of the set of real numbers $\mathbb{R}$ (nested interval lemma, finite covering, limit point).
3. Limit of a sequence and the Cauchy criterion for its existence. Tests for the existence of a limit of a monotonic sequence.
4. Infinite series and the sum of an infinite series. Geometric progressions. The Cauchy criterion and a necessary condition for the convergence of a series. The harmonic series. Absolute convergence.
5. A test for convergence of a series of nonnegative terms. The comparison theorem. The series $\zeta(s) = \sum_{n=1}^{\infty} n^{-s}$.
6. The limit of a function. The most important filter bases. Definition of the limit of a function over an arbitrary base and its decoding in specific cases. Infinitesimal functions and their properties. Comparison of the ultimate behavior of functions, asymptotic formulas, and the basic operations with the symbols $o(\cdot)$ and $O(\cdot)$.
7. The connection of passage to the limit with the algebraic operations and the order relation in $\mathbb{R}$. The limit of $\frac{\sin x}{x}$ as $x \to 0$.
8. The limit of a composite function and a monotonic function. The limit of $(1 + \frac{1}{x})^x$ as $x \to \infty$.
9. The Cauchy criterion for the existence of the limit of a function.
10. Continuity of a function at a point. Local properties of continuous functions (local boundedness, conservation of sign, arithmetic operations, continuity of a composite function). Continuity of polynomials, rational functions, and trigonometric functions.

V.A. Zorich, *Mathematical Analysis I*, Universitext,
DOI 10.1007/978-3-662-48792-1

11. Global properties of continuous functions (intermediate-value theorem, maxima, uniform continuity).
12. Discontinuities of monotonic functions. The inverse function theorem. Continuity of the inverse trigonometric functions.
13. The law of motion, displacement over a small interval of time, the instantaneous velocity vector, trajectories and their tangents. Definition of differentiability of a function at a point. The differential, its domain of definition and range of values. Uniqueness of the differential. The derivative of a real-valued function of a real variable and its geometric meaning. Differentiability of $\sin x$, $\cos x$, e^x, $\ln |x|$, and x^α.
14. Differentiability and the arithmetic operations. Differentiation of polynomials, rational functions, the tangent, and the cotangent.
15. The differential of a composite function and an inverse function. Derivatives of the inverse trigonometric functions.
16. Local extrema of a function. A necessary condition for an interior extremum of a differentiable function (Fermat's lemma).
17. Rolle's theorem. The finite-increment theorems of Lagrange and Cauchy (mean-value theorems).
18. Taylor's formula with the Cauchy and Lagrange forms of the remainder.
19. Taylor series. The Taylor expansions of e^x, $\cos x$, $\sin x$, $\ln(1+x)$, and $(1+x)^\alpha$ (Newton's binomial formula).
20. The local Taylor formula (Peano form of the remainder).
21. The connection between the type of monotonicity of a differentiable function and the sign of its derivative. Sufficient conditions for the presence or absence of a local extremum in terms of the first, second, and higher-order derivatives.
22. L'Hôpital rule.
23. Convex functions. Differential conditions for convexity. Location of the graph of a convex function relative to its tangent.
24. The general Jensen inequality for a convex function. Convexity (or concavity) of the logarithm. The classical inequalities of Cauchy, Young, Hölder, and Minkowski.
25. Legendre transform.
26. Complex numbers in algebraic and trigonometric notation. Convergence of a sequence of complex numbers and a series with complex terms. The Cauchy criterion. Absolute convergence and sufficient conditions for absolute convergence of a series with complex terms. The limit $\lim_{n\to\infty}(1+\frac{z}{n})^n$.
27. The disk of convergence and the radius of convergence of a power series. The definition of the functions e^z, $\cos z$, $\sin z$ ($z \in \mathbb{C}$). Euler's formula and the connections among the elementary functions.
28. Differential equations as a mathematical model of reality, examples. The method of undetermined coefficients and Euler's (polygonal) method.
29. Primitives and the basic methods of finding them (termwise integration of sums, integration by parts, change of variable). Primitives of the elementary functions.

2 Second Semester

2.1 Integration. Multivariable Differential Calculus

1. The Riemann integral on a closed interval. Upper and lower sums, their geometric meaning, the behavior under a refinement of the partition, and mutual estimates. Darboux's theorem, upper and lower Darboux's integrals and the criterion for Riemann integrability of real-valued functions on an interval (in terms of sums of oscillations). Examples of classes of integrable functions.
2. The Lebesgue criterion for Riemann integrability of a function (statement only). Sets of measure zero, their general properties, examples. The space of integrable functions and admissible operations on integrable functions.
3. Linearity, additivity and general evaluation of an integral.
4. Evaluating the integral of a real-valued function. The (first) mean-value theorem.
5. Integrals with a variable upper limit of integration, their properties. Existence of a primitive for a continuous function. The generalized primitive and its general form.
6. The Newton–Leibniz formula. Change of variable in an integral.
7. Integration by parts in a definite integral. Taylor's formula with integral remainder. The second mean-value theorem.
8. Additive (oriented) interval functions and integration. The general pattern in which integrals arise in applications, examples: length of a path (and its independence of parametrization), area of a curvilinear trapezoid (area under a curve), volume of a solid of revolution, work, energy.
9. The Riemann–Stieltjes integral. Conditions under which it can be reduced to the Riemann integral. Singularities and the Dirac delta-function. The concept of a generalized function.
10. The concept of an improper integral. Canonical integrals. The Cauchy criterion and the comparison theorem for studying the convergence of an improper integral. The integral test for convergence of a series.
11. Metric spaces, examples. Open and closed subsets. Neighborhoods of a point. The induced metric, subspaces. Topological spaces. Neighborhoods of a point, separation properties (the Hausdorff axiom). The induced topology on subsets. Closure of a set and description of relatively closed subsets.
12. Compact sets, their topological invariance. Closedness of a compact set and compactness of a closed subset of a compact set. Nested compact sets. Compact metric spaces, ε-grids. Criteria for a metric space to be compact and its specific form in $\mathbb{R}^n$.
13. Complete metric spaces. Completeness of $\mathbb{R}$, $\mathbb{C}$, $\mathbb{R}^n$, $\mathbb{C}^n$, and the space $C[a, b]$ of continuous functions under uniform convergence.
14. Criteria for continuity of a mapping between topological spaces. Preservation of compactness and connectedness under a continuous mapping. The classical theorems on boundedness, the maximum-value theorem, and the intermediate-value theorem for continuous functions. Uniform continuity on a compact metric space.

15. The norm (length, absolute value, modulus) of a vector in a vector space; the most important examples. The space $L(X, Y)$ of continuous linear transformations and the norm in it. Continuity of a linear transformation and finiteness of its norm.
16. Differentiability of a function at a point. The differential, its domain of definition and range of values. Coordinate expression of the differential of a mapping $f : \mathbb{R}^m \to \mathbb{R}^n$. The relation between differentiability, continuity, and the existence of partial derivatives.
17. Differentiation of a composite function and the inverse function. Coordinate expression of the resulting laws in application to different cases of the mapping $f : \mathbb{R}^m \to \mathbb{R}^n$.
18. Derivative along a vector and the gradient. Geometric and physical examples of the use of the gradient (level surfaces of functions, steepest descent, the tangent plane, the potential of a field, Euler's equation for the dynamics of an ideal fluid, Bernoulli's law, the work of a wing).
19. Homogeneous functions and the Euler relation. The dimension method.
20. The finite-increment theorem. Its geometric and physical meaning. Examples of applications (a sufficient condition for differentiability in terms of the partial derivatives; conditions for a function to be constant in a domain).
21. Higher-order derivatives and their symmetry.
22. Taylor's formula.
23. Extrema of functions (necessary and sufficient conditions for an interior extremum).
24. Contraction mappings. The Picard–Banach fixed-point principle.
25. The implicit function theorem.
26. The inverse function theorem. Curvilinear coordinates and rectification. Smooth k-dimensional surfaces in $\mathbb{R}^n$ and their tangent planes. Methods of defining a surface and the corresponding equations of the tangent space.
27. The rank theorem and functional dependence.
28. Local resolution of a diffeomorphism as the composition of elementary ones (diffeomorphisms changing only one coordinate).
29. Extrema with constraint (necessary condition). Geometric, algebraic, and physical interpretation of the method of Lagrange multipliers.
30. A sufficient condition for a constrained extremum.

Appendix A
Mathematical Analysis (Introductory Lecture)

A.1 Two Words About Mathematics

Mathematics is an abstract science. For example, it teaches how to count and add no matter of whether we count ravens, capital, or something else. Therefore, mathematics is one of the most universal and commonly used applied sciences. Mathematics as a science possesses a large number of features, and therefore it is usually treated with respect. For example, it teaches us to listen to arguments and appreciate the truth.

Lomonosov believed that mathematics leads the mind into order, and Galilei said, "The great book of nature is written in the language of mathematics." The evidence of this is obvious: anyone who wants to read this book, must have studied mathematics. Among them, there are representatives of natural sciences, technical professions, as well as humanities. As examples, there is the chair of mathematics at the Economics Faculty of Moscow State University, and also there is an Institute of Economical Mathematics in the system of the Russian Academy of Sciences. There is even a statement saying that a branch of science contains as much of science as there is mathematics inside of it. Although this point of view is too strong, it is in general a quite sharp observation.

Mathematics has the attributes of a language. However, it is clearly not merely a language (otherwise, it would have been studied by philologists). Mathematics not only can translate a question into mathematical language, but it usually provides the method for solving the formulated mathematical problem.

The capability to pose a question correctly is the great art of researchers in general and mathematicians in particular.

The great mathematician Henri Poincaré, whose influence is clearly seen in most courses in contemporary mathematics, remarked with humor, "Mathematics is the art of calling different things by the same name." For instance, a point is a barely visible particle under a microscope, an airplane on air traffic controls radar, a city on a map, a planet in the sky, and in general all those things, whose dimensions can be neglected on these scales.

V.A. Zorich, *Mathematical Analysis I*, Universitext,
DOI 10.1007/978-3-662-48792-1

Therefore, the abstract concepts of mathematics and their relationships, like number, are very useful in a tremendous sphere of specific phenomena and consistent patterns.

A.2 Number, Function, Law

The usual reaction to a miracle, "Is it really possible???" quickly and imperceptibly transforms into, "It could not be otherwise!!!"

We are so used to the fact that $2+3=5$, that we don't see any miracle there. But here it is not said that two apples and three apples will add up to five apples, it is said that this is the case for apples, elephants, and all other things. We already mentioned this above.

Next, we get used to the fact that $a+b=b+a$, where now the symbols a and b could mean 2, 3, or any other integer.

A function, or functional dependence, is the next mathematical miracle. It is relatively young as a scientific concept: just a little over three hundred years old. Nevertheless, we are confronted with it in nature and even in everyday life no less than with elephants or apples.

Every science or every field of human activity deals with a concrete area of objects and their relations. These relations, laws, or dependences, are described and studied by mathematics in an abstract and general way, relating the terms *function* or *functional dependence* $y=f(x)$ of the state (value) of one variable (y) to some other state (or value) of another variable (x).

It is especially important that now we are not dealing with constants, but with variables x and y and with the rule f relating them. A function is adapted to describe developing processes and phenomena, the nature of change of their states, and in general to describe the dependent variable.

Sometimes the rule f of a relation is known (given), e.g., by a government or by a technological process. We often try then, under the constraints of the acting rule f, to choose a strategy, i.e., some state (value) available to our choice of the independent variable x, in order to obtain the most favorable state (value) for us (in one way or another) of the variable y (given that $y=f(x)$).

In other cases (and this is even more exciting), we search for the law of nature relating certain phenomena. Though it is the job of experts in the corresponding specific branch of science, mathematics might be extremely helpful. Like Sherlock Holmes, it can deduce the new law f from very limited information accessible to the experts in this narrow domain. Following the parallel with Sherlock Holmes, mathematics uses a "deduction method" called *differential equations*, which were unknown to ancient mathematicians, and that emerged with the advent of differential and integral calculus at the seventeenth and eighteenth centuries thanks to the efforts of Newton, Leibniz, and their predecessors and successors.

So, let us begin a primer of modern mathematics.

A.3 Mathematical Model of a Phenomenon (Differential Equations, or We Learn How to Write)

One of the brightest and most long-lived impressions of school mathematics, of course, is that small miracle when you want to find something unknown to you, denoting it by the letter x or by the letters x, y, and then you write something like $a \cdot x = b$ or some system of equations

$$\begin{cases} 2x + y = 1, \\ x - y = 2. \end{cases}$$

After a bit of mathematical hocus-pocus, you discover what was unknown to you: $x = 1$, $y = -1$.

Let us try to learn how to write the equation in a new situation, when we do not have to find a number but instead the unknown rule relating two variables that are important for us, i.e., we look for the requisite function. Let us take a look at some examples.

In order to be more specific, we first shall talk about biology (proliferation of microorganisms, growth of biomass, ecological limitations, etc.). However, it will be clear that all this can be transferred to other areas, as for example the growth of capital, nuclear reactions, atmospheric pressure, and so on.

To warm up, consider the following playful problem:

Consider a primitive organism that replicates itself every second (doubling). Suppose it is put into an empty glass. After one minute, the glass is filled up. How long will it take to fill up the glass if instead of one organism, two of these organisms are put into an empty glass?

Now we get closer to our goal, and the promised examples.

Example 1 It is known that under favorable conditions, the reproduction rate of microorganisms, i.e., the biomass growth rate, is proportional (with a coefficient of proportionality k) to the current amount of biomass. We need to find the rule $x = x(t)$ of the change of biomass over time if the initial condition $x(0) = x_0$ is known.

We expect that if we knew the rule itself $x = x(t)$ of the variable x, we would know its rate of change at any time t. Without going too far into the discussion about how we calculate this rate of change from $x(t)$, we shall denote it by $x'(t)$. Since the function $x' = x'(t)$ is obtained from the function $x(t)$, in mathematics it is called the *derivative* of the function $x = x(t)$ (in order to find out how to calculate the derivative of a function and much more, it is necessary to learn differential calculus. This is coming).

Now we are able to write what is given to us:

$$x'(t) = k \cdot x(t), \tag{A.1}$$

where $x(0) = x_0$, and we want to find the dependence itself.

We just wrote down the first differential equation (A.1). In general, equations containing derivatives are called differential equations (some clarifications and exceptions are not yet relevant). It is worth noting that the independent variable is often omitted, in order to simplify the text while writing the equation. For instance, Eq. (A.1) is written in the form $x' = k \cdot x$. If the desired function is denoted by the letter f or u, then the same equation would have the form $f' = k \cdot f$ or $u' = k \cdot u$ respectively.

It is now clear that if we learn not only how to write, but also to solve or study differential equations, we will be able to predict and know many things. That is why Newton's sacramental phrase referring to the new calculus, was something like this: "It is useful to learn how to solve differential equations."

Problem 1 Write down an equation for the example above of reproduction in a glass. Which coefficient k, initial condition $x(0) = x_0$, and dependence $x = x(t)$ do we have?

Having succeeded with the first equation, let us try to encode by a differential equation several further natural phenomena.

Example 2 Assume now, as is always the case, that there is not infinitely much food and that the environment cannot support more than M individuals, i.e., that the biomass does not exceed the value M. Then the growth rate of the biomass will presumably decrease, for instance proportionally to the remaining conditions of the environment. As a measure of the remaining conditions, you can consider the difference $M - x(t)$, or even better consider the dimensionless value $1 - \frac{x(t)}{M}$. In this situation, instead of Eq. (A.1), we obviously have the equation

$$x' = k \cdot x \cdot \left(1 - \frac{x(t)}{M}\right), \tag{A.2}$$

which reduces to Eq. (A.1) at the stage when $x(t)$ is much less than M. Conversely, when $x(t)$ is very close to M, the growth rate becomes close to zero, i.e., the growth stops, which is natural. After having improved some skills, we shall find out later precisely how the rule $x = x(t)$ looks in this case.

Problem 2 A body having an initial temperature T_0 cools down in an environment that has a constant temperature C. Let $T = T(t)$ be the rule of variation of temperature of the body over time. Write down the equation that this function must satisfy, assuming that the cooling rate is proportional to the temperature difference between the body and the environment.

The velocity $v(t)$, which is the rate of change of the variable $x(t)$, is called the derivative of the function $x(t)$ and denoted by $x'(t)$.

The acceleration $a(t)$, as you may know, is the rate of change of the velocity $v(t)$. This means that $a(t) = v'(t) = (x')'(t)$, i.e., it is the derivative of the derivative of the initial function. It is called the second derivative of the initial function and it is frequently denoted by $x''(t)$ (some other notation will appear later). If we are able to find the first derivative, we may repeat the same process to calculate the derivative $x^{(n)}(t)$ of any order n of the initial function $x = x(t)$.

Example 3 Let $x = x(t)$ be the rule of motion of a point with mass m, i.e., coordinates for the position of the point as function of time. For simplicity, we may assume that the motion is along a straight line (horizontal or vertical); thus there is only one coordinate.

The classical Newton's law $m \cdot a = F$, relating the force acting on a point of mass m with acceleration caused by this action, can be written as

$$m \cdot x''(t) = F(t), \tag{A.3}$$

or in abbreviated form, $m \cdot x'' = F$.

If the acting force $F(t)$ is known, then the relation $m \cdot x'' = F$ can be considered a differential equation (of second order) with respect to the function $x(t)$.

For example, if F is the force of gravity on Earth's surface, then $F = mg$, where g is the acceleration at free fall. In this case, our equation has the form $x''(t) = g$. As you might know, Galileo discovered that in free fall, $x(t) = \frac{1}{2}gt^2 + v_0t + x_0$, where x_0 is the initial position and v_0 is the initial speed of the point.

In order to check that this function satisfies the equation, it is necessary to be able to differentiate a function, i.e., to find its derivative. In our case, we even need the second derivative.

Just below we shall give a small table of some functions and their derivatives. Its deduction shall be made later, in the systematic presentation of differential calculus. Now try yourself to do the following.

Problem 3 Write down the equation of free fall in the atmosphere. In this case, there will arise a resistance force. Consider it proportional to the first (or second) rate of change of the movement (the speed in free fall does not grow to infinity due to the presence of the force of resistance).

You should be convinced by now that it is worthwhile learning how to calculate derivatives.

A.4 Velocity, Derivative, Differentiation

First, let us consider a familiar situation in which we can trust our intuition (and we change the notation from $x(t)$ to $s(t)$).

Suppose that a point moves on the whole numerical real line, $s(t)$ is its coordinate at the moment t, and $v(t) = s'(t)$ is its velocity at the same moment t. After the time interval h, from its former position at the moment t, the point is located at the position $s(t+h)$. In our picture of velocity the quantity $s(t+h) - s(t)$, the distance traveled in the time interval h after the moment t, and its velocity at the moment t are related by the equation

$$s(t+h)-s(t)\approx v(t)\cdot h; \tag{A.4}$$

in other words, $v(t)\approx\frac{s(t+h)-s(t)}{h}$, and this approximation becomes closer to an equality as the interval h decreases its length.

Thus, we have to assume that

$$v(t):=\lim_{h\to 0}\frac{s(t+h)-s(t)}{h},$$

i.e., we define $v(t)$ as the limit of the quotient between the increment of the function and the increment of its argument as the latter approaches zero.

Now after this example, nothing impedes us from providing the usual definition for the value $f'(x)$ for the derivative f' of the function f at the point x:

$$f'(x):=\lim_{h\to 0}\frac{f(t+h)-f(t)}{h}, \tag{A.5}$$

i.e., $f'(x)$ is the limit of the quotient of increments $\Delta f/\Delta x$ as Δx approaches zero, where $\Delta f=f(x+h)-f(x)$ is the increment of the function with respect to the increment of its argument $\Delta x=(x+h)-x$.

Equation (A.5) can be rewritten in a similar form to (A.4) in the same convenient and useful form

$$f(t+h)-f(t)=f'(t)h+o(h), \tag{A.6}$$

where $o(h)$ is some error (of approximation), small in relation to h, as h approaches zero. (This means that the quotient $o(h)/h$ approaches zero as h goes to zero.)

Now we shall make some concrete calculations.

1. Let f be a constant function, i.e., $f(x)\equiv c$. Then it is clear that $\Delta f=f(x+h)-f(x)\equiv 0$ and $f'(x)\equiv 0$. This is natural, since the velocity of change is equal to zero if there is no change.
2. If $f(x)=x$, then $f(x+h)-f(x)=h$, and therefore $f'(x)\equiv 1$. If $f(x)=kx$, then $f(x+h)-f(x)=kh$ and $f'\equiv k$.
3. By the way, we can make two very common but extremely useful remarks: if the function f has its derivative f', then the function cf, where c is some arbitrary constant, has as derivative cf', i.e., $(cf)'=cf'$; in the same way, $(f+g)'=f'+g'$, i.e, the derivative of a sum is equal to the sum of the derivatives, if these derivatives are defined.
4. Let $f(x)=x^2$. Then $f(x+h)-f(x)=(x+h)^2-x^2=2xh+h^2=2xh+o(h)$; hence $f'(x)=2x$.
5. Analogously, if $f(x)=x^3$, then

$$f(x+h)-f(x)=(x+h)^3-x^3=3x^2h+3xh^2+h^3=3x^2h+o(h);$$

therefore, $f'(x)=3x$.

Table A.1

$f(x)$	$f'(x)$	$f''(x)$	...	$f^{(n)}(x)$
a^x	$a^x \ln a$	$a^x \ln^2 a$	...	$a^x \ln^n a$
e^x	e^x	e^x	...	e^x
$\sin x$	$\cos x$	$-\sin x$	...	$\sin(x + n\pi/2)$
$\cos x$	$-\sin x$	$-\cos x$	...	$\cos(x + n\pi/2)$
$(1+x)^\alpha$	$\alpha(1+x)^{\alpha-1}$	$\alpha(\alpha-1)(1+x)^{\alpha-2}$	...	?
x^α	$\alpha x^{\alpha-1}$	$\alpha(\alpha-1)x^{\alpha-2}$	...	?

6. Now it is clear that in general, if $f(x) = x^n$, then one has

$$f(x+h) - f(x) = (x+h)^n - x^n = nx^{n-1}h + o(h),$$

and therefore $f'(x) = nx^{n-1}$.
7. This implies that if we have a polynomial

$$P(x) = a_0x^n + a_1x^{n-1} + \cdots + a_{n-1}x + a_n,$$

then

$$P'(x) = na_0x^{n-1} + (n-1)a_1x^{n-2} + \cdots + a_{n-1}.$$

We calculated the above derivatives following the definitions. To develop and master the techniques of differentiation you have to practice. Now as examples and for your illustration we introduce Table A.1 with functions and their derivatives. Later we shall prove the results.

In Table A.1 e is a number (e = 2.7...), that appears everywhere in analysis, just like the number π in geometry. The logarithm with base e is frequently denoted by ln, instead of $\log_e$, and you can find it in the first row of the table in the second and third columns. The logarithm with this base is called the *natural logarithm* and it appears in many formulas.

Problem 4 Assuming that the formula of the first derivative f' is correct, verify the expression for $f^{(n)}$ and complete the table at the places where the question marks are. After that, calculate the value $f^{(n)}(0)$ in every case.

Problem 5 Try to find the derivative of the function $f(x) = e^{kx}$ and the solution of Eq. (A.1). Explain at what point in time the initial condition x_0 (capital, biomass, or something else represented by this equation) will double.

A.5 Higher Derivatives, What for?

A wonderful and very useful expansion of the central equation (A.6), which can be written as

$$f(x+h) = f(x) + f'(x)h + o(h), \tag{A.7}$$

turns into the following formula (Taylor's formula):

$$f(x+h) = f(x) + \frac{1}{1!} f'(x)h + \frac{1}{2!} f''(x)h^2 + \cdots + \frac{1}{n!} f^{(n)}(x)h^n + o(h^n). \quad \text{(A.8)}$$

If we set $x = 0$ and then replace the letter h by the letter x, we obtain

$$f(x) = f(0) + \frac{1}{1!} f'(0)x + \frac{1}{2!} f''(0)x^2 + \cdots + \frac{1}{n!} f^{(n)}(0)x^n + o(x^n). \quad \text{(A.9)}$$

For instance, if $f(x) = (1+x)^\alpha$, following Newton we find that

$$(1+x)^\alpha = 1 + \frac{\alpha}{1!}x + \frac{\alpha(\alpha-1)}{2!}x^2 + \cdots + \frac{\alpha(\alpha-1)\cdots(\alpha-n+1)}{n!}x^n + o(x^n). \quad \text{(A.10)}$$

Sometimes in the formula (A.9), it is possible to continue the sum to infinity, removing the reminder term, since it tends to zero as $n \to \infty$.

In particular, we have,

$$e^x = 1 + \frac{1}{1!}x + \frac{1}{2!}x^2 + \cdots + \frac{1}{n!}x^n + \cdots, \quad \text{(A.11)}$$

$$\cos x = 1 - \frac{1}{2!}x^2 + \cdots + (-1)^k \frac{1}{2k!}x^{2k} + \cdots, \quad \text{(A.12)}$$

$$\sin x = \frac{1}{1!}x - \frac{1}{3!}x^3 + \cdots + (-1)^k \frac{1}{(2k+1)!}x^{2k+1} + \cdots. \quad \text{(A.13)}$$

We have obtained a representation of relatively complex functions as a sum (an infinite sum, a *series*) of simple functions, which can be calculated with the usual arithmetic operations. The finite pieces of these sums are polynomials. They provide good approximations of the functions decomposed in such a series.

A.5.1 Again Toward Numbers

We have always tacitly assumed that we are dealing with functions defined on the set of real numbers. But the right-hand side of Eqs. (A.11), (A.12), and (A.13) make sense when x is replaced by a complex number $z = x + iy$. Then, we are able to say what the expressions e^z, $\cos z$, $\sin z$ mean.

Problem 6 Discover, after Euler, the following impressive formula: $e^{i\varphi} = \cos\varphi + i\sin\varphi$, linking these elementary functions and the remarkably beautiful equality $e^{i\pi} + 1 = 0$ resulting from this formula, which combines the basic constants of the mathematical sciences (arithmetic, algebra, analysis, geometry, and even logic).

A.5.2 *And What to Do Next?*

As we say in Russian, "with the fingers", without details and justifications, you have been given some idea of differential calculus, which is the core of a first-semester course in mathematical analysis. Step by step, we have become acquainted with the concepts of numbers, functions, limits, derivatives, series, which had been only superficially studied so far.

Now you know why it is necessary to take time to dive into a detailed and careful consideration of all of these concepts and objects. The understanding of them is necessary for a professional mathematician. For the average user, this is not required. Most people drive a car without even opening the hood. But that is reasonable, because someone well versed in engines designed a machine that works reliably.

Appendix B
Numerical Methods for Solving Equations (An Introduction)

B.1 Roots of Equations and Fixed Points of Mappings

Notice that the equation $f(x) = 0$ is obviously equivalent to the equation $\alpha(x) f(x) = 0$ if $\alpha(x) \neq 0$. The last equation, in turn, is equivalent to the relation $x = x - \alpha(x) f(x)$, where x can be interpreted as a fixed point of the mapping $\varphi(x) := x - \alpha(x) f(x)$.

Thus, finding roots of equations is equivalent to finding the fixed points of the corresponding mappings.

B.2 Contraction Mappings and Iterative Process

A mapping $\varphi : X \to X$ from the set $X \subset \mathbb{R}$ to itself is called a *contraction mapping* if there exists a number q, $0 \le q < 1$, such that for every pair of points x', x'', their images satisfy the inequality $|\varphi(x') - \varphi(x'')| \le q|x' - x''|$.

It is clear that this definition applies to arbitrary sets, without any changes, on which the distance $d(x', x'')$ between any two points is defined; in our case, $d(x', x'') = |x' - x''|$.

It is also obvious that a contraction mapping is continuous and cannot have more than one fixed point.

Let $\varphi : [a, b] \to [a, b]$ be a contraction mapping from the interval $[a, b]$ into itself. We shall show that the iteration process $x_{n+1} = \varphi(x_n)$, starting at any point x_0 from this interval, leads to the point $x = \lim_{n\to\infty} x_n$, the fixed point of the mapping φ.

We notice first that

$$|x_{n+1} - x_n| \le q|x_n - x_{n-1}| \le \cdots \le q^n|x_1 - x_0|.$$

V.A. Zorich, *Mathematical Analysis I*, Universitext,
DOI 10.1007/978-3-662-48792-1

Therefore, for all natural numbers m, n, with $m > n$, inserting intermediate points and using the triangle inequality, we get the estimate

$$|x_m - x_n| \le |x_m - x_{m-1}| + \cdots + |x_{n+1} - x_n|$$
$$\le (q^{m-1} + \cdots + q^n)|x_1 - x_0| < \frac{q^n}{1-q}|x_1 - x_0|,$$

and from this it follows that the sequence $\{x_n\}$ is fundamental (i.e., is a Cauchy sequence).

Hence, by the Cauchy criterion it converges to a point x in the interval $[a, b]$. This point is a fixed point of the mapping $\varphi : [a, b] \to [a, b]$, since by taking the limit $n \to \infty$ in the relation $x_{n+1} = \varphi(x_n)$, we obtain the equality $x = \varphi(x)$.

(Here we used the obvious fact that a contraction mapping is continuous; actually, it is even uniformly continuous.)

By passing to the limit $m \to \infty$ in the relation $|x_m - x_n| < \frac{q^n}{1-q}|x_1 - x_0|$, one gets the estimate

$$|x - x_n| < \frac{q^n}{1-q}|x_1 - x_0|$$

for the deviation value between the approaching point x_n and the fixed point x of the mapping φ.

B.3 The Method of Tangents (Newton's Method)

We presented a theorem stating that a continuous real-valued function from an interval taking values with different signs at the extremes of the interval has at least one zero in this interval (a point x where $f(x) = 0$). In its proof, we showed the simplest, but universal, algorithm for finding this point (splitting the interval in half). The order of the speed of convergence in this case is 2^{-n}.

In the case of a convex differentiable function, the method proposed by Newton can be much more efficient in terms of speed of convergence.

We draw a tangent to the graph of the given function f at some point $(x_0, f(x_0))$, where $x_0 \in [a, b]$. Next, we find the point x_1 where the tangent intersects the x-axis. By repeating this process, we obtain a sequence $\{x_n\}$ of points that quickly converges to the point x such that $f(x) = 0$. (It is possible to prove that each successive iteration leads to the doubling of the number of correct digits of x found at the previous step.)

It is easy to prove analytically (check it!) that the method of tangents reduces to an iteration process

$$x_{n+1} = x_n - \frac{f(x_n)}{f'(x_n)}.$$

For instance, the solution of the equation $x^m - a = 0$, i.e., the computation of $\sqrt[m]{a}$, according to the formula above, reduces to an iteration process

$$x_{n+1} = \frac{1}{m}\left((m-1)x_n - \frac{a}{x_n^{m-1}}\right).$$

In particular, for calculating $\sqrt{2}$ with the method of tangents, we obtain

$$x_{n+1} = \frac{1}{2}\left(x_n - \frac{a}{x_n}\right).$$

As can be seen from the formulas above, Newton's method looks for fixed points of the mapping $\varphi(x) = x - \frac{f(x)}{f'(x)}$. It is a special case of the mapping $\varphi(x) = x - \alpha f(x)$, discussed in the first section, and we obtain it by setting $\alpha(x) = \frac{1}{f'(x)}$.

Note that in general, the mapping $\varphi(x) = x - \alpha f(x)$ and even the mapping $\varphi(x) = x - \frac{f(x)}{f'(x)}$ involved in the method of tangents do not have to be contraction mappings. Moreover, as can be shown by simple examples, in the case of a general function f, the method of tangents does not always lead to a convergent iteration process.

If in the expression $\varphi(x) = x - \alpha f(x)$, the function $\alpha(x)$ can be chosen in the given interval so that $|\varphi'(x)| \leq q < 1$, then the mapping $\varphi : [a, b] \to [a, b]$, of course, will be a contraction.

In particular, if α can be taken as the constant $\frac{1}{f'(x_0)}$, then we obtain $\varphi(x) = x - \frac{f(x)}{f'(x_0)}$ and $\varphi'(x) = 1 - \frac{f'(x)}{f'(x_0)}$. If the derivative of the function f is at least continuous at x_0, then in some neighborhood of x_0, we have $|\varphi'(x)| = |1 - \frac{f'(x)}{f'(x_0)}| \leq q < 1$. If the function φ maps this neighborhood into itself (although this is not always the case), then the standard iterative process induced by the contraction mapping φ of this neighborhood will lead to a unique point of the mapping φ in this neighborhood at which the original function f vanishes.

Appendix C
The Legendre Transform (First Discussion)

C.1 Initial Definition of the Legendre Transform and the General Young Inequality

We call the *Legendre transform* of a function f with variable x a new function f^* with new variable x^*, defined by the relation

$$f^*(x^*) = \sup_x \bigl(x^* x - f(x)\bigr), \tag{C.1}$$

where the supremum is taken with respect to the variable x, for a fixed value of the variable x^*.

Problem 1

a) Check that the function f^* is convex in its domain of definition.

b) Draw a graph of the function f, the line $y = kx$ where $k = x^*$, and specify the geometric meaning of the value $f^*(x^*)$.

c) Find $f^*(x^*)$ for $f(x) = |x|$ and $f(x) = x^2$.

d) Prove that from the definition (C.1), it follows clearly that the inequality

$$x^* x \le f^*(x^*) + f(x) \tag{C.2}$$

is satisfied for all values of the arguments x, x^* in the domain of definition of the functions f and f^*, respectively.

Relation (C.2) is usually called the *general Young's inequality* or the *Fenchel–Young inequality*, and the function f^*, in convex analysis, for instance, is usually called the Young dual of the function f.

V.A. Zorich, *Mathematical Analysis I*, Universitext,
DOI 10.1007/978-3-662-48792-1

C.2 Specification of the Definition in the Case of Convex Functions

If the supremum involved in the definition (C.1) is attained at some inner point x from the domain of definition of the function f, and this function is smooth (or at least differentiable), then we find that

$$x^* = f'(x) \tag{C.3}$$

and therefore

$$f^*(x^*) = x^*x - f(x) = xf'(x) - f(x). \tag{C.4}$$

Thus in this case, the Legendre transform is specified in the form (C.3), giving the argument x^* and (C.4) providing the value $f^*(x^*)$ of the function f^*, i.e., the Legendre transform of the function f. (Notice that the operator $xf'(x) - f(x)$ was studied already by Euler.)

Moreover, if the function f is convex, then in the first place, the condition (C.3) will not only provide a local extremum, but also will be a local maximum (check it!), which in this case will be the global or absolute maximum.

In the second place, due to the monotonic increase of the derivative of a strictly convex function, Eq. (C.3) is uniquely solvable with respect to x, for such functions.

If Eq. (C.3) admits an explicit solution $x = x(x^*)$, then on substituting it in (C.4), we obtain an explicit expression for $f^*(x^*)$.

Problem 2

a) Find the Legendre transform of the function $\frac{1}{\alpha}x^\alpha$, for $\alpha > 1$, and obtain the classical Young's inequality

$$ab \le \frac{1}{\alpha}a^\alpha + \frac{1}{\beta}b^\beta, \tag{C.5}$$

where $\frac{1}{\alpha} + \frac{1}{\beta} = 1$.

b) What is the domain of definition of the Legendre transform of a smooth strictly convex function f having the lines ax and bx as asymptotes, for $x \to +\infty$ and $x \to -\infty$ respectively?

c) Find the Legendre transform of the function e^x and prove the inequality

$$xt \le \mathrm{e}^x + t \ln \frac{t}{\mathrm{e}}. \tag{C.6}$$

C.3 Involutivity of the Legendre Transform of a Function

As we already noted, Eq. (C.2), or equivalently the inequality

$$f(x) \ge xx^* - f^*(x^*), \tag{C.7}$$

holds for all values of the arguments x, x^* in the domains of definition of the functions f and f^*, respectively.

At the same time, as shown by formulas (C.3) and (C.4), if x and x^* are linked by the relation (C.3), then the last inequality, (C.7), becomes an equality, at least in the case of a smooth strictly convex function f. Recalling the definition (C.1) of the Legendre transformation, we conclude that in this case,

$$\left(f^*\right)^* = f. \tag{C.8}$$

So the Legendre transform of a smooth strictly convex function is involutive, i.e., the twofold application of this transform leads to the original function.

Problem 3

a) Is it true that $f^{**} = f$ for every smooth function f?

b) Is it true that $f^{***} = f^*$ for every smooth function f?

c) Differentiating Eq. (C.4), using (C.3) and provided that $f''(x) \neq 0$, show that $x = f^*(x^*)$ and therefore $f(x) = xx^* - f^*(x^*)$ (involutivity).

d) Check that at the corresponding points x, x^*, linked by Eq. (C.3), we have $f''(x) = 1/(f^*)''(x^*)$ and $f^{(3)} = -(f^*)^{(3)}(x^*)/((f^*)'')^2(x^*)$.

e) The family of lines $px + p^4$ depending on the parameter p is a family of tangents to a certain curve (the *envelope* of this family). Find the equation of this curve.

C.4 Concluding Remarks and Comments

As part of the discussion about convex functions, we gave the initial presentation of the Legendre transform at the level of functions of one variable. However, here we shall facilitate the perception of the Legendre transform and work with it in a number of important more general cases of applications, such as in theoretical mechanics, thermodynamics, equations of mathematical physics, calculus of variations, convex analysis, contact geometry, ..., and many more yet to be dealt with.

We shall analyze various details and possible developments of the concept of the Legendre transform. Here we add only the following remark. The argument of the Legendre transform is a derivative, or equivalently the differential of the original function, as is shown by Eq. (C.3).

If the argument x is, for instance, a vector of a linear space X equipped with a scalar product $\langle \cdot, \cdot \rangle$, then the generalization of the definition (C.1) is the equation

$$f^*(x^*) = \sup_x\left(\langle x^*, x\rangle - f(x)\right). \tag{C.9}$$

If we take x^* as a linear function on the space X, i.e., assume that x^* is an element of the dual space X^* and the action $x^*(x)$ of x^* on a vector x is denoted as before by $\langle x^*, x\rangle$, then Eq. (C.9) will continue to be meaningful, since the function f is defined on the space X, and therefore the Legendre transform f^* is defined in the space X^*, the dual of the space X.

Appendix D
The Euler–MacLaurin Formula

D.1 Bernoulli Numbers

Jacob Bernoulli found that $\sum_{n=1}^{N-1} n^k = \frac{1}{k+1} \sum_{m=0}^{k} C_{k+1}^{m} B_m N^{k+1-m}$, where $C_n^m = \frac{n!}{m!(n-m)!}$ are the binomial coefficients, and $B_0, B_1, B_2, \ldots$ are rational numbers, now called Bernoulli numbers. These numbers are met in different problems. They have the generating function $\frac{z}{e^z-1} = \sum_{n=0}^{\infty} \frac{B_n}{n!} z^n$, where they occur as coefficients in the Taylor expansion, and in this way they can be calculated.

These numbers can also be calculated with the following recurrence formula:

$$B_0 = 1, \qquad B_n = -\frac{1}{n+1} C_{n+1}^{k+1} B_{n-k}.$$

Problem 1 Find the first Bernoulli numbers and check that all Bernoulli numbers with odd indices, except B_1, are equal to zero, and that the signs alternate for the Bernoulli numbers with even indices. (The function $x/(e^x - 1) + x/2$ is even.)

Euler discovered the connection $B_n = -n\zeta(1-n)$ between the Bernoulli numbers and the Riemann ζ-function.

D.2 Bernoulli Polynomials

Bernoulli polynomials can be defined by various means. For example, Bernoulli polynomials are defined recursively by $B_0(x) \equiv 1$, $B_n'(x) = nB_{n-1}(x)$, with the condition that $\int_0^1 B_n(x)\,dx = 0$; they are also defined through the generating function $\frac{ze^{xz}}{e^z-1} = \sum_{n=0}^{\infty} \frac{B_n(x)}{n!} z^n$, through the formula $B_n(x) = \sum_{k=0}^{n} C_n^k B_{n-k} x^k$, or through $B_n(x) = \sum_{m=0}^{n} \frac{1}{m+1} \sum_{k=0}^{m} (-1)^k C_m^k (x+k)^n$.

V.A. Zorich, *Mathematical Analysis I*, Universitext,
DOI 10.1007/978-3-662-48792-1

Problem 2

a) Based on the different definitions, compute the first few Bernoulli polynomials, and check the coincidence between them and the fact that Bernoulli numbers are the values of the Bernoulli polynomials at $x = 0$.

b) By differentiating the generating function, show that the Bernoulli polynomials $B_n(x)$ defined through this function satisfy the recurrence relation above, which in turn means that

$$B_n(x) = B_n + n \int_0^x B_{n-1}(t)\,\mathrm{d}t.$$

D.3 Some Known Operators and Series of Operators

Problem 3

a) If A is an operator, then we employ the notation $\frac{1}{A}$, as is usual for numbers, for referring to the operator A^{-1}, the inverse operator of A.

The integration operator $\int$ is the inverse of the differentiation operator D (with a proper setting of the integration constant). Similarly, the sum operator $\sum$ is the inverse of the difference operator Δ, whose action is defined as $\Delta f(x) = f(x+1) - f(x)$. Specify exactly how to find $\sum f(x)$.

b) Do you agree with the fact that $B_n(x) = D(\mathrm{e}^D - 1)^{-1}x^n$?

c) According to Taylor's formula,

$$\Delta f(x) = f(x+1) - f(x) = \frac{f'(x)}{1!} + \frac{f''(x)}{2!} + \cdots = \left(\frac{D}{1!} + \frac{D^2}{2!} + \cdots\right) f(x);$$

therefore, $\Delta = \mathrm{e}^D - 1$ and $\sum = \Delta^{-1} = (\mathrm{e}^D - 1)^{-1}$, and since $\frac{z}{\mathrm{e}^z - 1} = \sum_{k=0}^{\infty} \frac{B_k}{k!} z^k$, then

$$\sum = \frac{B_0}{D} + \frac{B_1}{1!} + \frac{B_2}{2!} D + \frac{B_3}{3!} D^2 + \cdots = \int + \sum_{k=1}^{\infty} \frac{B_k}{k!} D^k.$$

D.4 Euler–MacLaurin Series and Formula

By applying this operator relation to the function $f(x)$, we guess the Euler–MacLaurin summation formula, more precisely, not the formula itself but the corresponding series

$$\sum_{a \le n < b} f(n) = \int_a^b f(x)\,\mathrm{d}x + \sum_{k=1}^{\infty} \frac{B_k}{k!} f^{(k-1)}(x)\bigg|_0^1,$$

where a, b, n are integers and k is a natural number.

These series differ in the way that the Taylor series is different from Taylor's formula, which is finite and contains certain information (remainder term) providing the possibility of estimating the remainder (the value of the approximation error).

When the sum is reduced to a single term, the simplest and at the same time basic Euler–MacLaurin formula with a remainder term has the form

$$f(0) = \int_0^1 f(x)\,\mathrm{d}x + \sum_{k=1}^{m} \frac{B_k}{k!} f^{(k-1)}(x)\Big|_0^1 + (-1)^{(m+1)} \int_0^1 \frac{B_m(x)}{m!} f^{(m)}(x)\,\mathrm{d}x.$$

It is assumed here that the original function f is smooth enough, for example, that it possesses continuous derivatives of the required order.

Problem 4 Using the formula of integration by parts, prove by induction the Euler–MacLaurin formula written above. (Recall that Taylor's formula with remainder term of integral type can also be obtained with a simple integration by parts.)

D.5 The General Euler–MacLaurin Formula

The general Euler–MacLaurin formula, providing the value of the sum $\sum_{a \le n < b} f(n)$, has the form

$$\sum_{a \le n < b} f(n) = \int_a^b f(x)\,\mathrm{d}x + \sum_{k=1}^{m} \frac{B_k}{k!} f^{(k-1)}(x)\Big|_a^b + R_m,$$

where a, b, n are integers and k, m are natural numbers,

$$R_m = (-1)^{(m+1)} \int_a^b \frac{B_m(\{x\})}{m!} f^{(m)}(x)\,\mathrm{d}x,$$

and $\{x\}$ is the fractional part of the number x.

Problem 5 Prove this formula, given that every interval $[a, b]$ whose endpoints are integers can be divided into unit intervals (with length 1) and each unit interval can be translated to the interval $[0, 1]$ with a shift.

D.6 Applications

Problem 6

a) Using the Euler–MacLaurin formula and setting $f(x) = x^n$, show that $\sum_{a \le k < b} k^{m-1} = \frac{1}{m} \sum_{k=0}^{m} C_m^k B_k (b^{m-k} - a^{m-k})$, and in particular, obtain Jacob Bernoulli's relation $\sum_{0 \le k < b} k^{m-1} = \frac{1}{m} \sum_{k=0}^{m} C_m^k B_k b^{m-k}$.

b) To calculate the asymptotic behavior of a sum or a series, usually the following kind of Euler–MacLaurin formula is used:

$$\sum_{n=a}^{b} f(n) \sim \int_a^b f(x)\,\mathrm{d}x + \frac{f(a)+f(b)}{2} + \sum_{k=1}^{\infty} \frac{B_{2k}}{(2k)!}\left(f^{(2k-1)}(b) - f^{(2k-1)}(a)\right),$$

where a, b are integers. The formula often remains valid for the extension from the interval $[a,b]$ to the whole line. In many cases, the integral on the right side can be calculated in terms of elementary functions, even if the sum on the left side cannot be expressed. Then all the terms of the asymptotic series can be expressed in terms of elementary functions. For example,

$$\sum_{s=0}^{+\infty} \frac{1}{(z+s)^2} \sim \int_0^{+\infty} \frac{1}{(z+s)^2}\,\mathrm{d}s + \frac{1}{2z^2} + \sum_{k=1}^{+\infty} \frac{B_{2k}}{z^{2k+1}}.$$

Moreover, in this case, the integral can be calculated and is equal to $\frac{1}{z}$.

c) Setting $f(x) = x^{-1}$, prove the asymptotic formula

$$\sum_{k=1}^{n} \frac{1}{k} = \ln n + \gamma + \frac{1}{2n} + \sum_{k=1}^{m} \frac{B_{2k}}{2kn^{2k}} - \theta_{m,n}\frac{B_{2m+2}}{(2m+2)n^{2m+2}},$$

where $0 < \theta_{m,n} < 1$, and γ is a constant (Euler's constant).

d) If we take $f(x) = \ln x$, show that

$$\sum_{k=1}^{n} \ln k = n\ln n - n + \sigma - \frac{1}{2}\ln n + \sum_{k=1}^{m} \frac{B_{2k}}{2k(2k-1)n^{2k-1}} - R_{m,n},$$

where

$$R_{m,n} = \phi_{m,n}\frac{B_{2m+2}}{(2m+1)(2m+2)n^{2m+1}},$$

$0 < \phi_{m,n} < 1$, and σ is a constant (in fact, equal to $\ln\sqrt{2\pi}$).

By raising to powers, we can obtain the asymptotic Stirling's formula for the values $n!$ as $n \to \infty$.

D.7 Again to the Actual Euler–MacLaurin Formula

Problem 7

a) If a and n are integers such that $a < n$ and f is a slowly changing function at the interval $[a,b]$, then the sum $S = \frac{1}{2}f(a) + f(a+1) + f(a+2) + \cdots + f(n-1) + \frac{1}{2}f(n)$ is a good approximation of the integral $I = \int_a^b f(x)\,\mathrm{d}x$.

Remember this by drawing a picture showing the geometric meaning of the quantities S and I, and at the same time recalling the numerical methods for calculating the integral.

b) If j is an integer, then integration by parts gives

$$\int_j^{j+1} f(x)\,\mathrm{d}x = \left(x - j - \frac{1}{2}\right) f(x)\Big|_j^{j+1} - \int_j^{j+1} \left(x - j - \frac{1}{2}\right) f'(x)\,\mathrm{d}x,$$

or

$$\frac{1}{2} f(j) + \frac{1}{2} f(j+1) = \int_j^{j+1} f(x)\,\mathrm{d}x + \int_j^{j+1} \omega_1(x) f'(x)\,\mathrm{d}x,$$

where $\omega_1(x) = x - [x] - \frac{1}{2} = \{x\} - \frac{1}{2}$.

(Recall that $[x]$ and $\{x\}$ are the integer and fractional parts of x, respectively.)

Summing these equalities for j from $j = a$ to $j = n - 1$, we obtain

$$S = I + \int_a^n \omega_1(x) f'(x)\,\mathrm{d}x.$$

Draw the graph of the function ω_1.

c) Now integrate the integral $\int_j^{j+1} \omega_1(x) f'(x)\,\mathrm{d}x$ by parts and obtain an expression with a new integral residue $\int_j^{j+1} \omega_2(x) f''(x)\,\mathrm{d}x$, where $\omega_2(x) = \int \omega_1(x)\,\mathrm{d}x$. Show the continuity of the function ω_2, given that $\int_j^{j+1} \omega_1(x)\,\mathrm{d}x = \int_0^1 (x - \frac{1}{2})\,\mathrm{d}x = 0$. Take the sum from $j = a$ to $j = n - 1$, as before, and obtain in one step a more advanced expression for the value of S with a new integral residue.

d) Continuing with this process, obtain the following Euler–MacLaurin formula:

$$S = I + \sum_{s=1}^{m-1} (-1)^{s+1} \omega_{s+1}(0) f^{(s)}(x)\Big|_a^n + (-1)^{m+1} \int_a^n \omega_m(x) f^{(m)}(x)\,\mathrm{d}x.$$

e) Check that $\omega_k(x) = \frac{1}{k!} B_k(\{x\})$ and compare the Euler–MacLaurin formula obtained with the one discussed previously.

Appendix E
Riemann–Stieltjes Integral, Delta Function, and the Concept of Generalized Functions

Basic Background

E.1 The Riemann–Stieltjes Integral

Specific Objective and Some Heuristic Arguments We have considered a number of examples of the effective use of the integral in the calculation of areas, volumes of solids of revolution, lengths of paths, work forces, energy .... We found the potential of the gravitational field and computed the escape velocity for Earth. Using the machinery of the integral calculus, we convinced ourselves of the fact that, for example, the path length does not depend on its parametrization. At the same time, we pointed out that certain calculations (e.g., the length of an ellipse) are associated with nonelementary functions (in this case elliptic functions).

All the quantities mentioned above (length, area, volume, work, ...) are additive like the Riemann integral. We know that every additive function $I[\alpha, \beta]$ of the oriented interval $[\alpha, \beta] \subset [a, b]$ has the form $I[\alpha, \beta] = F(\beta) - F(\alpha)$ if we set $F(x) = I[a, x] + C$. In particular, we can take an arbitrary function F and define an additive function $I[\alpha, \beta] = F(\beta) - F(\alpha)$ from it, considering $F(x) = I[a, x]$. If the function F is discontinuous on the interval $[a, b]$, then the function $I[a, x]$ is also discontinuous there. But then it cannot be represented as a Riemann integral $\int_a^x p(t)\,dt$ of any Riemann integrable function (a density p), because such an integral, as we know, is continuous at x.

Suppose, for instance, that the interval $[-1, 1]$ is a string having a bead of mass 1 in the middle. If $I[\alpha, \beta]$ is a mass within the interval $[\alpha, \beta] \subset [-1, 1]$, then the function $I[-1, x]$ is equal to zero for $1 \le x < 0$ and equal to 1 for $0 \le x \le 1$. If we tried to describe the distribution of the mass in the segment in terms of the density distribution (i.e., in terms of the limit of the ratio between the mass lying in a neighborhood of the point and the size of the neighborhood when the latter is contracted to a point), then we would have to define the density p as $p(x) = 0$ for $x \neq 0$ and $p(x) = +\infty$ for $x = 0$. Physicists and all scientists after Dirac called p a "function" (this distribution density). Today, we call p a "generalized function" or

V.A. Zorich, *Mathematical Analysis I*, Universitext,
DOI 10.1007/978-3-662-48792-1

"distribution". We denote it by δ and it is defined as $\int_\alpha^\beta \delta(x)\,\mathrm{d}x = 1$ if $\alpha < 0 < \beta$ and $\int_\alpha^\beta \delta(x)\,\mathrm{d}x = 0$ if $\alpha < \beta < 0$ or $0 < \alpha < \beta$, no matter what the numbers α and β are.

Of course, according to the traditional definition of the integral, e.g., the Riemann integral, this integral does not make sense (for the simple reason that the integrand is an unbounded "function"). We make liberal use of the integral symbol here. It is used only as a replacement of the additive function $I[\alpha, \beta]$, discussed above, when we considered the bead on a string.

Example 4 (Center of mass) Recall the fundamental equation $m\ddot{r} = F$ describing the motion of a point mass m due to the effect of a force F, where r is the radius vector of the point. If there is a system of n material points, then for each of them we have the equation $m_i\ddot{r}_i = F_i$. By summing all these equalities, we obtain the equation $\sum_{i=1}^n m_i\ddot{r}_i = \sum_{i=1}^n F_i$, which can be rewritten in the form $M\sum_{i=1}^n \frac{m_i}{M}\ddot{r}_i = \sum_{i=1}^n F_i$, or in the form $M\ddot{r}_M = F$, where $M = \sum_{i=1}^n m_i$, $F = \sum_{i=1}^n F_i$, and $\ddot{r}_M = \sum_{i=1}^n \frac{m_i}{M}\ddot{r}_i$. That means that if the total mass of the system is placed at some point in the space, with the radius vector being equal to $r_M = \sum_{i=1}^n \frac{m_i}{M} r_i$, and under the influence of the force $F = \sum_{i=1}^n F_i$, then the mass will move according to Newton's law, no matter how complex the mutual motion of the individual parts of the system is.

The point of the space we found with radius vector $\sum_{i=1}^n \frac{m_i}{M} r_i$ is called the *center of mass of the system of material points.*

Now suppose we have the task of finding the center of mass of a material body, i.e., a region D of space, in which a mass is somehow distributed. Let $\mathrm{d}v$ be the volume element, the concentrated mass $\mathrm{d}m$, and M the total mass of the body D. Then we suppose that $M = \int_D \mathrm{d}m$, and then the center of mass can be found with the formula $\frac{1}{M}\int_D r\,\mathrm{d}m$, where r is the radius vector of the mass element.

We still do not know how to integrate over domains in space, and therefore, we consider the one-dimensional case, which is also quite informative. Thus, instead of the region D, we consider the interval $[a, b]$ of the coordinate axis $\mathbb{R}$.

Then $M = \int_a^b \mathrm{d}m$, and the center of mass is found with the formula $\frac{1}{M}\int_a^b x\,\mathrm{d}m$, where x is the coordinate of the mass element $\mathrm{d}m$, which therefore can be written more precisely as $\mathrm{d}m(x)$.

The meaning of what we wrote evidently must be as follows. Take a partition P of the interval $[a, b]$, with some marked points $\xi_i \in [x_{i-1}, x_i]$. To the interval $[x_{i-1}, x_i]$ corresponds the mass Δm_i. We consider the sums $\sum_i \Delta m_i$, $\sum_i \xi_i \Delta m_i$, and passing to the limit as the parameter $\lambda(P)$ of the partition tends to zero, we obtain respectively what is denoted by $\int_a^b \mathrm{d}m$ and $\int_a^b x\,\mathrm{d}m$.

We arrive at the following generalization of the Riemann integral.

Definition of the Riemann–Stieltjes Integral[1] Let f, g be real-, complex-, or vector-valued functions defined over the interval $[a, b] \subset \mathbb{R}$. Let $(P, \xi) = (a =$

[1] T.J. Stieltjes (1856–1894) – Dutch mathematician.

$x_0 \le \xi_1 \le x_1 \le \cdots \le \xi_n \le x_n = b$) be a partition of this interval with marked points and parameter $\lambda(P)$. We consider the sum $\sum_{i=1}^{n} f(\xi_i)\Delta g_i$, where $\Delta g_i = g(x_i) - g(x_{i-1})$.

The *Riemann–Stieltjes integral of the function* f *with respect to the function* g *over the interval* $[a, b]$ is defined as the following value:

$$\int_a^b f(x)\,\mathrm{d}g(x) := \lim_{\lambda(P)\to 0} \sum_{i=1}^{n} f(\xi_i)\Delta g_i \tag{E.1}$$

if the above limit exists.

In particular, when $g(x) = x$, we return to the standard Riemann integral.

E.2 Case in Which the Riemann–Stieltjes Integral Reduces to the Riemann Integral

Note also that if the function g is smooth and f is a Riemann-integrable function over the interval $[a, b]$, then

$$\int_a^b f(x)\,\mathrm{d}g(x) = \int_a^b f(x)g'(x)\,\mathrm{d}x, \tag{E.2}$$

i.e., in this case, the computation of the Riemann–Stieltjes integral reduces to the computation of the Riemann integral of the function fg' on the same interval.

Indeed, using the smoothness of the function g and the mean value theorem, we can rewrite the sum at the right side of Eq. (E.1) in the following form:

$$\sum_{i=1}^{n} f(\xi_i)\Delta g_i = \sum_{i=1}^{n} f(\xi_i)\big(g(x_i) - g(x_{i-1})\big) = \sum_{i=1}^{n} f(\xi_i)g'(\tilde{\xi}_i)(x_i - x_{i-1}) =$$

$$= \sum_{i=1}^{n} f(\xi_i)g'(\xi_i)\Delta x_i + \sum_{i=1}^{n} f(\xi_i)\big(g'(\tilde{\xi}_i) - g'(\xi_i)\big)\Delta x_i .$$

By the uniform continuity of the function g' on the interval $[a, b]$ and the boundedness of the function f, the last sum approaches zero as $\lambda(P) \to 0$. The first sum is the usual integral sum for the integral, which appears at the right side of (E.2). By our assumptions about the functions f and g, the function fg' is Riemann integrable over the interval $[a, b]$. Thus, the sum above approaches the value of this integral for $\lambda(P) \to 0$, which completes the proof of Eq. (E.2).

Problem 1 We achieved the proof of this equality using the mean value theorem, which holds for real-valued functions. Using the general finite increment theorem, complete the proof for vector-valued functions (for instance, complex-valued functions).

E.3 Heaviside Function and an Example of a Riemann–Stieltjes Integral Computation

The *Heaviside step function* is defined through $H(x) = 0$ for $x < 0$ and $H(x) = 1$ for $0 \leq x$. Let us calculate the integral $\int_a^b f(x)\,\mathrm{d}H(x)$. Following the definition (E.1), we write the sum $\sum_{i=1}^n f(\xi_i)\Delta H_i = \sum_{i=1}^n f(\xi_i)(H(x_i) - H(x_{i-1}))$. Because of the definition of the Heaviside function, this sum is clearly equal to zero if the point 0 is not contained in the interval $[a, b]$, and equal to $f(x_i)$ if the point 0 falls into some of the intervals $[x_{i-1}, x_i]$ (more precisely, in the interior of it or at its endpoint x_i). In the first case, the integral is of course zero.

In the second case, under the limit $\lambda(P) \to 0$, the point $\xi_i \in [x_{i-1}, x_i]$ approaches 0. Therefore, if the function f is continuous at 0, then the limit of the sum above will be $f(0)$.

If the function f is discontinuous at 0, with small changes of the value of ξ_i it is possible to change essentially the value of $f(\xi_i)$, and thus the sums of integrals will not have a limit for $\lambda(P)$.

It is clear that the last calculation has a general nature, since the occurrence of joint points of discontinuity for the functions f, g involved in the Riemann–Stieltjes integral leads to the nonexistence of the limit if such a joint point occurs in the interior of the integration interval.

Therefore, the calculation above shows that if φ is, for instance, a function of class $C_0(\mathbb{R}, \mathbb{R})$, i.e., a function defined on the whole real line and continuous, that is identically zero outside of some bounded set, then

$$\int_{\mathbb{R}} \varphi(x)\,\mathrm{d}H(x) = \varphi(0). \tag{E.3}$$

E.4 Generalized Functions

E.4.1 Dirac's Delta Function. A Heuristic Description

As we previously remarked, physicists among other scientists use the delta function δ after its introduction by Dirac. This "function" is zero everywhere except at the origin, where its value is infinity. Along with this (and this is really important),

$$\int_\alpha^\beta \delta(x)\,\mathrm{d}x = 1 \quad \text{if } \alpha < 0 < \beta,$$

and

$$\int_\alpha^\beta \delta(x)\,\mathrm{d}x = 0 \quad \text{if } \alpha < \beta < 0 \text{ or if } 0 < \alpha < \beta,$$

for all real values α and β.

It is natural to assume that the multiplication of the integrating function by a number leads to the product of this number with the integral. But then, if a function φ is continuous at the origin, given that it is almost constant in some small neighborhood $U(0)$ of the origin and that $\int_{U(0)} \delta(x)\,\mathrm{d}x = 1$, we conclude that the following relation should hold:

$$\int_{\mathbb{R}} \varphi(x)\delta(x)\,\mathrm{d}x = \varphi(0). \tag{E.4}$$

By comparing Eqs. (E.2), (E.3), and (E.4) and continuing this chain of findings, we conclude that

$$H'(x) = \delta(x). \tag{E.5}$$

Of course, it does not fit into our classical setting. However, these considerations are very constructive, and if one were obliged to write a value $H'(x)$, then one would write what we now write, namely 0 if $x \neq 0$ and $+\infty$ if $x = 0$.

E.5 The Correspondence Between Functions and Functionals

One possible way out of this difficulty consists in the following idea of extension (generalization) of the concept of "function".

We shall look at the function through its interaction with other functions. (As usual, we are not interested in the internal structure of a device, such as a human, and we consider that we know an object if we know how the object responds to certain input actions or incoming questions.)

Take an integrable function f on the interval $[a, b]$, and consider the functional A_f (a function over functions) generated by f:

$$A_f(\varphi) = \int_a^b f(x)\varphi(x)\,\mathrm{d}x. \tag{E.6}$$

In order to simplify the technical difficulties, we shall consider smooth *test functions* even of class $C_0^\infty[a, b]$, i.e., infinitely differentiable functions vanishing in a neighborhood of the endpoints. It is even possible to continue both functions f and φ as zero outside the interval $[a, b]$, and instead of writing the integral over the interval, we can write the integral as

$$A_f(\varphi) = \int_{\mathbb{R}} f(x)\varphi(x)\,\mathrm{d}x. \tag{E.7}$$

Knowing the value of the functional A_f on test functions, we can find easily, if necessary, the value $f(x)$ of the function f at any point where this function is continuous.

Problem 2

a) Prove that the value $\frac{1}{2\varepsilon}\int_{x-\varepsilon}^{x+\varepsilon} f(t)\,\mathrm{d}t$ (integral average) tends to $f(x)$ for $\varepsilon \to 0$ at every point of continuity of the integrable function f.

b) Show that the step function $\overline{\delta}_\varepsilon$, equal to zero outside the interval $[\varepsilon, \varepsilon]$ and equal to $\frac{1}{2\varepsilon}$ inside this interval (the function $\overline{\delta}_\varepsilon$ imitates Dirac's δ function), can be approximated with smooth functions δ_ε with these properties: $\delta_\varepsilon(x) \geq 0$ in $\mathbb{R}$, $\delta_\varepsilon(x) = 0$ for $|x| \geq \varepsilon$, and $\int_{\mathbb{R}} \delta_\varepsilon(x)\,\mathrm{d}x = 1$.

c) Show that if $\varepsilon \to 0$, then $\int_{x-\varepsilon}^{x+\varepsilon} f(t)\delta_\varepsilon(x-t)\,\mathrm{d}t \to f(x)$ at every point of continuity of the integrating function f.

E.6 Functionals as Generalized Functions

Thus, an integrable function f provides a linear functional A_f (a linear function in the vector space of functions $C_0^\infty[a,b]$ or $C_0^\infty(\mathbb{R})$) defined through the formula (E.6) or (E.7), and moreover, with the use of the functional A_f, the integrable function f itself can be restored at all its points of continuity (i.e., almost everywhere). Therefore, the functional A_f can be thought of as a different encoding or interpretation of the function f considered in the mirror of functionals.

But in this mirror it is possible to find some other linear functionals that are not given through the integration of any function. As an example, we have the functional that we have studied before, $\int_{\mathbb{R}} \varphi(x)\,\mathrm{d}H(x) = \varphi(0)$, which we denote by A_δ (given that we would like to write $\delta(x)\,\mathrm{d}x$ instead of $\mathrm{d}H(x)$).

The functionals of the first type are called *regular*, while those of the second type are called *singular*.

We shall consider functionals as *generalized functions*. This set of functionals contains our usual functions as a subset, consisting of all regular functionals.

Thus, by relating the Riemann integral and its generalization the Riemann–Stieltjes integral, we gave an overview of the construction of generalized functions. We shall not delve into the details of the theory of generalized functions, which are related, for instance, to different spaces of test functions and the construction of linear functionals (generalized functions) on them. We prefer to prove the rule for differentiating generalized functions. As a final remark showing the usefulness of the Stieltjes integral, we would like to add here that on the space $C[a,b]$ of continuous functions φ on the interval $[a,b]$, every linear continuous functional (either regular or singular) can be represented as a Riemann–Stieltjes integral $\int_a^b \varphi(x)\,\mathrm{d}g(x)$ for some properly selected function g. (For instance, the singular functional A_δ representing the generalized function δ has the form $\int_{\mathbb{R}} \varphi(x)\,\mathrm{d}H(x)$, shown in Eq. (E.3).)

We began with an example in which we found the Stieltjes integral $\int_a^b x\,\mathrm{d}m(x)$ in the determination of the center of mass. The integral $M_n = \int_a^b x^n\,\mathrm{d}m(x)$ is called the *moment of order n* relative to a measure (e.g., a probability measure), mass, or charge distributed over the interval $[a,b]$. The moments M_0, M_1, M_2 are frequently met: M_0 is the total mass (or measure of charge); M_1/M_0 provides the center of mass in mechanics, and M_1 is the mathematical expectation (or expected value) of a random variable in probability theory; M_2 is the moment of inertia in mechanics and the scattering of a random variable with expected value $M_1 = 0$ in probability theory. One of the problems of the theory of moments is the restoration of a distribution through the computation of the moments.

E.7 Differentiation of Generalized Functions

Let A be a generalized function. What generalized function A' should be considered the derivative of A?

Let us consider first the derivative for regular generalized functions, i.e., for functionals A_f generated by a classical function f, in our case a smooth compactly supported function of class $C_0^{(1)}$. Then it is natural to consider $A_{f'}$ the derivative A'_f of A_f, generated by the function f', the derivative of the original function f.

Using integration by parts, we find that

$$A'_f(\varphi) := A_{f'}(\varphi) = \int_{\mathbb{R}} f'(x)\varphi(x)\,\mathrm{d}x = f(x)\varphi(x)\big|_{-\infty}^{\infty} - \int_{\mathbb{R}} f(x)\varphi'(x)\,\mathrm{d}x =$$
$$= -\int_{\mathbb{R}} f(x)\varphi'(x)\,\mathrm{d}x = A_f(\varphi').$$

Therefore, we find that in this case,

$$A'_f(\varphi) = -A_f(\varphi'). \tag{E.8}$$

This provides a reason to adopt the following definition of *derivative*:

$$A'(\varphi) := -A(\varphi'). \tag{E.9}$$

It is indicated here how the functional A' acts on a function $\varphi \in C_0^{(\infty)}$. Therefore, the functional A' is well defined.

The action of a linear functional on a function φ is frequently written in the form $\langle A, \varphi\rangle$, instead of $A(\varphi)$, recalling the scalar product, to emphasize that this product is linear in both of its variables.

With this notation if f is any generalized function, then according to Eq. (E.9), we have

$$\langle f', \varphi\rangle = -\langle f, \varphi'\rangle. \tag{E.10}$$

E.8 Derivatives of the Heaviside Function and the Delta Function

We shall compute the derivative of the Heaviside function, considering it a generalized function acting according to the usual rule of regular generalized functions

$$\langle H, \varphi\rangle = \int_{\mathbb{R}} H(x)\varphi(x)\,\mathrm{d}x.$$

Following the definitions (E.9) and (E.10), we have that

$$\langle H', \varphi\rangle := -\langle H, \varphi'\rangle := \int_{\mathbb{R}} H(x)\varphi'(x)\,\mathrm{d}x = -\int_0^{+\infty} \varphi'(x) =$$
$$= -\varphi(x)\big|_0^{+\infty} = \varphi(0).$$

We have shown that $\langle H', \varphi \rangle = \varphi(0)$. However, after the definition of the δ function we have $\langle \delta, \varphi \rangle = \varphi(0)$. Hence, we have proved that in terms of generalized functions, we have the equality

$$H' = \delta.$$

Let us compute, for example, δ' and δ'', i.e., we determine the action of the following functionals:

$$\left\langle \delta', \varphi \right\rangle := -\left\langle \delta, \varphi' \right\rangle := -\varphi'(0);$$
$$\left\langle \delta'', \varphi \right\rangle := -\left\langle \delta', \varphi' \right\rangle := \varphi''(0).$$

It is clear now that in general, we have $\langle \delta^{(n)}, \varphi \rangle = (-1)^n \varphi^{(n)}(0)$.

We realize that generalized functions are infinitely differentiable. This is their remarkable property, which has many consequences. This property allows operations that with usual functions are possible only under very special conditions.

To conclude, we would like to make the following remark of a general nature. Let X be a vector space and X^* its dual space, consisting of linear functions on X, and let X^{**} be the dual space of X^*. We shall write the value $x^*(x)$ of the function $x^* \in X^*$ at the vector $x \in X$ as a scalar product $\langle x^*, x \rangle$, as we did before. By fixing x, we obtain a linear function with respect to x^*. Thus every element of X can be interpreted as an element of X^{**}, i.e., we have an embedding $I : X \to X^{**}$. In the finite-dimensional case, all the spaces X, X^*, and X^{**} are isomorphic, and $I(X) = X^{**}$. In the general case, $I(X) \subsetneq X^{**}$, i.e., $I(X)$ is only a subset of the whole space X^{**}. This is what is observed in the transition from functions (corresponding to regular functionals) to generalized functions, which turned out to be a larger space.

Appendix F
The Implicit Function Theorem (An Alternative Presentation)

F.1 Formulation of the Problem

The formulation of the problem and the heuristic arguments are discussed, of course, in a course lecture; but we shall omit this here, since the relevant material can be read in Sect. 8.5 of Chap. 8.

We shall use a different approach for the proof of the implicit function theorem here, splendid and independent from that we presented in Sect. 8.5 of Chap. 8. This theorem assumes a somehow more advanced audience of readers, despite its conceptual simplicity, beauty, and generality. These readers are in general already more familiar with some general mathematical concepts, presented at the beginning of the second part of the textbook. In any case, all this information allows us to appreciate the real generality of the method, which we can show without loss of generality on simple visual examples in our familiar spaces.

F.2 Some Reminders of Numerical Methods to Solve Equations

By fixing one of the variables in the equation $F(x, y) = 0$, we obtain an equation in terms of the other variable. Therefore, it might be useful to remember how to solve equations $f(x) = 0$.

1) According to the properties of the given function f, one chooses the methods of solution.

For instance, if the function f is real-valued, continuous, and taking values with different signs at the endpoints of the interval $[a, b]$, then we know that in this interval there is at least one root of the equation $f(x) = 0$, and it is possible to find it through successive divisions of the interval. By dividing the interval in half, we get either the root or a half-interval where the function takes values with different signs at the endpoints. Continuing with this dividing process, we obtain a sequence (the endpoints of the intervals) converging to the root of this equation.

V.A. Zorich, *Mathematical Analysis I*, Universitext,
DOI 10.1007/978-3-662-48792-1

2) If f is a smooth convex function, then following Newton's method, it is possible to propose in this case a more efficient algorithm, in the sense of speed of convergence, to find a root.

Newton's method or the *method of tangents* works as follows. We build the tangent at the point x_0, we find the intersection point with the x-axis, and by repeating this process, we obtain a sequence of points with a recurrence relation

$$x_{n+1} = x_n - \left(f'(x_n)\right)^{-1} f(x_n) \tag{F.1}$$

converging rapidly to the root. (Estimate the velocity of convergence. Obtain the relation $x_{n+1} = \frac{1}{2}(x_n + a/x_n)$, allowing you to find the positive root of the equation $x^2 - a = 0$. Find $\sqrt{2}$ according to this formula with the desired accuracy and detect how many additional correct digits appear at each step.)

3) Equation (F.1) can be written in the form

$$x_{n+1} = g(x_n), \tag{F.2}$$

where $g(x) = x - (f'(x))^{-1} f(x)$. Thus, finding the roots of the equation for $f(x)$ reduces to finding a fixed point of the mapping g, i.e., a point such that

$$x = g(x). \tag{F.3}$$

This reduction, as we know, applies not only to Newton's method. In fact, the equation $f(x) = 0$ is equivalent to the equation $\lambda f(x) = 0$ (if λ^{-1} exists), and that is equivalent to the equation $x = x + \lambda f(x)$. Setting $g(x) = x + \lambda f(x)$ (λ can be a variable here), we arrive at Eq. (F.3).

The process of solution of (F.3), i.e., finding a fixed point of the mapping g in accordance with the recursive formula (F.2), is called an iterative process or method of iterations, as we already know. This means that the value found in the previous step becomes the argument or input of the function f in the next step. This cyclic process is suitable for implementation in a computer.

If the iteration process (F.2) is done in a region where $|g'(x)| \le q < 1$, then the sequence

$$\begin{aligned}
&x_0,\\
&x_1 = g(x_0),\\
&x_2 = g(x_1) = g^2(x_0),\\
&\vdots\\
&x_{n+1} = g(x_n) = g^n(x_0)
\end{aligned}$$

is always fundamental (or a Cauchy sequence). Indeed, by applying the mean value theorem, we have

$$|x_{n+1} - x_n| \le q|x_n - x_{n-1}| \le \cdots \le q^n |x_1 - x_0|. \tag{F.4}$$

We apply the triangle inequality to it, and we obtain

$$|x_{n+m} - x_n| \le |x_n - x_{n+1}| + \cdots + |x_{n+m-1} - x_{n+m}| \le$$
$$\le \left(q^n + \cdots + q^{n+m-1}\right)|x_1 - x_0| \le \frac{q^n}{1-q}|x_1 - x_0|. \qquad \text{(F.5)}$$

It is useful to remark that if we take the limit $m \to \infty$ in the last inequality, we obtain the estimate

$$|x - x_n| \le \frac{q^n}{1-q}|x_1 - x_0|, \qquad \text{(F.6)}$$

the deviation or evasion of x_n from the fixed point x.

Problem 1 Draw several variants of the curve $y = g(x)$ intersecting the line $y = x$ and a diagram simulating an iterative process $x_{n+1} = g(x_n)$ for finding a fixed point.

F.2.1 The Principle of the Fixed Point

The last arguments (relating formulas (F.3)–(F.6)) can obviously be applied in any metric space in which the Cauchy criterion is valid, i.e., where every fundamental sequence is convergent. Such metric spaces are called *complete metric spaces*. For instance, $\mathbb{R}$ is a complete metric space with respect to the standard distance $d(x', x'') = |x' - x''|$ between points $x', x'' \in \mathbb{R}$. The interval $I = \{x \in \mathbb{R} \mid |x| \le 1\}$ is also a complete metric space with respect to this metric. If we remove a point from $\mathbb{R}$ or I, then clearly, the resulting metric space will not be complete.

Problem 2

a) Prove the completeness of the spaces $\mathbb{R}^n$, $\mathbb{C}$, $\mathbb{C}^n$.

b) Show that the closed ball $B(a, r) = \{x \in X \mid d(a, x) \le r\}$ with radius r and center $a \in X$ in a complete metric space (X, d) is itself a complete metric space with respect to the induced metric d from the embedding $B \subset X$.

We recall now the following definition.

Definition 1 A mapping $g : X \to Y$ from a metric space (X, d_X) into another (Y, d_Y) is called a *contraction* if there exists a number $q \in [0, 1[$ such that for arbitrary points $x', x'' \in X$,

$$d_Y\left(g(x'), g(x'')\right) \le q d_X\left(x', x''\right).$$

For example, if $g : \mathbb{R} \to \mathbb{R}$ is a differentiable function with the property that everywhere $|g'(x)| \le q < 1$, then by the mean value theorem, we have $|g(x') - g(x'')| \le q|x' - x''|$, and therefore g is a contraction mapping. The same can be said about a differentiable mapping $g : B \to Y$ from a convex subset B of a normed space X (for instance, the ball $B \subset \mathbb{R}^n$) to a normed space Y if $\|g'(x)\| \le q$ at every point $x \in B$.

We are able to formulate now the following *fixed-point principle.*

A contraction mapping $g : X \to X$ of a complete metric space into itself has a unique fixed point x.

This point can be found through the iterative process $x_{n+1} = g(x_n)$, starting with any point $x_0 \in X$. The speed of convergence and the error estimate for the approximation are given by the inequality

$$d(x, x_n) \le \frac{q^n}{1-q} d(x_1, x_0). \tag{F.6$'$}$$

The proof of this fact was given above by the deduction of formulas (F.4)–(F.6), where instead of $|x' - x''|$ we have to write everywhere $d(x', x'')$.

In order to appreciate the usefulness and scope for generalizing this principle, consider the following important example.

Example 1 We look for the function $y = y(x)$ satisfying the differential equation $y' = f(x, y)$ and the initial condition $y(x_0) = y_0$.

Using the formula of Newton–Leibniz, we rewrite the problem in the form of the following integral equation for the unknown function $y(x)$:

$$y(x) = y_0 + \int_{x_0}^{x} f\bigl(t, y(t)\bigr)\,dt. \tag{F.7}$$

On the right-hand side there is a mapping g, which acts on the function $y(x)$, and we look for the fixed "point" of the mapping (action) g.

For example, let $f(x, y) = y$, $x_0 = 0$, and $y_0 = 1$. Then we deal with the solution of the equation $y' = y$ with the initial condition $y(0) = 1$, and Eq. (F.7) takes the form

$$y(x) = 1 + \int_0^x y(t)\,dt. \tag{F.8}$$

We then carry out the iterative process, starting with the function $y_0(x) \equiv 0$, and we successively obtain

$$\begin{aligned}
y_1(x) &= 1, \\
y_2(x) &= 1 + \int_0^x y_1(t)\,dt = 1 + x, \\
y_3(x) &= 1 + \int_0^x y_2(t)\,dt = 1 + \int_0^x (1+t)\,dt = 1 + x + \frac{1}{2}x^2, \\
&\vdots
\end{aligned}$$

$$y_n(x) = 1 + \frac{1}{1!}x + \cdots + \frac{1}{n!}x^n,$$

$$\vdots$$

It is clear, that we obtain the function $e^x = 1 + \frac{1}{1!}x + \cdots + \frac{1}{n!}x^n + \cdots$.

Problem 3 Show that if $\|f(x, y_1) - f(x, y_2)\| \le M\|y_1 - y_2\|$, then in a neighborhood of the point x_0 the iteration process is applicable in the case of the more general equation (F.7).

In this way, Picard (Émile Picard, 1856–1941) was looking for the solution of the differential equation $y'(x) = f(x, y(x))$, with the initial condition $y(x_0) = y_0$ as a fixed point of the mapping (F.7).

Banach (Stefan Banach, 1882–1945) formulated the fixed-point principle in the abstract form above, and in this form it is often called *Banach's fixed-point principle* or the *Banach–Picard principle*. However, its origins can be traced back to Newton, as we have seen.

F.3 The Implicit Function Theorem

F.3.1 Statement of the Theorem

We return now to the main object of our consideration and prove the implicit function theorem.

Theorem *Let X, Y, Z be normed spaces (for example, $\mathbb{R}^m, \mathbb{R}^n, \mathbb{R}^m$ or even $\mathbb{R}, \mathbb{R}, \mathbb{R}$), and suppose moreover that Y is a complete metric space with respect to the metric induced by the norm. Let $F : W \to Z$ be a mapping defined in a neighborhood W of the point $(x_0, y_0) \in X \times Y$, continuous at (x_0, y_0), together with the partial derivative $F'_y(x, y)$, which is supposed to exist in W. If $F(x_0, y_0) = 0$ and there exist $(F'_y(x_0, y_0))^{-1}$ and $\|(F'_y(x_0, y_0))^{-1}\| < \infty$, then there exist a neighborhood $U = U(x_0)$ of the point x_0 in X, a neighborhood $V = V(y_0)$ of the point y_0 in Y, and a function $f : U \to V$, continuous at x_0, such that $U \times V \subset W$ and*

$$\big(F(x, y) = 0 \text{ within } U \times V\big) \Leftrightarrow \big(y = f(x), x \in U\big). \tag{F.9}$$

In short, under the conditions of the theorem, the set determined by the relation $F(x, y) = 0$ within the neighborhood $U \times V$ is the graph of the function $y = f(x)$.

F.3.2 Proof of the Existence of an Implicit Function

Proof Without loss of generality and for brevity, we may assume that $(x_0, y_0) = (0, 0)$, which can always be achieved by the change of variables $x - x_0$ and $y - y_0$.

For a fixed x we shall solve the equation $F(x, y) = 0$ with respect to y. We look for the solution as the fixed point of the mapping

$$g_x(y) = y - \big(F'_y(0,0)\big)^{-1} F(x, y). \tag{F.10}$$

This is a simplified version of Newton's formula (F.1), where the coefficient λ is constant (see the paragraph following formula (F.3)). It is immediately clear that $F(x, y) = 0 \Leftrightarrow g_x(y) = y$.

The mapping (F.10) is a contraction if (x, y) is near $(0, 0) \in X \times Y$. Indeed,

$$\frac{\mathrm{d}g_x}{\mathrm{d}y}(y) = E - \big(F'_y(0,0)\big)^{-1} F'_y(x, y). \tag{F.11}$$

Here E is the identity (unitary) mapping, and since $F'_y(x, y)$ is continuous at the point $(0, 0)$, there exists a number $\Delta \in \mathbb{R}$ such that for $\|x\| < \Delta$ and $\|y\| < \Delta$,

$$\left\| \frac{\mathrm{d}g_x}{\mathrm{d}y} \right\| < \frac{1}{2}. \tag{F.12}$$

Finally, note that for every $\varepsilon \in\,]0, \Delta[$ there is $\delta \in\,]0, \Delta[$ such that if $\|x\| < \delta$, then the function g_x maps the interval (ball) $\|y\| \le \varepsilon$ into itself.

Indeed, because of $F(0, 0) = 0$ and from Eq. (F.10), we have $g_0(0) = 0$. In view of the continuity of F at the point $(0, 0)$, it follows from (F.10) that there is $\delta \in\,]0, \Delta[$ such that $\|g_x(0)\| < \frac{1}{2}\varepsilon$ for $\|x\| < \delta$.

Thus, for $\|x\| < \delta$, the mapping $g_x : B(\varepsilon) \to Y$ displaces the center of the interval $B(\varepsilon) = \{y \in Y \mid \|y\| \le \varepsilon\}$ no more than $\frac{1}{2}\varepsilon$. Therefore, by virtue of (F.12) it decreases $B(\varepsilon)$ by at least a factor of two. Hence, $g_x(B(\varepsilon)) \subset B(\varepsilon)$ for $\|x\| < \delta$.

By assumption, Y is a complete space, and therefore $B(\varepsilon) \subset Y$ is also a complete metric space (with respect to the induced metric).

Then by virtue of the fixed-point principle, there is a unique point $y = f(x) \in B(\varepsilon)$ that is fixed for the mapping $g_x : B(\varepsilon) \to B(\varepsilon)$.

Thus for every x with $\|x\| < \delta$, we have found a unique value $y = f(x)$ ($\|f(x)\| < \varepsilon$) in the neighborhood $B(\varepsilon)$ such that $F(x, f(x)) = 0$.

(The cross section of the domain $P = \{(x, y) \in X \times Y \mid \|x\| < \delta, \|y\| < \varepsilon\}$ passing through the point $(x, 0)$ is the interval (ball) $B(\varepsilon)$ in which lies the corresponding fixed point $y = f(x)$.)

Thus, we have shown that

$$\big(F(x, y) = 0 \text{ for } \|x\| < \delta \text{ and } \|y\| < \varepsilon\big) \Leftrightarrow \big(y = f(x) \text{ for } \|x\| < \delta\big). \tag{F.13}$$

Note that not only have we obtained the relation (F.9), but also, by virtue of the construction, for every $\varepsilon \in\,]0, \Delta[$ we can choose $\delta > 0$ such that (F.13) holds. Since the function f has been found already and is fixed, we have that $f(0)=0$ and f is continuous at $x=0$. □

The theorem just proved can be regarded as the existence theorem of the implicit function $y=f(x)$.

We shall see now what properties of the function F are inherited by the function f.

F.3.3 Continuity of an Implicit Function

If in addition to the conditions of the theorem, we know that the functions F and F'_y are continuous not only at the point (x_0, y_0) but also in some neighborhood of this point, then the implicit function is also continuous in some neighborhood of this point.

Proof Indeed, in this case, the conditions of the theorem will be fulfilled at all the points of the set $F(x,y)=0$ near (x_0,y_0), and each of them could be considered a starting point (x_0,y_0). The function f has been found already, and therefore is fixed.

Warning! Recall the exercise that if the mapping $A \mapsto A^{-1}$, where A is mapped to its inverse (for example for a matrix A) is defined on A, then it is defined on a neighborhood of A. □

F.3.4 Differentiability of an Implicit Function

If in addition to the conditions of the theorem, we know that the function F is differentiable at the point (x_0, y_0), then the implicit function f is also differentiable at the point x_0, and moreover,

$$f'(x_0) = -\big(F'_y(x_0,y_0)\big)^{-1} F'_x(x_0,y_0). \tag{F.14}$$

Proof Given the differentiability of F at the point (x_0,y_0), we can write

$$\begin{aligned} &F(x,y)-F(x_0,y_0)= \\ &\quad = F'_x(x_0,y_0)(x-x_0)+F'_y(x_0,y_0)(y-y_0)+o\big(|x-x_0|+|y-y_0|\big). \end{aligned}$$

Assuming for simplicity $(x_0,y_0)=(0,0)$ and considering that we are only moving along the curve $y=f(x)$, we obtain

$$0 = F'_x(0,0)x + F'_y(0,0)y + o\big(|x|+|y|\big),$$

or

$$y = -\left(F_y'(0,0)\right)^{-1} F_x'(0,0)x - \left(F_y'(0,0)\right)^{-1} o\left(|x| + |y|\right). \qquad \text{(F.15)}$$

Since $y = f(x) = f(x) - f(0)$, the formula (F.14) will be justified, if we can show that in the limit $x \to 0$, the second term on the right-hand side of (F.15) is $o(x)$.

But

$$\left|\left(F_y'(0,0)\right)^{-1} o\left(|x| + |y|\right)\right| \le \left\|\left(F_y'(0,0)\right)^{-1}\right\| \cdot \left|o\left(|x| + |y|\right)\right| = o\left(|x| + |y|\right).$$

Further,

$$\left\|\left(F_y'(0,0)\right)^{-1} F_x'(0,0)\right\| \le \left\|\left(F_y'(0,0)\right)^{-1}\right\| \cdot \left\|F_x'(0,0)\right\| = a < \infty.$$

Therefore, from (F.15) we obtain that $|y| \le a|x| + \alpha(|x| + |y|)$, where $y = f(x) \to 0$ and $\alpha = \alpha(x) \to 0$ for $x \to 0$. Hence,

$$|y| \le \frac{a + \alpha}{1 - \alpha}|x| < 2a|x|$$

for x sufficiently close to 0. Given this, for $x \to 0$ we obtain from (F.15) that

$$f(x) = -\left(F_y'(0,0)\right)^{-1} F_x'(0,0)x + o(x).$$

In view of $f(0) = 0$, we have (F.14). □

F.3.5 Continuous Differentiability of an Implicit Function

If in addition to the conditions of the theorem, we know that the functions F_x' and F_y' are defined and continuous in some neighborhood of the point (x_0, y_0), then the implicit function f is also continuously differentiable in some neighborhood of the point x_0.

In short, if $F \in C^{(1)}$, then f is also in $C^{(1)}$.

Proof In this case, the conditions of differentiability of f and (F.14) are fulfilled not only at (x_0, y_0) but at all points of the "curve" $F(x, y) = 0$ near (x_0, y_0) (see the above cautionary "Warning!"). Then, according to formula (F.14), in a neighborhood of the point x_0,

$$f'(x) = -\left(F_y'\left(x, f(x)\right)\right)^{-1} F_x'\left(x, f(x)\right), \qquad \text{(F.14')}$$

from which it is clear that f' is continuous. □

Warning! Recall that the mapping $A \mapsto A^{-1}$ is continuous.

F.3.6 Higher Derivatives of an Implicit Function

If in addition to the conditions of the theorem, we know that the function F is of class $C^{(k)}$ in some neighborhood of the point (x_0, y_0), then the implicit function f is also of class $C^{(k)}$ in a neighborhood of the point x_0.

Proof Suppose, for example, that F is of class $C^{(2)}$. Since f is of class $C^{(1)}$, the right-hand side of the equality (F.14′) can be differentiated according to the differentiation rule for composite functions (chain rule). We obtain then a formula for $f''(x)$, and from it the continuity of $f''(x)$ follows.

Moreover, as on the right-hand side of formula (F.14′), for $f'(x)$ the first partial derivatives of F and the function f itself (but not f') are involved; in the formula for $f''(x)$, the second partial derivatives of F, f, and f' are involved (but not f'').

Thus, if F is of class $C^{(3)}$, then we can differentiate $f''(x)$, and we arrive again at a formula for $f'''(x)$ in which are involved the third partial derivatives of F, and also the derivatives of the functions f (f, f', f'') of order less than three.

By induction, we obtain what we claimed. □

Warning! Recall that the mapping $A \mapsto A^{-1}$ is differentiable and even infinitely differentiable.

Problem 4

a) Find $f''(x)$ (write down the formula for the computation of $f''(x)(h_1, h_2)$ for given displacement vectors h_1, h_2).

b) What does the formula (simplified) for $f''(x)$ look like in the case that x, y, and $z = F(x, y)$ are real or complex variables?

Problem 5 (Method of undetermined coefficients). Suppose that we know the first (or all) coefficients of the Taylor series of the function F. Find the first (or all) coefficients of the Taylor series of the implicit function f.

Problem 6

a) Write in coordinate form the formulation of the implicit function theorem for the cases $F : \mathbb{R}^m \times \mathbb{R}^n \to \mathbb{R}^n$, when $m = n = 1$ and when $n > 1$.

b) Let $F : \mathbb{R}^m \to \mathbb{R}^n$ $(m > n)$ be a linear mapping with maximal rank $(= n)$. What is the dimension of the subspace $F^{-1}(0) \subset \mathbb{R}^m$ and what is its codimension? Let $F : \mathbb{R}^m \to \mathbb{R}^n$ $(m > n)$ be now an arbitrary smooth mapping, $F(0) = 0$ and $\operatorname{rank} F'(x) = n$. Answer the same questions ($\dim F^{-1}(0) = ?$, $\operatorname{codim} F^{-1}(0) = ?$) with respect to the set $F^{-1}(0)$.

References

1 Classic Works

1.1 Primary Sources

Newton, I.:

- a. (1687): Philosophiæ Naturalis Principia Mathematica. Jussu Societatis Regiæ ac typis Josephi Streati, London. English translation from the 3rd edition (1726): University of California Press, Berkeley, CA (1999).
- b. (1967–1981): The Mathematical Papers of Isaac Newton, D.T. Whiteside, ed., Cambridge University Press.

Leibniz, G.W. (1971): Mathematische Schriften. C.I. Gerhardt, ed., G. Olms, Hildesheim.

1.2 Major Comprehensive Expository Works

Euler, L.

- a. (1748): Introductio in Analysin Infinitorum. M.M. Bousquet, Lausanne. English translation: Springer-Verlag, Berlin – Heidelberg – New York (1988–1990).
- b. (1755): Institutiones Calculi Differentialis. Impensis Academiæ Imperialis Scientiarum, Petropoli. English translation: Springer, Berlin – Heidelberg – New York (2000).
- c. (1768–1770): Institutionum Calculi Integralis. Impensis Academiæ Imperialis Scientiarum, Petropoli.

Cauchy, A.-L.

- a. (1989): Analyse Algébrique. Jacques Gabay, Sceaux.
- b. (1840–1844): Leçons de Calcul Différential et de Calcul Intégral. Bachelier, Paris.

V.A. Zorich, *Mathematical Analysis I*, Universitext,
DOI 10.1007/978-3-662-48792-1

1.3 Classical Courses of Analysis from the First Half of the Twentieth Century

Courant, R. (1988): Differential and Integral Calculus. Translated from the German. Vol. 1 reprint of the second edition 1937. Vol. 2 reprint of the 1936 original. Wiley Classics Library. A Wiley-Interscience Publication. John Wiley & Sons, Inc., New York.

de la Vallée Poussin, Ch.-J. (1954, 1957): Cours d'Analyse Infinitésimale. (Tome 1 11 éd., Tome 2 9 éd., revue et augmentée avec la collaboration de Fernand Simonart.) Librairie universitaire, Louvain. English translation of an earlier edition: Dover Publications, New York (1946).

Goursat, É. (1992): Cours d'Analyse Mathématiques. (Vol. 1 reprint of the 4th ed. 1924, Vol. 2 reprint of the 4th ed. 1925) Les Grands Classiques Gauthier-Villars. Jacques Gabay, Sceaux. English translation: Dover Publ. Inc., New York (1959).

2 Textbooks[1]

Apostol, T.M. (1974): Mathematical Analysis. 2nd ed. World Student Series Edition. Addison-Wesley Publishing Co., Reading, Mass. – London – Don Mills, Ont.

Courant, R., John F. (1999): Introduction to Calculus and Analysis. Vol. I. Reprint of the 1989 edition. Classics in Mathematics. Springer-Verlag, Berlin.

Courant, R., John F. (1989): Introduction to Calculus and Analysis. Vol. II. With the assistance of Albert A. Blank and Alan Solomon. Reprint of the 1974 edition. Springer-Verlag, New York.

Nikolskii, S.M. (1990): A Course of Mathematical Analysis. Vols. 1, 2. Nauka, Moscow. English translation of an earlier addition: Mir, Moscow (1985).

Rudin, W. (1976): Principals of Mathematical Analysis. McGraw-Hill, New York.

Rudin, W. (1987): Real and Complex Analysis. 3rd ed., McGraw-Hill, New York.

Spivak, M. (1980): Calculus. 2nd ed. Publish Perish, Inc., Berkeley, CA.

Stewart, J. (1991): Calculus. 2nd ed. Thomson Information/Publishing Group. International Student Edition. Brooks/Cole Publishing Company, Pacific Grove, CA.

3 Classroom Materials

Biler, P., Witkowski, A. (1990): Problems in Mathematical Analysis. Monographs and Textbooks in Pure and Applied Mathematics, 132. Marcel Dekker, New York.

[1] For the convenience of the Western reader the bibliography of the English edition has substantially been revised.

Bressoud, D.M. (1994): A Radical Approach to Real Analysis. Classroom Resource Material Series, 2. Mathematical Association of America, Washington, DC.

Demidovich, B.P. (1990): A Collection of Problems and Exercises in Mathematical Analysis. Nauka, Moscow. English translation of an earlier edition: Gordon and Breach, New York – London – Paris (1969).

Gelbaum, B. (1982): Problems in Analysis. Problem Books in Mathematics. Springer-Verlag, New York – Berlin.

Gelbaum, B., Olmsted, J. (1964): Counterexamples in Analysis. Holden-Day, San Francisco.

Hahn, A.J. (1998): Basic Calculus. From Archimedes to Newton to its role in science. Textbooks in Mathematical Sciences. Springer, New York, NY.

Makarov, B.M., Goluzina, M.G., Lodkin, A.A., Podkorytov, A.N. (1992): Selected Problems in Real Analysis. Nauka, Moscow. English translation: Translations of Mathematical Monographs 107, American Mathematical Society, Providence (1992).

Pólya, G., Szegő, G. (1970/1971): Aufgaben und Lehrsätze aus der Analysis. Springer-Verlag, Berlin – Heidelberg – New York. English translation: Springer-Verlag, Berlin – Heidelberg – New York (1972–1976).

4 Further Reading

Arnol'd, V.I.

- a. (1989a): Huygens and Barrow, Newton and Hooke: Pioneers in Mathematical Analysis and Catastrophe Theory, from Evolvents to Quasicrystals. Nauka, Moscow. English translation: Birkhäuser, Boston (1990).
- b. (1989b): Mathematical Methods of Classical Mechanics. Nauka, Moscow. English translation: Springer-Verlag, Berlin – Heidelberg – New York (1997).

Avez, A. (1986): Differential Calculus. Translated from the French. A Wiley-Interscience Publication. John Wiley & Sons, Ltd., Chichester.

Bourbaki, N. (1969): Éléments d'Histoire des Mathématiques. 2e édition revue, corrigée, augmentée. Hermann, Paris. English translation: Springer-Verlag, Berlin – Heidelberg – New York (1994).

Cartan, H. (1977): Cours de Calcul Différentiel. Hermann, Paris. English translation of an earlier edition: Differential Calculus. Exercised by C. Buttin, F. Riedeau and J.L. Verley. Houghton Mifflin Co., Boston, Mass. (1971).

Dieudonné, J. (1969): Foundations of Modern Analysis. Enlarged and corrected printing. Academic Press, New York.

Dubrovin, B.A., Novikov, S.P., Fomenko, A.T. (1986): Modern Geometry – Methods and Applications. Nauka, Moscow. English translation: Springer-Verlag, Berlin – Heidelberg – New York (1992).

Einstein, A. (1982): Ideas and Opinions. Three Rivers Press, New York. Contains translations of the papers "Principles of Research" (original German title: "Motive des Forschens"), pp. 224–227, and "Physics and Reality," pp. 290–323.

Feynman, R., Leighton, R., Sands, M. (1963–1965): The Feynman Lectures on Physics, Vol. 1: Modern Natural Science. The Laws of Mechanics. Addison-Wesley, Reading, Mass.

Gel'fand, I.M. (1998): Lectures on Linear Algebra. Dobrosvet, Moscow. English translation of an earlier edition: Dover, New York (1989).

Halmos, P. (1974): Finite-Dimensional Vector Spaces. Springer-Verlag, Berlin – Heidelberg – New York.

Jost, J. (2003): Postmodern Analysis. 2nd ed. Universitext. Springer, Berlin.

Kirillov, A.A. (1993): What Is a Number? Nauka, Moscow (Russian).

Kolmogorov, A.N., Fomin, S.V. (1989): Elements of the Theory of Functions and Functional Analysis. 6th ed., revised, Nauka, Moscow. English translation of an earlier edition: Graylock Press, Rochester, New York (1957).

Kostrikin, A.I., Manin, Yu.I. (1986): Linear Algebra and Geometry. Nauka, Moscow. English translation: Gordon and Breach, New York (1989).

Landau, E. (1946): Grundlagen der Analysis. Chelsea, New York. English translation: Chelsea, New York (1951).

Lax, P.D., Burstein S.Z., Lax A. (1972): Calculus with Applications and Computing. Vol. I. Notes based on a course given at New York University. Courant Institute of Mathematical Sciences, New York University, New York.

Manin, Yu. I. (1979): Mathematics and Physics. Znanie, Moscow. English translation: Birkhäuser, Boston (1979).

Milnor, J. (1963): Morse Theory. Princeton University Press.

Poincaré, H. (1982): The Foundations of Science. Authorized translation by George Bruce Halstead, and an introduction by Josiah Royce, preface to the UPA edition by L. Pearce Williams. University Press of America, Washington, DC.

Pontryagin, L.S. (1974): Ordinary Differential Equations. Nauka, Moscow. English translation of an earlier edition: Addison-Wesley, Reading, Mass. (1962).

Spivak, M. (1965): Calculus on Manifolds: a Modern Approach to the Classical Theorems of Advanced Calculus. W.A. Benjamin, New York.

Uspenskii, V.A. (1987): What is Nonstandard Analysis? Nauka, Moscow (Russian).

Weyl, H. (1926): Die heutige Erkenntnislage in der Mathematik. Weltkreis-Verlag, Erlangen. Russian translations of eighteen of Weyl's essays, with an essay on Weyl: Mathematical Thought. Nauka, Moscow (1989).

Zel'dovich, Ya.B., Myshkis, A.D. (1967): Elements of Applied Mathematics. Nauka, Moscow. English translation: Mir, Moscow (1976).

Zorich, V.A. (2011): Mathematical Analysis of Problems in the Natural Sciences. Springer, Heidelberg.

Subject Index

V.A. Zorich, *Mathematical Analysis I*, Universitext,
DOI 10.1007/978-3-662-48792-1

Name Index

V.A. Zorich, *Mathematical Analysis I*, Universitext,
DOI 10.1007/978-3-662-48792-1

Printed in the United States
By Bookmasters

Printed in the United States
By Bookmasters